THE FLORA OF
DORSET

To all past and present botanists who have improved our knowledge of the Dorset Flora.

THE FLORA OF
DORSET

Humphry Bowen

piscespublications

First published by Pisces Publications 2000. Pisces Publications is the imprint of the **Nature**Bureau (formerly the Nature Conservation Bureau Limited).

British Library-in-Publication Data.
A catalogue record for this book is available from the British Library.

ISBN 1 874357 16 1

Designed and produced by the **Nature**Bureau, 36 Kingfisher Court, Hambridge Road, Newbury, Berkshire, RG14 5SJ. Tel: 01635 550380. Fax: 01635 550230.
E-mail: pisces@naturebureau.co.uk
Web site: http://www.naturebureau.co.uk

Printed by Information Press, Oxford.

CONTENTS

INTRODUCTION

This Flora is an account of the past and current status of plants in Dorset. Earlier Floras have restricted their coverage to vascular plants and ferns, but this one includes accounts of Mosses and Liverworts, Lichens, Fungi, Stoneworts and Seaweeds. It concentrates on those species which have been seen during the last ten years or so, a relatively short time period in the long history of plants in the county, but includes old records not recently confirmed. The area covered is more or less the Watsonian vice-county 9 (see Chapter 1) and does not cover Bournemouth and Christchurch. Limitations of space prevent this work from being a manual of identification.

The ensuing Chapters follow a conventional pattern. Chapter 1 summarises the physical background – topography, climate, geology, soils and river systems. A section on marine erosion suggests that at the end of the last ice-age Dorset extended seawards to more than twice its present size. Chapter 2 really deserves a volume on its own. It describes the main vegetational habitats, among which old woodland and coastal sites retain many botanical treasures, while the work of Thomas Hardy has made our heaths known to a wide audience. Chapter 3 attempts to describe the human effects on vegetation, starting with recent discoveries by archaeobotanists and continuing with sections on forestry, agriculture over the last five millenia, gardening since Roman times and conservation measures in the last five decades. Were more known, it might be possible to write a section on the effects of vegetation on man. For example Neolithic farmers exploited our Chalk uplands but were deterred by forests or swamps elsewhere; the forest cover north-east of the county probably delayed the displacement of the Celts by the Saxons until the middle of the dark ages.

Chapter 4 is an account of some of the deceased botanists who have laid the foundations of our present knowledge.

The Flora proper includes more than 2000 species of vascular plant that have been found here, some established and others transitory. It gives maps on a 2 km × 2 km tetrad basis, or localities for the rarer plants, and compares the current status of many species with that given in earlier Floras. Broadly speaking, the picture is one of decline of rarer native plants, especially those of wetlands (drained for agriculture), arable fields (sprayed with weedkillers) and sandy beaches (trampled by holiday-makers), while some foreign plants have spread dramatically fast. A few northern species may be declining for climatic reasons. Whether or not a given rarity is extinct is often a matter of opinion. Observant botanists are few in number, none of them is omniscient, and the longevity of buried seeds is rarely known.

Apart from the Lichens, which were covered by a Flora in 1976, no previous compilation of cryptogam records for the whole county have been published. They have been treated on a 10 km × 10 km hectad basis because of the lack of skilled recorders; this is glaringly obvious for the smaller fungi. The latest findings show that Dorset possesses many rare lichens and a few rare bryophytes. It is relatively rich in larger fungi and marine algae without many outstanding rarities, but much remains to be learnt about these groups.

ACKNOWLEDGEMENTS

I would particularly like to thank David Pearman, both for suggesting that this Flora be written and for constant help and encouragement, including letting me see his notes on manuscripts by J.H. Salter and R.D. Good. The authors of individual sections also deserve great praise, both for the excellence of their contributions and proof-reading, and for keeping to deadlines for delivery of manuscripts. The Flora could not have been compiled without the efforts of recorders too numerous to mention here, but whose initials appear frequently in the text: they have spent long hours in the field, often in botanically boring areas and in inclement weather. Especial help has been received from the Dorset Environmental Records Centre and from a number of institutes with herbaria, such as Bournemouth Natural Science Society (Mrs Sheila Mackintosh), Dorchester County Museum (Kate Hebditch), Liverpool Museum (John Edmondson), Nottingham Natural History Museum (G. Walley) and Reading University (Stephen Jury). Those who have kindly allowed me to see their unpublished work include R. Ashe (Flora of Portland), Mark Hill (Bryophytes), Martin Jenkinson (Orchidaceae) and N.F. Stewart (Charophytes). English Nature (at Furzebrook and Slepe Farm), the National Trust and the RSPB have let me see lists of records from their reserves. David Allen merits special thanks for his field and herbarium work on *Rubus*, for which he wrote the account here, as well as for compiling information on Dorset worthies in Chapter 4. Bryan Edwards and Vince Giavarini, both staunch workers on cryptogams in the field, have freely given their time in reading and emending the sections on bryophytes and lichens, as have John Keylock and Anne Leonard for the sections on larger and smaller fungi, M.J. Allen for the section on Archaeobotany and Mrs Kay Sanecki for the section on Gardens.

Experts on critical groups who have determined many Dorset specimens include John Akeroyd (*Rumex*), David Allen (*Rubus*), Jim Bevan (*Hieracium*), Eric Clement (Aliens), Brian Coppins (Lichens), I.K. Ferguson (*Salicornia*), Jeanette Fryer (*Cotoneaster*), Mark Hill (Bryophytes), Clive Jermy (Ferns), John Keylock (Fungi), Anne Leonard (Ascomycetes), C. Maggs (Marine Algae), Len Margetts (*Rubus*), R.D. Meikle (*Salix*), Jean Paton (Liverworts), David Pearman (Cyperaceae), Chris Preston (*Potamogeton*), A.L. Primavesi (*Rosa*), P.D. Sell (*Fumaria*), Mervyn Southam (*Oenanthe*), Clive Stace (Gramineae), N.F. Stewart (Stoneworts), Eric Watson (Bryophytes), P.F. Yeo (*Euphrasia*) and many others whose names appear in the text.

Tuition on computers has been given by Cameron Crook, Alison Stewart and Keith Woodhead, all of whom have been very patient with an aged tyro in this field. Dorset Environmental Records Centre staff have been consistently helpful in allowing the use of their records, and in many other ways. Photographs of Dorset worthies were provided by Dorset County Museum and the Bournemouth Natural Science Society. Line drawings of critical plants by Mrs Anita Pearman were specially commissioned. Colour photographs of plants and habitats were submitted by many people whose contributions are acknowledged, but especially by Bryan Edwards, Anne Horsfall, John Keylock and Tony Bates; the last named handled the transfer of pictures to a computer disc.

CHAPTER 1

PHYSICAL FEATURES

TOPOGRAPHY

Dorset is a southern county with an area of 2,521 km² (623,000 acres or 973 square miles). In shape it is an irregular lozenge with its longer east-west axis of 78 km (48 miles) and its north-south axis of 50 km (31 miles). Its southern boundary is the English Channel. To the west is South Devon (VC 3), and the artificial north-west border is with South Somerset (VC 5) and a little bit of North Somerset (VC 6). The north-east border with South Wiltshire (VC 8) and the eastern border with South Hampshire (VC 11) are mainly artificial, but include a massive earthwork, the Bokerly dyke.

In his Flora of 1948, Good subdivided the county into three botanical areas, each of which has its characteristic landscape determined by the underlying geology (see Figure 1, below (Good, 1948)).

Roughly half of the county is Chalk uplands. A chalk range with a north-facing scarp extends from Beaminster in the west to Melbury Down in the north-east, with the following notable heights: Pilsdon Pen, 277 m; Lewesdon Hill, 272 m; High Stoy, 262 m; Bulbarrow, 275 m; Okeford Hill, 225 m; Hambledon Hill, 190 m; Melbury Hill, 263 m. A much narrower chalk range stretches from

Beaminster to Ballard Head in the south-east, broken by a spectacular double gap at Corfe; vantage points include: Eggardon Hill, 252 m; Black Down, 199 m; Bincombe Hill, 162 m; Bindon Hill, 168 m; and Nine Barrow Down, 200 m. Much of the Chalk is now prairie-farmed, but some scarp slopes are wooded, and some south-facing slopes are covered with gorse. A few patches of chalk grassland remain in nature reserves or military enclaves, with fragments on ancient earthworks. The ridge of hard limestone outcropping at Portland, and from Durdle Door to Durlston Head, does not exceed 130 m in height, and is much quarried, with sea caves along the coast.

About a quarter of Dorset is on clay (Good's Border Vales), notably Marshwood and Blackmore vales in the west. Most are lowland, but outliers such as Swyre Head in the south-east rise to 192 m. The vales retain water better than the Chalk, and so are more fertile and have many small dairy farms, with a few small woods.

The third type of landscape is the sandy heathland of the Poole basin, a triangular outlier of the New Forest with apices at Cranborne, Dorchester and Studland. Much has been afforested with conifers, built over or fertilized for agriculture, but isolated fragments of low elevation remain, some large and many with associated bogs. Small

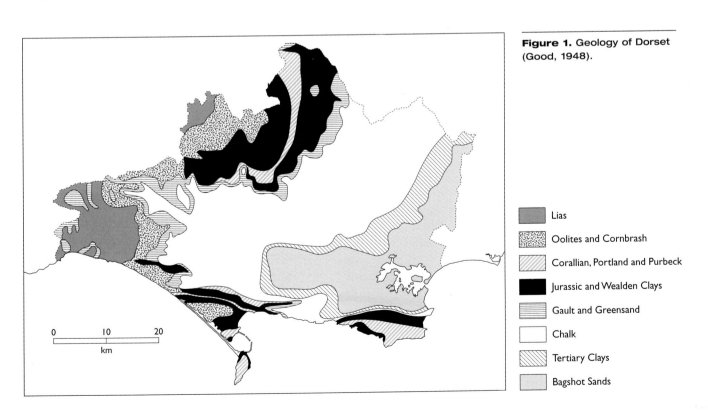

Figure 1. Geology of Dorset (Good, 1948).

Lias

Oolites and Cornbrash

Corallian, Portland and Purbeck

Jurassic and Wealden Clays

Gault and Greensand

Chalk

Tertiary Clays

Bagshot Sands

0   10   20
km

areas of heath also occur on hilltops in the west, as at Lamberts Castle, Hardown Hill, Holes Common and Black Down near Portesham. Where heath overlies chalk, dolines occur, but they are rare and small, e.g. Culpeppers dish near Affpuddle, and at Southover Heath (79W).

In his 1895 Flora, Mansel-Pleydell subdivided the county into six parts by its river basins (Mansel-Pleydell, 1895):

A: Axe and Brit in west Dorset
B: Yeo in north-west Dorset
C: Frome and Wey in central Dorset
D: Upper Stour, above Haywards bridge, in north Dorset
E: Lower Stour and Moors river in east Dorset
F: Corfe river in south-east Dorset (Purbeck)

The Stour (105 km long) rises in Wilts and flows in a generally south-east direction to its mouth at Christchurch, Hants. The Frome (56 km long) rises near Corscombe in the west, flows south to Dorchester and then turns eastward to its mouth in Poole Harbour. Here also emerges a parallel stream, the Piddle (40 km long), which rises near Alton Pancras. As these rivers cut across several types of landscape they do not define homogeneous botanical districts.

Dorset has some of the most striking and beautiful coastal scenery in Britain, perhaps because its 146 km of coastline is eroding rapidly. In the west, Lyme Regis landslip only just extends into the county. The highest headland is Golden Cap (188 m), and here the clay cliffs are constantly crumbling to produce bare habitats. Then come the orange, vertical cliffs of Burton Bradstock, which might not look out of place in the Sahara. This is where the Chesil Beach begins, a 29 km long storm beach, made of pebbles. East of Abbotsbury, the beach becomes a tombolo, separated from the mainland by the Fleet lagoon for 16 km. Where the beach ends, at Portland, the shingle is nearly 15 m above sea-level. The Isle of Portland is a triangular mass of hard limestone resting on a clay base, and projects 8 km from the rest of the coast into the Channel. The north end is 130 m high, and from there the island slopes downward to the Bill, which is 13 m above sea-level. Quarries are everywhere, and waste has been tipped over the vertical cliffs or Weares on both sides of the Isle.

Portland Harbour, in the lee of the Chesil bank, is the second largest artificial harbour in the world, created by the breakwaters which were made in the 1850s. Nearby Weymouth has an east-facing beach of fine sand backed by an artificial esplanade. Areas of salt-marsh at Radipole lake and Lodmoor have been partly filled in, but survive as bird reserves. Eastward, clay and limestone cliffs with rock ledges and reefs continue to Ringstead Bay, where the impressive mass of White Nothe (167 m) begins a stretch of superb chalk and limestone cliffs with shingle beaches, some only accessible by boat. Ancient geological convulsions, followed by selective erosion, have produced striking features such as Bats Head, Durdle Door, Stair Hole, Lulworth Cove and Worbarrow Bay. East of Worbarrow Tout the jagged

summits of Gad Cliff are followed by clay cliffs as far as the anchorage of Chapmans Pool. Then the limestone of St Aldhelm's head projects into the Channel, with vertical cliffs continuing to Durlston Head. Here the coastline bends northwards, via the sands of Swanage, to another stretch of chalk, with spectacular pinnacles between Ballard Head and Handfast point. North of this is Dorset's only accreting coast, between Studland and South Haven, a region of sand dunes and shallow lakes.

South Haven ferry crosses the entrance to Poole Harbour, which is the second largest natural harbour in the world; the rise in sea-level since Roman times has greatly extended its area. To walk round the harbour one would need to traverse about 144 km of reedbed and salt-marsh, but the depth of water only exceeds 2–3 m along the old bed of the Frome. Low islands within the harbour include Brownsea, Furzey, Green and Pergins Islands. To the north of Poole Harbour entrance, old dune systems and salt-marsh have been much affected by building. Some dune vegetation survives and is also present on the steep sandy cliffs between Flag Head and Branksome Chines; at the latter, VC 9 gives way to VC 11.

For the purposes of this Flora, the current county boundary has been modified in several places, shown by dotted lines on the species maps:

• In the west, about 45 km² around Hawkchurch and Wambrook, (SU39, ST20 and 30), strictly VC 9 but now in Devon or Somerset, have not been surveyed. In small areas in SU30 and 40 the modern county boundary has been followed rather than the VC boundary.

• South-east of Yeovil, the River Yeo is taken as the VC 9 boundary, although about 2 km² between the river and the railway line are administratively in Somerset. The modern county boundary is followed north of Sherborne in parts of ST51, 52, 61 and 62.

• In the south-east, the boundary of VC 9 lies in the built-up area between Poole and Bournemouth, which is often hard to locate in the field. About 84 km² of modern Dorset, including Bournemouth, Christchurch and Hengistbury, are in VC 11 and are not treated here, since they have been covered by recent Floras of Hampshire and the Christchurch Area. This is the only boundary change excluding several rare species of 'new' Dorset.

THE CLIMATE OF DORSET
by Dorothy Kerridge

The climate of Dorset is influenced by its location and by local topography. It is transitional between the wetter, maritime west and the drier, more continental east. The southern fringe of the county is affected by its proximity to the sea, and the great variety of topography inland has profound effects on both rainfall and temperature, with a resultant diversity of local climates.

The British Isles lie in a belt of westerly winds which bring a succession of depressions or cyclones from the Atlantic. Many of these depressions move northwards

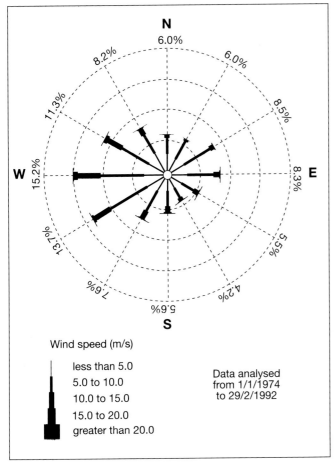

Figure 2. Wind rose for Portland.

Wind speed (m/s)

less than 5.0
5.0 to 10.0
10.0 to 15.0
15.0 to 20.0
greater than 20.0

Data analysed
from 1/1/1974
to 29/2/1992

16 vascular plant species, none of which are maritime, may be restricted to the coastal fringe because they need higher temperatures than the hinterland provides. Most of these are plants which are more at home in southern Europe, such as *Artemisia absinthium, Carduus tenuiflorus, Centaurium tenuiflorum, Medicago polymorpha, Pilosella peleteriana, Polycarpon tetraphyllum* and *Vicia bithynica.*

Sea breezes are particularly effective in reducing extremes of temperature in summer, while in winter frosts are infrequent and snow rarely lies for long. Onshore winds bring both moisture and salt, and the latter can scorch the leaves of vegetation. Plants of cliffs and undercliffs have to be able to tolerate salt, and trees and shrubs often lean away from prevailing winds. At Kimmeridge Bay, rock ledges exposed at low tide absorb heat and so warm the incoming sea-water that unusual marine algae grow there.

The rainfall map for 1995 (Figure 3) gives a good impression of the distribution of rain in the county. The areas of highest rainfall occur where air is forced to rise over the Chalk ridges. The grain of the relief also influences air movement and rainfall, as for example in the lines of showers which often follow the Purbeck ridge, dropping their rain on the north side but leaving dry the vale in which Swanage lies. The mean annual rainfall ranges from 636 mm at Portland to 1,010 mm at Winterbourne Abbas (1961–1990 figures): the average for the county is 883 mm for the same period, or 890 mm for the period 1856–1990. The wettest seasons are autumn and winter, as illustrated by monthly means in Swanage for the period 1972–1996 (Table 1).

High ground creates rain shadow areas on its lee side, usually to the east, even in a small region such as Portland. Large differences in rainfall can occur within short distances, as between a hilltop and a sheltered valley, or between windward and leeward slopes. Deep combes in the chalk can produce marked contrasts within a few kilometres, and cold air sinking down a slope often produces a frost-hollow at its foot. Summits are cool and windy, while valleys can be suntraps by day. East-facing sites are sheltered from the prevailing westerlies, but catch the full force of cold easterlies in winter. Cuttings for new roads have different vegetation on north-facing slopes from that on south-facing slopes. A small group of vascular plants may have their Dorset distribution determined by a requirement for high rainfall, among them *Carex strigosa, Chrysosplenium alternifolium, Colchicum autumnale, Gagea lutea, Paris quadrifolia* and *Stellaria neglecta.*

Soils and vegetation also affect local climates. Sandy soils warm up faster than clay soils in the spring, but are

along the west coast of Britain, and the weather fronts may have lost much of their activity by the time they reach Dorset. When centres of low pressure move up the Channel, the amount of rain received here can be considerable. Between depressions, ridges of high pressure bring quieter, brighter spells. When anticyclones become established they may produce heat waves in summer, and droughts at any season. Air movement is slow in anticyclones, and it is then that extremes of temperature occur. Sometimes anticyclones over Europe extend their influence over Britain, with prevailing easterly winds which can be cold and dry in winter. After a long spell of such cold weather the land surface may be so chilled that rain from incoming westerlies may fall as snow. Sometimes southerly winds are drawn in on the western flank of a European anticyclone, and arrive as a warm front with heavy thundery rain in summer. A wind-rose for Portland Bill (Figure 2) shows that the prevalent winds are south-west or west. At this spot the average number of days per year with winds of gale force is 20. Along the coastal belt the effects of winds on trees are marked (Plate 20).

The coastal fringe is affected by the temperature of the sea for some 3–4 km inland, varying from 5–7°C in winter to 17–18°C in summer. Land in the fringe is warmer in winter and cooler in summer than it is further inland. About

Table 1. Mean monthly rainfall in Swanage, 1972–1996.

Jan	89.3 mm	Jul	42.0 mm
Feb	68.8	Aug	48.8
Mar	70.1	Sep	70.3
Apr	43.7	Oct	90.3
May	47.3	Nov	90.3
Jun	46.8	Dec	98.5

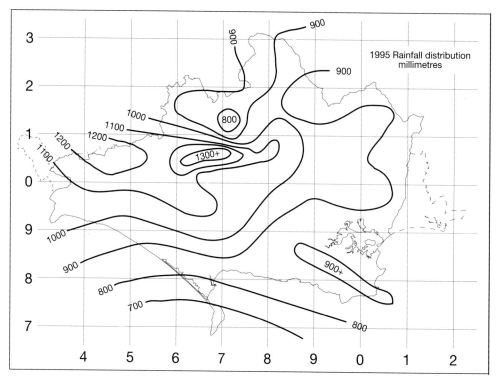

Figure 3. Dorset rainfall in 1995.

colder at night and during the winter. In the latter season radiation frosts are more frequent over a grass cover than when the soil is bare; a closed tree canopy protects against frost. Urban areas are warmer than rural ones, but Dorset towns are mostly too small for this to be significant. The growing season is believed to correspond with temperatures above 6°C, which occur between March and October. For the county as a whole the mean winter temperature is about 5°C, but it may be higher along the coastal fringe where plant growth may continue throughout the winter in favourable years. The mean summer temperature is 17–19°C. With the wide variations of climate we have, it is inadvisable to speculate on possible future effects of global warming. However, as will be seen, historical records show that a number of Dorset plants are retreating northwards, such as *Alchemilla filicaulis* and *Botrychium lunaria*, and some have become extinct this century, such as *Empetrum nigrum* and *Parnassia palustris*. Alien species currently invading the county, such as *Conyza sumatrensis* and *Echinochloa crusgalli*, are all from warmer climes.

DORSET GEOLOGY AND BUILDING STONES

There are many excellent texts on Dorset geology (Davies, 1935; Arkell, 1947; Wilson *et al.*, 1958; Perkins, 1977; Melville and Freshey, 1982), which it would be tedious to rephrase here. No less than seven major geological formations are named after Dorset sites where coastal erosion has exposed them. All Dorset rocks are sedimentary in origin. With a few gaps, they cover 200 million years of deposition, mostly in the Jurassic and Cretaceous periods when the region was submerged. The upper part of the Chalk is absent, but there are substantial Eocene deposits in the east. As far as botanists are concerned, the main kinds of rocks are calcareous (chalk and limestone), clays and sands. The geological map (Figure 1) shows that, apart from the Chalk, the distribution of these rocks is fragmented and complex, particularly so in west Dorset, the region north of Weymouth, and in Purbeck. Relatively few of the tetrads in the current survey cover a single geological formation, and most of these are on the Chalk. Needless to say, homogeneous geology is associated with monotonous botany.

Calcareous rocks; Limestones and Chalk

Areas of impure, iron-rich limestones in the north-west (parts of ST50, 61, 70, 71 and 72) include Inferior Oolite, Cornbrash, Forest Marble and Corallian, and were once much quarried. Ridges of the same rocks with Portland and Purbeck beds occur in the south-west (parts of SY58, 67 and 68). They are faulted in places, and rarely exposed, as at Coryates.

The Isle of Portland (SY66, 67 and 77) consists mostly of limestones, among them the hardest in the county: much is exposed in vertical sea cliffs and opencast quarries. South Purbeck (SY97 and SZ07) has Portland beds to the west of St Aldhelms Head and Purbeck beds to the east. Both are exposed in vertical sea cliffs, and inland are many underground mines with their spoilheaps.

The Upper, Middle and Lower Chalk, all very pure calcium carbonate, cover large parts of central Dorset, and also form the ridge between White Nothe and Ballard Head, with exposed cliffs at each end. Small chalkpits have been quarried here and there inland, but these revegetate rapidly.

The commonest tree of calcareous areas is the ash, both in old woods and as a colonist on steep slopes, as at Hambledon and Okeford Hills. Where woodland has been cleared, limestone grasslands develop with *Bromopsis erecta*, *Brachypodium pinnatum* or *Festuca ovina* as dominants and a characteristic flora of about 50 herbs (Chapter 2). Even in regions which are mostly arable, fragments of old grassland may survive on earthworks, lynchets, old quarries or verges. As far as Dorset is concerned, the flora of grassland on chalk differs little from that on limestone. However *Adiantum capillus-veneris*, *Cerastium pumilum*, *Hieracium leyanum*, *Ophrys sphegodes* and a few hepatics are only found on limestone, while *Carex humilis*, *Cephalanthera damasonium*, *Cirsium tuberosum*, *Euphrasia pseudokerneri*, *Herminium monorchis*, *Juniperus communis*, *Neotinea ustulata*, *Pulsatilla vulgaris*, *Tephroseris integrifolia* and *Verbascum nigrum* are only found on chalk.

Clays

Many different types of clay are found here; most are blue when freshly exposed, due to ferrous iron, and may contain nodules of iron pyrites. They soon weather to some shade of brown, due to ferric iron, while the pyrite forms sulphates. All are lowland, give rise to poorly drained gley soils, and have been much used locally for making bricks and tiles. Old brickpits flood and form ponds if not filled in. At Chickerell pits (SY67P) crystals of Selenite, or calcium sulphate, occur.

The calcareous Blue Lias covers parts of SY39 and 49 in the west. The calcareous Oxford Clay and Fullers Earth is widespread in scattered patches in the north and west, notably in the Blackmore and Wey vales (parts of SY48, 58, 59, 68, plus ST40, 50, 51, 60, 70, 71 and 72). Calcareous Kimmeridge Clay occurs close to Oxford Clay in many places, but also near the coasts of Portland, near Ringstead and Kimmeridge (SY78, 87 and 97) and extensively in the north (ST7l, 72, 73, 81 and 82). Gault clay forms a narrow belt along the northern edge of the chalk and more widely around Duncliffe Wood (ST70, 81 and 82). Acid London Clay occurs in an irregularly C-shaped arc between the southern edge of the Chalk and the heaths of the Poole basin. In north Purbeck it has been extensively mined for Ball Clay and China Clay; there is a huge opencast pit at Povington. When abandoned, the pits either flood, as at Blue Pool, or become bogs or sallow carr. The kaolinitic spoilheaps are infertile and slow to revegetate.

Clay areas were once dominated by oak forest. After clearance, grasslands or heaths form depending on the pH. Apart from the London Clay with a heathy vegetation, all the other clays develop a neutral grassland flora with grasses such as *Deschampsia caespitosa* and *Hordeum secalinum*, and herbs such as *Oenanthe pimpinelloides* and *Trifolium fragiferum*. In wetter places, rushes become abundant and *Carex otrubae* frequent.

Sands

Calcareous Bridport and Thorncombe sands cover parts of SY49 plus ST30, 51, 61 and 62 in the north and west. Wealden and Lower Greensand formations occur in narrow bands between the chalk and clays in the south (parts of SY78, 87, 88, 98, plus SZ07) and again in the north (part of ST81). The Lower Greensand is acidic and favoured by discriminating gardeners. Large areas of the Poole basin are covered by Eocene sands, mostly Bracklesham and Bagshot beds (SY78, 88, 89, 98, 99, plus SZ08, 09, plus SU00, 10 and 11). There are outliers in west Dorset, such as Black Down and Hardown Hill. Sand and gravel are extracted from huge opencast quarries near Binnegar and Warmwell, which develop an interesting heath or bog flora when worked out, if not used as tips.

Sandy soils will support oak forest, but pine, birch or rhododendron are more often seen on them today. Areas spared by farmers, builders and foresters are dominated by heathers (*Calluna vulgaris* and *Erica* spp.), gorses (*Ulex gallii* and/or *U.minor*) or Purple Moorgrass (*Molinia caerulea*), with many rarer species.

The Fossil Flora of Dorset

Most of the rich fossil biota of Dorset are animals. However, silicified trunks of trees (*Protocupressinoxylon purbeckensis*) are found in the Purbeck dirt beds where they grew about 150 million years ago: their trunks are up to 7 m long and 1.4 m in diameter (Francis, 1983). They are exposed as a fossil forest in cliffs east of Lulworth Cove, and have also been seen in quarries on Portland, south of Chalbury and near Poxwell. Freshwater deposits of Purbeckian age have chert nodules containing fossil *Chara* sp., while Roach beds of east Portland have fossils of a calcareous marine alga, *Solenopora*. About 10 million years later, in Cretaceous deposits of the Wealden formation at Swanage Bay, fossils of clubmosses, horsetails and cycads have been found. Much later, around 50 million years ago, the Bagshot beds near Studland provide fossil evidence of a tropical flora including trees such as *Aralia* and *Liquidambar*, fan-palms and ferns (Arkell, 1947). A rainforest flora, with trees of *Cinnamomum*, *Lindera* and *Neolitsea*, was present at Bournemouth in Eocene times (Bandulska, 1923, 1926, 1928).

Building stones of Dorset

Dorset, along with Somerset and Gloucestershire, has some of the best building stones in Britain (Clifton-Taylor, 1972). These are briefly discussed in view of their importance as a habitat for cryptogams, especially lichens.

Limestones Small churches, big houses, cottages, stone walls and tombstones are often built of stone dug locally. Examples are: Liassic Limestone near Lyme Regis and Bridport; Forest Marble at Long Burton; Limestone from the Oxford Clay at Melbury; Coral Rag at Marnhull and Sturminster Newton; Purbeck Oolite at Portesham and Sherborne. These may be grey, brown or golden in colour. Those with a high iron content are preferred by the lichen *Rinodina teichophila*.

Regionally or nationally important limestones include the golden Liassic Limestone from Ham Hill, Somerset, much used in north Dorset, e.g. at Sherborne Abbey; Purbeck roofing slates, made and used near Corfe since medieval times; Purbeck Marble, mostly used inside buildings since the Romans discovered it; and Portland stone. This last is the most used of all Dorset stones, e.g. for Rufus Castle, said to be built in 1080AD and still standing though roofless. Apart from many old houses on Portland and elsewhere in the county, this stone has been exported all over Britain. Arkell lists the uses of the many strata quarried on Portland (Arkell, 1947). All limestone exposures can be recognized from their lichen flora, and can be dated from the size of individual thalli. Sunny rocks have abundant white *Aspilicia calcarea*, black *Verrucaria nigra* and orange-yellow *Caloplaca* spp. Sheltered, north-facing sites have a different flora with grey *Dirina massiliensis* often dominant. Chalk is too soft to be used in many buildings, but cobb walls, made from chalk mud and protected by roofing tiles, are not infrequent. Their lichen flora is that of limestone.

Acid stone Most Greensand is soft, but hard grey-green blocks of Upper Greensand are used in churches, houses and walls around Shaftesbury and much of north Dorset, and also at Wimborne Minster. It is mildly acid and supports lichens such as *Caloplaca crenularia*, *Ochrolechia parella*, *Parmelia* spp. and *Tephromela atra* which are rare or absent on limestone. Flint and Chert are grey to black amorphous rocks, weathering white, used to face walls. Their smooth surfaces take a long time to become vegetated, but they can develop a rich acidiphilic lichen flora near the coast. Eocene Sandstone, dark reddish-brown and coarse grained, is used mixed with other local stones in many east Dorset churches. It has a poor lichen flora except when exposed for centuries, as on the 500 ton Agglestone near Studland. Sarsen stones, hard, close-grained mid-brown Eocene sandstones, occur more or less buried in the valley of the north Winterborne, near Corscombe and elsewhere. They were used in early megaliths and stone circles, and as foundation stones for medieval buildings in central and east Dorset. Few have been exposed for long enough to develop the interesting cryptogam flora found on such stones in Wiltshire. The puddingstones in the Valley of Stones near Portesham retain traces of this flora, including the moss *Hedwigia ciliata* and uncommon lichens.

Old bricks and tiles from the many Dorset brickpits, e.g. Blackmore Vale, Broadmayne, Chickerell and Motcombe, develop a typical acidiphilic lichen community after a century or so, if not shaded out by ivy. They can become hypereutrophic near farmyards, or mildly eutrophic under the drip of trees such as sycamores. Imported acid stones include blue-purple roofing slates from North Wales, and granites, both grey and pink, from Cornwall or North Britain. The former are extremely acid but support lichens such as *Parmelia verruculifera* and *Rhizocarpon geographicum*. The latter, used as memorials, have species such as *Polysporina simplex* which is rarely seen on other substrates.

SOILS OF DORSET
by Graham Colborne

Published maps and treatises on the soils of Dorset are few. Indeed, since KL Robinson's work (1948), only the Soil Survey of England and Wales has produced any study of significance, and then only at reconnaissance scale (Findlay and Colborne et al., 1984). Although the Soil Survey had intended to carry out a programme of detailed mapping in the county, only a few square km south of Yeovil were completed before the programme was ended (Colborne and Staines, 1987).

Soils and Geology
Over Southern England the absence of ice cover has ensured a fairly close relationship between soils and their underlying geology. For example, most soils overlying Kimmeridge Clay are heavy and poorly permeable, while most soils over limestones are shallow and brashy. There are exceptions: Gault Clay does not form a soil parent material here; Chalk is often covered by a shallow Clay-with-Flints soil, and in the vales some clays are masked by a loamy Head. Almost a quarter of Dorset soils form in a Head or Drift some 50 cm or more thick. In addition, the lithology of underlying strata is not always uniform: the Bridport Sands are not always sandy, nor are river Terrace Gravels always gravelly.

Soil Classification
The current soil classification in use in England and Wales (Avery, 1980) has made Robinson's terminology obsolete. The names of the Major Soil Groups (the highest level of the classification), and the approximate percentages of the county covered by them, are as follows, neglecting the 4% of urban land:

Terrestrial Raw and Raw Gley Soils <1
Podzolic Soils 10;
Lithomorphic Soils 21;
Surface-water Gley Soils 19;
Pedosols 4;
Ground water Gley and Peat Soils 6;
Brown Soils 36

Major Groups are subdivided successively into Groups, Subgroups and Soil Series. A 'Soil Association' is a convenient geographical grouping of soils similar to Robinson's "Local Soil Groups", and Soil Associations were mapped by the Soil Survey. Each Soil Association is likely to contain soils from two or more Major Groups. For example, the Sherborne Association is dominated by Sherborne Series (a Lithomorphic soil), but contains appreciable areas of Moreton Series (a Brown soil) and Evesham and Haselor Series (Pedosols). The position of a soil within the classification system, and a soil unit shown on a map, are not necessarily similar due to scale. Fifty Soil Associations occur in Dorset, and within these nearly 300 different Soil Series have been identified. Only on small-scale maps at field level can an individual Soil Series be shown as a pure mapping unit. At larger scales, units on a soil map include several distinct Soil Series. At present, botanists tend to refer to 'Chalk soils' or 'acid soils' without appreciating the wide range of soil types involved. As detailed soil maps showing the Soil Series in individual fields are rarely available, this is understandable, but when such maps are made there should be a great improvement in our appreciation of the interactions between soils and native plants. A broad outline of the characteristics of the Major Soil Groups may suggest some correlations with plant distribution.

Raw Soils

The chief feature of these soils is that they consist of little-altered mineral material. In Dorset they are either lithoskeletal gravels, as on the Chesil Beach, or soft estuarine muds as at Radipole Lake, Lodmoor or Poole Harbour. Both are saline, but some of the salt-marshes may contain pyrite (iron disulphide) which oxidizes to give a sulphate-rich horizon. Plant cover is halophytic, including many Chenopods and *Spartina* grass.

Lithomorphic soils

These are closely allied to their geological substrates, with shallow soil horizons to a depth of 30 cm. They include sand dunes at Studland, the limestones of Portland and Purbeck, and some parts of the Chalk. The Studland dunes are Sand Pararendzinas, including calcareous shell fragments. In fact only the dune ridge nearest to the sea is calcareous, while older dune ridges are acidic through leaching. There is a corresponding sharp change in vegetation from *Ammophila* to ericaceous plants in a traverse away from the sea.

Hard limestones generally give rise to well-drained clayey or loamy soils, whose fertility and calcareous content varies. Most are classed as Brown Rendzinas, particularly where ploughed, but some profiles under old grassland are neutral to slightly acid in their uppermost few centimetres. Near seepage springs, acidiphilic species such as *Eriophorum* may appear. Past and present management, slope and aspect provide a wide variety of plant habitats.

Lithomorphic soils overlying the Chalk include Brown, Grey and Humic Rendzinas, of which Brown types occupy about 12% of the county. Grey Rendzinas are similar, but are thinner and much chalkier, due to prolonged arable cultivation. Humic Rendzinas, which occupy about 2% of Dorset, occur where there are steep, unploughed slopes, and have both a high organic content and a rich flora. It is thought that at the time of woodland clearance in the Bronze Age, soils overlying the Chalk were 20–30 cm deeper than they are now (Avery, 1980; Bell, 1981; Staines, 1991 and Staines, 1993). Thousands of years of arable cultivation has depleted this soil resource. Archaeological evidence, such as excavation of barrows and ancient tree wind-throw holes, suggests that the original soils were neutral to slightly acid Brown Soils with a thin argillic subsoil enriched in clay. The eroded sediment has been washed downstream to form alluvium in river valleys or estuaries, or is present in the bottoms of dry valleys as colluvium.

Staines (1997) considers that prior to the impact of man, 40–75% of the Chalk near Maiden Castle and Dorchester was covered by Brown Soils, compared with the current 5%.

Rendzinas over Chalk are usually rich in calcium but deficient in magnesium, potassium and phosphorus. This may affect plant growth; for example, some lime-tolerant ericaceous plants will not grow where magnesium deficiency occurs (Jones and Roy, 1996).

Pedosols

These are clays with only slight seasonal waterlogging, and are all calcareous in Dorset, though the topsoil may be leached. They occur over Jurassic clays of the Forest Marble and Fullers Earth near Bridport, Weymouth and Sherborne, or on Drift originating from these strata. Tor Grass (*Brachypodium pinnatum*) is often frequent on them.

Brown Soils

These are defined by the presence of a weathered subsoil. They can be calcareous, neutral or slightly acid; clayey, loamy or sandy; be well-drained or not; rarely be rich in iron (as over the outcrop of iron ore at Abbotsbury). They occur on a wide variety of parent materials, for example Chalk, Greensand, some Jurassic beds such as Bridport and Yeovil Sands, Clay-with-Flints, Tertiary Beds, Plateau Drift, River terraces and alluvium. Given this variability, plant cover is equally variable, and most Brown Soils have been exploited for agriculture. Steep slopes on both Greensand and Bridport Sands support acid grassland locally.

Podzolic Soils

Podzolic soils are acid, infertile soils occurring mostly on coarse-textured parent materials. A dark, organic topsoil overlies a grey or white horizon from which almost all nutrients have been leached by percolating rainwater. Since their development is associated with the establishment of heath, which is thought to be brought about by human land-clearance, it is a matter of conjecture whether current podzols have always been so.

Typical Podzols occur in elevated sites on the oldest river terrace gravels; they are dry and well-drained. Such sites are either heaths, as at Canford Heath, or are afforested, as at Puddletown. More common are Gley-podzols in which the lower part of the sandy, permeable soil profile is regularly or irregularly waterlogged. They cover large areas of both wet and dry heath in the Poole basin, dominated by *Molinia* and ericaceous shrubs.

Surface-water Gley Soils

These are seasonally waterlogged, slowly permeable soils, usually grey or mottled in colour. The subsoil is coarsely structured and tightly packed, discouraging root development. They are extensive over the London, Kimmeridge, Oxford and Fullers Earth Clays, on the Lias in the Marshwood Vale, and on the Wealden Clay in Purbeck. Since they are not readily worked, they are often wooded; where cleared, a rich flora of damp grassland can develop, as at Lower Kingcombe and Corfe Common.

Ground-water Gley Soils and Peat Soils

These are also seasonally waterlogged due to a fluctuating level of groundwater, and are moderately permeable, allowing good root development for those species able to tolerate 'wet feet'. Alluvial gleys form water-meadows along all the major rivers and occupy about 5% of the county. They are variable in texture and in their contents of lime and organic matter, often over short distances. Where undisturbed, they have a rich flora and may develop patches of fen, reedswamp or sallow carr. Humic sandy gley soils and peat soils, which occupy only 1% of the county, form wet heaths and bogs around Poole Harbour and near the Moors river north of Ferndown. Peat in the bogs usually has a pH between 3.3 and 4.7, but when fed by less acid streams, as on Studland and Wytch Heaths, the topsoil may rise above pH5. Both wet heaths and bogs have a specialized flora which includes rarities such as *Erica ciliaris* and *Gentiana pneumonanthe*. Haskins (1980) provides a detailed vegetational history of peat bogs in Dorset.

Other soils

Areas which have been quarried for exploitation of their minerals include much of Portland, sands and gravels at Knighton, Crossways and Moreton, and Ball Clay pits, as between Corfe and Povington. More or less restored soils here cannot easily be classified, and are often very inhomogeneous. They do, however, provide a haven for many native and alien plants; for example, *Mentha pulegium* flourishes on Ball Clay spoil.

DORSET RIVERS AND THEIR PLANTS

In the late Pleistocene, the Dorset-Hampshire basin was drained by the Solent river, of which the Frome then formed the headwaters; this river spread flinty gravel and alluvium over much of the Poole basin, and entered the sea somewhere east of the Isle of Wight (Davies, 1935). The rivers Piddle, Stour and Avon were some of its tributaries. Present-day river topography in Dorset has been described (Mansel-Pleydell, 1895). The main rivers, which drain most of the north, east and centre, are the Stour (with tributaries Lidden, Cale, Divelish, Iwerne, Tarrant, N. Winterborne and Allen), the Frome (with tributaries Hooke, Sydling, Cerne and S. Winterbourne) and the Piddle. The Moors river, with its tributary the Crane, drains a part of east Dorset, the Axe and Yeo drain parts of the north-west, while in the south-west streams such as the Lim, Char, Brit, Bride and Wey have short courses to the coast.

Bulk flow-rates of some rivers are given as means in Table 2. Not only do these vary from year to year, they also have marked yearly cycles with maxima around February and minima in August (Casey & Norton, 1973; National Rivers Authority, 1995). Thus the Stour, which is sluggish in summer, has a winter flow-rate more than 10 times its summer one, so it is scoured in winter and has banks 2–3 m above its summer level. The Winterbornes rarely have any visible flow in summer, though this may continue 3–4 m below ground. Extraction of water for human use has

Table 2. pH, mean annual flow rate in cubic metre/s, and concentration of nutrients in mg/l, in Dorset streams.

Stream	pH	Flow rate	Ca	Mg	K	Na	P	Nitrate -N
Stour	8.0	5.0	94	4.4	5.5	17	0.21	7.1
Frome	8.1	5.2	84	2.5	1.9	12	0.13	2.8
Piddle	7.3		108	2.2	1.9	11	0.034	6.7
Cerne	8.3	0.39	87	1.9	0.8	7	0.042	2.1
Hooke	8.1	0.19	94	2.6	1.2	9	0.088	1.6
Moors	7.4	1.7					0.038	5.0
Crane	7.3		51	3.1	3.5	11	0.023	4.3
Furzebrook	6.9		22				0.016	0.41
Hartland	6.2		19	2.4	1.6	16	0.0006	<0.1
Luckford	6.4				16		<0.017	

reduced flow rates recently, notably in the Piddle and the Tarrant, but other human inflence is small. The Stour has a few weirs, but there are no locks, and very few boats above marine influence. It has suffered from Flood prevention schemes, which have cut down many trees and replaced the older ditches by underground pipes. Weeds, mostly *Ranunculus penicillatus*, are cut annually in parts of the Frome for anglers, and some channels are dredged. In earlier centuries the floodplains of the Frome, Piddle, Sydling and N.Winterborne had water-meadows with artificial ditches and leats for irrigation in spring, but maintenance of these has largely ceased.

The Chalk is a vast reservoir of water, and many springs occur at its base, as at Upwey and Millum Head near Bere; they are calcareous and alkaline. Some streams in the Poole basin, such as those draining Arne and Hartland Moor, differ in being acid and nutrient-poor. They contain much less calcium, nitrate and phosphate than do the larger rivers, but probably have higher concentrations of the plant-toxic elements aluminium, iron and manganese. From Table 2 we see that our main rivers have calcareous water and differ little as regards major nutrients. Annual cycles of nitrate (National Rivers Authority, 1995) and phosphate (Casey & Clarke, 1986), but not of silicate (Marker, Casey & Rother, 1984), occur in the Frome and presumably elsewhere, and are ascribed to the growth cycle of diatoms.

Holmes (Holmes, 1983) has classified the vegetation of British rivers as either Class A (Nutrient-rich lowland), B (Intermediate) or C/D (Nutrient-poor). His scheme only covers a selection of aquatic plants, of which roughly a third occur in all our rivers (Table 3). All large Dorset rivers and their tributaries are Class A, except for the Crane and Moors rivers which are Class B, and minor bog streams which are Class C.

Bolboschoenus maritimus and the Green Alga *Enteromorpha*, normally coastal, occur in many places along the Stour. This might be correlated with the high sodium content of Stour water (Table 2). One species, *Callitriche truncata*, is restricted to the River Axe here. Our limited understanding of the nutrient needs of aquatic plants is inadequate to explain these facts, nor can we account for the recent rapid spread of the aliens *Elodea nuttallii* and *Lemna minuta*, nor for the near extinction of natives such as *Hottonia palustris* and *Hydrocharis morsus-ranae*.

Future research may subdivide Holmes' Class A species into oligotrophs, which prefer low concentrations of nitrate and phosphate, and eutrophs, which can tolerate high concentrations of these nutrients. Calcareous springs are oligotrophic, and their vegetation deserves more study. Characteristic species include *Catabrosa aquatica, Hippuris vulgaris* and *Ranunculus penicillatus* subsp. *pseudofluitans*, the bryophytes *Aneura pinguis* and *Cratoneuron filicinum*, the lichens *Verrucaria elaeomelaena* and *V. rheitrophila* (Gilbert, 1996) and algae of the genus *Chara*. Many Dorset springs are used as watercress beds or trout farms which cause eutrophication downstream, but far more extensive eutrophication is caused by human and cattle sewage and washout from fertilizers. For example, in 1973 the concentrations of nitrate and phosphate in the Frome increased by factors of 1.5 and 2 respectively after passing Dorchester sewage works; farms currently contribute 80% of the input of nitrate into the Frome (Casey, Clarke & Smith, 1993). Field observations support the chemical data, but biological changes, with unpleasant smells and replacement of aquatic plants by sewage fungi, are only seen in ditches and streams close to farms. Eutrophic pollution favours *Glyceria maxima, Lemna* spp., *Myriophyllum spicatum* and the rare liverwort *Riccia fluitans*; it also causes *Apium nodiflorum* to become gross and succulent. Where aquatic weedkillers have been used by anglers, as at Tolpuddle, the moss *Fontinalis aquatica* may become dominant.

The vegetation of seasonally inundated areas needs more study; there are suitable sites along the upper Stour and North Winterborne. Flood-tolerant species include *Agrostis stolonifera, Mentha aquatica* and hybrids, *Phalaris arundinacea, Ranunculus aquatilis, Rorippa sylvestris* and *Tanacetum vulgare*. Several scarce bryophytes, e.g. *Cinclidotus fontinalis, Leskea polycarpa, Myrinia pulvinata, Orthotrichum rivulare* and *Tortula latifolia*, grow on seasonally flooded bark by the Stour. The flinty bed of the North Winterborne, scoured by winter torrents, has a rich flora of annual weeds in summer. At least 25 species occur, including *Chenopodium polyspermum, Lamium hybridum, Mercurialis annua* and *Urtica urens*, suggesting this habitat as a refuge for annuals before woodland clearance in Neolithic times.

EFFECTS OF MARINE EROSION AND DEPOSITION ON THE FLORA

About 20,000 years ago the sea was about 50 m lower relative to the land than at present, because so much water was locked up as ice (Wilson *et al.*, 1958). The Dorset coast would have been much further south than at present; about 27 km south of Lyme Regis and St Catherines Point, IOW, 8 km south of Portland Bill and 20 km south of St Aldhelm's Head. The area of the county would have been about 55% greater, with a southward extension of about 1,400 square kilometres. Subsequently the coastline receded northwards as a result of both rising sea-levels and erosion. The mean rate of recession has been between 0.4 and 1.35 m/yr, but most of it occurred before the Flandrian transgression in about 5000BC, when Britain was cut off from Europe.

At least nine lines of evidence confirm that both processes are continuing:
1. Submerged forest remains with peat and tree stumps near the mouths of the Char, Wey and Bourne (Arkell, 1947).
2. The existence of drowned valleys at Poole Harbour, Radipole Lake and Lodmoor, and the ponded-back alluvial floodplain of the Frome (Arkell, 1947). Freshwater peat deposits at a depth of 13 m near Hamworthy date from 7340BC (Perkins, 1977).
3. The Chesil Bank is slowly moving landward; before 3000BC it may have stretched from an extension of Portland Bill to a lost headland south of Beer in Devon (Arkell, 1947; Anon, 1994).
4. The seaward sides of the Iron Age camps at White Nothe and Flowers Barrow, built around 600BC, have been lost by erosion (Anon, 1994).
5. Excavations at Brownsea show that the sea-level has risen between 2.7 and 3.1 m since Roman times, when Poole Harbour may have been a quarter of its present size (Jarvis, 1993).
6. In the 16th century AD, the Fleet extended to the Mere at Portland (now a heliport), but has been overwhelmed by the Chesil bank (Arkell, 1947).

Table 3. Aquatic plants in Dorset faithful to Holmes' classification of vegetation in British rivers.

Class A	Class B
Berula erecta r	*Potamogeton* × *fluitans* o
Butomus umbellatus o	*Ranunculus fluitans* o
Lemna trisulca o	*Ranunculus omiophyllus* o
Myriophyllum spicatum f	
Nuphar lutea f	**Class C**
Oenanthe fluviatilis f	*Apium inundatum* r
Persicaria amphibia o	*Callitriche brutia* r
Potamogeton acutifolius r, FP	*Carex rostrata* o
Potamogeton friesii [F]	*Eleogiton fluitans* o
Potamogeton nodosus o,S	*Juncus bulbosus* f
Potamogeton perfoliatus o	*Littorella uniflora* r
P. × *salicifolius* r, F[S]	*Myriophyllum alterniflorum* r
P. × *schreberi* r,S	*Potamogeton obtusifolius* r
P. × *sudermanicus* r, FP	*Potamogeton polygonifolius* f
Ranunculus penicillatus a	*Sparganium natans* r
Sium latifolium [S]	*Utricularia intermedia* r
Spirodela polyrhiza o	*Utricularia minor* r
Veronica anagallis-aquatica o	**Bryophytes**
Veronica catenata o	*Drepanocladus fluitans* r
Veronica scutellata r	*Sphagnum auriculatum* + spp. o

F = Frome only; FP = Frome and Piddle only; S = Stour only; [ ] = extinct
a = abundant; f = frequent; o = occasional; r = rare

Table 4. Uncommon plants in Dorset Coastal Habitats.

Limy Cliffs (rock ledges)	Undercliffs/clifftops	Beaches; Sand/shingle	Salt-marsh/Mud
Adiantum capillus-veneris	Bromus hordeaceus ferronii	Asparagus prostratus	Alopecurus bulbosus
Asplenium marinum	Bupleurum tenuifolium	Crambe maritima	Althaea officinalis
Brassica oleracea	Centaurium tenuiflorum	Euphorbia paralias	Carex divisa
Hieracium leyanum	Chenopodium vulvaria	Glaucium flavum	Carex punctata
Inula crithmoides	Erodium maritimum	Lathyrus japonicus	Eleocharis parvula
Limonium dodartiforme	Euphorbia portlandica	Lavatera arborea	Hordeum marinum
Limonium 'obesifolium'	Frankenia laevis *	Mibora minima *	Parapholis strigosa
Limonium recurvum	Gastridium ventricosum	Ophioglossum azoricum	Polypogon monspeliensis
Matthiola incana *	Geranium purpureum	Orobanche purpurea	Puccinellia rupestris
	Marrubium vulgare	Polycarpon tetraphyllum	Suaeda vera
	Orobanche hederae	Rumex rupestris	
	Parapholis incurva	Trifolium suffocatum	
	Poa bulbosa	Vulpia ciliata ambigua	
	Puccinellia fasciculata	Vulpia fasciculata	
	Pyrola rotundifolia		
	Rubia peregrina		
	Sedum forsterianum		
	Silene nutans		
	Silybum marianum		
	Trifolium squamosum		
	Valerianella eriocarpa		
	Vicia bithynica		
	Vicia lutea		

N.B. * = recent record, probably introduced

---

7. Landslips are frequent, especially on cliffs between Lyme and Eype, on Portland, and at Osmington Mills, White Nothe, Houns Tout and St Aldhelm's Head (Arkell, 1947; Anon, 1994). In 1792, a block 2 km x 550 m fell vertically at north Portland (Hutchins, 1803). The huge landslip west of Lyme, which took place in 1839, is mostly in Devon (MacFadyen, 1970).
8. Between 1860 and 1890 the A353 road behind the beach at Lodmoor had to be set back 18 m, but was overwhelmed by pebbles after storms in 1899 and 1938 (Arkell, 1947).
9. Ordnance Survey maps show that the land has eroded at about 0.75 m/yr during the last 200 years at Hengistbury Head, VC11 (Cunliffe, 1978).

At the same time, at least one east-facing stretch of coast is accreting. Maps show that Little Sea did not exist before 1611, was an inlet of the sea between 1721 and 1811, after which it became a lake. The dunes north of Studland moved 73 m eastward between 1895 and 1924, an average of 2.5 metres per year (Diver, 1933).

As far as our flora is concerned, these rapid rates of change at the coast preclude agriculture and offer a wide range of habitats for plants to colonize. The habitats, especially undercliffs, act as refuges for thermophilic species driven northwards as the coast retreated. Some of these species are very rare, for example *Centaurium tenuiflorum* in west Dorset and the endemic *Limonium recurvum* and *Hieracium leyanum* on Portland. Table 4 includes the uncommon species of this kind, classified under habitat.

It is possible that some species recorded by R Pulteney and others in the 18th century may have been lost in coastal changes. Examples are *Corynephorus canescens*, *Corrigiola litoralis* and *Otanthus maritimus*, but herbarium evidence is often lacking, and human pressure on beaches is a cause of many such losses.

One can only speculate as to the kinds of habitat lost through marine erosion during the last 12,000 years. West of Portland there were probably extensive tracts of lowland clays, mostly calcareous, alternating with soft limestones. East of Portland there must have been a hard limestone ridge linking the island to St Aldhelm's Head, as well as Chalk hills and Wealden valleys. The coastline was presumably the scene of evolution of the *Limonium binervosum* complex.

CHAPTER 2
VEGETATION

OLD WOODLAND
by Anne Horsfall

Dorset is not well wooded. About 11% of the county is woodland, mostly plantations from the 18th century to the present day, and probably less than 2% of Dorset is old woodland. Old woods are mostly small and isolated, being the relics of clearance and human settlement since 3000–4000BC when most of Britain was forested.

The Dorset Domesday Book indicates that woodland cover had been reduced to 13% by 1086AD (Rackham, 1986). No woodland was recorded from the heathlands, little from the Chalk and none from the area south of Dorchester from Abbotsbury to Lulworth. The Domesday woods were mostly in the central and northern parts of the county, where fragments survive today (Fig. 4; Horsfall, 1996).

Old woodland can be identified from maps and other historical records. The most important single source is Isaac Taylor's 1765 map, before tree-planting became commonplace on Dorset estates (Taylor, 1765). His map shows 68 named woods and about a 130 others which can be located on modern maps. The larger woods are (west to east by Grid square): 39: Sleech. 59: Hooke and South Powerstock. 51: Clifton. 60: Batcombe, Holcombe and South Middlemarsh. 61: Honeycomb. 79: Athelhampton, Greys, Ilsington and Yellowham. 70: Armswell and Delcombe. 71: Brickles and Piddles. 88: Coombe Keynes, East Lulworth and Marley. 89: Bere and Great Coll. 80: Broadley. 82: Duncliffe. 98: Brenscombe and Creech. 99: Charborough and Lytchett. 91: Cranborne Chase. 00: Hinton Martell and Queen's Copse in Holt Forest. 01: Birches Copse, Boulsbury, Burwood, Edmondsham and Harley. Some, such as Beaulieu, have been grubbed up, or split up and partly replanted, but plants indicative of old woodland are still plentiful in all of them. Taylor's smaller woods are no less important. By using later maps up to and including the first Ordnance Survey map of 1811, a list of over 500 old woodland fragments has been compiled (Fig. 6; Spencer, 1988).

In Cranborne Chase, *Daphne mezereum*, *D. laureola*, *Helleborus viridis*, *Lathraea squamaria*, *Neottia nidus-avis*, *Ophrys insectifera* and *Polygonatum multiflorum* were among the rarities recorded by R Pulteney (1796). According to Chafin's account (1818), both species of *Daphne* were dug up and sold for horticulture. The Chase woods also provided such varied products as quickset plants of *Crataegus* for hedging, roots of *Valeriana officinalis* and truffles for the London market.

Woodland Classification
Six of the 25 types of natural woodland described in the National Vegetation Classification (Rodwell, 1991) occur in Dorset. Several types of scrub and carr are also represented, but these are transient and not typical of old woodland. The six types are: Oak-hazel (W10); Oak-birch (W16); Ash (W8); Yew (W13); Alder (W5); and wet Birch (W4).

Oak woodland with a hazel understorey is widespread on heavy, mildly acid soils, and is the commonest type of old woodland here. *Acer campestre* is common and *Sorbus torminalis* rare. A *Quercus-Pteridium-Rubus* community is frequent in drier places. The ground flora is often dominated by *Hyacinthoides* in spring. Scarce plants of oakwoods include *Epipactis purpurata* in the north; *Paris quadrifolia* in woods with substantial rainfall in central and west Dorset; *Pulmonaria longifolia* on acid soils in the southeast; and *Sedum telephium*, always in small quantity. These woods are rich in larger fungi.

Oak-birch woodland occurs on sandy, acid soils in the Poole basin; fragments on the south side of Poole Harbour may be relics of the type of forest cover in the Bronze Age. Trees include both native oaks, the two native birches, *Ilex aquifolium* and *Sorbus aucuparia*, and these support many epiphytic lichens. The ground flora may have *Conopodium majus*, *Melampyrum pratense* and *Solidago virgaurea*.

Typical ashwoods abound on the Chalk, especially on scarp slopes in central and north Dorset, and on the north slopes of the Purbeck hills. *Acer campestre*, *Corylus avellana* and *Crataegus monogyna* are common associates, while *Sorbus aria* and *Tilia cordata* are much more local. Sycamore spreads freely in these woods. The ground flora is dominated by patches of *Allium ursinum* and *Mercurialis perennis*, while rarer plants include *Gagea lutea* and *Lathraea squamaria*. Old ashes and maples are rich in epiphytic bryophytes and lichens.

Yew is a component of many estate woods, notably in north-east Dorset, where it is either bird-sown or a relic of old boundaries. The only pure yew wood in Dorset, on the southern flanks of Hambledon Hill, is of unknown origin. Woodland associates are few, but include *Clematis vitalba;* the ground flora is sparse but the lichen *Verrucaria dolosa* is dominant on exposed flints.

Alder grows in most parts of the county with a high water table, but alder woods are small. Place names such as Alderholt, Aldermore, Allers and Twisting Alders indicate some of the older ones. Woody shrubs such as *Frangula alnus*, *Salix* spp. and *Viburnum opulus* accompany the alders and *Ribes rubrum* is locally subdominant. The ground flora is rich in ferns, of which *Athyrium filix-femina*

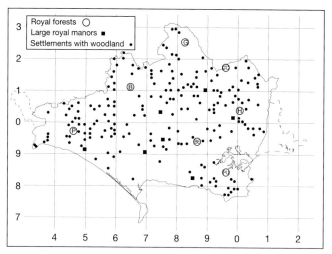

Figure 4. Distribution of woodland in Dorset, 1086.

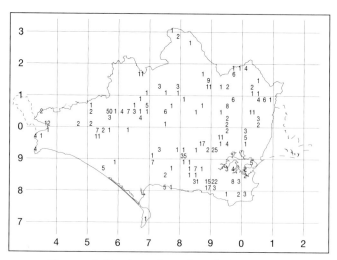

Figure 5. Number of species of old woodland lichens in Dorset tetrads: See Table 6.

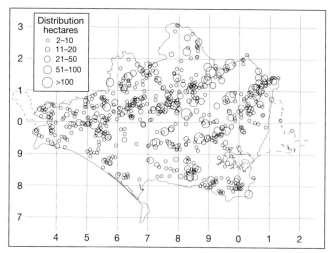

Figure 6. Current distribution of ancient woodland in Dorset.

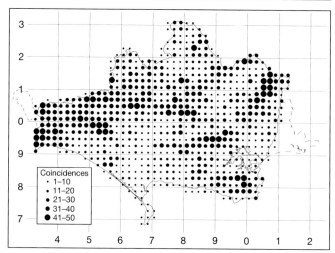

Figure 7. Current distribution of ancient woodland indicator plants: See Table 5.

is the commonest and *Dryopteris carthusiana* typical, also in sedges such as *Carex paniculata*. The edges of small streamlets have a rich flora with species of *Chrysosplenium*, and sometimes the bryophytes *Hookeria lucens* and *Trichocolea tomentella*. Less common plants are *Carex strigosa*, *Equisetum sylvaticum*, *Geum rivale* and *Viola palustris*.

Wet birchwoods are common on acid soils of the Dorset heaths. Some are undoubtedly long-established, but most are not, and changes in the water table can lead to their replacement by *Salix* carr. The dominant tree is *Betula pubescens*, often with an understorey of *Myrica gale*, while the ground flora is mostly *Molinia caerulea* with ferns such as *Dryopteris dilatata* and various rushes and sedges. Sphagnum mosses and their associated hepatics are frequent. *Lobelia urens* is very rare in this habitat.

Almost pure woods of Wych Elm (*Ulmus glabra*) used to occur in small valleys near the south coast, but have been destroyed by Dutch elm disease before they could be studied.

Individual woods often defy classification. Ashwoods merge into oakwoods at the base of chalk hills, and oakwoods merge into alderwoods along streams. Most old woods are mosaics developed after centuries of human exploitation. The diversity of woody species is commonly greater along the edges of old woods than within. The western edge of Clifton Wood in north Dorset is a fine example, though much of the wood has been replanted. As well as ash, hazel and oak, the margin has *Acer campestre*, *Crataegus monogyna*, *Ligustrum vulgare*, *Populus tremula*, *P.* × *canescens*, *Prunus spinosa*, *Rosa* spp., *Salix caprea* and *S. cinerea*, *Sambucus nigra*, *Viburnum lantana* as well as the climbers *Clematis vitalba* and *Tamus communis*. *Malus sylvestris*, *Sorbus torminalis* and *Tilia cordata* occur within, and there is documentary evidence of a pear tree felled in this wood in 1292 (Drew, MS in **DOR**). Trees such as *Sorbus torminalis* and *Tilia cordata*, surviving in hedges along former medieval boundaries, usually indicate former woodland.

Little remains of the woodland which made up Dorset's medieval and Royal forests (Horsfall, 1997). These provided the timbers for castles, churches, manor houses, barns and mills from the 12th century on. Stands of hazel were cut in rotation to yield a sustainable supply of rods, spars and brushwood. Most trees were coppiced, including oak for tanning and charcoal; hazel coppicing continues to this day in dozens of these woods. In Holt Forest, stands of tall oakwood remain among modern plantations, but much of the former forest is now wood pasture with holly undergrowth. Such wood pasture develops when natural regeneration is hindered by browsing animals, mostly deer or cattle, leaving only mature trees and dwarf grazed holly, as in parts of Oakers Wood. There was abundant wood pasture on common land in Dorset, as well as in the 90 medieval deer parks. Most of these are now farmland with relict woodland species on boundary banks. A few, such as Lulworth and Melbury Parks, are still parkland with venerable trees which support an exceptionally rich flora of bryophytes and lichens.

Figure 4 shows the distribution of woodland in Dorset in 1086AD, and Figure 5 the same in 1765. Figure 6 summarises the modern distribution, and Figure 7 is a composite map of the number of selected old woodland indicator plants found in each tetrad during the present survey.

Ferns and fern allies of old woods
See Table 5.

Vascular plants of old woods in Dorset
See Table 5. Tables 5 and 6 are subjectively chosen lists based on field observations. Other species which might have been included are *Ceratocapnos claviculata*, *Hypericum pulchrum*, *Potentilla sterilis*, *Stachys officinalis* and *Vaccinium myrtillus*, which turn up in hedgebanks and non-woodland habitats, as do many species in the list. *Aquilegia* is native in old woods, and a garden escape elsewhere; widely planted species such as *Helleborus foetidus*, *Myosotis sylvatica* and *Prunus avium* have been left out.

Bryophytes of old woods
See Table 6.

Lichens of old woods
See Table 6. A list of over 80 lichen species faithful to old woodland has been compiled by F Rose (1992). Seventy of these occur in Dorset, and a map showing the number of species in each tetrad (Figure 5) shows that the top 14 sites for old woodland species in the county are: Melbury Park (50), Oakers Wood (35), Lulworth Park (31), Morden Park (25), West Creech (22), Tyneham (17), Povington Wood (15), Bloxworth (17), Sadborow (12), Powerstock Common (11), Charborough Park (11), Handcocks Bottom (11), Holt Forest (11) and Sherborne Park (11).

Table 5. Ferns and fern allies and vascular plants of old woods in Dorset.

Ferns and fern allies

Athyrium filix-femina	D. affinis	Equisetum sylvaticum	Polystichum aculeatum
Blechnum spicant	D. carthusiana	Oreopteris limbosperma	P. setiferum
Dryopteris aemula	D. dilatata	Phyllitis scolopendrium	Thelypteris limbosperma

Vascular plants

Adoxa moschatellina	Elymus caninus	L. sylvatica	Primula vulgaris
Allium ursinum	Epipactis helleborine	Lysimachia nemorum	Pulmonaria longifolia
Anemone nemorosa	E. leptochila	Malus sylvestris	Ranunculus auricomus
Aquilegia vulgaris	E. phyllanthes	Melampyrum pratense	Ribes rubrum
Bromopsis ramosa	E. purpurata	Melica uniflora	Rosa arvensis
Campanula trachelium	Euphorbia amygdaloides	Melittis melissophyllum	Ruscus aculeatus
Carex laevigata	Festuca gigantea	Mercurialis perennis	Sanicula europaea
C. pallescens	Gagea lutea	Milium effusum	Scrophularia nodosa
C. pendula	Galium odoratum	Monotropa hypopitys	Sedum telephium
C. remota	Helleborus viridis	Narcissus pseudonarcissus	Sorbus torminalis
C. strigosa	Holcus mollis	Neottia nidus-avis	Tilia cordata
C. sylvatica	Hyacinthoides non-scripta	Ophrys insectifera	Veronica montana
Cephalanthera damasonium	Hypericum androsaemum	Orchis mascula	Vicia sepium
Chrysosplenium alternifolium	H. hirsutum	Oxalis acetosella	Vicia sylvatica
C. oppositifolium	Lamiastrum galeobdolon	Paris quadrifolia	Viola palustris
Colchicum autumnale	Lathraea squamaria	Platanthera chlorantha	Viola riviniana
Conopodium majus	Lathyrus linifolius	Poa nemoralis	
Daphne laureola	Luzula forsteri	Polygonatum multiflorum	
D. mezereum	L. pilosa	Populus tremula	

Table 6. Bryophytes and lichens of old woods in Dorset.

Bryophytes

On soil:

			On bark or dead wood:
Atrichum undulatum	Isopterygium elegans	P. nemorale	Dicranum tauricum
Brachythecium salebrosum	Isothecium myosuroides	P. succulentum	Frullania tamarisci
B. populeum	I. myurum	P. undulatum	Lejeunea ulicina
Chiloscyphus polyanthos	I. striatulum	Polytrichum formosum	Leucodon sciuroides
Cirriphyllum crassinervium	Leucobryum glaucum	P. longisetum	Metzgeria furcata
C. piliferum	L. juniperoideum	Rhizomnium punctatum	Neckera complanata
Dicranum majus	Mnium hornum	Rhytidiadelphus loreus	N. pumila
Diplophyllum albicans	M. stellare	Riccardia palmata	Nowellia curvifolia
Eurhynchium schleicheri	Plagiochila asplenioides	Scapania undulata	Orthotrichum lyellii
E. striatum	P. porelloides	Scleropodium cespitans	Pterogonium gracile
Heterocladium heteropterum	Plagiomnium rostratum	Thamnobryum alopecurum	Ulota crispa
Homalia trichomanoides	P. undulatum	Thuidium tamariscinum	Zygodon baumgartneri
Hookeria lucens	Plagiothecium curvifolium	Trichocolea tomentella	
Hylocomium brevirostre	P. denticulatum		

Lichens

Agonimia octospora	C. subflaccidum	Megalospora tuberculosa	P. inusta
Arthonia astroidea	Dimerella lutea	Micarea pycnidiophora	P. lyellii
A. ilicina	Enterographa sorediata	Nephroma laevigatum	Phyllopsora rosei
A. vinosa	Heterodermia obscurata	Ochrolechia inversa	Schismatomma niveum
Arthopyrenia antecellans	Lecanactis amylacea	Opegrapha corticola	S. quercicola
A. ranunculospora	L. lyncea	O. prosodea	Stenocybe septata
Bacidia biatorina	L. premnea	O. xerica	Sticta limbata
Bactrospora corticola	L. subabietina	Pachyphiale carneola	S. sylvatica
Biatorina sphaeroides	Lecanora jamesii	Pannaria conoplea	Strangospora ochrophora
Catillaria atropurpurea	L. quercicola	Parmelia crinita	Thelopsis rubella
Cetrelia olivacea	L. sublivescens	P. reddenda	Thelotrema lepadinum
Chaenotheca brunneola	Leptogium lichenoides	Parmeliella jamesii	Usnea ceratina
C. chrysocephala	L. teretiusculum	P. triptophylla	U. florida
C. hispidula	Lobaria amplissima	Peltigera collina	Wadeana dendrographa
C. trichialis	L. pulmonaria	P. horizontalis	Zamenhofia coralloidea
Cladonia caespiticia	L. scrobiculata	Pertusaria multipuncta	Z. rosei
C. parasitica	L. virens	P. velata	
Collema furfuracea	Loxospora elatina	Phaeographis dendritica	

Fungi of old woods

The vast majority of fungi are woodland species, and if more were known about them, a list of old woodland indicators might be made. From the present survey, based on 10 km Grid squares only, the squares with >300 species are SY59, 89, 98, ST50, SZ08 and SU01. Less than 60 species have been reported from SY49, 66, 67, 68, ST72 and 82, which more or less agrees with the distribution of old woods.

GRASSLANDS

by David Pearman and Bryan Edwards

In Dorset there are examples of calcicolous, neutral and acid grasslands, of which the former are the most important. Their history needs further research. Rackham (1986) interprets information from the Domesday Book to mean that 28% of the county was pasture or heath, while only 1% was meadow, in 1086. In medieval times much of the Chalk was sheepwalk, and by the 17th century water-meadows were widespread. Subsequent changes in management are summarized in Chapter 3; the most significant for plant life was the dramatic increase in arable, at the expense of grassland, in 1939.

Calcareous grassland

Dorset chalk grasslands have been surveyed three times recently (Jones, 1973; Wigginton, 1987; Edwards, 1997). All these surveys attempt to estimate the loss of downland, to delineate what remains, and to list plants from and categorise the remnants. For they are remnants. Dorset Chalk covers about 960 km² (Wigginton, 1987), but in the most recent survey (Edwards, 1997) just under 30 km² is described as unimproved or semi-improved. Most of this is either on steep slopes or on ancient earthworks.

Limestone grassland is much smaller in area, but equally fragmented; most of it is on Portland or near the coast, and is treated below. For differences between chalk and limestone plants see Chapter 1 above. Since these grasslands were created by man's livestock, continuous management is needed to maintain the habitat and its plant diversity. Recent changes in the abundance of rabbits, the ratio of cattle to sheep, and above all, undergrazing have led to substantial losses of plant species.

Six or seven of the types of calcicolous grassland recognized by the National Classification of Vegetation (Rodwell, 1992) occur in the county, but they intergrade so much that it is simpler to accept two main groups:

1. Open, heavily-grazed short turf, with *Festuca ovina* abundant to dominant and a rich assortment of dwarf shrubs, small herbs, bryophytes and sometimes lichens (CG1, 2 and 7).
2. Closed, lightly-grazed rank turf, with either *Bromopsis erecta* (CG 3) or *Brachypodium pinnatum* (CG 4) dominant, generally poor in herbs and cryptogams but with other grasses present.

Beginning in the north-east, Bokerly Dyke has suffered from growth of scrub. Its flora is poorer than on the Hampshire side, which has *Orchis anthropophora* and *Phyteuma orbiculare*, but still supports populations of *Carex humilis*, *Neotinea ustulata*, *Polygala calcarea*, *Tephroseris integrifolia* and *Thesium humifusum*, as well as a few plants of *Juniperus communis* and *Pulsatilla vulgaris*. Apart from

the considerable expanse of Pentridge Down (SU01I), with its huge population of *Carex humilis* and good bryophytes, other sites in this area are fragments dominated by *Bromopsis erecta*. *Filipendula vulgaris*, *Orobanche elatior* and *Verbascum nigrum* are frequent along verges. Sovell Down (ST91V) is one of the few Dorset sites for *Rosa agrestis*.

North and west of this region are large tracts of downland centred on Fontmell and Melbury Downs, where *Festuca ovina* and *Helictotrichon pratense* form extensive patches, with *Brachypodium sylvaticum* abundant in places. The rich associated flora includes large populations of *Gentianella anglica* (sporadic), *Hippocrepis comosa*, *Polygala calcarea* and *Primula veris*, together with smaller numbers of orchids such as *Gymnadenia conopsea*, *Dactylorhiza viridis*, *Neotinea ustulata* and *Spiranthes spiralis*. Oddly enough *Carex humilis* has not been found anywhere here, but the slopes of Melbury Downs are acid enough to support a few patches of *Calluna vulgaris*.

South-west of this are two outliers, Hod and Hambledon Hills, both of which are hillforts with mosaics of short and long grassland. They have rich floras, with much *Carex humilis*, and *Saxifraga granulata* is locally abundant in May. A nearby unusual site is Blandford Camp (90J). Here a First World War army camp is sited on a 19th century racecourse, thus preserving some old turf. *Carex humilis* is frequent, with abundant *Clinopodium acinos*, *Hippocrepis comosa*, *Polygala calcarea*, *Thesium humifusum* and *Thymus pulegioides*. Although there are thousands of *Carex humilis* plants in the resown and heavily mown lawns in

---

Table 7. Plants faithful to calcareous grassland in Dorset.

Vascular plants

Arabis hirsuta	Cerastium pumilum	Helictotrichon pratense	Pilosella peleteriana
Anacamptis pyramidalis	Cirsium acaule	Helictotrichon pubescens	Pimpinella saxifraga
Anthyllis vulneraria	Cirsium eriophorum	Herminium monorchis	Plantago media
Asperula cynanchica	Cirsium tuberosum	Hippocrepis comosa	Polygala calcarea
Blackstonia perfoliata	Dactylorhiza viridis	Inula conyza	Pulsatilla vulgaris
Brachypodium pinnatum	Euphrasia pseudokerneri	Koeleria macrantha	Ranunculus parviflorus
Briza media	Euphrasia tetraquetra	Leontodon hispidus	Reseda lutea
Bromopsis erecta	Festuca ovina	Marrubium vulgare	Sanguisorba minor
Campanula glomerata	Filipendula vulgaris	Neotinea ustulata	Scabiosa columbaria
Carduus nutans	Gentianella amarella	Ophrys apifera	Tephroseris integrifolia
Carex caryophyllea	Gentianella anglica	Ophrys sphegodes	Thesium humifusum
Carex humilis	Gymnadenia conopsea	Origanum vulgare	Verbascum nigrum
Carlina vulgaris	Helianthemum nummularium	Picris hieracioides	Viola hirta.
Centaurea scabiosa			

Bryophytes

Acaulon triquetrum	Entodon concinnus	Phascum curvicolle	Pterygoneurum ovatum
Barbula acuta	Fissidens adianthoides	Pleurochaete squarrosa	Rhodobryum roseum
Brachythecium glareosum	F. cristatus	Pottia bryoides	Scapania aspera
Campylium calcareum	Funaria muhlenbergii	P. caespitosa	Scorpiurium circinatum
C. chrysophyllum	F. pulchella	P. commutata	Southbya nigrella
C. stellatum var. protensum	Gymnostomum viridulum	P. crinita	Thuidium delicatulum
Ctenidium molluscum	Homalothecium lutescens	P. heimii	T. philibertii
Ditrichum flexicaule	Neckera crispa	P. wilsonii	Tortella tortuosa

this camp, it is not known whether they have seeded themselves. To the south is the very rich site of Badbury Rings (90R), with a similar flora on its ramparts, and less rich grasslands at Tarrant Rawston Cliff (90N) and Buzbury Rings (90C), the latter now partly a golf course.

Moving south-west again, the hills above Houghton (80C) have *Cirsium tuberosum*, a fairly recent record for *Neotinea ustulata*, and, surprisingly, *Pedicularis sylvatica*. Bulbarrow (70S) has chalk turf as well as leached patches with heather. Lyscombe Hill (70G) and other places on the scarp are of interest. The next major block of chalk turf lies around Cerne Abbas (60Q), where the composition of the grassland is somewhat changed. There is no *Carex humilis*, and *Filipendula vulgaris*, *Tephroseris integrifolia* and *Thesium humifusum* are very rare. Short turf here contains an abundance of *Leontodon hispidus*, *Scabiosa columbaria*, *Stachys officinalis* and *Succisa pratensis*. Many sites have an acid capping with *Galium saxatile*, *Teucrium scorodonia* and *Ulex gallii*. Batcombe Down (60H) is a rich site, but *Herminium monorchis* is now extremely rare. By Eggardon Hill (59H) the Chalk is near its westerly limit in Britain, apart from Beer in Devon. Rainfall is higher and the *Leontodon hispidus* etc. community is common; few of the rarer chalk plants occur, though there is an old record for *Carex humilis*. There is also abundant *Brachypodium pinnatum*, the scourge of the ungrazed downs of Purbeck, as well as those of Sussex and Kent.

Turning eastward, the stretch of chalk from Maiden Castle (68U) to Poxwell (78G) contains few plants of great interest apart from *Spiranthes spiralis* and *Thesium humifusum*. A rich herb flora begins where the Chalk meets the sea at White Nothe (78Q). This will be described below,

but one should note that *Brachypodium* grasslands dominate the whole ridge, except where grazing is heavy, as it is above Lulworth Cove (88F). Here scarcer herbs such as *Arabis hirsuta* and *Thesium humifusum* occur, but the specialities of the north-east Chalk are absent. Away from the cliffs as far as Lulworth, the long grass is poor in herbs, but in places leaching allows patches of acidiphilic species such as *Erica cinerea* and *Rumex acetosella*. From Lulworth to Mupe Bay (88K) the Chalk overshadows the coast, then it forms cliffs at Flowers Barrow (88Q) and a sharp ridge to Ballard Down (08K). Here the grasslands are dominated by *Brachypodium pinnatum*, with a good deal of invading scrub, mostly *Ulex europaeus* and *Crataegus*. The National Trust and other land-owners are beginning to clear some of this scrub, especially at the Ballard Down end. Much of the ridge is species-poor, but it is the headquarters of *Hypericum montanum* in Dorset. *Salvia verbenaca* and *Serratula tinctoria* are frequent herbs, and orchids such as *Anacamptis pyramidalis*, *A. morio* and *Ophrys apifera* are locally common. North-facing grassland above Church Knowle (98G) is particularly rich in bryophytes, while moderately rich lichen heaths occur on very shallow soils at Durdle Door (88A), Bindon Hill (88F) and at Ballard Down. Grassland Fungi are briefly discussed in Chapter 9.

The following sites have 30 to 40 species of vascular plant listed in Table 7: On the Chalk, Badbury Rings, Blandford Camp, Flowers Barrow, Hambledon and Hod Hills, Lulworth Cove, Pentridge area and White Nothe; on limestone, Durlston Country Park, Portland quarries and St Aldhelm's Head (see Figure 8, based on ten indicator species).

Lichens of Dorset chalk grassland have been investigated by Gilbert (Gilbert, 1993). Excluding

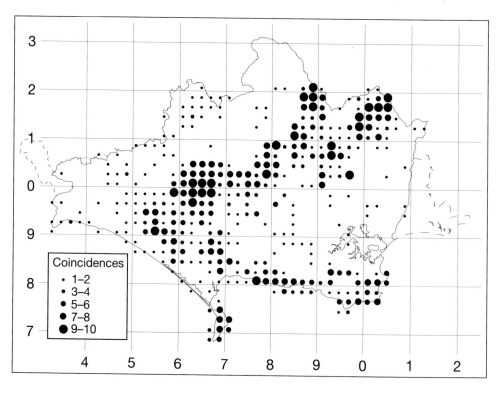

Figure 8. Location map of ten calcareous grassland indicators.
Indicator species;
Anacamptis pyramidalis,
Asperula cynanchica,
Campanula glomerata,
Dactylorhiza viridis,
Filipendula vulgaris,
Gentianella amarella,
Helianthemum nummularium,
Hippocrepis comosa,
Origanum vulgare,
Scabiosa columbaria.

Coincidences
· 1–2
· 3–4
• 5–6
● 7–8
● 9–10

species confined to exposed flints, they include: *Bacidia bagliettoana, Catapyrenium squamulosum, Cladonia foliacea, C. subrangiformis, Clauzadea immersa, C. metzleri, Hymenelia prevostii, Petractis clausa, Polyblastia gelatinosa, Rinodina bischoffii, Thelidium incavatum* and *Verrucaria simplex.*

Neutral Grasslands

These are confined to two main areas in Dorset, the northern and western vales and the Poole basin. Their total area has been estimated as 4 km² (Porley & Ulf-Hansen, 1991) or 22 km² (Edwards, 1997), but the majority are small, isolated fields, often with impeded drainage or on steep slopes.

In the western vales (SY39, 59 and ST50), despite great variations in soil, the drier, unimproved sites are all dominated by MG5 grassland, with *Cynosurus cristatus* and other grasses (Rodwell, 1992). Associated species include *Centaurea nigra, Gaudinia fragilis, Genista tinctoria, Oenanthe pimpinelloides, Ononis spinosa, Ophioglossum vulgatum, Stachys officinalis, Succisa pratensis* and *Trifolium medium.* Where the soil is slightly calcareous, as in parts of the Marshwood vale, *Anacamptis morio, Briza media, Leontodon hispidus,* and other calcicoles may appear. Slightly acidic soils, fairly extensive near Mapperton and Powerstock (SY59), often have *Conopodium majus, Galium saxatile, Pedicularis sylvatica* and *Potentilla erecta.* At Rampisham Radio Station (50K) there are more acid patches with *Calluna vulgaris, Deschampsia flexuosa, Ulex gallii* and the mosses *Dicranum scoparium* and *Pleurozium schreberi*; this type of vegetation may once have been widespread on leached soils overlying the Chalk, but has been lost to agriculture. Where waterlogging occurs, the vegetation turns either to fen (see below) or to rushy pasture (M23) with *Juncus acutiflorus* and *J. effusus.*

In the Blackmore vale (ST61, 62, 71, 81 and 82), neutral grassland sites are small and scattered. Breach Fields (82L) supports an MG5 community, while Burton Common (61K) is still grazed by horses and has species such as *Carex hostiana, Cirsium dissectum, Genista tinctoria* and *Oenanthe lachenalii.*

In the Poole basin (SY88, 98, 99, ST90, SU00) sites are equally small and dispersed. Some of the most varied are found on London Clay on the Army Ranges between Whiteway Farm and West Creech, also at Tyneham (both SY88). These are lightly grazed, but otherwise scarcely touched by modern agriculture, and have an MG5 sward with *Briza media, Hordeum secalinum, Ophioglossum vulgatum, Serratula tinctoria, Silaum silaus* and *Succisa pratensis.* The wetter flushes form *Juncus subnodulosus* fen-meadow (M22) with herbs such as *Achillea ptarmica, Anagallis tenella, Carex hostiana, C. pulicaris, Cirsium dissectum, Oenanthe lachenalii* and *Triglochin palustre.* Corfe Common (98Q) is an outstanding site with a long history of grazing. It is a mosaic of gorse, acid and neutral grassland, rush-pasture and fen-meadow with unique moss and sedge flushes. The many notable species include *Briza media, Carex hostiana, C. pulicaris,* an abundance of *Chamaemelum nobile, Cirsium dissectum, Eleocharis quinqueflora, Nardus stricta, Ophioglossum vulgatum, Scorzonera humilis, Serratula tinctoria, Succisa*

Table 8. Some typical plants of neutral grassland.

Achillea ptarmica	Leucanthemum vulgare
Botrychium lunaria	Lotus corniculatus
Bromus commutatus	Oenanthe pimpinelloides
B. hordeaceus	Ononis spinosa
B. lepidus	Ophioglossum vulgatum
B. racemosus	Pedicularis sylvatica
Carum verticillatum	Poa humilis
Centaurea nigra	Polygala vulgaris
Cirsium dissectum	Pulicaria dysenterica
Cynosurus cristatus	Rhinanthus minor
Danthonia decumbens	Scorzonera humilis
Euphrasia anglica	Serratula tinctoria
E. nemorosa	Silaum silaus
Gaudinia fragilis	Stachys officinalis
Genista tinctoria	Succisa pratensis
Hordeum secalinum	Trifolium fragiferum
Lathyrus nissolia	T. medium.
L. pratensis	

pratensis in several colour forms, *Trifolium fragiferum* and *Triglochin palustre.* There are one or two horse-grazed pastures near Corfe Mullen and Lytchett Matravers with a good flora and *Anacamptis morio* in quantity. *Wahlenbergia hederacea* is a very scarce plant in such grazed sites.

The Stour valley is almost entirely improved for dairy farms. A semi-improved meadow south-west of Wimborne (99Z) has an MG5 flora which must once have been typical; it has abundant *Hordeum secalinum* as well as *Oenanthe pimpinelloides* and *Silaum silaus.* Nearby, but along the River Allen (09E), is an unimproved meadow supporting the uncommon MG8 *Cynosurus cristatus-Caltha palustris* community with *Carex disticha, Filipendula ulmaria* and *Lysimachia nummularia.* The meadows of Pamphill Moor (90V) include both MG5 grassland and M23 rush-pasture, with plenty of *Chamaemelum nobile* in mown sites, *Dactylorhiza praetermissa* and *Oenanthe pimpinelloides* in quantity, and small amounts of *Botrychium lunaria* and *Spiranthes spiralis.*

Further north a similar mixture of neutral grass, fen-meadow and rush-pasture survives near Holt (00N), with much *Achillea ptarmica.* Holt Green (00G) is a traditional village green with a mosaic of acid, neutral and fen-meadow communities and *Cirsium dissectum, Nardus stricta, Oenanthe pimpinelloides* and *Pedicularis sylvatica.* Such sites must once have had *Pulicaria vulgaris,* now extinct in the county. In the north-east a few damp pastures have MG5 and M23 communities, e.g. the SSSIs at Bugden's Copse (00Z) and Sutton Holms (01K); the frequency of *Achillea ptarmica* here is noteworthy.

There are still examples of neutral grassland in old meadows and pastures all along the coast. The drier fields usually have *Hordeum secalinum, Linum bienne, Oenanthe pimpinelloides, Silaum silaus* and *Trifolium fragiferum.* Small areas of inundated grassland occur at West Bay (49Q), West Bexington (58I), Lodmoor (68V) and around Poole

Harbour. These are characterised by the presence of *Alopecurus bulbosus*, often with its hybrid *A.* × *plettkei* (Table 8). By no means are all these faithful, nor are there any characteristic cryptogams of this community, apart from some common fungi.

Acid grasslands

The distribution and composition of acid grasslands in Dorset has been inadequately studied, perhaps because so few fragments remain. At least three types can be distinguished, *Molinia* mires, *Agrostis curtisii* heaths and fallow or close-grazed heaths with more or less open vegetation.

Molinia mires (M25) cover wet, unimproved heathland in parts of the Poole basin with shallow, peaty soils, e.g. at Bovington Heath (88E). They have a monotonous vegetation, usually with *Potentilla erecta*, and can be hard to penetrate on foot. Their total area has been put at <0.2 km², but this is probably an underestimate (Porley & Ulf-Hansen, 1991).

Patches of *Agrostis curtisii*, often with *Festuca filiformis* and occasionally with *Nardus stricta*, occur on most of the drier Dorset heaths (see below). This *Agrostis* can colonise burnt areas faster than the dominant heathers.

Where heaths have been ploughed and left fallow, a transient assemblage of mainly annual herbs (U1) can develop. The commoner grasses, which are not dominant, include *Aira caryophyllea* and *A. praecox*, *Vulpia bromoides*, *V. ciliata* and *V. myuros*, while the abundant legumes on these nutrient-poor soils include *Lotus subbiflorus*, *Ornithopus perpusillus*, and at least eight species of *Trifolium*. This is the preferred habitat of some rare Dorset plants, such as *Anthriscus caucalis*, *Hypochoeris glabra*, *Lepidium heterophyllum*, *Moenchia erecta*, *Potentilla argentea*, *Stellaria pallida*, *Teesdalia nudicaulis* and *Vicia lathyroides*, and for the now extinct *Arnoseris minima*. The best examples can be found sporadically close to Poole Harbour, as at Wareham Common (98D), Studland Heath and South Haven (08),

or near Arne and Ower (98), but many heaths have fragments of this habitat. Species-poor outliers occur at Lambert's Castle (39U), Abbotsbury ironstone ridge (58S) with *Hordeum murinum* and *Vulpia bromoides* co-dominant, and on the top of Castle Hill, Cranborne (01L). Members of the same community can persist in modified heaths which have either been very closely grazed to lawns (MG6b), as at Blacknoll (88D) near Brownsea church (08I), or regularly mown, as at the RAOC depot at West Moors (00X). A particularly demanding community develops in slightly damp, rutted tracks in such places, with the rarities *Anagallis minima*, *Cicendia filiformis*, *Crassula tillaea*, *Illecebrum verticillatum* and *Radiola linoides*.

A unique natural community is found on fixed shingle between Small Mouth and the neck of Portland. Here *Festuca rubra* and *Armeria maritima* are the main dominants, with a long list of rare species such as *Anisantha madritensis*, *Asparagus officinalis* ssp. *prostratus*, a form of *Koeleria macrantha*, *Medicago polymorpha*, *Orobanche purpurea*, *Polycarpon tetraphyllum*, *Vulpia ciliata*, *V. fasciculata* and many maritime plants.

DORSET FENS
by Bryan Edwards and Humphry Bowen

Very little has been written about Dorset fens, perhaps because the habitat occupies such a small area, which has been estimated as <7 ha (Porley & Ulf-Hansen), but the actual area must depend on the definition of fen and its distinction from water-meadows. The latter are still fairly extensive, but other fens are restricted to tiny isolated patches. While the majority of fens here are dominated by *Juncus subnodulosus* (M22), there are a few examples of *Molinia caerulea* fen (M24) and *Filipendula ulmaria-Angelica sylvestris* tall-herb fen (M27; Rodwell, 1992; Rodwell, 1995). Because Dorset rivers are calcareous, but often traverse regions with acid soils, water-meadows tend to be atypical mosaics with a wide mixture of species.

In north and west Dorset patches of fen with dominant *Juncus subnodulosus* and/or *Molinia caerulea* occur at the junction of the Upper Greensand and Gault Clay. Such patches north of Beaminster, near Frome St Quintin and Rampisham may support *Anagallis tenella*, *Geum rivale* or *Valeriana dioica*. At Aunt Mary's Bottom (50L) both *Eleocharis quinqueflora* and *Eriophorum latifolium* survive. Gleyed soils at Powerstock Common (59N) have M24 fen with *Carex hostiana*, *C. panicea* and *C. pulicaris* plus acidiphiles such as *Erica tetralix*, *Dactylorhiza maculata*, *Genista anglica* and *Viola canina*, which is being managed. Other gleyed areas further north at Deadmoor Common, Lydlinch Common and Rooksmoor (all ST71) also have M24 vegetation and rich floras which include *Achillea ptarmica*, *Cirsium dissectum*, *Genista tinctoria* and *Oenanthe pimpinelloides*. Here scrub is being cleared to reduce shading. An isolated meadow near Hinton St Mary (81C) has a different species-rich flora with *Carex hostiana*, *Dactylorhiza incarnata*, *Thalictrum flavum* and *Triglochin palustre*.

Table 9. Some additional plants of dry, acid grassland.

Anchusa officinalis	T. ornithopodioides
Aphanes australis	T. scabrum
Centaurium pulchellum	T. striatum
Cerastium semidecandrum	T. subterraneum
Dianthus armeria	T. suffocatum
Erodium cicutarium	Veronica officinalis.
Filago minima	
F. vulgaris	**Moss**
Geranium pusillum	Brachythecium albicans
Himantoglossum hircinum	
Jasione montana	**Lichens**
Spergularia rubra	Cladonia cariosa
Thymus pulegioides	C. ciliata
Trifolium arvense	C. furcata
T. glomeratum	C. humilis
T. micranthum	C. portentosa

Table 10. Plants indicative of, but not always faithful to, fens or former fens.

Vascular plants		Bryophytes
Blysmus compressus	Eriophorum latifolium	Amblystegium riparium
Carex diandra	Groenlandia densa	Calliergon giganteum
C. disticha	Juncus subulatus	Campylium elodes
C. hostiana	[Leersia oryzoides]	C. stellatum
C. pulicaris	Lychnis flos-cuculi	Climacium dendroides
C. rostrata	Lysimachia vulgaris	Cratoneuron commutatum
C. vesicaria	Oenanthe fistulosa	Philonotis calcarea
C. viridula subsp. brachyrrhyncha	[Parnassia palustris]	Plagiomnium elatum
Carum verticillatum	Pedicularis palustris	Riccia fluitans
Catabrosa aquatica	Sanguisorba officinalis	Scorpidium scorpioides
Cladium mariscus	Scorzonera humilis	Sphagnum auriculatum
Cyperus longus	Stellaria palustris	S. subnitens
Dactylorhiza incarnata	Thalictrum flavum	
D. praetermissa	Triglochin palustre	
Eleocharis quinqueflora	Valeriana dioica	
Epipactis palustris	Veronica scutellata	

In the Poole basin, the edge of boggy heathland near Tadnoll (78Y) has some grazed M24 pastures with *Carex pulicaris*, *Potentilla palustris*, *Sanguisorba officinalis* and, sporadically, the bryophyte *Anthoceros punctatus*. Rather similar communities occur near the Bere stream (89L), and along the Piddle near Chamberlaynes and Turners Piddle (89G); the latter has *Carex viridula* subsp. *brachyrrhyncha* and a large colony of *Cyperus longus*. Both Piddle and Frome valleys have extensive unimproved water-meadows and grazed marshes and some *Phragmites* fen. Two outstanding fenny areas are at Wareham Common (98D) and The Moors (98N). Among the many rarities here are *Carex diandra*, *Carum verticillatum*, *Cladium mariscus*, the extinct grass *Leersia oryzoides*, *Schoenus nigricans*, *Scorzonera humilis* and *Thelypteris palustris*. There are small relict fens along the Sherford and Corfe rivers which may give some idea of the vegetation of our river basins before they were drained by man. The former has much *Cladium*, as well as *Epipactis palustris*, *Gymnadenia densiflora*, *Pedicularis palustris*, *Schoenus nigricans* and *Thelypteris palustris*, but is turning into *Salix* or *Alnus* carr. The latter has *Carex pulicaris*, *Eriophorum latifolium*, *Genista anglica*, *G. tinctoria*, *Lathyrus linifolius*, *Menyanthes trifoliata*, *Osmunda regalis*, *Pedicularis palustris* and patches of *Sphagnum*; it is being overgrown by *Myrica gale*. The extinct *Parnassia palustris* could have occurred in either of these sites.

A few fen-meadows (M23) can be found along the valleys of the Rivers Crane and Allen in the north-east. The best of these is the Bull meadow at Wimborne St Giles (01G), where *Blysmus compressus*, *Dactylorhiza incarnata*, *Geum rivale* and *Persicaria bistorta* occur in some quantity.

A hundred years ago there must have been small fenny sites by springs on calcareous cliffs in Purbeck, judging by old records of *Carex viridula* subsp. *brachyrrhyncha* and *Eriophorum latifolium* (Mansel-Pleydell, 1895). A few plants of *Cyperus longus* remain at Ulwell (08F), otherwise this habitat has gone. The small spring at Culver Well on Portland (66Z) might be considered as a surviving example.

HEATHS AND BOGS
by Steve Chapman, *et al.*

Large tracts of heathland developed in Neolithic times as man cleared the forest on sandy soils; these soils became podzols. Because heath was so infertile, it remained as waste land for millenia, with an area of about 400 km² when first accurately surveyed in 1811. Since that time, and especially since the 1940s, much heathland has been destroyed, so that only about 70 km² survive as fragments today (Moore, 1962; Webb, 1990). The main destructive agents have been agriculture, building and forestry, in order of area affected; the first two of these are virtually irreversible, but the third is not.

Heath is usually the product of a combination of fairly regular burning and light grazing. It was used in former centuries for grazing, and as a source of both peat and furze for fuel. These activities of man prevented the growth of scrub and trees, and so conserved the heath. Modern usage destroys the habitat, which is now rare in Europe, and poses problems for conserving what remains.

Despite the losses mentioned, Dorset still has some of the largest lowland heaths in Britain, a distinction it shares with Cornwall, Devon, Hampshire and the East Anglian Breckland. In the west there are a few small, isolated patches of plateau heath on hilltops, as at Hole Common, Lamberts Castle, Hardown Hill, Golden Cap and Blackdown near Portesham. Some of these have associated bogs near the springline. These have probably always been isolated, and are not floristically exciting apart from an old record for *Listera cordata* and a recent transient appearance of *Lycopodium clavatum*. A rich site at Champernhayes (39N) was lost to forestry in this century. The bulk of Dorset heathland has always been in the Poole basin, and a detailed breakdown of its area shows that about 20% is accounted for by scrub, carr, sandy tracks, firebreaks and open water (Webb, 1990).

The National Vegetation Classification (Rodwell, 1995) subdivides heaths and bogs into many communities, as follows:

Dry Heath

either H2 (*Calluna-Ulex minor*)

H8 (*Calluna-Ulex gallii*) in the west

H3 (*Ulex minor-Agrostis curtisii*)

or H4 (*Ulex gallii-Agrostis curtisii*).

Wet Heath

M16 (*Erica tetralix-Sphagnum compactum*)

Bogs or valley mires

either M21 (*Narthecium ossifragum-Sphagnum papillosum*)

or M14 (*Schoenus nigricans-Narthecium*),

with associated bog pools

either M1 (*Sphagnum auriculatum*)

M29 (*Hypericum elodes-Potamogeton polygonifolius*)

or M30 (*Hydrocotyle vulgaris-Baldellia ranunculoides*).

In this work, humid grassy heaths (M23, 24 and 25), lawns and greens (MG6B) and sandy grassland (U1) are treated as acid grassland above. However, most Dorset heaths actually consist of mosaics of many of these 15 abstract types which merge into one another. For example, dry heath (H2) is a common type which colours parts of the Poole basin purple and yellow in late summer. The type community is usually accompanied by patches where either bracken (*Pteridium aquilinum*) or Common Gorse (*Ulex europaeus*) are dominant.

The common heathers are dwarf shrubs which are not always dominant. Sparse *Calluna* occurs on a few superficials of the Chalk, as at Bulbarrow (70X), or in woodland rides. *Erica cinerea* is intolerant of shade but tolerant of drought, and is found in small colonies away from the main heath; there is a fairly recent record from Portland. *Erica tetralix* will grow on its own in wet fields, as in a relict M15 community near Coney's Castle (39T), or in cleared woods, as at Bere Wood (89S). All three species produce plenty of long-lived seed, as is shown by their regeneration when conifer plantations on old heathland are felled. The Forestry Commission has accepted a policy of clearing strips of their plantations to link up existing fragments of heath.

Dry heath, though beautiful, is floristically poor and its conservation is best justified by the specialised fauna which it supports. It commonly includes grasses such as *Agrostis curtisii, A. vinealis, Aira praecox, Deschampsia flexuosa,*

Festuca filiformis and (rarely) *Nardus stricta,* together with parasitic dodder *Cuscuta epithymum*. The more interesting herbs are often found in disturbed sites, lawns or edges of tracks. *Ulex gallii* has its eastern limit and *Ulex minor* its western limit in the county, but their precise ranges are uncertain because of possible misidentifications. Both grow together, along with *U. europaeus*, in a number of places such as Binnegar (88Y) and Great Oven's Hill (99F). Common bryophytes of dry heath include *Campylopus introflexus, Hypnum jutlandicum* and *Polytrichum juniperinum*, while there is a rich lichen flora of *Cladonia portentosa* and other *Cladonia* spp. on bare soil patches between young heather plants. *Absconditella lignicola* and *Hypogymnia physodes* colonise the stems of old heather bushes.

Wet heath (M16) is dominated by *Erica tetralix*, and its flora includes some scarce plants of which Dorset Heath (*Erica ciliaris*) is perhaps the most notable. An Atlantic species which has been in Dorset since the Bronze Age, *E. ciliaris* is locally abundant south of Poole Harbour and hybridises freely with *E.tetralix* (Rose, 1996; Chapman & Rose, 1994). There are isolated occurrences in many other heaths, and the plant is increasing its range; it colonises burnt or cleared areas vigorously, e.g. at Coombe Heath (98T). *Rhynchospora fusca* is another species whose largest British populations are in Dorset. Marsh Gentian (*Gentiana pneumonanthe*) is a striking perennial which often appears abundantly after a heath fire (Chapman, Rose & Clarke, 1989). Declining plants of wet heath include *Lycopodiella inundata* and *Platanthera bifolia*, which may have been affected by the cessation of peat cutting, and *Viola lactea*. Among bryophytes, *Sphagnum compactum* is always, and *S. tenellum* and *Odontoschisma denudatum* often, present, as is the lichen *Cladonia arbuscula*.

Shallow valley bogs develop in wet places where sphagnum moss decays to form peat, and the vegetation usually includes *Narthecium ossifragum*, often with *Drosera* spp., *Eriophorum angustifolium, Molinia caerulea, Rhynchospora alba* and *Trichophorum caespitosum*. Most Dorset bogs are M21 type, but a few are M14, dominated by *Schoenus nigricans*, as at Moigne Combe (78T). In some places nutrient-rich ground water emerges to produce a flush, with a rich vegetation of plants such as *Anagallis tenella, Cirsium dissectum, Dactylorhiza* spp., *Pinguicula lusitanica* and the liverwort *Aneura pinguis*. *Pinguicula vulgaris* is very rare, *Deschampsia setacea* slightly less so, and *Hammarbya paludosa* flowers sporadically on rafts of *Sphagnum*. The commonest bog-moss is probably *Sphagnum papillosum*, but about eight other species may be present including the uncommon, golden *S. pulchrum*. Interwoven with the Sphagna are liverworts such as *Cephalozia, Cladopodiella, Kurzia* and *Odontoschisma sphagni*. Where bog streamlets or pools of open water occur, they are soon filled with *Sphagnum auriculatum* or *S. subnitens*, and later with *Hypericum elodes, Juncus bulbosus* and *Potamogeton polygonifolius*. Even quaking bogs, as at Hartland Moor, are rarely dangerous to traverse in Dorset; they are preferred sites for rarities such as *Drosera longifolia, Pilularia globulifera, Utricularia intermedia, U. minor* and the charophyte *Nitella translucens*. The distribution of bog-

Table 11. Characteristic plants of dry heath, wet heath and bog.

Dry heath	Wet heath	Bog
Vascular plants		
Agrostis curtisii	Anagallis minima	Anagallis tenella
A.vinealis	Carex binervis	Apium inundatum
Aira caryophyllea	C. panicea	Baldellia ranunculoides
A. praecox	C. viridula oedocarpa	Carex dioica
Calluna vulgaris	Cicendia filiformis	C. lasiocarpa
Carex pilulifera	Dactylorhiza maculata	C. limosa
Crassula tillaea	Erica ciliaris	Deschampsia setacea
Cuscuta epithymum	Erica tetralix	Drosera intermedia
Deschampsia flexuosa	Gentiana pneumonanthe	D. longifolia
Erica cinerea	Lycopodiella inundata	D. rotundifolia
Festuca filiformis	Molinia caerulea	Eleocharis multicaulis
Nardus stricta	Radiola linoides	Eriophorum angustifolium
Sagina subulata	Rhynchospora fusca	Genista anglica
Ulex gallii	Trichophorum cespitosum	Hammarbya paludosa
U. minor	Viola lactea	Hydrocotyle vulgaris
Viola canina		Hypericum elodes
		Juncus bulbosus
		Littorella uniflora
		Menyanthes trifoliata
		Pilularia globulifera
		Pinguicula lusitanica
		P. vulgaris
		Potentilla palustris
		Rhynchospora alba
		Schoenus nigricans
		Utricularia intermedia
Bryophytes		
Campylopus brevipilus	Cephaloziella spp.	Aneura pinguis
C. fragilis	Fossombronia spp.	Aulacomnium palustre
C. introflexus	Gymnocolea inflata	Calypogeia spp.
C. paradoxus	Hypnum imponens	Campylium stellatum
Dicranum scoparium	Lophocolea bispinosa	Cephalozia spp.
Hypnum jutlandicum	Mylia anomala	Cladopodiella spp.
Nardia scalaris	Odontoschisma denudatum	Drepanocladus spp.
Pleurozium schreberi	Sphagnum compactum	Kurzia pauciflora
Polytrichum juniperinum	S.molle	Lophozia spp.
P. piliferum	S.tenellum	Odontoschisma sphagni
Tritomaria exsectiformis		Riccardia spp.
		Scorpidium scorpidioides
		Sphagnum spp.
Lichens		
Baeomyces spp	Cladonia arbuscula	Cladonia incrassata
Cladonia 13 spp.	C. strepsilis	Mycoglaena myricae
Coelocaulon spp.	C. uncialis	
Hypogymnia physodes	Icmadophila ericetorum	
Pycnothelia papillaria	Placynthiella uliginosa	
Trapeliopsis granulosa		

Table 12. Characteristic fungi of dry heath and bog.

Dry heath	Bog
Entoloma spp.	*Galerina paludosa*
Hygrocybe spp.	*Omphalina sphagnicola*
Lycoperdon foetidum	
Thelephora terrestris	

mosses in the Poole basin has been reviewed recently (Edwards, 1997).

The decline of many heath and bog plants has been investigated by Byfield and Pearman (1994) by comparing the present-day flora of 390 sites with Good's records in the 1930s. About 10% of the sites are no longer heathland, and almost all the scarcer plants have suffered dramatic declines. Since these declines have been less marked in the New Forest, the authors conclude that lack of grazing in Dorset is a major factor. An obvious example is the decline of *Chamaemelum nobile*, which only survives on grazed or regularly mown sites. Although the vegetation of so many heathland sites have been affected, species losses in the last 100 years have been relatively few. Probable extinctions include *Empetrum nigrum*, *Eriophorum gracile*, *Huperzia selago*, *Listera cordata* and *Simethis planifolia*, together with the bryophyte *Splachnum ampullaceum*, but some of these are northern species which may be declining for climatic reasons. If nothing is done, many more heath and bog species will be lost in the 21st century.

THE DORSET COAST AND ITS PLANTS
by David Pearman

The coast of Dorset has an underlying calcareous nature all the way from the Devon border to Studland, where acid tertiary beds outcrop. This long stretch of calcareous clay or limestone cliffs is broken by small areas of salt-marsh at West Bay and near Weymouth. Most of the cliffs are unstable, especially where they are largely clay, and undercliffs are frequent. These undercliffs have a flora of pioneer species, which can develop into a maquis of blackthorn, privet and sallow until the next cliff-fall opens up the habitat again. Unusual communities develop in fenny areas round temporary streams, with plants such as *Apium graveolens*, *Epipactis palustris* and *Samolus valerandi*, and also on the extreme edges of cliffs, where the open conditions suit *Bromus hordeaceus* subsp. *ferronii*, *Hyoscyamus niger* and *Silybum marianum*.

Beginning at Lyme Regis, the massive landslip extending west into Devon is now largely wooded. A hundred years ago Miss Lister recorded several lichens here which are no longer to be found; *Degelia atlantica*, *Pseudocyphellaria aurata* and *Teloschistes chrysophthalamus*. East of Lyme, at The Spittles, *Lathyrus sylvestris* is frequent on the undercliff and in small fields above, and *Isolepis cernua* occurs. The undercliff between Charmouth and Golden Cap is rapidly eroding and bare, but the strip

behind it is owned by the National Trust and has a rich vegetation, including *Moenchia erecta* in one of its few Dorset sites. The neutral grassland meadows west of St Gabriels (39W) are superb. East of Golden Cap, the cliff-top strip is narrow but the undercliffs are exciting as far as Eype Mouth. There are eight patches of *Centaurium tenuiflorum* in its only British location, as well as a new colony of *Pyrola rotundifolia*, *Epipactis palustris*, many legumes and large amounts of the bryophyte *Phaeoceros laevis*.

At Eype Mouth, and again at West Bay (49Q), the grasses *Parapholis incurva* and *Puccinellia rupestris* occur. At the latter site, with much sea defence work going on, *Frankenia laevis* and *Limonium hyblaeum* are found as aliens with *Puccinellia fasciculata*, and sea-front lawns consist of *Plantago coronopus*. In a brackish meadow behind West Bay, there is a large field with *Alopecurus bulbosus*, *Puccinellia rupestris* and *Trifolium ornithopodioides*. The largest British colony of *Chenopodium vulvaria* grows at the edge of vertical cliffs to the east, along with *Phleum arenarium*, *Trifolium scabrum* and *Torilis nodosa*, while *Atriplex longipes* and *Crambe maritima* occur below.

A semi-natural shingle beach community with much *Crambe* and *Glaucium flavum* is found at Cogden beach (58E) where the Chesil Bank begins; *Hordeum marinum* occurs, and just inland big populations of *Gastridium ventricosum* develop in favourable years. The seaward side of the beach is barren shingle, and the winter storms are such that even the marine algal flora is poor. At Abbotsbury the bank becomes a tombolo. *Lathyrus japonicus* forms colonies on the shingle here, while *Althaea officinalis*, *Carex divisa* and *Suaeda vera* grow at the edge of the Fleet lagoon, and *Allium ampeloprasum* appears to be native. The inner side of the Chesil bank has a rather poor but interesting flora, including Privet and Polypody. Within the Fleet there are large populations of *Ruppia cirrhosa*, *R. maritima*, *Zostera angustifolia*, *Z. noltei* and the rare charophyte *Lamprothamnium papillosum*, as well as more than 60 species of marine algae. Along its landward margin the low cliffs have been cultivated almost to the edge, but there are sites for *Bupleurum tenuissimum*, *Falcaria vulgaris*, *Lathyrus tuberosus* and *Trifolium squamosum*. Near Wyke these cliffs are mostly scrub, with patches of *Brachypodium pinnatum* grassland and scarce legumes such as *Lathyrus aphaca*, *Vicia bithynica* and *V. lutea*. Small Mouth near Ferrybridge (67S) is an especially rich site with *Calystegia soldanella*, *Phleum arenarium*, *Polycarpon tetraphyllum*, *Vulpia ciliata* and *V. fasciculata*. On the other side of the road linking Weymouth to Portland are *Asparagus officinalis* subsp. *prostratus*, *Eryngium maritimum*, *Orobanche purpurea* in an Armerietum, and a tiny salt-marsh with *Limonium dodartiforme*.

Portland deserves a section to itself. Its diversity has been modified by quarrying and development for centuries, a process which continues apace. Very little of the original grassland remains on top of the island, but near the Bill (66Z) there is a little *Bupleurum tenuissimum* and *Gentianella anglica*; the common *Arum* in waste places is *A. italicum*. The cliffs and undercliffs have plenty of the endemic *Limonium recurvum*, and the endemic *Hieracium leyanum* in two places. At Church Ope Cove there is much

Valerianella eriocarpa, with *V. dentata* later in the season, as well as *Polypodium australe* and *Sedum forsterianum*. Hard ground near the sea has *Parapholis incurva*, and a rich bryophyte flora with *Gymnostomum viridulum* and *Southbya nigrella*. More bryophytes occur in block scree, notably *Eurhynchium meridionale* in its only British site, and *Marchesinia mackaii*; exposed boulders have a rich lichen flora, mainly calciphiles like *Roccella phycopsis*, but also acidiphiles like *Lecanora aghardiana* and *Sclerophytum circumscriptum* on chert. Abandoned quarries have *Adiantum capillus-veneris*, and many aliens; *Polypogon monspeliensis* is established in one, and large Fig trees in others. Trees are scarce, but there is much undercliff scrub. The abandonment of strip cultivation north of the Bill has caused the loss or near loss of *Adonis annua* and *Scandix pecten-veneris*, but *Lathyrus aphaca*, *Medicago polymorpha* and *Vicia parviflora* are regularly present. Altogether the flora is more reminiscent of a mediterranean island like Majorca than it is of the rest of the county.

Eastwards from the mouth of the Fleet, we find *Anisantha madritensis*, *Geranium purpureum* and scarce legumes along the old railtrack. Portland Harbour is sheltered and has a rich algal flora, but the east-facing sandy beach at Weymouth is too densely peopled to support any plants, apart from a small colony of *Crithmum maritimum* on its retaining wall; it must once have been interesting. The mouth of the Wey, Radipole lake, is still a 'reedy pool' with dominant *Phragmites* and some plants of brackish sites. The nearby salt-marsh, Lodmoor, is now a reserve despite being partly overwhelmed by shingle and partly used as an urban tip. It has been invaded by *Spartina*, and at long intervals produces plants of *Rumex maritimus*.

East of Weymouth to Ringstead Bay (78K) are cliffs of calcareous clay, with sites for *Gastridium*, *Trifolium squamosum* and *Rumex rupestris*, though the last has not been seen for 20 years. Reefs covered by the tide shelter *Zostera marina* and many algae, and Bran Point has the marine lichen *Lichina pygmaea*. Beyond the Chalk headland of White Nothe both the flora and the scenery improve. Steep cliffs from here to Arish Mell (88K), and where the Chalk meets the sea at Ballard Head (08K), have plenty of *Brassica oleracea*, *Glaucium flavum*, *Inula crithmoides* and *Limonium dodartiforme*, with *Papaver somniferum* looking native. At the top of these majestic cliffs are scarce plants such as *Allium oleraceum*, *Marrubium vulgare*, *Pilosella peleteriana*, *Silene nutans* and *Valerianella eriocarpa*. The shingle beaches below, many inaccessible except by boat, are floristically poor with abundant *Atriplex glabriuscula*. Near Lulworth there are some rare bryophytes, including *Acaulon triquetrum* and *Pottia* spp, with the only British site for *Bartleya ohioensis* on dripping rocks near the sea.

From Lulworth Cove eastwards the cliffs are mostly clay and floristically less rich, though varied, and many beaches are difficult to reach. Gad Cliff (87Z) has a rich undercliff, with *Erodium maritimum*, *Lathyrus aphaca* and *Vicia parviflora*. Fields are cultivated very close to the cliffs round Kimmeridge (97E), but while the vascular flora is restricted, there are many thermophilic marine algae in the onshore nature reserve.

Between Chapmans Pool (97N) and Peveril Point (07P) is a seven mile stretch of limestone cliffs with few beaches, and undercliffs only where the coast faces east or west. At St Aldhelm's Head cultivated fields hold a diminishing population of *Adonis annua*, *Scandix pecten-veneris* and other rare weeds, with *Marrubium vulgare* along field borders. The bouldery undercliff has a cryptogam flora similar to that on Portland, with bryophytes such as *Cololejeunea rossettiana* and *Marchesinia* and many lichens including *Anaptychia runcinata* and *Ramalina siliquosa*. Much of this stretch has a buffer of unimproved grassland, notably in Durlston Country Park. The long list of rarities makes this a botanical Mecca, with *Arum italicum*, *Centaurium erythraea* var. *capitatum*, *Cerastium pumilum*, *Gastridium ventricosum*, *Gentianella anglica*, *Limonium binervosum agg*, *Ophrys sphegodes*, *Parapholis incurva*, *Rubia peregrina*, *Thesium humifusum* and *Valerianella eriocarpa*. There are also records for scarce cryptogams such as *Acaulon triquetrum*, *Pterygoneurum ovatum* and the lichen *Solenopsora vulturiensis*.

The sandy beach at Swanage has no botanical interest, and the low cliffs north of the town are scrubbed over, with some naturalised *Senecio cineraria*. Then comes Punfield Cove and the Chalk headland of Ballard Point, where the lichen *Heterodermia leucomelos* once occurred. At Handfast Point or Old Harry, there is a small area of good chalk grassland with *Marrubium vulgare* and *Silene nutans*, and a rich scrub including yew. Cultivated fields in 08L have *Trifolium glomeratum* and *Vicia lathyroides*.

Northwards one reaches the tertiary soils of Studland and the blown sand dunes of South Haven peninsula, intensively studied by Good and colleagues in the 1930s. The shore here has extensive beds of *Zostera marina* and some unusual algae. The outer dune ridge has much *Ammophila arenaria* and *Festuca arenaria*, with *Calystegia soldanella*, *Jasione montana*, and *Leymus arenarius*. Recently the alien *Acaena novae-zelandiae* has begun to spread here, and the grasses *Lagurus ovatus* and *Mibora minima* have appeared. The grass *Polypogon monspeliensis*, the bryophyte *Bryum knowltonii* and the lichen *Leptogium corniculatum* have no recent records. Dune slacks here tend to merge into a rich bog and heath vegetation with much *Osmunda regalis*.

The inner shores of Poole Harbour are either salt-marsh or dense *Phragmites* swamp. The best salt-marshes may be those fringing the larger islands (Brownsea, Furzey, Green and Pergins Island). These have *Carex extensa*, *Limonium vulgare*, *Sarcocornia perennis*, *Suaeda vera* and rarely *Atriplex laciniata* and *Scrophularia scorodonia*. The distribution of species of *Salicornia* and *Zostera* here needs further study. Beds of *Spartina anglica* are extensive, though reduced from their area of greatest spread in the 1920s. There is a single large clump of *Scirpoides holoschoenus* near Creekmoor, but neither *Eleocharis parvula* nor *Seriphidium maritimum* have been seen for some years.

North Haven continues the dunes of South Haven but is largely built over. *Cynodon dactylon* and ephemeral clovers persist on verges, as does *Polycarpon tetraphyllum* as a lawn and pavement weed. From Flaghead Chine to Branksome are sandy cliffs with *Ammophila, Carex arenaria* and heath plants, as well as many naturalised aliens which are still better represented on similar cliffs in Bournemouth (VC 11). *Simethis planifolia*, which may have been introduced with *Pinus pinaster*, grew at Branksome Chine from 1847 to 1914, but the Chine is now a public garden. A recent interesting find here has been the liverwort *Telaranea murphyae* under Rhododendron.

Although the National Vegetational Classification of maritime habitats has not yet appeared, we can say that Dorset has examples of sand dune, shingle beach, brackish meadow, salt-marsh and rocky cliff, as well as extensive undercliffs. There are four possible reasons why the coastal flora is so rich – much richer than in Hampshire, for example:

1. The cliffs and undercliffs are too unstable for cultivation
2. Salt spray delivers extra nutrients such as magnesium and sulphate, but requires salt-tolerant plants
3. The coastal strip has a higher mean temperature than the hinterland (see Chapter 1)
4. The rapid rates of coastal erosion may have driven populations of land plants northwards, to survive as relicts on undercliffs; accreting coasts, as at Studland, provide good habitats for colonists.

One feature of sandy beaches is their popularity with holiday makers. It will become clear in the main body of this Flora that many plants of this habitat have suffered from human pressure. They have either become extinct, as have *Otanthus* and *Polygonum maritimum*, or very rare, as have *Cakile, Rumex rupestris* and *Salsola kali*. Plants of shingle beaches and salt-marshes have declined less.

Plate 1. Ancient woodland, Yellowham Wood

Plate 2. Chalk landscape, Hog Cliff Hill

Plate 3. Chalk grassland, Fontmell Down

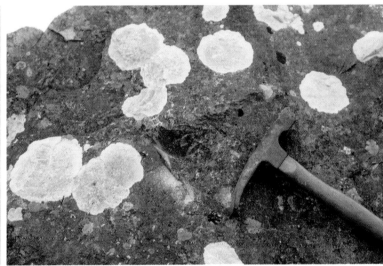

Plate 4. Lichens on limestone, Portland

Plate 5. Valley of stones, Portesham

Plate 6. Neutral grassland, Kingcombe

Plate 7. Heath landscape, Hardown Hill

Plate 8. Heath, Winfrith

Plate 9. Bog, Higher Hyde Heath

Plate 10. Shingle, Chesil beach, Abbotsbury

Plate 11. Coastal landslip, Black Ven

Plate 12. Coastal limestone, Lulworth

Plate 13. Coast, Mupe Bay; limestone
in foreground, chalk behind

Plate 14. Coast, Kimmeridge

Plate 16. Coastal sand dunes, Studland

Plate 15. Spring tide algae, Kimmeridge

Plate 17. Salt marsh, Poole Harbour

Plate 18. Fossil forest, Lulworth

Plate 19. Forestry, Woolsbarrow

Plate 20. Wind blown tree, Golden Cap

Plate 21. Ancient fields, Kingcombe

Plate 22. Old hedge, Kingcombe

Plate 23. Modern agriculture, Piddlehinton

Plate 24. Tout Quarry, Portland

Plate 25. Old oak 'Billy Wilkins', Melbury Park

Plate 26. *Limonium dodartiforme*
Rock Sea-lavender

Plate 27. *Limonium recurvum*
Portland Sea-lavender

Plate 28. *Erica ciliaris*
Dorset Heath

Plate 29. *Oenanthe pimpinelloides*
Corky-fruited Water-dropwort

Plate 30. *Centaurium tenuiflorum*
Slender Centaury

Plate 31. *Gentianella anglica*
Early Gentian

Plate 32. *Gentiana pneumonanthe*
Marsh Gentian

Plate 33. *Pulmonaria longifolia*
Narrow-leaved Lungwort

Plate 34. *Lobelia urens*
Heath Lobelia

Plate 35. *Orobanche purpurea*
Yarrow Broomrape

Plate 36. *Cirsium tuberosum*
Tuberous Thistle

Plate 37. *Hieracium leyanum*
Portland Hawkweed

Plate 38. *Scorzonera humilis*
Viper's-grass

Plate 39. *Rhynchospora fusca*
Brown Beak-sedge

Plate 40. *Leucojum vernum*
Spring Snowflake

Plate 41. *Himantoglossum hircinum*
Lizard Orchid

Plate 42. *Ophrys sphegodes*
Early Spider-orchid

Plate 43. *Lophocolea bispinosa*

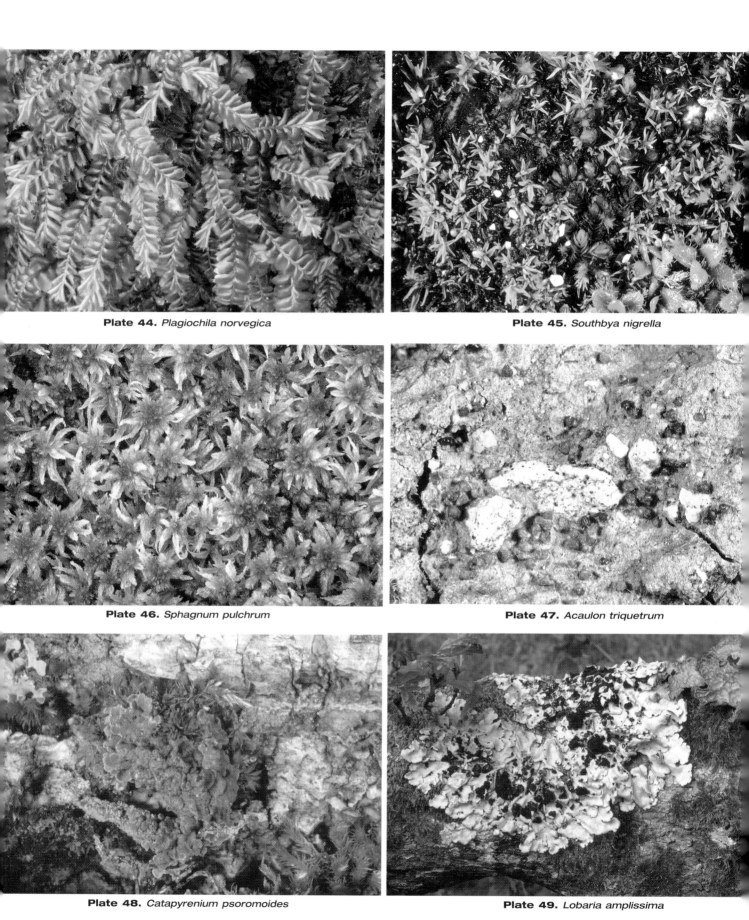

Plate 44. *Plagiochila norvegica*

Plate 45. *Southbya nigrella*

Plate 46. *Sphagnum pulchrum*

Plate 47. *Acaulon triquetrum*

Plate 48. *Catapyrenium psoromoides*

Plate 49. *Lobaria amplissima*

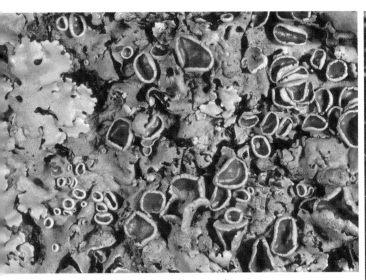

Plate 50. *Parmelia quercina*

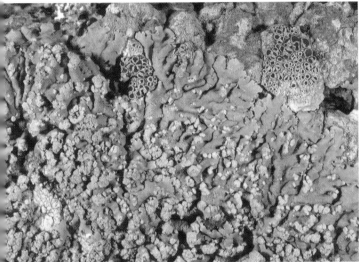

Plate 51. *Physcia tribacoides*

Plate 52. *Teloschistes flavicans*

Plate 53. *Hericium erinaceum*

Plate 54. *Pulcherricium caeruleum*

Plate 55. *Clathrus ruber*

Plate 56. *Cyathus olla*

Plate 58. *Enteromorpha* sp.

Plate 57. *Pisolitus tinctorius*

Plate 59. *Jania rubens*

CHAPTER 3

EFFECTS OF MAN

ARCHAEOBOTANY OF DORSET
by Pat Hinton

Archaeobotany is the study of plant remains recovered during the excavation of archaeological sites. The commonest state of recovered plant remains is in the form of charred fragments of durable parts such as seeds, cereal grains, fruit stones or charcoal. Other plant parts may survive for long periods in wet conditions when air is excluded, but few such cases have been found in Dorset. Most of the early evidence comes from pollen, which survives burial in peat and some buried soils. Where pollen is poorly preserved, as in calcareous soils, other evidence such as the molluscan fauna may illustrate ancient vegetation-types. The presence of plant remains is not proof that the plant was growing at the find-spot, nor does lack of a record indicate absence. Parts recovered are those which chanced to survive, and their condition often means that identification can be no closer than to genus.

When the Devensian glacial period ended, around 10,000BC, the coastline of what is now Dorset extended much further south, the land was covered by tundra or taiga and was sparsely peopled by hunting and fishing groups. After initial fluctuations, both climate amelioration and increasing human activity caused changes in the plant cover.

Early post-glacial (10,000–8500BC)
Much of the evidence of early vegetation comes from sites in the Poole basin where conditions are suitable for preserving pollen. Studies by Haskins (1980) and Scaife (1991) illustrate sequences of colonization, development and changes of forest cover. Late Devensian peat deposits at East Stoke (88T) and Morden (99B) show that birch was the dominant tree, with some pine and sallow and a high proportion of pollen from shrubs (*Calluna, Empetrum, Helianthemum, Hippophae* and *Juniperus*) as well as *Artemisia, Filipendula* and *Phragmites*. Alder pollen was also found at East Stoke, an early occurrence (3.2).

Mesolithic (8500–3500BC)
The pre-Boreal period, warmer and moister than before, began about 8500BC. Birch was still dominant, but a marked increase in pine pollen occurred at sites around Poole Harbour. Pine reached its maximum at about 8000BC in the earlier part of the warmer and drier Boreal period, but then was almost completely replaced by deciduous trees of alder, ash, elm and oak. Local stands of pine may have persisted here and elsewhere in southern England (Cameron & Scaife, 1988; Scaife, 1991). Sparse amounts of pine pollen from the Bronze Age peats near Poole Harbour, from Atlantic and later periods on the Isle of Wight, together with Bronze Age charcoal from Chick's Hill (88S), support this persistence. On Furzey Island (08D) pine pollen was found in buried soil under a Middle Iron Age enclosure bank.

Finds of flint tools, particularly in the south and north-east, are evidence of hunter-gatherers at this time. Pollen and molluscan studies indicate the clearance of some patches of forest, whether by gales, man or beavers is unknown, often followed by regrowth of forest. A study of the molluscs in a tufa deposit at Blashenwell (98H), first described by Mansel in 1857 (Mansel, 1886), indicate open marshy conditions at first, *c.*7000BC (Preece, 1980). This was followed by a shaded environment and then by partial clearance at the upper level, where a mesolithic midden was found. Impressions of leaves of elm, hazel and oak on the tufa surface were described by Reid (1896).

With the increasingly oceanic climate which marks the beginning of the Atlantic period around 5000BC, thermophilous species such as ash, beech, holly, hornbeam, lime and maple become prominent in pollen analyses; alder pollen also increased in all eight of the Poole basin sites studied (Haskins, 1980). A pollen profile from bog peat at Rimsmoor pond (89B) from the Boreal period to the present day shows an elm/ lime/ oak forest with alder, ash, birch, hazel and *Phragmites* at the transition from Boreal to Atlantic conditions (Walton & Barber, 1987).

Neolithic (3500–2000BC)
During the 4th millenium BC, changes in stone technology, pottery and cereal cultivation mark new human activities. The landscape was opened up and long barrows, causewayed enclosures and the Cursus in Cranborne Chase were constructed. Small areas were cleared to sow crops, but the first cereal cultivation seems to have been an addition to Mesolithic foraging rather than an abrupt innovation. At Rowden (68E) and Down Farm, Handley (01C) emmer *Triticum dicoccon* and barley *Hordeum vulgare* have been found, together with fruits and charcoal of apple, hawthorn and hazel. An early Neolithic site near Dorchester includes seeds of plants of disturbed soils,such as *Bromus* sp., *Fallopia convolvulus, Galium aparine, Veronica hederifolia* and *Vicia/Lathyrus* sp. A surprising find is a charred grape pip from Hambledon Hill (81L) with a date around 3400BC (Jones & Legge, 1987).

A decline in elm pollen occurred in this period, which may have been caused by the climate, disease (as the elm beetle *Scolytus scolytus* was present (Girling, 1988), or use of elm leaves as animal fodder (Robinson & Rasmussen, 1988); it coincides with an increase in herb and cereal

Table 13. Archaeobotanical records from Dorset.

Notes. Sites are given in approximate order of age. Genera are shortened to five letters, specific names (where known) to three letters. **BA** = Bronze Age;
C = Century; **e** = early; **IA** = Iron Age; **l** = late; **m** = middle; **Med** =Medieval; **Mes** = Mesolithic; **Neo** = Neolithic. This list is not fully inclusive, particularly for post-**IA**
sites; for fuller details the original references should be consulted.

Period	Site (Tetrad)	Taxa	Ref.
Mes	Blashenwell (98K)	*Coryl, Querc, Ulmus*	(Reid, 1896; Preece, 1980)
Mes	Gussage St Michael (91X)	*Pinus*	(Green & Allen, 1997)
Neo	Hambledon (81L)	*Bromu, Buxus sem, Coryl, Erica vag, Galiu apa, Horde vul, Lathy/Vicia, Triti dic, Vitis vin*	(Jones & Legge, 1987; in press)
Neo	Rowden (68E)	*Coryl, Crata, Fraxi, Horde, Querc, Triti dic*	(Carruthers, 1991b)
Neo	Thickthorn (91R)	*Coryl, Crata, Pinus, Pyrus, Querc*	(Maby, 1936)
Neo	Pamphill (90V)	*Crata, Prunu spi, Querc*	(Dimbleby, 1964)
Neo	Dorchester (68Z)	*Bromu, Fallo con, Galiu, Horde vul, Lathy/Vicia,Triti, Veron hed*	(Straker, 1997)
l Neo	Maiden Cas (68U)	*Coryl, Fraxi, Prunu, Querc*	(Gale, 1991)
Neo	Dorchester (69V)	*Horde, Prunu, Querc, Triti*	(Jones & Straker, 1993)
Neo	Maiden Cas (68U)	*Berbe, Cornu, Coryl, Fraxi, Prunu, Querc, Taxus*	(Gale, 1991)
Neo/BA	Pamphill (90R)	*Cheno, Triti*	(Ede, 1989)
l Neo	Handley (01C)	*Cheno, Coryl, Horde, Malus syl, Triti*	(Jones, 1980)
Neo/BA	Poundbury (69V)	*Horde, Persi lap, Triti spe?, Vicia*	(Monk, 1987)
EBA	Lyscombe (70K)	*Coryl, Fraxi, Querc*	(Morgan, 1980)
BA	East Stoke (88S)	*Alnus, Pinus syl, Querc*	(Ashbee & Dimbleby, 1958)
BA	Canford Heath (09C)	*Alnus, Betul, Callu, Coryl, Erica?, Polyp, Pteri aqu, Querc, Tilia, Ulmus, Verba nig*	(Green, 1980a)
BA	Maiden Cas (68U)	*Arrhe ela, Cheno, Fallo con, Galiu apa, Horde, Polyg avi, Rumex, Triti, Veron hed*	(Tomlinson, 1995)
BA	Poole (09A)	*Alnus, Betul, Coryl, Erica?, Fagus, Plant, Querc, Salix ?, Tilia, Ulmus*	(Dimbleby, 1952)
BA	Piddlehinton (69X)	*Fraxi?*	(Gale, 1990)
BA	Cowleaze (68E?)	*Acer, Coryl, Fraxi, Popul/Salix, Prunu, Querc*	(Carruthers, 1991b)
e/mBA	Tolpuddle Ball (89C)	*Alnus, Cornu, Coryl, Heder hel, Horde vul, Prunu, Querc, Triti dic*	(Hinton, 1999)
e/mBA	Wytch Farm (98S)	*Horde, Triti*	(Carruthers, 1991)
mBA	Wytch Farm (98S)	*Atrip, Avena, Bromu, Carex, Cheno alb, Cheno, Fallo con, Galiu, Horde vul, Lathy/Vicia, Persi lap, Plant lan, Prunu, Ranun, Rumex acetosel, Rumex, Sperg arv, Stell med, Trifo, Tripl ino, Triti dic, Triti spe*	
mBA	Dorchester (68U)	*Arrhe ela, Cheno alb, Horde vul, Triti, Valerianella, Veron hed*	(Carruthers, 1991)
mBA	Rowden (68E)	*Arrhe ela, Bromu, Coryl, Fallo con, Fraxi, Galiu, Horde vul, Lathy/Vicia, Polyg avi, Prunu, Querc, Sperg arv, Stell med, Triti dic, Triti spe, Vicia fab*	(Carruthers, 1991b)
lBA	Dorchester (68U)	*Aethu cyn, Atrip, Brass, Bromu, Cheno alb, Euphr/ Odont, Fallo con, Galiu apa, Horde vul, Linum cat, Medic lup, Papav, Persi mac, Plant lan, Polyg avi, Prune vul, Rumex, Sambu nig, Trifo, Tripl ino, Triti aes, Triti dic, Triti spe, Valer den*	(Carruthers, 1992)
eIA	Maiden Cas (68E)	*Anisa ste, Avena, Bromu, Horde vul, Persi lap, Secal cer, Triti aes, Triti dic, Triti spe*	(Helbaek, 1952)

Table 13. ...continued.

Period	Site (Tetrad)	Taxa	Ref.
eIA	Worth (97A)	*Triti spe*	(Helbaek, 1956)
eIA	Portland (67N)	*Avena, Bromu, Horde vul, Triti spe*	(Helbaek, 1952)
eIA W	Parley (09Y)	*Betul, Querc, Pteri aqu*	(Drew, 1929)
e/mIA	Sutton Poyntz (68W)	*Agros git, Anthe cot, Atrip, Avena, Bromu sec?, Carex acuti?, Cheno, Coryl, Euphr/Odont, Fallo con, Festu?, Galiu, Horde vul, Lapsa com, Litho arv, Loliu per?, Medic lup, Phleu?, Plant lan, Plant maj, Poa ann, Rumex, Stell, Thali fla?, Trifo pra?, T. rep?, Triti aes, T. dic, T. spe, Valer den, Vicia fab, Vicia hir?, Vicia sat, Vicia tet, Viola*	(Hinton, to be published)
e/mIA	Hengistbury (19Q)	*Atrip, Avena, Brass/Sinap, Callu vul, Capse bur, Carex, Cheno alb, Cheno fic, Cheno vul, Coryl, Eleoc pal, Fallo con, Horde vul, Juncu, Luzul, Persi mac?, Plant maj, Polyg avi, Poten, Rapha rap, Rumex acetosella, Rumex cri, Sperg arv, Stell gra, Stell med, Trifo, Triti aes, Triti dic, Veron ser*	(Nye & Jones, 1987)
l IA/Roman	Hengistbury (19Q)	as above + *Anthe arv, Artem?, Bolbo mar?, Calth pal, Cardam?, Cirsi, Crata, Eriop, Galiu, Genis ang, Hyosc nig, Lamiu, Linum cat, Linum usi, Malva?, Monti fon, Myosotis, Persi lap, Pisum, Plant lan, Ranun fla, Sambu nig, Sorbu auc, Stell uli, Tripl ino, Triti spe, Ulex, Valer dent, Vicia fab, Viola*	(Nye & Jones, 1987)
IA	Dorchester (78E)	*Acer, Coryl, Fagus, Popul, Querc, Taxus*	(Morgan, 1979)
m/l IA	Wytch Fm (98S)	*Avena, Bromu, Callu vul, Carex, Cheno alb, Chrys seg, Erica cin?, Erica tet?, Fallo con, Galiu apa, Horde, Monti fon, Scler ann, Trifo, Triti dic, Triti spe, Veron hed, Vicia fab*	(Carruthers, 1991c)
l IA	Gussage All Sts(01A)	*Agros git, Atrip pat, Avena, Bromu, Horde vul, Lens ?, Litho arv, Rumex cri, Triti aes, Triti spe, Vicia fab, Vicia tet*	(Evans & Jones, 1979)
l IA	Dorchester (68Z)	*Anisa ste, Arrhe ela, Avena, Bromu, Cheno fic, Danth dec, Eleoc, Fallo con, Galiu apa, Horde vul, Litho arv, Plant lan, Polyg avi, Ranun fla +, Rumex, Sangu min, Stell med, Trifo pra?, Tripl ino, Triti spe, Urtic dio, Urtic ure, Valer den*	(Straker, 1997)
l IA	Wytch Fm (98S)	*Atrip, Avena, Bromu, Fallo con, Persi mac?, Polyg avi, Prunu spi, Rumex acetosella, Scler ann, Secal cer, Sperg arv, Veron hed*	(Carruthers, 1991c)
IA	Maiden Cas (68T)	*Acer, Aira, Alnus, Anisa ste, Aphan arv, Arrhe ela, Atrip, Avena, Brass, Bromu, Callu?, Carex, Cheno alb, Cirsi?, Cornu, Danth dec, Empet nig, Erodi, Fallo con, Festu, Fraxi, Galiu, Hiera, Horde vul, Hyosc nig, Lens?, Leuca vul, Litho arv, Monti fon, Persi hyd?, Persi min, Phleum, Plant lan, Poa ann, Polyg avi, Prune vul, Prunu, Querc, Rumex obt +, Sambu, Senec, Shera arv, Silen, Sperg arv, Stell med, Thlas arv, Toril, Trifo, Tripl ino, Triti aes, Triti dic, Triti spe, Ulex, Veron, Vicia fab, Viola*	(Gale, 1991; Palmer & Jones, 1991)
IA	Pilsdon Pen (40A)	*Alnus, Coryl, Querc*	(Limbrey, 1977)
IA	Poundbury (69V)	*Avena?, Atrip, Avena, Bromu, Cheno alb, C. rub?, Fallo con, Fumar, Horde vul, Hyosc nig, Juncu, Loliu, Medic lup, Papav, Persi lap, Pisum, Plant lan, Polyg avi, Poten, Ranun, Rumex acetosella, Sambu, Silen, Solan nig, Sonch?, Stell gra, Stell med, Trifo, Tripl ino, Triti aes, Triti spe, Urtic dio, Vacci myr, Valer den, Valer rim?, Vicia fab*	(Monk, 1987)
IA	Tolpuddle (89C)	*Agrostis, Anisa ste, Arrhe ela, Atrip, Avena, Bromu sec?, Callu, Carex, Ceras fon, Cheno alb, Cheno pol, Cirsi, Coryl, Erica, Fallo con, Fraxi, Galiu apa, Horde vul, Litho arv, Loliu per, Luzul,*	(Hinton, 1999)

Table 13. ...continued.

Period	Site (Tetrad)	Taxa	Ref.
IA	Tolpuddle (89C)	*Lychn flo, Medic lup, Molin cae, Papav som, Plant lan, Poa ann, Polygala, Prunu, Pteri aqu, Querc, Ranun, Rumex acetosa, Rumex cri, Sambu, Shera arv, Stell gra, Stell med, Thymu, Trifo pra, Tripl ino, Triti aes, Triti dic, Triti spe, Ulex?, Valer den, Veron cha?, Vicia hir, Vicia sat?, Vicia tet*	(Hinton, 1999)
I IA	Worth (97U)	*Avena, Bromu, Callu, Carex, Cheno alb, Euphr?, Fallo con, Galiu apa, Galiu ver?, Horde, Linum bie, Malva syl, Medic lup, Plant lan, Poa, Polyg avi?, Ranun par, Rumex acetosella, Shera arv, Sisym off, Stell gra?, Stell med?, Trifo cam?, Trifo pra?, Triti, Urtic dio, Urtic ure, Veron ser?, Vicia fab, Vicia hir/tet*	(Woodward, 1991)
I IA/Roman	Wytch Fm (98S)	*Atrip, Avena, Bromu, Callu, Carex, Cheno alb, Coryl, Cytis sco, Erica cin?, Erica tet?, Eriop, Fallo con, Galiu apa, Horde, Hyosc nig, Monti fon, Myosoton aqu, Persi amph?, Plant lan, Plant maj, Pteri, Ranun sar +, Rapha rap, Rumex acetosella, Shera arv, Stell med, Tripl ino, Triti aes, Triti dic, Triti spe, Ulex, Valer den, Vicia fab*	(Carruthers, 1991c)
I IA/Roman	Worgret (98D)	*Coryl, Fagus, Prunu, Querc, Ulmus*	(Carruthers, 1991a)
I IA/Roman	Sutton Poyntz (68W)	*Anthe cot, Avena, Brass, Bromu sec?, Carex fla?, Carex ova, Coryl, Galiu apa, Horde vul, Loliu, Medic lup, Papav, Poa ann, Polyg avi, Prunu spi, Ranun par, Rumex, Trifo pra?, Triti aes, Triti dic, Triti spe, Vicia fab, Vicia hir, Vicia sat, Vicia tet, Viola*	(Hinton, to be published)
Roman	Maiden Cas (68U)	*Atrop bel, Danth dec, Litho arv, Papav som, Shera arv, Sparg, Valer den*	(Straker, 1997)
Roman	Dorchester (69V)	*Anisa ste, Triti aes, Vicia fab*	(Aitken, 1982)
Roman	Dorchester (69V)	*Anthe cot, Ceras?, Coniu mac, Ficus, Galeo tet, Hyosc nig Lapsa com, Malus/Pyrus, Prunu dom?, Rubus fru, Thlas*	(Ede, 1993)
Roman	Dorchester (69V)	*Anthe cot, Apium, Eleoc, Ficus, Lemna, Malus/Pyrus, Plant med, Resed luteola*	(Jones & Straker, 1993)
Roman	Poundbury (69V)	*Cheli maj, Conop mac, Fagus?, Pisum, Solan nig*	(Monk, 1987)
Roman	Dorchester (69V)	*Leuca vul, Papav som, Ranun fla, Rapha rap, Silen lat*	(Letts, 1997)
Roman	Poxwell (78H)	*Triti aes, Triti spe?, Horde vul*	(Jones, 1986)
Roman	Broadmayne (78I)	*Avena, Triti aes?, Triti dic, Triti spe*	(Green, 1980)
Roman	Marnhull (71U)	*Coryl, Fraxi, Ilex, Querc*	(Williams, 1950)
Roman	Tolpuddle (89C)	*Agrostis, Callu vul, Erica*	(Hinton, 1999)
Roman ?	Cann (82Q)	*Buxus sem*	(Gray, 1918)
Roman	Worgret (98D)	*Angel syl, Epilo, Fagus, Juncu, Monti fon, Poten, Ranun rep, Ranun sect. Bat, Rubus fru, Sagina, Sonch asp?*	(Ede, 1988a; Carruthers, 1991a)
Roman	Wareham (98I)	*Agros git, Callu vul, Linum usi, Malus, Pisum, Vacci*	(Green, 1978)
Roman	Bucknowle (98K)	*Canna sat, Malus, Prunu, Rubus, Sambu, Vitis vin*	(Green, 1980b)
Roman	Wytch Fm (98S)	*Callu vul, Chrys seg*	(Carruthers, 1991c)
I Roman	Brownsea (08J)	*Callu vul, Carex, Erica, Honck pep, Juncu art, Juncu squ, Suaed mar, Trigl + Sphagnum pap*	(Haskins, 1992)
Roman	W. Moors (00W)	*Ceras, Epilo, Erica?, Galeo tet, Hyosc nig, Lycop eur, Ranun sect. Bat, Rubus fru, Sagin, Silen, Sonch asp?*	(Ede, 1988)
Post-Roman	Poundbury (69V)	*Aethu cyn, Anthe cot, Avena fat, Buple rot?, Cheli maj, Chrys seg, Coniu mac, Galeop?, Hyosc nig, Papav rho, Picri ech, Prune vul, Sinap arv, Solan nig, Toril jap, Valer den*	(Monk, 1987)

Table 13. ...continued.

Period	Site (Tetrad)	Taxa	Ref.
m Saxon	Gillingham (82C)	*Acer, Agros git, Alnus, Anthe cot, Prunus, Ulmus*	(Ede, 1991)
Med	Dorchester (69V)	*Eleoc, Medic lup?, Raph rap, Ulex eur*	(Jones & Straker, 1993)
Med	Sutton Poyntz (68W)	*Agros git, Anthe cot, Bromu sec?, Carex, Papav som*	(Hinton, to be published)
Med	Wimborne (09E)	*Agros git, Anthe cot*	(Ede, 1992)
Med	Bere Regis (89H)	*Anthe cot, Centa cya?, Chrys seg, Rapha rap*	(Tomlinson, 1995)
Med, C12–13	Bere Regis (89H)	*Agros git, Callu vul, Erica tet, Malva, Trifo cam/dub*	(Hinton, to be published)
Med, C10–12	Wareham (98I)	*Carex?, Rubus fru*	(Monk, 1977)
Med, C12–13	Wareham (98I)	*Agrim?, Alche, Aphan arv, Buple rot, Callu, Chrys seg, Danth dec, Erica cin, Erica tet, Eupho hel, Isolep, Linum usi, Ranun fla, Rapha rap, Rubus ida, Rubus fru, Verbe off*	(Carruthers, 1995)
Med, C13	Wytch Fm (98S)	*Agros git, Anaga min?, Callu vul, Centa cya, Chrys seg, Erica cin?, Erica tet?, Ranun fla, Rapha rap*	(Carruthers, 1991c)
Med, C13–14	Rowden (68E)	*Coryl, Fraxi, Prunu, Querc*	(Carruthers, 1991b)
Med, pre C16	Christchurch (19R)	*Jugla reg, Prunu avi, Prunu dom, Rubus fru, Vitis vin*	(Green, 1983)
Med, C16–18	Christchurch (19R)	*Acer pse, Coryl ave, Prunu avi, Prunu dom*	(Green, 1983)
Post-Med	Wimborne (09E)	*Anthe cot, Bromu, Horde vul, Triti aes*	(Ede, 1992)

pollen. At Rimsmoor the decline in elm pollen occurred over some centuries within the range 4220–3790BC and 3640–3210BC (Watson & Barber, 1987; Woodward, 1991). Lime pollen, which appeared in the pre-Boreal and increased during the Atlantic period to dominance in the Sub-Boreal, declined as the climate cooled. At Rimsmoor the decline of lime pollen occurred between 2470 and 2040BC (Watson & Barber, 1987; Woodward, 1991). There are various factors affecting the interpretation of this finding, but the cause is likely to be anthropogenic rather than climatic.

On the Chalk south of Dorchester, where pollen is poorly preserved, work on snails shows that the landscape was changing (Allen, 1997). Evidence of primary woodland was found in the early Neolithic of Maiden Castle (68U) at *c*.4000BC, but snails of long grassland occurred in a buried turf of similar date at Dorchester. After *c*.3500BC snail evidence from a bank barrow and a circle of post pits suggested grazed grassland. Other sites illustrate woodland clearance, and by *c*.2500BC this area had become a patchwork of small arable fields in open pasture. Established, and undoubtedly managed, woodland was available around 3000BC for the palisades of henge monuments such as Mount Pleasant (78E), and the arc of pits marked in the basement of a car park at The Greyhound Yard, Dorchester once bore oak posts averaging 1 m in diameter.

Bronze Age (2000–600BC)
Valuable evidence of the environment prior to the construction of Bronze Age round barrows is revealed when buried land surfaces are examined. These confirm woodland clearance, at the end of the 5th millenium BC in the case of barrows on the South Dorset Ridgeway. At *c*.2000BC pollen of alder, birch, hazel, holly and oak, with herbs such as *Dianthus* sp.?, *Plantago coronopus*, *P. lanceolata*, *Ranunculus*, *Rumex* and *Taraxacum* and spores of bracken and polypody were found at Golden Cap (49B: Scaife, 1993). In the Poole basin hazel, ericaceous shrubs and herbs increased at this time. Podzolisation, possibly starting in the Mesolithic period, led to the formation of heath, with pollen of *Agrostis*, *Crassula tillaea*, *Rhynchospora* and *Ulex* and macrofossils of *Erica ciliaris* and *Gentiana pneumonanthe* at Rempstone (98S: Haskins, 1980).

Agriculture grew in importance during the later Bronze Age, and delimited fields associated with small round houses were constructed on ridges of the Chalk. Spelt (*Triticum spelta*), with emmer and barley, became major crops suited to the soils then present. Repeated tillage resulted in erosion of the Brown Earth soils on the slopes with deposition at their feet (Limbrey, 1975), and the higher parts of the Chalk downs became largely pasture.

Iron Age (600BC–43AD)
The construction of hillforts in this period may reflect increasing pressure on fertile land, and the need to safeguard farm animals and stored crops. Increasing trade with mainland Europe probably brought in new weedy species, and most of the commoner arable weeds and grassland plants were present in the country, if not yet found in Dorset, by this time. Cereal acreage increased, as storage pits and grain driers testify. *Avena strigosa* and *Secale cereale* may have occurred as weeds before they were selected for cultivation. Other crop-plants, such as *Pisum sativum*

and *Vicia faba* which were first found in the Neolithic, are found more frequently in this age. Henbane (*Hyoscyamus niger*), a ruderal drug-plant, has been found in Iron Age and Roman deposits at Dorchester, Maiden Castle and Wytch Farm (98S).

Roman (43–400AD)

During the Roman era, new roads were built, and towns and villas established, with small native farmsteads continuing alongside. Agriculture was intensive, and spelt the major wheat crop. There are more records of arable weed seeds from Roman sites than from any other period. Villas almost certainly had gardens, and it is conceivable that plants such as *Artemisia absinthium*, *Foeniculum vulgare* and *Vinca minor* are survivals from these (see Chapter 3D). However there are no modern records of plants which might be considered to be survivors at the main Roman town of Dorchester, nor along the main roads of this period. Leaves of box, *Buxus sempervirens*, first found in the Neolithic on Hambledon Hill, lined a lead coffin at Cann (82Q), while remains of hemp, *Cannabis sativa*, were found in a waterlogged deposit in a Roman bath-suite at Bucknowle (98K).

Post-Roman (400–1400AD)

There is little information from the Saxon period, but ovens in Gillingham (82D) contained charred grains of bread wheat *Triticum aestivum* and weeds such as *Agrostemma githago* and *Anthemis cotula*. These seeds, also found in 5th–6th century AD grain driers at Poundbury, became more common as heavy clay soils were cultivated using the deep plough with mould-boards.

Charred cereal and weed seeds from Medieval sites in Bere Regis, Dorchester, Sutton Poyntz, Tolpuddle, Wareham and Wimborne include new crop plants, especially fodder plants and legumes such as *Vicia sativa* subsp. *sativa*. Flax (*Linum usitatissimum*), cultivated since the Neolithic period for its oil-rich seeds, became important for netting and rope-making at Bridport from the early 13th century (Pahl, 1960). The investigation of a garderobe at Christchurch Priory revealed fruits of grape, walnut and sycamore. The last of these was in a level dated 1500–1700AD, about the time of the trees's introduction to Britain.

FORESTRY IN DORSET
by Roger McKinley

In the 'wildwood' before 5000BC Dorset probably had lime as the commonest tree, then hazel, oak and elm, with alder in the wetter areas (Rackham, 1986). This was much modified by human activities, and by medieval times the term 'forest' meant an area used by the king for hunting, whether it carried trees or not. Today it is convenient to use 'forestry' to define the art of growing and managing trees in large plantations, and 'woodland management' for managing smaller woods in a more traditional manner. The skills may be complimentary, but they are certainly not the same, and the latter were nearly lost in Britain during this century. The harvesting of woodland products has been practised by rural populations from prehistory until c.1914. The emphasis has varied, but one feature has remained timeless – that man was exploiting the surplus of a naturally renewable resource, leaving the resource to grow again for the future. Until Napoleonic times the underwood was often more valuable than timber trees, and woodland renewed itself after coppicing by natural regeneration. The trend this century has been to plant young trees as a timber resource, to fell this when mature and renew it by replanting. This is an efficient but highly artificial system.

In 1793, Claridge wrote of Dorset 'the county is extremely barren, both in timber and wood' (Claridge, 1793). In 1998 the picture is rather different, with trees covering about 11% of the county, of which 77% are conifers and 23% deciduous trees. Corsican pine is the most common conifer (53%), with Scots pine and Douglas fir at 15% and 13% respectively. Scots pine was once native, but all other conifers are introductions. Among the planted broadleaved trees, beech is the most common (45%), with mixed woodland (26%) and oak (10%) coming next. Since beech is probably not native in Dorset, existing woods and plantations consist mainly of introduced trees. New plantations brought a range of new problems, ranging from a huge increase in deer numbers to infestations by pests such as the great spruce bark beetle *Dendroctonus micans*.

The use of nursery-grown saplings for planting trees developed between 1760 and 1850, at a time when many alien tree species were brought to Britain (Jones, 1981). It had a profound effect on forestry and on large parks. Not only did this technique allow the creation of a major forest resource, but also it led to the degradation of many ancient woods, and, in Dorset, the 'coniferisation' of heathland. For example, in 1808 Dr Bain planted over 5,000 *Pinus pinaster* in the Bournemouth area, where the tree is now naturalised.

British woodlands have been destroyed by man on a huge scale, from the early Bronze Age when woodland cover was about 90% to the 1920s when it had been reduced to about 4%. Today the area stands at about 12%, making Britain the least wooded country in Europe, other than Ireland. The Forestry Commission was formed in 1919, initially to create a reserve of timber to sustain the nation during a three-year war, as well as to provide employment. Its massive afforestation policy continued to 1945. Its new forests were designed for efficiency – laid out and fenced in straight lines, and planted with the quickest growing species possible, these species invariably being introductions. Douglas Fir, for example, grows to timber size in less than 50 years, compared with oak which takes at least 150 years. The conifer also produces about four times as much timber per year from a given area of land. In economic terms, native species simply cannot compete – but good woodland management involves much more than producing timber.

On private land, management has always been more traditional, with shooting a by-product important to many owners. Both fine timber and many woodland products resulted, and the silvicultural input by resident woodland

Figure 9. Forestry statistics for Dorset (abstracted from Forest Enterprise information).

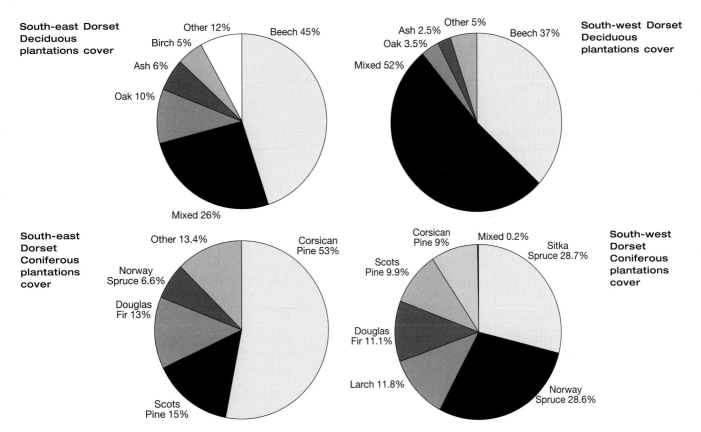

South-east Dorset Deciduous plantations cover

South-west Dorset Deciduous plantations cover

South-east Dorset Coniferous plantations cover

South-west Dorset Coniferous plantations cover

staff was impressive. It was usually on a small scale, however, and many private foresters formed the backbone of the early Forestry Commission. The instincts of foresters to grow the right tree on the right soil were not entirely subordinated to the demands of the Civil Service. This can be seen in many Forestry Commission plantations and woods on chalk or clay soils, especially in west Dorset. Here conifers were used as a nurse crop for young beeches or oaks, and to provide an early return from thinnings, while the broadleaved trees were intended as a final crop.

In the mid-1960s, national priorities began to change. Dense, serried ranks of conifers were said to be unfriendly to wildlife. Of course the pendulum of criticism swung too far, and of course the forest industry was slow to realize the need for change. However one should see this against the big picture, which aims to rebuild a woodland resource which had fallen to a critically low level. In the 1980s, the emphasis changed again. Ancient woodlands have better protection, and incentives are in place to grow more deciduous trees, to restore traditions such as hazel coppicing, and to leave more open glades in woodland for wildlife. There is a new policy of clearing conifers from corridors of land in order to connect up previously isolated fragments of heathland. Many foresters are learning a lesson that their forefathers could have taught them; that working with nature in a low-input – low-output way can be more effective than fighting it, and that plantations can be both pleasing to the eye and good for nature conservation.

In prehistoric times wood was used as fuel and to build houses, fences and fortifications. Military and Naval uses became of increasing importance from 1066 onwards, large buildings needed massive beams and peasants needed fuel. Today wood grown in Dorset is used for a range of purposes: in buildings, often in the form of reconstituted products such as chipboard, and for vast numbers of fence posts; for fuel, especially beech thinnings, and for making charcoal; beech is in great demand to make brown paper and cardboard. Coppice products are also making a small but welcome comeback.

AGRICULTURE IN DORSET

Hinton (above) summarized evidence for the beginnings of farming in Dorset. Around 3000BC the long-barrow people had stock animals which grazed in the woodlands on the Chalk, and grew emmer wheat and barley in small clearings. A similar pattern of farming was followed by the round-barrow people, who arrived about 2500BC, for many years. Around the 8th century BC there was a revolution in arable farming, caused by the introduction of spelt wheat which could be germinated in the autumn (Cunliffe,1974). Later, Iron Age invaders, a Celtic group who were called Durotriges by the Romans, increased the area under cultivation to an estimated 200 km^2 (50,000 acres), and were able to plough on slopes as steep as 24°. Many arable weeds were present. The Celts built at least 27 hill-forts in the county, the largest of which is Maiden Castle, and outlines of 'Celtic fields' are widespread on the Chalk.

After the Romans had conquered the Durotriges in the 1st century AD, the Celtic farmers did not change their methods of farming much, except on the large estates associated with the 20 Villas in the county, where new crops were grown using iron ploughs. During the 350 years of settled conditions, the human population increased by a factor of five. When the legions left in 406AD, the Villas were abandoned but peasant farming went on. A major change occurred when the Saxons invaded Dorset from the north-east, breaching the Bokerly Dyke and Combs Ditch, around 700AD. The Celts departed, and nearly all our place-names today are of Saxon origin. It is possible that Celtic impoverishment of upland chalk soils may have encouraged the Saxons to clear and plough the clay vales with their heavy ploughs. The pattern of small, hedged fields or assarts found in west Dorset today is a legacy of woodland clearance by the Saxons. After 900AD, monasteries such as Bindon, Cerne, Milton and Sherborne were founded and some became managers of large estates. With different owners, such large estates persist to the present day; in 1991, 27% of the county was held in estates exceeding 3 km² (750 acres).

The Norman conquest, followed by the compilation of Domesday Book in 1086, allows certain inferences. The human population was about 35,000, and settlements were uniformly distributed on chalk and clay soils, but were sparse on heathland. About 44% of the county was arable land, sheep were numerous, there were 276 watermills to grind corn, and orchards and vineyards had been planted. The pattern of large, feudal estates owned by Norman nobles, co-existent with many peasant farmers, continued as the population increased for three centuries. Common fields, regular fallowing and dunging by sheep, and strip cultivation (still seen in south Portland) were the rule. In 1348, the Black Death entered the country at Weymouth, causing a catastrophic drop in population and a decline in farming. Recovery was slow, with some villages permanently abandoned. Monastery records show that seven institutions kept a total of 25,000 sheep; however in the 16th century

monastic estates were sold by Henry VIII. Enclosure of underused land began then and continued until the early 19th century. More land was made available to farmers by disafforestation; for example the Forest of Gillingham, once a hunting precinct, was disafforested in 1628. The management of water-meadows, to produce spring pasturage, began in the 17th century and continued for 200 years until made obsolete by the advent of artificial fertilizers. In 1796 W Marshall noted weedy, waterlogged fields in west Dorset, with fallowing every fourth year, and much unenclosed chalk downland. At this time hemp and flax were locally important fibre crops near Bridport, together with woad for dyeing, as they had been for 400 years.

Only since the mid-19th century have annual statistics been kept on farmer's crops and stock animals (Simons, 1998; MOAFF, 1866–1966; HMSO, 1991; Bettey, 1974, 1980; Figure 11). During this period the human population has risen from 159,000 in 1831 to 647,000 in 1991. Between 500BC and 1800AD the area of grassland in the county exceeded that of arable land, except perhaps between 1100 and 1350. This reflects the dominant position of cows, sheep, and to a lesser extent pigs in its agricultural history. Since 1866, there have been marked fluctuations in the ratio of arable to grassland, corresponding to changes in the price of wheat, meat and dairy products. From 1866 to 1910 some arable reverted to grassland each year, in an agricultural depression referred to by Mansel-Pleydell in his *Flora* (1895). Despite some reversal of this trend during the 1914–1918 war, grassland increased at the expense of arable until 1939, when there was less arable land in Dorset than there had been for many centuries. Note that Good collected most of the data for his *Flora* during the period 1930–1939 (Good, 1948). The onset of the second world war in 1939 triggered a massive increase in arable, which continued until at least 1965. Since then some arable has returned to sown grassland. The number of sheep kept in the county confirms this fluctuating picture: 500,000 in 1850, 300,000 in 1900, 47,000 in 1947 and 202,000 in 1996.

The agricultural changes in the second half of the 20th century, when data for the present Flora have been

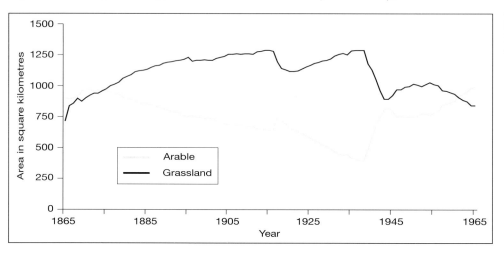

Figure 10. Areas of arable land and grassland in Dorset 1865–1965.

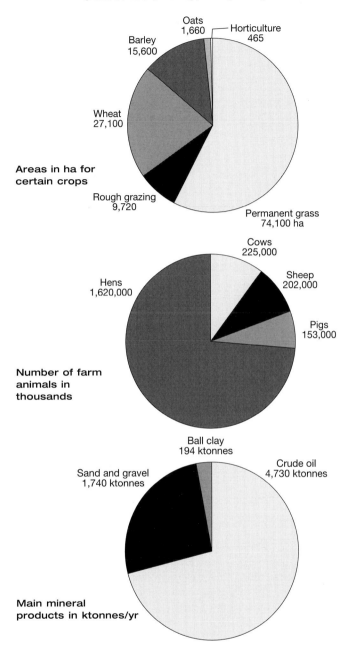

Figure 11. Agricultural statistics for Dorset in 1996 (HMSO, 1991; Bettey, 1974, 1980).

Areas in ha for certain crops

- Oats 1,660
- Barley 15,600
- Horticulture 465
- Wheat 27,100
- Rough grazing 9,720
- Permanent grass 74,100 ha

Number of farm animals in thousands

- Cows 225,000
- Hens 1,620,000
- Sheep 202,000
- Pigs 153,000

Main mineral products in ktonnes/yr

- Ball clay 194 ktonnes
- Sand and gravel 1,740 ktonnes
- Crude oil 4,730 ktonnes

with a monoculture of rye grass. Only rarely, as at Burton Common and Keysworth, is light grazing by cattle used; most grazing is intensive, and flowering and fruiting of any remaining native plants is discouraged. At the same time heathy commons in the Poole basin, once grazed by goats, geese or other stock, now suffer from undergrazing so that tiny ephemeral plants are squeezed out. Much of the wetter grassland has been drained, irrigation channels for water-meadows have been abandoned and allowed to silt up, while most field ponds on clay are now either dry for half the year or shaded out by willows.

The most boring fields for a botanist are those used for arable prairie farming on the Chalk. These are mostly monocultures of wheat or barley, less commonly of oilseed rape or flax, fed by chemical fertilizers and repeatedly sprayed to kill weeds and insects. The herbicides used, along with cleaner seed than formerly, has led to the decline or extinction of many familiar cornfield weeds, and their replacement by herbicide-resistant or nitrophilic species (see Table 14). Another factor is the speed of ploughing after harvest; scarcely has a group of combines removed the straw and grain before another posse of tractors arrives to disc-harrow, plough, spray and re-fertilize the stubbles, so that autumn-germinating annuals have little chance to flower or fruit. Set-aside fallow schemes have offered some respite. Of the many doom-laden predictions for arable farming of this kind, two could be serious. One is the compaction of soils by the huge weight of current farm machinery. The other is the long-term retention of poisons in the soil, notably the herbicide paraquat, which displaces potassium from the clay minerals which are the basis of soil fertility.

Table 14 gives the number of sites or tetrads from which arable weeds were recorded during the period indicated. 1850–99 records are due mostly to JC Mansel-Pleydell or EF Linton, who give their subjective views on frequency in their *Floras*. 1925–1940 records are based on RD Good's records and Handbook plus records by AW Graveson, D Meggison and others, and are probably underestimates.

In summary, of 66 less common arable weeds, nine appear to have increased recently; all nine are nitrophiles, but not all nitrophiles have increased. Of the 25 decreasing weeds, of which 13 appear to be extinct, seven began to decline before 1930 and 18 after 1940. The other 32 species cannot be said to have changed in abundance much from the limited evidence given here.

DORSET GARDENS AND THEIR PART IN PLANT INTRODUCTION

The first gardeners in Britain were probably the Romans. The courtyards of their villas of the 3rd and 4th centuries AD, as at Frampton, Halstock, Hinton St Mary and Tarrant Hinton, had formal gardens. Trees such as apple, plum, vine and walnut, and probably box, fig, laurel and plane, were grown, with smaller plants such as chervil, ground

collected, have been much more radical than mere statistics show. Hedges, once allowed to burgeon but layered at intervals, are now effectively castrated by mechanical flails each autumn. Their associated verges are mowed 'for tidiness', but the debris is never removed, thus ensuring that few herbs set seed and a dull sward of coarse grass becomes dominant. In arable districts many field hedges have been uprooted by bulldozers. Permanent grassland still occurs on some farms, as at Kingcombe, but it is exceptional to find grass fields which have not been 'improved' by ploughing, adding fertilizer and reseeding

Table 14. Summary of dated records of some less common arable weeds in Dorset.

Species	1850–1899	1925–1940	post-1985	Species	1850–1899	1925–1940	post-1985
Adonis annua	7,r	9,o	4,r#	*F. vaillantii*	1,r	0	1,r
Agrostemma githago	n,f	31,o	14, r#	*Galeopsis angustifolia*	17,f	11,f	12,r#
Alopecurus myosuroides	n,f	n,o	75,o	*Galium tricornutum*	10,o	8,o	0#
Alyssum alyssoides	3,r	1,r	0#	*Gastridium ventricosum*	19,r	10,o	10,r
Amaranthus retroflexus	1	0	34,o*	*Kickxia elatine*	24,f	11,f	107,o
Anchusa arvensis	23,f	5,f	85,o	*K. spuria*	39,f	2+,f	83,o
Anthemis arvensis	18,r	4,o	7,r#	*Lamium hybridum*	5,r	2,o	35,o*
A. cotula	n,f	9,f	53,o	*Legousia hybrida*	30,f	7,f	36,o
Apera spica-venti	2,r	2,r	2,r	*Lithospermum arvense*	23,f	10,o	9,r#
Arnoseris minima	4,r	7,r	0#	*Matricaria recutita*	3,o	n,o	122,o
Avena strigosa	4,r	l,r	0#	*Melampyrum arvense*	1,r	0	0#
Briza minor	15+,o	10,o	10,r	*Misopates orontium*	17,f	10,f	44,o
Bromus secalinus	13+,o	2,f	0#	*Myosurus minimus*	13,o	8,o	2,r#
Bupleurum rotundifolium	l,r	3?	0#	*Papaver argemone*	13,f	12,f	23,r
Caucalis platycarpa	1,r	2	0#	*P. hybridum*	9,r	11,o	28,o
Centaurea cyanus	16,o	7,o	24,r	*Petroselinum segetum*	24,r	6,o	32,r
Centaurea solstitialis	4,r	5,r	0#	*Polygonum rurivagum*	7,r	0	5,r
Chenopodium ficifolium	3,r	1,o	102,o*	*Ranunculus arvensis*	25,o	12,o	9,r#
C. glaucum	0	2,r	2,r	*Scandix pecten-veneris*	n,f	17,f	6,r#
C. hybridum `	15,r	8,o	27,r	*Scleranthus annuus*	n,f	10,f	15,r#
C. murale	16,r	12,f	23,r	*Sherardia arvensis*	n,f	n,f	191,o
C. rubrum	8,r	n,f	213,f *	*Silene gallica*	29,f	21,f	11,r#
C. urbicum	4,r	3,r	0#	*S. noctiflora*	0	5,r	7,r
Chrysanthemum segetum	n,a	n,f	87,f	*Solanum nigrum*	14,o	n,f	409,f *
Consolida ajacis	2,r	2,r	3,r	*Stachys arvensis*	28,o	6,f	66,o
Echinochloa crus-galli	1,r	11,r	21,o*	*Torilis arvensis*	15,o	7,o	0#
Erysimum cheiranthoides	9,r	4,o	6,r	*Urtica urens*	n,f	13,f	141,f *
Euphorbia exigua	n,f	1+,f	62,o	*Valerianella carinata*	2,r	7,o	114,o*
E. lathyris	4,r	11,o	93,o*	*V. dentata*	n,f	13,f	9,r#
E. platyphyllos	6,r	7,o	11,r	*V. rimosa*	20,r	9,o	1,r#
Filago lutea	2,r	0	0#	*Veronica agrestis*	2+,o	1+	30,o
F. pyramidata	6,r	1?	0#	*V. polita*	n,f	3+	133,o
Fumaria densiflora	10,r	4,r	4,r				
F. parviflora	4,r	1,r	3,r				

a = abundant, f = frequent, n = many, o = occasional, r = rare,
+ = underestimate, * = increasing, # = decreasing

elder, nettle, valerian and white deadnettle, and probably fennel, mint, parsley, periwinkle, sweet violet and wormwood (Putnam, W, 1984; Sanecki, 1994; Ryley, 1995). While some of these could have escaped when the Romans left around 400AD, they are not obvious relicts near the sites of villas. It is possible that fennel, fig, parsley and wormwood on Portland might once have grown in a Roman quarryman's garden.

Gardening records are scant during the next 900 years, during which the Celts left and were replaced by the Saxons. There are 17 Saxon place names related to the pear in Dorset, such as Parley, Parnham and Purcombe, suggesting that this tree was widespread then (Mills, 1986). By 816AD monastic gardeners grew alexanders, betony, catmint, chervil, clary, fennel, mint, opium poppy, pennyroyal, radish, tansy, white horehound and wormwood (Sanecki, 1994; Thacker, 1979).

After the Norman Conquest, Corfe Castle became a seat of power; its mound is still botanically interesting, with plants which could have escaped from early gardens, e.g. alexanders, black horehound, borage, ground elder, vervain and a large-flowered form of calamint. In the 15th century John Gardener lived at Lytchett Heath in Dorset, and in about 1440 he wrote a poem called 'The Feate of Gardening'. This mentions at least 78 species suited for gardens, including alexanders, borage, camomile, coriander, dittander, elecampane, fennel, garden orache, henbane, pellitory, periwinkle and white horehound (Amherst, 1896). Many more cultivated plants are described in the Herbals of Turner (1550) and Gerard (1597). From these and later sources we can establish the approximate dates of introduction to Britain, though not to Dorset, as listed in Table 15 (Mills, 1986; Amherst, 1896; Ilchester, 1899). Early introductions such as *Leucojum vernum* and *Tulipa sylvestris* survive in the wild here: the best known site of the latter plant is near the 15th century turf-maze at Leigh. A similar maze at Pimperne was ploughed up in 1730 (Thacker, 1979).

It is hard to say which garden has the longest continuous history in the county. Monasteries, which kept plants in cultivation for centuries, were dissolved in 1539. Surviving monastic buildings at Bindon, Cerne, Forde and Milton Abbey have redeveloped their gardens recently, while Sherborne Abbey is urbanized. Our two oldest gardens may be those of Cranborne Manor and Abbotsbury (Thacker, 1979). Cranborne was designed by John Tradescant in the 17th century. Abbotsbury was laid out in the same period, and by 1899 was growing an impressive number of plants, even including *Cypripedium calceolus* (Ilchester, 1899). Escapes here are numerous. *Petrorhagia saxifraga*, *Securigera varia* and *Teucrium chamaedrys* were established until this century, but are now lost; however *Acanthus spinosus*, *Allium roseum*, *Iris orientalis*, *Nectaroscordum siculum*, *Salpichroa origanifolia*, *Symphytum bulbosum* and *Tellima grandiflora* survive. Kingston Lacy and Melcombe Bingham also have ancient gardens, with Martagon Lily naturalized in both. The former has fine trees and an abundance of snowdrops, while the latter has a yew hedge planted *c.*1580.

Eighteenth century gardens of note include Sherborne Park and Milton Abbey, both laid out by Lancelot 'Capability' Brown (Paterson, 1978), Kingston Maurward and Minterne, all of which have fine specimen trees. Established aliens such as *Acorus calamus*, *Phuopsis stylosa* and *Symphytum tuberosum* occur at Sherborne, while Minterne has *Lathraea clandestina*, *Leucojum vernum*, *Symphytum bulbosum* and *Trachystemon orientale*. The first owner of Brownsea Island planted 10,000 saplings in 1722, and one of his successors spent £50,000 on his garden. More exotics were planted in the 1850s, and the most obvious escapes today are *Arbutus unedo*, *Rhododendron ponticum* and cultivars of *Narcissus* planted in 1925 (Battrick & Lawson, 1978). A nursery catalogue of 1782 (Galpine, 1983) lists over 200 hardy plants for sale at Blandford, including *Amelanchier lamarckii*, *Hypericum hircinum* and *Rhododendron ponticum*. In the old garden now forming part of Bryanston School grounds, plants such as *Buxus*, *Doronicum pardalianches*, *Duchesnea indica*, *Philadelphus coronarius*, *Sasa palmata* and *Vinca minor* are thoroughly naturalized, and may have survived two centuries since their introduction.

Although the gardens at Minterne were begun in the previous century, important introductions of Magnolias and Rhododendrons were made from expeditions by J D Hooker in the 19th century and by Farrer and Kingdon-Ward in the 20th (Paterson, 1978). Canford School (Hooker, 1993), Melbury Park and Milton Abbey School all have interesting arboreta planted during the past 150 years. Apart from the big estates, most villages have delightful cottage gardens. From these have spread many plants which like old walls, such as *Antirrhinum majus*, *Arabis caucasica*, *Aubrieta deltoidea*, *Buddleja davidii*, *Campanula* spp, *Centranthus ruber*, *Cerastium tomentosum*, *Cymbalaria muralis*, *Draba aizoides*, *Erigeron karvinskianus*, *Erinus alpinus*, *Erysimum cheiri*, *Linaria purpurea*, *Pseudofumaria lutea*, *Sedum* spp. and *Sempervivum tectorum*. Many of these, notably *Centranthus*, have spread to sea cliffs, as has *Matthiola incana* on Portland. The sandy cliffs west of Branksome also have many species of garden origin, such

Table 15.	Approximate dates of plants introduced to Britain.
Pre-1550	*Acer pse, Acoru cal, Antir maj, Armor rus, Asaru eur, Casta sat, Erysi cheiri, Inula hel, Isati tin, Melil alb, Papav som, Popul alb, Sapon off, Sempe tec, Smyrn olu, Vinca maj, Vinca min*
1550–1599	*Anaph mar, Centr rub, Cycla hed, Ficus car, Galan niv, Hespe mat, Leuco ver, Melis off, Pinus pin, Pseud lut, Syrin vul, Tulip syl, Vibur tin*
1600–1699	*Aescu hip, Carpo edu, Conyz can, Cymba mur, Hyper hirc, Larix dec, Lathy tub, Linar pur, Lonic cap, Oxali corn, Prunu lau, Robin pse, Trifo inc*
1700–1789	*Alliu ros, Amara ret, Aster novi-b, Ceras tom, Clayt sib, Coron did, Iris ori, Iris spu, Lyciu bar, Oenot gla, Pilos aur, Prune lac, Querc cer, Rhodo pon, Sisym ori, Solid gig, Symph alb, Trach ori*
1790–1899	*Buddl dav, Coton hor, Elode can, Epilo bru, Fallo bal, Galin par, Lepid dra, Lupin pol, Mahon aqu, Matri dis, Mimul gut, Petas fra, Symph × upl*
1900–2000	*Azoll fil, Cicer mac, Crass hel, Elode nut*

as *Erigeron glaucus*, *Escallonia macrantha*, *Lupinus arboreus* and *Senecio cineraria*. Garden introductions often found as patches on verges include *Fallopia japonica*, *Lonicera japonica*, *Petasites fragrans*, *Veronica filiformis* and bulbs such as *Crocus* spp., *Muscari armeniacum* and *Narcissus* spp. *Crocus vernus* at Forde Abbey and Studland, and *Eranthis hiemalis* at Milborne St Andrew and Tarrant Crawford, are so well established that they have probably been there longer than printed records show. *Pinus pinaster*, planted near Ringwood *c.*1801, is still spreading on heathland (Woodhead, 1994); the extinct *Simethis planifolia* may have come with it. *Rhododendron ponticum*, perhaps first planted near Moreton, is now locally dominant in acid woodland. *Quercus ilex* forms small woods at Durlston, and seedlings are frequent along the coast. *Cotoneaster integrifolius* was first reported on Portland in 1862, and is now abundant there. Few of these species were mentioned by Mansel-Pleydell in his *Flora* of 1895.

SOME MAN-MADE HABITATS: QUARRIES, HEDGES, VERGES OF ROADS AND RAILWAYS

Quarries

The county has many pits and quarries, which are often relict sites for a grassland or heath vegetation which has been destroyed around them. If not used as tips, the bared areas of abandoned pits revegetate quickly and show interesting stages of succession. Temporary or permanent pools in old pits are always worth inspection. Even when overgrown by scrub, some old quarries are SSSIs for their geological value.

The numerous limestone pits on Portland (Davies, 1935) have a rich vegetation with a surprising number of fig trees, *Adiantum capillus-veneris*, *Polypogon monspeliensis* and many cryptogams. South of Swanage the quarries are mines rather than pits, but there are good plants at Winspit (97T), Belle Vue (07E) and Tilly Whim (07I). Smaller limestone pits which are SSSIs are at Blind Lane, Abbotsbury (58S), Chalbury (68W), and on the Oolite at Peashill (49V), Horn Park (40L), Babylon Hill (51X), Trill (51W), Halfway House and Lows Hill (61D), Sandford Lane (61I), Frogden (61P), and Goathill (61T). Few of these are botanically exciting, but one has a colony of *Draba muralis*.

The largest working chalkpit is on the wooded scarp above Shillingstone (80J), with *Cephalanthera damasonium* nearby. There are abandoned pits at Maiden Newton (59Z), Upwey (68S), Godlingston (08A) and on the north side of the ridge between Cocknowle and Swanage. These, as well as many of the smaller pits, harbour chalk grassland species such as *Gentianella amarella* and *Ophrys apifera* with uncommon bryophytes. Sandstone pits near Shaftesbury are defunct (Chatwin, 1960). The Upper Greensand stone that was mined there favours *Ceterach* and the lichen *Caloplaca crenularia*.

Much clay has been dug in the past to produce kaolinite or bricks and coarse earthenware. Old brickpits have been mapped (Young 1972); most are either in the Jurassic of north-west Dorset or in the Poole basin. Old brickpits near Broadmayne (78I) have mostly been reclaimed; those near Chickerell (67P) on the Oxford Clay were working in 1970, but some are now part of a nursery for aquatic plants. The numerous pits on the Bagshot beds near Poole include a big one at Beacon Hill (99S). Many of these are now flooded or used as tips, as at Canford Heath (09H). Modern ballclay (China Clay) pits are underground, but there is a huge working opencast pit at West Creech (88W). Nineteenth century opencast pits between Corfe and Wareham are now either flooded, as at Blue Pool SSSI (98G), or overgrown with carr. *Pilularia* and *Stratiotes* occur in such pits, with *Mentha pulegium* on the infertile spoil.

Vast sand and gravel pits have been excavated near Knighton Heath (78P), Crossways (78U and Z), and between Hyde Heath and Binnegar (88Y–89K). Notable colonists at Knighton are *Epilobium brunnescens*, *Salix* × *pontederiana* and the alien liverwort *Lophocolea bispinosa*. Further east, Stokeford pits have our largest colony of *Lycopodiella inundata* with *Littorella uniflora* nearby. Binnegar pits were a tip in the 1990s with many casuals and the hybrid × *Conyzigeron*. Aquatics in flooded pits include *Hydrocharis*, *Potamogeton pusillus* and *P. trichoides*. Small gravel workings, now SSSIs, are at Broom (30G) and Blackdown (68D).

Hedges

There are two kinds of hedge in the county (Rackham, 1986; Good, 1966). The older kind follows a sinuous course, may have big coppice stools, and has a rich flora with many different shrubs and woodland-edge plants. This type is commonest on the Jurassic clays of north-west Dorset, and may date back to Saxon times; 6% of ancient charters for this region refer to hedges (Rackham, 1986). Shrubs include barberry, butcher's broom, crabapple, field

maple, hazel, holly, pear and spindle while prominent herbs are *Alliaria petiolata*, *Arum maculatum*, *Ballota nigra*, *Chaerophyllum temulum*, *Inula helenium*, *Silene dioica*, *Sison amomum*, *Stachys sylvatica*, *Stellaria neglecta*, *Torilis japonica* and many ferns. A variant is the double hedge, at least twice the width of a normal hedge, which acts a haven for woodland plants in arable terrain. Its origin is uncertain, but may involve parish boundaries.

The other kind of hedge probably originated at the time of the enclosures in the 18th century or later. Such hedges tend to be straight, to have few or no trees, and to be mainly of hawthorn, sometimes blackthorn, or gorse on heathland, and to have a poor ground flora. This kind is common on the Chalk. It may contain shrubs such as buckthorn, dogwood, grey willow or wayfaring tree, and herbs such as *Viola odorata*, but these are neither faithful nor characteristic. Only 1% of ancient charters on the Chalk refer to hedges (Rackham, 1986); Portland has never had hedges at all. The commonest lowland hedgerow tree was the elm, until all big trees succumbed to disease in the 1960s and 1970s. Elm suckers still appear, but are attacked by the fungus and killed after 10 years or so.

Few hedges are now managed by layering, but old layered branches are frequent in them. Current practice is to trim with a mechanical flail in early winter, which is not just unsightly, but destroys saplings and prevents flowering and fruiting of most shrubs.

Roadside verges

The widest verges are on the Clay, where tracks became morasses in winter before tarmac arrived in the 1930s (Good, 1966). Verges of unmetalled green lanes often have relict species. Apparently relict vegetation on road verges is atypical, since it has been perturbed by deposits of lead salts from exhaust fumes, nutrient-rich mud, salt from de-icing and spilt agricultural seed or garden rubbish. Lead scarcely affects vascular plants, but favours the spread of the lichen *Stereocaulon pileatum* on suburban walls. Nutrient enrichment is indicated by species such as *Anthriscus sylvestris*, *Atriplex prostrata*, *Chenopodium album*, *Conium maculatum*, *Lepidium draba*, *Sisymbrium officinale* and *Urtica dioica*. De-icing has allowed the recent spread of maritime plants such as *Armeria*, *Cerastium diffusum* and *Cochlearia danica* along verges and especially central reservations of major roads. Seed spills account for the presence of *Brassica napus*, *Linum usitatissimum* and cereal grasses at the junction of verge and tarmac. Garden outcasts can persist in clumps which spread, e.g. *Cicerbita macrophylla*, *Fallopia japonica*, *Geranium pratense*, *Lathyrus latifolius*, *Pentaglottis sempervirens*, *Petasites fragrans* and *Saponaria officinalis*. *Inula helenium* and *Smyrnium olusatrum* may be in this category, but have survived for centuries.

Denizens now widespread along verges, though absent from the hinterland, include *Artemisia vulgaris*, *Crepis vesicaria*, *Geranium pyrenaicum*, *Hordeum murinum* and *Lactuca serriola*. Rare aliens such as *Oenothera stricta* prefer sandy verges, especially where there is bare soil following disturbance. Species known to have been deliberately sown are *Anthyllis vulneraria* subsp. *polyphylla*, *Festuca brevipila*, *Leucanthemum vulgare*, *Lotus corniculatus* var. *sativus*, *Onobrychis viciifolia* and *Primula veris*. Too little effort is made to reseed disturbed verges with native plants; letting nature take its course is preferable to applying herbicides and later sowing ryegrass. While few rarities survive on Dorset verges, the wildlife trust has marked many sites of less common plants with blue posts, between which appropriate mowing regimes are specified. Notably rich verges can be found on the Chalk near Blandford Camp (90D), on clay at Mutton Street (39Z) and on sand at Povington Heath (88R), Toners Puddle Heath (89F) and Thrashers Heath (98R).

Railways

Apart from early tramways used for moving stone on Portland and China Clay in Purbeck, most railways were built between 1847 and 1865 (Lucking, 1982). In 1885 the county had 160 miles of track, which were searched by botanists such as EF Linton and WM Rogers. About half of this was abandoned between 1952 and 1972.

Railways provide a truer cross-section of the countryside than do roads. Botanically, cuttings are more interesting than embankments, though by defunct lines both are mainly scrub and young woodland. Embankments sometimes have associated ditches with wetland plants. Some plants favour open cinder tracks, e.g. *Chaenorhinum minus*, *Linaria repens*, *Senecio squalidus*, *S. viscosus*, *Valerianella locusta*, *Vulpia bromoides*, *V. myuros* and the liverwort *Marchantia polymorpha*; the latter is not killed by weedkillers. Among the many aliens which have colonized railway banks, *Buddleja* spp., *Cotoneaster* spp., *Fragaria ananassa*, *Hieracium spp*, *Leycesteria formosa*, *Lonicera japonica*, *Parthenocissus quinquefolia* and *Solidago* spp. are prominent.

The list below summarizes the habitats (in brackets), and items of botanical interest along specific stretches of railway; square brackets mean that the line is no longer used. Yeovil to Maiden Newton (Sand, clay and chalk): *Epilobium lanceolatum*, *Trifolium glomeratum* and *T. subterraneum* near Yeovil Junction. [Maiden Newton to West Bay (Clay and limestone)]: *Ophrys apifera* and calcicoles at Powerstock SSSI. Abbeyford to Gillingham (Clay and limestone): *Cerastium diffusum* and *Saxifraga tridactylites* at Gillingham. Maiden Newton to Dorchester (Chalk): *Prunus cerasus*, *Verbascum blattaria* and aliens at Dorchester. Dorchester to Weymouth (Chalk): *Clinopodium acinos* and *Orobanche elatior*, plus aliens at Weymouth Station. [Upwey to Abbotsbury (Clay)]. [Weymouth to Portland (Urban, coastal)]: *Geranium purpureum*, *Hypericum hircinum* and *Asparagus officinalis* subsp. *prostratus*. Dorchester to Wareham (heath and water meads): rich. [Wareham to Swanage (heath, chalk and clay, partly restored line)]: aliens and cryptogams at Corfe cutting. [Stalbridge to Blandford (clay and chalk)]: part a public footpath. [Blandford to Broadstone (Stour valley)]: old records from Bailey Gate. [Broadstone to Alderholt (heath)]:

Parentucellia viscosa and *Viola lactea* at Cranborne Common, old records from Daggons road. [Blue Pool to Ridge Wharf tramway (heath, marsh)]: rich area. [Norden to Middlebere tramway heath and bog)]: rich; *Hypochoeris glabra*, *Rosa pimpinellifolia* and *Cladonia* spp on the line. [Goathorn tramway (heath)]: rich; *Crassula tillaea*, *Lotus subbiflorus*, *Trifolium ornithopodioides*.

NATURE RESERVES AND CONSERVATION IN DORSET
by Joyce Bowcott and Humphry Bowen

Reserves and conservation are recent ideas. The National Trust was formed in 1895, primarily to conserve historic houses, but since the 1960s it has enlarged its role to conserve estates, beauty spots and their flora. It now owns two huge estates in the county – Golden Cap (772 ha) in the west and Kingston Lacy (3,559 ha) in the south-east. Its total holding of more than 5,500 ha dwarfs the area held by other organisations with conservation at heart. While the National Trust owns many important botanical sites, some of its holdings are of farmland. English Nature (EN; formerly the Nature Conservancy Council) is a government body founded in 1949. It began acquiring reserves in the 1980s, and now owns or manages 1,909 ha of National Nature Reserves (NNR) in Dorset, mostly in the south-east. It also has the responsibility for designating Sites of Special Scientific Interest (SSSIs), whose combined area in the county now totals nearly 18,000 ha, but which have little legal protection against determined developers, and cannot be called full reserves. Other owners of reserves include the Royal Society for the Protection of Birds (RSPB), with 881 ha, mostly heaths or reedbeds; Dorset

Table 16. Area in ha of reserves and SSSIs in Dorset in 1998.

Habitat	Reserves	SSSIs
Woodland	476	992
Calcareous grassland	947	1,754
Neutral grassland	415	864
Heath and bog	4,270	6,711
Coastal	1,090	7,427
Fen and reedbed	101	191

NB. There is also a marine reserve at Kimmeridge, and large SSSIs in the Fleet and Poole Harbour

Table 17. Approximate areas of reserves and SSSIs in 10 km grid squares in Dorset.

Area/ha	Reserves	SSSIs
>1,600	SY98, SZ08	SY88, 97, 98, SZ08
800–1,600	SY39	SU00, SY58, 99
400–800	ST90, SU00	ST81, SY49, 59, 89, SZ09
200–400	ST81, SY59, 78, 99	ST50, 91, SY67, 68, 69
<200	all others	all others

Table 18. Nature reserves larger than 1 ha in Dorset in 1998, ordered by 10 km Grid Square (GS).

GS	Name	Area/ha	Manager	Vegetation	GS	Name	Area/ha	Manager	Vegetation
ST40	Lewesdon Hill	11	NT	Wood	SY59	Eggardon Hill	19	NT	CGr
	Pilsdon Pen	15	NT	He		Kingcombe	152	DWT	NGr+
	Winyards Gap	9	NT	Pl		Loscombe	10	DWT	NGr
ST50	Bracket's Coppice	23	DWT	Wood+		Powerstock Common	115	DWT	Wood+
	Corscombe	19	DWT/PL	NGr	SY67	Broadcroft Quarry	7	B	CGr
ST60	Batcombe	15	DCC	CGr+	SY68	Black Down	21	NT	He
ST62	Holway Woods	16	DWT	Wood		Lodmoor	65	RSPB	SMa
ST70	Greenhill Down	12	DWT	CGr		Radipole Lake	87	RSPB	Rb
	Horse Close	17	WT	Wood	SY69	Hog Cliff NNR	88	EN	CGr+
	Woolland Hill	4	NDDC	Wood		Muckleford	3	DWT	CGr
ST71	Piddles Wood	24	DWT	Wood		Nuney mead	5	DWT	NGr
ST72	Fifehead Wood	20	WT	Wood	SY78	Ringstead	182	NT	CGr/Co
ST80	Turnworth	54	NT	Wood		Tadnoll	45	DWT	He
	Mill Down	15	NDDC	CGr	SY79	Thorncombe Wood	27	DCC	Wood
ST81	Fontmell Down	59	NT/DWT	CGr	SY88	Coombe Heath	40	DWT	He
	Hambledon Hill	74	EN	CGr+		East Stoke	5	DWT	Fen
	Hod Hill	54	NT	CGr		Winfrith Heath	103	DWT	He
ST82	Duncliffe Wood	86	WT	Wood	SY89	Higher Hyde Heath	54	DWT	He
	Kingsettle Wood	21	WT	Wood		Turners Puddle	14	DCC	He
ST90	Badbury Rings	23	NT	CGr+	SY97	Kimmeridge	–	DWT	Mar
	Kingston Lacy Park	101	NT	Pl/NGr		Spyway Farm	21	NT	Co+
	Pamphill	20	NT	NGr	SY98	Arne	498	RSPB	He/Rb+
ST91	Compton/Melbury					Arne reedbeds NNR	9	EN	Rb
	Downs	200	NT	CGr		Corfe Castle	10	NT	CGr
	Sovell Down	2	DWT	CGr		Corfe Common	98	NT	NGr+
SU00	Great Barrow,					Grange Heath	69	RSPB	He
	Ferndown	20	DCC/HCT	He		Green Pool	4	HCT	He
	Holt Heath NNR	488	NT/EN	He		Hartland Moor NNR	353	EN	He
	Slop bog	23	DCC	He		Stoborough Heath	128	RSPB	He
SU01	Garston Wood	34	RSPB	Wood	SY99	Ham Common	24	PDC	He/Co
SU11	Cranborne Common	45	DWT	He		Morden bog NNR	149	EN	He
SY39	Coneys/				SZ07	Bellevue Farm	26	NT	CGr
	Lamberts Castle	103	NT	He		Durlston	106	DCC	CGr/Co
	Golden Cap Estate	384	NT	Co+		Townsend	16	DWT	CGr
	Ware Cliff	12	NT	Co	SZ08	Ballard Down	90	NT	CGr
SY48	Burton Cliff	34	NT	Co		Brownsea Island	202	NT/DWT	He/Co
SY49	Allington Hill	13	WT	Wood		Old Harry	16	NT	CGr/Co
	Eype/St Gabriels	388	NT	Co+		Studland Heath NNR+	1214	NT/EN	He/Co
	Hardown Hill	10	NT	He	SZ09	Bourne valley	52	PDC	He
SY58	Cogden beach	105	NT	Co+		Canford Heath	62	PDC	He
	Limekiln Hill/					Corfe Hills	30	PDC	He
	Labour-in-vain	106	NT	CGr+		Hatch Pond	11	PDC	He
	West Bexington	20	DWT	Co/Rb					

For abbreviations, see text above +; B = Butterfly Conservation; CGr = Calcareous grass; Co = Coastal; He = Heath;
HCT = Herpetological Conservation Trust; Mar = Marine; NGr = Neutral grass; Pl = Plantation; PL = Plantlife; Rb = Reedbed; SMa = Salt-marsh

Wildlife Trust (DWT), initiated by Miss Helen Brotherton in 1961, which owns 760 ha (Bates, 1997); Dorset County Council (DCC), North Dorset District Council (NDDC) and Poole Borough Council (PBC), which between them protect 383 ha; the Woodland Trust (WT) with 157 ha of woodland, mostly in the north; and Plantlife, which has 15 ha of grassland. All these reserves, over 90 of very variable size, have been created since the 1960s.

Prospective visitors should note that access to many reserves is often more or less restricted. There are also over 100 SSSIs, (excluding quarries designated for their geological interest), which may coincide with or include reserves. In 1998, reserves covered 3.8% of the county, and SSSIs 7.1%.

With so many reserves, many containing several different habitats, generalisations are hard to make. Table

16 shows the approximate areas of different habitats in reserves and SSSIs.

It will be seen that calcareous grass, heath and coastal sites are relatively well protected, woods and fens are less so. The area of fen is small because this is such a rare habitat in the county, and is mainly reedbeds owned by the RSPB. Coastal reserves include much of the south-west and south-east coasts owned by the NT, as well as Brownsea Island. The area of coastal SSSIs is large because they include the south-west and south coasts, the Chesil Bank, the Fleet and the whole of Poole Harbour. This illustrates the difference between reserves and SSSIs. If the latter were full reserves, the coasts would not be disfigured by caravan parks, the concrete and metal litter on the Chesil Bank would be removed, and that part of Poole Harbour used as a major shipping route would surely be excluded from reserve status.

Table 17 shows that reserves and SSSIs are not uniformly distributed in the county.

By far the greatest areas of both reserves and SSSIs are in SY98, which has c.1,600 ha of reserves and c.6,000 ha of SSSIs; this square is the richest in the county, botanically. Otherwise well-protected grid squares are on the coast, in the Poole basin, on Portland (an SSSI on its own) or on the north-east Chalk.

Management of reserves is still in its infancy. Woodland needs relatively infrequent management, but may require deer control, or coppicing every few years to benefit the woodland flora. The RSPB are doing an excellent job at Garston Wood. Grassland and fen require either grazing, mowing or drastic removal of scrub to keep their diversity. DWT reserves at Kingcombe and Fontmell Down, and the NT reserve at Cogden beach, are successful examples of management. Heaths and bogs may benefit from light grazing, also from control of pine and gorse. Good diversity is maintained at the RSPB reserve at Arne, and at Warmwell Heath SSSI. English Nature have a team of professionals who are skilled at managing heaths, and of recreating the habitat from acid grassland. Reedbeds at Radipole are regularly cut by the RSPB to keep a balance between open water, young and old reeds. Coastal sites with undercliffs are best left wild, while lagoons such as the Fleet and the shallower parts of Poole Harbour have restrictions on access by power boats.

Some very rich sites in Dorset, only recently promoted to SSSI status, are rich because they they have been managed for many years in a manner favouring wildlife. Thus Melbury Park, which has been managed for deer for a long time, is among the best British sites for bark lichens. It has ancient trees and has never been drained, as has most agricultural land. Abbotsbury Swannery has been managed in the same way for centuries, and reeds are still cut there. Povington Heath presents some extraordinary features. It was taken over by the army in 1913 for artillery practice, and public access is forbidden, though stock graze there. As a result it preserves the farming landscape of almost a century ago, with a rich and varied flora, which includes many rare vascular plants and lichens.

The following very rare plants, of red data book status, occur in Dorset Reserves: *Centaurium tenuiflorum, Chenopodium vulvaria, Erica ciliaris, Gastridium ventricosum, Leucojum aestivum, Ophrys sphegodes*. However the other 20 red data plants in the county are not protected in reserves, though most are in SSSIs (Milton & Pearman, 1993). Very few of the county's red data cryptogams are in reserves, although most are in SSSIs such as Melbury Park, Oakers Wood, Povington or Portland.

CHAPTER 4

SOME EARLIER WORKERS ON THE DORSET FLORA

by David Allen

Dorset had to depend on visits by outsiders for the investigation of its flora until comparatively late. The first of these visitors that we know of was William Turner, the 'father of English Botany', who in 1550 made a holiday tour along the coast, taking in Portland, Poole and Purbeck. In his New Herball, he records finding sea-kale and scurvygrass at Portland and sea-purslane "beside the Ile of Purbeck". Hard Fern and Stinking Iris also owe their first mention as British plants as well as Dorset ones to Turner's trip, though he was vague about the county's limits, ascribing Salisbury Plain to Dorset. More than half a century later John Parkinson may have visited the county, if the records in his herbal of wild madder from Stourpaine and Wareham were truly his discoveries.

Weymouth, Portland and the Chesil Bank were attracting botanists before George III rendered the district fashionable in the late 18th century. Among them were Sir Thomas Browne, Sir Thomas Cullum, John Ray, William Stonestreet and William Hudson. Hudson also followed Turner to Purbeck, where he added bithynian vetch to the national list. Particularly promising was a young Edinburgh medical student, Thomas Yalden, who reported his finds in the area to the Rev. John Lightfoot. Unfortunately these were not published, and Yalden drowned at Venice in 1777, before starting his career.

By that time the county had acquired its first resident botanist, Dr Richard Pulteney (1730–1801). Born in Loughborough as one of 13 children, Pulteney was introduced to natural history in early boyhood by an uncle. After being apprenticed to an apothecary, he tried to establish himself as a doctor in Leicester, meanwhile recording nearly 600 species in that county's flora. His nonconformist views did not help his medical practice, but his scientific reputation led to a Fellowship of the Royal Society in 1762 and election to the Royal College of Physicians. In 1765 he moved to Blandford, where he amassed a large library, devoted his leisure to botany and conchology, and made a late but childless marriage. His extensive correspondence with most of the leading British naturalists, now preserved at the Linnean Society, kept him from writing a Flora of Dorset; however he did produce lists of Dorset birds, shells and rarer plants for the second edition of John Hutchins' History and Antiquities of the County of Dorset in 1799. He was revising a plate for this when his final illness intervened. He was buried at Langton Long, and there is a tablet to his memory in Blandford Church; his herbarium is now in **BM**.

At this time the first intensive study of the county's seaweeds was under way at the unlikely hands of a colonel in the Oxfordshire Militia, Thomas Velley (1748–1806). Resident in Bath for most of his life, Velley devoted much

time to scouring the south coast for algae, even taking time off while his troops were stationed there. Both Poole and the Weymouth area engaged his attention by 1786, his national *magnum opus* appearing in 1795. Six years later he fell from a runaway stage coach, suffered concussion and never recovered. Eight folios of his exquisitely preserved specimens are in **LIV**.

In the 1830s a 'home-grown' botanist at last emerged – only to be lost almost at once to the Isle of Wight. Thomas Bell Salter (1814–1858) was a product of the interlocking of medicine and natural history of that era. His father, maternal uncle and two brothers were, like him, all members of the medical profession and Fellows of the Linnean Society. Growing up in Poole, Bell Salter began in boyhood what was hailed at his death "as one of the most complete British Herbaria ever formed". The cream of his local finds appeared in print in 1839 as an appendix to J Sydenham's History of the Town and County of Poole, in which he announced his intention of producing a Dorset Flora, but in the same year he set up in practice in Ryde

Figure 12. Dr. Richard Pulteney (1730–1801).

Figure 13. John Clavell Mansel (1817–1902).

and never botanized in Dorset again. Some of his Dorset specimens survive in **BM** and **QMC**.

Some twenty years later John Clavell Mansel (1817–1902) began his career as the dominant 19th-century Dorset naturalist. The family home at Smedmore allowed him free access to the Purbeck countryside, where his interest in geology and natural history were encouraged by his mother and the local rector. Later, at Cambridge, he came under the influence of two teachers of botany, Babington and Henslow. Graduating in 1839, he then followed the life expected of a future estate owner, first reading for the bar and then serving as a cavalry officer. At that period he was elected to the Geological Society, but he did not join the Linnean Society until 1850. The recently-founded Purbeck Society published several of his papers and elected him Vice-President. It was probably in the 1860s that his thoughts turned to producing a Flora of the county, for herbarium specimens collected in that decade reveal that he was sending critical groups to national experts.

It was a period when county Floras were much in fashion, stimulated by HC Watson's work on plant distribution, and it was to Watson that the *Flora of Dorset* was dedicated ("by his sincere friend") when it appeared in 1874. About two years earlier Mansel had updated

Pulteney's list in a new edition of *Hutchins' History*, which probably helped to flush out many extra records. Nevertheless, though nine local botanists were major contributors to the Flora, far and away the greatest number of localized records were Mansel's own.

Singled out in the preface for a special tribute were the paintings of Mrs Mary Frampton of Moreton, who knew Pulteney; the paintings are still treasured by her descendants. Three other helpers merit notice here. Little is known of Susannah Mary Payne (1832–1899) other than that she was unmarried, lived in Weymouth and left a herbarium of over a thousand sheets in **DOR**. William Bowles Barrett (1833–1915) was a solicitor in Weymouth who compiled an unpublished Flora of the district, now in its Public Library with other manuscripts of his. In one of these he recorded the botanical reminiscences of TB Flower of an earlier generation, full of gossip which is a goldmine for historians. The third helper, James Buckman (1814–1884), was the largely self-educated son of a shoemaker. He displayed such prowess at Cheltenham as a botanist and geologist that he became a professor at the Royal Agricultural College at Cirencester. Resigning in 1863, he moved to Bradford Abbas, where he started a model farm and took in agricultural pupils. Since the local geology was the same as in the Cotswolds, he was able to continue his major study of its fossils for which he is now mainly remembered. His other interests were botany, entomology and archaeology.

Buckman was one of the two chief founders of the Dorset Field Club, which later became the Dorset Natural History and Archaeological Society. The other was the Rector of Holwell, Henry Hayton Wood (1825–1882), a former Oxford don and Dorset's first bryologist. Buckman's experience in running the successful Cotteswold Club made him the obvious choice as first Honorary Secretary and Editor when the Field Club began in 1875, tasks which he ably fulfilled for the rest of his life. The equally obvious choice as first President was Mansel, (who by then had changed his name to Mansel-Pleydell), and who was so devoted to the club that he was re-elected annually to the office for 27 years. He contributed numerous papers to the club's Proceedings and was a constant attender of its field meetings. It was not merely his social prominence that told; it was what one of his obituarists described as his "almost boyish ardour in his pursuit of nature". Such was his keenness that he never stopped to eat lunch, but ate as he walked. By a happy coincidence, the Dorset Field Club was born in the same year that Mansel-Pleydell's *Flora* was published and the author succeeded to two family estates. His father's estate in Purbeck was smaller than that of his mother's family at Whatcombe, the two together totalling 3,642 ha (9,000 acres). With these responsibilities came new public duties, culminating in his appointment as High Sheriff in 1876, but these scarcely affected his natural history output.

Just as in nearby Hampshire, the publication of a Flora stimulated so much new fieldwork that a second edition was called for. Though Mansel-Pleydell was nearing 80, the new edition of 1895 was substantially his work. It covered 123 more species than its predecessor, gave many new localities and improved the treatment of critical groups. However, the author did not long survive his second *Flora*, for it was on his way to an annual meeting of the Field Club at Dorchester that he suffered a fatal heart attack. He was buried in the family graveyard at Winterborne Clenston. A Mansel-Pleydell Prize to encourage the study of natural sciences was established by the Club, whose museum, which he had much enriched over the years, received his large herbarium of local and European plants.

The two decades that elapsed between the two editions of this *Flora* witnessed the arrival of two cryptogamic experts. Gulielma Lister (1860–1949) was the daughter of Arthur Lister, a world authority on mycetozoa, who had a summer residence at Lyme Regis. She never married, but had a passion for natural history and collected widely in both Essex and west Dorset. Highly regarded nationally, she received the exceptional accolades for a woman at that period of being twice elected President of the British Mycological Society and twice voted on to the council of the notoriously conservative Linnean Society. Her bryophyte herbarium is in **DOR**, while she also found such lichen rarities as *Teloschistes chrysophthalmus* and *Pseudocyphellaria aurata* near Lyme. By chance, the contemporary efforts of Edward Morell Holmes (1843–1930) in east Dorset complemented Miss Lister's work. His sole spell as a resident was a few years at Wimborne Grammar School, which he left at 14 to train as a pharmacist. When in business at Plymouth he began investigating the local algal, bryophyte and lichen floras, and in later years he continued fieldwork in Dorset and other southern counties. With a strong physique and excellent eyesight he was able to spot interesting cryptogams at a reputed pace of five miles an hour. In 1872, after failing to obtain a post at Kew, he became Curator of the teaching collections at the Pharmaceutical Society of Great Britain for what was to prove the next 50 years. By a macabre coincidence, in his later years Holmes, like Velley, was badly injured in a road accident; unlike Velley, he escaped with the loss of a foot. He published extensively, and among his papers at the **BM** is an unpublished Lichen Flora of Dorset intended for the Victoria County History. At his death, his herbarium was divided; his seaweeds are in **BIRM**, his mosses in **CGE**, his hepatics in **NMW** and his lichens in **DOR** and **NOT**.

Several contributors to the second edition of the *Flora* left the county for careers elsewhere. Lester Vallis Lester, later Lester-Garland, (1860–1944) was active from his boyhood in Langton Matravers and schooldays at Sherborne. His career as a schoolmaster took him to Jersey, where he published a Flora, and then to retirement in Bath. His large herbarium is in **K**. It was a similar story with Henry Nicholas Ridley (1855–1956) and William Fawcett (1851–1926), whose thorough working of the coast is acknowledged by Mansel-Pleydell. Ronald Good mentions receiving from Ridley confirmation of records made in 1876 when he was an Oxford undergraduate. Both joined the British Museum, and Fawcett departed

for Jamaica, where he was later co-author of the standard Flora. Ridley went to Malaya, where he was to play a key role in founding its rubber industry, and went on to become the first-known centenarian botanist.

As if in compensation, Dorset now began to receive those scholarly clergymen who dominated British field botany in the late-Victorian period. The first of these was William Moyle Rogers, who held a succession of livings, mostly in Devon but including one at Chetnole in 1879, before retiring to Bournemouth. He became an expert on *Rubus*, his collection of which is in **BM**, while the rest of his herbarium is in **LANC**. The next to arrive was Richard Paget Murray (1842–1908), a descendant of the dukes who once owned the Isle of Man, where he grew up. He held a living in Somerset, and collected material for a *Flora* of that county which came out between 1893 and 1896. In 1893 he became vicar at Shapwick, where he remained for 25 years and was finally buried. Despite his fervent attachment to entomology, and spending time abroad in the Canaries and elsewhere, he was active with the Dorset Field Club, where his genial personality and keen sense of humour were evident. He was another *Rubus* expert, who first found and described east Dorset's speciality, *Rubus durotrigum*; his herbarium is in **BM**. A third clerical collaborator on brambles was Edward Francis Linton (1848–1928), who moved to Bournemouth in 1888 for reasons of health. It was he who initiated the issuing in 1892–1895 of the three fascicles of the 'Set of British Rubi', which did so much to advance our knowledge of the group; many specimens came from Dorset. In 1900 Linton published his *Flora of Bournemouth*, with a mass of new localities of plants from both Vice Counties 9 and 11. One year later, his health restored, he became rector of Edmondsham. Two decades there led to a harvest of records published as supplements to his Flora in 1919 and 1925; in 1923 he was awarded the Morris Medal by the Bournemouth Natural Science Society. The first list of Dorset fungi was also published by Linton in 1916–1917. Soon after, he retired to east Bournemouth; his herbarium is now in **BM**. Three brambles, *Rubus lintonii, R. moylei* and *R. murrayi* commemorate the impressive clerical triumvirate.

Bournemouth was now becoming the chief source of new Dorset botanists. When Charles Baylis Green (1845–1918) retired to Swanage in 1910 after working for the LMS. railway in London, it was to the Bournemouth Natural Science Society that he chose to report his finds in Purbeck, and to its museum that he was to leave his herbarium, photographs and slides. That Society had a first existence from 1883 to 1893, and its refounding in 1903 was partly due to Richard Vowell Sherring (1847–1931). Sherring, who settled in Bournemouth on account of ill-health, hid a kindly heart behind a frosty reserve and a long beard which gave him the appearance of an Old Testament prophet. A keen naturalist from boyhood, he had done good work on the geology of the Mendips until ill-health sent him to the West Indies, where he collected the rich fern flora for Kew. He knew Green, and may have been responsible for the latter's interest in cultivating ferns. During his 30 years at Bournemouth, before retiring to Bristol in 1925, he charted the spread of *Spartina* in Poole Harbour, led field meetings, served as secretary of the botanical section of the BNSS and built up its herbarium. The society responded by making him the recipient of its first Morris Gold Medal in 1921.

Another collector was John Henry Salter (1862–1942), a Professor of Botany at Aberystwyth who published a *Flora of Cardiganshire* in 1935. In 1916 his wife's tuberculosis led him to settle for seven years at Verwood. He had no car, and relied on trains for his fieldwork, but his powers of endurance as a walker were legendary; even at the age of 80 he could cover 15 miles in a day. A tall, taciturn figure in knickerbockers and heavy boots, he added to his Shavian appearance an equally Shavian regime of total abstention. He was an all-round naturalist who also published a bird book and left a large insect collection. His herbarium is in **NMW**, and 24 volumes of natural history diaries in the National Library of Wales in **ABS** have been quarried by David Pearman for the rich Dorset material they contain.

Another Bournemouth botanist was Norman Douglas Simpson (1890–1974), who began forming a herbarium at the age of 12. After his graduation at Cambridge, the family moved to Bournemouth from Yorkshire, and Simpson became a professional taxonomist, mainly in North Africa. Private means enabled him to retire in his early forties when the Depression caused his government post to lapse. He inherited a spacious family home which allowed him to give full rein to his hobby of collecting books, especially local Floras, which led, in 1960, to his *Bibliographical Index of the British Flora*. That he published scarcely anything else resulted from a mixture of perfectionism, scepticism and diffidence. Apart from books and botany, his third passion was motoring, and for 40 years he drove his beloved Alvis far and wide in search of plants, often with congenial botanist friends who were fellow bachelors. Though not otherwise sociable, he seldom missed meetings of the BSBI, where his impish sense of humour and idiosyncratic laugh, developing from a wheeze to a chuckle, always stood out. His herbarium of 18,000 sheets is in **BM**. It includes part of the collection of Leslie Beeching Hall (1875–1945). Hall worked for the chemical industry in London until he retired to Parkstone after the First World War, and for six years was Curator of Botany for the London NHS, which now holds part of his herbarium and his Dorset notebooks. He acted as Curator and also as Chairman of the Bournemouth NSS until his death.

Three Dorset botanists of the inter-war years must be mentioned. Arthur William Graveson (1893–1979) was a Quaker. After gaining a double first at Cambridge, he became a schoolmaster, first at Blandford and then for 40 years at Beaminster Grammar School, (where his nickname was 'Weeds'). Much of his fieldwork was carried out by bicycle, and his extensive diaries are in the Dorset County Museum. He joined the Botanical Exchange Club in 1920 and later became a friend of JE Lousley, who learnt of many Dorset localities from him; his herbarium is in **BDK**. Dorothy Meggison (1897–1976) was the daughter

of a doctor in Dorchester, and is chiefly remembered for her work in the Red Cross blood transfusion service. Around 1923 she joined the Wild Flower Society and rose to become a star performer and a secretary of one of its senior branches for nearly 20 years. Her botanical notebooks are in the County Museum. She and her friend Mary J Andrews of Upwey contributed many records to the last of this trio, Ronald D'Oyly Good (1896–1992). Good's father was another Dorchester doctor, living near the County Museum, which his son visited regularly from the age of 10. After finding *Ranunculus ophioglossifolius* in 1914, Good enlisted in the 4th Dorsets for the First World War. After the war he graduated with a double first at Cambridge and took a post in the British Museum. On the strength of his publications on plant geography he became head of the Botany Department in the newly-established University College, Hull in 1928, becoming Professor in 1946, when his 'Geography of the Flowering Plants' brought him an international reputation. On retirement in 1959 he settled at Parkstone until his wife's death in 1975.

Yet Good had never fully left. Every university vacation was spent in Dorset in the 1930s, where his amazingly thorough survey of old roads, vascular plants and lost villages resulted in books published in 1940, 1948 and 1979 respectively. He also published a history of Weymouth in 1945. It is hard to believe that any one individual will produce such a massive amount of diverse scholarship on a single county. Good was no team man, and so reserved

Figure 14. Ronald D'Oyly Good (1896–1992).

that even close relatives found him difficult to understand. His *Geographical Handbook of the Dorset Flora* has two separate strands: a demonstration of how plant distributions can be correlated with external factors, and an updating of Mansel-Pleydell's *Flora* with a checklist of recent finds. The strings of localities traditional up till then in county Floras were mostly omitted. Good had little interest in aliens and critical taxonomy was not his forte. He grumbled at "all those horrible varieties" sent him by Simpson, but included them all indiscriminatingly. The main part of the book, with its novel distribution maps, was a feast by comparison and proved widely influential. There was then no National Grid, but Good invented a Dorset Grid, the squares of which were sub-divided into 16 parts. He amassed a quarter of a million records from 7,500 stands, which he could convert manually to dot-maps. The full set of species maps, extensive field notes and part of his herbarium are at **WRHM**, while further specimens are in **DOR**. They await computerization. The fact that Good did not join the Dorset Natural History Society until 1939 is not surprising, but once he returned to live in the county he played an active part in leading field meetings, editing the annual reports on botany and rainfall, serving as President and finally becoming a Trustee. In 1984 the Society reciprocated by subsidizing the publication of his *Concise Flora of Dorset*, revised and extended.

One source of Dorset records available to Good was an album of wildflower paintings by two sisters, the Misses Griffiths, who lived at Mells in Somerset in late Victorian times. They included some outstanding rarities such as man orchid, and Good felt it better to include these despite the lack of other evidence. Unknown to Good, some similar paintings from Sherborne by Miss DR Wilson *c.*1906 are kept in Sherborne Museum. Good also had problems with roses, especially those collected by AEA Dunston in the 1940s and now in **BM**. Dunston lived in Donhead St Mary, Wiltshire and was an all-round naturalist who published many Purbeck sites for *Sphagnum* mosses, with specimens now in **DOR**.

A final botanist who merits special mention is Elsie M Burrows (1913–1986). Educated in Leicestershire, she married an industrial chemist and moved to Liverpool, where she joined the University Botany Department and eventually became a Senior Lecturer and Chairperson. She specialized in the ecology of the larger seaweeds and was a founder member of the British Phycological Society. During her retirement in Dorset she made an important collection of algae, beautifully mounted and now in **DOR** and **LIV**. Sadly her later years were clouded by illness and personal tragedies, and she did not live to see the completion of her monograph on British Chlorophyta.

CHAPTER 5

INTRODUCTION TO THE PLANT LISTS

The twin aims of any Flora must be, first to inform readers where a given species may be sought, and second, to instruct future botanists whether their finds are original or refinds. This is why I have included many old records. Few people nowadays own copies of earlier Floras of Dorset, and Good gave few localities in his Handbook of 1948. Other old records are buried in herbaria which are rarely consulted, or in unpublished notebooks of earlier botanists. The latter include manuscript notes by AW Graveson and Miss D Meggison in **DOR**, and by JH Salter in **ABS**. My reasons for including a rather large number of alien species, apart from a personal interest, are twofold. So little of Dorset now consists of semi-natural habitats, that records from the large areas affected by man are needed to give an objective picture of its current flora. Of course, some attempt has been made to categorize the aliens as established, casual, planted and so on, even though such attempts are subjective ones. Secondly, while some purists prefer to ignore aliens, field-workers often come across them and others (for example entomologists or plant physiologists) may actually want to know where they can find certain non-native species.

Some conventions and abbreviations used are as follows. Latin and English names of vascular plants follow Stace's Flora (Stace, 1997), except for four orchid species, with the addition of some local English names collected by Anne Horsfall (Horsfall, 1991). The number given in most entries is the number of tetrads from which the plant has been recorded since 1984; the maximum is 739, and the number gives an objective indication of current frequency; numbers in square brackets indicate the extra number of tetrads with pre-1980 records. Place-names are spelt as in a gazetteer and its supplement (Shuttleworth, 1984; Lock, 1998); each entry starts with places in the south-west and ends with those in the north-east. A *record* is regarded as true and/or verifiable, but this is not the case for a *report*.

Directions such as north, west and so on, refer to the north or west of Dorset unless otherwise stated. Remarkable trees have dimensions given as H × G, where H is the height and G the girth in metres. Species or locations now extinct are enclosed in square brackets [ ], while species whose records are thought to be erroneous, or in another vice-county, are enclosed in round brackets (). Tetrad maps are given for most native species which are neither universal nor very rare, and for selected alien species; post-1984 records are given as black discs, and 19th century records as open circles. Records of intermediate date are given as grey circles; while most of these are 1930s records due to Good and his colleagues, most of Good's own records have yet to be put on a computer.

HOW GRID SQUARES ARE REFERRED TO IN THE TEXT

Parts of Dorset occur in four 100 × 100 km² National Grid Squares, ST, SU, SY and SZ. 10 × 10 km² Grid Squares are referred to as, for example, SY98 where 9 refers to Eastings and 8 to Northings. As there are no duplications of the two numbers within the county, the two letters are sometimes omitted.

2 × 2 km² Grid Squares, or tetrads, are referred to using the 10 × 10 km² Grid reference plus an additional capital letter, as in the table below.

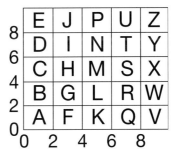

Figure 15. National Grid Square reference used in the text.

For example, SY98N could equally well be referred to as SY9486.

A NOTE ON METHODS OF COLLECTING RECORDS AND STATISTICS

All 739 tetrads were visited at least once, and almost all had three or more visits by more than one recorder. Recording was greatly helped by the use of printed cards with the names of 856 species in 'botanical shorthand', e.g. Achil mil for *Achillea millefolium*. Records were entered using Macprecs, and tetrad maps were produced using Dmap. Master cards are deposited at Dorset Environmental Records Centre.

Accessibility of tetrads was rarely a problem, as footpaths are numerous and most land-owners were sympathetic. The least accessible sites were parts of the Chesil bank, the island booms enclosing Portland Harbour, a few beaches and golf courses and some islets in Poole Harbour. Most of the military and naval bases in the county were visited under permit.

The mean number of species recorded per tetrad was 285, a figure which is certainly incomplete, but is quite respectable when one realises that 185 of our tetrads contain areas of sea, or of land belonging to other Vice

Counties. The actual number found in any tetrad depends partly on the zeal of the recorders, but also on the types of habitat represented there. High numbers, exceeding 400, were concentrated around Cranborne and in the lower Frome and Piddle basins, while low numbers, around 200, were mainly in the deforested area northwest of Weymouth and on the parts of the central Chalk ridge lacking both water and habitations. Numbers exceeding 500 were found for the tetrads including north Swanage, Wareham and Winfrith Heath.

An attempt was made to test whether the presence of sea coast in a tetrad increased its tally of species, but this proved inconclusive. Obviously no tetrads with coastline are complete, but one can compare incomplete coastal tetrads with incomplete marginal tetrads, arbitrarily subdivided into those with areas greater or less than 2 square km. The means, standard deviations (and numbers of tetrads involved) are as follows;

Coastal, >2 sq km	300 ± 131 (53)
Coastal, <2 sq km	206 ± 95 (37)
Marginal, >2 sq km	285 ± 56 (45)
Marginal, <2 sq km	181 ± 79 (50)

The large standard deviations make the larger numbers found in coastal tetrads statistically insignificant. Most coastal tetrads lack woods, while marginal tetrads need not do so. Similar attempts to compare numbers in tetrads with and without features such as woods, water or villages have not been carried out because of the large standard deviations observed.

List of abbreviations
agg. = aggregate
Br.Rub. = Set of British Rubi
c. = approximately
Ch = Church (yard)
CP = Country Park
Cv = Cultivar
d = old penny
DERC = Dorset Environmental Records Centre
det = determined by
et al. = other botanists
FP = Forest Park
Ho = House
HWM = High Water mark
Is = Island
LWM = Low Water mark
m = metre (s)
NHS = Natural History Society
NNR = National Nature Reserve
RSPB = Royal Society for the Protection of Birds
sp = species
s.s. = sensu stricto
Sta = Railway Station
Subsp. = subspecies
var. = Variety
VC = vice-county

WFS = Wild Flower Society
x = hybridized with
! = seen growing by the author

List of recorders or experts initials, other than herbaria
ACJ Jermy, ACL Leslie, AH Horsfall, AJB Byfield, AJCB Beddow, AJEL Lyon, AJR Richards, AJS Silverside, AL Leonard, A C, ALB Bull, ALP Primavesi, AM Mahon, AN Newton, APC Chamberlain, ARV Vickery, AWJ Jones, AWW Westrup, BAB Bowen, BAM Miles, BBS Brit. Bryol. Soc., BC Candy, BE Edwards, BJ, BJC Coppins, BLS Brit. Lichen Soc., BMS Brit. Mycol. Soc., BNHS Bryanston NHS, BNSS Bournemouth Nat. Sci. Soc., BSBI Bot. Soc. Br. Is., CBG Green, CCT Townsend, CDP Preston, CES Salmon, CHS Schofield, CIS Sandwith, CS Clayesmore School, CT Turner, DAP Pearman, DCL Leadbetter, DEA Allen, DFW Westlake, DG Green , DJG Godfrey, DM Meggison, DNHS Duncliffe NHS, DRS Seaward, DRW Wilson, DSR Ranwell, DWT Dorset Wildlife Trust, EAP Pratt, EBB Bishop, ECW Wallace, EFL Linton, EH Hodgson, EMB Burrows, EMH Holmes, EN Nimmo, ESE Edees, EWJ Jones, FAW Woodhead, FG Greenshields, FHH Haines, FR Rose, FWG Galpin, GAC Crouch, GAM Matthews, GCD Druce, GDF Field, GL Lister, HCP Prentice, HHH Haines, HHW Wood, HJB Brotherton, HAJ, HJG Goddard, HJR Riddelsdell, HJW Wadlow, HNR Ridley, HWP Pugsley, IAR Ricketts, IPG Green, JA Appleyard, JAG Gibson, JAL Larmuth, JAP Paton, JB Bowyer, JCMP Mansel-Pleydell, JDF Fryer, JEL Lousley, JEW Woodhead, JFA Archibald, JFGC Chapple, JGK Keylock, JHS Salter, JHSC Cox, JKH Hasler, JMB Bowcott, JN Nall, JO Ounsted, JPMB Brenan, JRA Akeroyd, JRW White, JWW White, KBR Rooke, KC Cook, KEB Bull, KG Gorringe, EK, LBH Hall, LC Cumming, LJM Margetts, LJW Ward, LMS Spalton, LVL Lester(-Garland), MAA Anderson, MB Blower, MF Frampton, MFVC Corley, MH Holden, MHL Lock, MI Ilchester, MJA Andrews, MJG Galliott, MJS Southam, MM Milnes-Smith, MNJ Jenkinson, MOH Hill, MP Porter, MPY Yule, MRH Hughes, NAS Sanderson, NCC Nature Conservancy, NDS Simpson, NFS Stewart, NHS Natural History Society, OLG Gilbert, PBLS Poole Borough Leisure Services, PDO Orton, PDS Sell, PET Toynton, PFY Yeo, PMH Hall, PRG Green, PWJ James, RA Ashe, RB Burt, RCP Palmer, RCS Stern, RDE English, RDG Good, RDM Meikle, RDR Randall, RF Fitzgerald, RHD Deakin, RJS Surry, RM McGibbon, RMB Burton, RMW Walls, RP Pulteney, RPM Murray, RSRF Fitter, RVS Sherring, RWD David, SBC Chapman, SFRG Southern Fungus Recording Group, SG Griffith, SMB Bovey, SME Eden, SMP Payne, SRD Davey, SS Sherborne School, TBR Ryves, TBS (Bell-)Salter, TCEW Wells, TCGR Rich, TH Hooker, TWC Chester, VJG Giavarini, VMS Scott, WAC Cocks, WB Borrer, WBB Barrett, WCRW Watson, WEN Nicholson, WF Fawcett, WGT Teagle, WHM Mills, WMR Rogers, WRL Linton, ZJE Edwards.

Initials used for Herbaria
ABS Aberystwyth, Univ. Coll. Wales, **ALT** Alton, Curtis museum, **BBSUK** British Bryological Society, **BDK** Baldock, N. Herts District Council Museum, specimens now

in **HITN**, Hitchin, **BEL** Belfast, Ulster Museum, **BIRA** Birmingham City Museum, **BIRM** Birmingham University, **BMH** Bournemouth, Museum of BNSS, **BOL** Bolton Museum, **BRISTM** Bristol University, **CGE** Cambridge University, **CLR** Colchester Museum Nat. Hist., **CLEY** Cley Museum, **CYN** Croydon Nat. Hist. Sci. Soc., **DOR** Dorchester Museum, **E** Edinburgh University, **FKE** Folkestone Museum, **GE** Genova University, **GL** Glasgow University, **GLAM** Glasgow Museum, **HDD** Huddersfield, Tolson Museum, **HIWNT** Hampshire & Is of Wight Nat. Trust, **HLU** Hull University, **K** Kew, **LANC** Lancaster University, **LDS** Leeds University, **LIV** Liverpool Museum, **LTR** Leicester University, **MANCH** Manchester Museum, **MDH** Middlesborough Museum, **NCH** Norwich Museum, **NOT** Nottingham, Natural History Museum, **NMW** National Museum of Wales, Cardiff, **NY** New York Botanic Garden, **OXF** Oxford University, **PMH** Portsmouth Museum, **PVT** Private, **QMC** Queen Mary College, London, **RDG** Reading Museum, **RNG** Reading University, **SDN** Swindon Museum, **SLBI** S. London Botanical Institute, **TLS** Tunbridge Wells Museum, **UEA** University of E. Anglia, Norwich, **WAR** Warwick University, **WRHM** Winfrith, Centre for Ecology and Hydrology, **WRN** Warrington Museum, **YRK** York University.

CHAPTER 6
THE VASCULAR PLANTS OF DORSET

LYCOPODIOPSIDA
CLUBMOSSES

A relict group of about 1,000 species whose leaves have midribs and ligules; some accumulate aluminium.

LYCOPODIACEAE

[*Huperzia selago* (L.) Bernh. ex Schrank & C. Martens
FIR CLUBMOSS
Extinct. 4. Old records from heaths, last seen near Chamberlaynes (89L, 1876, JCMP in **DOR**).]

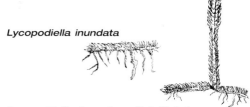

Lycopodiella inundata

Lycopodiella inundata (L.) Holub
MARSH CLUBMOSS
Very local and sporadic in damp, bare heaths and flooded sandpits or dune slacks. 27 + [20]. All recent records are in the Poole basin, lost from the west, and probably decreasing.

Lycopodium clavatum L.
STAG'S-HORN CLUBMOSS
A northern plant with four old records and one recent, but transient one. [5]. Pilsdon Pen (40A, 1969–75, DRS).

SELAGINELLACEAE

Selaginella kraussiana (Kunze) A. Braun
KRAUSS'S CLUBMOSS
A rare alien near glasshouses. 1 + [1]. Brownsea Is (08I, 1965, SW Limborn); Northleigh Ho, Colehill (00F, 1993, !).

ISOETACEAE

Isoetes echinospora Durieu
SPRING QUILLWORT
Rare on the floor of acid lakes or flooded pits. 1 + [4]. Moigne Combe (78T, 1951–70, !); Furzebrook (98G, 1957, !); Arne Heath (98, 1928, G Haines in **K**); Decoy

Heath (99F, 1958, RSRF); Little Sea, Studland Heath (08H, 1901, HJR, *et al.* in **BDK, BM, E, K, NMW, OXF, RNG** and **WRHM**, and 1995, BE and DAP). No British records further east.

EQUISETOPSIDA
HORSETAILS

A relict group of 30 species with jointed stems and whorled leaves; all accumulate silica.

EQUISETACEAE

Equisetum variegatum Schleicher
VARIEGATED HORSETAIL
Rare in damp sandy places. 1 + [1]. Worbarrow Bay cliffs (88Q, 1989, JRW and 1998, BE); Rempstone Heath (98X, 1966–78, CE Ollivant, !). Predominantly northern in Britain.

Equisetum fluviatile L.
WATER HORSETAIL
In acid ditches and pools. 68. Local in the Poole basin, scattered elsewhere but absent from the Chalk and most of the coastal strip.

Equisetum fluviatile × *arvense* = *E.* × *litorale*
Kuehl. ex Rupr.
Scarce, forming colonies on damp clay or by ponds. 7. Powerstock Common (59N, 1982, AJCB); Tadnoll (77Y, 1997, DAP); Encombe (97N, 1961, DFW); Norden pits (98L, 1993, !); Corfe Common (98Q, 1966–83, !).

Equisetum arvense L.
FIELD HORSETAIL
Frequent in hedgebanks, by ditches, in disturbed soil including gardens, on both acid and clay soils. 472. It avoids drier places on the Chalk and very acid sands.

Equisetum sylvaticum L.
WOOD HORSETAIL
A scarce northern plant forming colonies in shaded, damp acid sites. 16 + [4]. Rare in the north and west, with outliers at Lytchett Heath, Stour Row and Daggons; reports from Okeford Fitzpaine (80A), Hart's Copse (90V) and Swanage (07J) need confirmation.

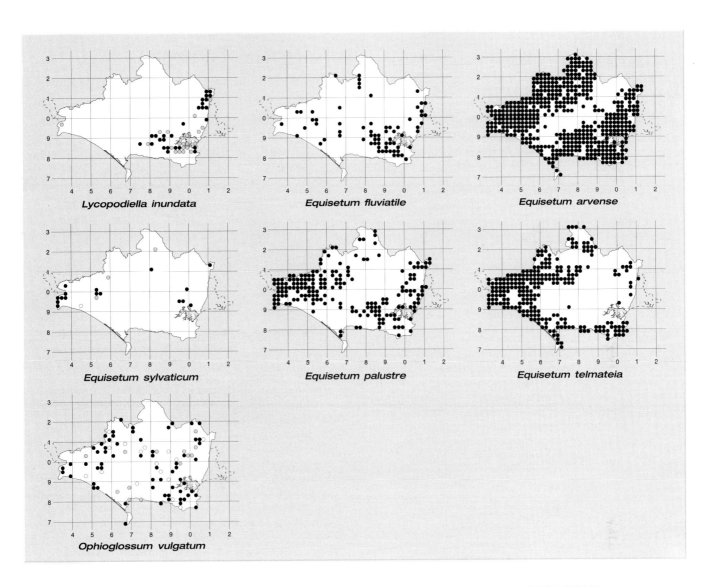

Lycopodiella inundata

Equisetum fluviatile

Equisetum arvense

Equisetum sylvaticum

Equisetum palustre

Equisetum telmateia

Ophioglossum vulgatum

Equisetum palustre L.
MARSH HORSETAIL
In wetter places than *E. arvense*, on waterlogged clay soils.
189. Absent from the drier Chalk.

Equisetum telmateia Ehrh.
GREAT HORSETAIL
In waterlogged clay soils, especially on undercliffs. 256.
Locally frequent along the coast, and in the north and
west but absent from the dry chalk ridges. Tolerant of
shade.

PTEROPSIDA
FERNS

**A group of about 10,000 diploid species producing
haploid spores.**

OPHIOGLOSSACEAE

Ophioglossum vulgatum L.
ADDER'S-TONGUE
An easily overlooked fern of woods, old grassland, fens
and heathy verges. 61 + [30]. Indifferent to soil, but
destroyed by ploughing.

Ophioglossum azoricum C. Presl.
SMALL ADDER'S-TONGUE
Once found on Studland Heath (08H, 1935, ECW in **RNG**);
also in Avon CP (10F, 1991–95, FAW) but in VC11. [1].

Botrychium lunaria (L.) Sw.
MOONWORT
A rare northern plant of acid soils, often under bracken,
much less common than a century ago. 3 + [20]. Recent
records: Kingcombe (59P, 1997, AO Chater); Black Down
(68D, 1970, !); Hartland Moor (98M, 1973, JB); Scotland
Heath (98S, 1959–65, JB, !); Pamphill (90V, 1993, M
Heath).

OSMUNDACEAE

Osmunda regalis L.
ROYAL FERN
Locally abundant in dune slacks, occasional in alder swamps and by acid ditches and pools in the Poole basin, also in the west. 67 + [11]. Tolerant of shade and marine influence, as on undercliffs at Cain's Folly and Branksome. Eaten by cows. Sporelings found on the Agglestone (08G, 1974, !). Planted at Abbotsbury Swannery, Melbury Park, Minterne Magna, Moigne Combe, Creech Grange and Green Is.

ADIANTACEAE

Adiantum capillus-veneris L.
MAIDENHAIR FERN
Occasional on Portland, usually in horizontal clefts of limestone cliffs which are hard to access, and where it has been known since 1864 (SMP in **DOR**). 9 + [4]. Also a transient colonist of old walls: Portesham (1936, DM); Moreton Ch (1936, DM); Creech Grange (1917, CBG); Steepleton Ho (81Q, 1994, !); Gold Hill, Shaftesbury (82R, 1999, VJG); Edmondsham Ho (01Q, 1990, !).

Adiantum pedatum L.
A self-sown clump persists in the wall of the Victorian fernery at Kingston Lacy (90V, 1993–96, !); it is also planted in Abbotsbury gardens. 1.

MARSILIACEAE

Pilularia globulifera L.
PILLWORT
Rare, but locally abundant in shallow, acid ponds and old claypits. 4 + [7]. Apart from an old record from Portland by H Groves, all records are from the Poole basin. Post-1900 records: Mare pond (88W, 1935, RDG, and 1999, BE) and Pool pond (88W, 1997, BE); Furzebrook (98G, 1986, ACJ, *et al.* in **BM, CGE** and **RNG** and 1997, RMW); Ridge (98I, 1913, CBG in **BMH**, *et al.*

in **BM, BRIST** and **NMW**); Norden (98L, 1912, LVL in **BRIST** and **NMW**, also 1915, CBG in **BMH**); Hartland Moor (98M, 1933, RDG in **WRHM** and 1991–98, AJB, DAP and RMW); Corfe Castle (98Q, 1916, A Holden in **TLS**); Slepe Ford (98S, 1933, RDG); Holt Heath (00M, 1932–37, FHH, DM); Crane Bridge, Verwood (00U, 1919, JHS); Verwood Lower Common (00X, 1920, AWG and 1938, RDG in **HLU** and **WRHM**); Daggons Road, Alderholt (11B, 1874, JCMP, 1892, FA Rogers in **LANC** and 1990, RMW).

POLYPODIACEAE

Polypodium vulgare agg.
POLYPODY
Not all recorders distinguished the segregates, which are widespread on shaded banks, old walls or as epiphytes in woods. 77 (agg.). Commonest in the wetter north and west. Var. *multifidocristatum* Moore, a monstrosity with asymmetric fronds, is planted on a wall in Corfe (98Q).

Polypodium vulgare L.
Frequent in the north and west, occasional elsewhere. 74.

Polypodium interjectum Shivas
Widespread, but rare on Portland, extending to shaded peaty banks at the edge of Poole Harbour. 411

Polypodium cambricum L.
Native among limestone rocks on the east side of Portland, scattered elsewhere on walls. 5. Abbotsbury Ch (58S, 1963–96, !); Yetminster (51V, 1990, ! in **RNG**); Church Ope Cove (67V, 1952, MCF Proctor in **CGE**, and 1997, !); Grove undercliff (77B, 1993, !); Sherborne Park (61N, 1970, ACL); Stalbridge (71J, 1993, BE).

DICKSONIACEAE

Dicksonia antarctica Labill. is planted in Abbotsbury gardens.

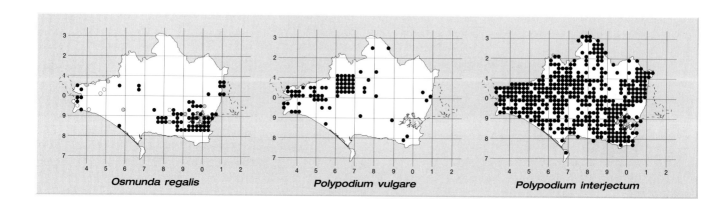

Osmunda regalis   Polypodium vulgare   Polypodium interjectum

DENNSTAEDTIACEAE

Pteridium aquilinum (L.) Kuhn.
BRACKEN
Locally dominant in woods, hedgebanks, on dry heaths, and leached soils overlying chalk, occasionally on walls, but absent from waterlogged soils. 584. Spores abundant in excavated soil of *c.*2000BC. Sometimes 3 m or more tall. The National Trust mow it to inhibit it from over-running heaths. At least 12 Dorset place names are derived from Fern, modified to Farn, Farr or Fur.

THELYPTERIDACEAE

Thelypteris palustris Schott
MARSH FERN
Local; a few colonies in wet woods or alder carr on acid soil, mostly in the Poole basin. 7 + [2]. Westwood Coppice (59N, 1986, RM); Mount Skippet (78T, 1991, DAP); Holme Priory (88Y, 1874, JCMP in **DOR**); Stoborough (98I, 1914, CBG in **BMH** and 1990, AJB); Between Wareham and Hamworthy junction (98N ?, 1910, LVL); The Moors (98N, 1937, RDG in **DOR** and 1990, AJB and DAP, !); Morden Decoy (90A, 1887, JCMP in **DOR**); Morden Mill and Whitefield Fen (90B & C, 1989, BE, !); Elder Moor (90M, 1988, G Marsh, !).

[*Phegopteris connectilis* (Michaux) Watt
BEECH FERN
Extinct or dubious, a northern plant. The report from Bickham Wood (20Y, 1975, TJ Wallace) is in Somerset, while that from Hawkchurch (30K, 1895, JCMP) may have been from Devon.]

Oreopteris limbosperma (Bellardi ex All.) Holub
LEMON-SCENTED FERN
Now rare in sheltered woods or hedgebanks on acid soils, formerly more frequent. 7 + [15]. White's Wood (60G, 1987, !); Higher Hyde (89K, 1991, BE); Coombe Heath (89S, 1992, BE); Stroud Bridge (89V, 1996, BE); Elder Moor (99M and S, 1989, ACJ *et al.*); St Leonard's Peats (10A, 1985, NCC).

ASPLENIACEAE

Phyllitis scolopendrium (L.) Newman
HART'S-TONGUE, HORSE-TONGUE
Locally dominant in sheltered ravines, especially in the north and west, and occasional in carr, among sheltered rocks, on walls and in wells. 617. Tolerant of shade, but absent from very acid soils, and confined to artificial habitats in Poole. Var. *crispum* Gray was found at Coombe Coppice (09D, 1917, CBG), and var. *multifidum* Gray at Portland (67, 1908, WBB).

Asplenium adiantum-nigrum L.
BLACK SPLEENWORT
Occasional in sandy hedgebanks and woods in the west, as far east as Hurst Heath (78Z) and Lee Wood (80F), and widespread but in small quantity on old mortared walls, often in churchyards or on railway bridges. 153 + [20]. It looks native among limestone boulders on the east Portland undercliffs.

[*Asplenium obovatum* Viv.
LANCEOLATE SPLEENWORT
Only one record of this mainly Cornubian fern, which grows in shaded hedgebanks. Near Corfe Castle (98, 1915, RVS in **BM**). The report from SY67 in the 1978 Fern Atlas was an error.]

Asplenium marinum L.
SEA SPLEENWORT
In small quantity in crevices of limestone cliffs, often high above the sea, or in adjacent quarries. 11 + [1]. Lyme Regis (39, 1874, E Newman); Portland (1864, SMP in **DOR**): recent records; Portland Bill and Freshwater Bay (66U and Z, 1981–82, AJB, !); Bowers Quarry (67Q, 1978–90, !); Blacknor (67V, 1999, DAP); Long Tout cutting (77A, 1982, AJB); Gad Cliff (87Z, 1996, BE); St Aldhelm's Head (97S, 1995, BE & !); Winspit (97T, 1925, JHS, *et al.*, and 1980, RS Cropper); Worth Matravers (97Y, 1984, FAW, also WGT and 1995, BE); Dancing Ledge and Tilly Whim (07D and I, 1895, JCMP, *et al.*, and 1981, AJB). Probably the source of erroneous reports of *Asplenium fontanum* L., *Pteris incompleta* Cav. and *P.serrulata* L.f. in 1852. It does not occur further east along the south coast.

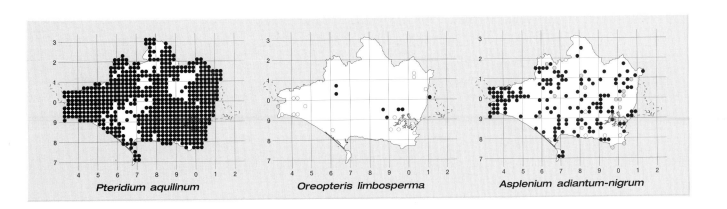

Pteridium aquilinum   Oreopteris limbosperma   Asplenium adiantum-nigrum

Asplenium trichomanes L.
MAIDENHAIR SPLEENWORT
In small quantity on exposed limestone of east Portland undercliffs. Otherwise widespread but local on old walls, commonest in the north and west, or on calcareous or sandstone rocks in railway cuttings. 192.

Asplenium ruta-muraria L.
WALL-RUE
In small quantity on exposed limestone near Abbotsbury

Castle (58T) and on Portland undercliffs. Widespread on limestone and mortared crevices of old brick walls elsewhere. 259. Least common in the east and on the Chalk.

Ceterach officinarum Willd.
RUSTYBACK
In small quantity on limestone undercliffs at East Portland (77B, 1996, BE) and Holworth (78Q, 1978, RHS and S Hatton). Also common on Greensand walls in the north-west, occasional on limestone walls, and rare in the east. 116 + [10].

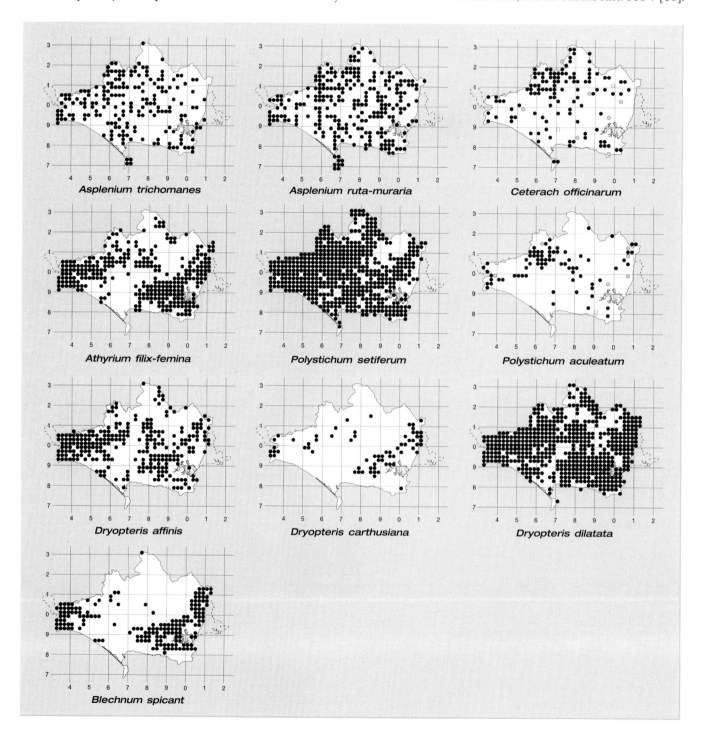

Asplenium trichomanes

Asplenium ruta-muraria

Ceterach officinarum

Athyrium filix-femina

Polystichum setiferum

Polystichum aculeatum

Dryopteris affinis

Dryopteris carthusiana

Dryopteris dilatata

Blechnum spicant

WOODSIACEAE

Matteucia struthiopteris (L.) Tod
OSTRICH FERN
Planted in a few wild gardens and more or less naturalised at Minterne (60S, 1984–96, !). 6. Also noted in 58S, 61I, 73Q, 98W and 00K.

Onoclea sensibilis L.
SENSITIVE FERN
Planted in wild gardens at Forde Abbey, Abbotsbury and Spetisbury Old Mill (90B, 1994), ± naturalised in the last. 3.

Athyrium filix-femina (L.) Roth
LADY-FERN
Locally abundant in wet, shaded ravines and alder woods, less frequent in marshes and drier woods. 296. It prefers acid, humus-rich soils and is absent from most of the chalk and limestone.

[*Cystopteris fragilis* (L.) Bernh.
BRITTLE BLADDER-FERN
Rare and transient on limestone rocks and walls. [2]. Greenhill, Sherborne (61I, 1958–81, CT *et al.*); Purbeck (1799, RP); Langton Matravers quarry (99M, 1895, LVL). The report from Bryanston in 1944 is doubtful.]

DRYOPTERIDACEAE

Polystichum setiferum (Forsskal) T. Moore ex Woynar
SOFT SHIELD-FERN
Frequent in sheltered woods and hedgebanks, commonest in the north and west, and absent or sparse in deforested areas, the north-east and east of Wimborne. 527. Var. *plumosum* Wollaston was reported from Hawkchurch (30K, 1876, J Bevis), perhaps in Devon.

Polystichum setiferum × *aculeatum* = *P.* × *bicknellii*
(Christ) Hahne
Single clumps are rarely seen where the parents meet. 4. Corscombe (50H, 1982, AJCB); Uphall (50L, 1993, LJM and LMS); Charborough Park (99I, 1992, BE); Honeybrook Copse (00B, 1993, BE).

Polystichum aculeatum (L.) Roth
HARD SHIELD-FERN
In similar habitats to *P. setiferum*, but much less common and mainly in the north and west. 74 + [8].

Cyrtomium falcatum (L.f.) C. Presl.
HOUSE HOLLY-FERN
On a limestone garden wall near the sea, Chiswell (67W, 1989–97, !); the wall was remortared in 1998.

Dryopteris filix-mas (L.) Schott
MALE-FERN
Frequent in woods, plantations and hedgebanks. 653. Commonest in the north and west and scarce in deforested areas and the north-east. Tolerant of sea-spray on Stair Hole cliffs (87J).

Dryopteris filix-mas × *affinis* = *D.* × *complexa*
Fraser-Jenkins
Charmouth (39R, 1937, AHG Alston in **BM**).

Dryopteris affinis (Lowe) Fraser-Jenkins
SCALY MALE-FERN
Occasional in woods, plantations and hedgebanks on sand, Greensand and clay. 264. Commonest in the west, and absent from deforested areas; usually in sites with either more shelter or higher rainfall than other ferns tolerate. First record: Bovington (88J, 1818, J Ladbrook in **BM**). Few recorders distinguished the subspecies, but subsp. *borreri* (Newman) Fraser-Jenkins is much more frequent than subsp. *affinis*, and subsp. *cambrensis*. Fraser-Jenkins has only been seen at Elder Moor (99M, 1990, ACJ). The forma *polydactyla* occurred at Rempstone (98W, 1980, M Baxter).

Dryopteris aemula (Aiton) Kuntze
HAY-SCENTED BUCKLER-FERN
Very local in sheltered woodland. 1 + [1]. Monkton Wyld (39I, 1866, ZJE, *et al.*, 1989, !); Whitchurch Canonicorum (39X, 1913, HHH in **LTR**). Reports from Clifton Wood (51R) and Afflington Wood (97U) need confirmation.

(*Dryopteris cristata* (L.) A. Gray
There are no specimens to support the three reported sites in JCMP's Flora, and they were believed to be errors by EFL and RPM.)

Dryopteris carthusiana (Villars) H.P. Fuchs.
NARROW BUCKLER-FERN
Local in alder swamps on acid and humus-rich soils. 68. Commonest in the west, absent from calcareous regions.

Dryopteris dilatata (Hoffm.) A. Gray
BROAD BUCKLER-FERN
Frequent to abundant in wet and dry woods, plantations and hedgebanks, but scarce in deforested areas. 521. It grows on some undercliffs, usually in a depauperate condition.

Dryopteris wallichiana and other hardy ferns have recently been planted in Abbotsbury gardens.

BLECHNACEAE

Blechnum spicant (L.) Roth
HARD-FERN
Locally frequent in woods and shaded ditch-banks, always in shelter and on ± acid soils. 172. Occurs in the west and in the Poole basin, with a few outliers as at Ashley Chase (58U) and on Greensand at Bourton (73Q), almost never on the Chalk.

Blechnum cordatum (Desv.) Hieron.
CHILEAN HARD-FERN
Planted in the grounds of Forde Abbey (30M, 1992, !). 1.

AZOLLACEAE

Azolla filiculoides Lam.
WATER-FERN
Sporadic in ponds, ditches, cress-beds and once in a salt-marsh. 8. Sold by aquarists. Chideock (49G, 1993, DAP); Abbotsbury (58S, 1995, DAP, !); Lewell Mill (78J, 1981, DAP and RMW); West Stafford (79F, 1981, EF Lenton and 1999, JHSC and WP Cox); Waddock (79V, 1977, V Jesty); Hinton St Mary (71X, 1989, !); R. Frome, Holme Bridge (88Y, 1999, !); Swineham (98I, 1996, DAP); Rempstone (98W, 1916, CBG in **BMH**); Ower (98X, 1935, BNSS); Herston (07E, 1977, WGT); Knitson Farm (08A, 1999,!); Rivendell Nursery (00N, 1994, !); Stephen's Castle (00Z, 1993, VMS).

PINOPSIDA
GYMNOSPERMS OR CONIFERS

A group of about 600 species which produce naked ovules and seeds, but lack flowers, and without vessels in their vascular tissues.

GINKGOACEAE

Ginkgo biloba L.
MAIDENHAIR TREE
Planted as single trees in Parks and churchyards, or as a pavement tree in Poole. 9. In 1970 the largest were at Sherborne Castle (25 m × 3 m) and Melbury Park (23.5 m × 3 m); also seen in 30S, 51Y, 61N, 79X, 80H, 99I and 09A.

PINACEAE

Abies alba Miller
EUROPEAN SILVER-FIR
Only seen in two plantations, at Nether Cerne (69U, !) and Canford School (09J, !). 2. Like most firs, it dislikes our Atlantic climate and is subject to attack by both insects and fungi.

Abies nordmanniana (Steven) Spach
CAUCASIAN FIR
Block planted at Burton Common (61K) and single trees at Bere Heath and Brownsea Is. 3.

Abies grandis (Douglas ex D. Don) Lindley
GIANT FIR
Occasionally block planted, forming dense shade, at Bowden Hill, Cole Wood, Sares Wood, Bere Heath, Ranksborough Gorse, Longthorns, Arne Big Wood (98U, 1954, N Moore), Charborough Park, Monmouth's Hill, Sovell Down and Ferndown Forest. 12.

Abies procera Rehder
NOBLE FIR
Block planted at Bere and Bloxworth Heaths, seedlings noted (89R, 1990–96, !).

Other species of Fir have been planted as specimen trees: *A. amabilis* Forbes and *A. homolepis* Siebold & Zucc. at Bere Heath (a big tree of the latter at Abbotsbury fell in 1980); *A. cephalonica* Loudon (36m × 4.6m) at Melbury Park; *A. lasiocarpa* (Hooker) Nutt. at Minterne; and *A. pinsapo* Boiss. in churchyards at Netherbury, Winterborne Kingston and Kingston Lacy; the last was at Charborough Park in 1914.

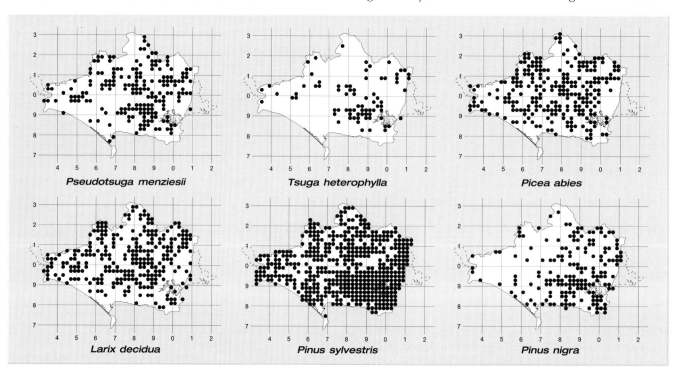

Pseudotsuga menziesii

Tsuga heterophylla

Picea abies

Larix decidua

Pinus sylvestris

Pinus nigra

Pseudotsuga menziesii (Mirbel) Franco
DOUGLAS FIR
An important forestry tree, block planted in woods on most soils. 190. Seedlings noted at Puddletown Forest (79L) and Broomhill Plantation (79W) on sand, and at Milton Park Wood (80B) on chalk.

Tsuga heterophylla (Antoine) Carriere
WESTERN HEMLOCK-SPRUCE
A forestry tree, block planted in woods on mostly acid soils. 77. Seedlings noted in Puddletown Forest and on Bere and Throop Heaths. It can succeed on chalk.

Picea sitchensis (Bong.) Carriere
SITKA SPRUCE
Block planted on acid soils, especially in the west. 32. Seedlings in wet heath, Morden Decoy (99A). There is a tree 38m × 3m at Melbury Park.

Picea omorika (Pancic) Purk.
SERBIAN SPRUCE
Block planted at Shillingstone Hill (80J) and near St Leonards (10A, HA Gillen). 2.

Picea abies (L.) Karsten
NORWAY SPRUCE
Block planted on acid soils, especially in the west, and in small amounts in plantations elsewhere. 244. Seedlings in Bere Wood and at Bloxworth Heath (89R and S). Sold at Blandford in 1782; the pollen is confined to the top 12 cm of soils, so the tree has never been a native. It looks sickly compared to trees in Norway.

Other species of **Picea** occur as planted specimen trees: *P. bicolor* (Maxim.) Mayr, 23m tall at Melbury Park; *P. breweriana* S Watson at Canford School and Brownsea Is; *P. glauca* Voss at Sherborne Park, Almer (99E) and Green Is; *P. orientalis* (L.) Carriere and *P. polita* (Siebold & Zucc.) Carriere at Canford School; *P. smithiana* (Wallich) Boiss. 33m tall at Melbury Park, also on Brownsea Is, and in Charborough Park in 1914; *P. spinulosa* (Griff.) Henry 26 m tall at Melbury Park.

Larix agg.
LARCH
Block planted in plantations on all soils, and in small amounts in parks etc. 269. Few recorders determined the species or hybrid they saw.

Larix decidua Miller
EUROPEAN LARCH
The least common larch in recent block plantings, but old trees occur singly in woods, and seedlings were seen in Bere Wood (89S).

Larix decidua × kaempferi = L. × marschlinsii Coaz
HYBRID LARCH
Commonly block planted, but under-recorded. 24+. On very acid soil at Godlingston Heath (08B) it fails to flourish as it does elsewhere. Seedlings seen at Bere Wood (89S).

Larix kaempferi (Lindley) Carriere
JAPANESE LARCH
Often block planted, but under-recorded. 19+. First record, Rempstone (98W, 1984, ! in **RNG**). Seedlings noted.

Cedrus deodara (Roxb. ex D. Don) Don
DEODAR
Rare as single trees in parks in 70V, 88L, 81F, 99I, 08I and 00Z. 7. In a plantation at Stubhampton Bottom (81Y, 1994, !).

Cedrus libani A. Rich.
CEDAR-OF-LEBANON
Specimen trees are planted in parks and churchyards. 34. The largest, 29 m × 6.7 m, at Ranston Ho, Bryanston was planted in 1680. Other imposing trees occur at Sherborne Castle, Chettle Ho, Bloxworth Ho and Spetisbury, and a tree in the avenue at Kingston Lacy was planted by the German Kaiser in 1907.

Cedrus atlantica (Endl.) Carriere
ATLAS CEDAR
Specimen trees rarely seen in Parks and churchyards. 10. Seen in 40Q, 61N, 70W, 71G, 99W, 90V and Y, 08I, 09J and W: no cedar seedlings have been reported.

Pinus sylvestris L.
SCOTS PINE
Abundant and self-sown on heaths, and block planted on sandy soils; less common on chalk and absent from waterlogged clays. 479. Although generally considered an alien in south Britain, it was here in the mesolithic and neolithic periods, and pine charcoal dating to 1500BC has been found at East Stoke (Ashbee & Dimbleby, 1958). Trees were planted at Hurn Court in VC11 in 1746, and were on sale at Blandford in 1782. Hilltop clumps of pine, as at Symondsbury (49L), Wolfeton (69X) and Elderton (89N) were planted as landmarks for cattle-drovers. Six Dorset place names are derived from 'pin', but some could have arisen from old English 'pinn' meaning peg. Larvae of several moth species eat the needles.

Pinus nigra J.F. Arnold
Subsp. *nigra* is rare. Subsp. *laricio* Maire – CORSICAN PINE – is extensively block planted on acid or leached soils, and is widespread as isolated trees elsewhere. 175. It tolerates salt-spray on the coast. A tree 34 m × 5 m is in Kingston Lacy Park.

Pinus muricata D. Don
BISHOP PINE
Block planted on Bloxworth Heath (89W), and perhaps mistaken for *P. nigra* elsewhere. 1.

Pinus contorta Douglas ex Loudon
LODGEPOLE PINE
Block planted on heaths in the Poole basin. 16. Self-sown in boggy places, but often fails.

Pinus pinaster Aiton
MARITIME OR BOURNEMOUTH PINE
Frequently self-sown on heaths and sandy cliffs in the Poole basin, and sometimes planted elsewhere. 83. It survives on chalk but prefers sand. Its introduction may date to 1801 at Ringwood, VC11.

Pinus radiata D. Don
MONTEREY PINE
Occasionally block planted on heaths, more often planted as single trees or clumps, especially along the coast. 65. This three-needle pine produces seedlings at Abbotsbury (58S), Bere Heath (89R), Rempstone (98W, EAP), Gore Heath (99F) and Godlingston Heath (08B). Large trees noted at Holworth Ho, Brownsea and Furzey Is.

Pinus strobus L.
WEYMOUTH PINE
Block planted at West Moors (00W, HA Gillen), subject to blister rust. 1. A five-needle pine. There was a large specimen at Charborough Park in 1914, with *P. cembra* L.

Pinus peuce Griseb.
MACEDONIAN PINE
Block planted at Uddens (00K, HA Gillen); another five-needle pine. 1.

Other **Pinus** species occur as planted specimen trees; two-needles: **P. pinea** L., 17 m × 3.4 m at Bryanston School; **P. thunbergii** Parl. 20m tall at Lytchett Heath; three-needles: At Bere Heath and Canford School, **P. jeffreyi** Murr. ; five-needles: **P. armandi** Franchet and **P. parviflora** Siebold & Zucc. ; **P. wallichiana** AB Jackson at Bere Heath, Melbury Park and Sherborne Park. **Pseudolarix amabilis** (Nelson) Rehder is planted at Canford School (09J, TH).

TAXODIACEAE

Cryptomeria japonica (L.f.) D. Don
JAPANESE RED-CEDAR
Single trees planted since before 1894 in plantations, parks and churchyards. 16. Seedlings seen at Marshalsea Farm (30V, IPG). A tree 27 m × 4 m at Melbury Park; another large tree at Charborough Park, first reported in 1914; other trees seen in 71J, 88L and N, 89M, R and S, 80A, E and J, 98E, 99D and S and 00Z.

Cunninghamia lanceolata (Lambert) Hook. is planted at Canford School (09J, TH).

Metasequoia glyptostroboides Hu & W.C. Cheng
Specimen trees seen in seven parks; 51Y, 69V, 61N, 73Q, 99W, 08U and 09J.

Sequoia sempervirens (D. Don.) Endl.
COASTAL REDWOOD
Occasional in plantations and parks. 15. A tree 41 m × 6.4 m is at Melbury Park.

Sequoiadendron giganteum (Lindley) Buchholz
WELLINGTONIA
Specimen trees planted in plantations, parks and churchyards. 28. Crichel Ho has one 34 m × 8.8 m.

Taxodium distichum (L.) Rich.
SWAMP CYPRESS
Specimen trees planted in parks, rarely in woodland, as at Shillingstone Hill (80J). 10. A large tree at Dean's Court, Wimborne was planted in the 17th century; others seen in 30M, 40Q, 50S, 51V, 78T, 70B, 80J, 99S and W, and 09J.

CUPRESSACEAE

Cupressus macrocarpa Hartweg ex Gordon
MONTEREY CYPRESS
Widely planted in hedges and grounds, and tolerant of lime. 68. Trees make quite a feature in the county, though subject to windblow and occasional dieback. Seedlings noted in Abbotsbury gardens (58S, SG) and Ferndown Forest (00N). Melbury Park has a tree 37 m × 5.8 m; another large tree is at Sturminster Marshall Ch (90K).

X Cupressocyparis leylandii (A.B. Jackson & Dallimore) Dallimore
LEYLAND CYPRESS
Much planted in hedges. 11+. First record away from gardens, Ryme Intrinseca (51Q, 1994, JAG).

Cupressus glabra Sudw.
SMOOTH ARIZONA CYPRESS
Planted at Glebeland estate, Studland (08F, 1998, EAP). 1.

Chamaecyparis lawsoniana (A. Murray) Parl.
LAWSON'S CYPRESS
Sometimes block planted, more often used for tall hedges, on sandy and calcareous soils. 290. Seedlings are often abundant in plantations, and occur in Dorchester.

Chamaecyparis pisifera (Siebold & Zucc.) Siebold & Zucc.
SAWARA CYPRESS
Rarely planted. 1. Big trees in Netherbury churchyard (49U, !).

Chamaecyparis nootkatensis (Lambert) Spach
NOOTKA CYPRESS
Rarely planted. 2. An avenue at Branksome Ch (09Q), also a tree at Canford School.

Thuja plicata Donn ex D. Don
WESTERN RED-CEDAR
Block planted on sandy and calcareous soils; occasionally in churchyards. 84. Seedlings seen at Bowden Hill (61J), Chetterwood (90T) and Colehill (00F).

Thuja occidentalis L.
AMERICAN ARBOR-VITAE
Rarely planted. 1. One at Carey Ho was 21 m × 2 m in 1968.

Platycladus orientalis (L.) Franco
CHINESE ARBOR-VITAE
Planted in a churchyard at Long Crichel (91Q). 1.

Juniperus communis L.
COMMON JUNIPER
Once scattered over the county on most types of soil, now almost destroyed by agriculture except for a few bushes in the north-east. 6 + [17]. Rarely planted, as at Little Wood, Frampton (69G) and Minterne (60S). No seedlings noted; natural regeneration is stimulated by removal of topsoil. Puncknowle Common (58J, 1905, WBB and 1956, J Warrington); Langton Herring Common (68B, 1963, JB); Bloxworth (89X, 1895, JCMP); Langton Matravers (97Z, 1895, E Bankes); Hartland bog (98M, burned in 1965, DSR); Bushey and Wytch Heaths (98S and W, 1912, CBG in **BMH**); Luccombe Down (80A, 1874, JCMP); Little Coll Wood (80Q, 1895, RPM); Hod Hill (81K, 1978, RHD); Hambledon Hill (81L, 1952, AS Thomas); Fontmell Down (81Y, 1979, DWT); King Down (90R, 1932, FHH); Crichel and Week Street Downs (91R, 1873, LKW); Gussage Hill (91X, 1873, LKW); Anvil Point (07I, 1980, RJH Murray); one bush, Ackling Dyke (01C, 1992, HCP); Bottlebush Down (01D, 1873, LKW); 11 bushes, St Giles Park (01F, 1990, DAP); single bushes, Bokerly Dyke (01J and P, 1982, AH, !)

Juniperus chinensis L.
CHINESE JUNIPER
Two old trees planted in Wimborne Minster churchyard (09E). 1.

Juniperus oxycedrus L.
PRICKLY JUNIPER
One bush planted at Rempstone (98W, 1974, ! in **RNG**). 1.

Calocedrus decurrens (Torrey) Florin
INCENSE CEDAR
Specimen trees planted in parks and churchyards. 7. Noted in 30M, 69J, 88L, 99S, 90Q, 08U and 09J.

Thujopsis dolabrata (L.f.) Siebold & Zucc.
HIBA ARBOR-VITAE
Specimen trees planted in parks and churchyards. 6. Large ones at Abbotsbury (18 m × 1.2 m) and Melbury Park (20 m × 1.2 m), others in 78I, 72W, 80C and 99W.

ARAUCARIACEAE

Araucaria araucana (Molina) K. Koch
MONKEY-PUZZLE
Odd trees persist well away from houses, or in churchyards. 9. On chalk at Milldown (80Y), and on sand at Brownsea Is (08J); also seen in 51V, 79I, 90V, 07J, 09F, G and W.

TAXACEAE

Taxus baccata L.
YEW
Native since neolithic times, this rarely forms pure woods, as at Hambledon Hill (80K) and Blackbush Down (01M). Frequent and seeding well in woods and plantations on the Chalk, even close to the sea as at Handfast Point (08L), but also found on sandy soils, as at Arne (98U and Z). 415. There are large trees in Lower Lodge plantation (70V), and in many churchyards, e.g.; Woolland (17 m × 8.8 m), Stoke Abbott (girth 5.5 m), Upwey, Trent, Sandford Orcas, Kington Magna, Turnworth, Gussage St Michael and

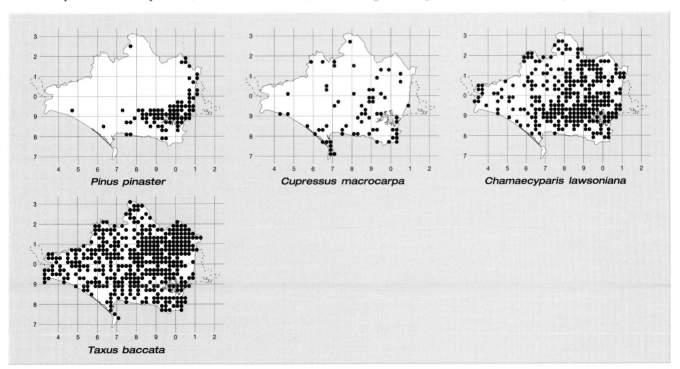

Pinus pinaster

Cupressus macrocarpa

Chamaecyparis lawsoniana

Taxus baccata

Knowlton. In the church porch at Lytchett St Mary the adjacent yew is certified as being 1,600 years old; as trees more than 400 years old become hollow, this is hard to establish. In the 18th century there was a tree in East Stour churchyard under which 1,500 men could stand. Yew is much used for hedges, as at Hardown Hill (49C); a hedge at Bingham's Melcombe dates from *c.*1580, and is 4.3 m tall by 5.5 m deep. Var. *fastigiata* Loud., the Irish Yew, is often planted in churchyards and survives in derelict gardens and roadsides. Of the seven Dorset place names derived from yew, the most notable is the River Iwerne and its villages. The fruit is eaten by all members of the thrush family (Wilson, Arroyo & Clark, 1997), and the wood is valued for furniture.

Podocarpus salignus D. Don makes a fair-sized tree at Abbotsbury gardens.

MAGNOLIOPSIDA
ANGIOSPERMS OR SEED PLANTS

A group of perhaps 300,000 diploid species, mostly with flowers, and all with haploid sex cells. They all form seeds, plus well developed roots and vascular tissue.

DICOTYLEDONS

LAURACEAE

Laurus nobilis L.
BAY
Widely planted in shrubberies and churchyards. 33. Self-sown on Lyme undercliffs (39F, AWJ), Abbotsbury gardens (58S) and Western Ledges (67Y). Big trees at Kingston Lacy (15 m × 1 m) and Green Is.

Laurus azorica (Senb.) Franco
AZORES BAY
Planted at Abbotsbury gardens (58S, SG), where there is a tree 10 m × 1 m and frequent seedlings. 1.

MAGNOLIACEAE

Liriodendron tulipifera L.
TULIP-TREE
Planted as specimen trees. 15. Large ones at Little Bredy (58Z), Furzey Is (08D) and Canford School (09J); also in 30M, 70V and W, 89M and S, 80A, H and P, 81U, 99W, 09G and X.

Magnolia spp. are planted in parks and churchyards. *M. campbelli* Hooker f. & Thoms, big trees at Abbotsbury and Minterne; *M. grandiflora* L. at Maiden Newton Ch and

East Holme Ch.; *M. × soulangeana* Soulange-Boudin at Netherbury Ch and Compton Abbas Ch ; *M. stellata* Maxim, in churchyards at Corscombe, Maiden Newton, Stourton Caundle and Yetminster.

ARISTOLOCHIACEAE

Asarum europaeum L.
ASARABACCA
Once used in pharmacy, introduced. 1 + [1]. Under hazel, far from houses, Corscombe (50I, 1929, CJ Troyle-Bullock, *et al.* in **BDK, OXF, RNG**, and 1982, AJB, !) – hard to find among Ramsons; Cranborne (01L, 1966–79, WEA Evans).

NYMPHACEAE

Nymphaea alba L.
WHITE WATER-LILY
Probably native in the Stour, and in ponds in the Poole basin, but planted or escaped in many artificial lakes, angler's ponds and old claypits. 67. Plants at Slop bog (00V, 1990, AJB) were recorded as subsp. *occidentalis* (Ostenf.) N. Hyl., which is a northern taxon. Some of the introductions fall under *N. × marliacea* Latour.-Marl., such as pale-yellow flowered plants at Broadmayne (78I), Breach plantation (98B) and New Swanage (08G), or pink-flowered plants at Parnham (40Q), Droop (70P), Moreton (88E), Furzebrook (98G) and Ham Common (99V).

Nuphar lutea (L.) Sm.
YELLOW WATER-LILY, CLOT
Native and frequent in the larger rivers as well as smaller streams such as the Allen, Axe, Lodden, Lydden and Yeo. 82. Tolerant of shade. Probably planted in some artificial lakes, but absent from the south and west. In several poems William Barnes refers to the "cloty Stour".

CERATOPHYLLACEAE

Ceratophyllum demersum L.
RIGID HORNWORT
Occasional in the Stour, local in lakes and pits and often in garden ponds. 29.

RANUNCULACEAE

A family showing great diversity in form of both flowers and fruits.

Caltha palustris L.
MARSH-MARIGOLD, BULL'S-EYES, MAY BLOB
Local in alder swamps, wet woods, reedbeds and as a relict in marshy fields or ditches by the larger rivers. 211. Absent from the chalk ridges and Portland. A double-flowered form is planted at Winterbourne Steepleton (68J).

Helleborus foetidus L.
STINKING HELLEBORE
Though perhaps native in a few old woods, most recent records are of introductions, since it seeds freely in gardens and escapes. 27. It may be native in three places: Milton Park Wood (80B, 1937, DM and 1997, !); woods near Gunville (91B, 1957, JAL); Rushmore (91P, 1874, JCMP in **DOR**, but perhaps in Wilts.).

Helleborus viridis L.
GREEN HELLEBORE
Scarce in old woods. 17 + [10]. It is rarely cultivated, and seeds sparingly, but is persistent. Marshwood (39Z, 1985, AWJ); Bettiscombe (49E, 1919, AWG, *et al.*, and 1999, IPG); Ashley Chase (58N, 1874, JCMP, *et al.*, 1996, IPG and PRG, !); Coombe Coppice (59K, 1992, BE); Toller Porcorum (59N, 1976, E Sykes); Bere Wood (89S, 1900, O Pickard-Cambridge and 1991, !); under Lewesdon Hill (40F, 1921–28, AWG); Weston Wood (50C and D, 1919, AWG and 1992, BE, !); Rampisham (50L, 1895, Miss Rooke); west of Bulbarrow (70T, 1993, MJG); Ibberton Ch (70V, 1895, JCMP, *et al.*, 1998, !); Buckhorn Weston (72M, 1895, H D'Aguilar); Lee Wood, Whatcombe (80F, 1895, JCMP, also 1938, DM and 1999, !); Bryanston (80T, 1966, BNHS); Iwerne Minster (81S, 1895, RPM); Ashy Coppice (81W, 1999, BE); Chettle Wood (91L, 1981–88, C Tandy); Tollard Royal (91N, 1799, RP and 1978, AH Dunn); Deanland (91Z, 1999, P Amies); High Hall (00B, 1895, JCMP and 1955, KG); Tarrant Gunville (00M, 1916, AWG); Dewlands Common (00U, 1918, JHS); Edmondsham (01R, 1905, EFL and 1991, BE); Castle Hill Wood (01S, 1959, KBR and 1993, GDF).

Helleborus argutifolius Viv.
CORSICAN HELLEBORE
Planted in a few churchyards or on waste ground. 4. Beaminster (40V, !); Dorchester Industrial estate (69V, !); Osmington Ch (78G, !); near Blue Pool (98G, !).

Helleborus orientalis Lam.
LENTEN ROSE
In abandoned garden, Bourton (73K, 1996, IPG and PRG); in a plantation, Middle Lodge, Bryanston (80T, 1999, !). 2.

Eranthis hyemalis (L.) Salisb.
WINTER ACONITE
Rarely naturalised in churchyards and plantations. 13. For over 100 years in a grazed copse, Milborne St Andrew (79Y, 1977, CA Whitby and 1997, !); verge east of Stair Hole (88A, BE & !); Manston Ch (81C, !); Tarrant Crawford (90G, 1961, HJB and 1996, !); also seen in 30M, 59Q, 50V, 60S, 70R, 89B and G, 90B and Q.

Nigella damascena L.
LOVE-IN-A MIST
Escaped and casual on disturbed verges and tips. 21. Fallen cliff, Sandsfoot (67T, 1986, ! in **RNG**).

Nigella arvensis L.
In a cultivated plot at New Barn, Bradford Peverell (69L, 1997, !). 1.

Aconitum napellus L.
MONK'S-HOOD
Probably native by streams such as the Bride and Yeo in

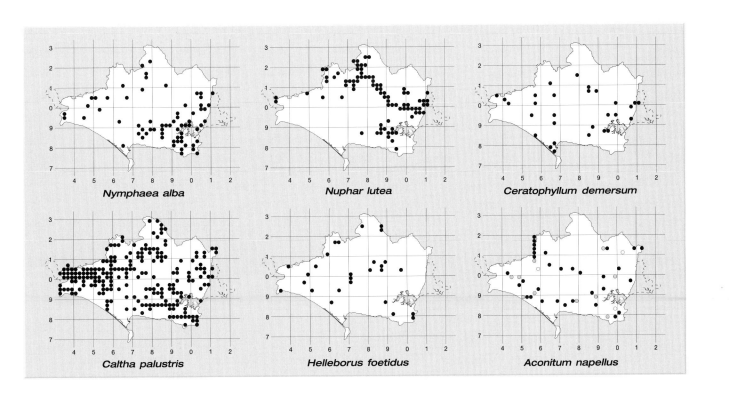

Nymphaea alba

Nuphar lutea

Ceratophyllum demersum

Caltha palustris

Helleborus foetidus

Aconitum napellus

the west and north. 32 + [13]. Other records may be garden escapes, which are persistent but localized to single clumps.

Consolida ajacis (L.) Schur
LARKSPUR
An occasional casual or garden escape on tips or disturbed ground by roads or railways. 3 + [10]. It persisted for many years in sandy arable west of Wareham (98D and E, 1915, CBG in **BMH** and 1955, !); other post-1950 records are: Tenantrees (78J, 1977, !); Wool railway (88N, 1989–93, DAP, !); Warren Heath (89K, 1977, !); Furzebrook (98H, 1957, PFH in **WRHM**); South Haven verge (08I, 1994, RSRF & !); Fleet corner tip (09B, 1977, !); Canford Magna (09J, 1953, KG).

Anemone nemorosa L.
WOOD ANEMONE, SNAKE-FLOWER
Locally abundant in old woodland and hedgebanks on all but the most acid soils. 324. Absent from deforested areas, including Portland, and built-up areas at Poole. Var. *purpurea* DC., with purplish petals, is present in most large populations.

Along with *Pulsatilla* and many terrestrial species of *Ranunculus*, this contains the toxin protoanemonine.

Anemone apennina L.
BLUE ANEMONE
Occasionally planted in churchyards or hedgebanks near houses. 14 + [1]. First record, Langton Matravers (97Z, 1889, LVL in **DOR**). With white flowers at Chalmington (50V); also seen in 39X, 30M and R, 49U, 40K and V, 59Q, 50Z, 69V, 79Q, 81S, 99X and 91L.

Anemone blanda Schott & Kotschy
BALKAN ANEMONE
Seen in a lane at Gussage St Michael (91W, 1996, GDF),and in churchyards at Stratton (69L) and Pimperne (90E). 3.

Anemone × hybrida Paxton
JAPANESE ANEMONE
A garden escape 5 + [1]. Verne (67W, 1979); Lodmoor tip (68V); Wareham (98I); Charborough Park (99D); Swanage (07J, EAP); Ferndown (09U).

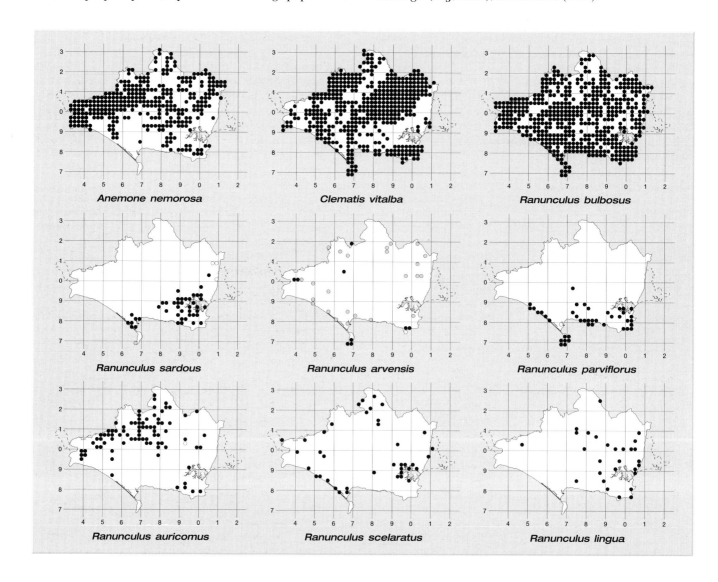

72

Other garden Anemones are planted very locally, e.g. *A. coronaria* L. at Witchampton Ch (90Y), *A.* × *fulgens* Gay at Friar Waddon (68M) and *A. payonina* Lam. at Clayton Meadow (O8F, EAP).

Pulsatilla vulgaris Miller
PASQUEFLOWER
1–2 plants in mature chalk turf on the Bokerly Dyke (01P, 1983, PET and 1998, FR, !), probably native. 1. It self-seeds in my garden at Winterborne Kingston, and in a nursery at Briantspuddle, but the ripe seed remains viable for a very short time.

Clematis vitalba L.
TRAVELLER'S-JOY, BEDWINE, OLD-MAN'S-BEARD
Woods, plantations and hedges, mostly on calcareous soils, and frequent in the old quarries of Portland and Purbeck. 432. It forms thick lianes in neglected plantations on chalk. Var. *integrata* DC., with integral leaves, is probably frequent – reported from Worbarrow Bay (87U, 1937, NDS) and Boveridge (01S, GDF).

Clematis flammula L.
VIRGIN'S BOWER
Established by the railway, Corfe Castle (98R, 1983, AJB and 1993, JB). 1.

Clematis tangutica (Maxim) Korsh
ORANGE-PEEL CLEMATIS
This may be the name of the straggler from a garden at Lewell Mill (78J); One survives at Canford tip (09I, 1999, !). 2.

Clematis viticella L.
PURPLE CLEMATIS
A garden escape. Abbotsbury (58S, 1953–89, !); Milborne St Andrew (89D, 1977–91, !). 2. It was on sale at Blandford in 1782, and grown at Abbotsbury in 1899 (MI).

Ranunculus acris L.
MEADOW BUTTERCUP, GILCUP
Frequent in old meadows, pastures and verges, on neutral to calcareous soils. 653. Flowers in May and again in early autumn if mown. Avoided by grazing stock. Whole fields coloured yellow by this plant, as by the Frome at Fordington (79A), are no longer common.

Ranunculus repens L.
CREEPING BUTTERCUP, RAM'S-CLAWS
Very common in woodland rides, meadows, pastures, verges, marshes, pond-margins and disturbed soil, but absent from heaths. It can colour damp meadows yellow in May. 732. Also a persistent lawn weed, resistant to trampling, which spreads by seed and runners. It varies greatly in size and luxuriance. Double-flowered forms occurred at Stockbridge Oak, (61F, 1998, !) and Stoborough (98C, 1999, RSRF & !).

Ranunculus bulbosus L.
BULBOUS BUTTERCUP, DILL-CUP
Locally frequent on dry, grassy banks, preferring calcareous soils. 492. A double-flowered form was found on a railway bank at Bailey Gate (99P, 1893, EFL), and plants with almost white flowers occurred on Slepe Heath (98M, 1999, !). Variable in stature and hairiness, but var. *dunensis* Druce has not been reported.

Ranunculus sardous L.
HAIRY BUTTERCUP
Local and sporadic near the coast since the late Iron age, preferring disturbed habitats, and only seen in quantity in fallow near Stoborough (98C). 43 + [4]. It occurs on both acid and calcareous soils and favours clay. There are few records west of Weymouth or in the north, e.g. Holt Heath (00L, 1995, GDF).

Ranunculus arvensis L.
CORN BUTTERCUP
Once a frequent cornfield weed, especially on well-drained calcareous soils, this has declined to near extinction. 9 + [26]. Good had it from five stands in the 1930s. Recent records are: Bettiscombe (30V and 40A, 1998, IPG); Portland (66U and Z, + 67V, 1984, RA and 1982, AJB); Lyon's Gate Farm (60M, 1988, AM); Crendle (61Z, 1993, JFA); Winspit (97T, 1988, M Dyke); Seacombe (97Y, 1983, AJB).

Ranunculus parviflorus L.
SMALL-FLOWERED BUTTERCUP
Occasional on dry banks, especially where bare or disturbed, near the coast. 42. Mostly on chalk, but also on sand at Moreton, Stoborough, Studland and Brownsea Is (08I). It germinates in autumn. First recorded in the late Iron Age, and historically in the late 18th century by W Withering.

Ranunculus auricomus L.
GOLDILOCKS BUTTERCUP
Local, never in quantity, in old woods and hedgebanks. 83. It prefers clay soils in the north-west, but occurs on Wealden soils in Purbeck.

Ranunculus sceleratus L.
CELERY-LEAVED BUTTERCUP
Rather uncommon on bare mud. 39. It tolerates some salinity, and occurs on the landward side of the Fleet, at Radipole Lake and around Poole Harbour. It is also found in the north-west, and in the upper Stour basin, but is rare near other Dorset rivers.

Ranunculus lingua L.
GREATER SPEARWORT
Once a rare plant of riversides and their ditches, now even rarer as a native, but widespread in ponds as an escape from water gardens. 28. It tolerates some shade. Now lost from the Frome, Allen and Moors Rivers, but probably native in three sites near the Stour: Spetisbury Old Mill(90B, 1994, !; in a wild garden, but not planted); Sturminster Marshall (90K, 1979, RA Leney); Cowgrove (90V, 1905, EFL, *et al.* in **BDK** and **BMH**, and 1993, DAP and M Heath). Probably extinct as a native as follows: [Redcliff, 1900; The Moors, 1937; Morden Park Lake, 1942; Hammoon, 1874;

Witchampton Mill, 1937; High Hall, 1909; Hinton Parva, 1895; Moors River, 1930]. A well-established alien at Bloxworth (89S, !), Lower Row (00M, DJG) and Woodlands Common (00U, GDF and JO).

Ranunculus flammula L.
LESSER SPEARWORT
Frequent in wet woodland rides, heathy ditches, flushed bogs and old wet pastures. 209. It prefers acid soils and is absent from the chalk, though a relict at Culverwell on Portland (66Z).

[*Ranunculus ophioglossifolius* Vill.
ADDER'S-TONGUE SPEARWORT
In a wet spot near Tincleton (79Q, 1914–17), found by a youthful Ronald Good. (Good, 1948).]

Ranunculus ficaria L.
LESSER CELANDINE, GILCUP
Common in woodland margins and hedgebanks, surviving in grassland, verges and as a garden weed. 683. Our common plant is subsp. *ficaria*. Subsp. *bulbifer* Lambinon is frequent in damp, shaded places, but is seldom recorded. Subsp. *ficariiformis* (F Schultz) Rouy & Fouc. has records from Portland (67V, 1990, ! in **RNG**) and Ringstead (78K, 1982, !), but does not look as distinct as it does in Portugal.

Ranunculus hederaceus L.
IVY-LEAVED CROWFOOT
Local in wet, bare muddy places and by springs or seeps. 57 + [9]. Tolerant of some salinity on cliffs, and of a wide range of pH, from wet heath to fen. Absent from the Chalk; on Portland it was last seen in 1912 by WBB.

Ranunculus omiophyllus Ten.
ROUND-LEAVED CROWFOOT
Local in acid streams, ditches and ponds, tolerant of shade. 24. Mostly in the Poole basin, also in the west, and absent from calcareous districts.

Ranunculus tripartitus DC.
THREE-LOBED CROWFOOT
Very rare or extinct, by acid ponds in the east. 1 + [4]. The Moors (98N, 1956, EM Burrows in **LIV**, and 1968, DSR); Studland shore (08G, 1856, WB); Parley (09Z, 1987, FAW); West Moors (00R, 1889, WMR in **DOR**); Verwood Lower Common (00Y, 1917, JHS in **NMW**). A mainly Cornubian plant in Britain.

Ranunculus baudotii Godron
BRACKISH WATER-CROWFOOT
Now very local in brackish pools near the sea, once more widespread. 10 + [5]. Some of JCMPs specimens in **DOR** are wrongly named. Recent records: Burton Mere (58E, 1989, DAP); West Bexington (58I, 1990, SME); Puncknowle (58J, 1990, SME); Abbotsbury (58S, 1971, RDE); Radipole Lake (68Q, 1924, RDG); Lodmoor (68V, 1994–99, DAP); below Borstal, Portland (77B, 1985, RA); Overcombe (78A, 1997, MJG); Southdown Farm (78Q,

1998, MJG); Encombe (97P, 1937, NDS in **BM** and JFG Chapple in **HDD**); Worth (97T, 1990, A Spink); Little Sea (08H, 1932, RDG).

Ranunculus trichophyllus Chaix
THREAD-LEAVED WATER-CROWFOOT
Uncommon in mature lakes and ponds, sometimes on undercliffs. 17 + [17]. Now found near Bexington and in the south-east, decreasing.

Ranunculus aquatilis L.
COMMON WATER-CROWFOOT
Occasional in ponds and lakes, winter streams and muddy pond margins. 41. Sometimes on undercliffs.

(*Ranunculus × bachii* Wirtgen: the origin of the report in Stace (Stace, ed. 1975) has not been traced.)

Ranunculus peltatus Schrank
POND WATER-CROWFOOT
Uncommon in streams and ponds, mostly in the north. 6. Confused with forms of *R. penicillatus*. Holwell Drove (61V, 1997, BE); Winterborne Kingston (89D, 1991, !); Fontmell Magna (81T, 1937, RDG in **WRHM**); Stour Row (82A, 1976, DWT); Almer (99E, 1995, !); Corfe Common (98Q, 1963–73, JB); Stanbridge Mill (00E, 1979, JN); Holt Wood (00I, 1991, DAP); Gussage All Saints (01A, 1982, SD Webster).

Ranunculus penicillatus Dumort.
STREAM WATER-CROWFOOT
Abundant in the larger rivers and their tributary streams. 151. It prefers clear water unpolluted by sewage, detergents or particulates. The amount cut or dredged to benefit cress-farmers and anglers was estimated as 880 ktonne/year from the R. Frome alone in 1968 (Westlake, 1968). Variable; named varieties and forms are well represented in **RNG**, coll. SD Webster. William Barnes wrote a poem about this plant.

Ranunculus circinatus Sibth.
FAN-LEAVED WATER-CROWFOOT
Rare in still waters. 1 + [9]. Not refound in old localities listed by JCMP, therefore decreasing. Chickerell (67P, 1997, CDP and DAP).

Ranunculus fluitans Lam.
RIVER WATER-CROWFOOT
Rare in flowing water. 2. All older reports are errors for *R. penicillatus*, except for those in Moors River (10A and B, 1991, NT Holmes and 1996, ! and RSRF).

Adonis annua L.
PHEASANT'S-EYE
Rare in arable fields, and lost from most of its old localities. 2 + [16]. It germinates in autumn, which does not help a weed under current farming practice. South Perrott (40T, 1928, AWG); Abbotsbury New Barn (58W, 1936, RDG in **WRHM**); Portland (1832, R Blunt in **WAR**, *et al.*, frequent in 1912, WBB, and 66U, 1964, !); Stoke Wake (70R, 1895,

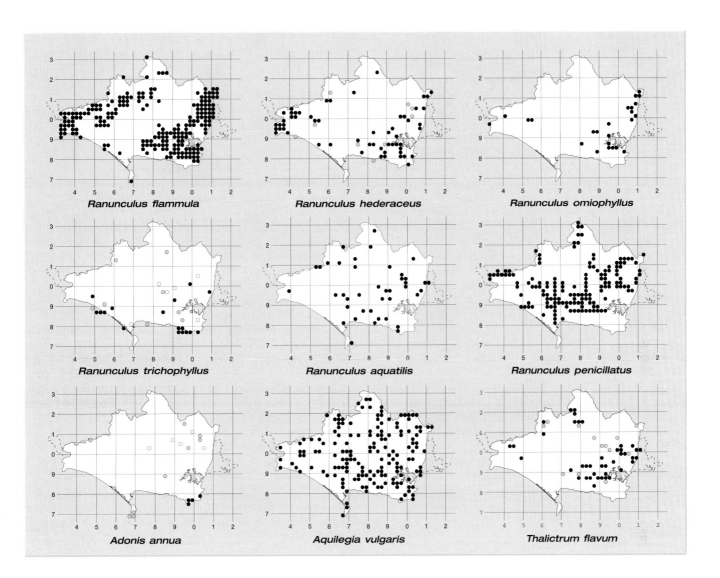

Ranunculus flammula

Ranunculus hederaceus

Ranunculus omiophyllus

Ranunculus trichophyllus

Ranunculus aquatilis

Ranunculus penicillatus

Adonis annua

Aquilegia vulgaris

Thalictrum flavum

WMR); Bourton (73Q, 1902, E Hannam in **LIV**); St Aldhelm's Head + Seacombe, Winspit and Worth (97S, T and Y, 1874, WB, *et al.*, and 97S, 1996–99, DAP + 97T, 1977, ME Barnsdale + 97Y, 1991, AH); King Down (90R, 1900, ES Marshall, *et al.*, 1975, HJB); Chettle Park (91M, 1938, RDG in **WRHM**); Swanage (07, 1928, AWG); North Farm, Horton (00J, 1922, JHS, *et al.*, and 1972, J Mason).

Myosurus minimus L.
MOUSETAIL
Once widespread, now scarce, in damp corners of arable fields on clay or alluvium. 2 + [19]. Its decline may be due to intensive ploughing of headlands and corners of arable. Post-1900 records: Stoke Abbott (40F, 1920–32, AWG in **BDK**); Warren Hill (59J, 1921–28, AWG); Lulworth (88, 1901, E Armitage in **CGE**); Hyde (89Q, 1935, DM); Stoborough (98H, 1914, CBG in **BMH)**; Wareham (98I, 1981, AH and 1997, !); Upton Park (99W, 1979, C Holt); Cowgrove (90V, 1937, DM); High Hall (00B, 1945, KG); Pig Oak (00G, 1991, AJB); Redmans Hill (00T, 1920, AWG); Edmondsham (01Q, 1905, EFL).

Aquilegia vulgaris L.
COLUMBINE
Occasional in small quantity in old woods, and much more frequent as a garden escape in village lanes and hedgebanks. 162. Records believed to be native are mostly in the north, away from the coast, as follows: 39N, 49P, 88H, 89A and R, 99C and I, 50D, 61F, 80G, P and V, 90U, 91C and Z, 01E, Q and R.

Thalictrum aquilegifolium L.
FRENCH MEADOW-RUE
A rare garden escape. 1 + [1]. Osmington (78G, 1973, !); Rempstone Heath (98W, 1995, !).

Thalictrum flavum L.
COMMON MEADOW-RUE
Local in wet pasture, by ditches and in sallow carr. 36. Although mainly in the Poole basin and the north-west, it does not grow in very acid soils, but in alluvial water-meads of the Stour, Frome, Piddle, Allen and other streams. A small population survives near Poole Yacht Club.

Thalictrum minus L.
LESSER MEADOW-RUE
A rare garden escape, scarcely naturalised. 10 + [1]. Above Dungy Head (88A, 1996, ! in **RNG**); Tarrant Gunville Ho (91G, 1957, JAL); Ashley Heath railway (10C, 1994, VMS), also in 49I and T, 67T, 68W, 81D and T, 82D and V and 08I.

BERBERIDACEAE

Berberis vulgaris L.
BARBERRY
Usually as single bushes in hedges, perhaps preferring calcareous soil. 30 + [20]. On sale in Blandford in 1782, but rarely planted today as it harbours wheat rust. Native, as remains have been found of neolithic age. Persistence is shown by the records from Oborne (61P, 1837, J Buckman, also 1958, CT and 1992, MJG). Large bushes occur at Brownsea Is (08I) and High Hall (00B).

Berberis thunbergii DC.
Sometimes planted. 5. Beaminster (40W, LJM and LMS); Holworth Chapel (78Q, !); Compton Abbas (81U, !); Studland (08G, !).

Berberis wilsoniae Hemsley
Two bushes on undercliff east of Lulworth Cove (88A, 1998, SP Chambers); Planted in Bryanston School grounds (80T, 1945, BNHS). 1 + [1].

Berberis gagnepainii C. Schneider
Planted at Giddy Green (88I) and by a pit near Huish (99D). 2.

[*Berberis julianae* C. Schneider
In a hedge at Bindon Abbey (88N, 1961, !)].

Berberis darwinii Hook.
Occasionally planted in hedges and churchyards. 14. Self-sown at Beaminster (40Q, LJM) and by the railway at Baiter (09F, !).

Berberis × stenophylla Lindley
Noted as planted in hedges or churchyards in 78G, 89F, 97T and 81U. 4.

Mahonia aquifolium (Pursh.) Nutt
OREGON-GRAPE
Planted and surviving, usually in small quantity, in woods, plantations, hedges and churchyards. 28. Indifferent to soil; said to be frequent at Worgret Heath (88Y, MHL).

Mahonia japonica (Thunb.) DC.
Planted as single bushes on verges or in churchyards in 67V, 78Q, 89E and M, 81U and 91G. 6.

Epimedium alpinum L.
BARREN-WORT
Colonies survive in three wild gardens at Minterne,

Sherborne Castle and Kingston Lacy; an *Epimedium*, perhaps *E. × versicolor*, also grows at Spettisbury Ch (90B). 3.

Podophyllum hexandrum Royle
A sterile colony grew at Charmouth (39R, 1958, WD Lang) but has not been refound. [1].

PAPAVERACEAE

Papaver pseudoorientale (Fedde) Medw.
ORIENTAL POPPY
A rare garden escape. 4. Blackdown (30W, 1998, IPG); naturalised by the railway near Dorchester West Sta (69V, 1992, !); casual at Herrison (69X, 1998, !); Fernbrook Farm (82H, 1999, !).

Papaver atlanticum (Ball) Cosson
ATLAS POPPY
Rarely established on waste ground or walls near houses. 9. Cerne (60Q, 1970, RDE); Upwey (68S, 1999, !); Dorchester (69V, 1994, !); Puddletown (79M, 1992–94, !); Tolpuddle Ball (89D, 1989–96, !); Turnworth (80I, 1999, !); Blandford (80Y, 1999, !); Lytchett Matravers (99M, 1998, !); Sturminster Marshall (99P and 90K, 1999, !).

Papaver somniferum L.
OPIUM POPPY
Perhaps native, as there are records from the Iron Age. Found on cliffs, beaches and nearby arable or fallow; also a common garden escape on verges, disturbed soil and tips. 242. When ancient turf at White Nothe (78Q) was ploughed in 1969, this poppy was dominant for a season; it has been known here since at least 1870 (RF Thompson in **DOR**). Most plants have entire pink and purple petals, but some escapes have petals scarlet or laciniate. The latex contains at least forty alkaloids, including the analgesic morphine.

Papaver rhoeas L.
COMMON POPPY
Widespread in arable and fallow, but less abundant than formerly due to weedkillers. 481. It colours new verges, exposed chalk and bare waste ground on all but the most acid soils. Plants with pale red flowers were at Badbury Rings (90R) and Fontmell Down (81Z), and with purplish flowers at Down Farm (91X).

Papaver dubium L.
LONG-HEADED POPPY
An occasional arable weed, also along verges and cliff-edges, preferring dry soils. 60. Never in quantity today, but JCMP found it rare in the 1890s and Good only had it from seven stands in the 1930s. Subsp. *lecoqii* (Lamotte) Syme is uncommon in arable, most often a garden weed. 15. It was first found at Southwell (67U, 1882, WBB) and was still there in 1982 and 1991; also seen in 67S, 69V, 61G,

78K, 70A and P, 71C and X, 72Y, 80H and Z, 97T and 90R.

Papaver hybridum L.
ROUGH POPPY
A scarce arable weed found mainly on chalk, rarely on sand. 31 + [3]. JCMP gives seven sites in the 1890s, and Good had it in 17 stands.

Papaver argemone L.
PRICKLY POPPY
Scarce in chalky arable, even scarcer on sand. 25 + [3]. JCMP gives 13 sites in the 1890s, and Good had it from six stands in the 1930s.

Meconopsis cambrica (L.) Viguier
WELSH POPPY
Occasional in a plantation at Durlston CP, and as a persistent wall or pavement weed in towns and villages. 23. It requires shelter, and tolerates shade. First record: La Lee Farm (80F, 1860, JCMP in **DOR**).

Glaucium flavum Crantz
YELLOW HORNED-POPPY
Scattered all along the coast on shingle beaches, clay undercliffs and calcareous clifftops. 31 + [3]. Uncommon on Portland and absent from dunes and salt-marshes. First record: W Turner, *c.* 1551.

Glaucium corniculatum (L.) Rudolph
RED HORNED-POPPY
A rare casual; the flowers are more orange than red. Charmouth (39R, 1948, WD Lang); Portland (*c.*1840, Mr Macgillivray in WBB); Radipole Lake (67U, 1926, MJA in **K**). [3].

Chelidonium majus L.
GREATER CELANDINE
Remains have been found in Roman material; now widespread and persistent in hedgebanks and lanes near villages, rarely far from houses. 186. Frequent in the plantations of Bryanston School (80X), but not a native woodland plant as in N. Europe. Var. *laciniatum* Mill. occurred at Parley (09Z, 1935, LBH), and a double form occurs at Zelston (89Y).

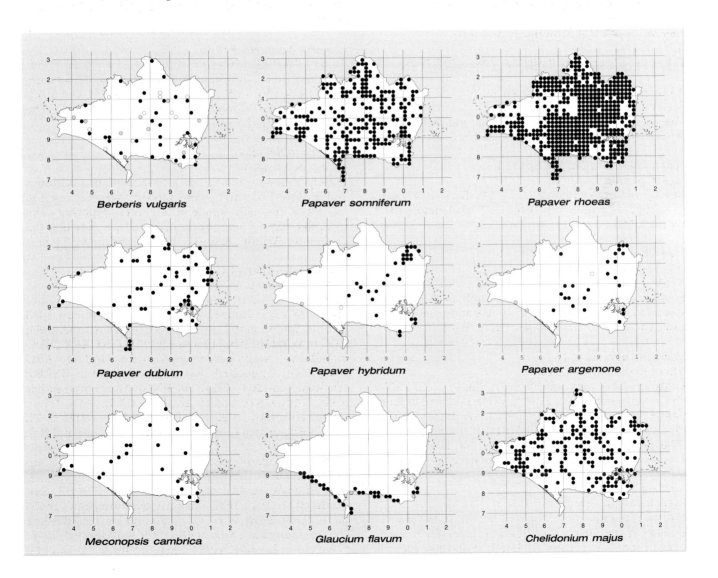

77

Stylophorum diphyllum (Michx.) Nutt.
WOOD-POPPY
Naturalised in the grounds of the Old Mill, Spetisbury (90B, 1994–99, !). 1.

Eschscholzia californica Cham.
CALIFORNIAN POPPY
A casual from gardens on disturbed verges, tips, building sites, old quarries and cliffs. 25. Sandsfoot cliffs (67T, 1986–87, !).

FUMARIACEAE

Dicentra formosa (Andrews) Walp.
BLEEDING-HEART
Established in a small plantation at Broadstone (99X, 1991, ! in **RNG**). 1.

Corydalis solida (L.) Clairv.
BIRD-IN-A BUSH
Rare in wild gardens. Long Bredy (59Q, 1993, !); Milborne St Andrew (89D, 1869, JCMP in **DOR**). 1 + [1].

Corydalis cava (L.) Schweigger & Koerte
HOLLOWROOT
A rare garden escape. Ballard Down (08F, 1977, BNSS). [1].

Pseudofumaria lutea (L.) Borkh.
YELLOW CORYDALIS
Widespread on old walls, often in shade and shelter, and a persistent pavement weed. 158. The seeds are spread by ants.

Pseudofumaria alba (Millen) Liden
PALE CORYDALIS
Rare on old walls in villages. 2 + [1]. Burton Bradstock (48Z, 1992, !); Upton Fort (78K, 1973, !); Ashmore (91D, 1981–99, MB).

Ceratocapnos claviculata (L.) Liden
CLIMBING CORYDALIS
Local in woods, pine plantations and heaths, or among bracken, on acid soils. 52 + [5]. Mainly in the extreme west and the Poole basin, with outliers at Encombe (AH), Alton Pancras (SB) and Piddles Wood (DWT).

Fumaria capreolata L.
WHITE RAMPING-FUMITORY
A rare casual of acid soils. 2 + [5]. Pilsdon (49E, 1930, AWG); Abbotsbury (58S, 1934, E Milne-Redhead in **K** and 1980, J Fryer); Carey (98D, 1955, JKH); Wareham (98I, 1901, FHH); Arne (98U, 1979, MRH); Piddles Wood (71W, 1988, RMW); Swanage (07J, 1875, BT Lowne in **K**). Reports from Portland and elsewhere need confirmation.

Fumaria bastardii Boreau
TALL RAMPING-FUMITORY
Rare and sporadic on acid soils. 3 + [10]. JCMP gives 28 localities in the 1890s; either these were errors for *F. muralis* or the plant has greatly declined. Post-1900 records:

Beaminster (40V, 1965, AWG in **BDK**); Sherborne (61N, 1968, ACL); Woodsford tip (78U, 1977, !); Ridge (98I, 1963, JB and 1989, ! + CDP in **RNG**, det. PFY); Pamphill (90V, 1988, RF); Chilbridge Farm (90W, 1982, HJB); Studland (08H, 1912, AWG in **BDK** and 1960, HJB); Fleets Corner tip (09B, 1977, !); Kinson (09T, 1964, MPY); Wimborne (00A, 1994–99, !); Verwood (00Z, 1919, AWG).

Fumaria reuteri Boiss.
MARTIN'S RAMPING-FUMITORY
Casual on a building site. [1]. Sherborne (61N, 1968, ACL and 1970, MM Webster in **E**).

Fumaria muralis Sonder ex Koch subsp. *boraei* (Jordan) Pugsley
COMMON RAMPING-FUMITORY
Occasional in sandy arable, mostly in disturbed verges, allotments and tips. 78. Commonest in the Poole basin, also in the west.

Fumaria muralis × *officinalis* = *F.* × *painteri* Pugsley
Once found at Wareham (98I, 1916, CBG in **BMH** det. HWP). [1].

Fumaria purpurea Pugsley
PURPLE RAMPING-FUMITORY
Rare and usually transient, but persistent at Abbotsbury (58S, 1895, LVL in **K**, 1915, J Groves in **BM** and 1971, ACL); casual at Wimborne (1919, LVL in **K**). [2]. The report from Sherborne was an error, fide PDS.

Fumaria officinalis L.
COMMON FUMITORY
Locally abundant but sporadic as a weed in calcareous arable, disturbed soil, gardens and tips. 178. Var. *wirtgeni* Haussk. is probably frequent, but rarely reported. The fruit, like that of other declining weeds, is eaten by the turtle dove (Wilson, Arroyo & Clark, 1997).

Fumaria densiflora DC.
DENSE-FLOWERED FUMITORY
Rare in arable on the north-east chalk, sporadic and rare elsewhere. 7 + [15]. Confused by JCMP in **DOR**. Post-1920 records: Bridport (49R, 1928, AWG); Dorchester (69, 1934, RDG in **WRHM**); Winterborne Kingston (89T, 1934, RDG in **WRHM**); Nothe gardens (67Z, 1988, DAP and ARG Mundell); Chebbard Farm (79U, 1998, RSPB); Witchampton (90T, 1997, ! in **RNG**); Sixpenny Handley (01D, 1996, MJG); Bottlebush Down (01E, 1996, MJG); Creech Hill (01G, 1918, JHS); Bowldish Pond (01H, 1999, J.Phillips); north of Cranborne (01M, 1933, GDF and JO).

Fumaria parviflora Lam.
FINE-LEAVED FUMITORY
Rare in chalky arable, once seen on sand. 3 + [6]. East Portland (77?, 1912, WBB); Milborne St Andrew (89D, 1866–90, JCMP in **DOR**, also 79V, 1976, MS Warren, needing confirmation); Little Wood (80L, 1995, !); Ridge (98I, 1956, G Williams); Keysworth (98J, 1961, !); Jubilee Wood (90M,

1997, ! in **RNG**); Badbury Rings (90R, 1904, RPM, *et al.*, and 1959, HJB); Bowldish Pond (01H, 1997, IPG).

Fumaria vaillantii Lois
FEW-FLOWERED FUMITORY
Rare in calcareous arable. 1 + [2]. Stalbridge (71J, 1868, Mrs Allen in **DOR**); Renscombe Farm (97T, 1979, D Welchman); Jubilee Wood (90M, 1997, ! in **RNG**); possibly this at Greenland Farm (08C, 1999, !).

PLATANACEAE

Platanus × *hispanica* Miller ex Muenchh.
LONDON PLANE
Often planted in small quantity in parks and plantations. 43. A tree at Bryanston, planted in 1740, was 47 m × 6 m in 1983 and is probably the tallest tree in Dorset; other large trees occur at The Cliff nearby and near Parnham Ho (40Q). An avenue tree in Dorchester.

Platanus orientalis L.
ORIENTAL PLANE
Rarely planted in parks. 6. A tree at Kingston Maurward (79A) was 26 m × 6 m in 1986, and one at Melbury Park (50T) was 23 m × 6 m in 1980; also noted in 30M, 80X and Y, 09J.

ULMACEAE

Ulmus glabra Hudson
WYCH ELM
Widespread and, prior to the epidemic of elm disease, often

dominant in small valley woods along the coast, where the leaves are often blackened by gales. Occasional in woods, hedgebanks and parks on the richer soils, but absent from acid soils or leached soils over chalk. 336. The only trees on Portland in 1543, according to J Leland. Most large trees are now dead, though the stools survive. There are six Dorset place names derived from elm, and remains date back to the mesolithic period.

Ulmus glabra × *minor* = *U.* × *vegeta* (Loudon) Ley
A large tree south of West Compton (59R, JAG); several near South Admiston Farm (79R, 1997, !); near Durlston CP (07I, 1999, EAP); also planted at Canford School (09J, TH). 4. A large hybrid elm survives at South Perrott (40T, BE).

Ulmus procera Salisb.
ENGLISH ELM, ELEM
Common on hedgebanks with nutrient-rich soil in valleys. 590. No large trees survived the attack of elm disease in the 1970s. Suckers from the stools grow until their bark is corky, and then mostly succumb. A young avenue is still at Herringston (68Z). William Barnes often referred to this tree and devoted a poem to it. This used to be the main foodplant of the White-letter Hairstreak, and many moths.

Ulmus minor Miller subsp. *angustifolia* (Weston) Stace
CORNISH ELM
Occasional, looking native, on undercliffs; rarely planted elsewhere. 16. Possibly native sites are: Beaminster (40V and W, 1921, AWG in **BDK**, and 1993, LJM); Portland Bill (66U, 1989, !); Osmington Mills cliffs (78F, 1960–93, !); Encombe (97N, 1937, NDS); Durlston Head (07I, 1977–

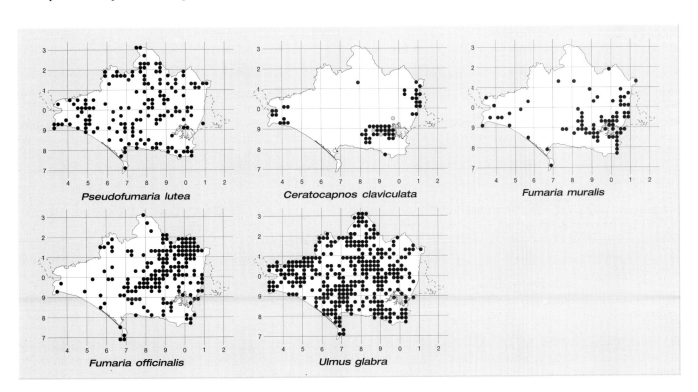

Pseudofumaria lutea — Ceratocapnos claviculata — Fumaria muralis — Fumaria officinalis — Ulmus glabra

97, !); West Orchard (81I, 1996, !). Trees 30 m tall, possibly this species, in Great Wood, Creech (98A, 2000, BE).

Zelkova serrata (Thunb.) Makino
Surviving in Little Wood, Creech (98B, 1973, ! in **RNG**); rarely planted, as at Canford School (TH). 2.

CANNABACEAE

Cannabis sativa L.
HEMP
It occurs in Roman remains, and was grown as a crop for rope-making near Bridport in the 17th and 18th centuries. 6. Now a rare crop or a casual on tips from bird-seed, or grown surreptitiously to smoke; seen in 68V, 88Y, 99K, 09I, 01K and L.

Humulus lupulus L.
HOP
In carr, at woodland edges and hedges on nutrient-rich soils, usually in valleys near water. Absent from deforested areas, including Portland, and from very acid soils. 317. The foliage was carved on capitals inside Woolland Ch (70T), probably in 1855. It may have been cultivated in the past, but not for the last 50 years; it grows in hedges near the brewery at Blandford. The bitter principles in it include humulone and lupulin, and pillows stuffed with dry hops are used by insomniacs.

MORACEAE

Morus nigra L.
BLACK MULBERRY
Rarely planted in parks. 6. Parnham Ho (40Q, !); Milton Abbas (70W, !), Stock Gaylard (71G), Green Is (08D) and Brownsea Is (08I and J).

Morus alba L.
WHITE MULBERRY
Rarely planted in parks. 3. Compton butterfly farm (51Y), to feed silkworms; Charborough Ho (99I); Canford School (09J, TH).

Ficus carica L.
FIG
Large trees established in Portland quarries (66U, 67V and W, 77B) and rarely introduced along the coastal strip, as at Abbotsbury, Godlingston, Green Is and Brownsea Is. 14.

URTICACEAE

Urtica dioica L.
NETTLE
Dominant in nutrient-rich soil of woods, under rookeries, in carr, abundant in hedgebanks, by rivers, frequent in rough grassland and able to colonise the upper parts of shingle beaches. 735. Tolerant of shade, where it can grow up to 4.8 m tall. Near the sea, the foliage can be blackened by salty gales. Its liking for potash and phosphate explains why nettle colonies often surround old, isolated cottages or barns, and for its predominance on mature river dredgings. The hairs contain histamine and acetylcholine. An important foodplant for several large Nymphalid butterflies, also some moths.

Urtica galeopsifolia Wierzb. ex Opiz
This controversial diploid lacks stinging hairs and has attenuate leaves, but is probably a variety of ordinary Nettle. It grows by streams or in fens, and is recorded from 39U, 49T, 61J, 88T and 09J.

Urtica urens L.
SMALL NETTLE
Locally abundant in dry arable soils of high nutrient status, in allotments, farmyards and gardens. 142. Least common in the predominantly grassland areas of the north-west, but indifferent to soil pH. First found in late Iron Age deposits; JCMP thought it common, Good had it from only 13 stands, and it has probably increased recently due to the heavy application of agricultural fertilizers. Eaten by sheep only when starving.

(*Urtica pilulifera* L.
One doubtful report as a painting, from Weymouth, *c.*1890, in Good's Handbook.)

Parietaria judaica L.
PELLITORY-OF-THE-WALL
Local among rocks, at the base of cliffs, on shingle beaches and especially on walls, or as a pavement weed in villages and on tips. 142. Widespread, including Portland, but not reported from SY69 or ST50.

Soleirolia soleirolii (Req.) Dandy
MIND-YOUR-OWN-BUSINESS
Widespread as an escape on damp, sheltered walls and as a pavement weed in villages and churchyards, sometimes by streams as at Portesham (68C), Winterbourne Abbas (69A) and Ringstead (78K). 107. First noted in 1935. The above-ground parts are killed by frost in hard winters.

JUGLANDACEAE

Juglans regia L.
WALNUT, WELSHNUT
Widely planted as single trees in parks, fields and hedges. 76. Regenerating at Sandsfoot (67T, !), Stallen (61D, IPG), Sherborne (61M, !), Long Crichel (91Q, !) and probably elsewhere.

Juglans nigra L. is planted at Kingston Lacy Park (90V), and at Canford School, where four other alien walnuts are planted (09J, TH).

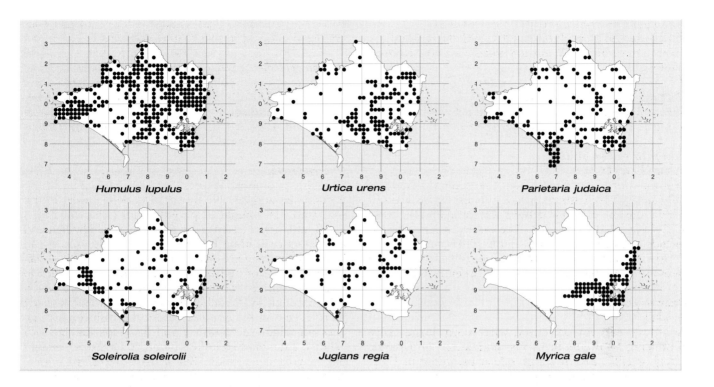

Humulus lupulus   Urtica urens   Parietaria judaica

Soleirolia soleirolii   Juglans regia   Myrica gale

Pterocarya fraxinifolia (Poiret) Spach
CAUCASIAN WINGNUT
Specimen trees are very rare outside parks, where they sucker and seed freely. 6. Wheelhouse Lane, in a hedge (30S, IPG). A tree 35 m × 5.5 m is at Melbury Park, and another 24 m × 4.9 m at Abbotsbury, probably planted *c.* 1780; also seen in 30M, 59K, 70V and 81T.

Carya **sp.**(Hickory); *C. ovata* (Miller) K Koch is planted at Canford School (09J), and big hickory trees were seen at Horton (00I, 1922, JHS).

MYRICACEAE

Myrica gale L.
BOG-MYRTLE
Locally abundant to dominant in alder-swamps, bogs and marshes, including some near the sea, but restricted to acid soils in the Poole basin with no records from the west. 82. A predominantly northern species in Britain, but showing no signs of decline here. Mature colonies can be 2 m tall, and are used as refuges by Roe deer. This may be because they are the source of an essential oil which repels insects.

FAGACEAE

Fagus sylvatica L.
BEECH
Although pollen has been found from Bronze Age times on, and the tree is undoubtedly native in the New Forest, this is considered to be an introduction. It is widespread, and often block planted, in plantations, parks and as hedges. 587. The absence of native beechwoods in Dorset means that some of the associated ground flora and fungi are absent or rare. Very large trees occur at Melbury Park, Sherborne Castle, Kingston Lacy, Morden Park and near Caesar's Camp (91H), but the largest is at Milton Abbas (70W; 36 m × 7.3 m in 1982). A notable avenue, planted in 1835, consists of 365 trees on either side of the B3087 near Badbury (90R); some trees were uprooted in the gales of 1987, but a row of saplings has been planted on both sides. There is another avenue, about 1 km long, at Bottlebush Down (01H), and a young avenue at Crichel Ho (90Z and 00E). Cv. *purpurea* is occasional and a specimen of Cv. *asplenifolia* is 27 m × 3 m in Melbury Park; other cultivars and *F. engleriana* Seemen are planted at Canford School (09J, TH). The place names Bockhampton and Bokers Farm could be derived from boc = beech. A number of moth larvae eat the leaves, and the mast, when fertile, is food for birds such as woodpigeons and hawfinches (Wilson, Arroyo & Clark, 1997).

Nothofagus obliqua (Mirbel) Blume
ROBLE
Occasionally planted, and fruiting, in plantations and parks. 9. Seen in 78T, 89R and S, 98J, 99C, 50S, 62A, 90Y and 09J.

Nothofagus nervosa (Philippi) Krasser
RAULI
Block planted on Bloxworth Heath (89W, ! in **RNG**); one seedling found. 12. Minor plantings, mostly on acid soils, in 30M, 50S, 79F, 73Q, 89R, T, V and W, 99I and W, 08F and 09J. Trees with leaves up to 20 cm long north of Swanage (08F, EAP) might be *N. alpina* (Poepp. & Endl.) Krasser, which is probably the correct name for this variable tree.

Nothofagus antarctica (Forst.f.) Oersted
Rarely planted on acid soils in 89W, 99C, 08D and 09J. 4.
Some of these may be *N. pumilio* (Poepp. & Endl.) Krasser.
Whitefield (99C, ! in **RNG**).

Nothofagus dombeyi (Mirbel) Blume
Rarely planted, mostly on acid soils in 60S, 79F and 70D.
3. A tree 20 m × 1.8 m in Minterne Park in 1967.

Castanea sativa Miller
SWEET CHESTNUT
Probably introduced; wood from the Roman period was
dug up at Woodcutts. Now commonest in woods and
plantations on acid soils in the Poole basin and the west.
271. Occasionally coppiced. The largest trees recorded
include one 21 m × 8.5 m at Gaunts (00H), and one with
a girth of 13.4 m at Canford School (09J); other large
trees are at Melbury Park (50S) and Green Is (08D).

Quercus cerris L.
TURKEY OAK
A widespread introduction, with many large trees
producing seedlings in woods, plantations and parks,
rarely on cliffs. 170. It was on sale at Blandford in 1782,
but was not mentioned by JCMP in the 1890s.

Quercus cerris × *ilex* ?
A tree which may be this keeps its leaves until December, in
an oak wood east of Red Bridge, Moreton (88E, 1990, ! in
RNG). 1.

Quercus cerris × *suber* = *Q.* × *crenata* Lam.
LUCOMBE OAK
There are large planted trees in parks or on verges at
Wootton Ho (39S), Bridport (49R), Abbotsbury gardens
(58S), Evershot (50S), Kingston Lacy (90V; 27 m ×
3.4 m), Durlston CP (07I) and Canford School (09J). 6.

Quercus ilex L.
EVERGREEN OAK
Much planted on both calcareous and sandy soils; seedlings
are frequent on cliffs, in calcareous grassland, heaths
and even marshes. 167. Notably most common in the
coastal belt, and dominant in places in Durlston CP (07I)
and Canford Cliffs (08U and Z), where it is resistant to
salt-spray. Recorded as planted in shelter belts at
Abbotsbury in 1899, where there are still large trees,
as there are in parks at Forde Abbey, Bridport, Lulworth
and Melbury.

Quercus canariensis Willd.
ALGERIAN OAK
A planted tree 30 m × 3.6 m at Melbury Park, also planted
at Canford School (09J, TH). 2.

Quercus petraea (Mattuschka) Liebl.
SESSILE OAK
An indicator of old woodland occurring in the west, the
north-east, near Oakers Wood and around the shores of

Poole Harbour. 50 + [5]. Probably under-recorded. More
restricted to dry, acid soils than *Q. robur*.

Quercus petraea × *robur* = *Q.* × *rosacea* Bechst.
Less common than either parent, but able to grow in wet
places on Poole Harbour shores. 5. Old Warren Hill (49M,
1994, P Macpherson); Pilsdon Pen (40A, 1980, !); Sares
Wood (89B, 1981, AJCB); Coombe (98T, 1994, !); Lytchett
Bay (99R, 1994, !).

Quercus robur L.
PEDUNCULATE OAK, WOAK
A dominant and variable tree of old woods on wet or dry
soils since the mesolithic period. Least common on the
Chalk and in deforested areas. Frequent in old hedges and
as isolated trees in parks, pasture and arable. 678. It can
be seen colonising grassland on clay at Lorton (68R). Oak
wood, charcoal or pollen have been found dating back to
mesolithic times; oak coffins have been found under tumuli
at Stoborough and Scrubbity Barrows. The timber has been
much used over the centuries, as for the 13th century oak
chest in Wimborne Minster, and for ships. In the 16th
century, a survey of Holt Forest found 6,697 oaks. Existing
oaks of great age include: 'Billy Wilkins', a pollard 15 m ×
12 m in Melbury Park (50S) and a nearby tree of var.
fastigiata, 27 m × 2.7 m; the Smuggler's oak in Sherborne
Park (61T), 9.6 m in girth; the Stockbridge oak (61F),
6.4 m in girth; a tree at Lytchett Matravers with graffiti
from 1849; the Remedy oak at Woodlands (01K), now
hollow, under whose shade Edward VI 'touched' scrofulous
persons in the mid-16th century. There are oak avenues at
Stock Gaylard (71G) and Pamphill Green (90V; planted in
1846). More or less extinct oaks of historic interest include:
the Silton oak (72Z) where Court leets were held; the
Boundary oak at Bloxworth (89S), now a dead stump but
said to have been planted in 1087; the Damory oak, (80Y,
29 m × 21 m), whose hollow was used as an inn during the
Civil war, felled in 1755 and the timber sold for £14 (some
is now in the pulpit of Okeford Fitzpaine Ch); the Burnham
oak at SZ 014800, so called after a man who was hanged
on it, lost in the 1920s; and the Mountjoy oak at Canford
School (09J), which died in 1992 and is now a shell. Oak,
as noake, ock or oke, occurs in 29 Dorset place names;
however, Oakers wood may be a corruption of Wulfgar's
wood. William Barnes devoted four poems to the 'woak'.
The leaves are an important food resource for many insects,
such as larvae of the Green Tortrix moth and the Purple
Hairstreak butterfly. Acorns are a winter food resource for
squirrels and jays, but have caused stomach upsets in horses.
Many fungi have mycorrhizal associations with the roots,
others feed off the living or dead wood. Once used as a
magic cure for toothache (Udal, 1889).

Quercus rubra L.
RED OAK
Commonly planted as specimen trees in woods, plantations
and parks, rarely block planted. 62. Seedlings seen at
Blacklawns Coppice (30M, IPG), Abbotsbury (58S) and Crichel
Park (90Z, RF). A tree 29 m × 5.2 m is in Melbury Park.

Many other oaks are planted as specimen trees in parks, but none has been shown to sow itself. Canford School has *Q. acutissima* Carruthers, *Q. coccinea* Muenchh., *Q. frainetto* Ten., *Q. macranthera* Fischer & Mayer, *Q. marilandica* Muenchh. and *Q. palustris* Muenchh. (TH), while Melbury Park has large trees of *Q. coccinea*, *Q. frainetto*, *Q. pubescens* Willd. and *Q. pyrenaica* Willd. *Q. suber* L. has been planted at Abbotsbury gardens and Swannery and Sherborne Castle, and *Q. velutina* Lam. at Forde Abbey.

BETULACEAE

Betula pendula Roth
SILVER BIRCH

Locally common, especially in felled oakwoods on sandy soils, but present in most woods in small amounts. It colonises wet heaths, and forms nearly pure woods there, but is absent from deforested areas, including Portland.

429. Pollen of neolithic age, and timber of Roman age, have been identified by archaeobotanists. A tree 19 m × 0.77 m grows at Athelhampton (79S). At least eight Dorset place names are derived from birch. Young trees may be defoliated by larvae of moths such as the Early Thorn.

Betula pubescens Ehrh.
DOWNY BIRCH

Less common than *B. pendula*, and more restricted to acid soils in the Poole basin and the north-west, rare on the Chalk. 233. Apart from its preference for acid soil, this occupies the same habitats as *B. pendula* and often grows with it. Both birches are regarded as trash by foresters. The hybrid *B.* × *aurata* Borkh. closely resembles *B. pubescens*, and has not been positively identified.

Other species of birch are planted as specimen trees, e.g. *B. costata* Trautv. at Colliton Park, Dorchester (69V), *B. maximowicziana* Regel at Bere Heath (89R, ! in **RNG)** and *B. papyrifera* Marsh. at Stourton Caundle (71C).

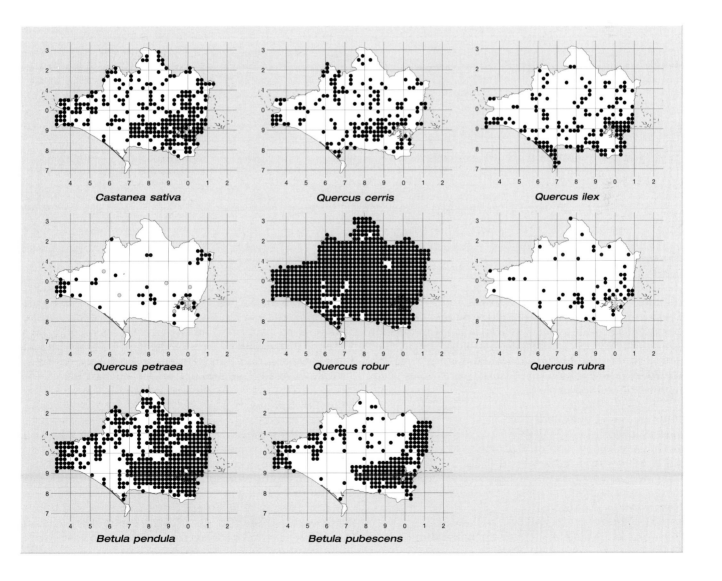

Castanea sativa

Quercus cerris

Quercus ilex

Quercus petraea

Quercus robur

Quercus rubra

Betula pendula

Betula pubescens

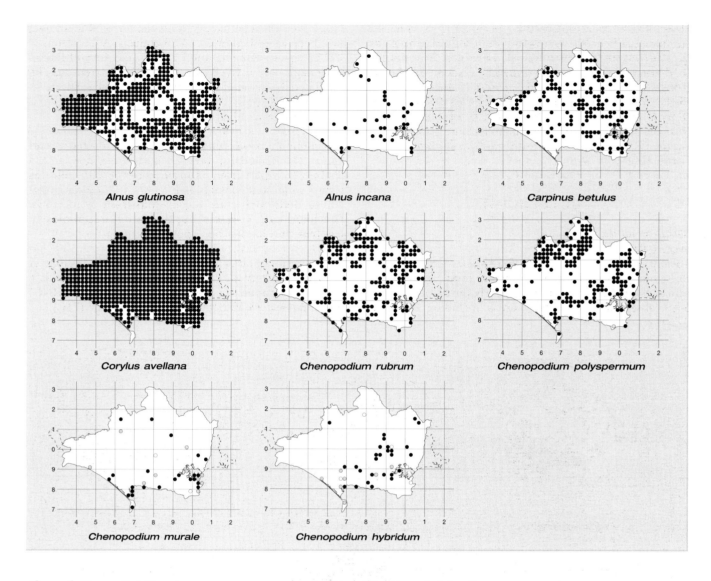

Alnus glutinosa (L.) Gaertner
ALDER, ALLER-TREE
Locally dominant in some valley swamps, as at Powerstock Common (59M), frequent along rivers and streams, but absent from deforested areas and from most of the Chalk; sometimes obviously planted. 439. Its pollen is frequent in excavations back to neolithic times. Its huge root nodules allow it to fix atmospheric nitrogen and act as a pioneer tree. Alder or aller is found in 15 Dorset place names. Many moth larvae feed on the leaves.

Alnus incana (L.) Moench.
GREY ALDER
Rarely block planted, more often planted in swamps, by rivers, and as hedges for fruit farms and fields. 36. First record: Radipole Lake (68Q, 1976, DT Ireland, also ! in **RNG)**.

Alnus cordata (Lois.) Duby
ITALIAN ALDER
Much planted recently by streams, ponds and in hedges. 26. First record: Tomson (89Y, 1989, !).

Carpinus betulus L.
HORNBEAM
Probably native in some woods, at least in the east, as pollen from the neolithic age has been found. 158. The tree tolerates chalk, clay and sandy soils, but any native distribution pattern is obscured by blocks or odd trees planted in plantations, parks and hedges. Large trees noted at Durweston (80P) and Pimperne Fox Warren (91A); seedlings locally common in Cranborne Chase (81X).

Corylus avellana L.
HAZEL, HAZZLE
An abundant understorey shrub in oak and ash woods on all soils, also a major component of hedges in wooded regions, known since the mesolithic age. 681. Absent from Portland and deforested areas round Weymouth. Its seedlings are frequent in woods, and it colonises grassland near old woods, as at Creech Great Wood (98A). Rarely a tall tree, as in Oakers Wood (89A). In central Dorset, hazel is still regularly coppiced by hand, with spectacular effects on the ground flora. The spars are used for hurdles and

thatching pegs, less often today for pea and beansticks or by dowsers. The nut crop is often good, but is now food for grey squirrels; it is nowhere harvested commercially, as it was in Cranborne Chase in the early 19th century, when the price was 6d a peck. Many moth larvae eat the leaves, and some fungi associate with the roots. Dorset place names include five derived from hazel and four from nut. William Barnes wrote a poem about the tree.

Corylus maxima Miller, and its *Cv.* **purpurea,** are planted on Green Is (08D) and Canford School (09J, TH).

PHYTOLACCACEAE

Phytolacca acinosa Roxb.
INDIAN POKEWEED
A rare casual, seen outside a garden at Piddletrenthide (70A, 1990, !), and earlier at Compton Ho (51Y, 1944, Col. Goodden) and Canford Cliffs (08P, 1944, WR Thompson). 1 + [2].

AIZOACEAE

Carpobrotus edulis (L.) N.E. Br.
HOTTENTOT-FIG
An alien temporarily established on a few warm cliffs, but killed by frost. 1 + [6]. Chesilton (67W, 1980, !); Lulworth Cove (88F, 1967, WEA Evans); between Studland and Swanage (08F, 1963, EC Payne); Parkstone (09K, 1933, H Phillips in **BDK**); Sandbanks (08N, 1960, !); Canford Cliffs (08U, 1930–61, RDG and 1992, RMW, !).

Tetragonia tetragonioides (Pallas) O.Kuntze
NEW ZEALAND SPINACH
A rare casual of ports and tips. Hamworthy quay (99V, 1956, AK Harding); Parkstone (09F, 1932, HJW). [2].

CHENOPODIACEAE

Chenopodium botrys L.
STICKY GOOSEFOOT
A very rare casual. 1. Greenland Farm (08C, 1999, DCL and EAP, det. EJ Clement in **PVT,** !).

Chenopodium capitatum (L.) Ascherson
STRAWBERRY-BLITE
A rare casual. [3]. Corscombe (50D, 1939–41, AWG); Rampisham (50L, 1931, AWG in **BDK**); Lydlinch (71L, 1969, RDG).

Chenopdium bonus-henricus L.
GOOD-KING-HENRY
Uncommon on grassy verges or near buildings. 6 + [17]. Old records suggest that this ancient herb of cultivation was once more common. Abbotsbury Tithe Barn (58S,

1937, DM and 1992, !); Leigh (60J, 1984, MP Hinton); Marnhull Ch, on verge (71Z, 1992, !); Melbury Down (91E, 1997, P Amies); Ferndown (00K, 1998, J Crewe); Moors River (00V, 1986, RMW).

Chenopodium glaucum L.
OAK-LEAVED GOOSEFOOT
A rare garden or pavement weed on sandy soil. 2 + [1]. Weymouth Backwater (68, 1927, MJA); Norden, 1 plant (98L, 1996, DAP and NFS); Upton CP (99W, 1991–97, !).

Chenopodium rubrum L.
RED GOOSEFOOT
Local on rich mud by ponds and by manure heaps, or in manured arable. 215. Like most of the family, this plant tolerates soils with high osmotic pressure, and its distribution pattern bears little relation to other edaphic factors. Variable in colour, size and habit.

Chenopodium polyspermum L.
MANY-SEEDED GOOSEFOOT
Locally common in manured arable, on mud by ponds, in farmyards, rich gardens and tips; remains have been found back to the Iron Age. 184. Absent from the Chalk ridge and scarce in the coastal strip.

Chenopodium vulvaria L.
STINKING GOOSEFOOT
Once fairly common along the coast, but now restricted to a stretch near West Bay, where it tolerates gull droppings. 2 + [12]. It converts excess nitrogen to trimethylamine, which is what makes it stink. Burton Bradstock (48U, 1895, JCMP, *et al.,* and 1988, DAP); West Bay cliffs (49Q, 1927, AWG, *et al.* in **RNG**, and 1998, DAP); Portland (1911, WBB); Sandsfoot (67T, 1860s, WBB); Weymouth (67, 1895, WBB and 1937, DM); Rodwell (68Z, 1927, MJA); Lodmoor tip (68V, 1950, !); Spettisbury railway (90F, 1975, VJG); Swanage (07J, 1895, JCMP); Studland (08G, 1889, EFL in **LIV**); Hamworthy (99V, 1975, CS); Poole (09A, 1830, TBS and 1922, JHS).

Chenopodium murale L.
NETTLE-LEAVED GOOSEFOOT
Scarce in disturbed, rich soil of farmyards, allotments, gardens and tips, mostly near the coast, and on the beach at Brownsea Is (08D). 23 + [13]. Persistent near Weymouth, Portland and Wareham, but becoming rarer.

Chenopodium hybridum L.
MAPLE-LEAVED GOOSEFOOT
Uncommon, sporadic and in small quantity in disturbed, rich, bare soil of gardens, allotments, verges and tips. 27 + [13]. It prefers sandy soils and there are few records from the north-west.

[***Chenopodium urbicum*** L.
UPRIGHT GOOSEFOOT
Extinct. Once occasional in rich, manured soils. [10]. Portland (67, 1874, HE Fox and 1912, WBB); Weymouth and Wyke (67T and U,

1860s, WBB); Rodwell (67U, 1926, MJA); Radipole Lake (68, 1936, MJA); Bere Regis (89M, 1895, JCMP); Ridge Farm (98N, 1914, AWG in **BDK**, also CBG in **BMH**, *et al.* over 50 years when it was a magnet for botanists, 1966, !); Westley Wood (90F, 1918, JHS).]

Chenopodium ficifolium Smith
FIG-LEAVED GOOSEFOOT
Occasional in manured arable, locally abundant with kale or maize crops, and in farm tracks, gardens, disturbed verges and tips. 102. Increasing, as JCMP only had it from three places in the 1890s, but it occurs in Iron Age deposits.

Chenopodium album L.
FAT HEN, BACONWEED, LAMB'S TONGUE, MUTTON-TOPS
Frequent in arable on all types of soils, especially in crops such as kale where weedkillers are not used, also in gardens and on tips, rarely on beaches. 698. Variable in leaf-form and pigmentation. The seeds are edible and may have been spread by neolithic men; they are important for flocks of small finches in autumn (Wilson, Arroyo & Clark, 1997).

[*Chenopodium leprophyllum* Dumort. agg.
SLIMLEAF GOOSEFOOT
Casual. Swanage Camp (08F, 1917, CBG in **BMH**, det. GCD as *C. pratericola*).]

Chenopodium giganteum D. Don
TREE SPINACH
Casual in arable, Stoke Abbott Mill (49I, 1976, AWG det. JG Dony); Down Farm (91X, 2000, !). 1 + [1].

Chenopodium probstii Aellen
PROBST'S GOOSEFOOT
Casual, Sherborne tip (61I, 1970, ACL). [1].

Chenopodium quinoa Willd.
QUINOA
Grown as a crop for game cover, Cutt Mill (71T, 1997, ! in **RNG**) and Turnworth (80!, 1999, !); among bird-seed aliens, Canford Tip (09I, 1999, !). 3. Mature plants may be orange, scarlet or crimson.

Bassia scoparia (L.) Voss
SUMMER-CYPRESS
Casual. Garden form at Lodmoor tip (68V, 1960, !); branched, open form at Tatchell's Pit (98E, 1993, ! in **RNG**). 1 + [1].

Atriplex hortensis L.
GARDEN ORACHE
A rare garden escape on disturbed verges or tips. 2 + [2]. Lodmoor tip (68V, 1977, !); Sherborne (61I, 1970, ACL); Lydlinch Common (71G, 1992, !); Canford Heath tip (09I, 1999, !).

Atriplex prostrata Boucher ex DC.
SPEAR-LEAVED ORACHE
Common in arable, gardens, disturbed verges and on

beaches, preferring bare, rich soil. 621. Variable in habit and leaf-form; on beaches this could be due to hybridisation.

Atriplex prostrata × *longipes* = *A.* × *gustafssoniana* Taschereau
Rare or overlooked on the coast. 1. West Bay (49Q, 1997, CDP, DAP and TD Dines in **CGE**).

Atriplex glabriuscula Edmonston
BABINGTON'S ORACHE
Locally frequent at the strand-line on pebbly or muddy beaches, rare on sandy beaches. 56. Often with *A. prostrata*, but less variable than that species.

Atriplex longipes Drejer
LONG-STALKED ORACHE
Rare or overlooked on muddy beaches. 2 + [1]. West Bay (49Q, 1988, DAP); behind Chesil bank, Abbotsbury (58S, 1988, RF and JRA); Brands Bay (08C, 1969, P Symmons in **RNG**).

Atriplex littoralis L.
GRASS-LEAVED ORACHE
Locally frequent on muddy beaches on both sides of the Fleet, on Portland and its breakwaters, and inside Poole Harbour. 22 + [1].

Atriplex patula L.
COMMON ORACHE
Less common than *A. prostrata*, but in similar places. 391.

Atriplex laciniata L.
FROSTED ORACHE
Rare and in small quantity on muddy shores or at the edge of salt-marshes. 4 + [10]. Sporadic around Poole Harbour and casual elsewhere. Charmouth (39R, 1966, AWG); Chesil bank (67, 1965, JB); Hobarrow Bay (87Z, 1975, WGT); Chapmans Pool (97N, 1963, JB); Durlston CP (07I, 1977, KC); Studland (08, 1911, CES, *et al.*, and 1960, WGT); South Haven and Shell Bay (08I, 1839, TBS, *et al.* in **BMH** and **DOR**, and 1994, BE, !); North Haven and Sandbanks (08I, 1839, TBS, *et al.*, and 1994, BE); Brownsea Is (08E, 1981, KC); Lilliput (08J, 1985–97, DAP); Poole (09A, 1911, CES); Canford Heath tip (09I, 1998, !).

Atriplex halimus L.
SHRUBBY ORACHE
Planted and established on sand or clay on a few undercliffs. 7. Lyme Regis (39, 1950, JPMB); Bexington (58I, 1961, !); Abbotsbury (58S, 1899, MI and 1978, !); Weymouth Sta (67U, 1961, !); The Nothe (67Z, 1977–95, !); Sandbanks (08N, 1943, BNSS and 1958–91, !); Flaghead Chine to Canford Cliffs (08P and U, 1992–97, !); north side of Poole Harbour (09, 1917, CBG in **BMH**); Baiter (09F, 1989, !).

Atriplex portulacoides L.
SEA-PURSLANE
Dominant or locally abundant on muddy shores in salt-marshes; on a cliff east of Lyme Regis, CDP and DAP. 33. First record: 1551, W Turner. On both sides of the

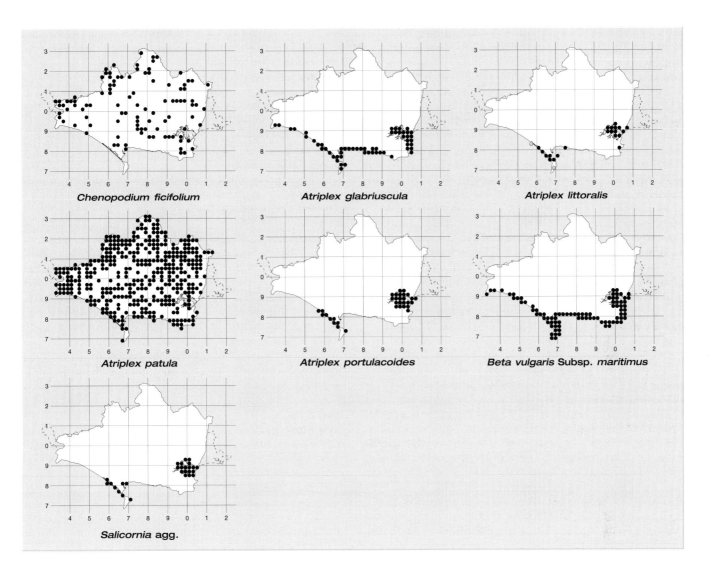

Chenopodium ficifolium

Atriplex glabriuscula

Atriplex littoralis

Atriplex patula

Atriplex portulacoides

Beta vulgaris Subsp. maritimus

Salicornia agg.

Fleet, east Portland and all round Poole Harbour inner shores. This and *Suaeda maritima* are the habitat for the rare mollusc *Truncatella subcylindrica*.

[*Atriplex pedunculata* (L.) L.
PEDUNCULATE SEA-PURSLANE
A rare salt-marsh annual now confined to Essex. Portland (19th Cent., G Donaldson in **BM**).]

Beta vulgaris L.
BEET
Subsp. *maritima* (L.) Arcang. is frequent all along the coast on pebble or muddy beaches, and on clay undercliffs and cliff-edges. 86. Common inside the Fleet and Poole Harbour. First record; 1801, W Withering. Subsp. *vulgaris* is a casual on tips and disturbed verges away from the coast, mostly as mangold, which is still grown as a crop, rarely as red beetroot. 15.

Sarcocornia perennis (Miller) A.J. Scott
PERENNIAL GLASSWORT
Locally frequent in salt-marshes around the Fleet, on

Portland and inside Poole Harbour. 8. Between Wyke and Portland (67, 1826, MJ Berkeley in **CGE**, *et al.* in **CGE, RNG** and **WRHM**, and 67S, 1999, !); Radipole Lake (68Q, 1785, J Lightfoot in **K** and 1924, RDG); Durdle Pier (77B, 1984, RA); Arne (98U and Z, 1936, RDG in **WRHM** and 1992, DAP and RMW); Wytch (98X, 1915, CBG in **BMH)**; Ower (98Y, 1900, EFL and 1915, CBG in **BMH)**; Hamworthy (99, 1915, AWG); Furzey Is and Green Is (08D, 1994–95, !); Mead and Sandy Points (08H, 1999, EAP); Brands Bay and Jerry's Point (08I, DSR in **UEA** and 1981, AJCB); Brownsea Is (08J, 1968–77, DSR); Poole and Sterte (09A, 1895, JCMP, also 1933, AJ Willmott in **BM**, and 1957, !).

Salicornia agg.
GLASSWORT
Locally abundant on mud in salt-marshes of the Fleet, east Portland, Radipole Lake and inside Poole Harbour. 27. Hard to name, and few experts have looked at Dorset plants; * = det. IK Ferguson. **S. pusilla** J Woods is reported from east of Weymouth (68, 1938, AJ Willmott in **BM)**;

East Weare, saline pools (77B, 1999, !*); Middlebere and Arne Bay (98, 1936, RDG in **WRHM** and 1934, AJ Willmott in **BM**); Hamworthy (99, 1912, EFL in **CGE**); Brands Bay and Goathorn (08, 1965, DSR); South Haven (08F, 1913–17, CBG in **BMH**); Brands Bay (08H, 1999, EAP); Poole (09, 1909, EFL, also 1913, CBG in **BMH** and 1933, AJ Willmott in **BM**). *S. pusilla* × *ramosissima* is claimed for Dorset by Stace (Stace, ed. 1975). *S. ramosissima* J Woods is probably the commonest taxon here. It has specimens in **BMH** and **WRHM** and post-1960 reports from 58V and W, 67S*, 68A*, Q and V*, 98N, P* and Z, 99W, 08C, D, H, I, J and P and 09A and B. *S. fragilis* PW Ball & Tutin occurs at Small Mouth (67S, 1999, !*), Lodmoor (68V, 1999, !*) and Patchins Point (98Z, 1999, !*), and probably elsewhere around Poole Harbour. *S. dolichostachya* Moss has records from six tetrads; Wyke (67, 1912, CBG in **BMH** and 1933, RDG in **WRHM**); Shag Looe Head (98P and U, 1999, JHSC*); Hamworthy (99, 1915, AWG); Brands Bay (08C and H, 1981, AJB and 1999, EAP).

Suaeda vera Forsskal ex J. Gmelin
SHRUBBY SEA-BLITE
Locally dominant in a strip along the landward side of the Chesil bank, occasional along the inner shores of the Fleet and also Poole Harbour, on undisturbed beaches. 18 + [5]. Rare elsewhere on the coast, as on Portland breakwater (67Y, 1996, !): Winspit (97T, 1966, RDE); Little Sea (08H, 1969, VM Trembath). First recorded by J Ray in 1670 on the Chesil bank, and by many later collectors in **CGE, DOR, HIWNT, RNG** etc. Poole Harbour records post-1900 are: Fitzworth (98Y, DAP); Arne (98U and Z, !); Rockley point (99Q, !); Goathorn, Green Is and Furzey Is (08D, !); Brownsea Is (08E, DAP); Bramble Bush Bay (08H, 1999, EAP); Salterns (08J, 1921, JHS). It does not occur west of Abbotsbury in Britain. Though sometimes inundated by high tides, its woody stems support several species of lichen; it regenerates well in the Fleet.

Suaeda maritima (L.) Dumort.
ANNUAL SEA-BLITE
Locally abundant in salt-marshes inside the Fleet, at Lodmoor and inside Poole Harbour, with outliers near West Bay. 36 + [1]. Variable in habit and colour. Pollen has been found in deposits from the Roman period on Brownsea Is, where the species still occurs.

Salsola kali L.
PRICKLY SALTWORT
Although there are old records from beaches from Charmouth to Canford Cliffs, this has markedly declined since the 1950s, probably from human pressure. 2 + [13]. Three plants occurred on the 'unspoilt' shore of Furzey Is (08D, 1995, !), and about 50 at Shell Bay (08I, 1999, EAP). Subsp. *ruthenica* (Iljin) Soo is a casual seen at Sherborne tip (61I, 1969, ACL) and Swanage Camp (08F, 1917, CBG in **BMH**). [2].

[*Axyris amaranthoides* L.
RUSSIAN PIGWEED
A casual at Beaminster (40Q, 1928, AWG). No specimen seen.]

AMARANTHACEAE

Amaranthus retroflexus L.
COMMON AMARANTH
Locally abundant in manured arable, maize fields, fruit farms and gardens on sandy or chalky soils. 35 + [6]. First record: Wareham (98, 1899, JCMP in **DOR**), since when it has spread widely.

(*Amaranthus hybridus* L.
No certain record, all reports prove to have been *A. retroflexus*.)

Amaranthus cruentus L.
PURPLE AMARANTH
A casual, doubtfully distinct from *A. hybridus*. Keyneston (90H, 1900, E Smith in **DOR**); Sherborne tip (61I, 1970, ACL). [2].

Amaranthus albus L.
WHITE PIGWEED
A casual. Sherborne (61I, 1970, ACL); Greenland Farm (08C, 1999, RSRF & !, det. TBR. 1 + [1].

[*Amaranthus blitum* L.
GUERNSEY PIGWEED
A casual. Netherbury (49U, 1930s, AWG). No specimen seen.]

Celosia cristata L.
COXCOMB
A garden escape. Sherborne tip (61I, 1970, LW Frost). [1].

PORTULACACEAE

Portulaca oleracea L.
COMMON PURSLANE
A garden weed, Lower Westport (98I, 1967, RC Trench and 1976, D McClintock); Parkstone (09F, 1930s, HJW). [2].

Claytonia perfoliata Donn ex Willd.
SPRINGBEAUTY
A persistent introduced weed of sandy soils, often in gardens, tolerating shade. 7 + [8]. Tizards Knap (49C, 1982, AWJ); Beaminster (40Q, 1937, DM); Langton Matravers (97Z, 1895, LVL); Wareham (98I, 1994, AH); Rempstone (98W, 1895, C Calcraft); Studland (08G, 1937, DM and 1996, !); North Haven (08I, 1999, !); Lilliput and Sandbanks (08N and P, 1900, EFL, *et al.* in **BMH**, and 1994, !); Branksome and Canford Cliffs (08U, 1891, JCMP, *et al.*, and 1992, RMW); Poole Park (09F, 1929, J Allner and 1977, H Thomas); Ferndown (00Q, 1932, FHH and 1960, RDG in **DOR**); West Moors (00R, 1945, BNSS). It germinates in autumn.

Claytonia sibirica L.
PINK PURSLANE
Widely scattered in woods, hedgebanks and gardens on sandy soils, as an introduction. 13 + [3]. Most frequent in the north-west. First record: Crab Coppice (49Z, 1920–47, AWG in **BDK**).

Montia fontana L.
BLINKS

In wet, bare muddy places, often in woodland rides on acid soils; sometimes in nursery gardens and once in a fen. 57 + [10]. The subsp. have no obvious ecological differences, and have only been determined in 17 localities; of these, nine were subsp. *amporitana* Sennen, seven subsp. *chondrosperma* (Fenzl) Walters, and one subsp. *variabilis* Walters.

Calandrinia ciliata (Ruiz Lopez & Pavon) DC.
RED-MAIDS

Casual, once found in a plantation at Keysworth (98J, 1978, AH and HJS Clark in **PVT** of EJ Clements). [1].

CARYOPHYLLACEAE

Arenaria serpyllifolia L.
THYME-LEAVED SANDWORT

Rather local in dry places, anthills, wall-tops and pavements, mostly on calcareous soils. 255. Absent from clayey or very acid soils. Subsp. *serpyllifolia* is commonest. Subsp. *leptoclados*

(Reichb.) Nyman has only seven records, mostly from walls, but is under-recorded. Subsp. *lloydii* (Jordan) Bonnier is a condensed form of cliff-edges and dunes: West Bay (49K, 1914, EBB in **BMH** and 1993, ! and RSRF); Smallmouth (67S, 1991, !); Winspit (97T, 1934, JEL in **RNG**); Studland (08L, 1991, ! and 08H, 1997, FAW). 3 + [1].

[*Arenaria balearica* L.
MOSSY SANDWORT

A garden escape. 2. Bowood (49P, 1925, AWG); Abbotsbury (58S, 1861, W Pamplin).]

Moehringia trinervia (L.) Clairv.
THREE-NERVED SANDWORT

Widespread, but rarely in quantity, in old woods and shaded hedgebanks. 375. It tolerates a wide pH range, but is absent from deforested areas, Portland, and built-up places.

Honckenya peploides (L.) Ehrh.
SEA SANDWORT

Occasional on low cliffs, or on beaches of fine shingle, sand

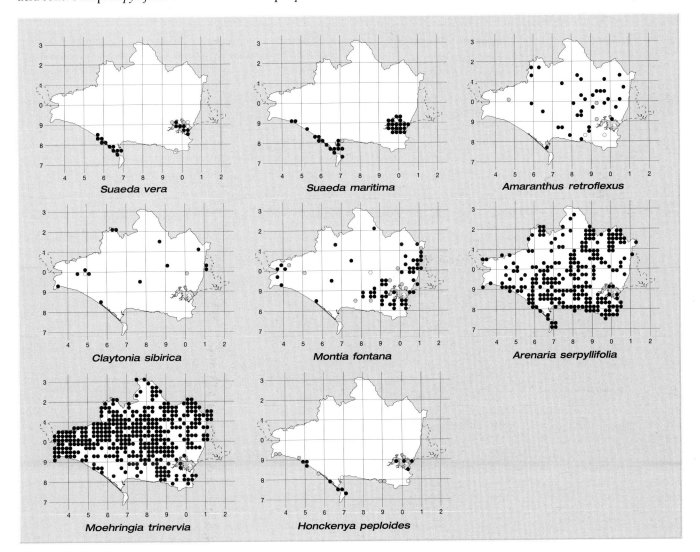

Suaeda vera

Suaeda maritima

Amaranthus retroflexus

Claytonia sibirica

Montia fontana

Arenaria serpyllifolia

Moehringia trinervia

Honckenya peploides

or mud. 11 + [10]. Its range extends all along the coast, but it has disappeared from most sandy beaches where humans congregate. Pollen from the Roman period has been found at Brownsea Is, where the plant still occurs. It also persists near the north end of Portland, where it was first noticed by M de L'Obel in 1576.

Minuartia hybrida (Villars) Schischkin
FINE-LEAVED SANDWORT

A rare casual of verges and railway tracks. First recorded by Pulteney in 1799, but the only records this century are: [Fleets Corner railtrack (now a road; 09B, 1977, !); Woodyates (01J, 1937, AWG).] There is a specimen from Whatcombe (80F, 1897, JCMP in **BM**). [5].

Stellaria media (L.) Villars
COMMON CHICKWEED

Very common in woodland rides, arable, cliff-tops, muddy beaches and gardens, always on rich soil; remains known from the Bronze Age. 726. It tolerates salt-spray, and is currently increasing in heavily fertilized sown grass leys. A useful foodplant for many moth larvae, and the seeds are taken by partridges and small birds (Wilson, Arroyo & Clark, 1997).

Stellaria pallida (Dumort) Crepin
LESSER CHICKWEED

A scarce ephemeral annual of bare, sandy soils, mostly near the coast. 21. Over-recorded by some, and probably by JCMP in his Flora. It germinates in December.

West Bay (49Q, 1999, DAP); Burton Mere (58D and E, 1993, LJM); Portland Bill (66Z, 1999, DAP); Small Mouth (67S, 1991, AH); Poyntington (62K, 1995, IPG and GAC); Tyneham Cap (87Z and 98A, 1991, AH); Swyre Head and Smedmore Hill (97J, 1999, BE and DAP); St Aldhelm's Head (97S and T, 1999, BE and DAP); Shipstal Point (98Z, 1999, !); Swanage Taxi rank (07J, 1999, EAP); Newton Bay (08C, 1996, MJG); Brownsea Is (08D and E, 1963–99, ! in **RNG**); Studland (08G, 1994, RF); Redhorn Quay (08H, 1930, LBH and 1991, AJB and DAP); Jerry's Point to South Haven (08I, 1991, RF, !); Sandbanks (08N, 1991, !); Poole Park (09F, 1991, RF).

Stellaria neglecta Weihe
GREATER CHICKWEED

Earlier Floras give an inadequate picture of the distribution of this plant, which is locally frequent by shaded streams, sheltered hedgebanks and edges of woods in the north-west, with few outliers. 140. It is prominent in May, but disappears from view after June.

Stellaria holostea L.
GREATER STITCHWORT, BATCHELOR'S BUTTONS, SNAPJACKS, THUNDERBOLTS

Frequent in woods and hedgebanks on dry soils of all kinds, preferring sandy ones, but absent from deforested areas and Portland. 529.

Stellaria palustris Retz
MARSH STITCHWORT

Rare in marshes or fens on richer soils in the north-east.

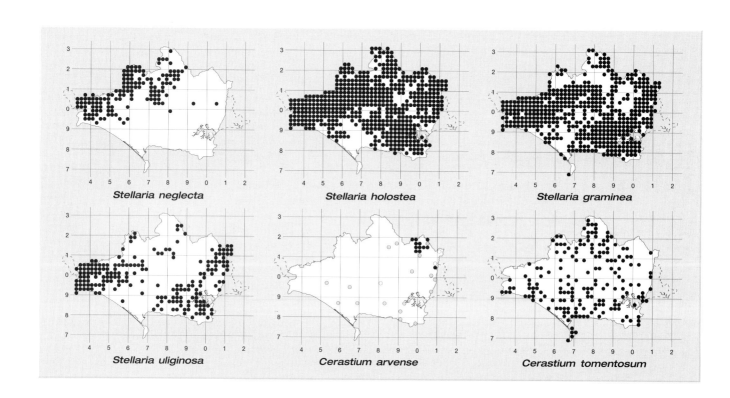

Stellaria neglecta
Stellaria holostea
Stellaria graminea
Stellaria uliginosa
Cerastium arvense
Cerastium tomentosum

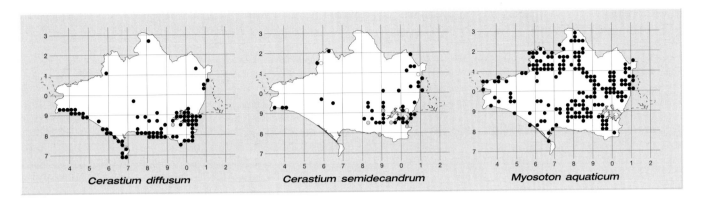

Cerastium diffusum | Cerastium semidecandrum | Myosoton aquaticum

1 + [13]. Often reported in error, but the plant has probably declined through drainage. The only reliable recent records are from Stanbridge Mill (00E, 1989, J Davis and 1994, JO). It may still persist by the Stour or Moors River, as it does by the Avon in 'New Dorset', VC11.

Stellaria graminea L.
LESSER STITCHWORT
Frequent in old grassland on all soils but the most acid, least common on the drier chalk. 485. It does not like shade, grows in less moist soils than does *S. uliginosa*, and recolonises these after disturbance.

Stellaria uliginosa Murray
BOG STITCHWORT
Occasional in marshes and damp grassland, especially in the Poole basin and the west, but mostly absent from the Chalk and Portland. 202.

Cerastium arvense L.
FIELD MOUSE-EAR
Once widespread in short calcareous turf, rare on heathland tracks. 12 + [20]. Its range has contracted to a small area on the north-eastern chalk, where the conversion of sheepwalk to arable has confined it to tracksides, verges and ancient earthworks. Only one outlier on heathland is known, at West Moors (00X, 1895, JCMP and 1990, DAP), but until recently it was at Talbot Heath (09R, 1959, KG), East Parley (09Z, 1955, BNSS), Ferndown (00Q, 1931, FHH) and Verwood (00Y, 1925, JHS).

Cerastium tomentosum L.
SNOW-IN-SUMMER
A frequent garden escape on verges and walls on all types of soil, rarely far from houses. 180. Invasive and persistent.

Cerastium fontanum Baumb.
COMMON MOUSE-EAR
Common in grassland and verges on all types of soil, also a garden weed. 726. It resists trampling, grazing and mowing.

Cerastium glomeratum Thuill.
STICKY MOUSE-EAR
Common in dry sandy or calcareous soils, preferring bare ground in arable, gardens or lawns. 609. Variable in stature.

Cerastium diffusum Pers.
SEA MOUSE-EAR
Common on cliff-tops, undercliffs and fixed shingle all along the coast, indifferent as to soil here. 80. It has recently spread to sandy roadside verges inland, to railway tracks at Gillingham (82D, 1995, ! and DAP), and to some pavements in towns.

Cerastium pumilum Curtis
DWARF MOUSE-EAR
A small ephemeral of dry, short, open grassland on limestone, rarely on chalk, mostly near the sea. 8. Eggardon Hill (59M, 1950, AR Clapham and CD Pigott); East Portland (67V and 77A, 1970, RMB and 1993, DAP and RF); Bat's Head (78V, 1993, BE); Bindon Hill (88K, 1997, BE); Spyway Farm, above Dancing Ledge (97Y, 1917, GCD, *et al.*, and 1997, BE and DAP); Langton Matravers (97Z, 1917, CE Britton in **K** and 1930, Anon in **RNG**); Blacker's Hole (07D, 1997, BE and DAP); Anvil Point and Durlston CP (07I, 1970, RMB, *et al.*, and 1994, AJB and DAP); Swanage quarries and Townsend reserve (07J, 1912, CBG in **BMH** and 1990–97, WGT and BE).

Cerastium semidecandrum L.
LITTLE MOUSE-EAR
An easily overlooked annual of bare, dry, sandy soils. Mostly on sandy cliffs or disturbed heaths, rarely on railway tracks and once seen on a tiled roof. 41 + [8]. Commonest in the Poole basin and the west, absent from the Chalk.

Myosoton aquaticum (L.) Moench
WATER CHICKWEED
Occasional in damp woods, more often along streams and ditches or by mature ponds, and colonizing dredged mud. 177. Absent from the Chalk (except on spoil) and from very acid sites.

Moenchia erecta (L.) Gaertner, Meyer. & Scherb.
UPRIGHT CHICKWEED
An annual, varying in abundance from year to year, which
has greatly declined over the last century. It prefers dry,
bare places on acid or sandy soils. 6 + [34]. Old records
are mostly from the Poole basin, where it is now very local.
Recent records are: Stonebarrow Hill (39R and W, 1985,
AWJ and 1993, MJG); Pilsdon Pen (40A, 1920, AWG and
1992, JGK, also IPG); Abbotsbury Castle (58S, 1937, DM
and 1960, !, also 58N, 1992, I Taylor); Blacknoll (88D,
1954, ! and 1980, RHD); Wareham Common (98D, 1930,
BNSS, *et al.*, and 1996, !).

Sagina nodosa (L.) Fenzl
KNOTTED PEARLWORT
An inconspicuous plant of disturbed heaths, tank tracks,
marshes and dune slacks, never in quantity. 11 + [25].
Much rarer than in the last century, when JCMP found it
on chalk downs. Recent records: Small Mouth (67S, 1911,
GCD and 1960, !); Portland (67, 1978, BSBI); Black Down
(68D, 1978, !); Winfrith Heath (88D, 1961, KBR and 1966, !);
Hyde Heath (88P, 1996, R Gibbons & N Spring); Five
Barrow Hill (88R, 1992, !); Bovington Heath (89F, 1947,
WAC and 89F and G, 1996, BE, !); Gallows Hill (89K,
1895, JCMP and 1990, AH); Corfe Common (98Q, 1900,
EFL, *et al.*, and 1992, AH); Wareham Common (98D, 1988,
RD Porley); Arne (98U, 1979, MRH); Studland Heath (08H,
1900, RPM, *et al.*, and 1935, RDG); Brownsea Is (08J, 1968–
77, KC).

Sagina subulata (Sw.) C. Presl.
HEATH PEARLWORT
Uncommon on bare, sandy soils of heath tracks, often with
small annuals such as *Radiola*. 25 + [5]. Mostly in the Poole
basin, with outliers at Champernhayes (39N, DAP), Coney's
Castle (39T, BE), Stonebarrow (39W, RJS) and Black Down
(60D, !).

Sagina procumbens L.
PROCUMBENT PEARLWORT
Widespread on most types of soils; in woodland rides,
between tussocks in fens, and especially on bare tracks and
as a pavement weed in gardens and built-up areas. 471.

Sagina apetala Ard.
ANNUAL PEARLWORT
Occasional to frequent on dry, bare soil, heathy tracks,
gravel paths and shingle beaches above the strandline,
also on walls and tips. 163. Absent from heavy clays, and
mostly in the Poole basin or along the coast. Few recorders
distinguished the two subspecies, but subsp. *apetala* is
probably commoner than subsp. *erecta* F. Herm.; the latter
has only 11 certain records.

Sagina maritima G. Don
SEA PEARLWORT
Frequent in bare, rocky places or hard soil near the sea, all
along the coast, and sometimes as a pavement weed. 39 +
[7]. Inland, it has been seen on the ironstone ridge at

Abbotsbury (58S, 1978, !) and at Longham (09T, 1994,
TCGR).

Scleranthus annuus L.
ANNUAL KNAWEL
Although known in Iron Age deposits, frequent in the
1890s, and common in the Poole basin in the 1930s, it is
uncommon now. 15 + [17]. It occurs in dry arable on sand,
less commonly on chalk, and sometimes as a weed of gravel
paths and verges. Recent records: Lower Kingcombe
(JGK, !); Frome St Quintin (DAP); Hurst Heath (1978, !);
Bovington (BE and DAP); Whiteway (BE); Grange Arch;
Wareham by-pass (1980, AH & !); East Hill, Corfe;
Greenland (AH and DAP); north of Arne (!); Kinson (FAW);
Castle Hill Wood (1979, !); Ashley Heath (VS).

[*Corrigiola litoralis* L.
Casual?. Blandford (80Y, 1885, RPM in **NMW**).
Reported from a painting at Weymouth in the 1890s in Good's
Handbook, but never confirmed.]

[*Herniaria hirsuta* L.
HAIRY RUPTUREWORT
A casual alien in a garden at Beaminster (40, 1939, AWG in **BDK**).]

Illecebrum verticillatum L.
CORAL-NECKLACE
Very rare on bare, wet heaths or by acid ponds. 2 + [1].
[Small Mouth (67S, 1792, W Sole)]; Gundry's Enclosure
(00W and X, 1981, HCP in **CGE** and 1990, DAP); (one
plant at Barnsfield Heath (10G, 1994, BE) but in VC11).
Essentially southern in Britain.

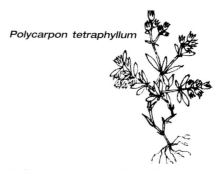

Polycarpon tetraphyllum

Polycarpon tetraphyllum (L.) L.
FOUR-LEAVED ALLSEED
This inconspicuous Cornubian annual, so common on
Guernsey, is rare here and occurs on fixed shingle near
the sea, or as a pavement weed. 3. It has been known near
Weymouth for more than two centuries. Small Mouth (67S,
1774, J Lightfoot in **K**, *et al.*, e.g 17th Century, JS Baly in
WAR, 1876, JCMP in **DOR**, 1904, AS Montgomery in **LIV**,
and 1997, SME, !); Pavement weed, North Haven (08I,
1999, !) and Sandbanks (08N, 1979, RMB in **RNG** and
1991, AJB).

Spergula arvensis L.
CORN SPURREY
Locally frequent in dry arable, mostly on sandy soil, since

the Bronze Age. 148. Occasionally on beach sand near the sea and on fire-breaks and rabbit scrapes, or on bare chalky soil, rare on clay.

Spergularia rupicola Lebel ex Le Jolis
ROCK SEA-SPURREY.
Local on limestone, clay and Wealden cliffs and undercliffs, sometimes at cliff-edges or on sea walls. 37 + [4]. All along the coast except for the Chesil Bank.

[*Spergularia rupicola × marina*
A sterile hybrid once found on a sea wall at Lyme Regis (39L, 1911, HWP).]

Spergularia media (L.) C. Presl
GREATER SEA-SPURREY
Local on muddy shores and in salt-marshes, especially inside the Fleet and Poole Harbour, including the islands. 31 + [3].

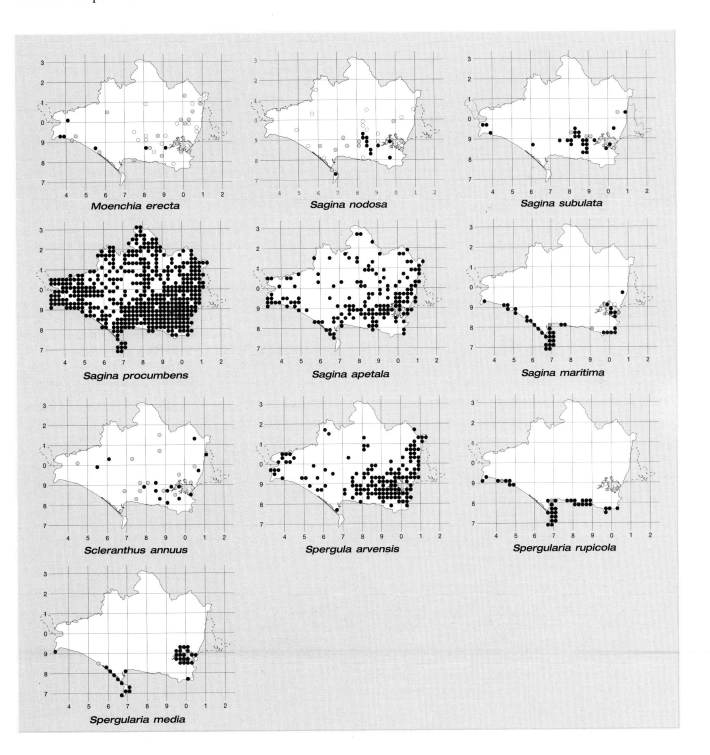

Moenchia erecta

Sagina nodosa

Sagina subulata

Sagina procumbens

Sagina apetala

Sagina maritima

Scleranthus annuus

Spergula arvensis

Spergularia rupicola

Spergularia media

Spergularia marina (L.) Griseb.
LESSER SEA-SPURREY
In the same habitats as *S. media*. 38 + [2]. Not yet reported from salted verges.

Spergularia rubra (L.) J.S. Presl & C. Presl
SAND SPURREY
Local on dry sandy or gravelly tracks, in bare places in old gravel-pits, on rail tracks, firebreaks in heaths and bare, sandy lawns. 71 + [5]. Mostly in the Poole basin, in the west and near Sherborne.

Lychnis coronaria (L.) Murray
ROSE CAMPION
A garden escape, sometimes established for a few years, in village lanes, on small tips, or far from houses. 24. It prefers dry soils, either calcareous or sandy.

Lychnis flos-cuculi L.
RAGGED-ROBIN, ROBIN HOOD
A plant of wet pastures and marshes, tolerant of shade but then not flowering. 244. It is absent from the drier parts of the Chalk and from Portland.

Agrostemma githago L.
CORNCOCKLE
A common cornfield weed from the Iron Age to Victorian times, now rare. 14 + [42]. Possibly introduced with impure seeds, as by the 1930s it was infrequent and decreasing; Good had it from 23 stands. Since the 1960s it has been a casual, mostly on verges or in gardens, and is a constituent of 'wild flower' seed mixtures. It was, and is, found on calcareous or sandy soils, and still appears when old

hedgebanks or verges are disturbed: garden records are omitted below. Bettiscombe (39V and 40A, 1998, IPG); Uphill (50L, 1993, !); Sandsfoot undercliff (67T, 1986–87, DAP, ! in **RNG**); Milborne St Andrew (89D, 1992, BE, !); Bloxworth Ch (89X, 1992, !); thousands in arable, Combe Bottom (80D, 1991, !); Durlston CP (07I, 1990, AL); verge, Holes Bay (09A, 1989, AM); Muscliff (09X, 1989–91, FAW).

Silene italica (L.) Pers.
ITALIAN CATCHFLY
Introduced on a wall, Sydling St Nicholas (69J, 1981, AJCB and 1982, ! in **RNG**; lost by 1999). 1.

Silene nutans L.
NOTTINGHAM CATCHFLY
Very local at the edge of calcareous cliffs, but easily overlooked when not in flower. 6 + [1]. Rousden Cliff, perhaps in Devon (39F, 1975, J Fairley); White Nothe to Bats Head (78Q and V, 1971, A Lack, *et al.*, and 1998, !); Newlands Warren (80A, 1984, NCC); Arish Mell (88K, 1992, DAP & !); Corfe Castle (98L, 1972, C Jones); Ballard Point (08K, 1917, CBG, *et al.* in **DOR, K, RNG**, and 1995, WGT); Old Harry (08L, 1891, HNR in **DOR** and 1953–93, !).

Silene vulgaris Garcke
BLADDER CAMPION, BLETHERWEED
Occasional, never in quantity, in arable or on calcareous verges. 182. Var. *pubescens* DC. used to be found but has not been seen recently.

(*Silene vulgaris* × *uniflora* was reported by Stace (Stace, ed. 1975), but no specimens have been seen. The parents rarely grow together, except at Cogden beach (58I).)

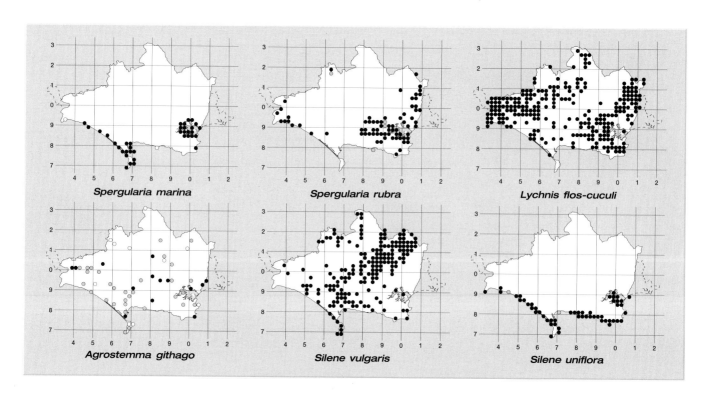

Spergularia marina

Spergularia rubra

Lychnis flos-cuculi

Agrostemma githago

Silene vulgaris

Silene uniflora

Silene uniflora Roth.
SEA CAMPION
Locally frequent on shingle beaches, also on clay, chalk and limestone cliffs, all along the coast. 52 + [3]. Less common in muddy salt-marshes. Conspicuous on the landward side of the Chesil Bank and at Houn's-tout Cliff.

Silene armeria L.
SWEET-WILLIAM CATCHFLY
A garden escape on waste ground at Puddletown (79M, 1992, !) and North Haven (08I, 2000, !). 2.

Silene noctiflora L.
NIGHT-FLOWERING CATCHFLY
A scarce weed of arable, first found by EFL but not by JCMP, nor found by Good in any of his stands. 8 + [19]. Now only on the north-east Chalk. Recent records: Combs Ditch (80Q, !); Fox Grounds (80S, !); Gunville Down (91B, !); Chettle Down (91M, !); Sovell Down (91V, !); Sixpenny Handley (91Y, 1980, AH); Wyke Down (01C, GDF); frequent near Ackling Dyke (01D, GDF); Bowldish Pond (O1H, 1999, J. Phillips).

Silene latifolia Poiret
WHITE CAMPION, MILKMAIDS, ROBIN WHITE
A frequent weed of arable and grassy verges, mostly on calcareous soils; first noted in Roman deposits. 301. Var. *macrocalycinum* Rouy occurred at Mupe Bay (87P, 1937, NDS).

Silene latifolia × *dioica* = *S.* × *hampeana* Meusel & K. Werner
Isolated plants are common, usually with one or both parents on verges and waste ground. 137. Increasing, as

Good only had one record in the 1930s. First record: Abbotsbury (58S, 1908, HJR).

Silene dioica (L.) Clairv.
RED CAMPION, RED GRANFER GREGGLE, ROBIN HOOD
Common in woods and hedgebanks on all soils but the most acid ones. 657. Commonest in the west, least common in the east, on the Chalk ridge and in deforested areas. Female plants have larger flowers and remain in flower late in the year. White-flowered plants are rare; Lewesdon Hill (40F, 1996, IPG and PRG); Monmouth Hill (70V, 1997, !). Rabbits rarely eat this plant.

Silene gallica L.
SMALL-FLOWERED CATCHFLY
Frequent in the Poole basin in the 1890s, and the 1930s when Good found it in 63 stands, this is now a rare and sporadic weed of sandy arable and disturbed places. The decline is due either to selective weedkillers or to increased usage of fertilizers. 11 + [35]. Recent records: Lower Kingcombe (59P, 1994, SME); Portland (67W, 1983, M Leicester); Whitcombe Barn (78D, 1993, !); White Nothe top (78Q, 1984, MH Collier); Hurst Heath (78Z, 1990, BE); Stoke Heath (88P, 1991, !); Winterborne Kingston (89P, 1989, !); Carey (98E, 1997, RB); Wareham and its by-pass (98H and I, 1980, AH and 1987, MHL); Fitzworth (98Y, 1998, BE and DAP). The forma *quinquevulnera* is a rare casual seen at Branksome in 1902, Wimborne in 1946 and in a garden at Motcombe (82M, 1994, !) from bird-seed.

[*Silene conica* L.
SAND CATCHFLY
Extinct. 2. In a heathy field at Parkstone from 1886 to 1902 (WH

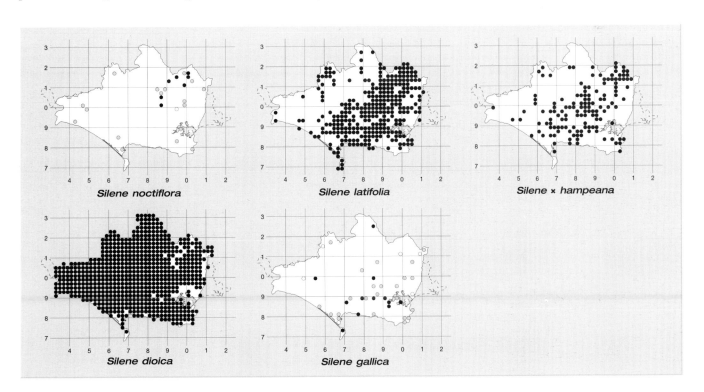

Silene noctiflora

Silene latifolia

Silene × hampeana

Silene dioica

Silene gallica

Moberly in **DOR**, also EFL in **BM**, **DOR**, **LIV** and **OXF**); Sherborne (61, 1906, DRW).]

Other *Silene* species have occurred as casuals or garden escapes, but not since 1960; *S. coeli-rosa* (L.) Godron on Grove cliffs (77B, 1959, !); *S. cretica* L. at Godlingston (08A, 1917, CBG); *S. laeta* (Aiton) Godron in arable at Stoborough (98I, 1916, CIS); *S. pendula* L. at Northport (98J, 1898, EFL) ; and *S. schafta* C. Gmel. ex Hohen. on a wall at Abbotsbury (58S, 1951, !) and at Portland (67, 1927, J Rayner).

Cucubalus baccifer L.
BERRY CATCHFLY
Tenuously established in a shaded hedgebank at Portesham (68C, 1971, EA Higgins and 1988, RJS, !).

Saponaria officinalis L.
SOAPWORT
Small colonies of this persistent alien are established on verges, railway banks and in village lanes. 47. Once used as a substitute for soap; the *flore pleno* forms are garden throwouts.

Vaccaria hispanica (Miller) Rauschert
COWHERB
Single plants of this casual alien are occasionally found near buildings or on tips, but not lately. [11] Compton Abbas (1894, EFL); Portland and Radipole Lake (1926, MJA); Weymouth (1930s, NDS); Lodmoor tip (1963, !); Leigh (1970, SS); Sherborne tip (1968, ACL); Melbury Abbas (1895–1904, EFL in **DOR**); Wareham (1930, BNSS); Swanage Camp (1917, CBG in **BMH**); Parkstone (1933, HJW).

[*Petrorhagia saxifraga* (L.) Link
TUNICFLOWER
A rare garden escape, erroneously recorded as *P. prolifera* by some. 1. On a wall near Abbotsbury gardens (58S, 1893, T Benbow, *et al.*, and 1932, MAA in **RNG**).]

Dianthus gratianopolitanus Villars.
CHEDDAR PINK
Planted on a wall at Upwey (68S, 1999, !), and a casual on a tip at East Weare, Portland (77B, 1959, !). 1 + [1].

Dianthus caryophyllus L.
CLOVE PINK
A persistent denizen on the walls of Old Sherborne Castle (61N, 1876, JCMP in **DOR**, *et al.*, and 1992, DAP); planted on a garden wall at Strode Manor (49P, 1972, !). 1 + [1].

Dianthus plumarius L.
PINK
Persistent on old walls at Powerstock (59C, 1983, D Griffiths) and Hooke (50F, 1961–98, !, also D Harris). 2.

Dianthus deltoides L.
MAIDEN PINK
A rare garden escape in dry grassland and on tips. 1 + [2]. Sherborne School and tip (61, 1969, ACL); Bryanston (80T, 1941, DM); Parley Common (09Z, 1995, FAW; five plants, one white-flowered).

Dianthus barbatus L.
SWEET-WILLIAM
Commonly grown in gardens, rarely escaping but not

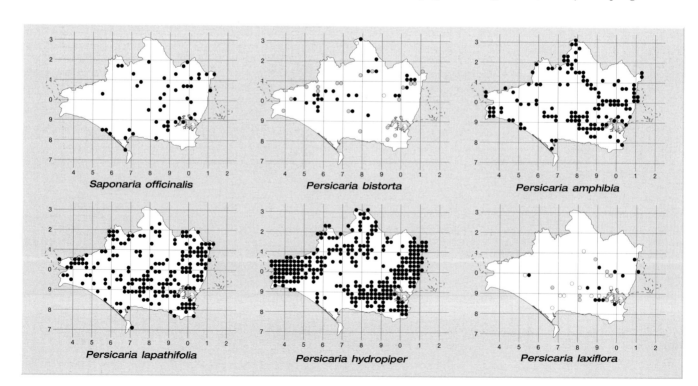

Saponaria officinalis   Persicaria bistorta   Persicaria amphibia

Persicaria lapathifolia   Persicaria hydropiper   Persicaria laxiflora

persisting long in village lanes and on tips. 2 + [6]. Litton Cheney (59K, 1981, AJCB); Portland Bill (66U, 1959, !); Grove cliffs (67W, 1960, EH in **LANC**); Lodmoor tip (68V, 1951, !); Stallen (61D, 1955, SS); Sherborne tip (61I, 1970, ACL); Knighton pits (78P, 1998, !); Puddletown (79M, 1992, !). The seeds are eaten by larvae of the moth *Hadena bicruris*.

Dianthus armeria L.
DEPTFORD PINK
Rare on dry banks with open soil. 1 + [4]. Whitchurch Canonicorum (39X, 1930s, G Cole); On the ironstone ridge and in the gardens, Abbotsbury (58S, 1893, N Richardson in **DOR**, also 1928, AWG and 1936, RDG in **WRHM**); Caundle Holt (71, 1846, T Digby); Hethfelton (88N, 1938, RDG in **DOR** and **WRHM**, and 1998, !).

Gypsophila muralis L.
ANNUAL GYPSOPHILA
A casual reported from Greys Wood (79G, 1976, MS Warren), perhaps from pheasant food; Wareham pavement (98I, 1998, ! in **RNG**). 1 + [1].

POLYGONACEAE

Persicaria campanulata (J.D. Hook.) Ronse Decraene
LESSER KNOTWEED
A rare but persistent garden escape. 2 + [1]. Old railway, Uplyme (39G, 1985, RCP in **OXF**); planted and established in Abbotsbury gardens (58S, 1984–93, !); Rempstone (98W, 1969, DG Hewett in **WRHM**).

Persicaria wallichii Greuter & Burdet
HIMALAYAN KNOTWEED
An occasional denizen of verges and abandoned gardens. 4 + [2]. Lyme Regis (39G, 1984, AWJ); Hewood (30L, 1998, IPG); Birdsmoor Gate (30V, 1994, GM Kay and JE Hawksford); Oborne (61P, 1971, ACL); Bourton (73K, 1996, IPG and PRG); Hammoon (81H, 1983, AJCB); Shaftesbury (82R, 1983, AJCB); Swanage Camp (08F, 1917, CBG in **BMH**); St Leonards Peats (10A, 1984, RMW).

Persicaria bistorta (L.) Samp.
COMMON BISTORT
Uncommon in wet meadows, mostly in central and north Dorset. 27 + [21]. It has probably decreased due to 'improvement' of such meadows. Easily confused with *P. amplexicaulis* when sterile.

Persicaria amplexicaulis (D. Don) Ronse Decraene
RED BISTORT
A garden escape sometimes established on verges, usually near houses. 18. Distinct from *P. bistorta* in its deep pink flowers, and in its aerial shoots being cut down by frost. At Charminster (69R) it is slowly colonising the banks of the Cerne.

Persicaria amphibia (L.) Gray
AMPHIBIOUS BISTORT
Frequent, either floating in lakes and rivers, or on river banks and marshes, sometimes a weed of damp arable, and near the shore inside Poole Harbour. 137. Most records are from the valleys of the Stour, Frome, Piddle and Moors Rivers.

Persicaria nepalensis (Meissner) Gross
NEPAL PERSICARIA
A rare weed of nurseries. 3 + [1]. Abbotsbury gardens (58S, planted in 1899, MI and a weed, 1984–91, !); Moreton Nursery (88E, 1989–92, !); Blandford pavement (80Y, 1999, !); Stapehill Nursery (00K, 1968, E Thorne in **RNG**).

Persicaria capitata (Buch.-Ham. ex D Don) H. Gross
A rare escape from hanging baskets, as a pavement weed. 2. Lulworth Cove (88F, 1998, RJ Swindells); Wareham (98I, 1996, !).

Persicaria maculosa Gray
REDSHANK
A common and variable weed of arable, also in green lanes, disturbed verges, waste places, gardens and glasshouses. 597. It prefers moist, sandy, rich soils where it often appears in abundance and matures fast. The seeds are eaten by partridges and small birds (Wilson, Arroyo & Clark, 1997).

[*Persicaria maculosa* × *minor* = *P.* × *brauniana* (F. Schultz) Sojak
JCMP recorded this at Puncknowle Mill (58J, 1891) but I have not seen a specimen. [1].]

Persicaria lapathifolia (L.) Gray
PALE PERSICARIA
A frequent weed of damp, nutrient-rich arable fields, river dredgings or on mud by ponds, first evident in the neolithic period. 193. The flower colour varies from greenish-white to pink. A hairy form, var. *incana* Lej. & Court, was found at Tricketts Cross (00V, 1930s, NDS).

Persicaria pensylvanica (L.) M. Gomez
PINKWEED
Casual at Sherborne tip (61H, 1966, ACL).

Persicaria hydropiper (L.) Spach
WATER-PEPPER
Rather local in damp, muddy places in woodland rides, heathy tracks and by rivers. 293. It contains insecticidal substances. The var. *densiflorum* A Braun occurred at Blackney (49J, 1993, LJM and LMS).

Persicaria laxiflora (Weihe) Opiz
TASTELESS WATER-PEPPER
Very local and sporadic in wet, muddy places near the larger rivers, or by ponds. 12 + [18]. There is an outlier at Kingcombe (59P, SME) not far from JCMP's record from Puncknowle Mill (59J). Specimens are in **ALT, BDK, BMH, K** and **WRHM**.

Persicaria minor (Hudson) Opiz
SMALL WATER-PEPPER
Rare and sporadic in wet, muddy places in water-meadows
and by ponds. 6 + [25]. Much less frequent than formerly,
when it occurred mostly in the Poole basin, in the valleys
of the Piddle and Moors Rivers. Recent records: Hunters
Bridge (61Q, 1992, JO and LJM); Bryanston grounds
(80T, 1999, CDP and DAP); Morden Park Lake (99B,
1991, BE); Mannington (00S, 1993, GDF and JO);
Dewlands Common (00U, 1991, AJB and DAP); West
Moors (00X, 1992, DAP).

Fagopyrum esculentum Moench
BUCKWHEAT
Rarely grown as a crop today, as at Chamberlaynes (89L,
1989, !). 11. Otherwise a casual on tips, in arable, gardens
or disturbed soil.

(*Fagopyrum tataricum* (L.) Gaertner
GREEN BUCKWHEAT
Reported from Bowood (49P, 1930s, AWG). No specimen
seen.)

(*Polygonum maritimum* L.
The recent arrival of a strong colony on the shore at Hengistbury
Head (19V, 1997, !) leads one to hope that it will spread from
VC11 to VC9.)

Polygonum oxyspermum C. Meyer & Bunge ex Ledeb.
subsp. *raii* (Bab.) Webb & Chater
RAY'S KNOTGRASS
Rare on sandy beaches. Post-1930 records: Charmouth
(39R, 1937, FK Makins in **K** and 1999, !); West Bay

(49Q, 1895, JCMP, *et al.* in **BDK** and **K**, and 1961, !);
Portland (67, 1934, HB Redgrove); Studland Bay (08G,
1935, RDG). Its decline is due to human pressure.
1 + [10].

Polygonum arenastrum Boreau
EQUAL-LEAVED KNOTGRASS
Very common in arable and in hard, bare soil of tracks
and verges, resistant to trampling. 581.

Polygonum aviculare L.
KNOTGRASS
Very common in the same habitats as *P. arenastrum*, also
on shingle beaches. 669.

Polygonum rurivagum Jordan ex Boreau
CORNFIELD KNOTGRASS
Rare, but easily overlooked, in calcareous arable or on
verges; old records may not be trustworthy. 5 + [9]. Sheep
Down (68E, 1993, !); south-east of Cerne (69U, 1973,
RDE); Bell Hill (80E, 1989, !); Blandford St Mary (80X,
1994, !); Badbury Rings (90R, 1995, FAW); sandy beach,
south of Jerry's Point (08H, 1999, ! det. JRA).

Fallopia japonica (Houtt.) Ronse Decraene
JAPANESE KNOTWEED
Widespread, forming small colonies on verges, railway
banks and tips. 236. Most records are post-1970, probably
rejects from frustrated gardeners; first record, Beaminster
(40Q, 1926, AWG). Young shoots are often killed by late
frosts; it flowers well, but fruit have not been found.
Some efforts are now being made to control its spread by
mowing.

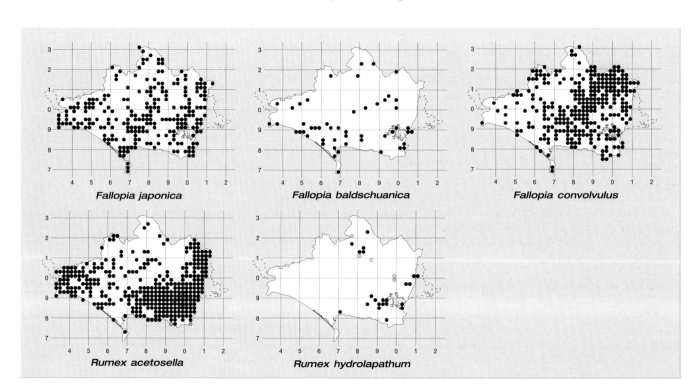

Fallopia japonica

Fallopia baldschuanica

Fallopia convolvulus

Rumex acetosella

Rumex hydrolapathum

Fallopia japonica × sachalinensis = F. × bohemica
(Chrtek & Chrtkova) J. Bailey
Reported from Lyme Regis (39G, 1991, AWJ) and perhaps
overlooked elsewhere. 1.

Fallopia sachalinensis (F. Schmidt ex Maxim) Ronse
Decraene
GIANT KNOTWEED
Planted in Abbotsbury gardens in 1899 (MI), now only
found in grounds of large estates. 4. Forde Abbey (30M,
1992, !); Over Compton (51Y, 1996, !); Milton Abbas
(80A, 1989–99, !); Green Is (08D, 1993, !).

Fallopia baldschuanica (Regel) Holub
RUSSIAN-VINE
A vigorous climber escaping from gardens, and usually
seen overgrowing hedges on all types of soil. 51. Its
habit makes it hard to tell whether the plant originates
inside or outside a garden, but it is established on
undercliffs at Sandsfoot (67T) and Kimmeridge (97E).
First record: Winterborne Whitechurch (80F, 1980, !).

Fallopia convolvulus (L.) A. Love
BLACK-BINDWEED
Frequent in arable, especially on calcareous soil, also on sandy
beaches and tips; known since neolithic times. 294. Var.
subalatum Lej. & Court, confused with *F. dumetorum*, is rare:
Portland (67, 1912, WBB); Durdle Door caravan site (88A,
1999, !); Wallis Down (09S, 1900, EFL); Stanbridge Mill
(00E, 1996, !). Partridges eat the seeds (Wilson, Arroyo &
Clark, 1997).

Fallopia dumetorum (L.) Holub
COPSE-BINDWEED
A rare and sporadic climber along woodland margins and
old hedges on sandy soils, probably restricted to the north-
east. 2 + [5]. Ensbury (09T, 1874, JH Austen); Horton
(00I, 1937, DM); Lower Mannington (00S, 1928, FHH and
1993, GDF and JO); between Hill Farm and Perry Copse
(11G, 1991, JO, also HCP & ! in **RNG**). Reports from
Sherborne Park, Corfe and Colehill are best treated as
doubtful.

Muehlenbeckia complexa (Cunn.) Meissner
WIREPLANT
Locally dominant as a garden escape on sandy cliffs at
Branksome Chine (08U and 09Q, 1992–96, RMW, !); also
further east on similar cliffs in VC11. 3. Planted in
Abbotsbury gardens in 1899 (MI) and still there in 1993.

Rheum × hybridum Murray
RHUBARB
Much grown in gardens, but only once seen outside them.
Jerusalem Hill (61T, 1970, ACL). [1].

Rumex acetosella L.
SHEEP'S SORREL
Common in heaths, sandy grassland, acid fallow and
arable. 334. Occasional in leached grassland overlying

chalk, but absent from clay regions. Subsp. *acetosella* is the
commonest form, rarely as var. *tenuifolius* (Wallr.) Love on
very acid heaths in the Poole basin. Subsp. *pyrenaicus*
(Pourr.) Akeroyd may be widespread but rarely noticed,
e.g. Scotland Heath (98S, 1953, CCT in **RNG**); Parkstone
(09F, 1927, LBH in **RNG**).

Rumex acetosa L.
COMMON SORREL
Common in old grassland, sometimes giving a whole field
a red tinge. 683. Usually in richer turf than *R. acetosella*,
and absent from very acid soils.

[***Rumex salicifolius*** Weinm.
WILLOW-LEAVED DOCK
A casual, once found at Swanage Camp (08F, 1916–17, CBG in
BMH and **OXF**).]

Rumex hydrolapathum Hudson
WATER DOCK
Rather local in ditches and wet places in the valleys of the
Stour, Frome, Piddle and Moors Rivers, also in brackish
dune slack pools at Shell Bay. 22 + [5]. It has decreased
or disappeared from the Stour between Shillingstone and
Wimborne.

Rumex hydrolapathum × crispus = R. × schreberi
Hausskn.
Rare in ditches or cliffs. 2 + [1]. Redcliff (98I, 1989, ! in
RNG); Rockley sands (99Q, 1999, ! det. JRA); St Leonards
(10B, 1958, AK Harding).

Rumex hydrolapathum × obtusifolius = R. × lingulatus
Jungn.
Rare in wet places. [3]. Culeaze (89L, 1930s, RDG);
Stoborough (98I, 1911, CES); dredgings north of Wareham
(98I, 1980, AH & !); Swanage and Little Sea (07J and 08H,
1915, CBG in **BMH**).

Rumex crispus L.
CURLED DOCK
Very common in disturbed grassland, verges, waste places
and tips. 719. Subsp. *litoreus* (J Hardy) Akeroyd is common
on some beaches such as the landward side of the Chesil
Bank and inside Poole Harbour. Subsp. *uliginosus* (Le Gall.)
Akeroyd should be sought on muddy beaches. As with all
docks, the leaves are a source of oxalic acid, while the root
contains anthraquinones and has been used to alleviate
constipation.

Rumex crispus × conglomeratus = R. × schulzei Hausskn.
Only once recorded; Crab Orchard (00Y, 1990, AJB and
DAP). 1.

Rumex crispus × sanguineus = R. × sagorskii Hausskn.
Occasional in waste places. 6. Snipe Gate (68G, 1995, !);
Redcliff Point (78A, 1999, ! det. JRA); Knighton pits (78P,
1997, ! det. JRA); Scratchy Bottom and Wool (88A and N,
1999, !); Pamphill Common (90V, 1989, AJB).

Rumex crispus × pulcher = R. × pseudopulcher Hausskn.
Rare in waste places near the sea. 1 + [1]. Abbotsbury (58S, 1971, AWG in **BDK**); Fleet (68F, 1995, !).

Rumex crispus × obtusifolius = R. × pratensis
Mert. & Koch
A fairly common, but overlooked hybrid of waste ground and tips. 25 + [10]. JCMP gave 10 localities in the 1890s, and recent records show where experienced recorders, such as IPG, have been.

Rumex conglomeratus Murray
CLUSTERED DOCK
Frequent in disturbed grassland and verges, often in marshy places and on mud by ponds. 488. It is less common than *R. sanguineus* and less tolerant of shade.

Rumex conglomeratus × pulcher = R. × muretii Hausskn.
Rare in Purbeck. 1 + [2]. Chapmans Pool (97N, 1980, AH & !); Corfe (98R, 1911, CES); Langton Matravers (07E, 1990, ! in **RNG**).

Rumex conglomeratus × obtusifolius = R. × abortivus
Ruhmer
Rare. 2 + [1]. Wool (88N, 1997, CDP and DAP, det. JRA); Challow Farm (98R, 1943, AEA Dunston in **RNG**); Greenland Farm (08C, 1999, ! det. JRA).

Rumex sanguineus L.
WOOD DOCK
Very common in woods, plantations, hedgebanks and verges. 673. Var. *sanguineus*, Blood-veined Dock, has been a garden plant since 1782 and occasionally escapes, e.g.

Wareham (98I, 1929, LBH); Swanage (07J, 1915, CBG in **BMH**); West Moors (00S, 1995, !).

Rumex sanguineus × obtusifolius = R. × dufftii Hausskn.
Rare but probably under-recorded. 2 + [1]. Birdsmoor Gate (30V, IPG); Higher Halstock Leigh (50D, IPG and PRG); Corfe (98, 1943, AEA Dunston in **RNG**).

Rumex rupestris Le Gall
SHORE DOCK
Very rare, sporadic and perhaps extinct on the strandline of shingle beaches. 2 + [3]. Lyme Regis (39L, 1923, AWG in **BDK**); West Bay (49Q, 1949, AWG in **BDK**); Ringstead Bay (78Q, 1914, CBG in **BMH**, *et al.*, and 1985, PJ Wilson, !); Durdle Door (88A, 1963, ! and 1985, PJ Wilson). Its decline must be ascribed to human pressure on beaches; its British distribution is Cornubian.

Rumex pulcher L.
FIDDLE DOCK
Local in old grassland, or on verges, especially along the coastal strip including Portland. 49 + [14].

Rumex pulcher × obtusifolius = R. × ogulinensis Borbas
Rare in the coastal strip. [3]. Burton Bradstock (48Z, 1949, AWG in **BDK**); Chapmans Pool (97N, 1980, !); Corfe (98R, 1943, AEA Dunston in **RNG**).

Rumex obtusifolius L.
BROAD-LEAVED DOCK
The commonest Dorset Dock, found in disturbed grassland, hedgebanks, verges and tips. 727. Its leaves are often discoloured by the fungus *Venturia rumicis*, and it is

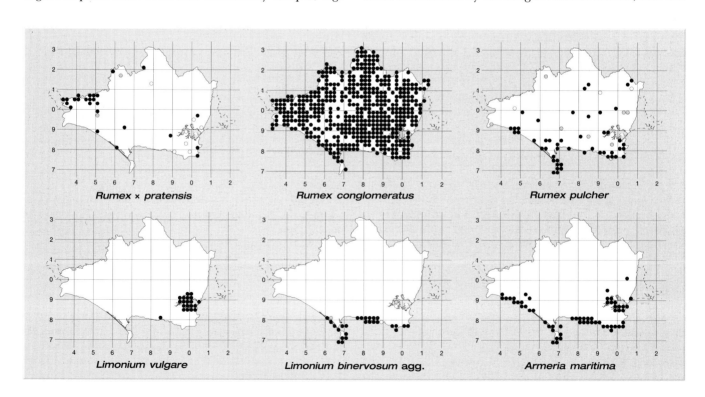

Rumex × pratensis

Rumex conglomeratus

Rumex pulcher

Limonium vulgare

Limonium binervosum agg.

Armeria maritima

a foodplant for many moths. William Barnes wrote a poem on the dock.

Rumex maritimus L.
GOLDEN DOCK
Sporadic in brackish marshes and ditches, appearing at long intervals after disturbance. 1 + [5]. Bexington (58, 1937, DM); Radipole Lake (68Q, 1926, MJA, *et al.*, and 1960, RGB Roe, !); Weymouth (67, 1832, R.Blunt in **WAR**); Lodmoor (68V, 1883, JCMP in **DOR**, *et al.*, and 1999, CDP, !); Parkstone (09, 1929, LBH).

PLUMBAGINACEAE

Limonium vulgare Miller
COMMON SEA-LAVENDER
At present this is locally frequent to abundant in muddy salt-marshes inside Poole Harbour and its islands. 21 + [5]. Very rare elsewhere, but it may once have been more common on beaches, as old reports suggest: Abbotsbury Swannery (58S, 1840s, MF); Weymouth (67, 1799, RP and 1840s, Miss Cotton in **NCH**); Osmington Mills (78F, pre-1990, CA Stace); Lulworth Cove (88, 1799, RP and 1953, AWW); Arish Mell (88K, 1992, DAP & !); Swanage (07J, 1799, RP and 1954, SJ Langford). A white-flowered form occurs near Jerry's Point (08I). Dead flower-heads are sold as 'white heather' in Dorchester by gypsies.

(**Limonium humile** Miller has been reported several times from inside Poole Harbour, but never confirmed; it is hard to tell from **L. vulgare**.)

Limonium hyblaeum Brullo
ROTTINGDEAN SEA-LAVENDER
Introduced by gulls from a garden in the 1980s, now established on a sea wall and clay cliffs at West Bay (49K, 1992–98, ! in **RNG**). 1.

Limonium binervosum (G.E. Sm.) C.E. Salmon agg.
ROCK SEA-LAVENDER
Locally frequent on chalk and limestone cliffs, concrete sea walls and salt-marshes. The three subtaxa, two of them endemic, are hard to distinguish.

Limonium dodartiforme Ingrouille
This occurs on shingle on the landward side of the Chesil bank, and on cliffs between White Nothe and Durlston CP. 19. First record: Lulworth Cove (87J, 1848, JWoods). Specimens are in **BM, DOR, LTR** and **RNG**. In a salt-marsh near Portland Harbour it grows with *L. recurvum*.

Limonium recurvum C.E. Salmon
PORTLAND SEA-LAVENDER
Locally frequent on Portland and its harbour breakwaters, on limestone cliffs, also by brackish ponds below East Weares. 8. First collected by T Velley (1790 in **LIV**), and first distinguished as 'L. dodartii' by Babington in 1862. Specimens in **BDK, BM, BMH, DOR, LIV, LTR, RNG**,

WAR, WRHM etc. According to Stace & Ingrouille, two subpecies occur. Subsp. *recurvum* Salmon is endemic to Portland Bill. Subsp. *portlandicum* Ingrouille includes the salt-marsh plants and those on Portland breakwaters, and is also found in Kerry.

Limonium 'obesifolium' sp. nov.
Recently detected in material collected by CDP and DAP from limestone cliffs at Durlston CP (07D) and Tilly Whim (07I). 2. There are older reports of cliff plants from this stretch of coast, e.g. Seacombe (97Y, 1970s, R Jennings); Dancing Ledge (07D, 1900, EFL and 1981, RB); Durlston Head (07I, 1995, IPG and PRG), which may be this or *L. dodartiforme*. The *Limonium binervosum* specimens from Arne bay salt-marshes (98U, 1895, JCMP in **DOR** and 1937, RDG in **WRHM**) and Middlebere (98T, 1936, RDG in **WRHM**) have not been critically examined.

Armeria maritima (Miller) Willd.
THRIFT, LADY'S CUSHION
Locally frequent to dominant along the coast on cliffs, cliff-edges, fixed shingle and salt-marshes. 59. It makes an impressive mass of colour at Small Mouth in May, where a few clumps have white flowers. Rare on dry hilltops and verges inland: Ironstone ridge, Abbotsbury (58S, 1978, !); B3157 at Burton Common (58E, 1995, IPG and PRG); A35 at Askerswell (59G and K, 1990, !); Ferndown by-pass (00K, 1997, !). Variable in habit.

Armeria arenaria (Pers.) Schult. agg.
JERSEY THRIFT
Reported from dunes at Studland (08G, 1985, CM Lovatt), but no specimen seen. (An *Armeria* with large pink or white flowers, referred to *A. latifolia* hort. by Stace, is frequent on sandy cliffs at Bournemouth (09V, 19F, !) but in VC11).

Ceratostigma plumbaginoides Bunge
An escape on a wall at Portesham Ch (68C, 1991, ! in **RNG**). 1.

PAEONIACEAE

Paeonia officinalis L.
GARDEN PEONY
A garden escape, or planted and persisting. 5. Castleton Ch (61N, !); Cowherd Shute Pond (82S, !); Kingston (97P, !); Glebelands estate (08F, !); Woodlands (00P, GDF).

ELATINACEAE

Elatine hexandra (Lapierre) DC.
SIX-STAMENED WATERWORT
Very local in shallow pools with sandy or peaty bottoms. 3 + [1]. Corfe Common (98Q, 1985–86, DAP and WGT); Little Sea (08H, 1928, LBH in **CGE** and **RNG**, *et al.* in **BDK, RNG, NMW** and **WRHM**, and 1998, AH, also 08G, 1999, JHSC); Eastern lake (08H, 1991, AJB); Poole (09, 1937, DM).

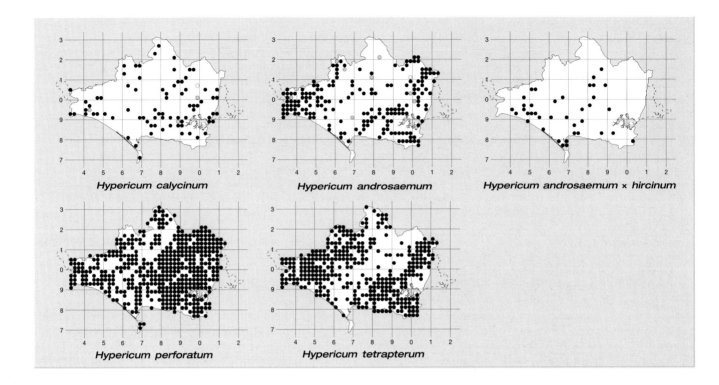

Hypericum calycinum

Hypericum androsaemum

Hypericum androsaemum × hircinum

Hypericum perforatum

Hypericum tetrapterum

(*Elatine alsinastrum* L., reported from Wytch in 1985, was probably *Crassula helmsii*.)

CLUSIACEAE

Hypericum calycinum L.
ROSE-OF-SHARON
Often planted on verges, in village lanes, and sometimes on undercliffs, but rarely naturalised. 76 + [4]. A persistent garden plant which tolerates shade, drought and the drips from trees, which has certainly increased since its first record in 1874.

Hypericum calycinum × *cyathiflorum* Cv. Hidcote
Much grown, but rarely outside gardens; in plantations. 5. Dottery (49M, !); Giddy Green (88I, !); Milborne St Andrew (89D, !); Bere Heath (89K, 1990, ! in **RNG**); Lodden Lakes (82C, 1999, !).

Hypericum androsaemum L.
TUTSAN
Widespread but rarely in quantity in old woods and hedgebanks. 180 + [9]. It is sometimes found in wet woods or carr, and tolerates calcareous and acid soils. Absent from deforested areas and the drier parts of the chalk ridge.

Hypericum androsaemum × *hircinum* = *H.* × *inodorum* Miller
TALL TUTSAN
Naturalised in village lanes, large estates, wild gardens and churchyards. 36. The fruit is sought out and eaten by pheasants, which may help to spread the plant.

Hypericum hircinum L.
STINKING TUTSAN
An introduced shrub, long established on shaded banks and stonework by the River Wey, also along the old railway at East Weares, rarely elsewhere. 10. On sale in Blandford in 1782. Old railway, Sandsfoot to Small Mouth (67T, 1991, !); North Portland (67W, 1980, !); Portesham (68C, 1952, AWW and 1981, J Fryer); Upwey (68S, 1864, SMP in **DOR**, *et al.*, and 1960–98, ! in **RNG**); also Broadwey, Nottington and Radipole (68L, R and Q); Weymouth (68V, 1980, !); Grove (77B, 1973, RCP and 1974–94, ! in **RNG**); one bush, Lake (99V, 1995, !).

Hypericum perforatum L.
PERFORATE ST JOHN'S-WORT
Widespread and common in woodland clearings, hedgebanks, dry grassland and verges. 461. Var. *angustifolium* Gaud. is noted occasionally, but is little more than a form induced by poor nutrition. The plant contains hypericin and is poisonous to farm stock; this does not prevent its recommendation and sale by some herbalists.

Hypericum perforatum × *maculatum* = *H.* × *desetangsii* Lamotte
Near Sherborne Sta (61N, 1999, ! conf. NKB Robson).

Hypericum maculatum Crantz subsp. *obtusiusculum* (Tourlet) Hayek
IMPERFORATE ST JOHN'S-WORT
Rare and decreasing in damp woodland rides and marshes. 8 + [17]. Some records need confirming. Bowshot and Prime Coppices (39M and V, 1968, JB and D Kennedy);

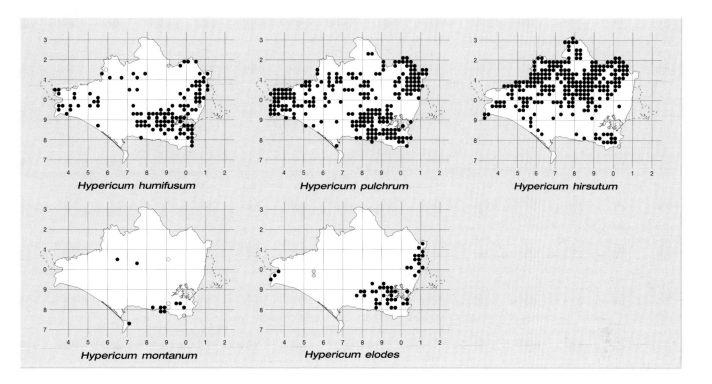

Hypericum humifusum

Hypericum pulchrum

Hypericum hirsutum

Hypericum montanum

Hypericum elodes

Parnham (40Q, 1979, D Griffith); Edmundcombe (40V, 1947–63, AWG in **BDK**); Came Down (68X, 1977, JKH); Warmwell (78T, 1997, FAW); Worbarrow Tout and Mupe Bay (87P and U, 1932, RDG); Corfe Common (98K, 1960, BNSS); The Moors (98N, 1987, AH, !);Woolgarston Wood (98Q, 1992, JB); Corfe Sta (98R, 1991, AH); Simmons Bottom (99H, 1968, AH Dunn); Spetisbury (90A, 1988, BM Hotchkiss); Pimperne (90D, 1934, RDG in **WRHM**); West Moors (00R, 1989, FAW); Denbose Wood (01E, 1991, GDF); Noon Hill (10E, 1997, FAW); Daggons Road (11B, 1891, TRA Briggs).

Hypericum tetrapterum Fr.
SQUARE-STALKED ST JOHN'S-WORT
Frequent in wet woodland rides, marshes and by ditches and streams. 361. Absent from the dry chalk ridge and Portland.

Hypericum humifusum L.
TRAILING ST JOHN'S-WORT
Occasional, rarely in quantity, in bare, sandy soil of woodland clearings or heath tracks, or in bare, wet places on clay. 119 + [2]. Sometimes a weed in Nursery gardens, or by rail tracks. Mostly in the Poole basin and the north-west, absent from the chalk and deforested areas.

Hypericum pulchrum L.
SLENDER ST JOHN'S-WORT
Characteristic of dry, shaded banks at the edges of woods or on heaths, but also in mature, leached chalk grassland in the coastal belt east of White Nothe. 230. Least common on the Chalk and not on Portland.

Hypericum hirsutum L.
HAIRY ST JOHN'S-WORT
Common in dry woodland rides and hedgebanks on the chalk and calcareous clays. 259. It does not grow on acid soils, appears to avoid the coast, and is scarce in deforested areas.

Hypericum montanum L.
PALE ST JOHN'S-WORT
Scarce and local on dry calcareous soils, often in scrub as it tolerates some shade. 12 + [4]. The headquarters of this plant have always been on Portland and near Corfe; reports from the north-central Chalk are doubtful. It has not been seen lately at three old sites: Seacombe (97Y, 1933, RDG in **WRHM**); Creech Grange Wood (98A or B, 1799, RP, *et al.*, and 1929, AWG); Old Harry (08L, 1911, CES).

Hypericum elodes L.
MARSH ST JOHN'S-WORT
Very locally frequent to dominant in bog pools, acid streamlets and ditches and old claypits. 48 + [2]. Not uncommon in the Poole basin, rare in the west and not seen since 1968 in the Kingcombe/Powerstock Common region.

TERNSTROEMIACEAE

Camellia japonica L.
Occasionally planted in big estates and surviving among semi-native vegetation on acid soils. 3. Charborough Park (99I, BE & !); Branksome Chine (08U, !); Heathy How (00Y, BE). There is a commercial nursery at Hampreston.

TILIACEAE

Tilia platyphyllos Scop.
LARGE-LEAVED LIME

Although probably present in neolithic times, this is usually seen today as single trees in plantations, where it is rare. 14 + [1]. Two records may represent native trees: one coppiced tree among ash and hazel, Charborough Park (99I, 1997, BE); Oakhills Coppice (90Z, 1988, NCC). An avenue was planted at Kingston Lacy (90Q) in the 17th century. There are large planted trees at Melbury Park (50S; 22 m × 3 m), Turnworth (80D), Bryanston (80T), Upton CP (99W) and Eastbury Ho (91G).

Tilia platyphyllos × *cordata* = *T.* × *europaea* L.
LIME, LINDEN

Commonly planted in plantations and parks, rarely in old woods, e.g. Broadley Wood (80M). 291. There are avenues of old trees at Child Okeford (81G) and Canford School (09J), but the avenue at Kingston Maurward has mostly died, and avenues in Dorchester are disfigured by pollarding; other avenues are at Rogers Hill Farm (89H) and Langbourne (90E). A tree dwarfed by exposure to wind grows in the old churchyard at Church Ope Cove (67V), and there is a tree 30 m × 6.4 m at Ranston Ho (81R); another large tree is planted at The Cross, Winterborne Stickland (80H). A foodplant for larvae of the White-letter Hairstreak and several moths. "Linden Lea" is perhaps William Barnes' most famous poem.

Tilia cordata L.
SMALL-LEAVED LIME

Local, sometimes coppiced, in old woods and hedges, mostly in the north-east. 54 (22 + native). Pollen is known from 5000 to 1900BC from Rimsmoor Pond etc. The tree was found in Cranborne Chase (81, 1691, J Aubrey) and Holt Forest (00, 1595, J Pointer); it survives in both. Recently, many saplings have been planted in plantations, parks and reserves. At Forde Abbey (30M) is a tree 24 m × 1.65 m, and other large trees are at Birkin Ho and Kingston Maurward (79A) and Burwood (01S). Four Dorset place names may be derived from 'lind', though confusion with 'lin' = flax is possible.

Tilia tomentosa Moench
SILVER-LIME

Occasionally planted in parks and churchyards, usually as Cv. *petiolaris*. 22. There is a short avenue of big trees at Almer (99E), and large trees have been seen at Little Bredy (58Z), Dorchester (69V), Herringston (69X), Stinsford (79A), Melbury Park (50S) and Kingston Lacy (90Q).

MALVACEAE

Malva moschata L.
MUSK-MALLOW

Widespread, rarely in quantity, in dry woodland rides and verges. 125. Tolerant of chalk soils, but not confined to them, absent from deforested areas and infrequent near the sea. Often planted in gardens. White-flowered forms were seen at Red Bridge, Moreton (78U), Corfe Mullen (99U) and Canford Heath tip (09D and I).

Malva sylvestris L.
COMMON MALLOW, CHEESES

Common by farm buildings, in village lanes and hedgebanks and on cliffs. 467. It prefers disturbed sites on dry, rich soils and is tolerant of salt-spray. Var. *lasiocarpa* Druce, with hairy seeds, was found at Worth Matravers (97T, 1930s, NDS).

Malva parviflora L.
LEAST MALLOW

Rare and mostly casual by farmyards or on tips. 1 + [4]. Lamberts Castle (39U, 1928, AWG); Chickerell tip (68K, 1980, !); Middle Farm, Dorchester (69Q, 1955, !); Hinton St Mary (71Y, 1989, !); Poole (09, 1920s, JHS).

Malva pusilla Smith
SMALL MALLOW

A rare casual of disturbed soil. 2 + [9]. Some recorders may have confused this with *M. neglecta*. Golden Cap and Beaminster (49B and 40, 1961, AWG in **BDK**); Netherbury (49U, 1928, M Moores); Melbury Bubb (50Y, 1978, P Symons in **RNG**); Small Mouth verge (67S, 1987, !); Dorchester (69Q, 1955, !); Swanage Camp (08F, 1917, CBG in **BMH**); Greenland Farm (08C, 1999, !); Goathorn

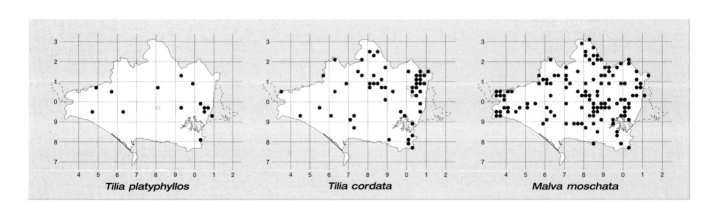

Tilia platyphyllos   Tilia cordata   Malva moschata

(08D, 1917, CBG); Lilliput (08P, 1930s, NDS); Edmondsham (01Q, 1909, EFL).

Malva neglecta Wallr.
DWARF MALLOW
Rather uncommon in small quantity in rich, disturbed soil by farmyards and on verges. 27 + [18]. Mostly in the Poole basin and scarce on the Chalk.

[Malva verticillata L.
CHINESE MALLOW
A casual once found at Gotham (01V, 1925, JHS).]

Lavatera arborea L.
TREE-MALLOW
Occasional all along the coast on hard soil of beaches and cliffs near the sea, where it is sporadic. 45 + [6]. Inland it is a rare casual on tips or a garden escape. First record: Chesilton (67W, 1670, J Ray) where it still occurs. Seedlings, and even mature plants, are killed by hard frosts.

Lavatera cretica L.
SMALLER TREE-MALLOW
A rare denizen with a Cornubian distribution, flowering much earlier than *Malva sylvestris*, which it resembles. 1 + [1]. Near buildings, Rodwell (67Z, 1992–96, !); between Wareham and Stoborough (98I, 1884, JCMP in **DOR**).

Lavatera olbia × thuringiaca
GARDEN TREE-MALLOW
This floriferous, but short-lived garden plant has been seen increasingly on verges near houses since 1991. 19. It is usually propagated from cuttings, as the seedlings do not come true, suggesting hybridity.

Lavatera trimestris L.
ROYAL MALLOW
A rare, showy, casual garden escape. 1 + [1]. Beaminster (40V, 1921, AWG in **BDK** as *Malope hispida*); by new hedge, Milborne St Andrew, with pink and white flowers (89D, 1992, !)

Althaea officinalis L.
MARSH-MALLOW
Very local in brackish marshes and ditches along the landward side of the Fleet. 7. In the 18th century, Pulteney reported it from the north side of Poole Harbour, and it still occurs to the east in Stanpit Marsh (19; VC11). Abbotsbury Swannery (58S, 1895, SMP in **DOR**, *et al.*, and 1989, DAP, !); West Fleet (58W, 1994, SME); East Fleet (67J, 1989, DAP); Wyke Regis (67N, 1992, !); north of Ferry Bridge (67T, 1985, RA); Langton Herring, Herbury and Rodden Hive (68A and B, 1895, JCMP, *et al.*, and 1991, AJB, !). An extract from the root was once used as a cough medicine.

[Althaea hirsuta L.
ROUGH MARSH-MALLOW
Casual, but not seen this century. 2. Weymouth railway (67, 1892, JCMP in **DOR**); arable, Dunbury (80G, 1889, JCMP in **DOR**).]

Alcea rosea L.
HOLLYHOCK
An occasional garden escape, persistent for a few years on verges and tips. 8 + [1]. Lodmoor tip (68V, 1978 and 1997, !); Dorchester by-pass (68U, !); Dancing Hill (61H, !); Caundle Marsh (61R, 1968, ACL); Corfe (98K, !); Post Green (99L, !); France Firs (80Z, !); Okeford Fitzpaine (81A, !); Alderholt (11G, GDF).

Abutilon theophrasti Medikus
VELVETLEAF
A casual of maize or buckwheat fields, or on tips. 4. Mosterton (40M, 1990, JGK); Puncknowle (58J, 1989, F Woodhouse in **RNG**); Chamberlaynes (89L, 1989, ! in **RNG**); Tatchells Pit (98E, 1994, !).

Hibiscus trionum L.
BLADDER KETMIA
A rare casual. 2 + [1]. Sherborne tip (61I, 1970, ACL); Greenland Farm (08C, 1999, EAP); Ferndown (00Q, 1995, PD Stogden).

SARRACENIACEAE

Sarracenia purpurea L.
PITCHERPLANT
Planted in a bog at Hyde Heath (89K, 1998, PH Sterling). 1. Spreading, over 80 clumps present.

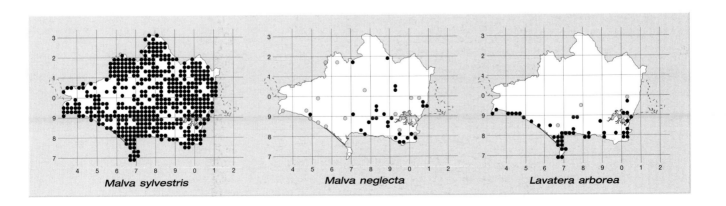

Malva sylvestris   Malva neglecta   Lavatera arborea

DROSERACEAE

Drosera rotundifolia L.
ROUND-LEAVED SUNDEW
Locally frequent in open bogs on peat or *Sphagnum*, or by wet sandy tracks. 97 + [1]. Mostly in the Poole basin, with outliers at Charmouth Forest (39), Pilsdon Pen (40A), Abbotsbury Castle (58N, P Stevens) and Mount Pleasant (59P, SME). It benefits from light grazing by deer, which help to prevent its habitat become overgrown by *Molinia*.

Drosera rotundifolia × intermedia = D. × beleziana G. Camus.
In a bog near Agglestone (08G, 1999, DAP). 1.

Drosera rotundifolia × longifolia = D. × obovata Mert. & Koch.
A rare hybrid in boggy pools, now only in the Poole basin in very small numbers. 4 + [1]. [Champernhayes (39N, 1930, AWG in **BDK**)]; Mare Pond (88W, 1994, EDV Prendergast); Great Knoll, Hartland (98M, 1990, AJB *et al.*); Arne Heath (98U, 1993, B Pickess); Little Sea and Spur Bog, (08H, 1895, HNR and WF, *et al.*, and 1987, AH).

Drosera longifolia

Drosera longifolia L.
GREAT SUNDEW
In small quantity in very wet bogs, rare in *Cladium* fen, and colonising mature flooded claypits in the Poole basin; extinct in the west. 12 + [7]. Some old records may be confused with *D. intermedia*. There are specimens in **BDK**, **DOR** and **WRHM**.

Drosera intermedia Hayne
OBLONG-LEAVED SUNDEW
Locally frequent on bare, wet peat of heaths and in *Sphagnum* bogs, sometimes surviving in wet rides in conifer plantations. 80 + [5]. Confined to the Poole basin except for a single outlier in the west: Charmouth Forest (39U, 1988, NCC). Favoured by light grazing or disturbance.

CISTACEAE

Helianthemum nummularium (L.) Miller
COMMON ROCK-ROSE
Rather local in mature calcareous grassland, or at the edges of woods on chalk. 147 + [2]. Occasionally dominant or abundant, as at Turnworth (80J) and Stubhampton Bottom (81Y). Secretions from this shrub inhibit germination of other plants. Rare on a heathy verge north of Bovington Camp (88J), which also has *Blackstonia*. Hairy plants have been seen above Ringstead (78L, 1960, !) and at Corfe (98R, 1930s, NDS). An orange-flowered form occurred at Corfe (98L, 1980, AJS).

Several species of *Cistus* are planted on verges and churchyards, or persist in abandoned gardens. Most are frost-resistant hybrids such as **C. × corbariensis** at Sydling St Nicholas (69J) and Blue Pool (98G) or **C. × skanbergii** at Abbotsbury (58S) and Milborne St Andrew (89D, 1980–99). [**C. albidus** L. was at Sandbanks (08N, 1958, !), now built over.]

VIOLACEAE

Viola odorata L.
SWEET VIOLET
Widespread and frequent in plantations, hedgebanks, village lanes and churchyards on rich soils that are none too acid. 374. It is probably native, though uncommon in old woods, and a local practice was to plant this when installing a new gate or fence posts. The white form, var. *dumetorum* Jordan, is common, while the purple (var. *typica* G.Beck) and dull pink (var. *subcarnea* Jordan) forms are scarcer.

Viola odorata × hirta = V. × scabra F. Braun
Probably overlooked, though the parents do not often meet. 2 + [6]. Recently seen at Beaminster (40V, 1993, LJM) and Dancing Hill (61H, 1983, AJCB); specimens in **BDK** and **DOR** came from Marshwood (39U), Bettiscombe (39Z), Muckleford (69L) and Higher Waterston (79H).

Viola hirta L.
HAIRY VIOLET
Frequent in short calcareous turf. 191. The forma *calcarea*

(Bab.) EF Warburg, with starved flowers, has not been noticed recently, but there are many old records.

Viola hirta Reichb.
COMMON DOG-VIOLET
Our commonest violet, found in most old woods and hedgebanks, and in open grassland in the west; a foodplant for several Fritillary butterflies. 564.

Viola riviniana × reichenbachiana = V. × bavarica
Schrank
In light shade at Badbury Rings (90R, 1998, SP Chambers); first found by Good in the 1930s.

Viola riviniana × canina = V. × intersita G. Beck
EF Linton found this in three heathland sites in 1900; since then there have been three reports; Maiden Castle (68U, 1938, SA Taylor); Hartland Moor (98M, 1960, HJB); Pentridge Hill (01I, 1933, PMH). [6].

Viola riviniana × lactea
Two records from the Poole basin; Winfrith Heath (88D, 1960, HJB); Stoborough Heath (98H, 1933–34, AWG in **BDK** and JEL in **RNG**). [2].

Viola reichenbachiana Jordan ex Boreau
EARLY DOG-VIOLET
Fairly common in old woods and hedgebanks, often with *V. riviniana* but flowering earlier. 225. Absent from deforested areas, except in certain churchyards, where it is locally abundant under yew.

Viola canina L. subsp. canina
HEATH DOG-VIOLET
Occasional in tracks, firebreaks and bare, grassy places on heaths; rarely in short chalk turf on steep slopes. 30 + [12]. Old records were much confused with *V. riviniana*, and different botanists may still interpret these species in different ways.

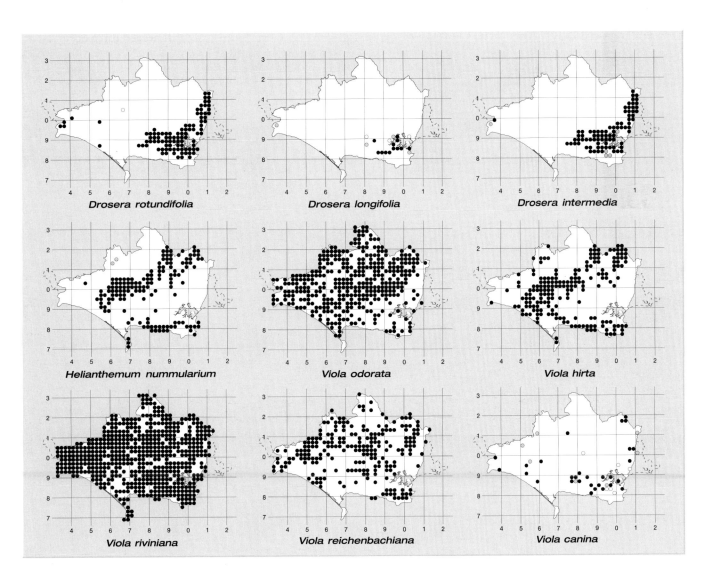

Drosera rotundifolia — Drosera longifolia — Drosera intermedia — Helianthemum nummularium — Viola odorata — Viola hirta — Viola riviniana — Viola reichenbachiana — Viola canina

(*Viola canina* × *lactea* is doubtfully present, as old reports were probably *V. riviniana* × *lactea*.)

Viola lactea Smith
PALE DOG-VIOLET
Sporadic, rare and in small quantity in dry, grassy heaths, often after disturbance. 14 + [12]. Mostly in the Poole basin with outliers at Clarkes Gorse (50C, 1933, AJ Holloway in **BDK**); Holnest (61K, 1960, CT); Cranborne Common (01V

and 11A, 1960, KBR in **DOR** and 1996, BE). Its British distribution is mainly Cornubian.

Viola palustris L. subsp. *palustris*
MARSH VIOLET
Local in alder-swamps, sallow carr and sheltered marshes. 35 + [10]. As noted by Good, this has a disjunct distribution pattern with colonies in the west and the Poole basin, but none on the Chalk.

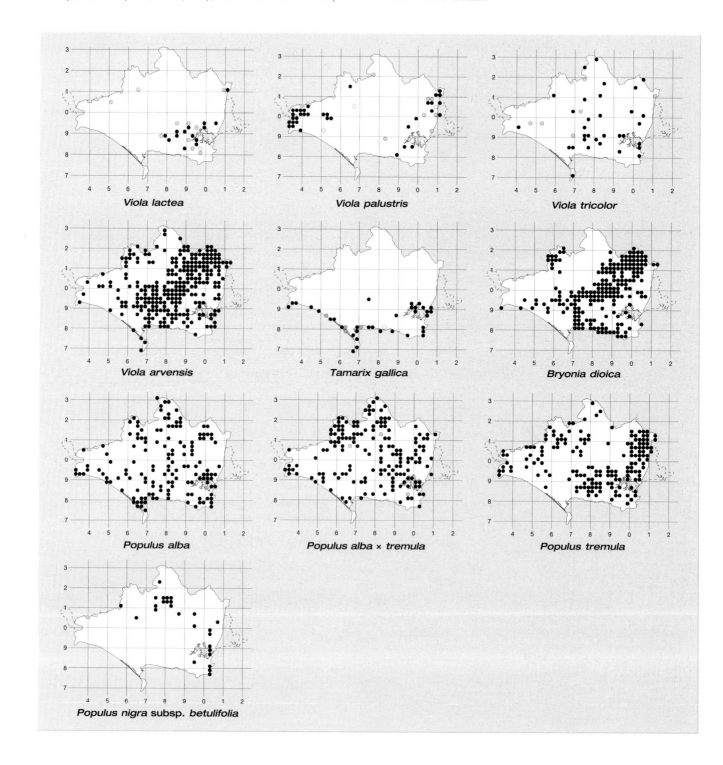

Viola × wittrockiana Gams ex Kappert
PANSY
A casual escape from gardens, where it is much grown, on waste ground or tips. 13.

Viola tricolor L.
WILD PANSY
An occasional arable weed, more often a casual in allotments or gardens and tips. 33 + [6]. It grows on both sandy and chalky soils, and in fallow fields. Good did not distinguish this from *V. arvensis*.

V. tricolor × arvensis = V. × contempta Jordan
Once found in a heathy field near Morden Decoy (99A, 1900, EFL), but easy to overlook. [1].

Viola arvensis Murray
FIELD PANSY
Frequent in dry arable fields, especially on the Chalk, but also on sandy soils. 265. Good had *V. tricolor* agg. in 139 stands, so it has not decreased. Variable in flower size and also in colour, which is predominantly pale yellow with different patterns; cream and purple flowers can occur on the same plant.

Viola labradorica Schrank
A garden plant which is likely to spread. 1. Abundant in wild garden, Litton Cheney (59K, !). The leaves are dark purple underneath.

TAMARICACEAE

Tamarix gallica L.
TAMARISK
More or less obviously planted all along the coast, including parts of the Chesil bank, the landward shores of the Fleet and inside Poole Harbour on mud, on shingle beaches, clay and sandy cliffs. 34 + [5]. Inland it occurs on waste ground on tips, probably from cuttings, which root easily.

FRANKENIACEAE

Frankenia laevis L.
SEA-HEATH
Recently introduced to the coast. 2. Well established on clay undercliff by gulls who use it for nesting, West Bay (49K, 1992, LMS, !); planted in a sandy rockery, Sandbanks (08N, 1990, !).

CUCURBITACEAE

Bryonia dioica Jacq.
WHITE BRYONY, MANDRAKE
Fairly common in woods, hedges and among block scree, largely restricted to calcareous soils but absent from Portland. 206. The roots, which can grow into large, suggestive shapes, were once used as a fertility charm for horses.

Citrullus lanatus (Thunb.) Matsum. & Nakai
WATER MELON
Casual at Oborne tip (61P, 1971, ACL). [1].

Cucumis melo L.
MELON
Occasional on tips, where it has been found with ripe fruit. 3 + [1]. Sherborne (61H, 1970, ACL and 1989, !); Crossways (78P, 1999, !); Binnegar (88Y, 1994, !); Stourpaine (80U, 1989, !).

Cucumis sativus L.
CUCUMBER
A rare casual on tips or a pavement weed. 3. Crossways tip (78P, 1999, !); Binnegar tip (88Y, 1993–97, !); Canford Cliffs (08U, 1997, !).

Cucurbita pepo L.
MARROW
A casual on tips. 2 + [3]. Lodmoor (68V, 1977, !); Sherborne (61H, 1970, ACL); Crossways tip (78P, 1999, !); Binnegar (88Y, 1992 and 1996, !); Fleets Corner (09B, 1977, !).

Cucurbita maxima Duchesne ex Lam.
PUMPKIN
Seen on Binnegar tip (88Y, 1994, ! in **RNG)** and Canford tip (09I, 1999, !) 2.

SALICACEAE

Populus alba L.
WHITE POPLAR
Widespread in small numbers, as a planted tree in parks, plantations, hedgebanks, by streams, and on cliffs, as it tolerates salt. 144. All poplars are foodplants for the larvae of Poplar Hawk, Eyed Hawk and other moths.

Populus alba × tremula = P. × canescens (Aiton) Smith
GREY POPLAR
Occasional as single trees in wet woods, more often in plantations or by roads. 162 + [3].

Populus tremula L.
ASPEN
Local in wet woods, where it forms colonies by suckering. 164. Absent from deforested areas, also from most of the Chalk and uncommon near the sea.

Populus nigra L. subsp. **betulifolia** (Pursh.) W. Wettst.
BLACK-POPLAR
A relict native tree found in very small numbers, most often near the upper reaches of the River Stour. 30. Female trees outnumber males, and are sometimes pollarded; no fertile seed is recorded. Recent interest has led to this being planted more often, sometimes as a street tree. With galls of *Pemphigus spirothecae* at New Swanage (08F, EAP). Var. *italica* Muenchh. LOMBARDY POPLAR is a conspicuous planted tree. 20+.

Populus nigra × deltoides = P. × canadensis Moench
HYBRID BLACK-POPLAR
Sometimes block planted in wet places, and frequent in plantations and hedges, often as large trees. 278. Possibly the Poplar written about by William Barnes.

Populus trichocarpa Torrey & A. Gray ex Hook.
WESTERN BALSAM-POPLAR
Rarely planted by streams or in parks. 9. Broad Oak (49I, JAG); Round Hill (59X, JAG); Poyntington (61P, JAG); Southover Ho (79W, !); Tomson (89Y, ! in **RNG**); near Durlston CP and New Swanage, 07I and 08F, EAP); Canford School (09J, BNSS); Milham's Brook (09T, FAW); Gotham (01W, GDF).

Populus × jackii Sarg.
BALM-OF-GILEAD, BALSAM-POPLAR
Occasionally planted in large estates and churchyards, mostly in the north. 27 + [2]. On sale at Blandford in 1782, and growing at Abbotsbury in 1899 (MI), first recorded 'wild' by AWG in 1928. It grows on cliffs at Charmouth (39L, LJM).

Other poplars, such as **P. lasiocarpa** Oliver and **P. wilsonii** Schneid, are planted at Canford School (09J, TH).

Salix pentandra L.
BAY WILLOW
As a native this does not occur further south than mid-Wales, but it is sometimes planted in wild gardens from which it escapes as single bushes. 4 + [5]. In wet woods or by rivers as follows: Tadnoll (78Y, 1928, FHH, *et al.*, and 1998, DAP, !); East Burton (88I, 1914, CBG in **BMH)**; Coombe Heath (88S, 1993, DAP and SME); Moreton Heath (89A, 1919, FHH and 1932–37, BNSS). Also in wild gardens at Loscombe (59E); Norden (98L, 1968, MT Horwood); Lytchett Heath (99S); High Hall (00B, 1932, FHH); and Verwood Lower Common (00Y, 1920, AWG in **BDK)**.

Salix fragilis L.
CRACK-WILLOW
Occasional in wet woods, and frequent by rivers, ditches and ponds. 384. Often planted, but rarely pollarded here. Seedlings are frequent on tips. A large tree was seen at Athelhampton (79S).

Salix fragilis × alba = S. × rubens Schrank
Occasionally planted by streams and ditches, once seen in a dune slack. 29.

[*Salix fragilis × triandra = S. × alopecuroides* Tausch
Tarrant Keyneston (90H, 1897, EFL in **LIV**); two trees at Kinson (09T, 1890, EFL in **LIV**).]

Salix alba L.
WHITE WILLOW
Occasional in wet woods, by rivers and streams, and planted in drier sites on verges and railway banks. 197. Big trees seen at Westham (67U) and Bere Regis (89M). Var. *britzensis* has been found near Beaminster (40W, LJM), and the red-twigged var. *vitellina* (L.) Stokes is not uncommon, and conspicuous in winter. Var. *caerulea* (Sm.)

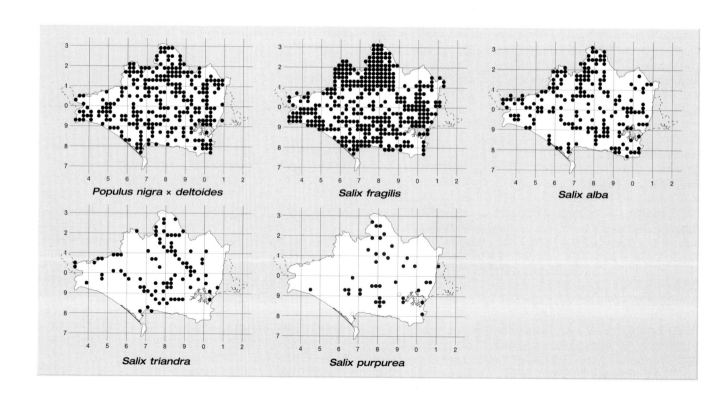

Populus nigra × deltoides

Salix fragilis

Salix alba

Salix triandra

Salix purpurea

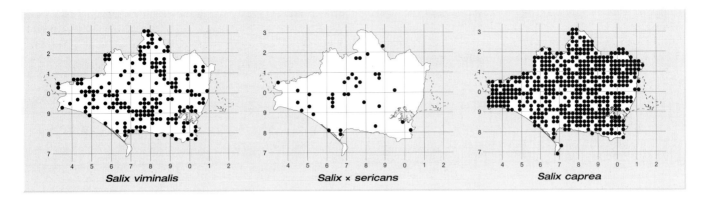

Salix viminalis   Salix × sericans   Salix caprea

Koch, with olive twigs, is probably often planted: Swanage (07J, EAP, det. RDM). As with all willows, the bark contains salicylates.

Salix alba × babylonica = S. × sepulchralis Simonkai
WEEPING WILLOW
Planted in parks, by rivers, mill ponds and lakes, mostly as Cv. *chrysocoma*. 19. In the 1930s a graveyard near Beaminster had a tree said to be grown from one planted on Napoleon's tomb in St Helena.

Salix triandra L.
ALMOND WILLOW
Occasional by rivers, streams, ponds and flooded pits, especially by the Stour. 88. Var. *hoffmanniana* (Sm.) Bab. is recorded from the Frome and Piddle valleys, e.g. Wool (88S, 1960, RDM).

Salix triandra × viminalis = S. × mollissima Hoffm. ex Elwert
Only seen by the Stour at Muscliff (09X and Y, 1992, FAW). 2.

Salix purpurea L.
PURPLE WILLOW
Occasional by rivers, streams and ponds, especially in the valleys of the Stour, Allen and Frome, but not in the west. 41. Sometimes in damp hedgebanks.

Salix purpurea × cinerea = S. × pontederiana Willd.
One bush of this rare hybrid at Knighton Pits (78P, 1997, ! det. RDM). 1.

[**Salix purpurea × viminalis = S. × rubra** Hudson
Though said to be frequent by JCMP, he only gave two sites and it has not been found since.]

[**Salix × forbyana** Smith
A rare triple hybrid. [2]. There are old records by EFL in **LIV** from Trigon Farm (88Z) and Tarrant Crawford Mill (90H).]

Salix daphnoides Villars
EUROPEAN VIOLET-WILLOW
Rarely planted by streams, lakes or as a street tree, first recorded by AWG at Verwood Lower Common (00Y, 1920).

8 + [1]. Recent records are from 67U, 78P, 79X, 89M, 80X, 98E, 08F and 00K.

Salix viminalis L.
OSIER
Frequent by rivers, streams and ditches, also planted in wet woods or withy beds and on wet undercliffs. 197. No longer used for making baskets, and rarely to make lobster pots.

Salix viminalis × caprea = S. × sericans Tausch ex A. Kerner
BROAD-LEAVED OSIER
The commonest sallow hybrid here, but only as single bushes in wet places or hedges. 36.

[**Salix × calodendron** Wimmer
A rare triple-hybrid reported from Spetisbury (as *S. acuminata*) by JCMP (90B, 1874).]

Salix viminalis × cinerea = S. × smithiana Willd.
There are seven 19th century records, but it is rare today. 6 + [9]. Post-1910 records are: Winsham Bridge (30T, IPG); Chapel Court Farm (40N, IPG); Lecher Water (40T, IPG); Holway Copse (61J, 1970, ACL); Sandford Orcas (62F, IPG and PRG); Wool (88N, 1932, FHH); Bloxworth (89X, !); Stourpaine (80U, !); Wareham (98I, 1915, CBG in **BMH**); Langton Matravers (07E, 1914, CBG in **BMH**).

Salix viminalis × repens = S. × friesiana Andersson
Corfe Sta (98R, 1984, ACL det. RDM). 1.

Salix caprea L.
GOAT WILLOW
Present as a few bushes in most old woods, also by hedges, rivers, streams and ponds. 450. It can grow close to the sea, as at Portland and Worbarrow Tout (87U). A tree 2.1 m in girth grows in the car park of Parnham Ho (10Q). An important source of nectar for early moths, as well as a foodplant for many larvae.

Salix caprea × aurita = S. × capreola J. Kerner ex Andersson
One bush near Gillingham (72Y, 1998, ! det. RDM). 1.

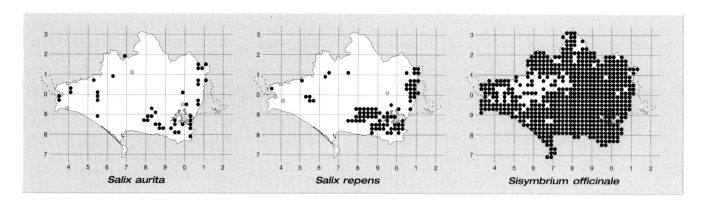

Salix aurita   Salix repens   Sisymbrium officinale

Salix caprea × cinerea = S. × reichardtii A. Kerner
Probably overlooked. Beaminster (40V, 1993, LJM and
LMS); Muston (89T,1999, !); Brownsea Is (08B, 1999, M
Keith-Lucas & !). 3+.

Salix cinerea L.
GREY WILLOW, COMMON SALLOW
By far the commonest *Salix* in Dorset, in hedges, edges of
woods and by water, also on undercliffs as it tolerates salt-
spray. 680. Seedlings are freely produced in bare ground.
Our common plant is subsp. *oleifolia* Macreight, but the
subsp. *cinerea* occurs in hedges near the fen at Wytch (98R,
1997, ! det. RDM). The twigs are used to make whistles.

Salix cinerea × aurita = S. × multinervis Doell.
Rare, though recorded from the Poole basin by EFL in
1900. 5 + [8]. The only later records are: Pilsdon Pen and
Lewesdon Hill (40A, B and F, IPG and PRG); Horn Park
(40R, IPG); Batcombe (60H, 1981, AJCB); Ulwell (08F, 1916,
CBG in **BMH**); Holt Forest (00H, 1981, AH & ! in **RNG**).

Salix cinerea × repens = S. × subsericea Doell
On wet heaths, probably overlooked. 1. Southover Heath (79W,
1998, ! det. RDM), perhaps also on Corfe Common (98).

Salix aurita L.
EARED WILLOW
Occasional, perhaps overlooked, in woods, old pastures,
heaths and by streams on acid soil. 43 + [3]. Found in the
Poole basin and the north-west, but not on the Chalk. It
usually forms a bush, but var. *nemorosa* was a tree 3 m tall
near Kingcombe (50K, 1995, ! det. RDM).

Salix aurita × repens = S. × ambigua Ehrh.
Rare on damp heaths. 2 + [2]. Stoke Heath (89K, 1991, !);
South Haven peninsula (08, 1896–1904, EFL); Holt Heath
(00S, 1980, HCP).

(***Salix phylicifolia*** L. was reported as planted near Wimborne (1937,
DM); no specimens seen.)

Salix repens L.
CREEPING WILLOW
Locally frequent in wet heaths and bogs, including grazed
areas. 83 + [3]. Rare in old pastures or verges on acid soil

and in dune slacks. Restricted to the Poole basin and the
north and west, apart from the floor of an old chalkpit near
Whitemill Bridge (90K). Here it survived for many years
until the pit was filled in, before 1996. Var. *argentea* Sm. has
only been seen planted on a verge at Holes Bay (09B, 1997, !
det. RDM).

Salix elaeagnos Scop.
OLIVE WILLOW
Occasionally planted in parks, verges, shrubberies or by rivers.
7. Recorded from 61N, 88P, 98A, 99P and W, 08D and 09X.

Salix udensis Trautv. & C.A. Mey.
SACHALIN WILLOW
Planted near Parnham Ho (40Q, !) and Durweston Mill
(80P). 2.

BRASSICACEAE

[***Sisymbrium irio*** L.
LONDON ROCKET
A casual, recorded pictorially near Hurst bridge during rebuilding
(79Z, 1800s, MF). Later reports appear to be errors.]

Sisymbrium altissimum L.
TALL ROCKET
A casual, recorded three times from Weymouth (67, 1890,
SMP in **DOR**, 1932, FHH and 1951, !), also from Monkton
(68U, 1937, DM), Wareham (98I, 1930, BNSS), Corfe Sta
(98R, 1959, !) and Poole (09A, 1958, KG). [4].

Sisymbrium orientale L.
EASTERN ROCKET
An alien, sporadic in big towns and ports and casual in
bare, dry places. 8 + [9]. Lyme Regis (39, 1950, JPMB);
Wootton Fitzpaine (39S, 1970, JB); Burton Bradstock (48Z,
1919–28, AWG); West Bay (49Q, 1932, AWG); Beaminster
(40V, 1938, Anon in **RNG**); Weymouth Sta (67U, 1926,
MJA, *et al.* in **OXF**, and 1951–97, !); Dorchester (69V,
1926, MJA, *et al.*, and 1978, !); Bedmill Farm (61C, 1956,
JL Gilbert); Lulworth Castle (88L, 1991, AH); Lower
Hamworthy (08E, 1977–99, !); Swanage Camp (08F, 1915,
CBG in **BMH**); Sandbanks (08N, 1946, KG, *et al.*, and
1969–99, !); Poole (09A, 1927, LBH in **OXF** and 1995–

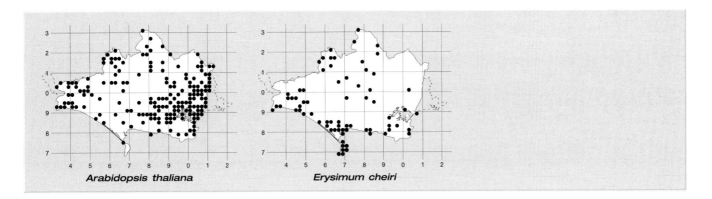

Arabidopsis thaliana   Erysimum cheiri

99, !); Holes Bay (09B, 1991, !); Canford Heath tip (09I, 1999, !); Wimborne (00, 1930, FHH).

Sisymbrium officinale (L.) Stock
HEDGE MUSTARD
Common everywhere except on clay soils in the north-west. 610. Mostly in disturbed hedgebanks, verges and waste ground, but sometimes frequent in dry arable. Var. *leiocarpum* DC., with hairless pods, was found at Parkstone (09, 1937, NDS), near Bindon Abbey (88T, 1992, !) and at Greenland Farm (08C, 1999, !).

Descurainia sophia (L.) Webb ex Prantl
FLIXWEED
Scarce and probably always transient, in sandy arable or on tips. 2 + [11]. First record: Ham railway (99V, 1869, JCMP in **DOR**). Post 1950 records: Stoke Abbott (40K, 1966, AWG in **BDK**); near Yeovil (51Q, 1956, IAR) ; Sherborne (61I, 1962, CT); Durweston Mill (80P, 1905, EFL in **LIV** and nearby, 1984, AH); Paradise Farm (81W, 1996, !); Bear Cross (09M, 1965, MPY).

Alliaria petiolata (M. Bieb.) Cavara & Grande
GARLIC MUSTARD
Common at the edges of woods and in hedgebanks, also seen on a sandy beach at Studland (08G). 622. A form with cuspidate leaf-teeth occurred at Netherbury (49U). This is an important foodplant for the Orange-tip butterfly.

Arabidopsis thaliana (L.) Heynh.
THALE CRESS
Rather local on bare, dry, sandy soil of old walls, gardens, pavements and tips. 179. Uncommon in arable, and scarce on the Chalk. Its winter rosettes are familiar in town gardens, where it has the shortest life-cycle of any British plant.

Isatis tinctoria L.
WOAD
Occasionally grown in gardens, as at Stoke Mill (88T, 1991, FH Dawson, !) and Bryants Puddle (89B, 1987, P Bowell). 3. Many thousands appeared on chalk spoil-heaps near Down Farm (01C, 1992–98, O Linford, !). The cultivation of this plant in past centuries as a source of indigo for dyeing is reflected in names such as Waddon Hill (49X and 40K), Waddon (68H) and Waddon Ho (81E).

Bunias orientalis L.
WARTY-CABBAGE
A casual alien, not established here, though perennial. [2]. Bailey Gate (99P, 1909, RPM); Weymouth (67, 1963, JB).

Erysimum cheiranthoides L.
TREACLE-MUSTARD
A scarce and often casual arable weed on dry, sandy or chalky soils. 6 + [14]. Post-1930 records are: Nettlecombe (1965, 59C, DWT); Weymouth (67U, 1979, R Harris); Portland (67W, 1979, CE Richards); Thornford nursery (61B, 1989, IPG and PRG); Bishops Caundle (71B, 1989, IPG and PRG); Trigon (88Z, 1961, !); Hyde (89, 1937, DM); Pimperne Wood (91A, 1997, !); Holes Bay (09A, 1977, !); Woodlands (00P, 1996, GDF); Verwood (00Z, 1961, KG).

Erysimum cheiri (L.) Crantz
WALLFLOWER
Frequent on cliffs, old walls, and rocky places, and occasional on tips. 71. Commonest along the coastal belt and in the north-west, but rare in the Poole basin. Most naturalised plants have yellow flowers.

Hesperis matronalis L.
DAME'S-VIOLET
Occasionally established in hedgebanks, village lanes or shaded waste places. 29 + [10]. On sale as a garden plant since 1782; flowers white or pink.

Malcolmia maritima (L.) R. Br.
VIRGINIA STOCK
A rare, casual, garden escape on tips or a pavement weed, preferring rich, bare soil in the coastal belt. 5 + [2]. Chickerell (68K, 1995, !); Lodmoor tip (68V, 1961, KG); Ringstead cliffs (78K, 1950, !); Binnegar tip (88Y, 1997, !); Tatchells Pit (98J, 1992, !); Swanage (07J, 1991, WGT); Poole (09, 1917, CBG); Canford tip (09I, 1998–99, !).

Matthiola incana (L.) R. Br.
HOARY STOCK
Established on a few cliffs of limestone or sand, rarely on old walls or on tips. 5 + [4]. West Bay (49Q, 1993, LJM); Small Mouth (67S, 1909, CD Day); wall, Weymouth (67Z, 1969, !); East Weare cliffs (77B, 1958–97, !); north of

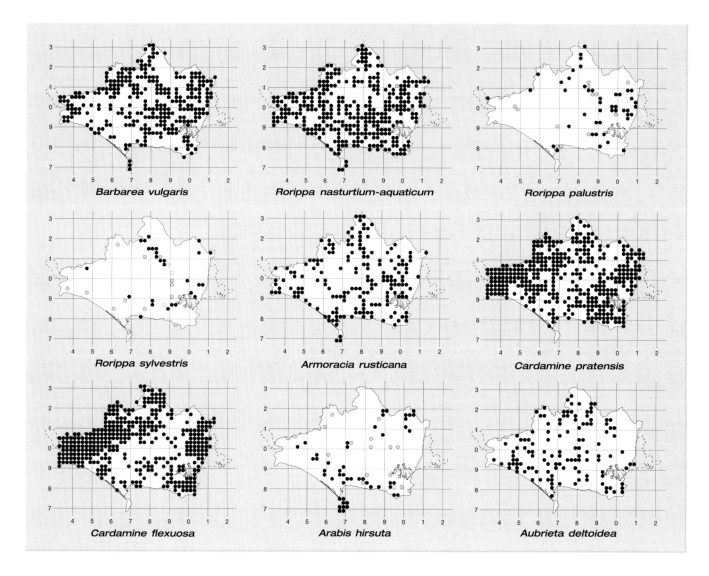

Barbarea vulgaris

Rorippa nasturtium-aquaticum

Rorippa palustris

Rorippa sylvestris

Armoracia rusticana

Cardamine pratensis

Cardamine flexuosa

Arabis hirsuta

Aubrieta deltoidea

Swanage (07J and 08F, 1915, AWG and 1991, !); Canford Cliffs (08U, 1992, RMW). Var. *annua* was at Sherborne tip (61I, 1969, ACL). Flowers mostly white.

Matthiola sinuata (L.) R. Br.
SEA STOCK
Reported from Brownsea Is (08, 1798) and Flag Head Chine (08P, 1931); no specimens seen. [2].

Matthiola longipetala (Vent.) DC.
NIGHT-SCENTED STOCK
A rare casual or pavement weed. 2 + [2]. Nothe Gardens (67Z, 1977, !); Dorchester (69V, 1992, !); Sherborne tip (61I, 1970, ACL); Sturminster Newton (71X, 1997, !); in arable south of Winfrith Newburgh (88B, 1984, P Bowell).

Barbarea vulgaris R. Br.
WINTER-CRESS
Common, rarely abundant, by rivers and ponds, in damp hedgebanks and waste ground, often appearing after disturbance. 290. Commonest in the Stour valley.

Barbarea stricta Andrz.
SMALL-FLOWERED WINTER-CRESS
Rare and sporadic in damp places. 2 + [1]. Seaborough (40I, 1993, JGK); Hammoon (81H, 1981–83, AJCB); Bear Cross (09M, 1965, MPY).

Barbarea intermedia Boreau
MEDIUM-FLOWERED WINTER-CRESS
Scarce, but perhaps increasing, usually as a few plants on verges. 6 + [5]. A critical plant easily confused with *B. vulgaris*; * = det. TCGR or EJ Clement. Fern Hill (39M, 1985, AWJ); Mosterton (40M, 1993, JGK *); Lower Kingcombe (59P, 1993, SME *); Sherborne (61I, 1970, ACL *); West Stafford (78P, 1987, DAP*); Cutt Mill (71T, 1979, WAC); Sturminster Newton (71X, 1900, JCMP); Wool (88N, 1947, KG); between Bindon and West Holme (88T, 1968, JB); Swanage (07J, 1999, EAP); Alderholt (11B, 1993, JO *).

Barbarea verna (Miller) Asch.
AMERICAN WINTER-CRESS
An uncommon escape on verges, made ground and tips. 4

+ [11]. Persistent at Portland, but not refound in any of JCMPs sites. Whitchurch Canonicorum (39X, 1996, !); South Perrott (40T, 1928, AWG); Portland (67W, 1912, WBB and 1989–97, !); Lodmoor tip (68V, 1960, !); Sherborne (61I, 1981, AJCB and 1994, IPG); Bere Heath (89K, 1970, RMB); Tatchells pit (98E, 1995, !); Wareham (98I, 1934, JEL in **RNG**).

Rorippa nasturtium-aquaticum (L.) Hayek
WATER-CRESS
Widespread in clear, chalk streams and ponds, and much cultivated where springs emerge from the base of the Chalk. 347 + [5]. The local name Kersey, cognate with cress, could have given rise to the name of Kershay Farm (49N). Sold locally, as at Bere Regis, and on a large scale for supermarkets.

Rorippa nasturtium-aquaticum × *microphylla* = *R.* × *sterilis* Airy Shaw
Rare or overlooked. Lyme Regis (39,1973); Wimborne (09, 1892). [2].

Rorippa microphylla (Boenn.) N. Hylander ex A. & D. Love
NARROW-FRUITED WATER-CRESS
Scarce in wet meadows, by ditches, streams and ponds. 11 + [3]. Beaminster (40V, 1967, AWG in **BDK**); Sherborne Lake (61N, 1980, AJCB); Waddock (79V, 1981, AJCB); Hinton meads (71X and 81C, 1994, !); Lower Street (89P, 1991, !); Black Ven Farm (82G, 1999, !); Swyre Head (97J, 1996, !); Holton Heath (99K, 1996, !); Tarrant Hinton (91F, 1957, F.Partridge); N Herston and S Swanage (07 E and I, 1999, EAP); Studland, (08G, 1998, EAP); Stanbridge Mill (00E, 1979, JN).

Rorippa palustris (L.) Besser
MARSH YELLOW-CRESS
Occasional on mud by ponds and streams, also as a garden weed. 46 + [15].

Rorippa sylvestris (L.) Besser
CREEPING YELLOW-CRESS
Occasional on mud in the upper Stour valley as a native, also more or less casual as a garden weed or on tips elsewhere. 27 + [17].

Rorippa amphibia (L.) Besser
GREAT YELLOW-CRESS
This riverside plant is either very rare in Dorset, or has been misidentified. 2?. In the absence of specimens, all records need confirming. Tadnoll (78Y, 1981–83, RC Stern); R. Stour, Hammoon (81C, 1962, RGB Roe and 1982, AJCB).

Armoracia rusticana P. Gaertner, Meyer & Scherb.
HORSE-RADISH
A persistent, deep-rooted plant of old grassland or verges near villages. 182. The strong taste and lachrymatory properties of the grated roots are partly due to allyl isothiocyanate, and they are used as a condiment.

(*Cardamine amara* L. Some 10 reports have not been substantiated, and are thought to be errors for *C. pratensis* var. *dentata*.)

Cardamine pratensis L.
CUCKOOFLOWER
Locally frequent in wet woodland rides, water-meadows, damp lawns and churchyards. 387. Scarce on the drier parts of the Chalk, and not found on Portland. Var. *dentata* Schultes, which is tall with large white petals, is frequent in the north-west and has puzzled many botanists.

[*Cardamine pratensis* × *flexuosa* = *C.* × *fringsii* Wirtg. f.
Once found on an island in the Stour near Shapwick (90F, 1900, WRL).]

Cardamine flexuosa With.
WAVY BITTER-CRESS
Frequent in alder-swamps, wet woods and hedgebanks, and along shaded streams. 381. Absent from Portland and most of the Chalk.

Cardamine hirsuta L.
HAIRY BITTER-CRESS
A common garden weed of open soils, bare places in lawns, pavements and waste ground, sometimes in fallow. 568. An albino form from Glanvilles Wootton (60U, IPG) may have been due to weedkiller.

Cardamine asarifolia L.
Naturalised in small quantity in the wild garden at Minterne Magna (60S, 1984–87, ! in **RNG**). 1.

[*Arabis glabra* (L.) Bernh.
TOWER MUSTARD
Extinct, on sandy banks. Bridport (1800, J Sowerby) and Parkstone (1898, AE Hudson). I have not seen specimens; Pulteney's earlier record was an error.]

Arabis caucasica Willd. ex Schldl.
GARDEN ARABIS
Occasionally self-sown or planted on old walls. 12. Chideock (49F); Little Bredy (58Z); Mapperton (59E), Hooke (59J and 50F); Corscombe (50H); Rampisham (50R); Osmington (78G); Marnhull (71U); Iwerne Minster (81S); Sturminster Marshall (90K); Witchampton (90Y).

Arabis hirsuta (L.) Scop.
HAIRY ROCK-CRESS
Local, in small quantity, in short, dry, calcareous turf. 49 + [15]. Found in old quarries, railway cuttings, undercliffs and places where the soil is shallow. Most frequent on Portland; the few records from areas with acid soils are from old limestone walls or lime-enriched verges.

Aubrieta deltoidea (L.) DC.
Planted and frequently self-sown on old walls or dry banks in almost every village. 125.

Lunaria annua L.

HONESTY

A garden plant, on sale in Blandford in 1782, and frequently escaping to shaded verges, village lanes, hedgebanks and tips. 177. Flowers deep pink, rarely white.

[**Alyssum alyssoides** (L.) L.

SMALL ALISON

A casual in arable, not seen in the last 70 years. 4. Radipole Lake (67U, 1926, MJA); Badbury Rings (90R, 1900, EFL); Kinson (09T, 1892, WRL in **LIV**); Bourne Valley (09, 1891, H Fisher).]

Alyssum saxatile L.

GOLDEN ALYSSUM

A rock-garden plant, on sale in Blandford in 1782. 9. Seen on old walls at Little Bredy (58Z); Long Bredy (59Q); Upwey (68S); Sutton Poyntz (78B); East Chaldon (78W); Stourton Caundle (71C); Buckhorn Weston (72M); Silton (72Z); Gussage All Saints (91V) and New Swanage (08F, EAP).

[**Berteroa incana** (L.) DC.

HOARY ALISON

Reported from Weymouth (67, 1766, RP).]

Lobularia maritima (L.) Desv.

SWEET ALISON

A garden escape established on cliffs, waste ground and pavements along the coastal belt, also on tips. 63.

Draba aizoides L.

YELLOW WHITLOWGRASS

A few clumps established on limestone walls at Charmouth (39R). 1. The original wall, found by Mr Templeman in 1921, was destroyed in the 1960s, but a second site was found by KEB in 1985, where the plant survives, RCP, !. It was introduced from Pennard Castle in south Wales (1931, MAA in **LIV**).

Draba muralis L.

WALL WHITLOWGRASS

Very local in bare places in limestone quarries, or on old walls. 1 + [4]. Chart Knowle (40K, 1921, AWG in **BDK** and **RNG**, *et al.*, and 1994, P Brough); Chedington (40X, 1928,

AWG); Chickerell (68L, 1924, IM Roper); Oborne (61P, 1937, DM). More frequent across the Somerset border.

Erophila verna (L.) Chevall

COMMON WHITLOWGRASS

A locally frequent ephemeral in bare places on rocks and cliffs, along dry paths and as a wall-top or pavement weed; once seen on sandy rifle-butts. 126. It occurs on both limestone and sandy cliffs. There are old reports by EFL of var. *praecox* (Steven) Diklic, which has not been seen recently.

E. glabrescens Jord.

Occurs in a heathy ride at Tadnoll (78Y) and has a herbarium record in SY98.

E. majuscula Jord. has a herbarium record in SU00.

Cochlearia anglica L.

ENGLISH SCURVYGRASS

Occasional or locally frequent in the upper parts of salt-marshes or on muddy shores. 23. Found on the eastern shores of the Fleet and inside Poole Harbour including its islands.

(**Cochlearia officinalis** L.

No certain records. Early reports were probably *C. anglica* or *C. danica*; the distinction between this and *C. anglica* is critical. It does not occur along the south coast of Hampshire, but is found in north Somerset where it has spread to the M5 motorway.)

Cochlearia anglica

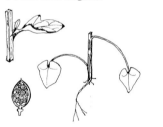 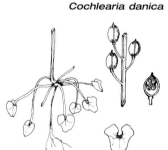

Cochlearia danica

Cochlearia danica L.

DANISH SCURVYGRASS

Locally frequent on cliff-tops, rocky, sandy and shingle

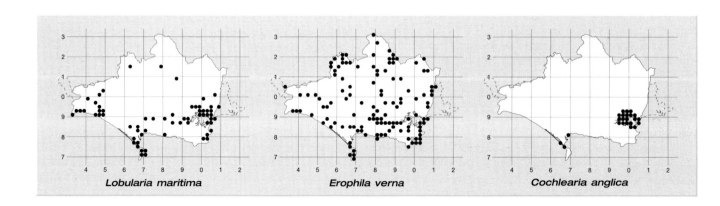

Lobularia maritima   Erophila verna   Cochlearia anglica

beaches, sea walls and pavements near the coast. 63. It has recently spread along the central reservation of double-track roads (A31, A35, A3049 and A338) in both west and east Dorset. Not usually found in salt-marshes, but large plants on beaches can mimic *C. officinalis*. Probably this referred to by W Turner in 1550.

Camelina sativa (L.) Crantz
GOLD-OF-PLEASURE
Once a weed of flax fields, but not seen in them today, when it is a rare casual from bird-seed. 1 + [5]. Lyme Regis (39, 1652, W Hudson); Bridport (49, 1652, W Hudson and 1920, AWG); Beaminster (40Q, 1919, AWG); Ferry Bridge, in a garden (67T, 1991, !); La Lee Farm (80F, 1874, JCMP); Swanage Camp (08F, 1915, CBG).

[*Neslia paniculata* (L.) Desv.
BALL MUSTARD
Once found at Radipole Lake (67U, 1926, MJA det. **K**).]

Capsella bursa-pastoris (L.) Medikus
SHEPHERD'S-PURSE
A very common weed of bare, dry soil in arable, gardens, waste ground and paths. 706.

Teesdalia nudicaulis (L.) R. Br.
SHEPHERD'S CRESS
Very rare in bare, dry sandy soil in the east. 1 + [2]. Sandbanks (08N, 1954, KG & MM Webster); Talbot Heath (09R, 1980, C Thomas); Ensbury (09T, 1863, JH Austen and 1900, EFL); Ebbslake (10D, 1994, VMS).

Ionopsidium acaule (Desf.) Reichenb.
Casual, Wimborne Minster (09E, 1973, Mrs M Parish in **BM**, det. TCGR).

Thlaspi arvense L.
FIELD PENNY-CRESS
A locally frequent weed of arable, new verges, gardens and tips on calcareous or sandy soil, known since Iron Age times. 93. Its odd distribution pattern could reflect its requirement for sulphur, as in many crucifers.

(*Thlaspi alliaceum* L. should be looked for, as it occurs at Winkton (19T, FAW) in VC11.)

Thlaspi macrophyllum Hoffm.
CAUCASIAN PENNY-CRESS
A perennial more or less established in a wild garden at Spetisbury Old Mill (90B, 1994, !). 1.

Iberis sempervirens L.
PERENNIAL CANDYTUFT
A garden plant, rarely found on old walls or in churchyards. 4. Beaminster (40W), Bincombe (68X), Shillingstone (81G) and Kingston (97P). A 19th century record of *I. amara* from walls at Sutton Poyntz (78B, WBB) could have been this.

[*Iberis amara* L.
WILD CANDYTUFT
Extinct for 90 years as a casual. 4. Portland, south of the prison (67W, 1857, WBB); Shitterton (89H, 1874, JCMP); Wareham Sta (98E, 1893, JCMP in **DOR**); Edmondsham (01, 1908, EFL).]

Iberis umbellata L.
GARDEN CANDYTUFT
An annual garden escape found on tips or as a pavement weed. 6. Lodmoor tip (68V, !); Sherborne tip (61H, ACL and 1989, !)); Giddy Green (88I, !); Wareham (98I, !); Canford tip (09I, !); Ferndown (00Q, !).

Lepidium sativum L.
GARDEN CRESS
A rare casual on tips or in flax fields. 1 + [7]. Bridport quarries and West Bay (49G and Q, 1920s, AWG); Weymouth (67U, 1865, SMP in **DOR**); Lodmoor tip (68V, 1960-63, !); Sherborne tip (61H, 1970, ACL and 1981, AJCB); Swanage Camp (08F, 1915, CBG in **BMH**); Parkstone (09, 1937, NDS).

Lepidium campestre (L.) R. Br.
FIELD PEPPERWORT
Sporadic, often as a few plants, on dry banks, verges and rail tracks. 22 + [12]. Recent records are from the Poole basin or the north-west, but it used to occur near Weymouth.

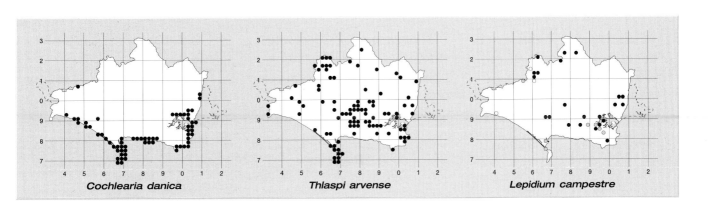

Cochlearia danica   Thlaspi arvense   Lepidium campestre

Lepidium heterophyllum Benth.
SMITH'S PEPPERWORT
Sporadic and in small quantity, on dry banks and along railways, in fallow fields, sandy shores and on tips. 14 + [22]. First record: Ham shore (99, 1696, W Stonestreet); also seen near here by DM in 1937, and near Poole Yacht Club (09A, 1999, !). The plant is mostly found not far from Poole Harbour.

[*Lepidium virginicum* L.
LEAST PEPPERWORT
A rare casual. 4. Weymouth (67, 1917); Dorchester (69, 1919); Swanage Camp (08F, 1915–17, CBG in **BMH**); Tricketts Cross (00V, 1946, NDS).]

Lepidium ruderale L.
NARROW-LEAVED PEPPERWORT
A rare casual on tips. 2 + [10]. There are about 10 records pre-1960, also: Lodmoor tip (68V, 1960, !); Waddock (78Z, 1983, AJCB); Binnegar tip (88T, 1993, !).

(*Lepidium latifolium* L. has been seen at Bournemouth (19B, 1962, MPY) but in VC11.)

Lepidium draba L.
HOARY CRESS
Introduced and sometimes locally abundant on cliffs, along rail tracks, the upper part of beaches or on tips, where it persists. 31 + [11]. Most common in the coastal belt and around Weymouth and Portland. First report: Moreton (88E, 1830s, MF).

[*Lepidium perfoliatum* L.
PERFOLIATE PEPPERWORT
Casual, once seen at Bowood (49P, 1930, AWG).]

Coronopus squamatus (Forsskal) Asch.
SWINE-CRESS
Widespread on compacted soils near gates, farmyards, pond margins, tips and rarely on beaches. 415. On chalk or clay, but not on very acid soil.

Coronopus didymus (L.) Smith
LESSER SWINE-CRESS
An introduced plant now locally frequent in gardens, disturbed soil, tips and beaches. 251. It prefers acid soil, but can occur on chalk though absent from the higher downs. It is often detected by its scent of cress when bruised. First record: Poole quay (09, 1839, TBS); JCMP only knew it from eight sites in 1895, and Good found it infrequent in the 1930s, so it is increasing.

Diplotaxis tenuifolia (L.) DC.
PERENNIAL WALL-ROCKET
Uncommon in waste places along the coastal belt, commonest near Weymouth and Portland, and sometimes casual elsewhere. 12 + [10].

Diplotaxis muralis (L.) DC.
ANNUAL WALL-ROCKET
Appearing native on calcareous cliffs, more often as a weed of bare, disturbed soil on chalk or sand. 55 + [7]. Especially common along rail tracks, on tips and in the coastal belt.

Diplotaxis erucoides (L.) DC.
WHITE WALL-ROCKET
Once reported from Durlston CP (07I, 1985, AM); no specimen seen. The petals are white with pink veins.

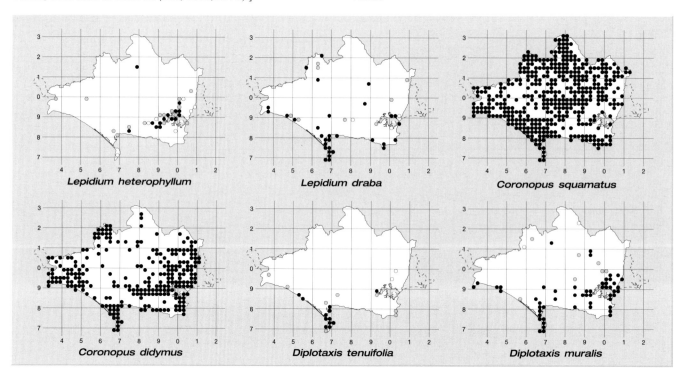

Lepidium heterophyllum — Lepidium draba — Coronopus squamatus — Coronopus didymus — Diplotaxis tenuifolia — Diplotaxis muralis

Brassica oleracea L.
CABBAGE
Locally frequent, rarely dominant, on calcareous cliffs, undercliffs and cliff-tops between Portland and Handfast Point. 26 + 13 introd. + [2]. Coastal records from further east and inland records are introductions in tips or waste places, usually the var. *viridis* L. (Kale), which is a common crop. The native plant flowers early and attracts immigrant cabbage white butterflies.

Brassica napus L.
RAPE
Much grown as a crop, and a frequent but under-recorded escape in arable, fallow, verges, waste ground and tips. 74+.

Brassica rapa L.
TURNIP-RAPE
A frequent casual, easily confused with *B. napus*. 88. It was much grown as a crop in the last few centuries, but seldom today.

Brassica juncea (L.) Czernj.
CHINESE MUSTARD
An alien recently found on tips. 3. Lodmoor (68V, !); Tatchells Pit (98E, !); Greenland Farm (08C, !). The pods are almost torulose.

Brassica nigra (L.) Koch
BLACK MUSTARD
Locally frequent on clay cliffs, riverbanks and dredgings, sometimes in arable, waste places and tips. 153. It prefers clay soils and is most frequent along the coastal belt.

Sinapis arvensis L.
CHARLOCK
An arable weed since post-Roman times, which was once common, but is now much less so on account of weedkiller use. 433. It can become dominant in fallow or on new verges, and is also present on tips. The seeds were once important for linnets, but these birds have now adapted to oil-seed rape (Wilson, Arroyo & Clark, 1997).

Sinapis alba L.
WHITE MUSTARD
An arable weed which has declined through the use of weedkillers, always on calcareous soils. 96 + [4]. It is found in odd corners of arable fields which have escaped spraying, or in newly disturbed ground. Rarely grown today for its seeds, an ingredient of mustard.

Erucastrum gallicum (Willd.) O.E. Schulz
HAIRY ROCKET
Casual, Swanage (07J, 1956, HW Border in **BM**).

Coincya monensis (L.) Greuter & Burdet
WALLFLOWER CABBAGE
A casual, once reported from Sandbanks (08N, 1976, V Follett); no specimen seen.

Hirschfeldia incana (L.) Lagr.-Fossat
A rare alien. Sandbanks (08N, 2000, !); Seldown, Poole (09A, 1997, FAW), spreading from VC11. 2.

Cakile maritima Scop.
SEA ROCKET
Sporadic, usually in small numbers along the strand-line of sandy or shingle beaches. 19 + [6]. Its disappearance

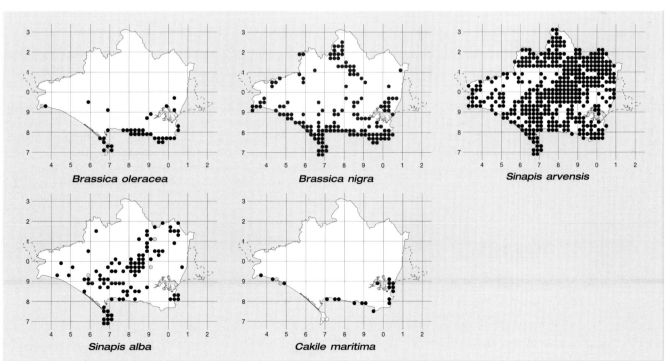

Brassica oleracea   Brassica nigra   Sinapis arvensis

Sinapis alba   Cakile maritima

from most beaches in the west is probably due to human pressure.

Rapistrum rugosum (L.) Bergeret
BASTARD CABBAGE
A casual, associated with the old Weymouth to Portland railway, rarely on waste ground or tips elsewhere. 4 + [3]. Portland (67, 1960, HJB); Weymouth (67, 1931, FHH); Small Mouth (67T, 1988, !); Weymouth Sta (67U, 1996, !); The Nothe (67Z, 1988, ARG Mundell); Lodmoor tip (68V, 1958, KG and 1963, !); Sherborne tip (61I, 1970, ACL); Greenland Farm (08C, 1999, EAP).

Crambe maritima L.
SEA-KALE
Fairly frequent on shingle beaches, or on cliffs at West Bay, between Charmouth and the west end of the Chesil bank, and sporadic on beaches elsewhere. 21 + [7]. Subject to storm damage and human pressure. First record: Portland (1576, M L'Obel); it still occurs on the Island.

Crambe cordifolia Steven
GREATER SEA-KALE
A garden throw-out in a lane at Hinton St Mary (71Y, 1989, !); propagated by root cuttings. 1.

Raphanus raphanistrum L.
WILD RADISH
A locally frequent arable weed, preferring the sandy soils of the Poole basin, but not confined to them; first evident in Iron Age deposits. 109. Petals mostly white, sometimes yellow. Subsp. *maritimus* (Smith) Thell., which most often has yellow flowers here, is local on shingle beaches and undercliffs. 14.

Raphanus sativus L.
GARDEN RADISH
An escape from cultivation. 1 + [2]. Radipole Lake (67U, 1926, MJA); Herringston (68Z, !); Swanage (07J, 1915, CBG in **BMH**).

Conringia orientalis (L.) Dumort
HARE'S-EAR MUSTARD
A rare casual. [5]. Beaminster (40Q, 1927, AWG); Sherborne (61I, 1970, ACL); Wareham (98I, 1917, CBG); Swanage Camp (08F, 1915, CBG in **BMH**); Poole (09, 1898, AE Hudson and 1917, CBG).

[*Erucaria hispanica* (L.) Druce
 A casual on the railway between Poole and Lytchett (99W, 1873, JCMP in **DOR**).]

RESEDACEAE

Reseda luteola L.
WELD
A widespread biennial of woodland margins, bare chalk, rabbit warrens, cliff-tops and arable headlands. 136. It prefers chalky soils but is not confined to them, is sporadic and requires occasional disturbance of its habitat.

Reseda alba L.
WHITE MIGNONETTE
A rare casual of verges and railway banks. 2 + [5]. Ridgeway Hill (68S, 1927, M. Moores); Came Wood (68X, 1925, CW Hewgill); Durweston (80P, 1952, BNHS); Swanage (07J, 1895, WMR); Poole (09A, 1998, I Davenport); Holes Bay (09B, 1930, S Haines and 1991, BE); Poole park (09F, 1900, EFL).

Reseda lutea L.
WILD MIGNONETTE
Occasional in dry, disturbed soils, in arable, on verges and by rabbit warrens, mostly on the Chalk. 126.

Reseda odorata L.
GARDEN MIGNONETTE
A garden escape once seen on Lodmoor tip (68V, 1955, !). [1].

CLETHRACEAE

Clethra alnifolia L. and **C. delavayi** Franch are grown in wild gardens on Greensand. A fine specimen of the latter in Minterne Magna gardens is mentioned by Bean.

EMPETRACEAE

[*Empetrum nigrum* L.
CROWBERRY
This was present in the Iron Age at Maiden Castle, and survived near the mouth of Poole Harbour until 1900; its nearest site is now Exmoor. 2. East side of Little Sea (08H, 1893, W Mitten and 1900, JCMP in **DOR**); Sandbanks (08N, 1892, CB Clarke).]

ERICACEAE

Rhododendron ponticum L.
RHODODENDRON
An introduced shrub which seeds freely and can rapidly come to dominate the understorey of woods or plantations on wet or dry, acid soils; bushes planted on the Chalk survive in a chlorotic state but do not spread. 230. Introduced between 1796 and 1820 at Hurn (VC11; FAW), not mentioned by JCMP in 1895 but abundant in the 1870s near Bournemouth, *fide* J.M. Faulkner. It now has the distribution pattern of a native plant, and is spectacularly abundant at Hurst and Warmwell Heaths (78), Puddletown Forest (79) where there is a Rhododendron mile, Bere Wood (89), Arne (98), and Brownsea and Green Is (08). Efforts to control it are ineffective.

Rhododendron luteum L.
YELLOW AZALEA
Introduced and seeding itself in private grounds on acid soils. 14. Locally frequent near Carey (98E) and in Charborough Park (99I).

Many other Rhododendron species or hybrids are planted in churchyards or estates, but none are spreading naturally. *R. ferrugineum* and *R. 'maximum'* were on sale at Blandford in 1782. Sites where specimens with putative names have been seen are: *R. japonicum* agg. at Netherbury Ch (49U) and Green Is (08D); *R. arboreum* W Smith & Waterers hybrids at Melbury Park (50S); *R. falconeri* Hook.f.? and *R. thomsonii* Hook. f. at Bere Heath (89R) and Charborough Park (99I); *R. arboreum*? at Little Wood (98B); *R. × praecox*? at Brownsea Is (08J); *R. indicum* Sweet at Branksome Ch (09Q). There are several commercial Rhododendron nurseries near Three-legged Cross.

Daboecia cantabrica (Hudson) K. Koch
ST DABEOC'S HEATH
Planted and surviving in plantations or wild gardens on acid soils. 3 + [1]. Trigon (88Z, 1930s, DM); Carey Heath (98E, planted *c.*1883, CBG, *et al.* in **BDK** and **BMH**, and 1990–97, !, with a white-flowered form); Lytchett Heath (99S, 1978. AJB and 1989, !); Upton Park (99W, 1992, !).

Gaultheria shallon Pursh
SHALLON
Sometimes planted in shaded plantations or wild gardens on acid soils, and then spreading on its own to become locally abundant. 14 + [3]. First noticed by RDG in the 1930s. An oil from this is rich in methyl salicylate.

Gaultheria mucronata (L.f.) Hook. & Arn.
PRICKLY HEATH
Odd bushes have been found on acid soils, but this is nowhere naturalised. 4. Forde Abbey (30M, 1); Bere Heath (89R, 1981–88, !); Studland (08G, !); Branksome Chine (08U, !).

Arbutus unedo L.
STRAWBERRY-TREE
Occasionally planted in churchyards and verges, surviving on all soils. 20. Around Poole Harbour it is naturalised in plantations, cliffs and railway cuttings on sandy soil.

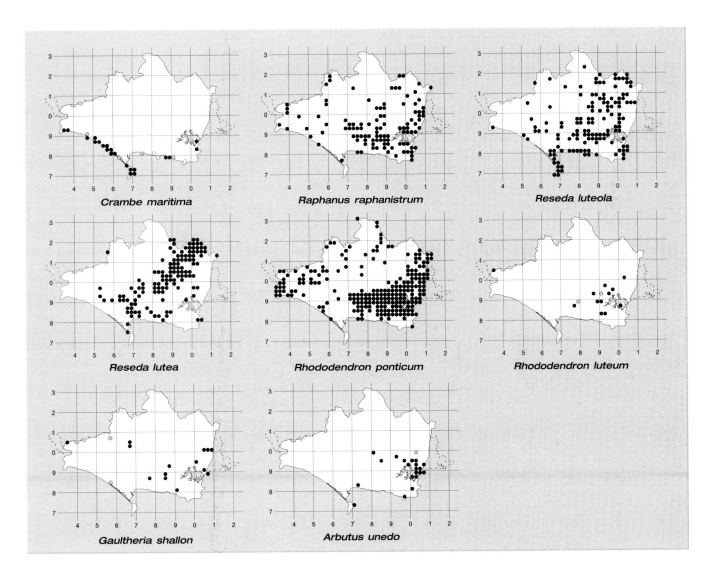

Crambe maritima

Raphanus raphanistrum

Reseda luteola

Reseda lutea

Rhododendron ponticum

Rhododendron luteum

Gaultheria shallon

Arbutus unedo

Other *Arbutus* sp. are rare but notable features of gardens, e.g. *A. menziesii* Pursh at Portland Ho, Rodwell (67Z) and Arbutus Close, Dorchester (69V), and *A.* × *andrachnoides* Link at Lytchett Heath (99S), where a tree planted in 1877 was 10 m × 1.2 m in 1970.

Calluna vulgaris (L.) Hull
HEATHER, LING
Locally dominant and producing a mass of colour on dry, sandy heaths in the Poole basin and the west. 190 + [4]. It tolerates light shade and is well able to recolonise disturbed areas such as cleared plantations, gravel or kaolin workings and burnt ground. Scarce among grass in leached soils over the Chalk in central and north-east Dorset. A foodplant for the Silver-studded Blue and many moth larvae, and beloved by beekeepers. The hairy var. *incana* Reichb. is occasional, and white-flowered bushes are rare.

Erica ciliaris L.
DORSET HEATH
Locally abundant in wet heathland in the Poole basin. 30 + [17]. Well represented in herbaria (e.g. **BDK, BM, DOR, LIV, MDH, RNG, WAR**). First record: Corfe and Wareham Heaths (98, 1833, R Blunt in **WAR**). Its distribution has been mapped (Chapman, 1975), and since then some sites have been lost to forestry while isolated bushes have spread to heaths east of the headquarters of the plant in north Purbeck. While the plant has been here since early times, some could have been introduced with *Pinus pinaster* from west France *c*.1800; it is Cornubian in its British distribution, but reports from west Dorset are errors.

Erica × *watsonii*

Erica ciliaris × *tetralix* = *E.* × *watsonii* Benth.
Frequent on wet heaths wherever *E. ciliaris* is common (Chapman, 1975). 13. First recorded from between Stoborough and Arne (98N, 1891, JCMP and 1998, !). Specimens are in **BMH** and **LANC**.

Erica tetralix L.
CROSS-LEAVED HEATH
Locally frequent or abundant in wet heaths and bogs in the Poole basin and the west. 141 + [3]. It is not very tolerant

of shade, but accepts some eutrophication in fenny sites, and is able to compete with *Molinia*. Some outliers have been lost to cultivation. Var. *fissa*, with flowers replaced by coloured bracts, was found by LBH at Moreton (78Z, 1932), Studland Heath (08, 1928) and in the Bourne valley (09, 1928). White- and pale-pink-flowered bushes are rare.

Erica terminalis Salisb.
CORSICAN HEATH
Two bushes seen planted at Lytchett Heath (99S, 1978, AJB). 1.

Erica cinerea L.
BELL HEATHER
Locally abundant, and a colourful sight in flower, on dry heaths in the Poole basin and the west. 168 + [6]. Rare in bogs, occasional in dry leached grassland over Chalk, and once seen on north Portland (67X, 1960, RGB Roe). Var. *rendlei* LB Hall, with aborted flowers, has been found at Upton Moor (99X, 1983, Mrs Maginnis); Parkstone (09, 1930, LBH in **RNG**) and Parley Common (09Z, 1984, MB in **RNG**). White-flowered plants are rare, e.g. Studland (08G, EAP).

Erica lusitanica Rudolphi
PORTUGUESE HEATH
Occasionally planted, and naturalised in at least two places on acid soils. 5. Uplyme railway (39G, 1972, KEB and 1985, RCP); Trigon (88Z, 1912, CBG in **BMH**); Creech Grange (98B, 1973, !); Lytchett Heath (99S, planted in 1876, Lord Eustace Cecil, *et al.* in **DOR, LANC** and 1989, !); Knoll Ho, Studland (08G, 1996, !); Luscombe valley (08P and 09Z, 1980, AJB and 1995, !).

Erica × *darleyensis* Bean
DARLEY DALE HEATH
Planted in churchyards and rarely on roadside banks. 13. It tolerates calcareous soil. Seen in 39R, 49U, 51V, 78I, 79M, 70Z, 81U, 98I, 90B, 91G, 08F and U, and 01L.

Erica vagans L.
CORNISH HEATH
Planted and naturalised in a few places. 6 + [3]. Trigon (88Z, 1912, CBG in **BMH**, *et al.* in **BDK** and **BM**, and 1950, JKH); Canford Heath (09H, 1995, BPLS); Parkstone (09, 1935, M Richmond in **LIV**). Obviously planted in 49U, 88D, 98E and H, 99S and W and 08U. The record from neolithic remains on Hambledon Hill is intriguing.

Erica arborea L.
TREE HEATH
Rarely planted on acid soil, forming large bushes. 3. Charborough Park (99I, 1992, BE & !); Upton Park (99W, 1992, !); Green Is (08D, 1994, !).

Erica scoparia L.
A few bushes of this greenish-flowered heath are planted at Clouds Hill (89F, 1992, !). 1.

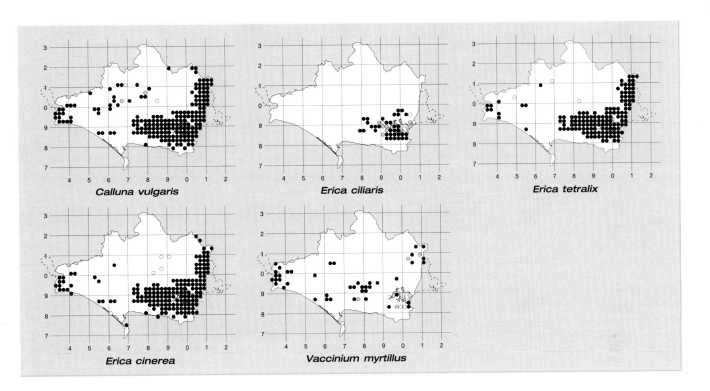

Calluna vulgaris   Erica ciliaris   Erica tetralix

Erica cinerea   Vaccinium myrtillus

Vaccinium macrocarpum Aiton
AMERICAN CRANBERRY
Very locally frequent in a *Sphagnum* bog on the Hampshire border, between Ashley Heath and Lions Hill (10B and C, 1979, JRW, *et al.* in **RNG**, and 1994, VMS, !); how introduced unknown. 1.

Vaccinium myrtillus L.
BILBERRY
Local in old woods and on dry ridges or banks on acid soils. 43 + [7]. Good had it from 30 sites in the west but only five in the east. It fruits here too sparingly to be worth gathering.

Vaccinium corymbosum L.
BLUEBERRY
Cultivated for its fruit in a few plantations on acid soils, near which bird-sown seedlings are frequent. 6. Skippet Heath (78T, 1985, APC); Boswells Plantation (79K, 1990, !); Bere Heath (89R, 1985, G Tuley and 1991–98, ! in **RNG**); Longham (09U, 1993, JO); Ferndown (00Q, 1989, !); Ashley Heath (10C, 1993, FAW, perhaps in VC11).

Enkianthus campanulatus Nichols
A planted shrub. 2. Bere Heath (89R, 1992–98, !); Green Is (08D, 1994, !).

Oxydendron arboreum (L.) DC.
A planted tree in parks. 2. Lytchett Heath (99S, 1989, !); Canford School (09J, TH).

Pieris formosa (Wallich) D. Don
Rarely planted, not naturalised, in parks or churchyards

on acid soils. 5. Minterne Magna (60S, !); Trigon (88Z, !); Compton Abbas (81U, !); Rempstone (98W, 1974, ! in RNG); Charborough Park (99I, !).

Pieris japonica (Thunb.) D. Don
Rarely planted. 2. Minterne Magna (60S, !); Charborough Park (99I, !).

PYROLACEAE

Pyrola minor L.
COMMON WINTERGREEN
Rare in plantations on acid soils, but not seen recently. [4]. Mainly northern in Britain, and found neither by JCMP nor RDG. Halstock (50I, 1965, WG Latham); Hurst Heath (78Z, 1961–78, ! in **RNG**); Waddock Copse (79V, 1976, MS Warren); Broomhill Plantation (79W,1953, WAC det. SME).

Pyrola rotundifolia L. subsp. **maritima** (Kenyon) E. Warb.
ROUND-LEAVED WINTERGREEN
Very rare on the coast. 1 + [1]. Thorncombe Beacon undercliff (49F, 1996, FAW in **BM**); Studland Heath (08H, 1973–75, JB Hope-Simpson, also KM Rushton and R Cox).

MONOTROPACEAE

Monotropa hypopitys L.
YELLOW BIRD'S-NEST
Rare, in small quantity and probably declining in plantations. 3 + [23]. Found under beech or pine. Post-1930 records: Bride Head drive (58Z, 1937, RDG in

WRHM); Langton Herring (68G, 1976, JKH); Well Bottom (69K, 1936, RDG); Blagdon Copse (70X, 1936, RDG); Bryanston Hangings (80T, 1954, BNHS); Hod Hill (81K, 1930–50, AWG); Iwerne (81S, 1942, BNHS); Kingston Lacy (90Q, 1936, DM); Melbury Down Wood (91E, 1997, P Amies); Cashmoor (91R, 1937, DM); Chase Wood (91Z, 1989, BE); Monkton Up Wimborne (01C, 1936, RDG in WRHM); Wimborne St Giles (01F, 1936, DM); Cranborne (01M, 1937, DM); East lake, Brownsea Is (08I, 1982, KC).

PRIMULACEAE

Primula vulgaris Hudson
PRIMROSE
In woods, hedgebanks, sheltered undercliffs and churchyards. 613. Found almost everywhere, but most common in the north and west, and least common in deforested areas, on very acid soils and in built-up regions. Var. *minor*, with small flowers, was in Yellowham Wood (79G, 1914, CBG in BMH). Forms with pink flowers with a yellow eye are not rare, especially in churchyards, and a form with brick-red flowers occurred at Godlingston Heath (08B, 1914, CBG in BMH). The flower is carved on the pulpit at North Poorton (built 1862). There is a Primrose Hill in 97Z.

Primula vulgaris × veris = P. × polyantha Miller
FALSE OXLIP
Widespread but in small numbers in woodland rides, scrub, hedgebanks, old grassland or edges of heaths, rarely in churchyards. 84. Individual plants are long-lived. Pink-flowered forms occur away from houses at Toller Fratrum (59T), also in churchyards as at Arne (98U) and Sixpenny Handley (91Y).

Primula vulgaris × juliae = P. × pruhonicensis Zemann ex Bergens
This is taken to be the name of the bright pink-flowered plant seen in churchyards at Thorncombe (30R), Chickerell (68K), Warmwell (78M), Woodsford (79Q), and Turners Piddle (89G). 5.

Primula veris L.
COWSLIP, CREWEL, HOLROD
Occasional to abundant in old calcareous or clay grassland, cliff-tops, sometimes in woodland rides or water-meadows, never in heathland. 380. Large populations still occur, as on the Lulworth ranges (87J) and near Compton Abbas (81U), though intensive farming has destroyed others. Sometimes sown on verges, as at Winterbourne Abbas (69F), where some plants have red flowers; this form has also been seen at Buckland Ripers Ch (68L). The seed remains for some months in its capsule. The plant benefits from grazing or mowing to reduce competition with tall grasses. It is the main foodplant for the Duke of Burgundy butterfly. 'Cowslip Balls' were once used by village maidens to reveal their future lovers (Udal, 1889).

Primula florindae Kingdon-Ward
TIBETAN COWSLIP
Persistent in wild gardens on wet, acid soil, as at Melbury Park (50T) and Knoll Ho, Studland (08G).

Primula japonica A. Gray
JAPANESE COWSLIP
Naturalised in one wet wood. West Field Coppice (98J, 1978, ! and 1984, RM). Seedlings are freely produced in many bog gardens elsewhere; some may be hybrids with *P. pulverulenta*. 1.

Primula pulverulenta Duthie
Established by a pond in Little Wood (98B, 1973, !), and often planted in bog gardens. 1.

Hottonia palustris L.
WATER-VIOLET
Rare, sporadic and probably introduced. Pulteney gave three localities in the east in the 18th century, since when there have been three records: Stalbridge (71, 1895, Mrs Allen in DOR); planted in farm pond near Swanage (07J, 1957, WB Alexander); ditch, Studland (08, 1956, RM Harley). [3].

Cyclamen hederifolium Aiton
SOWBREAD
Well naturalised in or near churchyards at Portesham (68C), Chetnole (60E, MP Hinton) and Tarrant Keyneston (90H). A few corms are planted in many other churchyards, or in village hedgebanks. 49. White-flowered forms were seen at Chetnole (60E), Arne (98U) and elsewhere.

Cyclamen coum Miller
EASTERN SOWBREAD
Naturalised at Forde Abbey (30S), Colehill Ch (00F), and a few plants seen at Durweston Ch (80P) and at Shaftesbury Abbey ruins (82R). 5. Also naturalised with *C. repandum* under beech in a garden at Long Bredy (59Q), where white-flowered plants are present.

Cyclamen repandum Sibth. & Smith
SPRING SOWBREAD
Naturalised at Colehill Ch (01F), and a few plants at Stratton Ch (69L), Anderson (89X), Middle Lodge, Bryanston (80T) and Steepleton Ho (81Q). 5.

Lysimachia nemorum L.
YELLOW PIMPERNEL
Occasional in woodland rides, preferring slightly acid and damp soils. 224. Least common on the Chalk, and absent from deforested and built-up areas.

Lysimachia nummularia L.
CREEPING-JENNY
Local in wet woodland rides and marshes on clay soils; also an escape from cultivation in a few churchyards and village lanes. 94. Commonest in the north and west, and along the upper Stour.

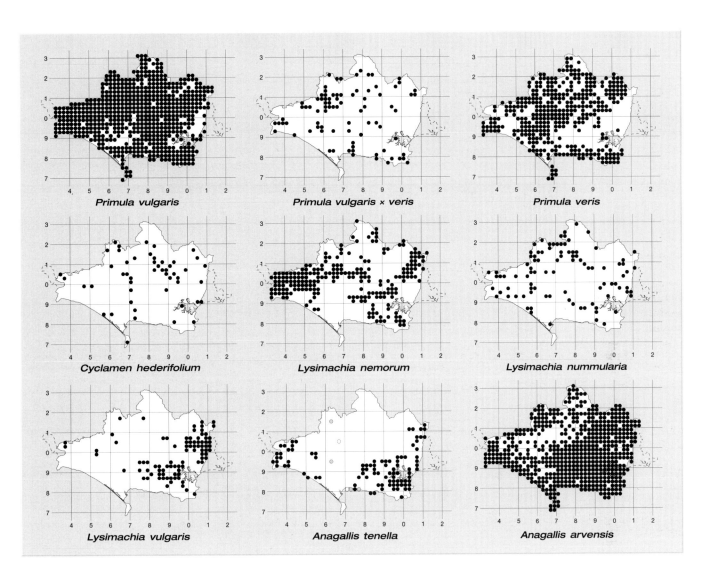

Primula vulgaris | Primula vulgaris × veris | Primula veris

Cyclamen hederifolium | Lysimachia nemorum | Lysimachia nummularia

Lysimachia vulgaris | Anagallis tenella | Anagallis arvensis

Lysimachia vulgaris L.
YELLOW LOOSESTRIFE
Locally frequent in tall vegetation by the larger rivers and streams, occasional in alder-swamps, wet woods, flushed bogs or reed beds. 87. Records outside the Poole basin are few and scattered, perhaps because wet habitats are less frequent; sometimes confused with *L. punctata*.

Lysimachia punctata L.
DOTTED LOOSESTRIFE
A garden escape or throw-out found as isolated patches on verges, old tips and in churchyards. 22 + [3]. It tolerates any soil, and while it is not usually found in wet places, it can be killed by drought. Some records might be the newly distinguished *L. verticillaris* Sprengel.

Anagallis tenella L.
BOG PIMPERNEL
This small, creeping plant of wet places is local in the Poole basin, the west and along the coastal belt. 94 + [7]. It has quite a range of habitats here; 45% bogs or wet heaths,

18% marshes, 11% flushes, 7% fen, and 19% divided between shaded carr, brackish marsh, rail tracks and urban lawns. Its survival is improved by moderate grazing, mowing or trampling.

Anagallis arvensis L.
SCARLET PIMPERNEL
A frequent, arable weed of light soils, in gardens, quarries, heathy tracks, dunes and upper strandlines of beaches. 566. Var. *carnea* Schrank, with pale pink flowers, is occasional near the coast. Var. *caerulea*, Blue Pimpernel, is so hard to tell from subsp. *caerulea* Hartman that records are lumped here: Forde Abbey (30M, 1992, !); Seatown (49F, 1961, !); Burton Mere (58D, 1927, AWG); Chilfrome (59Z, 1989, S Philp); Portesham (68C, 1941, RDG); Kingston Maurward (79F, 1996, MJG); Winfrith Heath (88D, 1932, FHH); Houghton (80E, 1867, JCMP in **DOR**); Iwerne (81, 1942, BNHS); Stoborough (98H, 1981, AH); Lytchett Matravers (99M, 1874, JCMP and 1958, AK Harding); Swanage (07, 1915, CBG in **BMH**).

Anagallis minima (L.) E.H. Krause
CHAFFWEED
Very local, scarce and hard to find in rutted tracks in heaths or woods, at the edges of sandy pools or in dune slacks. 14 + [11]. Restricted to the Poole basin and the west, where it often grows with *Radiola*. Post-1920 records: Champernhayes (39N, 1928, AWG and 1991, DAP); Pilsdon Pen (40A, 1928, AWG); Gilliards Coppice (78T, 1997, BE); Moreton (78Z, 1978, ! in **RNG** and 1991, DAP); Blacknoll (88D, 1964–90,!); Woolbridge Heath (88P, 1990, !); Whitehall (88W, 1998, BE); Oakers Wood and Affpuddle Heath (89A and B, 1930s, RDG in **DOR** and **WRHM**); Drinking Barrow (98B, 1998, BE); Stoborough (98H, 1992, DAP); Slepe Heath (98N, 1990, AJB); Wytch Heath (98S, 1977, ! in **RNG**); Godlingston Heath (08B, 1996, !); Queen's Copse (00N, 1981, AH); Crab Orchard (00Y, 1990, AJB); Gotham (01V, 1991, !).

Glaux maritima L.
SEA-MILKWORT
Locally frequent in salt-marshes and on muddy shores inside the Fleet lagoon and Poole Harbour, and sparse on beaches further west. 35 + [4].

Samolus valerandi L.
BROOKWEED
This plant, sacred to the Druids according to Tacitus, is found in small quantities in salt-marshes and in wet flushes on clay undercliffs along the coast. 40 + [13]. Inland it is rare in marshes or at the edge of flooded pits; Holdscroft Farm (39Z, 1999, DAP); Loscombe (59E, 1996, !); Brackets Coppice (50D, 1995, P Oswald); Halstock (50J, 1995, IPG); Honeycomb Wood (61H, 1962, CT); Moreton Heath (78Z, 1982, !); Bindon Abbey (88N, 1961, !); Turners Piddle (89G, 1978, AH); Wareham Common (98D, 1988, APC).

PITTOSPORACEAE

Pittosporum tenuifolium Gaertner
KOHUHU
Planted in churchyards, wild gardens or dunes. 6. A tree was 11 m tall in 1957 at Abbotsbury (58S, SG) where seedlings appear, as they do at North Haven (08I, !). Used

for hedging at Branksome Chine (08U), and also seen in 68X, 78Q, 99S and 08N.

Pittosporum tobira (Thunb.) Aiton f.
A rarely used hedging plant seen at Beaminster cemetery (40V) and Nothe gardens (67Z). 2.

HYDRANGEACEAE

Philadelphus coronarius L.
MOCK-ORANGE
Occasional in old plantations and woods, sometimes naturalised. 23. Established at Abbotsbury (58S), Dewlish (79T), Bryanston School grounds (80X, 1969–98) and probably elsewhere.

Philadelphus × *virginalis* Rehder
In plantations, easily confused with *P. coronarius*. 5. Seen in 70V, 81Q, 99K, 07I and 09I.

Philadelphus microphyllus Gray
Planted in hedges at Piddlehinton Camp (79I, 1992, ! in **RNG**), probably surviving from the 1940s.

Deutzia scabra Thunb.
Surviving in private grounds or abandoned gardens. 8. In a wood, Holt Forest (00H, 1994, !), also in 68W, 72S, 79I, 88V and Z, 80F and 99L.

Hydrangea macrophylla DC.
Abundantly planted in a shaded ravine at Abbotsbury gardens (58S, 1950–97), but not naturalised. 4. Also seen in a plantation at Chideock (49G, 1994, AWJ) and in churchyards at Eype (49K), Compton Abbas (81U) and Branksome (09Q).

Hydrangea serrata DC. is planted on Green Is (08D). 1.

GROSSULARIACEAE

Escallonia macrantha Hook. & Arn.
Occasionally established on sandy cliffs, frequently planted and escaping in hedges, plantations and

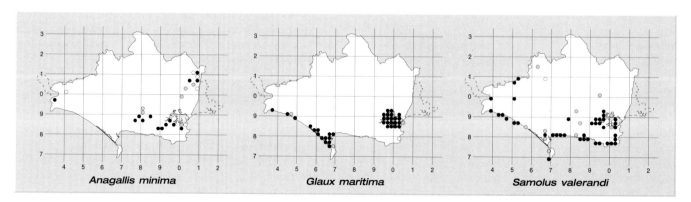

Anagallis minima   Glaux maritima   Samolus valerandi

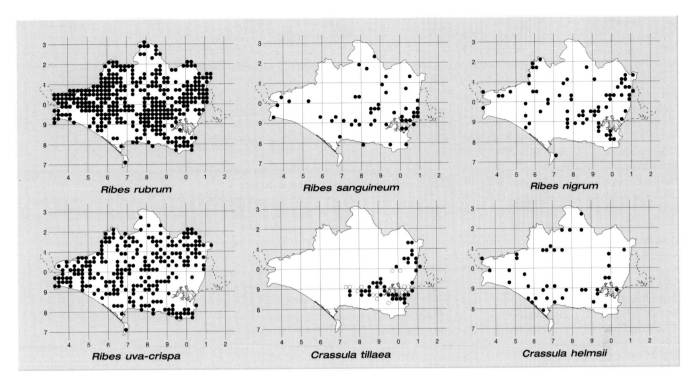

Ribes rubrum Ribes sanguineum Ribes nigrum

Ribes uva-crispa Crassula tillaea Crassula helmsii

churchyards, especially near the sea. 23. Naturalised on cliffs between Sandbanks and Branksome Chine, and further east in VC11 (08P, 1992, RMW, !). Moths love the flowers.

Escallonia × langleyensis Veitch
A pale-pink flowered shrub sometimes planted in hedges or churchyards. 11. Seen in 30M, 59Y, 78Q, 79I, 89K and M, 97E, 99L, 08D and G and 01Q.

Ribes rubrum L.
RED CURRANT
Common in most wet woods or by shaded streams, sometimes in hedges. 375. It fruits well and appears to be native. Frequently grown in cottage gardens and commercially in fruit farms.

[Ribes spicatum Robson
DOWNY CURRANT
A northern species recorded from Mapperton (59E, 1938, JEL in **RNG**) and Corfe Mullen (99U, 1908, EFL). Very close to *R. rubrum*.]

Ribes nigrum L.
BLACK CURRANT
Widespread by ponds or in wet woods, but never in quantity. 70. Perhaps mostly bird-sown from cultivated bushes. The leaf-scent is characteristic.

Ribes sanguineum Pursh
FLOWERING CURRANT
A garden shrub found increasingly as odd bushes in hedges, plantations and churchyards, and as seedlings on tips. 49. Mentioned by neither JCMP nor RDG. First record: Durlston

Head (07I, 1913, EBB). Well naturalised at Lamberts Castle (39U), Dungy Head cliffs (87E), Green Is (08D), Branksome Chine (08U) and Parkstone cutting (09K).

Ribes odoratum H.L. Wendl.
BUFFALO CURRANT
A yellow-flowered bush planted at Nether Cerne (69U, 1995, !) but not naturalised. 1.

Ribes uva-crispa L.
GOOSEBERRY
Widespread as isolated bushes in hedges and plantations on all soils but the most acid. 249. Probably an escape from cultivation, though long known; grown commercially in fruit farms.

CRASSULACEAE

Crassula tillaea Lester-Garland
MOSSY STONECROP
Very local on sandy tracks, old sandpits and cliffs with bare soil in the Poole basin. 38 + [17]. Somewhat sporadic, it occasionally colonises sandy fallow fields or nursery gardens in abundance, colouring the ground red.

Crassula helmsii (T. Kirk) Cockayne
NEW ZEALAND PIGMYWEED
In ponds, especially in gardens, and flooded pits. 41. First record: Ferndown (00V, 1962, MM Webster). Since then it has spread as an aquarists throw-out, and is still sold to unwary gardeners. It completely clogs ponds at Fishponds (39U), Admiral Digby Plantation (60P) and Redbridge Lane (78Z), but has not yet threatened native aquatics.

Umbilicus rupestris (Salisb.) Dandy
NAVELWORT, WALL PENNYWORT, CUPS-AND-SAUCERS
Local but persistent on old mortared walls in the south and west. 128 + [9]. Scarce on the Chalk. In the extreme west it becomes a plant of hedgebanks and sheltered woods, as it tolerates shade. It is at the eastern edge of its British range in Dorset. Seen as an epiphyte on ash at Fryer Mayne (78N), and on sandy cliffs at Redend Point (08G).

Sempervivum tectorum L.
HOUSE-LEEK, SELGREN
Planted on old walls and roofs, mostly in the north, also on a cliff-top boulder at Dungy Head (87E). 22 + [12]. In earlier days it was supposed to ward off fires.

Aeonium arboreum (L.) Webb & Berthel.
A tender garden escape once seen at Langton Matravers (97Z, 1997, !). 1.

Sedum spectabile Boreau
BUTTERFLY STONECROP
Common in gardens, escaping to verges outside and to waste heaps. 9. First noticed in 1994. Seen in 49E, 67V, 68B, 78B, 72M, 80H and W, 82S and 09I.

Sedum telephium L.
ORPINE
Very local and in small numbers in old woods and hedgebanks. 19 + [24]. Its distribution pattern is anomalous, with records from the west and the south-east, and it can occur on both acid and calcareous soils. It has an outlier at Piddles Wood (71W, 1895, JCMP and 1987, AH). Subsp. *telephium* is more often recorded than is subsp. *fabaria* (Koch) Kirschl.

Sedum spurium M. Bieb.
CAUCASIAN-STONECROP
A widespread planted alien or garden outcast on village walls and lanes, in churchyards and on tips. 25. First record: Castletown tip (67X, 1955, !).

Sedum stoloniferum S. Gmelin
LESSER CAUCASIAN-STONECROP
In similar habitats to *S. spurium*. 20. First record: On earth bank, Broadmayne (78I, 1993, !).

Sedum rupestre L.
REFLEXED STONECROP
Widespread on village walls. 102 + [13]. Rare in semi-native situations, e.g. Cliffs between Lyme Regis and Charmouth (39L, 1895, HNR and WF, *et al.*, and 1937, MAA in **WRHM**); rocks near Renscombe Farm (97T, 1980, !); Corfe railway cutting (98R, 1989, !). It also occurs on Portland (67, 1960, EH in **LANC**). Herbarium material is often inadequate to tell this from *S. forsterianum*.

Sedum forsterianum Smith
ROCK STONECROP
Frequent on undercliffs and in old limestone quarries on Portland (67 and 77, 1878, SMP in **DOR**, *et al.* in BM, and 1968–97, !). 6 + [7]. Other records are either old, or from village walls and presumably planted, as at Beaminster, Godmanstone, Long Burton, Trigon and Stoborough. There are no records further east along the south coast; reports from Lyme Regis cliffs need confirmation.

Sedum acre L.
BITING STONECROP
Native in fixed shingle near the coast. 212. Introduced and persistent on roofs, walls, rail tracks, pavements and disused asphalt.

Sedum sexangulare L.
TASTELESS STONECROP
A rare introduction on old walls. 3 + [1]. Abbotsbury (58S, 1908, HJR); Minterne Magna (60M, 1961–97, !); Lillington (61G, 1992, CT & !); Shaftesbury (82R, 1993, !).

Sedum album L.
WHITE STONECROP
Native, or well naturalised, on cliffs and screes at Church Ope Cove (67V) and Dungy Head (87E and 88A). 182 + [11]. Most records are from old walls, edges of drives or pavements in villages where the plant has been introduced for many years.

Sedum anglicum Hudson
ENGLISH STONECROP
Very local in dry, sandy places or on low cliffs, in the south-west and inside Poole Harbour. 16 + [11]. It also occurs in a gravel-pit at Black Hill (89G, 1987, RM) and along the

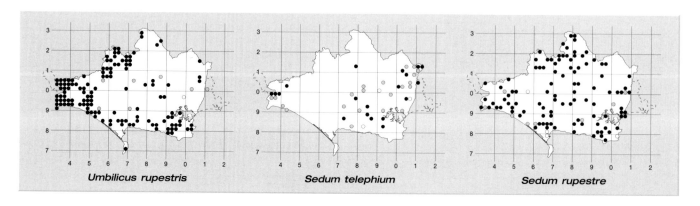

Umbilicus rupestris   Sedum telephium   Sedum rupestre

old tramline on Hartland Moor (98), and is abundant on Green Is and Furzey Is. It may be decreasing.

[Sedum dasyphyllum L.
THICK-LEAVED STONECROP
Introduced on walls, but not seen since 1938. 4. Burton Bradstock (48Z, 1937, DM); Upwey (68S, 1936, MJA); Blandford (80Y, 1893, JCMP in **DOR** and 1909, EFL); Corfe (98R, 1927, MJA); Swanage (07J, 1937, DM).]

Sedum hispanicum L.
SPANISH STONECROP
Once seen outside a garden near Broadmayne (78H, 1993, !). 1.

Sedum hybridum L.
SIBERIAN STONECROP
A yellow-flowered plant which is rarely introduced on old walls. 2. A35 near Askerswell (59B, 1994, !); Buckhorn Weston Ch (72M, 1995, !).

Sedum kamtschaticum Fischer & C. Meyer
A yellow-flowered introduction which is invasive and persistent. 2. Charmouth Ch (39R, 1994, !); Slepe Heath (98S, 1981, DG).

SAXIFRAGACEAE

Bergenia crassifolia (L.) Fritsch
ELEPHANT-EARS
A persistent garden escape in churchyards and on tips. 6 + [2]. Salway Ash Ch (49N); Abbotsbury (58M, 1961); Portland (67V); Portesham (68D); East Weare tip (77B, 1980); Black Hill sandpit, Bere (89H); Branksome Chine (08U); Merley Ho (09E).

Bergenia cordifolia (Haw.) Sternb. from Hooke Park (59J, 1993, LJM) and *B. ligulata* Engler from Sherborne tip (61I, 1970, ACL) are similar rare escapes, often hard to name.

Darmera peltata (Torrey ex Benth.) Voss ex Post & Kuntze
INDIAN-RHUBARB
Surviving in small numbers in wild gardens, in or by ponds and streams. 5 + [1]. Forde Abbey (30M); Abbotsbury (58S,

1985, C Jarratt, !); Minterne Magna (60R),; Spring Bottom (78K, 1961–78, !); Burleston (79S, 1961, RDG & G Williams); Winterborne Stickland (80H, 1982–96, !); Steeple (97A, !).

Saxifraga cymbalaria L.
CELANDINE SAXIFRAGE
A rock-garden annual, escaping but not established. [3]. Charmouth (39R, 1964, WD Lang); Langton Matravers (97Z, 1968, AB Ellis in **BM**); Parkstone (09F, 1964, KG). There are unconfirmed reports from Studland.

Saxifraga × urbium D. Webb
LONDON PRIDE
Rarely escaped in shaded sites. 3. Stoke Abbott (40K, 1991, DAP); Buckhorn Weston (72M, 1995, !); Pimperne Ch (90J, 1994, !).

Saxifraga granulata L.
MEADOW SAXIFRAGE
Very local in grassland or grassy woodland rides on chalk in the north. 11 + [11]. It is easily overlooked except during its short flowering season in May. Post-1920 records: Rampisham (50R, 1937, DM); Melbury Park (50T, 1921, AWG); Poyntington Down (62K, 1995, IPG and GAC); Bryanston (80T, 1966, BNHS); France Firs (80Z, 1991, !); Hambledon Hill (81L, 1936, RDG, *et al.*, and 1998, !); Melbury Hill (81U, 1953, CS); Ashmore Down (81Z, 1984, NCC); Zigzag Hill (82V, 1937, DM); Tarrant Hinton (91K, 1969, !); New Town (91Z, 1978, RB); Verwood (00Z, 1971, BNSS); Harley Down (01B, 1994, GDF); Garston Wood (01E, 1991, RJS); Pentridge (01I, 1984, NCC); Castle Hill (01L and R, 1904, EFL, *et al.*, and 1994, GDF, !); Whitey Top (01N, 1992, HCP). The double-flowered form was on sale at Blandford in 1782, and has been seen at Castleton (61S, 1970, ACL) and Manston Ch (81D, !).

Saxifraga hypnoides L.
MOSSY SAXIFRAGE
Garden forms of this, with white or deep-pink flowers, are planted and persist on walls or in churchyards. 6 + [1]. Charmouth (39R); Beaminster (40V, 1928, AWG and 1995, !); Sherborne (61, 1911, Anon in **HbSS**); Buckhorn Weston (72M); Milton-on-Stour (82E); Kingston (97P); Branksome (09Q).

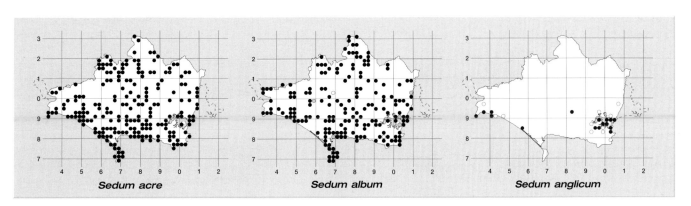

Sedum acre   Sedum album   Sedum anglicum

Saxifraga tridactylites L.
RUE-LEAVED SAXIFRAGE
Occasional amongst limestone boulders on undercliffs of Portland and the Purbeck coast. 70 + [11]. Elsewhere on walls, roofs, bridges, tombs, pavements, bare soil and rail tracks, variable in abundance from year to year.

Heuchera sanguinea Engelm.
CORALBELLS
Once seen on a bank outside gardens. Lyme Regis (39L, 1993, LJM and LMS). 1.

Tolmeia menziesii (Pursh) Torrey & A. Gray
PICK-A-BACK-PLANT
Rarely naturalised in churchyards or wild gardens in shade. 2. Litton Cheney (59K); Turners Piddle Ch (89G).

Tellima grandiflora (Pursh) Douglas ex Lindley
FRINGECUPS
Well naturalised in shaded plantations, wild gardens, village lanes or by streams, sometimes dominating the herb layer. 14. Forde Abbey (30M); Abbotsbury gardens (58S, 1984–96, !); Litton Cheney (59K); Melbury Park (50T, 1971, !); Minterne Magna (60R and S); Ambrose Hill (61J); Athelhampton (79S); Claysemore School (81S, 1969, PJN Lette and 1996, !); Corfe (98K); Knitson Farm (08A, 1999, !); Studland (08G, *c.*1945, *et al.*, and 1991–96, !); Kinson Common (09T, 1991, MA Stewart); Gaunts (00C, 1981, SPM Roberts and 1995, GDF).

Chrysosplenium oppositifolium L.
OPPOSITE-LEAVED GOLDEN-SAXIFRAGE
Locally frequent to dominant by springs and small streams in alder and other woods, sometimes in bare, open Greensand flushes. 175 + [4]. Common in the west, rather sparse in the Poole basin, and absent from deforested areas and the Chalk, except at Up Sydling (60F).

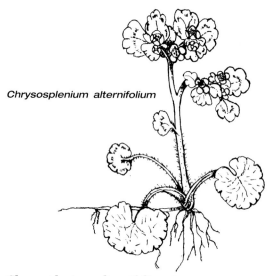
Chrysosplenium alternifolium

Chrysosplenium alternifolium L.
ALTERNATE-LEAVED GOLDEN-SAXIFRAGE
Local in alder-swamps, wet woods or by shaded streams in

the north and west, never in the open. 47 + [4]. Often with *C. oppositifolium* but much less common.

[*Parnassia palustris* L.
GRASS-OF-PARNASSUS
A northern plant, preferring flushed bogs or fens. 2. Wareham Heath (98, 1799,RP and 1893, RPM in **LIV**); Corfe bog, Norden Heath (98L, 1849, WM Heath in **DOR** and 1884, JCMP in **DOR**).]

ROSACEAE

Sorbaria sorbifolia (L.) A. Braun
A planted shrub which escapes by suckering to form colonies. 5. Winterbourne Abbas (69F); Sherborne Park Lake (61N); Old Park Wood (99H); Gussage St Andrew (91S); Wyke Down (01C, 1996, GDF).

Sorbaria tomentosa (Lindley) Rehder
Similar to *S. sorbifolia*. 2. Maiden Newton Ch (59Y); cliff below Holworth Ho (78Q).

Physocarpus opulifolius (L.) Maxim.
NINEBARK
Rarely planted in churchyards or escaped into hedges. 2 + [1]. Tinkers Barrow (78T, 1978, !); Bere Regis (89M); Compton Abbas Ch. (81U).

Spiraea salicifolia × *douglasii* = *S.* × *pseudosalicifolia*
Silverside
CONFUSED BRIDEWORT
Escaped from cultivation and forming colonies in hedges, plantations and waste ground, mostly on acid soil. 28 + [4]. All reports of *S. salicifolia* appear to be this. First record: between Broadstone and Poole (09C, 1889, WMR in **DOR**).

Spiraea douglasii Hook.
STEEPLE-BUSH
Similar to *S.* × *pseudosalicifolia* in status and habitat. 12 + [1]. First record: Creech claypits (98B, 1917, CBG in **BMH**).

Other cultivated *Spiraea* sp. have recent records as follows: *S.* × *arguta* Zabel Eype Ch (49K) and Sandford Orcas (62A, !) ; *S. canescens* D Don on tip, Woodlands (00P, GDF); *S. japonica* L.f. Giddy Green (88I) and old pit, Higher Hyde (89K, !); *S.* × *vanhouttei* (Briot) Carriere Southlands (67T), Bincombe Ch (68X), Tolpuddle fishing lakes (79X) and Winterborne Kingston Ch (89T); *S. veitchii* Hemsl. planted south of Canford Heath (09C).

Filipendula vulgaris Moench
DROPWORT
Frequent to rare in long calcareous grassland on the Northern Chalk, but absent from the chalk ridge in Purbeck. 80 + [6]. The distribution pattern is confused by isolated records of garden escapes on verges, sometimes in heathland, at Lyme Regis, Lodmoor, Hyde Heath, Arne, Durlston CP and elsewhere. The status of the plant on

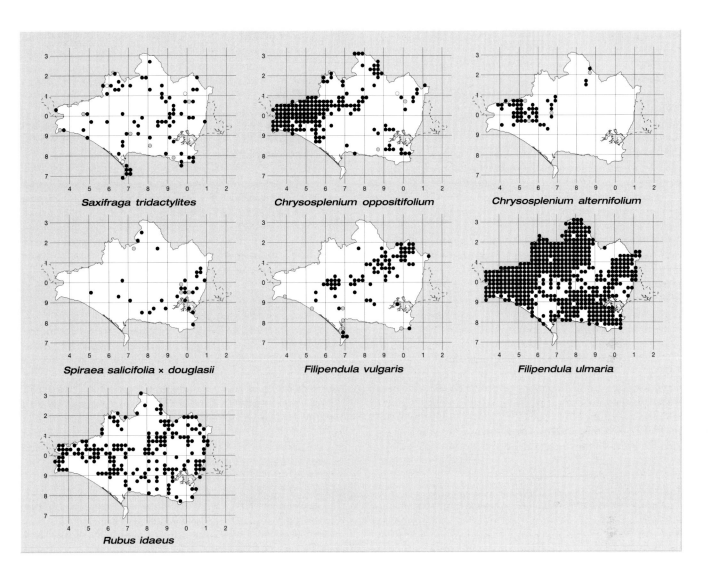

Saxifraga tridactylites

Chrysosplenium oppositifolium

Chrysosplenium alternifolium

Spiraea salicifolia × douglasii

Filipendula vulgaris

Filipendula ulmaria

Rubus idaeus

Portland is unclear; it was recorded there in 1912 by WBB, but today it is only seen as single plants which might be escapes.

Filipendula ulmaria (L.) Maxim.
MEADOWSWEET
Common in fens, wet pastures and verges, edges of woods and by rivers or streams. 499. It prefers clay soils, and can grow close to the sea, but is sparse on the dry chalk ridge. The plant was sacred to the druids, perhaps because of its healing action from the salicylates it secretes.

Kerria japonica (L.) DC.
Rarely planted in churchyards or on verges, usually as the double-flowered form. 4. Coles Farm (89E); Bere Regis (89M); Shapwick Ch (90F); Ferndown (00V, 1995, GDF).

Rubus tricolor Focke
CHINESE BRAMBLE
An invasive scrambler planted or escaped on banks

and verges since 1993. 7. Litton Cheney (59K); Sutton Poyntz (78B); Broadmayne (78I); West Melbury (82Q); Bailey Gate (99P); Broadstone (09C); Talbot Heath (09R, FAW).

Rubus parviflorus Nutt.
THIMBLEBERRY
Planted at Lower Lodge plantation (70V, 1994, !). 1.

Rubus idaeus L.
RASPBERRY
Widespread as colonies at the edges of woods, in hedgebanks and on railway banks. 193. It grows on both acid and calcareous soils, but avoids the coastal belt. Most wild clones have smaller fruits than modern cultivars, but the Cv. 'Lloyd George' originated from a wood near Corfe Castle.

Rubus phoenicolasius Maxim.
JAPANESE WINEBERRY
Rarely established. 1 + [1]. In wild garden, Litton Cheney (59K); Merley, old railway (09E, 1979, MB).

[*Rubus spectabilis* Pursh
SALMONBERRY
In a shrubbery, North Bowood (49P, 1925, AWG in **BDK**, also F Gale in **RNG**).]

Rubus fruticosus agg.
BRAMBLE
Very common in woods and plantations, woodland margins and clearings, hedges, scrubby heaths and waste ground, especially on sandy or gravelly soils, with a few microspecies on chalk or clay. 738. The berries, besides being gathered for human food, are important for small mammals and birds, while the leaves are eaten by the larvae of many moths. About 100 microspecies occur here, for two distinct *Rubus* florulas meet in the county, one Cornubian and the other from the Bournemouth region. Extensive collections of Dorset *Rubi* were made in the late 19th century by EFL, RPM, WMR and JCMP; some were circulated to herbaria as 'Set of British Rubi' (= Br. Rub. below). Good was no batologist, and recent work on the group has been done by LJM in the west, ESE in the south-east, and AN and DEA county-wide. The account below by DE Allen owes much to past batologists, and is the product of a critical culling of records from herbaria and literature; Allen's specimens are in **BM**. Dates are omitted, tetrads or six-figure grid references are given for some taxa, and the number of 10 × 10 km grid squares from which each microspecies is recorded is given. There is insufficient data to make tetrad maps worthwhile.

Subsect. *Rubus*
R. arrheniiformis W.C.R. Watson. Very rare. 1. Lambert's Castle (369988, LJM). A scarce near-endemic of England and Wales.
R. bertramii G. Braun. Very rare. 2. Rushy field, Lambert's Castle (39T, BSBI); Wareham (98, 1891, TS Lea in **BM**). Western.
R. briggsianus (Rogers) Rogers. Rare. 7. Powerstock Common (59, AN in **BM**); Halfway Ho Copse, now Wilkswood (97Z, JCMP and WMR in **BM** and **CGE**); Corfe (98L, LC in **OXF**); Daggons Road (11B, EFL, RPM and WMR in **BM, CGE** and **GL**). Endemic to the south-west British Isles.
R divaricatus P.J. Mueller. Rare. 1. West Moors (00R, AH Wolley-Dod in **BM** and WHM in **CGE**). Frequent in south-west Hampshire.
R. fissus Lindley. Very rare. 1. Golden Cap (49B, AN in **BM**). Mostly northern in Britain.
R. nessensis W. Hall. Rare. 4. Hardown Hill (49C, DEA); Yellowham Wood (79G, JCMP in **DOR**); Bere Wood (89S, EFL and RPM in **BM** and **DOR**); Alderholt (11B, EFL in **BM**). Mostly southern in Britain.
R. nobilissimus (W.C.R. Watson) Pearsall. Very local in wet places. 4. Old records from the northern Poole basin, recently from Kinson (068962) and Wools Bridge (101048; both BAM in **CGE**).
R. plicatus Weihe & Nees. Formerly local in the west and south-east, now rare. 7. Hardown Hill (49C, DEA); West Moors (00R, WHM in **CGE**).

R. scissus W.C.R. Watson. Uncommon. 5. Lambert's Castle (39U, BSBI); Worgret Heath (88Y, AN); East of Wareham Sta (98J, EFL, also NDS in **BM**); Queen's Copse (00I, EFL and WRL in **BM**); Holt Heath (00M, AN); West Moors (00R, RPM, also NDS in **BM,** WHM in **CGE**); near Fordingbridge (11B, EFL in **BM**).
R. sulcatus Vest. Rare. 3. Dullar Wood (99N, RPM in Br. Rub., also ESE in **NMW**); Westley Wood (90F, EFL); Alderholt (11B, EFL in **BM, CGE** and **LIV**).
R. vigorosus P.J. Mueller & Wirtgen. Formerly frequent in the south-east, now less so. 8. Worgret Heath (88Y, AN); east of Wareham Sta (98J, NDS in **BM**); Corfe Common (98K, DEA); Great Ovens Hill (99F, DEA); Turlin Moor (99Q, BAM in **CGE**).

Subsect. *Hiemales* EHL Krause.
 Ser. *Sylvatici* (P.J. Mueller) Focke.
R. albionis W.C.R. Watson. Frequent to locally abundant in the west, and on Okeford Hill (80E and J, DEA), rare elsewhere. 13. A widely western endemic.
R. boulayi (Sudre) W.C.R. Watson. Very local. 3. Puddletown Forest (79L, DEA); frequent near Bournemouth, e.g. Branksome Park (08U in Br. Rub. as *R. macrophyllus*). A trans-Channel species; elsewhere in the Hampshire basin and Normandy.
[**R. calvatus** Lees ex Bloxam. Extinct ?. 1. West Moors (00R, 1892, AHW in **BM**). Reported from South Haven (08I, 1935, NDS & AH Evans), but no specimen found. A mainly northern endemic.]
[**R. cambrensis** W.C.R. Watson. Extinct ?. 1. Piddles Wood (71W, 1889, TRA Briggs and WMR in **CGE**). Endemic to the south-west Midlands.]
R. errabundus W.C.R. Watson. Locally common. 6. Widespread in the Poole basin as a micromorph, notably on Brownsea Is, with outliers in the west at Wootton Hill (39N, LJM) and Abbotsbury Castle (58N, DEA). A mainly northern endemic.
R. imbricatus F.J.A. Hort. Uncommon in Purbeck and north-central Dorset. 8. Bound Lane, Chetnole (60, WMR in **CGE**); Rolf's Wood (81W, DEA); Crossways (773874, DEA); abundant, Halfway Ho Copse, now Wilkswood (97Z, WMR in **BM** and 993796, BAM in **CGE**, *et al.*); Studland Heath (016845, DEA).
R. laciniatus Willd. A garden escape. 5. Recorded from 80, 98, 08, 09 and 00, well established on Green Is (08D, !).
R. leucandriformis Edees & Newton. Local. 10. Mostly in the east, from Piddles Wood (71W), Cole Wood (88N), Corfe Common (88Q) and Verwood (00U). Endemic to the southern England.
R. lindleianus Lees. Once locally abundant, now frequent at most. 15.
R. macrophyllus Weihe & Nees. Rare, mostly as single bushes or patches. 6. All recent records are from the Poole basin, and early records are mostly errors. A wide-European species reaching its western limit in Britain here.
R. mollissimus Rogers. Scarce on heaths in the Poole basin. 5. Warmwell Heath (755872, DEA); Piddles Wood

(794127, BAM in **CGE**); Langton Matravers (97Z, WMR, Br. Rub. no.85); Rempstone Heath (98W, LC in **BM** and **OXF**); North of Verwood Sta (01V, EFL in **BM**, *et al.* in **CGE** and **SLBI**). Endemic to south England.

R. oxyanchus Sudre. Local. 10. Mainly in the Poole basin, also in Piddles Wood (71W). A trans-Channel species; elsewhere mainly in the Hampshire basin and Normandy.

R. perdigitatus Newton. Very rare. 1. Wootton Hill (356975, LJM). Endemic to south-west Wales and west Wessex.

R. pullifolius W.C.R. Watson. Local in heaths and open woods in the Poole basin. 9. Endemic otherwise to the Hampshire basin.

R. purbeckensis W.C. Barton & Riddelsd. Local in the west and the Poole basin. 9. Hardown Hill (49C); Pilsdon Pen (40A); Okeford Hill (80E) and Purbeck. The holotype in **BM** is from Parkstone (09). A S. Wessex-Irish near-endemic.

R. pyramidalis Kaltenb. Very local in the west and the Poole basin, with a micromorph, var. *parvifolius* Frid. & Gelert. 9. Stonebarrow Hill (39W, BSBI); Powerstock Common (59I, AN); Black Down (68D, DEA); Batcombe Hill (60Z, AN); Okeford Common (71V, LC in **BM**); Black Hill (89H, DEA) are outliers.

R. questieri Lef. & P.J. Mueller. Local in woodland, once locally dominant. 7. Kingcombe (50K, AN); Winfrith Heath (88D, DEA); Black Hill (89H, DEA); frequent in the Sturminster Marshall–Upton–Winterborne Zelston triangle (99); west of Studland (08, MA Rogers in **BM**). Common in north-west France, extending to south-west British Isles.

R. riparius Newton. Rare. 4. Lambert's Castle and Wootton Hill (39N and U, LJM); Hardown Hill (49C, DEA); Woolbridge Heath (88P, DEA); Sutton Holms (00P, BAM in **CGE**). A taxon of north Wales, south Wessex and Normandy.

R. subintegribasis Druce. Local in the Poole basin, west to Empool Heath (78N). 9. Endemic otherwise to the Hampshire basin.

Ser. *Rhamnifolii* (Bab.) Focke.

R. altiarcuatus W.C. Barton & Riddelsd. Local in the west and the Poole basin. 8. Lambert's Castle (39U, LJM); Abbotsbury Castle (58N, DEA); Fifehead Wood (72Q, DEA); Empool Heath (78N, DEA); Cole Wood (88N, DEA); Black Hill (89H, DEA); Seven Barrows (98E, BAM in **CGE**); between Corfe and Swanage (97Z, WMR in **BM** and ESE in **NMW**). A western taxon plentiful in north Devon and south Wales.

R. amplificatus Lees. Rare. 5. Lambert's Castle (39U, RDR); between Corfe and Swanage (97Z, HJR in **BM**); Henbury (99P, RPM in **CGE**); Woodlands (00D, AN); Sutton Holms (00P and U, 01K, WHM in **CGE** and **BAM**).

R. boudiccae A.L. Bull & Edees. Very rare. 1. Woodbury Hill (89M, M Porter). Also in east Wessex.

R. cardiophyllus Lef. & P.J. Mueller. Widespread and locally common, on acid and clay soils. 17. Var. *fallax* WCR Watson has been found in three places west of Bournemouth.

R. cissburiensis W.C. Barton & Riddelsd. Rare. 2. Woodbury Hill (89M, M Porter); Branksome Park (09K, AN). Spreading westwards from south-east England.

R. curvispinosus Newton & Edees. Very local. 5. Moreton Sta (780895, DEA); around Sturminster Newton, on clay (71); commonest in the east. Otherwise endemic to the Hampshire basin.

R. davisii D.E. Allen. Locally abundant in the Poole basin. 5. Winfrith Heath (88D, DEA), Bere Wood (89S, ESE) and heaths east of these sites. Known since the 1890s but only recently described.

R. dumnoniensis Bab. Local on heaths in the Poole basin. 5. A western taxon.

R. elegantispinosus (A. Schumach.) H.E. Weber. A naturalised garden escape. 1. Broadley Wood (850054, DEA); probably this on Brownsea Is (031877, DEA).

R. incurvatus Bab. Very local in the west and around Studland (08). 3. Abundant at Lambert's Castle (39U, LJM); most reports from the Poole basin belong to *R. davisii*. A western near-endemic.

R. lasiodermis Sudre. Local in the Poole basin. 4. It extends west to Black Hill (89H) and north to Verwood (00Z). An endemic of the Hampshire basin.

R. nemoralis P.J. Mueller. Occasional but widespread. 20. It occurs mainly on acid heathland.

R. pampinosus Lees. Locally abundant. 9. In woods and plantations in the Dorchester–Bere Regis–Wareham triangle, also at Piddles Wood (71W, RPM, also DEA) and Brownsea Is (08I, DEA). Br. Rub. no. 86 came from Bere Wood (89S). A western endemic.

R. pistoris W.C. Barton & Riddelsd. Rare and mostly as single bushes. 3. Black Hill (89H, DEA); Corfe Common (98K, DEA); Alderholt (11B, JCMP). Mainly a northern endemic, which also occurs in the New Forest.

R. polyanthemus Lindeb. Widespread, rarely abundant locally. 18.

R. prolongatus Boulay & Letendre. Locally abundant in the west and south-east. 14.

R. ramosus Bloxam ex Briggs. Very rare. 1. Dodpen Hill (353983, LJM in **BM**). A Cornubian endemic extending to east Devon.

R. riddelsdellii Rilstone. Rare. 3. Stonebarrow Hill (39W, BSBI); between Swanage and Corfe (97Z, HJR in **BM**, also LC in **BM** and **OXF**, det. AN); Corfe Common (98Q, AN). A mainly Cornubian endemic, here at its eastern limit.

R. rubritinctus W.C.R. Watson. Local in the west and the southern Poole basin. 8.

R. septentrionalis W.C.R. Watson. Very rare. 1. Branksome Park (09K, WMR, Br. Rub. no. 95). A northern species, elsewhere in south England only in Hampshire.

R. subinermoides Druce. Uncommon in hedgebanks and wood margins on clay. 5. Crossways (78N, DEA); Yellowham Wood (79G, AN); Fifehead Wood (72Q, DEA); Highwood (88S, DEA); West Moors (00R, WMR *et al.*, Br. Rub. no. 81). A near-endemic in south England, just extending to Devon.

Ser. *Sprengeliani* Focke.
R. sprengelii Weihe. Local in the far west, and in the Poole basin. 8.

Ser. *Discolores* (P.J. Mueller) Focke.
R. armeniacus Focke Cv. Himalayan Giant. An aggressive garden escape, which is spreading fast, as it tolerates all soils. 10. First reported in the 1970s.
R. armipotens W.C. Barton ex Newton. Very local, mainly in the east. 8. Warmwell Heath (78N); Fifehead Wood (72Q); Wareham (98); Verwood (00Z, all DEA); also in 49, 59, 50 (AN). Endemic to south-east England and spreading westwards, first recorded in 1964.
R. rossensis Newton. Locally common in the south-west, west of Abbotsbury (58) and Powerstock (59). 5. Also at Redbridge (78Z, AN). Endemic to south-west England and Wexford.
R. stenopetalus Lef. & P.J. Mueller. A rare calcicole in the north-east. 3. Fifehead Wood (72Q, DEA); Preston and Rolf's Woods (81W, DEA); Farnham Wood (91I, DEA). Elsewhere in Britain only in Hampshire and Wiltshire.
R. ulmifolius Schott. Very common on chalk and clay soils. 24+. A diploid species particularly prone to hybridise. Cv. *bellidiflorus*, with double flowers, was seen at Tarrant Monkton (90P, !).

Ser. *Vestiti* (Focke) Focke.
R. adscitus Genev. Rare; only a few bushes have been seen recently, in the west. 3. Lambert's Castle (39U, BSBI); Pilsdon Pen (40A, ALB and AN); Powerstock Common (59N, DEA). A western taxon, common in Devon and parts of Hampshire.
R. bartonii Newton. Very rare. 1. Cole Wood (885857, DEA); Worgret Heath (88Y, AN). A mainly Welsh taxon, also on the Mendips in Somerset.
R. corbieri Boulay ex Corbiere. Local, aggressively spreading, in the Poole basin; first found in 1892. 5. South of Moreton Sta. (78T and Z, DEA); Highwood Farm (88S, DEA); Wilkswood (97Z; Corfe Common (98K and Q, LC in **BM, CGE, NMW, OXF** etc, also DEA and SME); between Corfe and Swanage (98V?, EFL in **BM** and **LIV**); Rempstone Heath (98X, BAM in **CGE**); Studland Heath (08H, DEA and SME). A trans-Channel plant, found in the Cherbourg peninsula, the Channel Is and only Dorset in Britain.
R. criniger (E.F. Linton) Rogers. Very rare. 1. A patch on the Wessex ridgeway at Rolf's Wood (81W, DEA). An endemic of central and east England, also in west Hampshire.
R. lanaticaulis Edees & Newton. Uncommon. 6. Champernhayes (39N, HJR in **BM**, also BSBI); Pilsdon Pen (40A, DEA); above Valley of Stones (68D, DEA); Piddles Wood (71W, BAM in **CGE**, *et al.*); Whatcombe (80F, JCMP in **DOR** and RPM in **BM** and **CGE**). Endemic in the West of the British Is.
R. longus (Rogers & Ley) Newton. Rare in the north and west. 4. Stonebarrow Hill (39R and W, BSBI); Golden Cap (49B, AN); Batcombe Hill (60B, AN);

Chaffeymoor (73K, DEA). Endemic, mostly in south Wales.
R. orbus W.C.R. Watson. Rare in the north. 2. France Firs (80Z, DEA); Preston and Rolf's Woods (81W, DEA). A mainly Cornubian endemic.
R. vestitus Weihe. Locally abundant in woods on clay. 15. Absent from heaths and much of the west.

Ser. *Mucronati* (Focke) HE Weber.
R. cinerosiformis Rilstone. Very local in the west. 2. Champernhayes (39N, BSBI); Lambert's Castle (39U, LJM in **BM**); Pilsdon Pen (40A, DEA, *et al.*). Endemic to the south-west British Is.
R. mucronatiformis (Sudre) W.C.R. Watson. Locally abundant in the south-east, also in the south-west. 17. An outlier at Chaffeymoor (73K) on Greensand, near a Somerset population.

Ser. *Micantes* Sudre ex Bouvet.
(**R. glareosus** Rogers. Wood near Charmouth (39, 1937, FK Makins in **K**). Very likely to occur, but needs confirmation.)
R. heterobelus Sudre. Very local in the Bere Regis–Dorchester–Wareham triangle in the Poole basin. 4. First found in 1892 (Br. Rub. no. 15, as *R. praeruptorum*), in hedges west of Wareham. Almost endemic to Dorset: the lectotype is in **MANCH**.
R. leightonii Lees ex Leighton. Locally common in the west and south-east. 14. Also near Evershot (50, WMR) and Iwerne Minster (81X, DEA).
R. melanodermis Focke ex Rogers. Locally abundant, attaining its greatest profusion in Dorset. 16. The lectotype from Yellowham Wood and Puddletown Heath is in **CGE**.
R. micans Godron. Frequent. 15. Especially plentiful at Evershot (50). Predominantly south-western in the British Is.
R. moylei W.C. Barton & Riddelsd. Rare in the north. 2. Farnham Wood (91I, DEA); near Fordingbridge Sta. (11H, RPM in **BM**).
R. percrispus D.E. Allen & R.D. Randall. Very rare. 1. Hedge near Foxholes Wood (950984, EFL in **BM**, *et al.*, and DEA). So far known from south England and the Channel Is (Allen & Randall, 1995).
R. raduloides (Rogers) Dudre. Rare in central Dorset. 3. Bulbarrow (70S, AN); Piddles Wood (71W, TRA Briggs in **BM**, *et al.*, Br. Rub. no. 196, also DEA); Milton Abbas (80B, AN). Mainly a Cotswold plant.

Ser. *Anisacanthi* HE Weber.
R. anglofuscus Edees. Very rare. 1. Warmwell Heath (752866, DEA).
R. dentatifolius (Briggs) W.C.R. Watson. Locally common on acid soils of the west and south-east. 17.
R. hibernicus (Rogers) Rogers. Very rare. 1. In open plantation, Okeford Hill (80E, DEA). An endemic of west Britain and north Ireland, not found nearer than the Mendips.
R. leyanus Rogers. Rare. 6. Lambert's Castle (39U,

BSBI); Black Down and above Valley of Stones (68D, DEA); Yellowham Wood and Puddletown Forest (79G and L, DEA); Northport Heath (98E, M.Porter, also Carey Heath, DEA). A south-western taxon.

Ser. **Radula** (Focke) Focke.
R. bloxamii Lees. Locally abundant in the west and the north-eastern Poole basin. 12. Absent from heaths south of Lytchett Matravers, but extending to Melcombe Park (70S) and Cranborne Chase (91). In the west it has red styles and may be a distinct taxon, *R. multifidus* Boulay & Malbranche.
R. echinatus Lindley. Frequent, and tolerant of basic soils. 10.
R. euryanthemus W.C.R. Watson. Very rare. 1. Creech Heath (98B, ESE in **NMW**). Widespread in Britain but not found further west.
R. flexuosus P.J. Mueller & Lef. Local in south-east and central Dorset. 9. It is prone to hybridise.
R. insectifolius Lef. & P.J. Mueller. Scarce, and mostly in the Poole basin. 6. There are old records near Bournemouth by EFL, and others from Puddletown Forest (79K and L, DEA also ALB and AN); Black Hill (89G, DEA); Longthorns (89J, JCMP in **DOR**); Arne (98U, LC, *et al.*, and DEA).
(*R. longithyrsiger* Lees ex Focke is likely to be found, as it occurs in E. Devon, Somerset and Hampshire, and in the western extension of VC9 not covered by this Flora.)
(*R. radula* Weihe ex Boenn. Uncertain. 1. Probably 1 bush on clay, Fifehead Wood (72Q, DEA). Older reports are errors.)
R. sectiramus W.C.R. Watson. Very local. 4. Warmwell Heath (78N, DEA); Delcombe Wood (70X, DEA); Piddles Wood (71W, DEA); Broadley Wood (80M,DEA). Endemic to S.E. England; white-flowered in Dorset.
R. sempernitens D.E. Allen & L.J. Margetts. Very rare. 1. 1 bush at Monkton Wyld (331961, LJM). Elsewhere only in Devon and the Cherbourg peninsula.

Ser. **Hystrix** Focke.
R. asperidens Sudre ex Bouvet. Rare. 4. Golden Cap (49B, AN); Evershot (50S, AN); Piddles Wood (71W, many botanists in **CGE** etc, under various names.); Bere Wood (89S, AN).
R. atrebatum Newton. Very rare. 1. Pilsdon Pen (413011, DEA). Far from its headquarters in west Surrey.
R. dasyphyllus (Rogers) E.S. Marshall. Rare, often on chalk summits, and probably a recent invader. 6. Dogbury Gate (60M); Puddletown Forest (79L, DEA); Delcombe Wood (70X); Houghton North Down (80C); south of Hammoon (81B, MA Rogers in **BM**); Fontmell Down (81T); between Wareham and Corfe (98, HJR in **BM**).
R. durotrigum R.P. Murray. Locally abundant in, and largely endemic to, the Poole basin. 6. Br. Rub. no. 24; the lectotype is from Spetisbury (90B, RJM). It extends to the Chalk at Bulbarrow and south of Blandford, with an outlier at Alderholt (11B, JCMP in **DOR**).
R. marshallii Focke & Rogers. Very rare, and perhaps

introduced. 1. Black Down (68D, DEA). Endemic and mostly in west Surrey.
R. murrayi Sudre. Local. 4. Longthorns (89J, JCMP); also around Edmondsham and West Moors in the east. It does not occur further west in Britain.
R. phaeocarpus W.C.R. Watson. Local, probably invading from the east. 3. East of Evershot (50X, DEA); between Broadmayne and Crossways (78N); Bindon Abbey (88N); Highwood (88S).
R. pseudoplinthostylus W.C.R. Watson. Very rare, and too restricted in range to merit status as a microspecies. 2. Foxholes Wood (99N, RPM, *et al.*, and AN; the holotype is in **BM**); Rolf's Wood (886138,DEA)
R. rilstonei W.C. Barton & Riddelsd. Rare in the west. 1. Rides in Champernhayes forest (39N, LJM in **BM**). A Cornubian taxon.
R. scabripes Genev. Uncommon, mainly in the west. 4. Champernhayes (39N, BSBI); Birdsmoor Gate (30V, ALB and AN); Golden Cap (40B, AN); Seven Barrows plantation (69L, WMR in **BM**); Hurst Heath (78Z, DEA). A mainly Cornubian taxon.
R. venetorum D.E. Allen. Very rare. 1. One clump near Redbridge (776885, DEA). Found in west Britain and Brittany.
R. vigursii Rilstone. Very rare in the west. 2. Pilsdon Pen (40A, ALB and AN); Black Down (1 bush, 612876, DEA). A Cornubian endemic.

Sect. **Corylifolii** Lindley (intermediate between *R. fruticosus* agg. and *R. caesius*).
R. conjungens (Bab.) Rogers. Frequent on mostly basic soils. 10.
R. nemorosus Hayne & Willd. Local, mainly in the Poole basin, and often misdetermined. 9. Puddletown Forest (79L, DEA); Northport (99B, DEA); between Hamworthy junction and Bailey Gate (99M?, RPM in **CGE**); between Sutton Holms and Woodlands (00P, WHM in **CGE**); also in the west, AN.
R. pruinosus Arrh. Locally common, mostly in the west and south. 19.
R. transmarinus D.E. Allen. Locally common round Bournemouth, but too recently described to have many records. 2.
R. tuberculatus Bab. Locally common in the south-east, especially near the coast. 8.

Rubi innominati
There are a few locally-distributed brambles which cannot be matched with any named species, three of which are:
a. *R. leucostachys* var. *angustifolius* Rogers occurs in the east between Wimborne and Alderholt.
b. A member of Ser. **Sylvatici** abundant on Okeford Hill (80E), which resembles the Wealden taxon 'false *orbifolius*' but has pilose anthers and often clasping sepals.
c. A pink-flowered member of Ser. **Rhamnifolii** simulating *R. cissburiensis*, common in S. Devon, occurs at Lambert's Castle (39U, BSBI).

Subsect. *Caesii* Lej. & Courtois.
Rubus caesius L.
DEWBERRY
Common at the edges of woods, hedgebanks and heathland verges, but perhaps confused with some of its hybrids. 372. It flowers earlier than *R. fruticosus*.

Potentilla fruticosa L.
SHRUBBY CINQUEFOIL
Widely cultivated. 4. Bincombe Ch (68X, 1998, !); a hortal form on nursery waste, Lytchett (99S); self-sown, Corfe Mullen Ch (99U, !); a white-flowered form planted at Horton Ch (00I).

Potentilla palustris (L.) Scop.
MARSH CINQUEFOIL
Local in bogs, fens, dune slacks, ditches, edges of lakes or flooded pits in acid places. 30 + [11]. Restricted to the Poole basin and Powerstock Common (59I and P, 1850s, E Fox, *et al.*, and 1994, SME). A report from Lodmoor (68V) needs confirmation.

Potentilla anserina L.
SILVERWEED
Common in marshes, compacted soil of verges and wet paths, edges of ponds, in disturbed soils and locally abundant at the upper level of some beaches. 678. William Barnes devotes a poem to it.

Potentilla argentea L.
HOARY CINQUEFOIL
Very local in dry, sandy fields, verges or pits, preferring bare, disturbed ground. 13 + [13]. Confined to the Poole basin, where it may be decreasing; it is no longer seen on rail tracks. Grown in some gardens, e.g. at Compton Acres (08P). Post-1960 records: Bovington (88N, 1991, BE); Woolbridge Heath (88P, 1990, !); Gallows Hill (89K, 1998, BE and DAP); Worgret pit (98D, 1964, JB and 1993, MHL); Carey (98E, 1997, RB); Sandford (98J, 1980, AJB); Middlebere and Scotland (98S, 1990, AJB and DAP); abundant at Fitzworth (98Y, 1991–98, BE and DAP); Marsh Br, Almer (99E, 1999, !); Greenland (08H, 1991, AH and DAP); Castle Hill (01L and R, 1979, !).

Potentilla recta L.
SULPHUR CINQUEFOIL
Rarely established on verges. 9. Tizards Knap (49C, 1982, AWJ); Pilsdon Pen (40A, 1960, JKH); near Yeovil (51T, 1956, IAR); West Holme (88X, 1992, !); Milton Park Wood 980F, 1989, !); Compton Abbas (81U, 1996, !); Corfe Castle mound (98L, 1953, PJ Wanstall in **QMC**, *et al.*, and 1968–91, !); Hogstock (90N, 1981, AH); New Swanage (08F, 1998, EAP); Holes Bay (09B, 1991, !); Stephen's Castle (00Z, 1993, VS); Alderholt (11B, 1960, KBR).

[*Potentilla intermedia* L.
RUSSIAN CINQUEFOIL
A rare casual. 2. Stoke Abbott (40K, 1927, AWG in **BDK**); Weymouth Sta (67U, 1934, NDS).]

Potentilla norvegica L.
TERNATE-LEAVED CINQUEFOIL
A rare casual. 1 + [2]. Weymouth Sta (67U, 1935, NDS); New Swanage (08F, 1999, EAP); Branksome (09K, 1947, KG).

Potentilla erecta (L.) Rausch.
TORMENTIL
Frequent on acid soils in woodland rides, old grassland and heaths, including leached turf overlying Chalk or limestone. 374. Least common on the Chalk and on clays in the north; only once recorded from Portland (67X, 1981, D Griffith). Subsp. *strictissimum* (Zimm.) A Richards is found in Hewood Bottom (30R, 1993, LJM &LMS) and may have been under-recorded. A form with small, strongly emarginate petals occurs on Canford Heath (09C and H, !).

Potentilla erecta × *anglica* = *P.* × *suberecta* Zimm.
Hard to name. 2. Reports from Rye Water Farm (61Q, 1993, SME) and Holton Lee (99K, 1994, DG).

[*Potentilla erecta* × *reptans* = *P.* × *italica* Lehm
By the railway north of Verwood Sta (00Z, 1900-08, EFL); both railway and station are defunct.]

Potentilla anglica Laich.
TRAILING TORMENTIL
Locally frequent in grassy rides of old woods on both acid and calcareous soils. 128 + [9]. Absent from deforested areas and least common on the Chalk. Plants on Brownsea Is may be this or hybrids.

Potentilla anglica × *reptans* = *P.* × *mixta* Nolte ex Reichb.
Probably under-recorded, on disturbed sandy soils. 3 + [4]. Lyme Regis (39G, 1974, OM Stewart in **E**); Knighton pits (78P, 1996, BE, DAP and DG); Sturminster Newton (71X, 1890, RPM in **LIV**); Lytchett Matravers (99M, 1893, RPM); Swanage (07J, 1999, EAP); Kinson (09T, 1900, EFL); West Moors (00W, 1995, GDF and JO).

Potentilla reptans L.
CREEPING CINQUEFOIL
Frequent in more or less disturbed grassland, verges, undercliffs, lawns and gardens, perhaps preferring clay soils. 686. Var. *microphylla* Trutt, a small-leaved form, has a record from Weymouth (67, 1908). A robust, large-leaved form, as seen in the Mediterranean, is frequent on Portland; this may represent a form without introgression from other species.

Potentilla sterilis (L.) Garcke
BARREN STRAWBERRY
Locally frequent in old woods and dry hedgebanks. 392. Commonest in the west, and absent from deforested areas. It occurs in shelter on east Portland undercliffs.

Fragaria vesca L.
WILD STRAWBERRY
Frequent in open woods, hedgebanks, village lanes, verges, railway banks and churchyards. 401. Scarce in deforested areas and on very acid soils in the east.

Fragaria moschata (Duchesne) Duchesne
HAUTBOIS STRAWBERRY
A garden plant occasionally found in village lanes or churchyards, probably decreasing. 6 + [6]. Puncknowle (58J, 1883, JCMP in **DOR**); Litton Cheney (59K, 1995, !); Sydling St Nicholas Ch (69J, 1982, !); Whatcombe (80F, 1896, JCMP in **DOR**); Kingston Ch (97P, 1996, !); Rempstone (98W, 1895, JCMP in **DOR**); Studland (08G, 1998, !); Branksome Park (09K, 1993, FAW); Cranborne (01C, 1989, ACL).

Fragaria ananassa (Duchesne) Duchesne
GARDEN STRAWBERRY
Much planted for its fruit. Escaped colonies are commonest on railway banks, as they have been for a century. 10 + [7]. West Bay (49Q, 1993, LJM); Clifton Wood (51R, 1978, AE Newton); Ferry Bridge (67T, 1992, !); Stratton (69L, 1978, C Sargent); Dorchester W. Sta (69V, 1992, !); Sherborne railway (61N, 1970, ACL); Sutton Poyntz (78B, 1993, !); Tolpuddle (79X, 1997, !); Wareham Sta (98E and J, 1978–92, !); Bailey Gate (99P, 1900, EFL and 1995, !); Higher Row (00L, 1993, DJG); Ferndown (00Q, 1989, !).

Duchesnea indica (Andrews) Focke
YELLOW-FLOWERED STRAWBERRY
Well naturalised in plantations and churchyards, especially near Blandford. 5 + [1]. Corscombe (50C, 1950, AWG in **BDK**); Little Monington (80T, 1968, MJC Scoular, *et al.*, and 1990, !); The Cliff, Bryanston (80T and X, 1999, !); Blandford Ch (80Y, 1971, A Lack, and 1977–99, !); verge near Morden Decoy (99B, 1994, !); Gillingham Ch (82D, 1995, !).

Geum rivale L.
WATER AVENS
Very local in wet woodland rides, water-meads and in wet flushes on heaths; occasionally grown in gardens, but not reported as an escape. 36 + [7]. In the west and the Poole basin only.

Geum rivale × urbanum = G. × intermedium Ehrh.
In small numbers within the range of *G. rivale*. 7. Post-1925 records: Shipton Gorge (49W, 1977, EJ Chaplin): Meerhay (40W, 1928, AWG); Wynford Wood (59T, 1937,

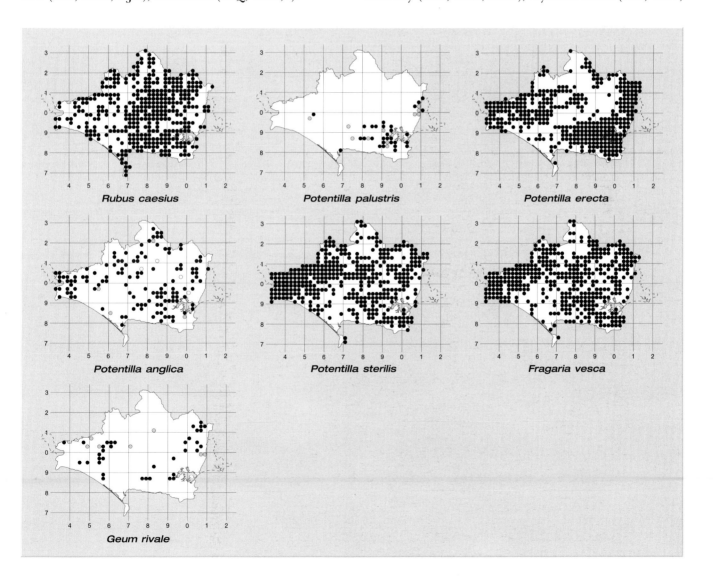

Rubus caesius

Potentilla palustris

Potentilla erecta

Potentilla anglica

Potentilla sterilis

Fragaria vesca

Geum rivale

DM); Benville (50L, 1928, AWG); Melbury Park (50X, 2000, JAG); Came Wood (68X, 1955, JKH); Batcombe (60B, 1969, F Newbould); Wareham meads (98D and J, 1981, AJB); West Moors (00R, 1991, BE); Burwood (01S, 1992, ACP).

Geum urbanum L.
WOOD AVENS
Very common in woods, hedgebanks and shaded lanes. 681.

Agrimonia eupatoria L.
AGRIMONY
Common, but rarely in quantity, in woodland rides, verges and old meadows. 514. Scarce in built-up areas round Poole and north of Weymouth.

Agrimonia procera Wallr.
FRAGRANT AGRIMONY
Scarce, in similar habitats to *A. eupatoria*. 24 + [12]. Commonest in the Poole basin, but tolerant of a wide range

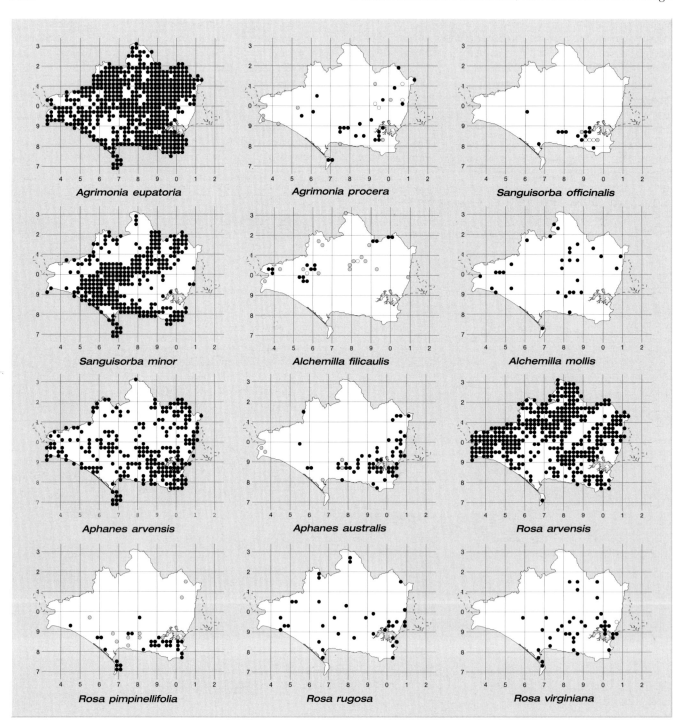

Agrimonia eupatoria

Agrimonia procera

Sanguisorba officinalis

Sanguisorba minor

Alchemilla filicaulis

Alchemilla mollis

Aphanes arvensis

Aphanes australis

Rosa arvensis

Rosa pimpinellifolia

Rosa rugosa

Rosa virginiana

of soils. It occurs on Portland, but is absent from most of the Chalk and the north.

Sanguisorba officinalis L.
GREAT BURNET
Local in water-meads or flushes in heaths, mostly in the lower Frome valley. 11 + [5]. JCMP had it from six sites, RDG from 11 stands. West Radipole Lake (68Q, 1977, D Ireland); Watercombe (78T, 1978, E.Pearse and 1991, DAP); Tadnoll (78Y, 1966–93, !); Winfrith Heath (88D, 1960–97, !); Povington ranges (88S and W, 1991, DAP, !); East Holme (88Y, 1937, DM); The Plantation (97P, 1987, AH and JMB); Stoborough (98C, D and I, 1980, AH, *et al.*, and 1993, JB, !); north east of Wareham (98J, 1987, RM, !).

Sanguisorba minor Scop.
SALAD BURNET
Frequent in calcareous grassland, sometimes on calcareous clay, and scarce on eutrophiated heathy banks, as at Abbotsbury Castle (58S). 306. Subsp. *muricata* (Gremli) Briq. is a scarce introduction, seen recently on sown turf, verges or railway banks at Beaminster (40V, IPG), Melbury Sampford (50X, IPG and PRG), Small Mouth Halt (67T, !), West Stafford (78J, !); Piddlehinton Camp (79I, !), Hinton St Mary (71X, !), Stourpaine (80U, !) and Baiter (09F, !). 8.

Acaena novae-zelandiae Kirk
PIRRI-PIRRI-BUR
An invasive alien in sandy soil. 6. Moreton caravan park (78Z, 1993, AH); Stoborough (98 H, 2000, R. J. Swindells; Swanage (07J, 2000, EAP); near Knoll Ho, Studland (08G, 1949, JO, *et al.*, spreading into dunes in 08H and I by 1998, !).

Alchemilla filicaulis Buser subsp. vestita (Buser) Bradshaw
LADY'S-MANTLE
Local and in small numbers in woodland rides, hedgerows and verges. 15 + [18]. A northern species, now only in the north and west of the county, but formerly more widespread. Pulteney saw it at Grange in south Purbeck in the 18th century.

Alchemilla mollis (Buser) Rothm.
GARDEN LADY'S-MANTLE
In small amounts, but seeding freely, on verges, in churchyards, wild gardens and tips; increasing. 31. First record: Sherborne tip (61I, 1970, ACL).

Aphanes arvensis L.
PARSLEY-PIERT
Frequent in dry, arable fields, disturbed grassland, on anthills and as a vernal weed of lawns. 249. It tolerates calcareous and sandy soils, and sometimes accompanies *A. australis*.

Aphanes australis Rydb.
SLENDER PARSLEY-PIERT
Occasional in dry, sandy turf, on banks, in fallow, by tracks, old railways, sandpits and sandy cliffs. 55 + [5]. Probably under-recorded, mainly in the Poole basin, absent from the Chalk.

Rosa multiflora Thunb ex Murray
MANY-FLOWERED ROSE
Rare in hedges on a variety of soils, first noted in 1990. 10. Burton Bradstock (48Z, ! in **RNG**); Beaminster (40Q, LJM and LMS); Winterbourne Abbas (68E, LJM and LMS); Bloxworth (89X, !); Iwerne Minster (81S, !); Blue Pool (98G, !); Lytchett Matravers (99M, !); Talbot Heath (09R, !); Ferndown (00V, GDF); Verwood (00Z, VMS).

Rosa arvensis Hudson
FIELD-ROSE
Frequent at the edge of woods and in old hedges. 381. Absent from deforested areas and parts of the Chalk. A record from Portland (67V, M Leicester) needs confirmation.

Rosa arvensis × stylosa = R. × pseudorusticana Crepin ex Rogers
Perhaps overlooked. 1 + [1]. Thorncombe (30R, 1992, CDP in **CGE**); Chetnole (60E, 1878, WMR in **OXF**).

[*Rosa arvensis* × *gallica* = *R.* × *alba* L.
One old record from Verwood (00Z, 1895, EFL).]

[*Rosa arvensis* × *micrantha* = *R.* × *inelegans* Wolley-Dod
Reported from Studland (08, 1920, GA Boulenger); no specimen seen.]

Rosa pimpinellifolia L.
BURNET ROSE
Local in dry, sandy heaths and on old tramlines in Purbeck, and in shallow turf over limestone on Portland. 30 + [11]. It is also a scarce garden escape inland, certainly so in 49L, 78P, 88F, 89D and 91W.

Rosa rugosa Thunb. ex Murray
JAPANESE ROSE
Frequently planted or escaping in hedges, verges, railway banks and churchyards. 35. First record: Warmwell (78N, 1956, !). Well established at Ham Common beach (99V), South Haven dunes (08I) and Canford Cliffs (08P). With white flowers at Corscombe (50H, LJM) and Anderson (89Y).

Rosa glauca Pourret
RED-LEAVED ROSE
Bushes escaped on heathland at Blue Pool (98G), Ham Common (99Q), Sandbanks (08N) and near Ferndown (00Q, 1998, ! det. ALP).

Rosa virginiana Herrm.
VIRGINIA ROSE
Planted or escaped in hedges, railway banks, churchyards, waste ground and tips, but not well naturalised. 39 + [2]. First record: Wallis Down (09S, 1891, EFL).

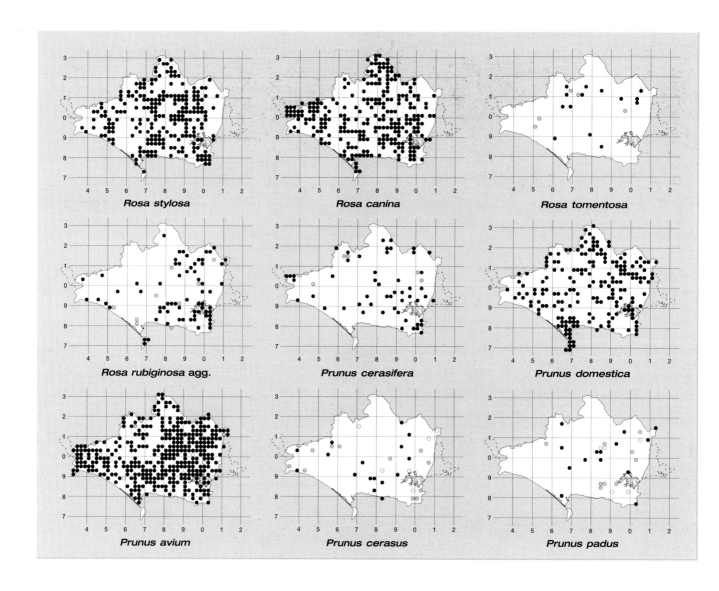

Rosa stylosa · Rosa canina · Rosa tomentosa

Rosa rubiginosa agg. · Prunus cerasifera · Prunus domestica

Prunus avium · Prunus cerasus · Prunus padus

Rosa stylosa Desv.
SHORT-STYLED FIELD-ROSE
Frequent, at the edges of woods and in hedges. 225. It prefers calcareous soils but is not confined to them. The flowers are often deep pink.

Rosa stylosa × *canina* = *R.* × *andegavensis* Bast.
Frequent as single bushes in hedges and scrub, probably under-recorded. 26.

Rosa canina L.
DOG-ROSE
Common in hedges and scrub. 260 (s.s.). When sterile, this and *R. stylosa* were often recorded as *R. canina* agg. 690 (agg.). Pure white-flowered forms were seen at Cuckoo Pound and Ballard Down (07D and 08F, EAP) and are probably frequent. Fruiting of hedgerow roses is much affected by flailing. The hips are eaten by winter birds, while the leaves are eaten by many moth larvae.

Rosa canina × *obtusifolia* = *R.* × *dumetorum* Thuill.
Rare. 1. On chalk, Compton Down (81Z, 1997, CDP and DAP det. ALP).

Rosa canina × *sherardii* = *R.* × *rothschildii* Druce
Rare. 1. Three bushes in a hedge, Verwood (00Z, 1993, VMS det. ALP).

Rosa canina × *micrantha* = *R.* × *toddiae* Wolley-Dod
Rare. 1. On chalk at Compton Down (81Z, 1997, CDP and DAP det. ALP).

[*Rosa caesia* Smith
GLAUCOUS DOG-ROSE
A northern species, once found at Chetnole (60E, 1879, WMR in **BM**) and reported as *R. obtusifolia*.]

Rosa tomentosa Smith
HARSH DOG-ROSE
Scarce at the edges of woods and in hedges, mostly in the

north. 14 + [6]. Since JCMP recorded it from about 40 sites in 1895, it has decreased.

Rosa sherardii Davies
SHERARD'S DOWNY-ROSE
Rare here and in southern England. 2. Uploders (59B, 1993, LJM); south-east of Stalbridge (71N, 1992, LJM and JO); also in Purbeck (98, 1944, EBB).

Rosa rubiginosa L. agg.
SWEET-BRIAR
Occasional in scrub, rarely in hedges. Of 25 recent reports giving habitat, 15 were from heath and 10 from calcareous grassland; it also occurs on clay. 68 + [10]. JCMP thought it rare in 1895.

Rosa micrantha Borrer ex Smith
SMALL-FLOWERED SWEET-BRIAR
Uncommon and in small quantity at the edges of woods and in scrub, on both heathy and calcareous soils. 11 + [6]. Post-1950 records: Waytown (49T, 1979, M James); Bracket's Coppice (50D, 1992, JGK); Hunters Bridge (61Q, 1992, JO and LJM); Owermoigne (78S, 1967, JB); Moreton (78Z, 1981, AJCB); Puddletown (79, 1954, AWW); Trigon (88Z, 1968, BNSS); Enford Bottom (80P, 1999, BE); Hambledon Hill (81L, 1999, !); Melbury Hill (81U, 1997, CDP and DAP); St Aldhelm's undercliff (97S, 1993, !); Wareham (98, 1954, AWW); Upton CP (99W, 1983, Anon); Blandford Camp (90I and J, 1999, BE); Studland (08G, 1968, JB); Shell Bay (08M, 1999, EAP); Verwood (00Z, 1993, VMS).

[Rosa micrantha × agrestis = R. × bishopii Wolley-Dod
A rare hybrid. 1. Kinson (09T, 1891, MA Rogers in **BM**, det. GG Graham).]

Rosa agrestis Savi
SMALL-LEAVED SWEET-BRIAR
Rare in chalk scrub in the north-east. 3 + [2]. Bere Wood (89S, 1895, RPM); Melbury Hill (81U, 1895, EFL and 1997, CDP and DAP); Fontmell Down (81Y, 1993, DWT); Sovell Down (91V, 1974, DJG and 1998, !). Protected and producing seedlings in the last site.

Alien planted roses include **R. moyesii** Hemsl. & Wils. at Tarrant Gunville Ch (91G) and probably **R. sericea** Lindley on Bere Heath (89R, RSRF & !).

Prunus persica (L.) Batsch.
PEACH
Seedlings or saplings are rare, where peach stones have been thrown. 5. Five Bridges (72K, JM Newton); Scratchy Bottom (88A, !); Tatchells Pit (98E, !); cliffs at Anvil Point (07I, !). 41 varieties of peach were on sale in Blandford in 1782, but it is rarely grown today.

Prunus cerasifera Ehrh.
CHERRY PLUM
Widespread as isolated trees in hedges or churchyards,

mostly as the Cv. *pissardii* with dark red leaves. 52 + [5]. First 'wild' record: Holt (00G, 1932, FHH). The green-leaved form sporadically yields edible red fruits.

Prunus spinosa L.
BLACKTHORN, SLOE
Very common in hedges, thickets, edges of woods and undercliffs, especially on clay but also on sandy and calcareous soils. 719. Fruits abundantly, but the sloes remain sour until January. The leaves are eaten by many moth larvae, but can be blackened by salty gales. Two Blackbush names and Sloes Hill refer to this plant; 'thorn' names are ambiguous. Once used to cure warts (Udal, 1889).

Prunus spinosa × domestica = P. × fruticans Weihe
Probably frequent but overlooked. South Perrott (40T, IPG). The fruit stone shape is intermediate between those of the parents.

Prunus domestica L.
WILD PLUM
Occasional, in small quantity, in hedges, usually close to cottages or their ruins, as in Piddles Wood (71W). 186. Twenty-eight varieties were on sale at Blandford in 1782. Fruits are mostly bluish-purple or red, rarely yellow as at Sandsfoot (67T). Subsp. *insititia* (L.) Bonnier & Layens, the BULLACE, was thought to be native by JCMP, but never looks so today, and it has only three modern records.

Prunus avium (L.) L.
WILD CHERRY, MERRY-TREE
Occasional to frequent in woods and hedges, commonest on calcareous soils and absent from very acid areas. 372. It is so often planted that the natural distribution pattern is obscured. Large trees, *c.* 100 years old, in Broadley Wood, were felled in 1997; they, and surviving big trees at Netherbury (49U) and High Hall (00B), were or are about 20 m tall. The fruit is eaten by many birds.

Prunus cerasus L.
DWARF CHERRY
Scattered in hedges, railway banks, verges and scrub in small numbers. 17 + [17].

Prunus padus L.
BIRD CHERRY
Single trees of this northern cherry are found in hedges and copses, often obviously planted. 14 + [14]. Some records might be *P. serotina*.

Prunus serotina L.
RUM CHERRY
Introduced and spreading rapidly by seed in one wood on acid soil. 4. Coltclose Corner (88D, 1997, !); North Oakers and Sares Woods (89A and B, 1970–97, ! in **RNG**); planted, Dudsbury (09T, 1984, FAW). The ripe fruit attracts many birds.

Prunus lusitanica L.
PORTUGAL LAUREL
An evergreen tree, frequently planted in plantations or shrubberies. 55. Seedlings have been found at Beaminster (40Q) and Tricketts Cross (00V). First record: Brownsea Is (08E, 1968, DWT).

Prunus laurocerasus L.
CHERRY LAUREL
Commonly planted in woods and plantations, where it sows itself and can become abundant in the understorey. 325. On sale, with *P. lusitanica*, at Blandford in 1782.

Many **Prunus** species are cultivated and occur in hedges, verges and churchyards, where they are hard to name when not in flower. Thus **P. serrulata** Lindl., Japanese Cherry, has been recorded from 78M and U, 81U, 07E and 09W, **P. serrula** Franch. from Huish Manor (99D) and **P. subhirtella** Miq. from 79U, 70Y, 80A and T, 81J, 98Z and 00Q, but none are naturalised. Canford School (09J, TH) has a particularly rich collection.

Cydonia oblonga Miller
QUINCE
Two planted bushes survive in the grounds of Spyway Ho (97Y, 1991–97).

Chaenomeles speciosa (Sweet) Nakai
CHINESE QUINCE
Planted, usually in churchyards, at Beaminster (40V), Osmington (78G), Stock Gaylard (71G) and Okeford Fitzpaine (81A); Brownsea Is (08I); a white-flowered form is at Wareham Ch (98I). 6.

Chaenomeles japonica (Thunb.) Spach.
JAPANESE QUINCE
Planted at Wareham Ch (98I). 1.

Pyrus communis L.
PEAR
Isolated trees are found on verges, railway banks, old gardens, also in quarries and undercliffs on Portland. 39 + [2]. They do not appear to be native, but remains are

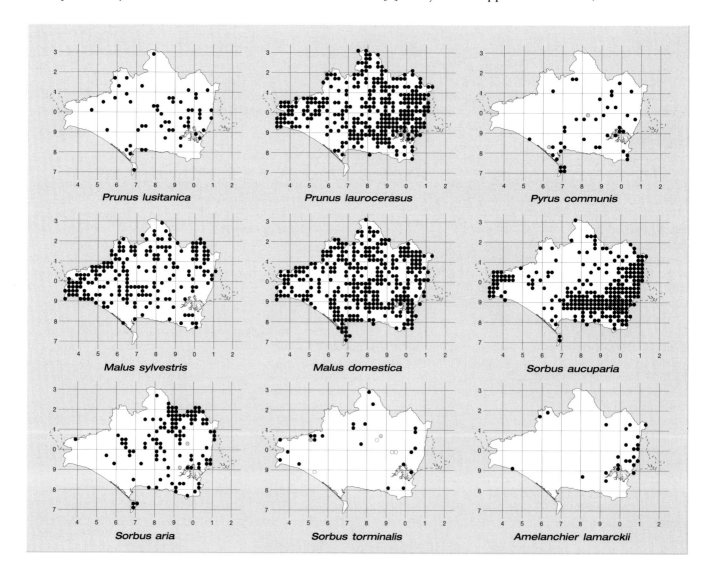

142

found in neolithic deposits. It has long been cultivated, and 38 cultivars were on sale at Blandford in 1782. Some 17 Dorset place names of Saxon origin have names derived from pear, e.g. Parley, Parnham, Perry, Purcombe and Purewell, so the tree may have been wild or much grown around the 8th century AD. Victorian folklore referred to a pear with small fruits (Udal, 1889).

Pyrus nivalis L.
SNOW PEAR
Two old, planted trees survive in Stutcombe Bottom (50S). 1.

Pyrus salicifolia Pallas
WILLOW-LEAVED PEAR
Planted at Bere Regis (89M) and Carey (90E). 2.

Malus sylvestris (L.) Miller
CRAB APPLE, GRIBBLE
Widespread in small quantity in woods and hedges, but scarce or absent in deforested areas such as Portland and near the sea; remains have ben found of neolithic age. 203. Five Dorset place-names have the 'crab' root.

Malus domestica Borkh.
APPLE
Widespread as isolated trees, all thought to have sprung from discarded pips. 340. 52 cultivars, including the local 'Piddle Redstreak', were on sale at Blandford in 1782. Planted orchards, mostly recent, are uncommon, but there is a museum of cider-making at Moigne Combe. A dense thicket of apple saplings grows on an old railway bank at Southlands (67U).

Other *Malus* taxa are planted on verges or as street trees. These include **M. baccata** (L.) Borkh. at Durweston (80U), Morden (99D) and Great Barrow (09U); **M. floribunda** Sieb. ex Van Houtte at Okeford Fitzpaine (81A); **M. × purpurea** (Barbier) Rehder at Naked Cross (99S), Green Is (08D) and Gussage All Saints (01A); and **M. sargentii** Rehder at Priors Down (71P, JO). More can be seen at Canford School (09J, TH).

Sorbus domestica L.
SERVICE-TREE
A rare tree reported from Bridport (49R, 1799, RP) and recently at Nottington (68R, 1990, DAP). 1. There is a big tree in the garden of Admiral Hardy's old house at Portesham (68C).

Sorbus aucuparia L.
ROWAN
Frequent in woods on acid soils or in scrub on heaths. 247. Rare on chalk or limestone, where it may be planted as a street tree, as on Portland. Mostly in the west and the Poole basin, perhaps increasing as JCMP thought it 'not common' in 1895. Rowan has a history of use as whip-handles, and was planted to ward off witches. The fruit is relished by birds.

Sorbus intermedia (Ehrh.) Pers.
SWEDISH WHITEBEAM
Much planted recently in woods, plantations, hedges and churchyards. 62. About 100 trees are planted in an avenue at Moor Crichel (90Z).

Sorbus aria (L.) Crantz
COMMON WHITEBEAM
Occasional in woods and scrub on Chalk, rare on acid soils and sometimes planted in plantations or on verges. 114 + [2]. This is commonest in the north-east, where large trees occur, as at Stubhampton Bottom (81Y); also native at Great Wood, Creech in Purbeck (98A). Stunted trees from east-facing limestone on Portland have been called subsp. *elegans*, but do not merit a new name (67V and W,1924, AWG in **BDK** and 1979, ! in **BRIST**).

Sorbus torminalis (L.) Crantz
WILD SERVICE-TREE
Uncommon in old woods and hedges in the north and west, where it fruits well and produces saplings; rarely planted elsewhere, as in churchyards. 15 + 6p + [6]. Probably native in Purbeck at Harmans Cross (98V, M Martin).

Other *Sorbus* species are planted. A tree in Sherborne Park (61S) is probably *S. croceocarpa* L., and one with hairless leaves in the avenue to Huish Manor (99D) may be *S. megalocarpa* Rehd. *S.* Cv. 'Joseph Rock', with yellow berries, has been planted on a verge at Gussage All Saints (91V), and other species are at Canford School (09J, TH).

Aronia arbutifolia (L.) Pers., Red Chokeberry, is planted at Forde Abbey (30M), while **A. melanocarpa** (Michaux) Elliott, Black Chokeberry, survives at Bere Heath (89R, 1991-98, ! in **RNG**).

Amelanchier lamarckii F.-G. Schroeder
JUNEBERRY
Planted and suckering or self-sown on acid soils in the Poole basin, elsewhere as a planted tree. 24. First record: Upton (99W, 1956, !); it is now naturalised by the shore here.

Cotoneaster frigidus Wallich ex Lindley
TREE COTONEASTER
Often planted on verges and railway banks, reproducing from seed. 16. Planted in quantity in a shelter-belt at Dorsetshire Gap (70G). Birds do not readily eat the berries.

Cotoneaster frigidus × salicifolius = C. × watereri Exell
Planted in churchyards and an escape on waste ground. 6. Hardown Hill (49C); Little Bredy (58Z); Waddock (79V); Todber pits (71Z, CT & ! in **RNG**); Tarrant Gunville (91G); Wyke Down (01C, GDF).

Cotoneaster salicifolius Franchet
WILLOW-LEAVED COTONEASTER
Rarely planted. 2. Shillingstone Ch (81F); Shaftesbury (82R).

Cotoneaster salicifolius × *horizontalis*
Cv. Valkenburg occurs on a railway bank at Lambs Green (99Z, 1997, ! det. JDF). 1.

Cotoneaster lacteus W. Smith
LATE COTONEASTER
On an old railway bank, Rodwell (67T, 1987, ! in **RNG** det JDF); Higher Kingston (79B, 1999, ! det. JDF). 2.

Cotoneaster integrifolius (Roxb.) Klotz
ENTIRE-LEAVED COTONEASTER
Long known, and now abundant, in quarries and undercliffs all over Portland; occasional in small quantity in calcareous grassland elsewhere. 31. First record: Rufus Castle (67V, 1862, SMP in **DOR**).

Cotoneaster horizontalis Decne.
WALL COTONEASTER
Widespread, mostly bird-sown, on walls, cliffs, calcareous banks or among limestone rocks, also on tips. 126. Usually as single bushes, but locally frequent on Portland.

Cotoneaster conspicuus Marquand
One bush on waste ground at Milborne St Andrew (89D, 1999, !); others at Lake (99V, 1999, ! det. JDF) and Holes Bay (09B, 1997, RSRF & ! in **RNG** det. JDF). 3.

Cotoneaster divaricatus Rehder & E. Wilson
SPREADING COTONEASTER
Blackdown (30W, 1998, IPG det. JDF); planted at Maiden Newton (59Y, 1995, ! det. JDF). 2.

Cotoneaster villosulus (Rehder & E. Wilson) Flinck & Hylmo
LLEYN COTONEASTER
Probably this in a plantation at Carey (98E, 1990, ! in **RNG**).

Cotoneaster simonsii Baker
HIMALAYAN COTONEASTER
Abundant in forestry plantations on acid soils, and occasional on walls, verges, railway banks and calcareous scrub, as on Portland. 142. Once seen as an epiphyte on elm by ACL. First record: Wareham Forest (1923, GCD), where it is now self-sown and common.

Cotoneaster bullatus Bois
HOLLYBERRY COTONEASTER
Self-sown on walls from gardens, and planted in churchyards or verges. 20. First record: Stourpaine railway bridge (80U, 1981, KEB). 'Wild' in Milborne Wood (79Y).

Cotoneaster rehderi Pojark
BULLATE COTONEASTER
Planted on verges and in churchyards and escaping. 14. First record: Naked Cross (99X, 1991, ! in **RNG** det. JDF).

Cotoneaster boisianus G. Klotz
Planted on a verge at Bere Regis (89M, 1995, ! det. JDF). 1.

Cotoneaster dielsianus E. Pritzel ex Diels
DIEL'S COTONEASTER
Scarce on railway banks or tips. 6. Sandsfoot (67T, 1991, ! in **RNG** det. JDF); Sherborne (61N); Stoke Heath (88P); Corfe (98R, 1982, GS Joyce); New Swanage (07J, EAP det. JDF); Poole quay (09A, 1999, ! det. JDF).

Cotoneaster sternianus (Turrill) Boom
STERN'S COTONEASTER
Planted in churchyards and on verges and sometimes self-sown, as at Chalbury Hill (68W). 16. First record: Lulworth (88F, 1963, D McClintock). It prefers calcareous soils.

Cotoneaster mairei Lev.
A scarce relict. 2. Piddlehinton Camp (79I, 1997, !); Woodville (82A, 1997, ! det. JDF).

Cotoneaster serotinus Hutch.
Whistling Copse (30R, 1998, IPG det. JDF). 1.

Pyracantha coccinea M. Roemer
FIRETHORN
Planted, and sometimes self-sown, on verges and sandy cliffs. 19. The fruits are eaten by pheasants, greenfinches and other birds.

Pyracantha rogersiana (A.B. Jackson) Coltman-Rogers
ASIAN FIRETHORN
Planted in rough grass, West Bexington (58I, 1992, SME); Blandford old railway (80Y, !) and probably elsewhere. 2.

Mespilus germanica L.
MEDLAR
Not seen wild, but there are large trees at Mapperton Ho (59E) and Stour Ho, Blandford (80X).

Crataegus submollis Sarg.
HAIRY COCKSPURTHORN
Probably this, planted as a street tree in Bridport (49R). 1.

Crataegus pedicellata Sarg.
PEAR-FRUITED COCKSPURTHORN
In a hedge at Hanover Wood (61T, 1991, JFA).

Crataegus persimilis Sarg.
BROAD-LEAVED COCKSPURTHORN
Planted in hedges, parks and churchyards in small numbers, but not naturalised. 21. A very large tree grows in a dell at Okeford Fitzpaine (81A).

Crataegus crus-galli L.
COCKSPURTHORN
All recent reports of this are *C. persimilis*. Old records from Moreton (78Z, 1917, T. H. Leach) and Stoke Road, Beaminster (40Q, 1920, AWG in **BDK**) may be correct.

Crataegus monogyna Jacq.
HAWTHORN, BLOODY BONES, MAY-TREE
Very common at the edges of woods, in both ancient and

recently planted hedges, and as a major component of scrub on chalk or clay soils. 727. The fruit, which vary in abundance from year to year, are an important source of food for small mammals and migrant birds; the practice of flailing hedges in autumn destroys this food. The leaves are eaten by many moth larvae, but can be blackened by salt gales. The flower-colour varies from white to pink, and a plant with variegated foliage was collected at Upcerne (60L, 1700, C DuBois in **OXF**). Twenty-two Dorset place names include 'thorn', but some of these, such as Turnworth, may refer to blackthorn. The flower-tops contain trimethylamine, and the berries have been used to treat angina.

Crataegus monogyna × laevigata = C. × macrocarpa Hegetschw.
Rare; probably this in a plantation north-east of Moreton (88E).

Crataegus laevigata (Poiret) DC.
MIDLAND HAWTHORN
Rare, and probably always introduced, in plantations and hedgebanks, though it might be native at Piddles Wood. 6 + [2]. Garden forms often have deep pink flowers. West Bay (49Q); hedge, Netherbury (49U, !); Powerstock Common (59I, 1974, AWG); above Bran Point (78K, 1961-93, !); Piddles Wood (71W, 1936, RDG in **WRHM**, *et al.*, and 1997, BE and FG); Tyneham Ho ruins (88Q); Talbot Heath (09R, 1980, C Thomas); Stanbridge Mill (00E, 1979, JN).

Crataegus laciniata Ucria
ORIENTAL HAWTHORN
Rare, planted in parks or as a street tree. 1 + [2].

Chedington (40X, 1928, AWG); Sherborne Park and streets (61N, 1990-95, !); Moreton (78Z, 1918, TH Leach).

Crataegus phaenopyrum Medikus
Possibly this at Castle Hill, Buckland Newton (60Y, 1865, SMP in **DOR**).

Many species of *Crataegus* and the hybrid × ***Crataegomespilus grandiflora*** (Smith) E.G. Camus were planted at Canford School (09J, TH), but may have been damaged by recent building work.

Eucryphia × nymansayensis Bausch
Trees have been planted in hedges at Bloxworth Ho (89X) and in Charborough Park (99I). 2.

CAESALPINIACEAE

Cercis siliquastrum L.
JUDAS-TREE
Rarely planted in churchyards or in large estates. 3 + [1]. It fruits well but no seedlings have been found. Bridport Ch (49R); Spring Bottom (78K); Bryanston (80T, 1944, BNHS); Charminster (09X, FAW).

FABACEAE

Robinia pseudacacia L.
FALSE-ACACIA
Planted in a few woods, plantations and churchyards, usually on sandy soil. 49. Commonest in the east, and often

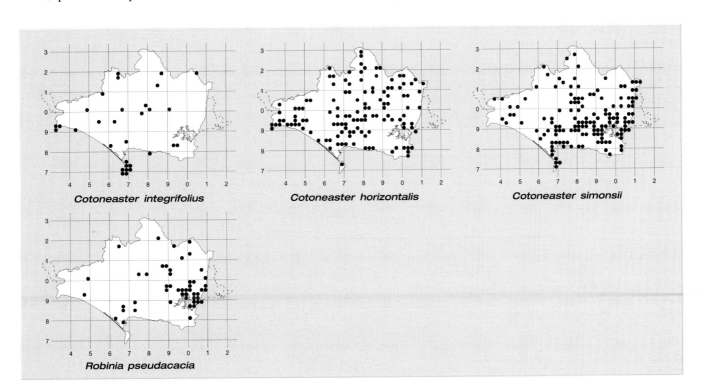

Cotoneaster integrifolius

Cotoneaster horizontalis

Cotoneaster simonsii

Robinia pseudacacia

self-sown, as on railway banks at Parkstone (09K). On sale at Blandford in 1782, but big trees were present earlier, e.g. one 18 m × 3.2 m at East Stour Ch (72W, 1797).

Psoralea americana L.
SCURFY PEA
A tender casual. Sherborne tip (61H, 1969, ACL).

Galega officinalis L.
GOAT'S-RUE
A garden escape, but usually short-lived. 3 + [3]. Burton Common (58E, 1994, DAP); Abbotsbury (58X, 1979, W. HA Picton); Weston Manor (50D, 1981, AJCB); West Stafford (78E, !); Swanage (07J, 1953, CCT and 08F, 1998, EAP); Ameys Ford (00R, 1975, VJG). Flowers mauve or white.

Colutea arborescens L.
BLADDER-SENNA
A rare garden relict. 2. Nottington (68R, 1981, AJCB); Okeford Fitzpaine (81A, 1996, !).

Astragalus glycyphyllos L.
WILD LIQUORICE
A rare native on calcareous soil, at the western end of its range in Britain. 1 + [2]. Bryanston (80T, 1941, DM); Scrubbity Barrows (91U, 1992, GDF, confirming a record near here made by T Bury in 1874); south of Swanage (07, 1912, HWP).

Onobrychis viciifolia Scop.
SAINFOIN, FRENCH-GRASS
Local on cliffs, rare in old chalk turf, also on railway banks and verges where it is sometimes sown, as at Baiter (09F). 36 + [7]. Mainly in Chalk and limestone areas, probably a denizen.

Anthyllis vulneraria L.
KIDNEY VETCH
Local, forming colonies in calcareous turf, especially in disturbed sites, quarries and cuttings for roads or railways which face south. 141 + [7]. Other habitats include calcareous clay cliffs and undercliffs, enriched heathy verges and railway banks, and fixed shingle at Small Mouth (67S),

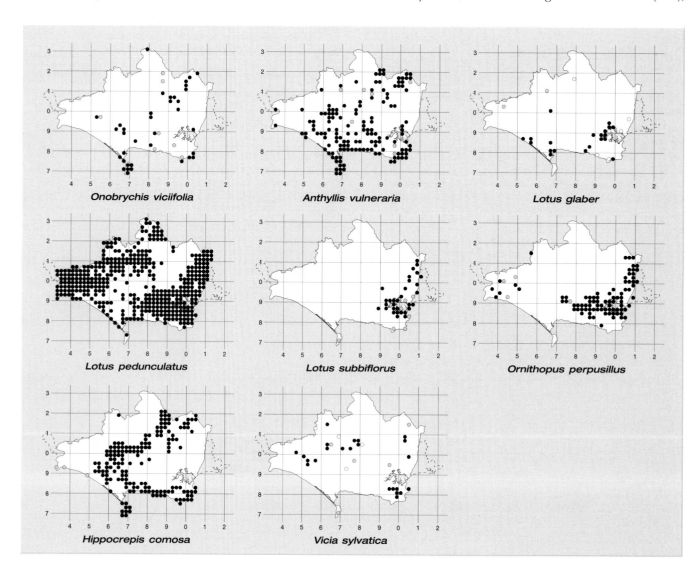

Onobrychis viciifolia

Anthyllis vulneraria

Lotus glaber

Lotus pedunculatus

Lotus subbiflorus

Ornithopus perpusillus

Hippocrepis comosa

Vicia sylvatica

where there are plants with pale yellow flowers. Subsp. *vulneraria* is common. Subsp. *corbieri* (CE Salmon & Travis) Cullen, with succulent leaves, is on chalk cliffs at Fortuneswell (67W, 1987, SL Jury & MF Watson in **RNG)** and Ballard Down (08K and L, 1959, WS Catling in **RNG** and 1990, !). Subsp. *polyphylla* (Ser.) Nyman, a tall, erect plant, has been planted on verges at Briantspuddle (89C, 1962, T Woodisse in **BM**) and Bere Regis (89M, 1989–98, ! in **RNG**). Subsp. *carpatica* (Pant.) Nyman var. *pseudovulneraria* Sag. occurs as a prostrate form at Poole port (09A, 1999, ! det. JRA).

Dorycnium hirsutum L.
CANARY CLOVER
This survives, planted, on Green Is (08D), and seeds itself in gardens elsewhere.

Lotus glaber Miller
NARROW-LEAVED BIRD'S-FOOT TREFOIL
Locally frequent in the upper parts of salt-marshes and coastal grasslands, rarely introduced on verges inland. 21 + [9]. Once seen as a casual weed in lucerne at Broadwindsor (40G, 1921, AWG).

Lotus corniculatus L.
COMMON BIRD'S-FOOT TREFOIL
Frequent in dry grassland, verges, railway banks, undercliffs and fixed shingle. 664. Variable. Var. *incanus* Gray is a hairy plant found on chalk cliffs at Mupe Bay (87P and 88K). Var. *sativus* Chrtkova is a tall, erect plant which has been found sown on new roadside banks. 9. f. *crassifolius* Pers. is a coastal form with succulent leaves. Other forms have narrow leaflets resembling *L. glaber*, while in unfavourable sites the pedicels bear single flowers.

Lotus pedunculatus Cav.
GREATER BIRD'S-FOOT TREFOIL
Frequent in wet woodland rides, marshes and wet grassland, often with rushes where few other legumes compete. 406. Scarce on the dry Chalk ridge. It varies much in hairiness.

Lotus subbiflorus Cav.
HAIRY BIRD'S-FOOT TREFOIL
Local and sporadic in bare, dry, disturbed places or fallow fields in the Poole basin. 32 + [7]. While it prefers sandy soils, it occurs on calcareous sites by the coast in Portland and South Purbeck. There are specimens in **BM, BMH, K** and **WRHM**. Cornubian in range in Britain.

(*Lotus angustissimus* L. has no confirmed records, but occurs at Wick (19L, 1992) in VC11.)

[*Tetragonolobus maritimus* (L.) Roth
DRAGON'S TEETH
Once reported from Ulwell Golf Course (08A, 1900, LVL).]

Ornithopus compressus L.
YELLOW SERRADELLA
A rare casual found in quantity on a new sandy verge at Holes Bay (09A and B, 1991, ! in **RNG**). 2.

[*Ornithopus sativus* Brot.
SERRADELLA
A casual in derelict arable at Parkstone (09, 1927, LBH in **BM**).]

Ornithopus perpusillus L.
BIRD'S-FOOT
Locally abundant in bare, dry, disturbed, sandy soil, heathy tracks or arable. 86 + [4]. Scattered in the west, frequent in the Poole basin and absent from chalk, limestone and clay soils.

Coronilla scorpioides (L.) Koch
ANNUAL SCORPION-VETCH
A rare casual. [2]. West Bay Sta (49Q, 1921, AWG); Sherborne (61I, 1970, ACL).

Hippocrepis emerus (L.) Lassen
SCORPION SENNA
A shrub planted in an orchard hedge, Gussage St Andrew (01A, 1994, !). 1.

Hippocrepis comosa L.
HORSESHOE VETCH
Locally frequent in short, dry, calcareous turf, especially on south-facing banks. 159 + [3]. Variable in stature; robust forms occur on the undercliffs of Portland and St Aldhelms Head, and perfume the air when in flower. A pale yellow flowered form occurred at Winspit (97T, 1930s, RDG) and Blacknor (67Q, !).

Securigera varia (L.) Lassen
CROWN VETCH
Rarely established for a time on railway banks or on waste ground. 1+ [4]. Escaping from a cottage garden at Symondsbury (49L, 1999, !); Abbotsbury (58M, 1960, !); Bradford Abbas (51X, 1876, RDS Stephens in **DOR**); by railway, Rodwell (67U, 1914, RDG and 1920, MJA); Lake (99V, 1955, EN); abandoned garden, Greenland Farm (08C, 1999, !).

Scorpiurus muricatus L.
CATERPILLAR-PLANT
A casual alien from bird-seed. [1]. Thorncombe (30R, 1974, M Roper).

Vicia cracca L.
TUFTED VETCH
Frequent in woodland rides, hedges, verges and marshes on all but the most acid soils. 565.

[*Vicia tenuifolia* Roth
FINE-LEAVED VETCH
A casual once found in a garden at Parkstone (09, 1916, Anon in **BMH**).]

Vicia sylvatica L.
WOOD VETCH
Local and sporadic in old woods, often springing up after disturbance. 23 + [8]. JCMP had it from 15 sites in 1895, RDG from only five in the 1930s. It has a strange

distribution, mostly along the northern scarp of the Chalk in both central Dorset and Purbeck, with outliers on Portland undercliffs (67, 1912, WBB), Dyett's Coppice (99M, 1981, I Cross) and Little Canford pits (09P, 1995, DAP). Specimens are in **BDK, DOR, RNG** and **WRHM**.

[*Vicia villosa* Roth
FODDER VETCH
Once found as a casual; Grid Square SU01, 1927, Anon in **BM**. NDS found it in SZ19, VC11 in 1924.]

Vicia hirsuta (L.) Gray
HAIRY TARE
Locally frequent on dry, grassy railway banks or verges. 192. Tolerant of most soils, but least common along the Chalk ridge.

Vicia parviflora Cav.
SLENDER TARE
Frequent on Portland, elsewhere uncommon on undercliffs or disturbed waste ground where it is sporadic. 19 + [11].

Mostly along the coastal belt, or in the north-west near Sherborne or Yetminster. Specimens are in **BM, CGE, DOR, HIWNT, RNG, WRHM** and **YRK**.

Vicia tetrasperma (L.) Schreber
SMOOTH TARE
Scattered in woodland rides, undercliffs, verges, railway banks and rough grassland, once seen in fen at Hartland Moor (98M). 76 + [2]. Often confused with *V. parviflora*, but both species occur on Portland. RDG thought it commonest in the west, which is not so today.

Vicia sepium L.
BUSH VETCH
Frequent at the edges of woods and in old hedgebanks. 474. Found on Portland, but absent from other deforested areas, also from very acid soils and built-up regions.

Vicia sativa L.
COMMON VETCH
First records are from Iron Age deposits. Subsp. *nigra* (L.)

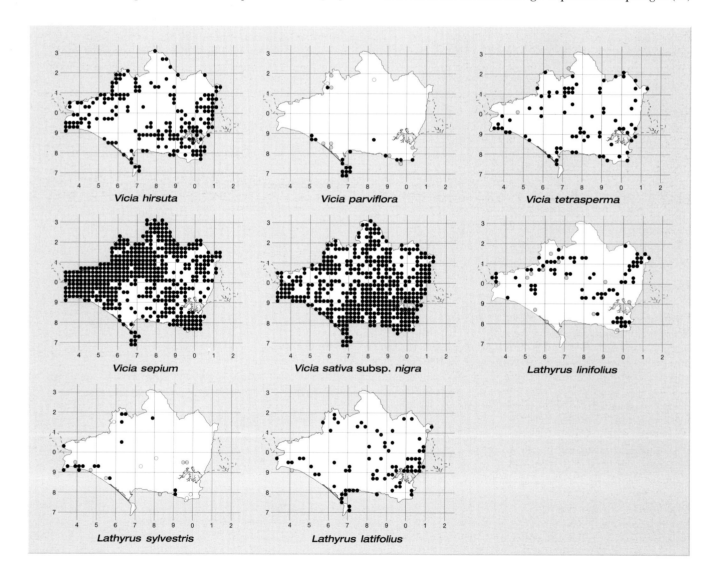

Vicia hirsuta

Vicia parviflora

Vicia tetrasperma

Vicia sepium

Vicia sativa subsp. nigra

Lathyrus linifolius

Lathyrus sylvestris

Lathyrus latifolius

Ehrh. is common in grassland, verges, railway banks and disturbed sites. 462. Subsp. *sativa* and subsp. *segetalis* (Thuill.) Gaudin were lumped in the present survey. They are common, but often transient, in sown fields, field-edges or disturbed sites near farmyards. 340. White-flowered plants were found by RDG in 1960, and at Eype (49K, 1998, !).

Vicia lathyroides L.
SPRING VETCH
Rare and sporadic in dry sandy places. 5 + [7]. Ever since Pulteney's time, this has been reported in error for *V. sativa*. Mostly around Poole Harbour, very local elsewhere. Ironstone ridge, Abbotsbury (58S, 1959, !); Bradford Abbas (51S, 1837, J Buckman); Weymouth (67, 1895, TG Cullum); Rempstone (98W, 1980, EA Parkin); Fitzworth (98X, 1991, BE); Punfield (08F, 1882, HNR and WF); Studland Heath and South Haven (08H and I, 1935, RDG, *et al.* in **RNG**, and 1998, MJG, !); Brownsea Is (08J, 1977, DWT); Old Harry (08L, 1991, !); Branksome cliffs (09Q, 1993, BE and RMW, but mostly in VC11); Castle Hill (01R, 1994, !).

Vicia lutea L.
YELLOW-VETCH
Local, rare and sporadic on cliffs or sunny banks along the coast, very rare inland. 3 + [8]. Abbotsbury (58S, 1928 AM Harris in **BDK** and **BM**, *et al.* in **DOR** and **WRHM**, and 1990, SME); Loders (59B, 1980, ER Sykes); Bradford Abbas (51S, 1838, J Buckman and 1884, RPM in **BM**); Portland (67V, 1924, DM and 1983, M Leicester); Weymouth (67, 1778, W Hudson, *et al.* in **BM, DOR, K** and **WAR**, and 1953, Miss Pearson in **DOR**); Wyke cliffs (67T, 1885, WBB and 1988, !); Sandsfoot (67T, 1895, GS Gibson, *et al.*, and 1987, DAP, !); Radipole Lake (68U, 1895, JCMP, *et al.* in **DOR**, and 1926–39, MJA); Lodmoor (68V, 1774, J Lightfoot in **BM**, *et al.*, and 1926, MJA); Preston beach (78A, 1935, MJA); Bailey Gate Sta (99P, 1893, EFL and RPM in **BM**); Canford Cliffs (08P, 1953, KG); Parkstone, from bird-seed, (09F, 1955, ER Banham). Many old recorders called this *V. laevigata* Smith.

Vicia bithynica (L.) L.
BITHYNIAN VETCH
Local on cliffs, railway banks and rough, clay grassland, along the coastal belt with one inland site. 10 + [3]. East and south-west of Lyme Regis (39F and L, 1880, HT Mennell in **BM**, *et al.*, and 1999, !); Eype (49F, 1997, DAP); Wyke cliffs (67S, 1896, CES in **BM**, *et al.*, and 1988, !); Sandsfoot (67T, 1801, J Stackhouse, *et al.*, and 1991, BE, !); Weymouth (67, 1885, JS Gale in **OXF**); Portland (66Z and 67V and W, 1937, DM and 1960–97, !); Radipole (68, 1833, M Athersoll in **BM**); Osmington Mills (78K, 1977, ! in **RNG**); Rooksmoor (71F, 1975–97, DAP); Purbeck (97?, 1762, W Hudson); Poole (09A, 1930, LBH).

Vicia faba L.
BROAD BEAN, HORSE BEAN
A frequent casual on tips, verges, or self-sown in arable after a crop. 14. The form with small seeds has been grown here as a crop since the middle Bronze Age.

[*Vicia melanops* Sibth. & Smith
BLACK-EYED VETCH
A casual from the Mediterranean: Poole (09A, 1930, LBH).]

[*Vicia narbonensis* L.
NARBONNE VETCH
A casual. Poole (09A, 1930, LBH in **BM**).]

Lathyrus japonicus Willd. subsp. *maritimus* (L.) P. Ball
SEA PEA
Once locally abundant on shingle beaches, but lost from many sites. 7 + [3]. Charmouth (39R, 1939, WD Lang); East Bexington (58M, 1895, JCMP); in patches along the landward side of the Chesil bank between Abbotsbury and Wyke (58, 67 and 68, 1801, J Stackhouse, *et al.* in **BDK, CGE** and **DOR** and 1994, SME, !); Abbotsbury (58S, 1999, J Dunn & M Palfrey); [east of Poole (09, 1874, Dr Argent)].

Lathyrus linifolius (Reichard) Baessler
BITTER-VETCH
Local in old woods and hedgebanks, railway banks and once in fen (98R). 74 + [15]. It has a strange distribution, mainly in the west and north, in south Purbeck and in the east along the junction of the Chalk and acid soils.

Lathyrus pratensis L.
MEADOW VETCHLING
Frequent in old meadows and pastures, mature grassy verges and railway banks. 677. Absent from heathland and 'improved' grassland.

Lathyrus tuberosus L.
TUBEROUS PEA
Very local and sporadic or transient on verges. 1 + [7]. East Fleet (67J, 1955, RDG in **DOR** and 1988–98, DAP); Middle Farm, Dorchester (69Q, 1938, DM); Owermoigne (78S, 1966–73, JB); Hurst Heath (78Z, H Bury, *et al.*, and 1960–80, !); Wool (88N, 1979, C Holt); Duncliffe (82G, 1947, BNHS); Beacon Hill, Lytchett (99S, 1975, DE Coombe); Godlingston (08A, 1926–36, MJA).

[*Lathyrus grandiflorus* Sibth. & Smith
TWO-FLOWERED EVERLASTING-PEA
Once established by Abbotsbury ruins (58M, 1960, !), now covered by tipped soil.]

Lathyrus sylvestris L.
NARROW-LEAVED EVERLASTING-PEA
Local at edges of woods, in scrub on undercliffs, or on roadside banks. 17 + [12]. It prefers clay soils and is mostly found in the north and west, with outliers at Kimmeridge and Worth. First recorded by RP, but not refound in his localities. Some reports have been rejected as probable errors for *L. latifolius*.

Lathyrus latifolius L.
BROAD-LEAVED EVERLASTING-PEA
Introduced on waste ground, undercliffs, verges and especially on railway banks. 79 + [2]. More tolerant of soil-

type and exposure than *L. sylvestris*. It was grown in Abbotsbury gardens in 1899 (MI), but its first 'wild' record is from Portland cliffs (67, 1908, GCD), where it still flourishes. White- and pale pink-flowered forms occur at Broadoak (71W).

Lathyrus hirsutus L.
HAIRY VETCHLING
A rare casual. [2]. Undercliff east of Lyme Regis (39L, 1964, J Cusden); Parkstone, in a garden (09F, 1955, ER Banham).

Lathyrus nissolia L.
GRASS VETCHLING
Local and sporadic in tall grassland on disturbed clay soils, on undercliffs, verges and railway banks, rarely on tips. 52 + [11]. Mainly in the coastal belt, with a few records in the north-west. A white-flowered form occurred at Wyke Regis (67T, 1988, M Dyke).

Lathyrus aphaca L.
YELLOW VETCHLING
Local, often sporadic, in calcareous grassland of sunny undercliffs, old quarries and verges. 24 + [13]. Frequent on Portland, otherwise mostly along the coastal belt; the old records scattered inland were probably casual. Specimens are in **CGE, DOR** and **WAR.**

[*Lathyrus inconspicuus* L.
A casual alien from bird-seed. [1]. Parkstone (09F, 1955, ER Banham).]

Pisum sativum L.
GARDEN PEA
An occasional relict of cultivation since the late Iron Age in arable fields, disturbed verges and tips. 8. Flowers normally white, but subsp. *elatius* M. Bieb. with bicoloured pink flowers is a rare arable weed (! in **RNG**). f. *arvense* L., with much-branched tendrils, is grown for animal food and may sow itself in fields or tips. In gardens, the seeds are attacked by larvae of the moth *Cydia nigricana*.

[*Cicer arietinum* L.
CHICK PEA
A rare casual. 2. Radipole Lake (67U, 1926, MJA); Poole and Parkstone (09, 1930s, HJW).]

Ononis spinosa L.
SPINY RESTHARROW
Local, in small quantity and declining, in fragments of old grassland on clay. 21 + [15]. Commonest in the west and north, rare on chalk or sandy soils. Likely to be confused with *O. repens* var. *horrida*.

Ononis repens L.
COMMON RESTHARROW, CAMMOCK
Locally frequent in old grassland on chalk or clay, on undercliffs, verges and railway banks. 231. Commonest along the coastal belt and on Chalk and limestone. The spiny var. *horrida* Lange was recorded from four sites along the coast by JCMP and recently at Five Barrow Hill (88S).

Melilotus altissimus Thuill.
TALL MELILOT
Widespread on cliffs and upper beaches, where it could be native. 117 + [3]. More commonly seen on all types of soil on verges, quarries, waste ground and tips. Starved plants occurred on the artificial breakwater of Portland Harbour (77C). Seeds were on sale in Blandford in 1782, so it may be a relict of cultivation, though it is not grown today.

Melilotus albus Medikus
WHITE MELILOT
Locally frequent but sporadic on waste ground and tips near Poole, Weymouth and elsewhere, or a rare casual. 20 + [16]. First record: Ham Quay (09A, 1839, TBS), probably arriving in ballast, and still found there.

Melilotus officinalis (L.) Pallas
RIBBED MELILOT
In the same habitats as *M. altissimus*, with which it can be confused, but less common and often casual. 26.

Melilotus indicus (L.) All.
SMALL MELILOT
Always casual, on urban waste land or on tips. 5 + [13]. First record: Parkstone (09F, 1888, WMR in **DOR**). Post-1970 records are; Portland (67W, 1978, J Fairley); Chickerell tip (68K, 1980, !); Radipole (68Q, 1977, !); Lodmoor tip (68V, 1963–77, !); Sherborne tip (61I, 1970, ACL); Crossways tip (78P, 1999, !); Blandford (80Y, 1989, !); Arne (98U, 1977, VJG); Lake (99V, 1995, !).

Melilotus sulcatus Desf.
FURROWED MELILOT
A rare casual resembling *M. indicus*. [1]. Sherborne tip (61I, 1970, ACL).

Trigonella corniculata (L.) L.
SICKLE-FRUITED FENUGREEK
A rare casual on a building site. 1. Lake (99V, 1995, ! in **RNG**).

Medicago lupulina L.
BLACK MEDICK
Common in short, calcareous turf, on verges, railway banks and as a garden weed. 683. It prefers dry, nutrient-rich soil. Var. *willdenowiana* Koch, with glandular pods, has not been reported since 1908.

Medicago sativa L.
LUCERNE
A deep-rooted plant occasionally planted for cattle fodder, as it has been for centuries, and a persistent relict on field borders and verges. 67. Subsp. *falcata* (L.) Arcangeli was first found at Poole (09A, 1839, TBS), and in the 1930s at Radipole and Branksome; its only recent record is from a tip in a Portland Quarry (77A, 1996, BE). 1 + [5]. Subsp. *varia* (T. Martin) Arcangeli was seen at Goathill (61T, 1996, FAW).

Medicago minima (L.) L.
BUR MEDICK
A rare casual. 1 + [2]. In a garden, Beaminster (40Q, 1959, AWG in **BDK**); Weymouth Backwater (68Q, 1938, DM and MJA); Branksome Chine (08U, 1992–95, RMW). Other reports are errors.

Medicago polymorpha L.
TOOTHED MEDICK
Locally plentiful on Portland and near Weymouth in

ruderal sites, and sporadic in small numbers in cliff-top grassland along the coastal belt. 21 + [16]. It likes bare, dry, sunny places. Collected on Portland in 1837 (**WAR**), 1862 (**DOR**), and 1932 (**RNG**), and still there.

Medicago arabica (L.) Hudson
SPOTTED MEDICK
Frequent in grassland, disturbed verges and village lanes in the south, and scattered in the north. 148. Although Good thought it frequent in the 1930s, he only found it in

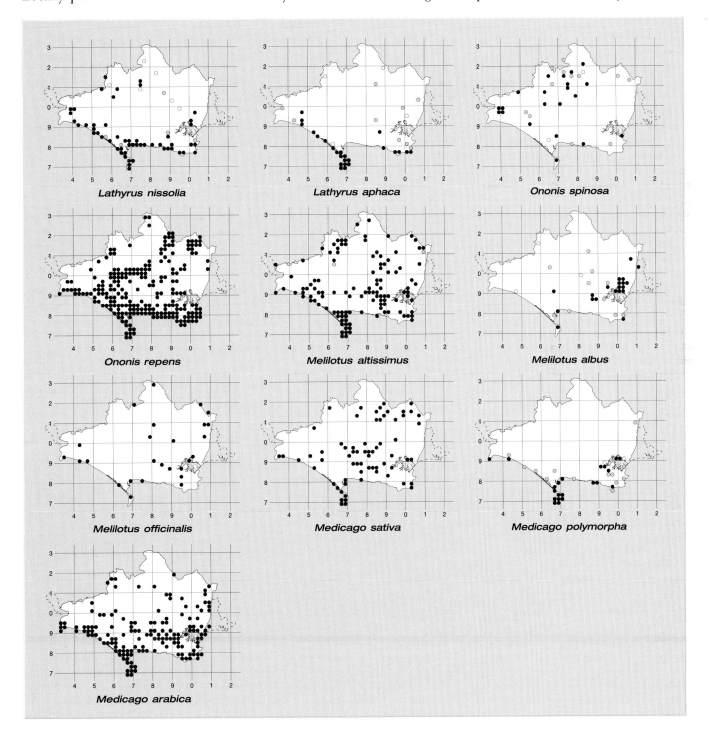

Lathyrus nissolia   Lathyrus aphaca   Ononis spinosa

Ononis repens   Melilotus altissimus   Melilotus albus

Melilotus officinalis   Medicago sativa   Medicago polymorpha

Medicago arabica

three stands, suggesting that then, as now, it does not grow in semi-native habitats. It prefers nutrient-rich soils and is locally dominant on Portland.

Trifolium ornithopodioides L.
BIRD'S-FOOT CLOVER
Local but sporadic in bare, dry places such as heath tracks, undercliffs and beaches in the coastal belt. 32 + [10]. It prefers sandy soils, but occurred on clay at Holworth cliffs and on limestone at Portland and Durlston CP. Specimens are in **BM, CLEY, DOR** and **WRHM**.

Trifolium repens L.
WHITE CLOVER
Abundant in closed turf, verges, undercliffs and fixed shingle, also often planted in leys. 735. Variable in size and leaf-pattern, and probably in its content of cyanogenic glucosides. It is a foodplant for the Common Blue and many moths. Forms with pale pink flowers occur near the sea. Var. *phyllanthes* Seringe, with a leafy calyx, is a common form in wet summers.

Trifolium hybridum L.
ALSIKE CLOVER
Scattered as a relic of cultivation in pastures, edges of fields and newly sown grass. 40. [Subsp. *elegans* (Savi) Asch. & Graebner was reported from Wareham and East Parley in 1900 by EFL, but has not been seen since.]

Trifolium glomeratum L.
CLUSTERED CLOVER
Very local and sporadic in bare, dry places along the coastal belt and in the Poole basin. 19 + [9]. There is an outlier at Yeovil Junction (51S); it does not occur further north in Britain. Specimens are in **BDK, DOR, LANC** and **WRHM.**

Trifolium suffocatum L.
SUFFOCATED CLOVER
Very local and in small numbers in bare, dry, sandy places in the Poole basin. 8 + [7]. It does not occur further north in Britain. Tincleton (79Q, 1962, RDG); East Burton Heath (88I, 1932, RDG); Wareham Sta (98I, 1884, WMR in **BM**); near Hartland Moor (98M and S, 1999, BE); Greenland (08C, 1948, WAC); Studland (08G, 1884, WMR, *et al.*, and 1938, DM); South Haven (08I, 1951, GA Swan, *et al.*, and 1999, EAP, !); Brownsea Is (08I, 1979, RMB); Old Harry (08L, 1900, EFL); Sandbanks (08N, 1958–91, !); Canford Cliffs (08P and U, 1992, RMW); Parkstone (09F, 1900, EFL); Turbary Common (09S, 1994, RMW); Castle Hill (01L, 1976, DE Coombe and 1993, RMW). Specimens are in **BDK, BM** and **DOR.**

Trifolium fragiferum L.
STRAWBERRY CLOVER
Locally frequent, but easily overlooked early in the season, in old pastures on alluvium or clay. 115 + [10]. Mostly in the coastal belt, in the north-west, and in the upper Stour valley. Scarce on the Chalk and on very acid soils.

Trifolium resupinatum L.
REVERSED CLOVER
A rare casual which can persist for a few years. 1 + [4]. West Bay (49Q, 1998, D Patrick); Beaminster (40V, 1929–35, AWG in **BDK**); Sherborne (61, 1937, DM); Ham Quay (99V, 1831, TBS); Poole Quay (09A, 1955, AF Chapman).

(**Trifolium aureum** Pollich was reported from Bagmans Copse (00J, 1986); no specimen seen.)

Trifolium campestre Schreber
HOP TREFOIL
In grassland, on verges and railway banks, frequent in the south but rare in the north. 224. It tolerates most soils in dry conditions.

Trifolium dubium Sibth.
LESSER TREFOIL
Common in short turf and a frequent weed in lawns and disturbed places. 580.

Trifolium micranthum Viv.
SLENDER TREFOIL
Occasional in very short turf, sandy lawns, golf courses, by heath tracks or edges of tarmac. 71 + [18]. Mostly in the Poole basin and absent from the Chalk, but found in some shallow, leached soils over limestone.

Trifolium pratense L.
RED CLOVER, BROAD CLOVER
Common in grassland on all types of soil, least common in nutrient-poor conditions. 731. Often planted in leys, and variable in stature and flower colour. Var. *sativum* Schreb. is a robust form, said to originate in Dorset, sold by seedsmen; forms with white, cream and pale pink flowers have been seen recently. Var. *parviflorum* Bab. was reported from Weymouth (67, 1926); a decumbent, pale pink-flowered form grows on the edge of the cliffs at White Nothe (78Q).

Trifolium medium L.
ZIGZAG CLOVER
Local and in small numbers in grassy rides, tracks or verges on clay, also in unimproved meadows, rarely on undercliffs. 90 + [17]. Commonest in the north-west, absent from the Chalk and from acid heathland.

(**Trifolium ochroleucon** Hudson has been erroneously reported from Poole by RP and from Portland by WBB.)

Trifolium incarnatum L.
CRIMSON CLOVER
Occasional in sown leys earlier this century, and then a striking sight in flower. 4 + [27]. Last seen *en masse* in the 1950s, but seeds were still on sale in Dorchester in 1992. First record: Almer (99E, 1865, JCMP in **DOR**), but JCMP omits it from his Floras, implying he considered it to be a crop plant. Since then the number of records per decade peaked in the 1930s; recent records are of single casual plants in kale fields or on new verges: Zelston (89Z, 1989, !

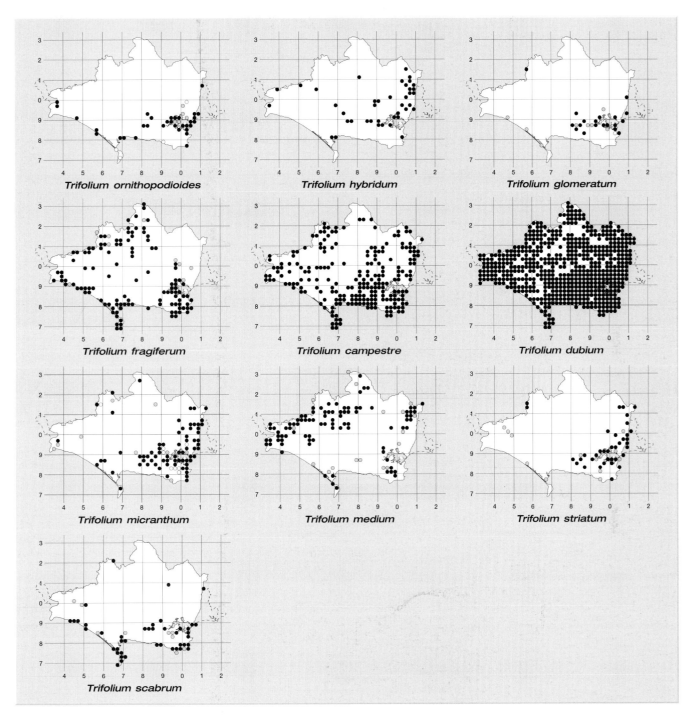

Trifolium ornithopodioides

Trifolium hybridum

Trifolium glomeratum

Trifolium fragiferum

Trifolium campestre

Trifolium dubium

Trifolium micranthum

Trifolium medium

Trifolium striatum

Trifolium scabrum

in **RNG**); Gillingham (82D, 1995, ! and AG Hobson); Duncliffe Wood (82G, 1983, Anon); Caesars Camp (91H, 1997, !). [Subsp. *molinerii* (Balbis ex Hornem.) Cesati, Long-headed clover, was a rare casual last century; Wareham (98I, 1884, HNR and WF in **BM**) and Poole (09A, 1898, HWP)].

Trifolium striatum L.
KNOTTED CLOVER
Local, mostly in small numbers, on bare, dry, sandy tracks, fallows, verges, rail tracks and cliff-tops. 37 + [12]. Most

common in the Poole basin, with outliers round Yeovil junction in the north-west and some old records from the south-west. Reports from chalk turf at Crincombe (70W, NCC) and Hod Hill (81K, BNSS) are excluded pending confirmation.

Trifolium scabrum L.
ROUGH CLOVER
Local, sometimes in quantity, on bare, dry soils of tracks, verges, old quarries and fixed shingle. 43 + [10]. It prefers acid soils, but tolerates limestone cliffs in Portland and

Purbeck, and is commonest along the coastal belt. The outlier at Blandford Camp (90J) was on sand imported for rifle-butts. WMR found it abundant in Purbeck in 1895, since when it may have declined.

Trifolium arvense L.
HARE'S-FOOT CLOVER

Local on bare, dry soils, heaths, sand dunes, sandy verges or rail tracks. 50 + [7]. Most recent records are from the Poole basin, or along the landward side of the Chesil bank. It was frequent near Sherborne in the 1950s, and there are isolated records from calcareous soils at Portland (67, RA), Melbury Hill (81Z, AH) and Martin Down (01P, JH Lavender, perhaps in VC11).

Trifolium squamosum L.
SEA CLOVER

Rare and sporadic on clay undercliffs between Seatown and Lulworth, on fixed shingle at Small Mouth (67S) and occasional in fields on Portland. 18 + [5]. Specimens are in **BDK, BMH, DOR** and **WAR.**

Trifolium subterraneum L.
SUBTERRANEAN CLOVER

Local in disturbed, dry, sandy fields, fallows, verges, golf courses, sandy lawns and cliffs. 60 + [10]. Mainly in the Poole basin, with occasional outliers along the coastal belt and near Yeovil junction (51T, 1875, RDS Stephens and 1994, MJG); it also grows in central Dorchester.

[*Trifolium diffusum* Ehrh. was a casual at Kingswood Farm (08A, 1913, CBG in **BMH**).]

[*Trifolium echinatum* M. Bieb.
HEDGEHOG CLOVER
Casual in felled woodland, Sherborne Park (61I, 1937, RDG in **RNG** and **WRHM**).]

Lupinus arboreus Sims
TREE LUPIN

Sometimes planted on verges, escaped on waste ground and tips, and self-sown on sandy undercliffs and dunes. 20 + [1]. Grown at Abbotsbury in 1899 (MI), but first

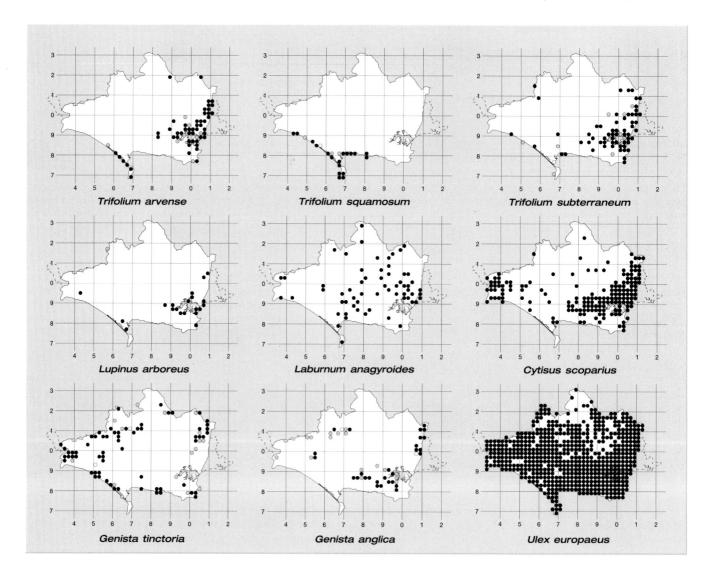

Trifolium arvense

Trifolium squamosum

Trifolium subterraneum

Lupinus arboreus

Laburnum anagyroides

Cytisus scoparius

Genista tinctoria

Genista anglica

Ulex europaeus

recorded 'in the wild' at Fleets Corner (09B, 1937, WAC), and still there despite major landscaping.

Lupinus arboreus × polyphyllus = L. × regalis Bergmans.
RUSSELL LUPIN
Naturalised in grass at Bovington (88J, 1992, !), otherwise transient in tips or waste places. 4 + [3].

Laburnum anagyroides Medikus
LABURNUM, GOLDEN CHAINTREE
Frequently planted in churchyards and on verges, rarely in woods, and sowing itself in many places. 61. On sale in Blandford in 1782. The leaves and seeds contain cytisine and are toxic.

Laburnum anagyroides × alpinum = L. × watereri
(Wettst.) Dippel
Planted at Swanage (07J, DCL) and perhaps elsewhere. 1.

Laburnum alpinum (Miller) Bercht & J.S. Presl
SCOTTISH LABURNUM
An old tree in Melbury Park (50S, 1971–97, ! det. AO Chater); also at Canford School (09J, TH).

The hybrid × ***Laburnocytisus adamii*** (Poiteau) Schneid. is planted at Chettle Ho (91L).

Cytisus multiflorus (L'Her. ex Aiton) Sweet
WHITE BROOM
Rarely planted but escaped and self-sown on a few verges and railway banks, mostly on acid soils in the Poole basin. 6 + [2]. Bovington Camp (88J, !); Wareham Common (98D, !); Wytch Farm (98S, !); Newton Farm (99G, !); Hamworthy (99V, 1979, KEB); Sandbanks (08N, 1958, !); self-sown at Merley (09E, ! in **RNG**); Ferndown by-pass (00Q, !).

Cytisus striatus (Hill) Rothm.
HAIRY-FRUITED BROOM
Occasionally planted on verges. 8. Bovington Camp (88J, !); Tatchells Pit (98E, !); Newton Farm (99G, 1988–99, ! in **RNG**); Ham Common (99Q, !); Lytchett Minster (99R, !); Upton Heath (99W, !); Merley (09E, !); Poole by-pass (09G, !).

Cytisus scoparius (L.) Link
BROOM
Frequent on disturbed or burnt heaths, sandy verges, railway banks and cliffs. 167. Mainly in the west and throughout the Poole basin, scarce or absent on chalk and clay soils. A foodplant for the Green Hairsteak and many moths, though an extract from the twigs has diuretic effects on mammals. Var. *andreanus*, with reddish wings, is reported from Golden Cap (49A, DG) and Verwood (00Z, VMS).

Spartium junceum L.
SPANISH BROOM
Planted on verges, railway banks and in churchyards, and

self-sown on Portland and on dunes at Sandbanks (08N). 8. On sale in Blandford in 1782. One of the few brooms to tolerate calcareous soils, but it also grows on sand. Seen in 67V and X, 78B, 98J, 99Q and W, 08N and 09C. It contains the alkaloid sparteine, with purging and emetic properties.

Genista monspessulana (L.) L. Johnson
MONTPELLIER BROOM
A rare garden escape which sows itself but is cut down by hard frosts. 2 + [1]. Overcombe (68V, 1982, ! in **RNG**); between Hyde and Wareham (89, 1915, Mrs Drummond); Northport (98J, 1993, !).

Genista tinctoria L.
DYER'S GREENWEED, WOODWEX
A local dwarf shrub of old grassland, tolerant of light grazing but destroyed by ploughing. 64 + [18]. Characteristic of ill-drained clay pastures in the north-west, but it will grow in both calcareous and acid grass along the coastal belt and in the north-east. It has declined since the 1930s due to the mechanization of farming. Subsp. *littoralis* (Corbiere) Rothm. may be the right name for prostrate forms on cliffs at Wyke Regis (67N, 1930, LBH) and east of Lulworth Cove (87J, 1895, WMR and 1984, FAW, !).

Genista anglica L.
PETTY WHIN
Local and in small numbers in bogs, wet heaths and acid marshes. 28 + [18]. Now mainly in the Poole basin; lost from many sites in the north and west owing to drainage.

Genista hispanica L.
SPANISH GORSE
Planted in churchyards and a few verges or dunes, established on sandy cliffs. 5. Netherbury Ch (49U, !); Moreton Ch (88E, !); Bryanston School (80T); Sandbanks (08N, !); Canford Cliffs (08U, RMW).

[***Genista florida*** L.
Planted, but destroyed with other brooms, on a verge at Merley (09E, 1991, ! in **RNG** det CA Stace). It flowers in July.]

[***Genista linifolia*** L.
NEEDLE-LEAVED BROOM
A garden relict, Sandbanks (08N, 1958, ! in **RNG**).]

Genista radiata (L.) Scop.
SOUTHERN GREENWEED
A garden escape at Poole (09, 1961, RDG). [1].

Ulex europaeus L.
GORSE, FURZE, VUZZ
Abundant and locally dominant on heaths, also on the summits and south-facing slopes of Chalk downs and limestone cliffs. 586. Widespread at the edges of woods, in sandy waste ground, and sometimes trimmed into hedges, as at Hurst Heath (78U) and Woodsford (79Q). Least common on clay in the north-west. Resistant to salty gales.

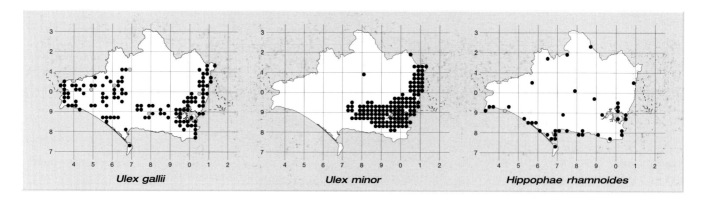

Ulex gallii   *Ulex minor*   *Hippophae rhamnoides*

When colonies are burned, both rootstocks and many seedlings regenerate. Old stems support epiphytic lichens and fungi, and provide shelter for birds and many spiders. At the RSPB reserve at Arne, gorse is cut in rotation to provide bushes 6–15 years old, which are favoured by Dartford Warblers. In earlier centuries, furze was cut for fuel, especially for bakeries.

Ulex europaeus × gallii
Although the parents often grow together, they flower at different seasons, so this hybrid is rare. 3. Stonebarrow Hill (39W, 1992, LJM det PM Benoit); Lewesdon Hill (40F, IPG); Winterbourne Abbas (68E, 1993, LJM); Holt Heath (00, 1982, HCP det. MCF Proctor).

Ulex gallii Planchon
WESTERN GORSE
Frequent to locally dominant in the west, in heaths and bogs or rarely in leached grassland overlying Chalk. Also frequent to abundant in the east, on mature heaths and their boundary banks, or in cleared conifer plantations and on sandy cliffs; it has recently colonised new by-pass verges north of Weymouth. 122 + [5]. Absent from most of the Chalk, and very rare on Portland. Sometimes confused with *U. minor*, as the two species differ mainly in spikiness, chromosome number and the lengths of calyces and flowers. In west Hampshire *U. gallii* is very rare, but in the Poole basin the ranges of *U. minor* and *U. gallii* overlap. All three *Ulex* species grow close together at Binnegar (88Y), Great Ovens Hill (99F), Greenland Farm (08C, EAP), Ferndown Common (09U) and Holt Heath (00L).

Ulex gallii × minor
A single plant closely resembling *U. gallii* but with chromosome number 2n = 48 was found at Gore Heath (99F, 1998, JM Bullock *et al.*). It needs further study but is probably this hybrid (Bullock, Connor, Carrington & Edwards, 1998).

Ulex minor Roth
DWARF GORSE
Locally frequent in dry heathland or mown rides in conifer plantations, and on sandy cliffs. 131. It colonises burnt or cleared heathland along with heathers. Its distribution pattern shows a sharp cut-off in mid Dorset; it grows

nowhere further west, as the few recent records there are errors for *U. gallii*. Two outliers are on lynchets at Ringmoor (80E, !, det. PM Benoit) and at Martin Down (01P, PET, perhaps not in VC9).

Phaseolus coccineus L.
RUNNER BEAN
A casual on tips, widely grown but frost-tender. 2 + [1]. Sherborne tip (61I, 1970, ACL); Crossways tip (78P, 1999, !); Tatchells Pit (98E, 1993, !).

ELAEAGNACEAE

Hippophae rhamnoides L.
SEA BUCKTHORN
Planted all along the coast, and occasionally inland in parks, on verges and waste ground. 39 + [1]. Naturalised by seed or suckers on cliffs at Charmouth (39F and L, 1969, AWG and 1985, RCP in **OXF**); Abbotsbury (58S, 1931, R Meinertzhagen in **BM** and 1998, !); Dungy Head (87E, !); above Lulworth Cove (87J, 1908, CW Dale and 1982–97, !); Flaghead Chine (08P, !). On sale at Blandford in 1782.

Elaeagnus spp. (Oleasters) are planted as hedges, in churchyards, verges and private grounds as follows: *E. argentea* Pursh on Green Is (08D); *E. multiflora* Thunb. at Hamworthy (09A, 1977, ! in **RNG**); *E. pungens* Thunb. in 67N, 78Q, 88D, 89E and M, 80U, 81U and 08G; *E. umbellata* Thunb. at Stourpaine (80U,! in **RNG**), Branksome Chine (08U) and Holes Bay (09B); The only species beginning to spread naturally is *E.* × *ebbingei* Door., by the railway at Southlands (67T), at Ham Common (99Q) and Baiter (09F), and on cliffs west of Bournemouth (08N); specimens in **RNG**, det. A.C. Whiteley.

HALORAGACEAE

(*Myriophyllum verticillatum* L. was reported from Wareham (98, 1799, RP) but no specimen has been seen.)

Myriophyllum aquaticum (Vell. Conc.) Verdc.
PARROT'S-FEATHER
An aquatic alien which is somewhat frost-tender, but is

spreading in garden and field ponds and ditches. 22. First record: Arne (98N, 1990, GM Kay, also ! in **RNG**).

Myriophyllum spicatum L.
SPIKED WATER-MILFOIL
Frequent in parts of the Rivers Yeo, Frome and Stour, occasional in ponds and flooded pits. 53 + [10]. It is tolerant of eutrophic pollution, as at Gillingham and Madjeston.

Myriophyllum alterniflorum DC.
ALTERNATE WATER-MILFOIL
Rare in oligotrophic water of ponds and lakes, also in the Moors River last century. 4 + [10]. Only in the Poole basin apart from a report from Holnest (61K, 1958, SS) which needs confirmation. Coombe Keynes Lake (88R, 1996, FAW); East Holme (88Y, 1955, !); Hartland Moor (98M, 1936, RDG); The Moors (98N, 1982, !); Arne (98P, 1968, DSR and 1979, MRH); Little Sea (08H, 1900, EFL, *et al.*, and 1997, JHSC, with a terrestrial form in 08G).

GUNNERACEAE

Gunnera tinctoria (Molina) Mirbel
GIANT-RHUBARB
Scarcely naturalised, in wet places in wild gardens or by ponds near houses. 15. Once seen on Canford tip (09H).

Gunnera manicata Linden ex Andre
BRAZILIAN GIANT-RHUBARB
Planted in wet wild gardens at Abbotsbury (58S), Mapperton (59E), Litton Cheney (59K) and Nottington (68R). 4.

LYTHRACEAE

Lythrum salicaria L.
PURPLE-LOOSESTRIFE
Frequent at the edge of rivers, streams, ditches and ponds or in water-meadows. 217. Absent from Portland and most of the Chalk.

Lythrum junceum Banks & Solander
FALSE GRASS-POLY
A rare casual, found in a dry pond near Beaminster (40Q, 1921–22, AWG and ML Moores in **BDK**); a record from near Bournemouth (09, 1962, Miss Penrose) may have been in VC11.

Lythrum hyssopifolium L.
GRASS-POLY
Very rare, and varying in abundance from year to year, at the edge of a damp arable field near Charborough Park (99I, 1987–98, BE); once from a garden and tip at Grange Road, Parkstone (09C, 1924, HJW det. J Fraser). Other reports of casual plants, which might be misidentifications, are from Chickerell (67P, 1992, JV Carrington), glasshouses at Canford Magna (09J, 1955, DM) and Verwood (00Z, 1973, D.Parrish). 1 + [4].

Lythrum portula (L.) D. Webb
WATER-PURSLANE
Rather local along wet tracks, or at the edges of ponds and ditches. 58 + [11]. It prefers acid soils and its only record from the Chalk is at Bonsley Pond (80J). Widespread in the Poole basin and scattered in the north and west.

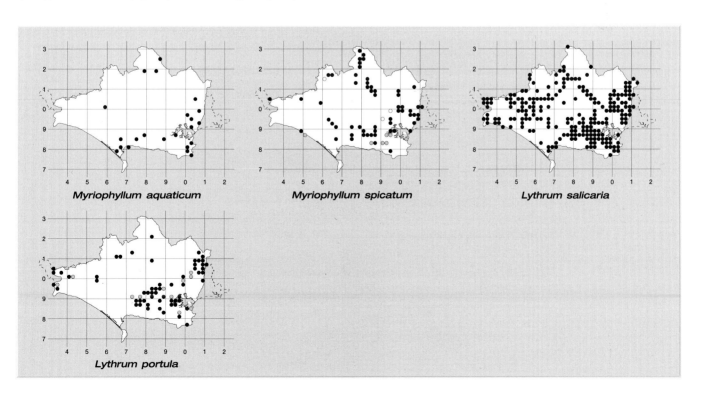

Myriophyllum aquaticum

Myriophyllum spicatum

Lythrum salicaria

Lythrum portula

THYMELEACEAE

Daphne mezereum L.
MEZEREON

A rare, short-lived shrub found in small numbers in calcareous woods or carr. 3 + [7]. On sale in Blandford in 1782 and commonly grown, so that the digging up of wild bushes may have been compensated by birds such as greenfinches spreading seed from gardens. It germinates in autumn and is toxic. Hooke Park (59J, pre-1874, E Fox); Came Wood (68Y, 1995, !); Owermoigne (78S, 1939, DM, also MJA in **RNG** and VM Leather in **OXF**); planted, Bryanston (80T, 1945, BNHS); Langton Copse (80X, 1799, RP); Morden (99C, 1990, BE, !); Well Bottom (91D, 1997, Anon); Eastbury (91G, 1799, RP); New Lane, Verwood (00J, 1922, JHS); Deer Park Farm (01K, 1920, JHS and 1921, AWG).

Daphne laureola L.
SPURGE-LAUREL

A short-lived shrub found in old woods, plantations and old hedges. 84 + [10]. It is very tolerant of shade, and favours calcareous clay soils, but is absent from deforested areas, Portland and heaths. It contains the acrid substance mezerein, which causes blisters.

Daphne odora Thunb.
Rarely planted in churchyards, also in urban Dorchester. 2. Turners Piddle Ch (89G) and Tarrant Keyneston Ch (90H).

HAMAMELIDACEAE

Liquidambar styraciflua L.
SWEET GUM

Occasionally planted as single trees in parks and churchyards. 13. It prefers acid soils and suckers freely. Seen in 78T, 79Q, 89M and Y, 80Y, 82M, 98E, 99S and W, 08D, 09J, N and X. On sale in 1782.

Parrotia persica (DC.) CA Meyer
IRON-TREE

A winter-flowering tree, planted at Abbotsbury before

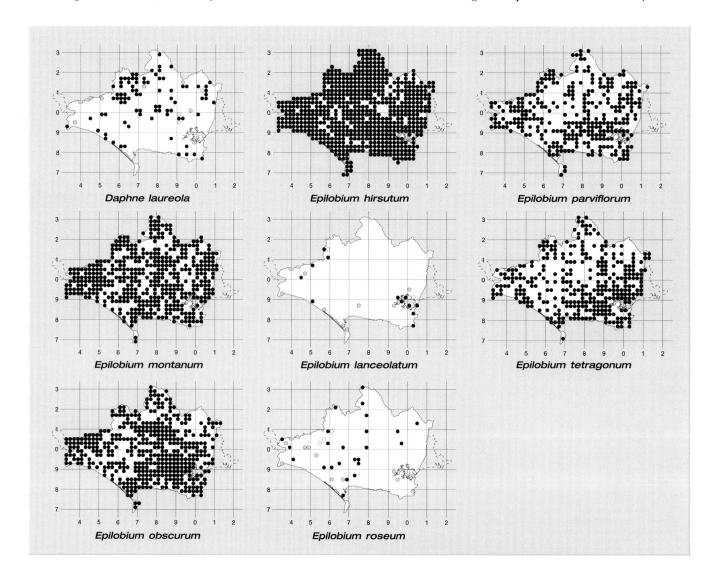

Daphne laureola

Epilobium hirsutum

Epilobium parviflorum

Epilobium montanum

Epilobium lanceolatum

Epilobium tetragonum

Epilobium obscurum

Epilobium roseum

1899; a tree there was the tallest in Britain (15 m) until it blew down in 1980. 11. Occasionally seen in plantations, parks and churchyards, as at Hyde (89K, ! in **RNG**), also in 39S, 40Q, 58S, 50S, 78T, 79A, 71G, 88Z, 08D and 09J.

Hamamelis mollis Oliv. is probably the name of bushes planted in Charborough Park (99, !) and Green Is (08D, !).

MYRTACEAE

Eucalyptus gunnii Hook. f. is occasionally planted outside gardens and fruits, but no seedlings have been seen; Coles Farm (89E), Northport Heath (98E), Green Is (08D) and Winton (09W, FAW).

Other eucalypts, such as *E. globulus* Labill., *E. nicholii* Maiden & Blakeley and *E. viminalis* Labill., are fair-sized at Abbotsbury. Also at Abbotsbury are trees of *Luma apiculata* (DC.) Burret, up to 12 m × 1 m, planted before 1899 for their dramatic bark; seedlings occur here (SG). *Myrtus communis* L. has once been seen planted on a verge at Gaunt's Common (00I).

Callistemon citrinus Stapf
RED BOTTLEBRUSH TREE
One bush planted by pheasant pen, Muston Down (89U), far from houses.

ONAGRACEAE

Epilobium hirsutum L.
GREAT WILLOWHERB, CHERRY PIE, CODLINS-AND-CREAM
Widespread along rivers, streams and ditches or by ponds. 644. Less common on dry verges or waste ground, where it can be an invasive weed. A form with white flowers occurred at Winfrith Newburgh (88C, 1938–60, !).

(*E. hirsutum* × *montanum* = *E.* × *erroneum* Hausskn. is said to have occurred (Stace, ed. 1975).)

E. hirsutum × *tetragonum* = *E.* × *brevipilum* Hausskn.
Brownsea Is (08I, 1979, RMB in **RNG**).

E. hirsutum × *ciliatum* = *E.* × *novae-civitatis* Smejkal.
Grimstone (69H, 1993, LJM and LMS); Whistley Copse (72Z, 1994, RM Veall & !); both det. CA Stace. 2.

Epilobium parviflorum Schreber
HOARY WILLOWHERB
Frequent in open, damp places such as winter streams, ditches and waste ground, sometimes in fallow. With white flowers at Moreton (78U). 311.

[*E. parviflorum* × *obscurum* = *E.* × *dacicum* Borbas. Old records from Ridge (98I, 1935, RDG), Lytchett Minster (99R, 1900, EFL) and Horton (00I, 1935, RDG).]

E. parviflorum × *roseum* = *E.* × *persicinum* Reichb.
Reported from Sherborne School (61I, 1971, SS).

E. parviflorum × *ciliatum*
Rare. North Poorton (59E, 1993, LJM and LMS); Bryanston (80T, 1945, NDS det. GM Ash); Durlston CP (07J, 1977, KC); New Swanage (08F, 1998, EAP). 2 + [2].

E. parviflorum × *tetragonum* = *E.* × *palatinum*
F.W. Schultz
Rare or overlooked. 1. With parents in fallow, near Marsh Br (99E, 1999, !).

Epilobium montanum L.
BROAD-LEAVED WILLOWHERB
A frequent weed in disturbed woodland, hedgebanks, gardens and waste places. 464. Variable in size. A form with white flowers grew at Bryanston (80X, !).

[*E. montanum* × *obscurum* = *E.* × *aggregatum* Celak
Bilshay Farm (49M, 1935) and Ranksborough Gorse (70J, 1937), both RDG in **WRHM** det. GM Ash.]

[*E. montanum* × *roseum* = *E.* × *mutabile* Boiss. & Reuter
Evershot (50S, 1928, CES); Wareham (98, 1940, LBH); Branksome (09K, 1945, NDS).]

Epilobium lanceolatum Sebast. & Mauri
SPEAR-LEAVED WILLOWHERB
A rare weed of gardens or on walls. 13 + [1]. Mostly along the coastal belt or in the north-west, much rarer than in Devon. A white-flowered form was seen at Holton Heath (99K, 1998, AJ Showler).

Epilobium tetragonum L.
SQUARE-STALKED WILLOWHERB
Frequent on disturbed verges, ditch banks, fallow, gardens, pavements and waste ground. 323.

(*E. tetragonum* × *palustre* = *E.* × *semiobscurum* Borbas is said to have occurred (Stace, 1975).)

Epilobium obscurum Schroeder
SHORT-FRUITED WILLOWHERB
Common in woodland rides, hedgebanks, verges, ditches, gardens and waste ground. 438.

Epilobium obscurum × *brunnescens* = *E.* × *obscurescens*
Kitchener & McKean
With parents, Knighton pits (78P, 1999) ! det. T.D. Pennington).

[*E. obscurum* × *palustre* = *E.* × *schmidtianum* Rostkov
East of Little Sea (08H, 1913, CBG in **BMH** det. HWP). [1].]

Epilobium roseum Schreber
PALE WILLOWHERB
An occasional weed with a liking for churchyards, also in shaded hedgebanks, lanes and gardens. 24 + [14]. It avoids

acid soils, and has not been reported from semi-native habitats.

Epilobium ciliatum Raf.
AMERICAN WILLOWHERB
Now common and widespread in woodland rides, gardens, pavements, waste places and tips. 379. First noticed in 1936, at Bloxworth (89X) and Charborough Park (99J, RDG in **WRHM**).

Epilobium palustre L.
MARSH WILLOWHERB
Local in marshes, wet woods, flushed bogs and water-meadows. 63 + [5]. Unlike most species in this genus it grows in closed vegetation. Scattered in the north-west and the Poole basin, absent from the Chalk and other dry regions.

Epilobium brunnescens (Cockayne) Raven & Engelhorn
NEW ZEALAND WILLOWHERB
Very local and often transient. 1 + [2]. Established on wet floor of old sandpit, Knighton (78P, 1992–99, DAP and

RMW, !); conifer nursery, Wareham (98, 1951, BA Harland); West Moors (00R, 1929, Miss Firbank in **K**).

Chamerion angustifolium (L.) Holub
ROSEBAY WILLOWHERB
Frequent in cleared woodland or burnt heath, occasional on verges, gardens and urban sites. 502. Widespread and much increased since 1895. White-flowered plants have only been seen in gardens.

Ludwigia palustris (L.) Elliott
HAMPSHIRE-PURSLANE
Very rare in an acid pond. 1. Near Edmondsham (01K, 1996, JHSC, !).

Oenothera glazoviana Micheli ex C. Martius
LARGE-FLOWERED EVENING-PRIMROSE
Occasional in bare, disturbed soil on cliffs, verges, railway banks, old pits or tips. 135 + [11]. Not found by JCMP. First record: Portland (67, 1912, WBB). It prefers sandy soils but is not restricted to them.

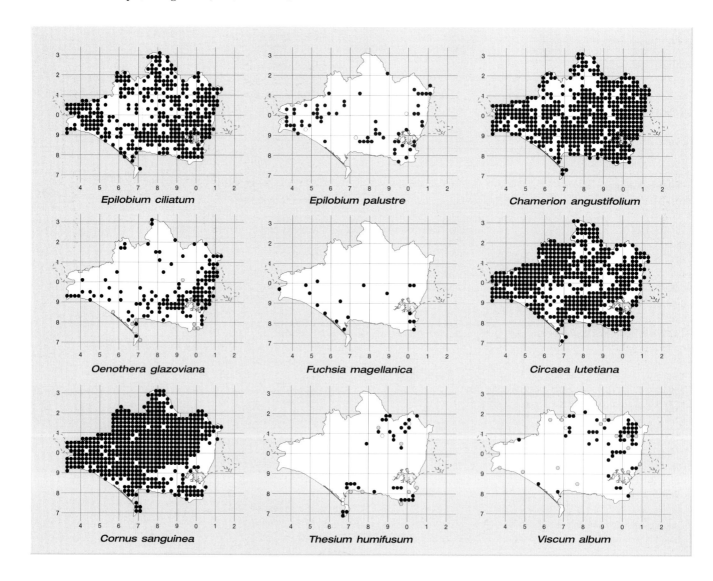

Epilobium ciliatum   Epilobium palustre   Chamerion angustifolium

Oenothera glazoviana   Fuchsia magellanica   Circaea lutetiana

Cornus sanguinea   Thesium humifusum   Viscum album

Oenothera × fallax Renner
INTERMEDIATE EVENING-PRIMROSE
Reported from Mosterton (40M, 1993, JGK), and perhaps overlooked, but it has only one recent record from Somerset.

Oenothera biennis L.
COMMON EVENING-PRIMROSE
Uncommon in bare, dry, disturbed soils in verges, railway banks or tips. 10 + [11]. First certain record: Studland (08, 1911, HE Fox in **OXF** det. K Rostanski). Mostly on sandy soils, but seen on chalk among kale at Tarrant Hinton (91F) and as a garden weed at Chettle Ho (91L).

Oenothera cambrica Rostanski
SMALL-FLOWERED EVENING-PRIMROSE
Doubtfully distinct from *O. biennis*, in similar habitats and on dunes. 7 + [4]. First report: South Haven (08I, 1888, JCMP in **DOR** det. !, also 1992, !).

Oenothera stricta Ledeb. ex Link
FRAGRANT EVENING-PRIMROSE
A scarce garden escape, established on sandy cliffs, verges and railway banks. 6 + [5]. Mainly near the coast and in the Poole basin. Lyme Regis (39, 1879, JCMP in **DOR** and 1954, BNHS); Abbotsbury (58S, 1981, !); East Knighton (88C, 1991, !); Holmebridge (88Y, 1955, !); Wareham (98I and J, WF in **BM**, *et al.* in **LIV**, and 1998, !)); Arne (98U, 1979, MRH); North Haven (08I and N, BNSS and 1969–99, !); Parkstone (09F, 1895, WMR).

Fuchsia magellanica Lam.
FUCHSIA
Planted in a few churchyards and hedges, and escaping to old quarries, railway banks and waste places. 18. Mostly transient, and cut down by frosts, but seedlings occur at Abbotsbury (58S, SG). First record: Compton Valence (59W, 1960, EH in **LANC**). Cv. *molinae*, with pale flowers, was seen by the Stour at Canford School (09J, !). It likes a higher rainfall than Dorset provides, but there is a Fuchsia nursery at Three-legged Cross.

Circaea lutetiana L.
ENCHANTER'S-NIGHTSHADE
Frequent in woods, plantations, hedgebanks and in shaded gardens. 503.

Clarkia amoena (Lehm.) Nelson & Macbr.
GODETIA
A pavement weed at Wareham (98I, 1998, !). 1.

CORNACEAE

Cornus sanguinea L.
DOGWOOD
Frequent at the edges of woods, in scrub and hedges on calcareous soils; known since the neolithic age. 546. Absent, unless planted, from acid heathland in the Poole basin. Starlings eat the fruit (Wilson, Arroyo & Clark, 1997).

Cornus sericea L.
RED-OSIER DOGWOOD
Planted in clumps by lakes and rivers in parks. 21. First record: Wimborne St Giles (01F, 1990, !).

Cornus alba L.
WHITE DOGWOOD
Less commonly planted than *C. sericea*, more often on verges, by car parks or in towns. 11. Seen in 49L, 40Q, 78J, 71G, 73Q, 89M, 80Z, 91G, 07I, 09D and 00E.

Cornus mas L.
CORNELIAN CHERRY
Rarely planted as single trees, and fruiting sparsely, but not naturalised. 6. A large tree near Pulteney's old house in Blandford was felled in 1995. A bush with variegated leaves is at Charminster Ch (69R), and a tree at West Stafford Ch (78J); several large trees grow in a spinney at Alderney (09M).

Other **Cornus** spp. are planted, as at Canford School (09J, TH), including **C. controversa** Hemsl., which produces seedlings at Abbotsbury (58S, SG).

Aucuba japonica Thunb.
SPOTTED-LAUREL
Occasionally planted in parks, plantations and church-yards. 21. Seedlings seen at Bradpole (49X) and Kingston (97P); shrubs noted in 39L, 49R, 40G, 58Z, 68R and S, 60S, 78J, 79A and Q, 80T, 81H, 98A, 99I, 90Q, 91G, 07I, 08D and L.

Griselinia littoralis (Raoul) Raoul
NEW ZEALAND BROADLEAF
Planted for hedging, or in parks, all along the coast. 12. Self-sown on sandy cliffs at Flaghead Chine (08P).

SANTALACEAE

Thesium humifusum DC.
BASTARD-TOADFLAX
Very local in short calcareous turf, often on ancient earthworks. 33 + [12]. It can grow near the sea, and was once found on fixed shingle at Small Mouth (67S, 1895, JCMP). Now found on the limestone of Portland and Purbeck, and on the Chalk in the south-east and north-east, notably at Badbury Rings and Fontmell Down. Specimens are in **ALT, BM, CGE** and **DOR**.

VISCACEAE

Viscum album L.
MISTLETOE
The native records for this parasite are mostly in the north-east. 50 + [15]. It is widely planted elsewhere, mostly on apple (19 records). Other hosts, where reported, are *Acer campestre* (7), *Crataegus* (5), *Populus* (10), *Tilia* (14), *Cotoneaster* (2) + 1 record each from *Acer pseudoplatanus*, *Fremontodendron*, *Prunus*, *Quercus* and *Sorbus*. It may be

increasing. There is a Mistleberry Wood in Cranborne Chase, and the berries are spread by Mistle Thrushes. All parts are poisonous to man, as they contain polypeptide viscotoxins which may be anticarcinogenic.

CELASTRACEAE

Euonymus europaeus L.
SPINDLE, SKIVER-WOOD
Occasional, in small numbers, at the edges of woods, in scrub and old hedges. 417. It prefers calcareous soils, and does not grow in most of the Poole basin. Large trees occur near Badbury Rings (90R), and a young scandent form was seen in a thicket at Higher Nyland (72L). The wood, though poisonous, was once used for skewers by butchers. The fruit is toxic to sheep, but robins eat it (Wilson, Arroyo & Clark, 1997).

Euonymus latifolius (L.) Miller
LARGE-LEAVED SPINDLE
Rarely planted, though on sale at Blandford in 1782. 3. Self-sown in a hedge at Stickland (80H, 1991, ! in **RNG**), also near Sycamore Down Farm (80H, 1999); planted at Holworth Ho (78Q) and Kingston Lacy (90Q).

Euonymus japonicus L. f.
EVERGREEN SPINDLE
Widely planted in hedges and shelter-belts near the sea, and inland in churchyards. 51. Well established on limestone undercliffs at Church Ope Cove (67V) and Lulworth (87J), and on sandy cliffs at Flaghead Chine (08P). Large trees occur at Abbotsbury (58M and S), Durlston (07I) and North Haven (08N).

AQUIFOLIACEAE

Ilex aquifolium L.
HOLLY
Frequent to abundant in oak woods, where it is browsed by deer, less common in ash and alder woods, extending to the edge of Poole Harbour. 688. Common in old hedges and along green lanes. Thirty cultivars were on sale at Blandford in 1782. Trees with yellow fruit occur in a hedge at Monkton Wyld (39I, LJM), and are planted near Charborough Ho (99I). *f. pendula* grows at Corfe Mullen Ch. (99Z). The white wood is valued by turners, and sucker shoots were once used as whip handles. The leaves are eaten by the Holly Blue, and the berries by wintering thrushes (Wilson, Arroyo & Clark, 1997). Despite being an evergreen, severe gales can defoliate exposed trees, while mild gales do not even blacken the leaves. Eighteen place-names based on holly (hollis, holm, holn) are found in the county.

Ilex aquifolium × *perado* = *I.* × *altaclarensis*
(hort. ex Loudon) Dallimore
HIGHCLERE HOLLY
Rarely planted in churchyards or parks. 5. Buckland

Newton (60X), Kingston Lacy Ch (90V), Canford School (09J) and Branksome Chine (08U and 09Q).

Other *Ilex* sp. are planted in Abbotsbury gardens, where *I. dipyrena* Walt. seeds itself (58S, SG), and Canford School (09J, TH).

BUXACEAE

Buxus sempervirens L.
BOX
Long cultivated in calcareous plantations, often seeding itself like a native, as at Chedington (40X), Came Wood (68X), Woolland (70T) and Bryanston (80T); it was here in the neolithic age. 173. Trees at France Firs (80Z, EN) were cut for their valued timber in 1963. Frequently planted in churchyards or as hedges in villages. The name Bexington might be derived from box, though it is not found there now.

Buxus balearica Lam.
BALEARIC BOX
Rarely planted in churchyards or cemeteries. 3. Over Compton Ch (51Y, !); Fortuneswell (67X, !); Brownsea Is (08I, !).

EUPHORBIACEAE

Mercurialis perennis L.
DOG'S MERCURY
Dominant in the drier parts of woods, frequent in old hedgebanks and sheltered undercliffs. 550. Rare on Portland, and scarce or absent in deforested areas and acid heaths of the Poole basin. Rabbits will not eat it, but bullfinches will take the fruit (Wilson, Arroyo & Clark, 1997). Where woods are eutrophiated, as under rookeries or at the base of steep slopes, mercury is displaced by nettle.

Mercurialis annua L.
ANNUAL MERCURY
Occasional in arable, more often in gardens, allotments, pavements or tips; on beaches at Brownsea Is. 180. It occurs on both acid and calcareous soils when they are nutrient-rich, and likes warm sites. The seeds are dispersed by the ant *Messor structor*.

[*Euphorbia peplis* L.
PURPLE SPURGE
Once found on sandy shores. [1]. West Bay (49Q, 1805, J Sims and 1875, Miss Clark).]

Euphorbia oblongata Griseb.
Rarely seen near gardens, though seeding freely there. 6. Shipton Gorge (49V, 1999, !); Elwell (68S, 1999, !); pavement, Dorchester (69V, !); Bere Regis (89M, 1995, !); Canford Heath tip, a hairy form (09I, 1999, ! det EJ Clement); Woolsbridge racing circuit (10C, 1999, GDF).

Euphorbia dulcis L.
SWEET SPURGE
A rare casual. 1 + [1]. In a garden, Houghton (80H, 1977, T Norman); on verge, Talbot village (09S, 1997, FAW).

Euphorbia platyphyllos L.
BROAD-LEAVED SPURGE
Very local and sporadic, but persistent in dry arable, gardens and tips. 11 + [18]. It has always been scarce, and Good did not find it in any of his stands in the 1930s. It does not occur further west in Britain, but survives around Beaminster, Sherborne, south Purbeck and Edmondsham.

Euphorbia serrulata Thuill.
UPRIGHT SPURGE
A rare casual in a garden at Stoborough (98H, 1998, AH). 1.

Euphorbia helioscopia L.
SUN SPURGE
Occasional in dry arable on chalk or sand, more common in allotments and gardens in villages. 288. Least common on very acid soils of the Poole basin, as it prefers nutrient-rich sites; variable in size.

Euphorbia lathyris L.
CAPER SPURGE
Widespread but sporadic in village lanes and verges, gardens, car parks, waste ground and tips. 93 + [12]. JCMP gave only two localities in 1874. The seeds have a burning taste and should not be eaten. The latex is an purgative emulsion containing more than 40 terpenoids, including the toxic euphorbin, eye irritants and probable carcinogens. The use of this plant as a mole-repellent is based on hearsay.

Euphorbia exigua L.
DWARF SPURGE
Local in dry, calcareous arable, often with *Kickxia* spp. 64 + [4]. Restricted to the chalk, the coastal belt and the north-western limestone. JCMP found it common, and RDG had it from 84 stands, so it is declining, either from excessive use of fertilizers or early ploughing of stubbles.

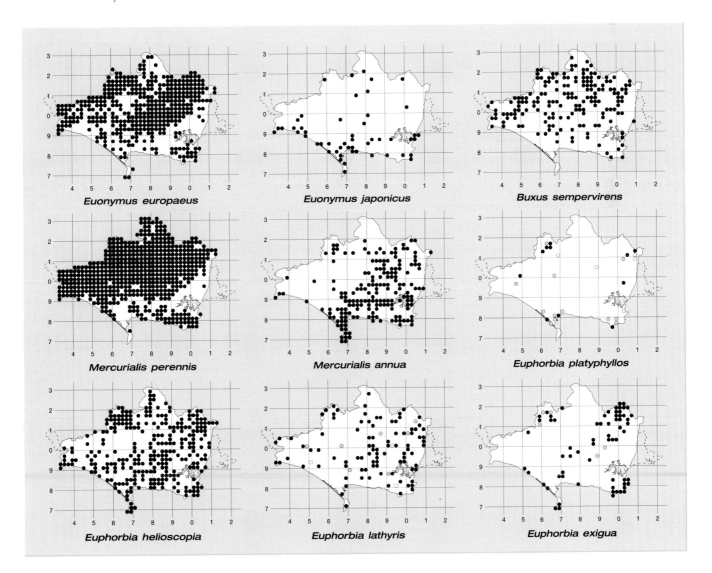

Euonymus europaeus

Euonymus japonicus

Buxus sempervirens

Mercurialis perennis

Mercurialis annua

Euphorbia platyphyllos

Euphorbia helioscopia

Euphorbia lathyris

Euphorbia exigua

Euphorbia peplus L.
PETTY SPURGE

Occasional in dry arable, much more common in allotments, gardens and as an urban weed. 401. It germinates early in winter, and needs the extra warmth associated with built-up areas to flourish. Its latex has caused rashes.

Euphorbia portlandica L.
PORTLAND SPURGE

Locally frequent on undercliffs, sometimes on shingle beaches, along the coast. 21 + [3]. Its main strongholds are Portland and the cliffs between Ringstead Bay and Kimmeridge. It is rare or in decline west of Portland, but there are few records on the south coast east of Dorset. An early report from Small Mouth (67S, 1570, M L'Obel) is ambiguous and may refer to *E. paralias*. Specimens are in **BDK, CGE, HIWNT, MDH** and **RNG**.

Euphorbia paralias L.
SEA SPURGE

Once scattered on sand and shingle beaches or clay cliffs, now very rare. 2 + [7]. Its decline is due to human pressure

on beaches. Burton Bradstock (48Z, 1961, RDG in **DOR**); Fleet (67, 1977, Anon in **WRHM**); Small Mouth (67S, 1799, RP, *et al.*, and 1996, DAP, !); East Portland (77, 1960, BSBI); Overcombe cliffs (69V, 1960, !); Osmington Mills (78F, 1895, FWG); Swanage (07J, 1799, RP); Sandbanks (08N, 2000, !); Poole (09, 1846, JC Dale).

Euphorbia × pseudovirgata (Schur.) Soo
TWIGGY SPURGE

A rare but persistent colonist of verges. 4 + [4]. Powerstock (59D, 1982, D Griffith); Portland (67, 1957, LE Cobb in **RNG**); Westham (67U, 1930s, DM in **WRHM**); below Came Down (68Y, 1936, RDG in **WRHM**, *et al.*, and 1960–95, !); Cale bridge (71P, 1919, Lady Douie, and 1934, RDG in **WRHM**); Lower Farm, Fifehead Magdalen (72Q, 1965, RDG in **DOR**); Blandford Camp road (90D, 1990, DAP, !); Pimperne (90E, 1996, !); N. Swanage (08F, 1999, EAP).

Euphorbia cyparissias L.
CYPRESS SPURGE

A garden escape, mostly transient, on verges and tips, but perhaps naturalised at Holt Forest (00H, 1999, BE) and

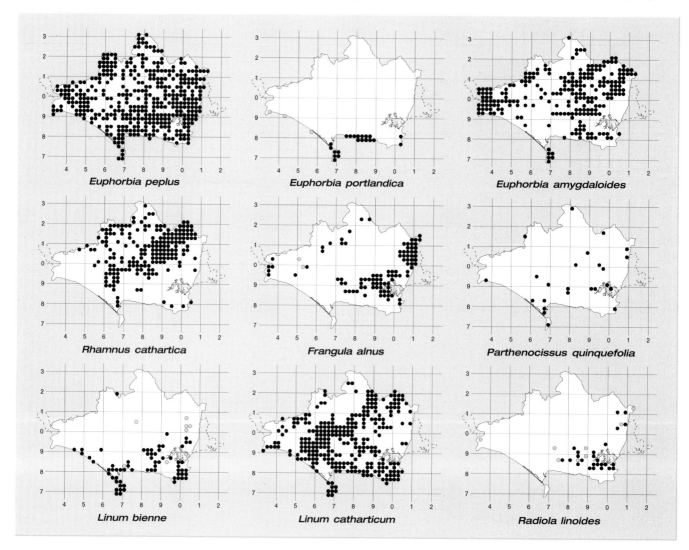

Euphorbia peplus

Euphorbia portlandica

Euphorbia amygdaloides

Rhamnus cathartica

Frangula alnus

Parthenocissus quinquefolia

Linum bienne

Linum catharticum

Radiola linoides

Crab Orchard (00Y, 1993, VS). 18 + [2]. First record: Netherbury (49U, 1935, AWG).

Euphorbia amygdaloides L.
WOOD SPURGE
Locally frequent in, and characteristic of, old woods, also on sheltered undercliffs at Portland and White Nothe. 235. Sometimes abundant and colouring large patches in clearings. Absent from deforested areas and very acid soils. Subsp. *robbiae* (Turrill) Stace occurs rarely as an escape in plantations, wild gardens and churchyards, often in deep shade, in 40Q, 68W, 69L, 60S, 79U, 89Q, 80H, 07J and 08G. 9.

Euphorbia characias L.
MEDITERRANEAN SPURGE
Rarely self-sown on sandy cliffs or on sunny verges outside gardens. 9. Naturalised at Flaghead Chine (08P, 1992, RMW and 1994–97, !), also seen in 40T, 68H, 78G, 81B, 82Q and V, 98I and 01A.

RHAMNACEAE

Rhamnus cathartica L.
BUCKTHORN
Occasional in hedges and scrub on calcareous soils, usually in dry places, but among reeds in Radipole Lake. 194. Mostly on the Chalk in central and north-east Dorset, and on limestone in the north-west; scarce on the Purbeck chalk ridge, and absent from Portland, the west and the Poole basin. Large trees seen in Jubilee Wood (90M). A foodplant of the Brimstone butterfly, with a purgative fruit. The inner bark is orange.

Frangula alnus Miller
ALDER BUCKTHORN
Occasional to frequent in acid woods, carr, wet heaths and bogs. 95 + [2]. Mostly in the Poole basin, local in the north and west, and absent from calcareous regions. It rarely grows with *Rhamnus catharticus*, but the species co-exist at Lydlinch Common (71G). It contains 3–7% of the purgative anthraquinones emodin and frangulin, as well as other compounds which can induce vomiting.

Ceanothus spp. are increasingly planted, e.g. at Turners Puddle Ch (89G).

VITACEAE

Vitis vinifera L.
GRAPE-VINE
A rare escape from cultivation. 4. Charlestown brickpits (67U, 1998, !); Sherborne (61H, 1970, ACL and 1989, !); Wool (88I, 1991–98, !); Wareham (98I, 1993, JB). Neolithic remains have been found on Hambledon Hill. The Domesday Book mentions a three-acre vineyard at 'Dervinestone' or Durweston in 1086. Twelve cultivars were on sale in Blandford

in 1782. Modern vineyards are at Tarrant Crawford (90B), Langton Long (90D) and Horton (00I).

Parthenocissus quinquefolia (L.) Planchon
VIRGINIA-CREEPER
An occasional garden escape climbing into plantations, tall hedges or railway banks, and on tips. 24.

Parthenocissus tricuspidata (Siebold & Zucc.) Planchon
BOSTON-IVY
A rare garden escape on walls. 2 + [1]. Abbotsbury (58S, 1960, EH in **LANC**); Sandford Orcas (62F, 1982, AJCB); Swanage (07J, 1991, WGT).

LINACEAE

Linum bienne Miller
PALE FLAX
Locally frequent in dry calcareous grassland, often on cliffs, also occasionally on calcareous or dry, nutrient-enriched verges through heathland. 66 + [8]. Mostly in the south, and absent from the north apart from an outlier at Crackmore (61U, JAG). Present since the Iron Age.

Linum usitatissimum L.
FLAX
Much grown around Bridport for fibre, from the late Iron Age and medieval times to 1816. Since 1992, subsidies have made this a profitable oil-crop, even if left unharvested. Both blue- and white-flowered forms are grown, notably Cvs. Barbara and Norlin. Now a common relict in stubbles and verges, also a casual from bird-seed in gardens and tips. 42+.

Linum catharticum L.
FAIRY FLAX
Frequent in well-drained soils with short turf, mostly calcareous, but also along heath tracks. 282. Common on chalk and limestone, occasional in the Poole basin, and mostly absent from clay. Var. *dunense* Druce, a dwarf form, grows on cliffs at Church Ope Cove (67V, !) and Old Harry (08L, 1957, VM Leather in **RNG**).

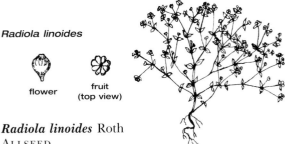

Radiola linoides

flower   fruit (top view)

Radiola linoides Roth
ALLSEED
Very local in rutted tracks of acid plantations or heaths, in grazed heaths or their mown firebreaks, rarely at the back of dunes. 25 + [8]. All recent records are from the Poole basin, where it is decreasing, but it was once found in the west, at Champernhayes (39N, 1928, AWG).

POLYGALACEAE

Polygala vulgaris L.
COMMON MILKWORT
Occasional in short grass on calcareous or neutral soils, scarce on acid soils. 192. It has decreased because so much of its habitat has been ploughed up. The petals are mostly blue, sometimes pink and least commonly white; all three colour-forms occur at Eastfield Hill (69P), Mupe Bay (88K) and Zigzag Hill (82V). Subsp. *collina* (Reichb.) Borbas is found at Cat Range Head (88R, 1998, P Selby).

Polygala vulgaris × *calcarea* was once found at Seacombe (97Y, 1935, NDS in **BM**). [1].

Polygala serpyllifolia Hose
HEATH MILKWORT
Occasional in acid rides of woods and plantations, or among heather. 112. Frequent in the Poole basin, scattered in the north and west, and absent from the Chalk. Forms

with pink flowers occur at Winfrith Heath (88D) and elsewhere, and with white flowers at Black Hill (89H) and on Brownsea Is (08D).

Polygala vulgaris
Sepal showing veins

× 4

Polygala calcarea
Sepal showing veins

Polygala calcarea F. Schultz
CHALK MILKWORT
Locally frequent in short calcareous turf, mostly confined to steep banks and earthworks which have escaped the plough. 79 + [9]. Some records from central and west Dorset may be errors for *P. vulgaris*, but the two can co-exist. The flowers are usually bright blue, sometimes pink and least commonly white; all three colour-forms occur at Badbury Rings (90R). At Cheselbourne (70K, 1976, JKH) this was picked to decorate the church at Rogation.

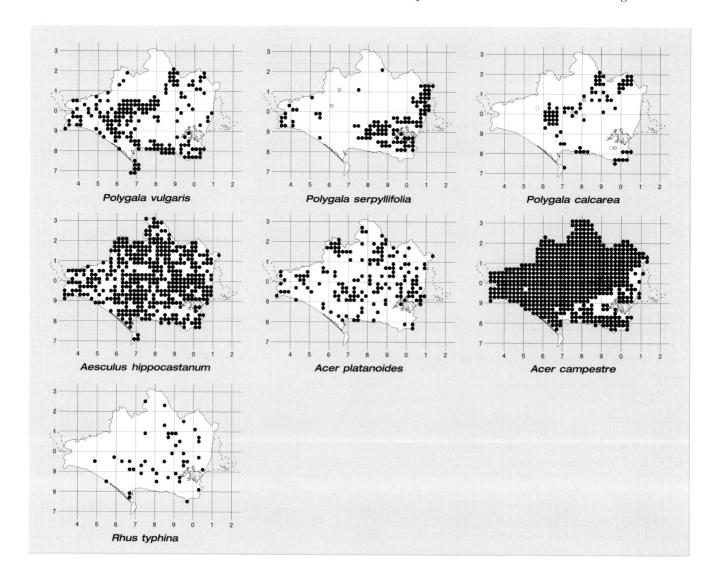

Polygala vulgaris

Polygala serpyllifolia

Polygala calcarea

Aesculus hippocastanum

Acer platanoides

Acer campestre

Rhus typhina

STAPHYLEACEAE

Staphylea pinnata L.
BLADDERNUT
Several trees grow in a plantation at Nether Cerne (69U, 1995, ! in **RNG**).

SAPINDACEAE

Xanthoceras sorbifolia Bunge was reported as planted at Bryanston (80T, 1945, BNHS).

HIPPOCASTANACEAE

Aesculus hippocastanum L.
HORSE-CHESTNUT
Frequently planted as one or a few trees in plantations, parks and village greens, or as avenues in Dorchester; there is a line of 12 trees near Newton farm (99G). 436. It seeds freely, and seedlings are frequent in gardens, urban sites and tips. The seeds, which are toxic, are sought by boys for the game of conkers. The tree was on sale at Blandford in 1782. A tree at Kingston Lacy (90V) was 35 m × 5.5 m in 1983.

Aesculus carnea Zeyher
RED HORSE-CHESTNUT
Planted in similar places to the last, but no seedlings have been reported. 37. On sale in 1782.

Other *Aesculus* spp. are planted, as at Canford School (09D, TH). *A. flava* Schrader at Minterne Magna (60S); *A. indica* (Cambess.) Hook. at Uplyme (39G) and Brownsea Is (08I); *A. turbinata* Blume at Netherbury (49U); and *A. wilsonii* Rehder at Melbury Park (50T, 21 m × 1.5 m in 1980).

ACERACEAE

Acer platanoides L.
NORWAY MAPLE
Occasional in plantations and parks, with cultivars often planted as street trees in villages. 165. Many seedlings noted. Colourful when flowering in spring and for its autumn leaves. On sale at Blandford in 1782, but not recorded 'in the wild' until 1968.

Acer campestre L.
FIELD MAPLE
A frequent understorey tree in old woods and hedges, and much planted today. 626. Common on chalk and clay soils, but scarce in deforested areas and the more acid soils of the Poole basin. Not known on Portland. Large trees seen in Up Cerne Wood (60M), Sherborne Park (61T), Duncliffe Wood (82G), Hogstock Wood (90N) and Cranborne Chase, where they support a rich lichen flora. The wood was used by the Romans for artefacts dug up on Brownsea Is, and

was favoured by the Saxons for making harps. William Barnes devotes a poem to the 'meaple'. About six place names, such as Mapperton, are derived from this tree.

Acer pseudoplatanus L.
SYCAMORE
Frequent in plantations, often self-sown in woods, and in hedges far from houses. 697. Introduced before the 13th century. There are sycamore avenues at Stinsford (79A), Ilsington (79L) and Anderson (89Y). Notable trees include the 'Posy Tree' at Mapperton (49Z), the scene of confrontation between Mapperton and Netherbury men in 1582, a plague year; a tree at Sherborne Park (61N) which was 30 m × 1 m in 1978; and the Tolpuddle Martyrs' tree (79X), a 12 m × 5.5 m hollow pollard in 1991. Although disliked by many, the sycamore is a good host for lichens and insects, and the wood is used for furniture.

Acer saccharinum L.
SUGAR MAPLE
Occasionally planted in parks or as a street tree. 11. Seen in 30S, 61N, 79A, 81N, 98U, 09E,P and X, 00I and Q and 01F.

Acer negundo L.
ASH-LEAF MAPLE
Rarely planted in parks or as a street tree. 9. Noted in 67U, 61I and N, 71W, 88J, 89T, 90V, 09E and J.

Other *Acer* spp. are planted in parks, as at Canford School (09J, TH). *Acer cappadocicum* Gleditsch with seedlings in Melbury Park (50X, IPG and PRG); *A. franchetii* Pax, a tree 10 m × 0.6 m at Abbotsbury (58S); *A. griseum* (Franch.) Pax at Holworth Ho (78Q); *A. palmatum* Thunb. at Sherborne Castle (61S); *A. rubrum* L., 18 m × 2.5 m at Melbury Park (58S), also at Stock Gaylard Park (71G).

ANACARDIACEAE

Rhus typhina L.
STAG'S-HORN SUMACH
A persistent garden escape, suckering on verges and railway banks, waste ground and tips. 43. On sale at Blandford in 1782. On St Aldhelm's Head undercliff, a relict of the radar station in the 1940s (97S). Its latex contains phenols which could affect sensitive skins.

Cotinus coggygria Scop. is planted in Sherborne Castle grounds (61S).

SIMAROUBIACEAE

Ailanthus altissima (Miller) Swingle
TREE-OF-HEAVEN
Planted trees seen at Forde Abbey (30M, !), Sherborne (61M, JAG), Bere Heath (89R, ! in **RNG**), Charborough Park (99I, first reported in 1914), Corfe Mullen Ch (99U) and Canford School (09J, TH); it suckers freely. 6.

RUTACEAE

Choisya ternata Kunth
MEXICAN ORANGE
Rarely planted in churchyards or parks. 5. Over Compton Ch (51Y); Sherborne Park (61N); Bryanston (80X); Kingston Lacy (90Q); Alderholt Ch (11B). The young shoots are blackened by frost.

Ptelea trifoliata L. is planted in Lower Lodge Plantation (70V).

Ruta graveolens L.
RUE
Naturalised on a wall at Upwey (68S, 1999, !), casual at West Lulworth (88F, 1981, !) and planted at Cranborne Ch (01L). 2 + [1].

Zanthoxylum piperitum (L.) DC. is planted near Sherborne Old Castle (61S).

Skimmia japonica Thunb. is planted at Alderholt Ch (11B).

OXALIDACEAE

Oxalis corniculata L.
PROCUMBENT YELLOW-SORREL
Widespread in warm, artificial habitats such as gardens, pavements and walls, never in arable. 154 + [12]. The purple-leaved var. *atropurpurea* Van Houtt ex Planchon is frequent.

Oxalis exilis Cunn.
LEAST YELLOW-SORREL
In the same habitats as *O. corniculata* but rarer. 4. Loders (49X, 1995, LJM and LMS); Winterborne Kingston (89P, 1989–99, !); Lake (99V, 1995, !); Branksome (09Q, 1996, PC Hall).

Oxalis stricta L.
UPRIGHT YELLOW-SORREL
Occasional on railway banks, or in warm sites near buildings or walls. 32 + [11]. Increasing. Specimens are in **BDK, RNG** and **WRHM**.

Oxalis articulata Savigny
PINK-SORREL
A garden plant but often found outside gardens in village lanes, waste places, tips and on sandy cliffs at Flaghead Chine (08P). 90. Planted in Abbotsbury gardens in 1899, first record 'in the wild': Redcliff (98I, 1951, !), where it persists. A form with white flowers is at North Haven (08I)

Oxalis acetosella L.
WOOD-SORREL, CUCKOO'S-BREAD, SLEEPING BEAUTY
Occasional in both dry and wet woods, preferring neutral to mildly acid soils. 227. Most frequent in the west, absent from deforested areas. Locally abundant in deep shade in Broomhill Plantation (79W).

Oxalis debilis Kunth
LARGE-FLOWERED PINK-SORREL
A scarce but persistent garden weed. 2 + [1]. Upwey (68S, !); Cerne Abbas (60Q, 1971, RDE); Upton CP (99W, !).

Oxalis latifolia Kunth
GARDEN PINK-SORREL
A weed infesting some allotments, nursery gardens and parks. 7 + [4]. Little Bredy (58Z, 1953, !); Abbotsbury (58Z, 1958, !); Wyke Regis (67N, 1960, PJO Trist); Sandsfoot (67T, !); The Nothe (67Z, !); Blandford (80Y, 1977, !); Steeple (97A, !); Smedmore Ho (97J,1998, EAP); Wareham (98I, !); Holton Heath (99K, !); Stapehill (00K, !).

Oxalis tetraphylla Cav.
FOUR-LEAVED PINK-SORREL
Reported as a weed at Nettlecombe (59C, 1979, Anon in Wild Fl. Mag.).

(**Oxalis lasiandra** Zucc. is a weed at Christchurch (19W, 1971, D McClintock in **BM**), but in VC11.)

Oxalis incarnata L.
PALE PINK-SORREL
A delicate weed of sheltered sites in gardens, under walls or in shade, which escapes to village lanes or shaded stream-banks. 15. It was sold at Blandford in 1782 as a tender plant. Seen in 39G and L, 48Z, 58S, 67T, 68B and C, 78B, 72Z, 81G, 99W, 90Y, 07J and 08G and I.

GERANIACEAE

Geraniaceae, together with Rutaceae and Myrtaceae, are some of the few families whose members secrete essential oils, which render the leaves resistant to insect and mammalian attacks.

Geranium endressii Gay
FRENCH CRANE'S-BILL
An occasional escape from gardens on verges, waste ground or tips, and in churchyards. 20 + [3]. First record: Parkstone (09, 1924, LBH in **BM**). Some records may be errors for *G.* × *oxonianum*.

Geranium endressii × versicolor = G. × oxonianum Yeo
An uncommon or under-recorded escape. 8. First record: Netherbury (49U, 1930, AWG). Low Down Farm (40B, IPG); Corscombe (50H, 1957, AWG in **BDK**, and 1997, IPG); Abbotsbury Swannery (58S, !); Osehill Green (60U, !); South Down Farm (78L, !); Ryeclose (78T, !); Buckhorn Weston (72M, !); New Swanage (08F, EAP); verge west of Canford Heath (09D, !).

Geranium versicolor L.
PENCILLED CRANE'S-BILL
An uncommon but well-established garden escape on

sheltered verges. 5 + [6]. Broadoak (49M, 1978, E Pearse); Waytown and Netherbury (49T and U, 1928, AWG); Abbotsbury (58S, 1989–97,); Magiston (69I, 1977, J May); Bere Regis (89M, 1894, EFL); Arne (98U, 1979, MRH); Eddy Green Farm (99M, 1954, H Bury, *et al.*, and 1997, !); Canford Magna (09J, 1978, RHD).

[*Geranium nodosum* L.
KNOTTED CRANE'S-BILL
Bere Regis (89M, 1876, N Bond in **DOR**).]

Geranium rotundifolium L.
ROUND-LEAVED CRANE'S-BILL
Fairly frequent on Portland and by the railway near Weymouth. 21 + [3]. Spreading elsewhere, scattered by old railways, on allotments or waste ground. First record: Portland (67, 1878, WBB).

Geranium pratense L.
MEADOW CRANE'S-BILL
Uncommon in grassy woodland rides, and in patches on verges and railway banks. 56. Mainly on calcareous soils, where some colonies are native but others may be garden escapes. It has been known here for 200 years, but Dorset is its western limit in Britain.

Geranium pratense × *himalayense* is planted in churchyards, usually as Cv. Johnson's Blue. Mappowder Ch (70I, !); Silton (72Z, RM Veall & !). 2.

Geranium himalayense Klotzsch
HIMALAYAN CRANE'S-BILL
A garden escape in a lane in Wimborne suburbs (09E, !). 1.

Geranium sanguineum L.
BLOODY CRANE'S-BILL
An uncommon garden escape on verges or tips. 6 + [1]. Winfrith Heath (88D, 1978, JKH); Muston (89T, !); Morden (99D, !); Swanage (07J, DCL); Talbot Village and Parley Common (09E and Z, FAW); Woodlands (00P, GDF).

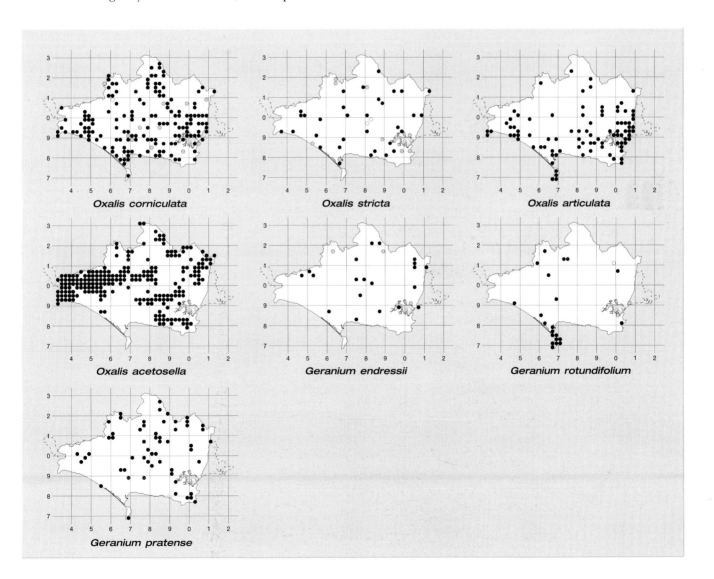

Oxalis corniculata

Oxalis stricta

Oxalis articulata

Oxalis acetosella

Geranium endressii

Geranium rotundifolium

Geranium pratense

Geranium columbinum L.
LONG-STALKED CRANE'S-BILL
Occasional, in small numbers, at the edges of woods or along rides, in warm hedgebanks, verges and railway banks. 114 + [25]. It occurs on dry calcareous and sandy soils, but is absent from most clays, wet soils and heaths.

Geranium dissectum L.
CUT-LEAVED CRANE'S-BILL
A common weed of grassland, lawns, verges, bare soil and gardens. 664. It germinates in autumn. A form with pale pink flowers occurred at Pimperne (90E).

Geranium ibericum × platypetalum = G. × magnificum N. Hylander
PURPLE CRANE'S-BILL
A garden escape on verges at Kington Magna (72R, 1999, !), Deanend (91T, 1996, !) and North Haven (08J, !). 3.

Geranium pyrenaicum Burman f.
HEDGEROW CRANE'S-BILL
Widespread on verges and railway banks among long grass.

181. Mostly on calcareous soil, but colonising acid verges in the east. JCMP found it rare in 1895, so it is increasing. The flowers vary in colour from dull to bright pink.

Geranium pusillum L.
SMALL-FLOWERED CRANE'S-BILL
A weed of bare, dry places, disturbed verges, sandy arable and tips. 102 + [10]. Commonest in the Poole basin, occasional on the Chalk and on Portland, rare in the west.

Geranium molle L.
DOVE'S-FOOT CRANE'S-BILL
A weed of short grassland, arable and gardens. 481. It occurs on both acid and calcareous soils, and germinates in both spring and autumn. Forms with pale pink or white flowers are not uncommon.

Geranium macrorrhizum L.
ROCK CRANE'S-BILL
Planted in churchyards, or a persistent garden escape. 4. Charmouth (39R, !); Minterne Magna (60S, !); Verwood (00Z, VMS); Ashley Heath (10C, VMS).

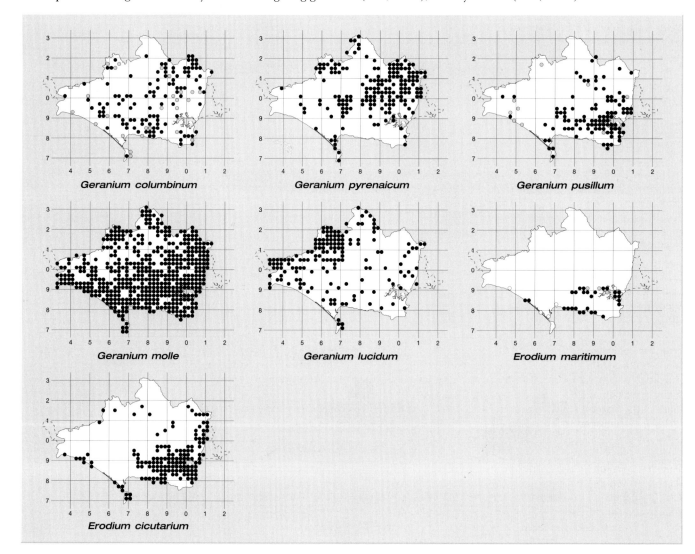

Geranium columbinum

Geranium pyrenaicum

Geranium pusillum

Geranium molle

Geranium lucidum

Erodium maritimum

Erodium cicutarium

Geranium lucidum L.
SHINING CRANE'S-BILL
Locally frequent among limestone rocks, on walls, in dry hedgebanks, village lanes and gardens. 161. It germinates in both spring and autumn, and tolerates shade. Commonest in the north and west and scattered elsewhere.

Geranium robertianum L.
HERB-ROBERT
Common in dry woods, shaded hedgebanks, on shingle beaches and as a garden weed. 711. The distinct-looking shingle beach form is common along the landward side of the Chesil Bank. White-flowered plants are rare: Champernhayes (39N); Wootton Fitzpaine (39S); Broadstone (09D); Canford Heath tip (09I); Ashley Heath (10C, FAW).

Geranium purpureum Vill.
LITTLE-ROBIN
Very local on shingle beaches, stone-heaps, old railways and waste places. 3 + [5]. Often confused with *G. robertianum*. Sandsfoot and Small Mouth Halt (67T, 1936, JD Grose in **RNG**, *et al.*, and 1997, !); Church Ope Cove (67V, 1966, JB); West Weares (67W, 1911, CES and 1953, WAC); Weymouth (67Z, 1985, AH); Charminster (69R, 1991, RCP in **OXF**); Dorchester and Fordington (69V, 1883, WBB and 1964, !); Swanage (07, 1724, W. Sherard, *et al.*, and 1917, CBG); Ulwell and Godlingston (08A, 1882, HNR and WF in **BM**, also JCMP in **DOR**). Reports from Abbotsbury and the Sherborne area need confirmation.

Geranium phaeum L.
DUSKY CRANE'S-BILL
A garden escape on verges, tolerant of shade. 3 + [5]. Fishpond (39U, !); Powerstock Castle (59H, 1874, E Fox, *et al.*, and 1980, DRS); Benville (50G, 1942, AWG in **BDK**); Chalmington (50V, !); Ryme Intrinseca (51V, 1930, AWG); Ambrose Hill (61J, !); Crossways (78U, 1937, DM); Melcombe Bingham (70R, 1846, CW Bingham); Gussage St Michael (91, 1887, LJW).

Geranium psilostemon Ledeb.
ARMENIAN CRANE'S-BILL
Becoming naturalised in a wild garden at Parnham Ho (40Q).

Erodium maritimum (L.) L'Her.
SEA STORK'S-BILL
Sporadic and variable in abundance in bare, dry habitats in the south-east. 28 + [9]. It grows on crumbling clay or chalk cliffs, also on sandy tracks in heaths, old cinder beds and tips. Once seen at West Bay (49Q, 1874, J Jones), also at Abbotsbury (58, !). It is abundant on Brownsea Is.

Erodium moschatum (L.) L'Her.
MUSK STORK'S-BILL
Rare, sporadic and in small numbers, but perhaps established on cliff-tops near Poole. 2 + [4]. Bridport (49Q, 1871, J Jones, *et al.* in **BDK**, and 1939, JEL in **RNG**); Portland (67, 1907, HJR); Small Mouth (67S, 1964, AJCB,

et al., and 1994, DCL); Chetnole (60E, 1895, WMR); Houghton (80D, 1895, JCMP); Northport (98J, 1980, AJCB); Flag Head Chine (08P, 1992, RMW, !). It has a mainly Cornubian distribution in Britain.

Erodium cicutarium (L.) L'Her.
COMMON STORK'S-BILL
Locally abundant in bare, dry arable, fallow or dunes on sandy soil. 148. Much less common on chalk or limestone except along the coastal belt or on new verges, and mainly in the Poole basin. Variable. The dune ecotype has sticky glandular hairs and is hard to tell from *E. lebelii*; it has been seen west of Bridport (49K), at Small Mouth (67S), Studland and South Haven (08I and L).

Erodium lebelii Jordan
STICKY STORK'S-BILL
Doubtfully distinct from *E. cicutarium*. Very local on sandy cliffs or dunes. 1 + [3]. West of Bridport (49Q, 1930, AWG and JEL); Small Mouth (67S, 1953, ! det. A Melderis); Shipstal Point (98Z, 1960, JFM Cannon in **BM**); South Haven (08I, 1930, LBH in **BM** and 1981, AJCB).

LIMNANTHACEAE

Limnanthes douglasii R. Br.
MEADOW-FOAM
A garden plant seen on a tip near Woodlands (00P, 1996, GDF). 1. It seeds freely in gardens.

TROPAEOLACEAE

Tropaeolum majus L.
NASTURTIUM
A tender garden escape. 14. Seen as a casual in 39F, 68V, 61H and V, 78P, 88Y, 82B, 98I and X, 90K, 08C, 09I, 00Q and X.

BALSAMINACEAE

[Impatiens noli-tangere L.
TOUCH-ME-NOT BALSAM
Dean Court, Wimborne (09E, 1799, RP), also painted by MF *c.*1800 from near Wareham, RP. All later reports are errors or uncorroborated.]

Impatiens capensis Meerb.
ORANGE BALSAM
This colonised the banks of the Stour *c.*1927, and has been introduced elsewhere, but is not common now and may be declining; it prefers ditches or carr to river banks. 8 + [11]. Weymouth (67, 1927, AWG); carr in Bryanston grounds (80T, 1999, CDP and DAP, !); Steepleton (81Q, 1994, !); Steeple Manor (98A, 1992, !); Spetisbury (90A and B, 1994, !); Shapwick (90F, 1927, LBH); Julian's Br, Wimborne (09E, 1999, !); Kinson & Ensbury (09J, 1994, FAW).

Impatiens parviflora DC.
SMALL BALSAM
A rare weed of shaded lanes, gardens or pavements. 1 + [1]. Sydling St Nicholas (69J, 1981, AJCB, and 1981–99, !); Dorchester (69V, 1937, DM and MJA, and 1960–76, JKH, !).

Impatiens glandulifera Royle
INDIAN BALSAM
An aggressive weed of river banks and damp waste places. 146 + [8]. First record: R. Brit, Netherbury (49U, 1924, AWG). It had spread to the Stour by 1933, the Frome and Piddle by 1937, and to the Moors River by 1977. It is commonest in the west and extreme east, where it is locally abundant, and is still spreading, helped by deluded gardeners, as along the North Winterborne. It prefers rich, somewhat acid mud, and germinates in March. A form with white flowers was seen by the R. Brit at Parnham (40Q, !) and at New Swanage (08F, EAP).

Impatiens balfourii Hook. f.
A tender plant which seeds freely and is likely to spread. 2. Upton CP (99W, 1996, !); Knoll gardens (00K, 1996, !).

ARALIACEAE

Hedera helix L.
IVY
Abundant to dominant on the ground in plantations, frequent in woods and hedges, and among rocks on undercliffs and old quarries. 733. It often drapes trees or stumps, and may kill trees which are already moribund. Germination occurs in autumn. Its flowers are an important source of nectar for insects in late autumn; its leaves are eaten by larvae of the Holly Blue, and its berries are eaten by birds in spring (Wilson, Arroyo & Clark, 1997). The leaves are resistant to salty gales. Much is gathered for decorating churches. Six Dorset place names are derived from Ivy, e.g. Ivers and Ivest Wood. Subsp. *hibernica* (Kirchner) D McClint. is not often recorded, but is probably the common plant of undercliffs and ravines near the sea. Three cultivars of ivy were on sale at Blandford in 1782, and one is dominant on the site of the old garden at Bryanston (80T).

APIACEAE or UMBELLIFERAE

Hydrocotyle vulgaris L.
MARSH PENNYWORT
Locally common in alder-swamps, marshes, fens and dune slacks, rarely in wet lawns. 121 + [3]. Tolerant of shade and of light grazing. Mostly in acid soils in the west and the Poole basin, with a few outliers such as Culverwell on Portland (66Z).

Sanicula europaea L.
SANICLE
Found in most old woods and adjacent hedgebanks in shade. 194 + [2]. It tolerates all but the most acid soils, but is absent from deforested areas, Portland and heaths in the Poole basin.

Eryngium maritimum L.
SEA-HOLLY
Scarce on sand or shingle beaches, and declining because of human pressure. 4 + [9]. Charmouth (39R, 1937, DM); Burton Bradstock (48Z, 1895, JCMP, *et al.*, and 1988, RJS); Eype (49K, 1930, AWG); West Bay (49Q, 1914, EBB and 1961, !); Burton or Cogden Mere (58D, 1930, AWG and 1991, J Pyett); West Bexington (58D, 1895, JCMP, *et al.*, and 1994, SME); Abbotsbury (58R, 1930s, DM and RDG); Portland (67, 1833, R Blunt in **WAR**); Small Mouth (67S, 1799, RP, *et al.*, and 1997, !); Kimmeridge (97E, 1874, JCMP); Winspit (97U, 1930s, E.Sykes); Swanage (07, 1874, JCMP); Studland (08H, 1799, RP, and 1964, BNSS); north shore of Poole Harbour (09, 1895, JCMP).

[*Eryngium* × *oliverianum* Delaroche was a garden escape at Lyme Regis (39, 1929, MAA det. J Fraser).]

Eryngium planum L.
BLUE ERYNGO
1 plant, Blandford by-pass (80Z, 2000, P.H. Sterling, !).

Chaerophyllum aureum L.
GOLDEN CHERVIL
A casual in nursery gardens, Cranborne (01L, 1978, E Balfour-Browne). [1].

Chaerophyllum temulum L.
ROUGH CHERVIL
Occasional to frequent in hedgebanks on calcareous or clay soils, but absent from sands and scarce in the west. 594. It flowers when *Anthriscus sylvestris* is in seed.

Anthriscus sylvestris (L.) Hoffm.
COW PARSLEY, GYPSY-LACE, QUEEN ANNE'S-LACE
Occasional in nutrient-rich woods or damp grassland, and locally abundant along verges and hedgebanks on all but the most acid soils. 707. The cut-leaved var. *angustisecta* Druce has been recorded from Charmouth (39R, 1928) and Parkstone (09, 1927, LBH in **RNG**).

[*Anthriscus cerefolium* (L.) Hoffm.
GARDEN CHERVIL
Sometimes cultivated in gardens as a herb, and a casual at Melbury Park (50, 1926, AWG).]

Anthriscus caucalis M. Bieb.
BUR CHERVIL
Scarce in bare, dry, sandy soils of hedgebanks, verges and the backs of dunes. 5 + [15]. Declining; only post 1910 records given: Near Yeovil (51T, 1956, IAR); Ferry Bridge (67S, 1997, CA Stace); Mount Skippet (78T, 1977, !); East Burton and Wool (88I and N, 1932, FHH); Northport and Kesworth (98J, 1895, WMR, *et al.*, and 1995, !); Wytch Farm (98S, 1992, !); Durlston Bay (07J, 1997, FAW); Studland (08G, 1912, CBG in **BMH,** also 08H, 1990, !);

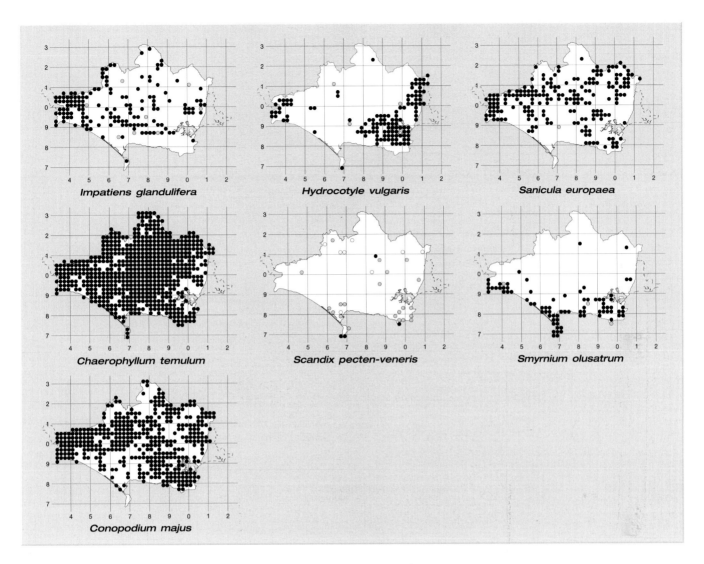

Impatiens glandulifera

Hydrocotyle vulgaris

Sanicula europaea

Chaerophyllum temulum

Scandix pecten-veneris

Smyrnium olusatrum

Conopodium majus

South Haven (08I, 1895, JCMP and 1999, EAP, !); Parkstone (09K, 1895, WMR and 1937, DM); Lower Mannington (00S, 1974, RHS Hatton).

Scandix pecten-veneris L.
SHEPHERD'S-NEEDLE
Once a frequent weed, now very rare, in calcareous arable. 4 + [32]. JCMP found it too common to give localities, while in the 1930s Good had it in 25 of his stands. Post-1960 records are: Portland Bill (66U, 1990, DAP and 66Z, 1993, J Pyett, !); Southwell allotments (67V, 1993, RSPB); North of Sherborne (61I, 1970, ACL); North-west of Bere Wood (89S, 1961, EN); Travellers Rest (80P, 1994, A Hosford); St Aldhelm's Head (97S, 1978, RHD, *et al.*, and 1986–96, DAP); Afflington Farm (98Q, 1965, JB). Old specimens are in **BM, BMH** and **DOR**.

Myrrhis odorata (L.) Scop.
SWEET CICELY
A rarely established garden escape from herb gardens. 2. Cheselbourne (70G, 1974, RHS Hatton); Bloxworth (89X, 1989, ! in **RNG**); Cranborne (01L, 1992, !).

Coriandrum sativum L.
CORIANDER
Sown in herb gardens since 1782, and a rare casual on new verges or tips. 2 + [5]. Near Yeovil (51T, 1956, IAR); Weymouth (67, 1930, AWG in **BDK** and 1935, NDS); Lodmoor tip (68V, 1960, EH in **LANC**); Sherborne tip (61I, 1970, ACL); Canford Magna (09J, 1953, DM and 1961, KG and 09I, 2000, !); Holes Bay (09B, 1997, !).

Smyrnium olusatrum L.
ALEXANDERS, HEALROOT
Long cultivated as a pot-herb, now well established on verges and waste ground in the coastal belt, much less common inland. 78 + [4]. It germinates in autumn, but seedlings die in hard winters.

Conopodium majus (Gouan) Loret
PIGNUT, HARE-NUT
Common in woods, also locally abundant in old pastures and parks. 412. Scarce or absent in deforested areas such as Portland, and in suburbs.

[*Pimpinella major* (L.) Hudson
GREATER BURNET-SAXIFRAGE
Reported, as f. *rosea*, from Wareham (98, 1938, LBH), but no specimens seen; it likes woods on clay.]

Pimpinella saxifraga L.
BURNET-SAXIFRAGE
Common in grassland on calcareous or calcareous clay soils. 333. Absent from acid soils of the Poole basin. Variable in size and leaf dissection, and often reported as *P. major*.

Aegopodium podagraria L.
GROUND-ELDER
Locally dominant in plantations, verges and rough grassland, and a bad weed in gardens. 462. It prefers nutrient-rich soils and tolerates a wide range of pH, but when found in old woods it is usually near houses or ruins. Cultivated in early times; the form with variegated leaves is still sold to unwary gardeners.

[*Sium latifolium* L.
GREATER WATER-PARSNIP
The only reliable records are from the Stour; most, if not all, recent reports are errors for *Apium nodiflorum* or *Oenanthe* sp., and Pulteney's 18th century records from the Frome are doubtful. [5]. Whitecliff Mill (80T, 1799, RP); Between Henbury and Sturminster Marshall (99P, 1890, JCMP in **DOR**); Clapcott's Farm (90B, 1977, DWT – unconfirmed); Shapwick (90K, 1890, RPM in **DOR**, EFL in **BM** and 1930, BNSS); Longham (09T, 1931, HHH in **LTR** and 1937, DM). It likes backwaters with tall vegetation on rich mud, and may have been lost through intense agriculture or flood-prevention work.]

Berula erecta (Hudson) Cov.
LESSER WATER-PARSNIP
Local in ditches draining rich marshes, fens or water-meadows. 30 + [19]. Now mostly in the valleys of the Frome, Piddle and Allen, as it has declined by the Stour. Reports from west Dorset are believed to be errors for *Apium nodiflorum*. Specimens are in **BDK** and **DOR**.

Crithmum maritimum L.
ROCK SAMPHIRE
Frequent all along the coast on rocky or clay cliffs and shingle beaches. 64 + [1]. It occurs along the inner shore of the Chesil bank, and occasionally on muddy shores of the Fleet and inside Poole Harbour. It has colonised man-made sea walls and concrete slipways at Lyme Regis, Weymouth beach, Ferry Bridge, Overcombe and Sandbanks, and was seen on timber at West Bay. The leaves have a strong taste, but are not used as food items.

Oenanthe fistulosa L.
TUBULAR WATER-DROPWORT
Local and scarce in marshes, fens, water-meadows and ditches, or by ponds. 22 + [20]. Commonest in the lower valleys of the Frome and Piddle, decreasing and rare near the Stour, and scattered elsewhere. Its decline is due to drainage and agricultural 'improvements'.

Oenanthe silaifolia M. Bieb.
NARROW-LEAVED WATER-DROPWORT
A plant of inland fens and water-meadows, which flowers in June, before *O. lachenalii* and *O. pimpinelloides*. MJ Southam has examined much Dorset material reported as this, and concludes that it may not occur here; there are

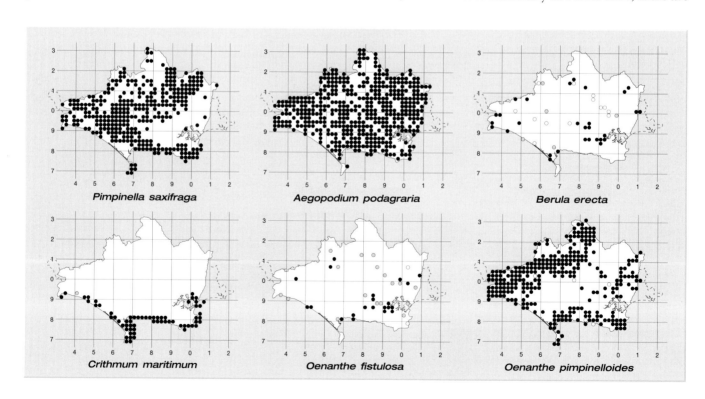

Pimpinella saxifraga

Aegopodium podagraria

Berula erecta

Crithmum maritimum

Oenanthe fistulosa

Oenanthe pimpinelloides

no good records from Hampshire or Somerset and only one from the Is of Wight. Specimens which may be correctly named are: Waddock (79V, 1935, RDG in **WRHM** det. PM Hall); Ridge (98S, 1933, AL Still in **RNG**); Lytchett Minster (99L, 1969, M Service det. NM Hamilton).

Oenanthe pimpinelloides L.
CORKY-FRUITED WATER-DROPWORT
Locally frequent in old pastures on calcareous clay, perhaps more frequent here than anywhere else in Britain. 286 + [5]. Commonest in the north and west and the coastal belt between Portland and Swanage. Absent from the Chalk except along the course of the Stour, and scattered in the Poole basin. It prefers damp soils, and tolerates salt-spray, as well as limited agricultural 'improvement'.

Oenanthe lachenalii C. Gmelin
PARSLEY WATER-DROPWORT
Occasional in tall salt-marsh vegetation along the coast, and scattered in marshes and fens inland. 35 + [10]. It has been much confused with other species in this critical genus. Inland records are: Bailey Ridge Farm (60J, 1888, WMR in **DOR**, refound in this tetrad in 1997, MP Hinton det. MJS); Tadnoll (78Y, 1995, BE); Deadmoor Common (71K, 1996, MJG); Winfrith Heath (88D, 1963, RDG in **DOR** and 1988, BE, !); Povington (88Q, 1996, BE det. MJS); Whitefield fen (99C, 1989, BE det. MJS, !); Stanbridge Mill (00E, 1979, JN); Edmondsham (01Q, 1909, EFL).

Oenanthe crocata L.
HEMLOCK WATER-DROPWORT
Common in wet woods, by rivers, streams and ditches and rarely on dry verges. 570. Absent from some of the dry

chalk uplands. At Radipole Lake (67U) and East Knighton (88C), many plants are suffused with pink pigment, including the flowers. The roots have poisoned cattle.

Oenanthe fluviatilis (Bab.) Coleman
RIVER WATER-DROPWORT
Locally frequent in the Stour below Blandford, the Frome below Dorchester and the Piddle below Trigon. 33 + [6]. Also reported from the Sydling brook, the Allen, and Sherford and Moors Rivers. The aerial parts become bloated in the larger rivers. It may have been introduced in the Frome, where it is now well established.

[Oenanthe aquatica (L.) Poiret
FINE-LEAVED WATER-DROPWORT
Very rare or extinct on river banks. [1]. By R. Stour, Kinson (09T, 1964, MPY). It has also disappeared from south Hampshire, but is still frequent on the Somerset levels. Most, if not all, early reports refer to O. fluviatilis.]

Aethusa cynapium L.
FOOL'S PARSLEY
Occasional in arable, gardens and tips on dry soils, both acid and calcareous, first evident in the late Bronze Age. 198. Subsp. agrestis (Wallr.) Dostal is a short form, rarely recorded, but has been seen at East Chaldon (78W, !) and Badbury Rings (90R, 1935, NDS).

Foeniculum vulgare L.
FENNEL
Occasional, mostly in small numbers, in verges, waste ground, pavements and tips. 103 + [4]. It prefers warm, dry sites, and it is commonest around Weymouth, Portland

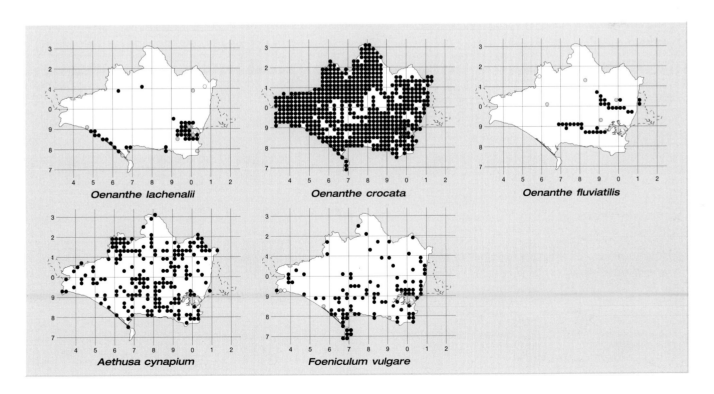

Oenanthe lachenalii   Oenanthe crocata   Oenanthe fluviatilis

Aethusa cynapium   Foeniculum vulgare

and Poole. Well established in chalk grassland at East Hill, Corfe (98R). First record: Purbeck (1799, RP). It contains fenipentol and its strong smell is due to anethol.

Silaum silaus L.
PEPPER-SAXIFRAGE
Local in, and characteristic of, old grassland on clay. 120 + [12]. Mostly in the coastal belt or the north-west, scarce or absent on the Chalk, but still present on clay in Portland, Purbeck and the north-east. Relict on verges or in churchyards even when all old grassland has been ploughed.

Conium maculatum L.
HEMLOCK
Widespread in hedgebanks, by ditches, often on dredged mud, disturbed verges and tips. 373. It prefers rich soils and avoids acid sites, and remains have been found in Roman deposits. It is highly toxic from its content of coniine (2-propyl piperidine).

Bupleurum fruticosum L.
SHRUBBY HARE'S-EAR
Planted in shade in Abbotsbury gardens in 1899 (58S, MI and still there in 1991, !). 1.

Bupleurum tenuissimum L.
SLENDER HARE'S-EAR
Very local and in small numbers in short turf near the coast, where it is declining. 3 + [7]. Lyme Regis (39, 1801, JE Smith); Portland Bill (66U, 1876, WBB, et al., and 1996, DAP, !); West Cliff, Portland (67, 1872, HE Fox in **BM**, also JCMP in **DOR**); Tidmoor Rifle range (67P, 1998, M King); Weymouth (67, 1868, J Irvine in **BM** and **CGE**, and

1894, FWG in **BM**); Lodmoor (67V, 1862, SMP in **DOR**, et al. in **BM**, **BMH** and **CGE**, and 1926, MJA); Langton Hive (68A, 1991, AJB); Chafey's Lake (68Q, 1874, GS Gibson and 1937, DM); Wareham (98, 1800s, MF); Tilly Whim (07I, 1896, LVL in **BM**).

Bupleurum subovatum Link ex Sprengel
FALSE THOROW-WAX
Probably this, confused with *B. rotundifolium*, as a casual bird-seed alien in allotments or gardens. [7]. Verne Common (67W, 1979, CE Richards); Puddletown (79M, 1973, JKH); Briantspuddle (89B, 1965, HW Cook); Bere Regis (89M, 1976, HGA Bates); Bryanston (80T, 1971, A Lack); Brownsea Is (08J, 1963, A Walton); Canford Magna 09J, 1978, RHD).

[Bupleurum rotundifolium L.
THOROW-WAX
A weed of arable since post-Roman and medieval times. Beaminster (40Q, 1920, AWG); Greenhill, Weymouth (68X, 1926, MJA); Sherborne (61, 1937, DM); Alderholt Ch (11B, 1998, !, from a florist's wreath) are the only records this century.]

Apium graveolens L.
WILD CELERY
Occasional by salt-marsh ditches, or by wet seeps on clay undercliffs all along the coast. 33 + [19]. In the 19th century it occurred in inland marshes in the west, as it does on the Somerset levels today. Var. *dulce* (Miller) DC., the cultivated celery, is a rare casual; Sherborne tip (61H, 1970, ACL); Wimborne (1937, DM). The plant produces an oil containing limonene, and its taste is due to other substances such as apiin and bergapten.

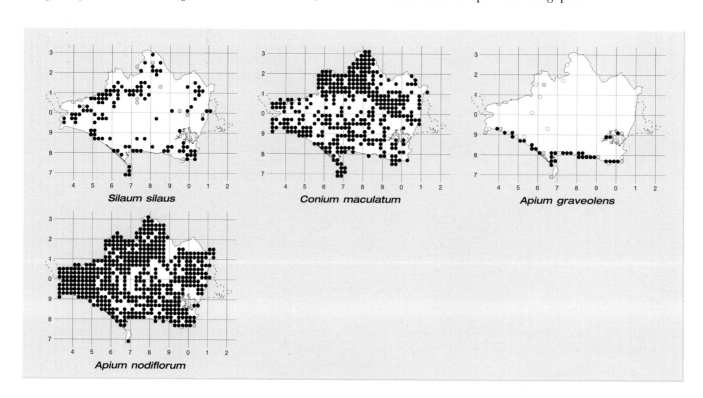

Silaum silaus

Conium maculatum

Apium graveolens

Apium nodiflorum

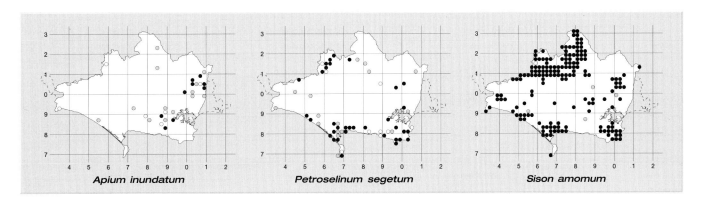

Apium inundatum   Petroselinum segetum   Sison amomum

Apium nodiflorum (L.) Lagasca
FOOL'S-WATER-CRESS
Common in wet places, in or by rivers, streams, ditches, ponds and marshes. 513. Var. *ochreatum* (DC.) Druce, which roots at every node, is frequent, and on rich mud produces grossly swollen forms which have been mistaken for other species, as near the lake at Kingston Lacy (79A).

Apium inundatum (L.) Reichb. f.
LESSER MARSHWORT
Once uncommon, now rare at the edges of acid ponds in the Poole basin. 8 + [21]. Its decline coincides with the filling-in or shading over of old cow ponds. Post-1960 records: Trigon (88U, 1991, DAP); Mare and Pool ponds (88W, 1991–97, BE and DAP, !); Piddle marshes, Wareham (98I, 1965, DSR); Stoborough meads (98I, 1980, AJB); Pamphill (90V, 1989, AJB); Studland (08, 1964, BNSS); Parley (09Z, 1961, MPY); Holt (00H and I, 1991, DAP); Dewlands Common (00U, 1991, AJB and DAP); West Moors (00W, 1995, GDF and JO); Verwood Lower Common (00X, 1990, AJB and DAP). There are earlier outliers in the west and north.

Petroselinum crispum (Miller) Nyman ex AW Hill
PARSLEY
Rare, perhaps native on cliffs near the sea. 2 + [3]. West Bay (49Q, 1830, TBS and 1930, AWG); Sandsfoot (67T, 1920, MJA and 1938, DM); North Portland (67W, 1935, MJA); Grove (77B, 1979–99, !); St Aldhelm's Head (97S, 1997, MJG). The garden plant, with crisped leaves, is sometimes found as a casual on walls or in dry waste places: 2 + [5]. Whitchurch Canonicorum (39X, 1961, !); Stoke Abbott (40K, 1930, AWG); Abbotsbury (58S, 1958, !); Monkton (67W, 1937, DM); Portland (67X, 1959–84, !; Blandford (80Y, 1991, !); Shaftesbury (82R, 1978, RHD). Both forms were on sale at Blandford in 1782, and have the same characteristic scent of apiole and myristicin when crushed.

Petroselinum segetum (L.) Koch
CORN PARSLEY
Very local and in small numbers, in disturbed grassland or arable, and on tips. 34 + [26]. Mostly on calcareous clay in the coastal belt, or on limestone soils in the north-

west. JCMP thought it rare, Good had it from 14 stands, and JHS was excited to find it in the 1920s.

Sison amomum L.
STONE PARSLEY
Locally frequent in hedgebanks or verges. 165 + [5]. Commonest on clay soils, especially in the north-west, between Abbotsbury and Weymouth and in Purbeck; tolerant of salt-spray near the sea.

Ammi majus L.
BULLWORT
A rare casual from bird-seed in gardens. 2. Motcombe (82M, 1994, !); Corfe (98Q, 1987, JB).

Ammi visnaga (L.) Lam.
TOOTHPICK-PLANT
A rare casual. [1]. Sherborne (61, 1970, ACL).

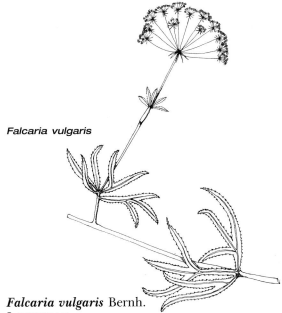

Falcaria vulgaris

Falcaria vulgaris Bernh.
LONGLEAF
A rare but persistent introduction in waste grassland. 2 + [1]. Grove (67W, 1995, B Collins); Admiralty Research Station, Portland (67, 1966, JB); East Fleet (68F, 1953, RDG in **DOR**, *et al.*, and 1986–96, DAP).

Carum carvi L.
CARAWAY
A rare casual on waste ground, railway sidings, allotments or tips. 1 + [3]. Yeovil junction (51T, 1956, IAR); Weymouth (67, 1929, AWG in **BDK** and 1938, DM); Longfleet (09A, 1900, EFL); Baiter (09A, 1983, MB).

Carum verticillatum (L.) Koch
WHORLED CARAWAY
Very local in wet pastures. 2. Wareham Common (98D, 1985, AH, *et al.*, and 1993, !); Ridge Farm (98N, 1914, TH Green, *et al.* in **BDK, DOR** and **LANC**, and 1996, !). It does not occur further east in Britain.

Angelica sylvestris L.
WILD ANGELICA
Frequent in wet woods, marshes and by rivers. 479. Absent from the drier parts of the Chalk, Portland and built-up areas. It contains beta phellandrene.

Angelica archangelica L.
GARDEN ANGELICA
Grown in a wild garden by the Stour at Spetisbury (90G), and sowing itself there. 1.

Levisticum officinale Koch
LOVAGE
A casual garden throw-out on a tip at Three-legged Cross (00X, 1982, ! in **RNG**).

(**Peucedanum ostruthium** (L.) Koch was reported from Studland (08, 1883, FC King), but seems unlikely and there are no supporting specimens.)

Peucedanum austriacum (Jacq.) Koch
A casual at Broadstone (09C, 1986, D MacGuiness in **BM**). 1.

Pastinaca sativa L.
WILD PARSNIP
Common in dry grassland, cliffs, verges, railway banks and waste land, mainly on calcareous soils. 250. In the Poole basin it is restricted to the south, on eutrophiated verges or tips. Absent from most of the north-west, except on the limestone west of Sherborne.

Heracleum sphondylium L.
HOGWEED, BROADWEED, ELTROT
Very common in woodland rides, tall, neglected grassland and on verges, on all soil-types but not in wetlands. 724. The flowers are sometimes pinkish. Young leaves are eaten by rabbits and deer, while the hollows in the dead stems are refuges for insects in winter. Var. *angustifolium* Hudson, with narrow leaf-segments, has been seen at Frome Whitfield (69V) and Pimperne (90E).

Heracleum mantegazzianum Sommier & Levier
GIANT HOGWEED
Rarely planted in gardens, but escaped and established along streams in the west, rare elsewhere. 18. First record,

without locality, in the 1930s by HJW. Especially common along the River Char at Charmouth and the Brit at Beaminster. When handled, the plant releases furocoumarins which may cause a skin reaction when exposed to sunlight.

Torilis japonica (Houtt.) DC.
UPRIGHT HEDGE-PARSLEY
Occasional in woodland rides, old hedgebanks and undercliffs, most often on clay. 397. It is the last of the Hedge-parsleys to come into flower.

Torilis arvensis (Hudson) Link
SPREADING HEDGE-PARSLEY
An arable weed of calcareous soils, which may now be extinct. [19]. JCMP had it from 13 localities, and RDG from only two stands. Post-1920 records: Weymouth (67, 1929, DM); Bincombe (68X, 1927, MJA); Dorchester (69V, 1935, RDG); Preston (78C, 1926, MJA); White Nothe top (78Q, 1960, !); Corfe (98, 1937, DM); Badbury Rings (90R, 1948, Anon in **BM**); North Farm, Horton (00J, 1922, JHS).

Torilis nodosa (L.) Gaertner
KNOTTED HEDGE-PARSLEY
Local, on hard ground of paths, cliffs, field-edges and walls, always in dry places, sporadic and in small numbers except in some mown suburban verges. 47 + [17]. Commonest along the coastal belt, less common on calcareous soil inland, as near Beaminster, Gillingham and Sherborne. Var. *pedunculata* (Rouy & Fouc.) Druce, with long peduncles, was seen at Renscombe (97T, 1935, NDS) and there is an unlocalised specimen in **DOR**.

Daucus carota L.
WILD CARROT
Frequent in dry calcareous grassland and undercliffs, occasional in sandy grassland or fixed dunes. 342. Most common along the coastal belt, and absent from parts of the clay grasslands in the north-west. A coastal form often has a single dark purple floret at the centre of the umbel. Subsp. *sativus* (Hoffm.) Arcang., Garden Carrot, grew on Lodmoor tip (68V, 1997, !); Subsp. *gummifer* (Syme) J.D. Hook., has succulent leaves and grows very close to the sea, all along the coast, on calcareous cliffs or shingle beaches. 26. The characteristic scent of carrot is caused by a mixture of asarone, carotol, limonene and pinene.

[**Caucalis platycarpos** L. has occurred in arable, as a casual. [3]. Mosterton (40M, RDG); Beaminster (40Q, 1934, AWG in **BDK**); Bradford Abbas (51S, 1837, J Buckman).]

[**Turgenia latifolia** (L.) Hoffm. was reported from arable at Dorchester (69V, 1883, J Sims), but no specimen has been seen.]

Ciclospermum leptophyllum (Pers.) Britton & E. Wilson
A rare casual of fallow at Merley (09E, 1995–6, MB in **RNG**); Miss Blower died before her specimen of this was named.

178

GENTIANACEAE

Cicendia filiformis (L.) Delarbre
YELLOW CENTAURY

Very rare, but locally abundant in a few grazed heaths, rutted tracks, firebreaks, mud by ponds or dune hollows. 7 + [19]. An inconspicuous Cornubian annual confined to the Poole basin, whose petals unfold in bright sunlight. First record: Moreton Heath (89, 1799, RP). Records extra to those given in Victorian Floras: Grange Heath (88W, 1925, FHH); Bloxworth Heath (89W?, 1931, DM); Creech Heath (98G, 1908, B King); Stoborough (98H, 1929, AWG, *et al.*, and 1996, DAP); Wareham (98I, 1905, Lady Davy in **RNG**, *et al.*, and 1937, DM); Sandford (98J, 1999, DCL); Slepe Heath (98N, 1935, DM, *et al.*, and 1996, RSRF & !); Threshers and Wytch Heaths (98R and S, 1911, CES, *et al.*, and 1997, !); Great Ovens Hill (99F, 1992, SME, !); Ham Heath (99Q, 19—, Anon in **CGE**); Poor Common (99S, 1915, AWG); Little Sea (08H, 1881, JWW in **CGE**, *et al.* in **DOR**, and 1937, DM); Holt Heath (00M, 1932, FHH); Mount Pleasant, Verwood (00U, 1937, RDG in

WRHM); Verwood Lower Common (00Y, 1919, AWG); East of Verwood Station (00Z, 1891, EFL in **CGE**); Gotham (01V, 1990, RD Porley, !).

Centaurium erythraea Rafn.
COMMON CENTAURY

Fairly common in woodland rides, short turf, tracksides and undercliffs on nutrient-poor soils. 328. Scarce or absent from rich soils on the Chalk and from wetland. Too bitter for rabbits to eat. The dwarf, congested form *capitatum* has long been known from Durlston CP and elsewhere along the coast: West Bexington (58I), Portand Bill (66), Holworth undercliff (78Q), Durlston (07I, in **BDK, BMH, CGE** and **RNG**) and Ballard Head (08K, in **RNG**). Forms with white flowers have been seen at Toot Hill (88N), Clouds Hill (89F) and Handfast Point (08L).

(*Centaurium littorale* (Turner ex Smith) Gilmour was reported from Encombe and Kimmeridge (97, 1937, NDS in **BM**), but there are no accepted records south of Wales.)

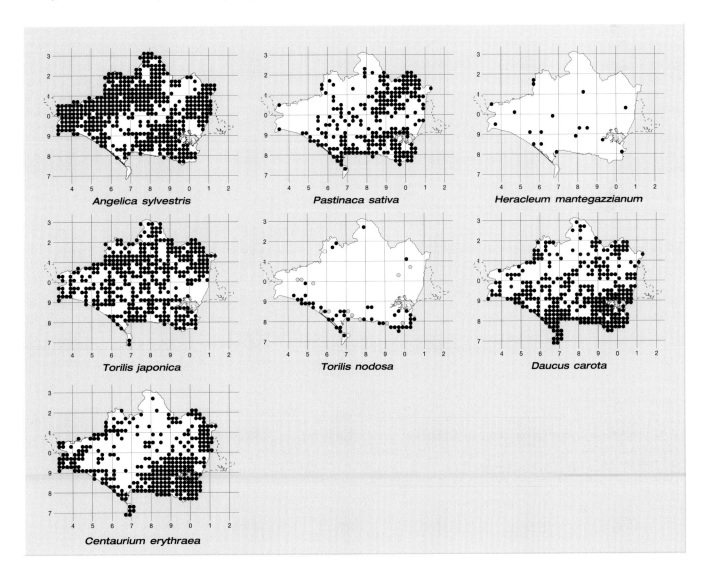

Angelica sylvestris

Pastinaca sativa

Heracleum mantegazzianum

Torilis japonica

Torilis nodosa

Daucus carota

Centaurium erythraea

Centaurium pulchellum (Sw.) Druce
LESSER CENTAURY

Occasional in woodland rides, short turf, scrapes, heathy tracks, undercliffs and salt-marshes on most soils, but especially on clays in dry or temporarily wet places. 87 + [10]. Most frequent along the coastal belt, also in both the north-west and the north-east. It often grows with *C. erythraea*.

Centaurium tenuiflorum (Hoffsgg. & Link) Fritsch.
SLENDER CENTAURY

Very local on damp, unstable clay undercliffs. 3. A Mediterranean species on the edge of its range, formerly found in the Isle of Wight, now only in west Dorset. First record: 1935, RDG. Our plant has white, or less often, pink, flowers. East of St Gabriel (39W, 1935, RDG in **WRHM** and 1967, !); Seatown to Golden Cap (49A, 1935, RDG in **WRHM**, *et al.* in **RNG**, and 1996, DAP, !); Thorncombe Beacon (49F, 1937, AWG in **BDK**, *et al.*, and 1994, MJG); Eype (49K, 1936, ECW in **LIV** and **RNG**, *et al.*, and 1994, DAP, !).

Blackstonia perfoliata (L.) Hudson
YELLOW-WORT

Rather local in short grassland or undercliffs on chalk, limestone or calcareous clay. 131 + [7]. It is tolerant of grazing but not of shade. Most frequent along the coastal belt and the north-western limestone, scattered on the Chalk, and confined to eutrophiated verges or tracks in the Poole basin. It may have a high magnesium requirement.

Gentianella campestris (L.) Borner
FIELD GENTIAN

A northern species, once rare, now perhaps extinct. [13]. It grew in short calcareous turf or on enriched verges of sandy tracks. In Somerset, it has not been seen since 1940, and it has also declined in the New Forest. Post-1900 records: Hooke (59J, 1930–36, AWG in **BDK**); Ballast Knap (78T, 1933, RDG in **WRHM** and 1953, RDG in **DOR**); Tadnoll (78Y, 1972, !); Winfrith Heath (88D, 1960, KBR, !); Swanage Quarries (07J, 1936, MJA); Blackbush Down (01M, 1947, MH Collier).

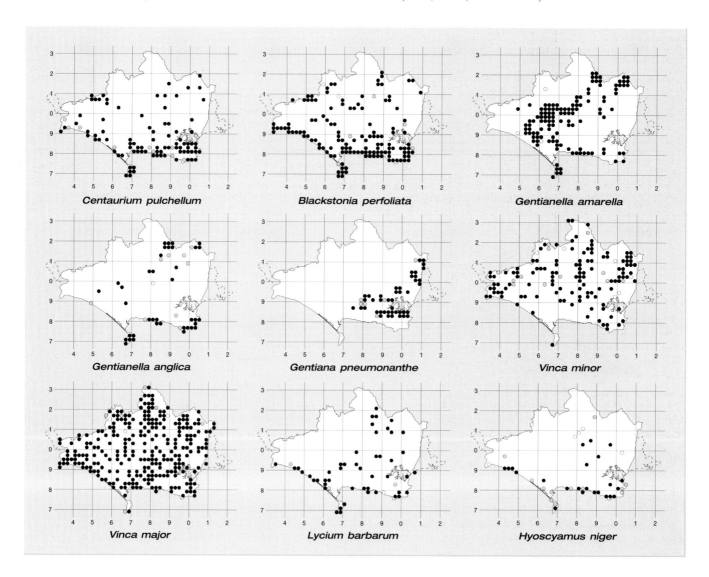

Centaurium pulchellum   Blackstonia perfoliata   Gentianella amarella

Gentianella anglica   Gentiana pneumonanthe   Vinca minor

Vinca major   Lycium barbarum   Hyoscyamus niger

Gentianella amarella (L.) Borner
AUTUMN GENTIAN
Occasional, locally frequent, in short, calcareous turf. 121 + [3]. Now usually in relict sites such as steep banks, ancient earthworks, old verges or railway cuttings. First record: 1551, W Turner. The leaves are too bitter for rabbits to eat. White-flowered forms have been seen at Abbotsbury (58X), The Warren (78V) and Hambledon Hill (81L), and pink-flowered forms at Badbury Rings (90R).

Gentianella amarella × *anglica* = *G.* × *davidiana*
T.C.G. Rich
In chalk turf at Bindon Hill (88K, 1998, BE, det. TCGR in **NMW**); Stonehill Down (98G, 1999, BE). 2.

Gentianella anglica (Pugsley) E. Warb.
EARLY GENTIAN
Local in short, calcareous turf, flowering earlier than *G. amarella* but often in the same site. 38 + [10]. Variable in abundance from year to year (Edwards, 1997). Restricted to the coastal belt and the Chalk, especially in the north-east, where it is known from 28 sites. First record: Durlston (07I, 1878, F Stratton). Dorset is a stronghold of this endemic species, with a population of about 60,000 in 1998. There are specimens in **BDK, CGE, DOR, OXF** and **RNG**.

Gentiana pneumonanthe L.
MARSH GENTIAN
Local, in wet heaths and at the margins of bogs. 43 + [6]. Somewhat sporadic, often more frequent after heath fires. Restricted to the Poole basin which, with the New Forest, is its main stronghold in Britain; it does not occur further west (Chapman, Rose & Clarke, 1989). White- and pale blue-flowered plants appear occasionally at Studland Heath (08H). I have seen bunches used to decorate café tables. Botanists have no need to pick it, as specimens are in **BDK, BEL, DOR, HIWNT, WRHM** and **YRK**.

APOCYNACEAE

Vinca minor L.
LESSER PERIWINKLE, ST CANDIDA'S EYES
Well established and locally dominant in a few woods, plantations and wild gardens. 120 + [20]. Otherwise a more or less obvious garden escape in hedgebanks, village lanes and verges. Perhaps introduced as far back as Roman times, but its distribution today is unlike that of any native plant. Sites where it is well naturalised include Winyards Gap (40X), Oakers Wood (89A), Bere Wood (89S), Milton Abbas (80B), Hod Wood (81K), and woods at Creech Hill and Wimborne St Giles (01K and L). Forms with white flowers occur at Nether Cerne (69U), West Stour (72W) and Huish (99D).

Vinca major L.
GREATER PERIWINKLE
Widely planted or escaped in plantations, hedgebanks and village lanes, but never looking more than a persistent introduction. 265 + [12]. Var. *oxyloba* Stearn, with narrow

petals, occurs at Whitchurch Canonicorum (39X), Beaminster (40Q, LJM), Caundle Wake (61W) and Bloxworth (89S).

SOLANACEAE

Nicandra physalodes (L.) Gaertner
APPLE-OF-PERU
An occasional casual of nutrient-rich gardens or tips, not seen before 1969. 15. Burton Bradstock (48Z, RJS); West Bexington (58I, SME); Chesilton (67S, 1973, WGT, !); Westham allotments (67U,!); Lodmoor tip (68V, 1978, !); Sherborne tip (61H, 1969, ACL); Crossways (78P, !); Tadnoll (78Y, !); Piddletrenthide (70A, !); Binnegar tip (88Y, !); Wareham (98I, 1999, JMB); New Barn Farm (07E, EAP); Parkstone (09K, !); Wimborne (00A, !); Three-Legged Cross (00X, 1982, ! in **RNG**); Cranborne (01R, GDF).

Lycium barbarum L.
DUKE OF ARGYLL'S TEAPLANT
A persistent garden-hedge plant, escaped in waste places, especially along the coastal belt. 48 + [5]. First records: Weymouth (67, 1866) and Lulworth (88,1867), both SMP in **DOR**.

Atropa belladonna L.
DEADLY NIGHTSHADE
Very local, sporadic and in small numbers in clearings in woods and plantations, or by old buildings. 5 + [10]. Sherborne (61I, 1970, ACL); Lulworth Cove (87J, 1840s, JC Dale); Melcombe Bingham (70R, 1956, C Woodhouse); Girdlers Copse (71W, 1983, DWT); Bryanston (80T, 1937, DM and 1977, CA Hutchinson); France Firs (80Z, 1951, EN); Great Yews (81, 1953, CS); St Aldhelm's Head (97S, 1900, E Bankes); Corfe (98, 1936, Anon in **ALT**); East Morden (99C, 1895, WM Heath); Charborough Park (99I and J, 1908, J Cross, *et al.*, and 1998, !); Spetisbury (90B, 1933, BNSS); Badbury Rings (90R, 1981, FAW); Pamphill (90V, 1990, RA Leney); Chettle Ch (91L, 1799, RP). It contains atropine and hyoscamine, and is toxic.

Hyoscyamus niger L.
HENBANE
Sporadic in eutrophic sites on cliffs and beaches, by farms, with fodder beans in arable or a pavement weed; known since the late Iron Age, when it may have been cultivated. The seeds are very long-lived. 25 + [19]. Mostly along the coast, and casual inland. It contains the toxic alkaloid hyoscine, small amounts of which act as a sedative. In the 19th century, Weymouth pharmacists obtained their supplies from Portland, according to WBB.

Salpichroa origanifolia (Lam.) Thell.
COCK'S-EGGS
Rare, but locally abundant in a plantation and an old garden at Abbotsbury (58S, 1937, Mrs Davies and 1960, EH in **LANC**, also 58M, 1960–98, !). 2.

Physalis alkekengi L.
JAPANESE-LANTERN
A rare but persistent garden escape in hedgebanks and verges. 4. Stallen (61D, JAG); Almshouse Wood (61H, JAG); Winterborne Whitechurch (80F, !); Farnham (91M, !).

Physalis peruviana L.
CAPE-GOOSEBERRY
A rare, tender casual, whose flowers often face the soil. 4. Crossways tip (78P, 1999, !); Binnegar tip (88T, 1993, ! in **RNG**); Greenland Farm (08C, 1999, !); Canford Heath tip, (09I, 1998, !).

Physalis ixocarpa Brot. ex Hornem.
TOMATILLO
Casual. Sherborne tip (61J, 1966, ACL).

Capsicum annuum L.
SWEET PEPPER
A rare casual which germinates on tips. 2 + [1]. Sherborne (61H, 1989, !); Oborne (61P, 1971, ACL); Crossways tip (78P, 1999, !).

Lycopersicon esculentum Miller
TOMATO
A tender casual, often self-sown and fruiting on tips, disturbed verges or urban sites, rarely on the strandline of beaches. 20. Seeds were sold at Blandford in 1782.

Solanum nigrum L.
BLACK NIGHTSHADE
Common in eutropicated arable, especially in maize crops, in gardens and on tips. 416. On beaches the prostrate var.

marinum Bab. has been seen at West Bexington (58I, 1960, !), South Haven (08I, 1993, RSRF & !) and Parkstone salterns (09, 1900, EFL). First evident in post-Roman deposits; JCMP found it 'not common', and it has probably increased in arable with increased fertilizer usage, though not in the stock-raising regions of the Blackmore and Marshwood Vales. Its berries are not toxic.

Solanum physalifolium Rusby
GREEN NIGHTSHADE
A rare but persistent weed of bare, sandy soil. 2. Blacknoll (88D, 1978–97, ! in **RNG**); Canford tip (09I, !). This was previously recorded as S. sarachoides.

Solanum dulcamara L.
BITTERSWEET
Common in fens, hedgebanks, riverbanks, by ponds and behind beaches. 691. It prefers eutrophic soils and varies much in habit and degree of hairiness, as along the inside of the Chesil bank. A form with black fruit was found at Bredy Farm (59A, 1986, G Greatham); Good reported a form with white flowers in the 1930s.

Solanum tuberosum L.
POTATO
A relic of cultivation, often found on tips, but able to persist for at least 10 years in rough grassland. 15+. Rarely grown as an agricultural crop, but common in allotments and gardens.

Solanum rostratum Dunal.
BUFFALO-BUR
A rare casual of disturbed soil. 5. Moonfleet (68A, 1999,

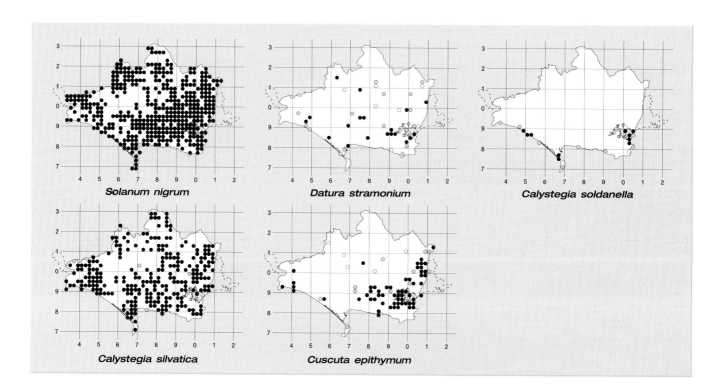

MR Chandler, !); Frampton (69G, 1988, DAP in **RNG**); Briantspuddle (89B, 1987, P Bowell); Pamphill (90V, 1995, RA Leney); Greenland Farm (08C, 1999, EAP, !).

[*Solanum fontanesianum* Dunal. was casual at Swanage (08, 1917, IM Roper).]

[*Solanum villosum* Miller
A rare casual. Sandbanks (08N, 1929, LBH); Poole (09, 1910, Anon).]

Datura stramonium L.
THORN-APPLE
A scarce and sporadic casual of disturbed verges, allotments, gardens and tips, rarely on beaches. 19 + [24]. It fruits well and the seeds remain viable for many years; the plant contains toxic alkaloids. Var. *chalybaea* WDJ Koch, with mauve flowers, occurred at Stoke Mill (49I, 1976, AWG), Greenland Farm (08C, 1999, EAP, !) and Parkstone (09, 1935, HJW).

Datura ferox L.
ANGEL'S-TRUMPETS
A casual in a chicken run at Frome St Quintin (50J, 1995, DAP). 1.

Nicotiana × *sanderae* Will. Watson
TOBACCO-PLANT
A casual garden escape found on disturbed verges, building-sites, waste ground and tips. 14. Portland Bill (66U, RF, !); Lodmoor tip (68V, 1978 and 1998, !); Sherborne tip (61H, 1970, ACL and 1989, !); Crossways (78P, 1999, !); Puddletown (79M, 1992–98, !); Binnegar tip (88Y, !); Blandford (80Y, !); Fiddleford (81B, !); Shillingstone (81F, !); Studland (08G, !). The flowers are mostly white or dull pink, but dull red forms occurred at Lodmoor, Binnegar and Shillingstone tips and at New Swanage (08F, EAP). These may be *N. forgetiana* Hemsley, but the correct name for garden hybrids is unsettled.

Petunia × *hybrida* Vilm.
A casual garden escape on building-sites, tips and pavements (from hanging baskets). 5 + [2]. Sherborne (61H, 1970, ACL); Creech Bottom (98H, !); Wareham (98I, !); Godlingston (08A, DCL); Greenland Farm (08C, EAP); Fleets Corner tip (09B, 1977, !); Canford tip (09I, !).

Schizanthus pinnatus Ruiz Lopez & Pavon
Once seen as a pavement weed in Dorchester (69V, 1999, !).

CONVOLVULACEAE

Convolvulus arvensis L.
FIELD BINDWEED, DEVIL'S-GUTS
Common in grassland, on verges and railway banks, and a persistent weed in gardens on all types of soil. 647. The flower-colour varies; pink flowers with a central white star are frequent.

Calystegia soldanella (L.) R. Br.
SEA BINDWEED
Local, rare and often sporadic on sandy foredunes or shingle beaches, rarely in salt-marsh. 10 + [10]. Colonising hollow concrete blocks of a slipway at Ferry Bridge (67T). Frequent in dunes north of Studland, with a white-flowered form (08G, H and I); [salt-marsh on the NW shore of Brownsea Is, 08J, 1966–77, KC].

Calystegia sepium (L.) R. Br.
HEDGE BINDWEED, SHIMMY-AND-SHIRT, WITHYWINE
Common in wet woods, fen carr and reed-swamps, or in hedges on rich soils. 625.

Calystegia sepium × *silvatica* = *C.* × *lucana* (Ten.) Don
Rare or overlooked in hedges. 2. Headstock Road (30M, 1998, IPG); Halscombe Bridge (30V, 1998, IPG).

Calystegia pulchra Brummitt & Heyw.
HAIRY BINDWEED
A rare introduction in villages and towns. 2. Dewlish (79U, 1991, !); Blandford (80Y, 1990, ! in **RNG**).

Calystegia silvatica (Kit.) Griseb.
LARGE BINDWEED
Widespread and well naturalised in hedges and gardens, usually near villages, rarely on riverbanks. 257 + [5]. First record: Parkstone (09, 1930s, NDS); Good gives only one locality, so its spread has been rapid.

[*Cuscuta europaea* L.
GREATER DODDER
Parasitic on nettle. Lyme Regis (39, 1895, FT Richards); Blandford (80, 1799, RP). No specimens seen.]

Cuscuta epithymum (L.) L.
DODDER
A local and sporadic parasite, mostly on *Calluna, Erica* or *Ulex*. 59 + [17]. Widespread on heaths in the Poole basin, also on summit heaths in the west. It also parasitizes *Centaurea, Galium, Genista, Lotus* and *Trifolium* in calcareous turf along the coast or inland; Portland Bill (66U, 1887, JCMP in **DOR**); Portland (67W, 1866, SMP in **DOR**, *et al.*, 1960, !); Glanvilles Wootton (60U, 1878, SMP in **DOR**); Bulbarrow (70S, 1984, NCC); Mupe Bay (87P and 88K, 1997, BE, !); Whatcombe (80F, 1874, JCMP in **DOR**); Bryanston (80T, 1966, BNHS). It was probably more widespread inland, away from the heaths, in the 19th century.

[*Cuscuta epilinum* Weihe
FLAX DODDER
Parasitic on flax. 3. Near Bridport (49, 1920, AWG in **BDK**); Netherbury (49U, 1922, AWG); Bradford Abbas (51S, 1874, J Buckman).]

MENYANTHACEAE

Menyanthes trifoliata L.
BOGBEAN
Local in alder-swamps, bogs, sedge fens, marshes, water-meadows and by acid ponds. 53 + [10]. Certainly planted in 07E, 07I and 08P. Mostly in the Poole basin and around Powerstock in the west, with an outlier at Millum Head (89I). It may be declining through drainage. Once used by herbalists, it contains bitter iridoids.

Nymphoides peltata Kuntze
FRINGED WATER-LILY
Extinct at Loders (49X, 1846, J Jones), but since 1981 introduced in ponds or lakes, rarely in rivers. 10. Chideock (49G, DAP); Mangerton Mill (49X, LJM and LMS); Littlewindsor (40H, 1981, AJCB); Abbotsbury Mill pond (58S, !); R. Lydden and R. Stour, Cutt Mill (71T, DAP); Kingston Magna (72R, JO, !); Fiddleford Mill (81B, DAP and RF, !); Hatch Pond (09B, !); Deans Court, Wimborne (09E, !); Talbot Heath pond (09R, !).

POLEMONIACEAE

Polemonium caeruleum L.
JACOB'S-LADDER
A garden escape, sometimes established for a few years on verges or waste ground. 5 + [6]. Chardstock (30G, 1886, JCM in **DOR**); Thorncombe Thorn (30Q, 1986, A Brook and C Coryell); North Bowood (49P, 1922, AWG); Radipole (68, 1862, SMP in **DOR**); Cruxton (69D, 1846, CW Bingham); Buckhorn Weston (72M, 1995, !); Winterborne Whitechurch (80F, 1945, Clifton Coll. NHS); Hanford (81K, 1982, GD Harthan); Iwerne Minster (81S, 1957, JAL); Blue Pool (98G, 1993–95, !); Sovell Down (91V, 1979, JN); Canford Heath tip (09I, 1999, !).

HYDROPHYLLACEAE

Phacelia tanacetifolia Benth.
A rare casual from bird-seed in gardens, arable fields or sown verges. 4. Below Pilsdon Pen (40A, 1984, JGK);

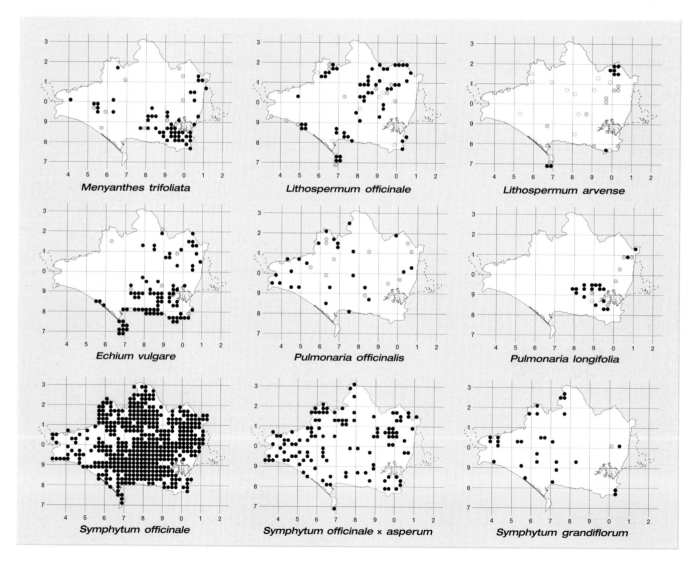

Kingcombe (59U, 1996, RJ Swindells); Down Farm (91X, 2000, !); Canford Cliffs (08P, 1999, OM Linford).

Phacelia campanularia A. Gray
CALIFORNIA-BLUEBELL
A rare casual at Sherborne tip (61H, 1970, ACL). [1].

BORAGINACEAE

Lithospermum purpureocaeruleum L.
PURPLE GROMWELL
A garden escape in shrubberies, waste ground or on walls, not established. 2 + [2]. Litton Cheney (59K, 1992, !); Sherborne Castle (61N, 1972, !); Affpuddle (89B, 1959, JG Lawn); Hod Hill (81K, 1982, FAW).

Lithospermum officinale L.
COMMON GROMWELL
Local in woodland rides and clearings, among scrub, in mature calcareous grassland and on undercliffs. 56 + [18]. It does not flower in shaded sites. Largely restricted to the Chalk, especially near the coast, and the limestone of Portland and the north-west.

Lithospermum arvense L.
FIELD GROMWELL
Now a rare weed in calcareous arable. 9 + [25]. JCMP found it frequent in the 1890s, but in the 1930s Good only found it in four stands. Only seen recently in Portland, Purbeck, and on the north-east Chalk. It probably dislikes over-fertilization or 'chemical farming', or it may not have time to set seed after harvest. Seeds have been found in late Iron Age excavations, and in 12th century material from Wareham.

Echium vulgare L.
VIPER'S-BUGLOSS
This biennial likes disturbed sites in sunny places on dry sandy or calcareous soils. 86 + [7]. It grows at the edge of grassland or heathland, in coppiced woods, on verges, railway banks and undercliffs, and is commonest in the coastal belt, but is often sporadic. Forms with small flowers and with pale-coloured flowers are found on cliffs at Church Ope Cove (77A).

Echium plantagineum L.
PURPLE VIPER'S-BUGLOSS
A rare garden escape on verges or tips. 2 + [1]. Bridport (49,1999, !); Worgret (98D, 1944, DM); Wareham and Redcliff (98I, 1989–93, ! in **RNG**, also RJS).

Echium pininana Webb & Berth.
GIANT VIPER'S-BUGLOSS
Planted and sowing itself as a pavement weed at Upton CP (99W, 1996, !). 1. Also grown in Abbotsbury gardens.

Pulmonaria officinalis L.
LUNGWORT
Escaped or planted in small numbers in plantations, hedgebanks, churchyards or verges, mostly close to houses. 25 + [12]. First record: Portland (67, 1890, Anon). Often misidentified.

Pulmonaria rubra Schott
RED LUNGWORT
Much planted in the grounds of Kingston Lacy Ho (90Q). 1.

Pulmonaria longifolia (Bast.) Boreau
NARROW-LEAVED LUNGWORT, MARY'S-TEARS
Local in old woods and hedgebanks on acid soils, rarely in artificial habitats such as railway banks. 18 + [8]. Restricted to the Hampshire basin and the Is of Wight in Britain. In deep shade it fails to flower, but revives after coppicing or clearance.

Symphytum officinale L.
COMMON COMFREY, KNIT-BONE
Common in wet woods, marshes, edges of rivers, streams and ditches, and occasionally on dry verges or edges of fields. 455. Its range extends to east Portland, and it is least common in west Dorset. Var. *patens* Sibth., with dull purple flowers, is as common as the type, which has cream flowers. Once used to make an ointment for sores (Udal, 1889).

Symphytum officinale × *asperum* = *S.* × *uplandicum* Nyman
RUSSIAN COMFREY
Once grown for forage, now occasionally established in patches on verges or in old fields. 119. Not mentioned by JCMP and noted as occasional by Good. First record: Abbotsbury (58S, 1908, B King), where it is still frequent. While this plant makes a good green manure and regrows vigorously when cut, its root contains poisonous pyrrolizidine alkaloids. A white-flowered form was seen at Loders (49X, !).

Symphytum asperum Lepechin
ROUGH COMFREY
A rare alien. [2]. Shorts Lane, Beaminster (40Q, 1963, AWG in **BDK**); Cranborne (01L, 1961, NDS). Sometimes reported in error for its hybrid, *S.* × *uplandicum*.

Symphytum tuberosum L.
TUBEROUS COMFREY
A scarce denizen of plantations or verges. 3 + [3]. Melbury Park (50X, DAP); Folke (61L, 1997, JAG); Sherborne Park (61N, 1899, JCMP in **DOR**, *et al.*, and 1972, !). All reports from Abbotsbury are errors for *S. bulbosum*, and the following need confirmation: Fern Hill (39M, 1938, G Cole and 1985, AWJ); Furzehill, Wimborne (00B, 1978, RHD); between Alderholt and Cranborne (01R, 1981, B Robinson).

Symphytum grandiflorum DC.
CREEPING COMFREY
Garden escapes are established in patches in verges and village lanes, and may dominate wild gardens. 30. First report: Benville (50G and L, 1880s, Anon) and still there. Increasing.

Symphytum Cv 'Hidcote Blue' hort. ex G. Thomas
HIDCOTE COMFREY
An invasive hybrid of uncertain origin, in the same habitats as *S. grandiflorum*. 9. First record: Lillington (61G, 1992, !), and later in 49U, 59A, 61A, 70P, 73T, 89G, 80H, 09I and 00N. Plants on a verge at Morcombelake (49B) may be *Symphytum* Cv. 'Hidcote Pink' hort.

Symphytum orientale L.
WHITE COMFREY
A denizen or garden escape in village lanes or urban sites, sometimes reported in error as *S. tauricum*. 12 + [4]. Lyme Regis (39G, AWJ); Beaminster (40Q, LJM and LMS, also IPG); Abbotsbury (58M and S, !); Corscombe (50D and H, 1930s, AWG, *et al.* in **RNG**, and 1993, LJM, !); Chelborough (50M, 1952, AWG in **BDK**); Adber (52V, JAG); Wrackleford (69R, 1967, JB); Dorchester (69V !); Cheselbourne (79U, !); East Lulworth (88L, 1967, J Hughes); Blandford (80Y, 1954, AWW); Studland (08G, 1999, EAP); Canford Heath tip (09I, !).

Symphytum bulbosum C. Schimper
BULBOUS COMFREY
Planted in churchyards and wild gardens where it becomes locally dominant in spring, but disappears by midsummer. 7 + [1]. Abbotsbury (58M and S, 1893, J Hawkins in **DOR**, *et al.* in **BDK**, and 1996, !); Long Bredy (59Q, !); Chalmington (50V, PRG); Pennsylvania Castle (67V, 1981, RA and 1994, !); Portesham (68C, !); Upwey (68S, !). See *S. tuberosum*.

Brunnera macrophylla (Adams) I.M. Johnston
GREAT FORGET-ME-NOT
A persistent but not invasive garden escape on verges in village lanes. 14. First record: Netherbury (49U, 1972–97, !). Also found in 39R, 59K, Q and T, 52V, 68C, 61X, 79U, 70W, 71C, 73Q, 08F and 09E.

Anchusa officinalis L.
ALKANET
Rarely established for a short time on verges or tips. 5 + [3]. Hurst (79V, 1986, MHL); Worgret pits (88V and 98D, 1949, NDS in **BM**, *et al.* in **BDK**, and 1987, MHL); Bryanston (80T, 1954, BNHS); Grange Arch (98A, 1991, ! – Cv. 'Peter Pan'); Swineham (98J, 1986, MHL); Poole Salterns (08P, 1926, L Baker in **BM**).

Anchusa arvensis L.
BUGLOSS
Rather local in sandy arable, fallow and disturbed heaths or new verges. 85 + [15]. Mostly in the Poole basin, scattered elsewhere. JCMP thought it frequent, Good had it from 33 stands.

Nonea lutea (Desr.) DC.
A garden escape at Witchampton (90Y, !). 1.

Pentaglottis sempervirens (L.) Tausch ex L. Bailey
GREEN ALKANET
An occasional, persistent denizen in hedgebanks and

verges, plantations or tips. 139 + [8]. It was rare in the 19th century and uncommon in the 1930s, so it is increasing. The roots were once used to produce a red dye.

Borago officinalis L.
BORAGE
Sporadic in disturbed waste ground, gardens and tips, usually near houses. 40 + [11]. Persistent in some places, notably on Portland (67, 1864, SMP in **DOR**, *et al.* in 1912 and 1937, and 1994, !) and Corfe Castle Mound (98L, 1895, JCMP, *et al.* in 1926, 1935 and 1965, and 1993, !). A form with white flowers was found near Blynfield Farm (82F, 1997, A Ferguson).

Borago pygmaea (DC.) Chater & Greuter
SLENDER BORAGE
Rarely planted, but self-sown in a wild garden at Litton Cheney (59K). 1.

Trachystemon orientalis (L.) Dum.
ABRAHAM, ISAAC AND JACOB
Occasionally planted in wild gardens or on shaded verges, and spreading to become locally dominant. 5 + [1]. Compton Valence (59W, 1960, TBR and 1998, JAG, !); Ryme Intrinseca (51V, 1998, JAG); Cerne Abbas (60Q, 1966, AWG in **BDK**); Minterne Magna (60R and S, 1971, RDE and 1984–96, !); Bourton (73Q, 1997, !).

[*Amsinckia lycopsoides* (Lehm.) Lehm.
SCARCE FIDDLENECK
Reported as an alien from chicken food, but no specimen has been seen. Bowood (49P, 1935, AWG); Netherbury (49U, 1929, AWG); Parkstone (09, 1933, HJW).]

Amsinckia micrantha Suksd.
COMMON FIDDLENECK
A rare weed at West Moors (00W, 1995, GDF and JO). 1. It prefers sandy soils.

Myosotis scorpioides L.
WATER FORGET-ME-NOT
Frequent on rich mud by rivers, streams and ponds, also in marshes in river basins. 228.

Myosotis secunda A Murray
CREEPING FORGET-ME-NOT
Occasional in wet, peaty places in the west, and less common in the Poole basin. 32 + [6]. Perhaps confused with *M. laxa* by some.

Myosotis laxa Lehm.
TUFTED FORGET-ME-NOT
Frequent in marshes, fens, wet meadows, by streams and flooded pits. 127 + [6]. It is absent from most of the Chalk and also the Stour basin.

Myosotis sylvatica Hoffm.
WOOD FORGET-ME-NOT
Possibly native in Cranborne Chase (91, 1931, AK Hervey),

but found by neither JCMP nor RDG. 171 + [13]. All recent records are garden escapes in plantations, hedgebanks and village lanes. It is classed as an introduction in both Somerset and south Hampshire. A white-flowered form was seen at Bryanston (80T), and pink-flowered forms persist in gardens.

Myosotis arvensis (L.) Hill
FIELD FORGET-ME-NOT
Occasional to frequent in dry soils of woodland rides, disturbed grassland, arable, village lanes and waste ground. 616. It also grows near rabbit warrens.

Myosotis ramosissima Rochel.
EARLY FORGET-ME-NOT
Occasional in dry, bare places on shallow calcareous soils, also in open sandy turf, by rabbit scrapes, on shingle beaches, rarely on walls. 100 + [15]. Most common along the coastal belt and scattered inland, absent from heavy clay soils.

Myosotis discolor Pers.
CHANGING FORGET-ME-NOT
Occasional in dry, open grassland, quarries, sandpits or rabbit-grazed lawns. 138 + [10]. Most common in the north-west and the Poole basin. Although scarce on the Chalk, it occurs on the leached soil of anthills. Var. *balbisiana* Jord., with white or pale yellow flowers, was found at Stonebarrow Hill (39W, 1880, HNR and WF), Upwey (68S, 1910, Anon) and Hurst Heath (78Z, 1999, !).

[Lappula squarrosa (Retz.) Dumort.
BUR FORGET-ME-NOT
A scarce casual. 6. West Bay (49Q, 1921, AWG); Netherbury (49U, 1927, AWG); Beaminster (40Q, 1919, AWG); Radipole Lake (67U, 1926, MJA); Corfe (98, 1915, CBG in **BMH**); Swanage (08F, 1917, CBG).]

Omphalodes verna Moench
BLUE-EYED MARY
Established in a few plantations or churchyards. 3 + [2].

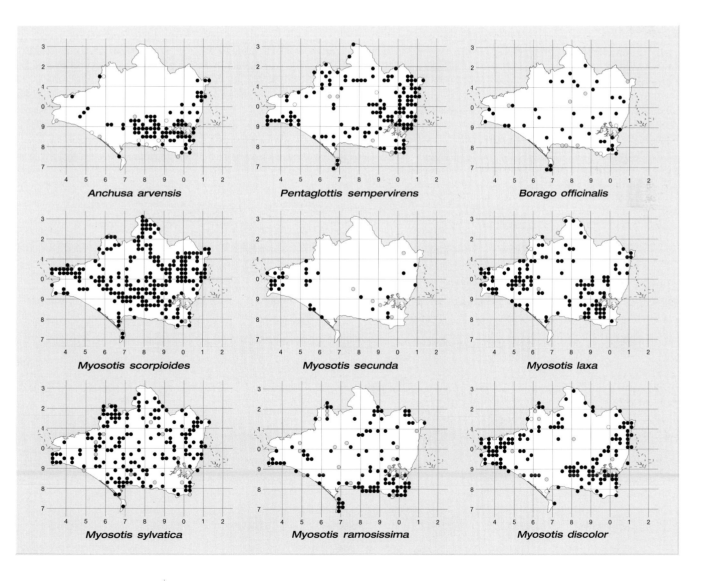

Anchusa arvensis

Pentaglottis sempervirens

Borago officinalis

Myosotis scorpioides

Myosotis secunda

Myosotis laxa

Myosotis sylvatica

Myosotis ramosissima

Myosotis discolor

Sandpit Farm, Drimpton (40H, 1925–35, AWG); Holwell (61V, 1993, AH, !); Crichel Park and Witchampton (90Y, 1919, JHS and 1965, AWG in **BDK**); Charborough Park (99I, 1936, RDG in **WRHM** and 1992, BE); Cranborne Ch (01L, 1991, !).

Cynoglossum officinale L.
HOUND'S-TONGUE
Occasional, in small numbers, in calcareous grassland, especially near rabbit warrens, rarely in disturbed heaths or on beaches. 60 + [26]. Commonest along the coastal belt and in the north-east. It has probably decreased through intensive agriculture. The leaves secrete acetamide which give them their characteristic smell of mice.

[*Alkanna lutea* DC. was a casual at Bindon Mill (88N, 1929, LBH in **BM**).]

Heliotropium europaeum L. occurred in arable at Stoke Mill (49I, 1976, AWG). [1].

VERBENACEAE

Verbena officinalis L.
VERVAIN
Rather local and sporadic, rarely in quantity, in dry, sunny places such as woodland rides, verges and railtracks, old pits and undercliffs, on both calcareous and sandy soils. 74 + [39]. Least common in the north and west. The plant attracts insects, and was sacred to the druids; it contains

bitter iridoids such as verbenin, which can cause vomiting.

Verbena bonariensis L.
ARGENTINIAN VERVAIN
A garden escape at a small tip near Redbridge (78Z, 1999, !). and Greenland Farm (08C, 2000, EAP). 2.

Clerodendron trichotomum Thunb.
A small flowering tree, rarely planted in private grounds. 2. Over Compton (51Y, !); three trees at Amen Corner (01A, !).

LAMIACEAE or LABIATAE

Most species contain essential oils.

Stachys officinalis (L.) Trev. St Leon
BETONY
Frequent in unimproved grassland, sunny woodland rides, along green lanes and old hedgebanks. 324. It favours leached clay soils, but also grows in mature calcareous grassland and in grazed (and so eutrophiated) heaths. It is least common where buildings or prairie farming have destroyed all natural vegetation, as on parts of the chalk. Var. *nana* Druce, a dwarf ecotype, has been seen at Winterbourne Abbas (69A, !), Hog Cliff (69D, 1977, CA Hutchinson), Flowers Barrow (88Q, 1916, CBG in **BMH**) and West Hill, Corfe (97N, !). A form with pale pink flowers occurs at The Warren (78V, !).

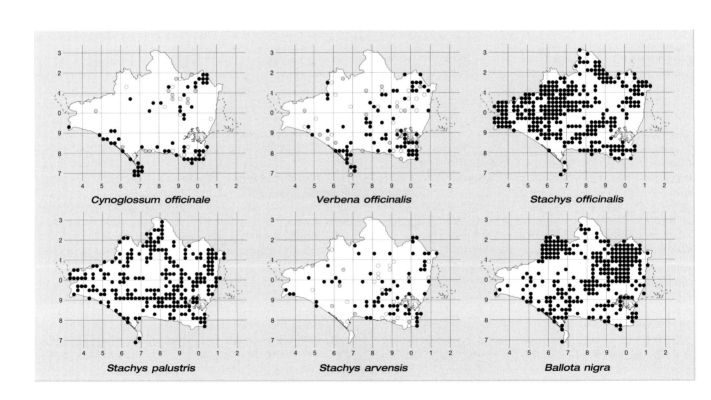

Cynoglossum officinale · Verbena officinalis · Stachys officinalis · Stachys palustris · Stachys arvensis · Ballota nigra

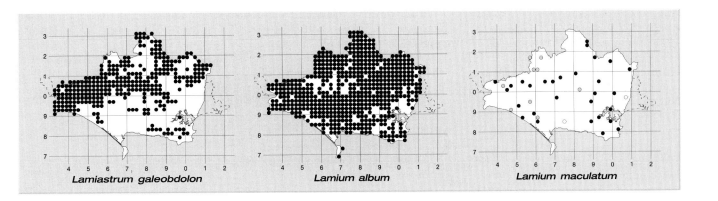

Lamiastrum galeobdolon   Lamium album   Lamium maculatum

Stachys byzantina K. Koch
LAMB'S-EAR
Planted in churchyards, or a garden escape on verges. 5.
Portland Bill (66, RA); The Nothe (67Z, !); Mappowder
Ch (70I, !); Giddy Green (88I, !); Milborne St Andrew
(89D, !).

Stachys sylvatica L.
HEDGE WOUNDWORT
Common at the edges of woods, in hedgebanks and on
verges. 677. The leaves have a rank scent.

Stachys sylvatica × palustris = S. × ambigua Smith
Very local, in patches along ditches or on verges, persisting
for some time, though lost from many old sites. 12 + [16].
First record: Sturminster Marshall (99P) and Wimborne,
1696, W. Stonestreet. Noted recently in 49H, 40H, 58Z,
59D, 50H, 70C, 71T and V, 88D, 98B, 90V and 08F.
Specimens are in **BMH** and **RNG**.

Stachys palustris L.
MARSH WOUNDWORT
Frequent along rivers, streams and ditches, and
occasionally an arable weed in drier places. 250. With white
flowers at Blandford (80Y).

Stachys arvensis (L.) L.
FIELD WOUNDWORT
Occasional, in small numbers, as a weed in arable,
disturbed verges or tips. 66 + [29]. Most frequent in sandy
soils in the Poole basin, but sometimes in chalky fields.
JCMP found it 'not common', and RDG had it from 37
stands.

[**Stachys annua** (L.) L.
ANNUAL YELLOW-WOUNDWORT
Casual at The Park, Weymouth (67, 1890, WBB in **DOR**).]

Ballota nigra L.
BLACK HOREHOUND
Frequent in 'unnatural' habitats such as hedgebanks,
ditches, disturbed grassland, verges, allotments and tips.
274. Most frequent on calcareous soils, though rare on
Portland, and least common in the west and the Poole

basin. The flowers are mostly dull pink, but have been
seen white at Pimperne (90E) and Hinton Parva (90X),
pale pink at Blandford allotments (80Y) and the mauve
shade of *Clinopodium ascendens* at Barnsley Farm (90W).

[**Leonurus cardiaca** L.
MOTHERWORT
An extinct denizen of waste places, recorded by JCMP before 1895
from Dorchester, Church Knowle and Studland. 3. The nearest
recent report is from a tip at Iford Lane (19G, 1978, RHD) in
VC11.]

Lamiastrum galeobdolon (L.) Ehrend & Polatschek
YELLOW ARCHANGEL
Subsp. *montanum* (Pers.) Ehrend & Polatschek is locally
abundant in old woods and hedgebanks. 320. It prefers
humus-rich clay soils, tolerates chalk, but is absent from
acid sands in the Poole basin, and also from deforested
areas north of Weymouth and Portland. Subsp. *argentatum*
(Smejkal) Stace, with white-striped leaves, is an aggressive
garden escape or throw-out, in plantations, churchyards,
verges and village lanes. 117. First recorded in 1990, it is
spreading very fast.

Lamium album L.
WHITE DEAD-NETTLE
Common in 'unnatural' habitats such as hedgebanks,
village lanes and verges, mostly near houses. 569.

Lamium maculatum L.
SPOTTED DEAD-NETTLE
An occasional garden escape in churchyards, hedgebanks
and verges. 29 + [12]. First record: Poxwell (78M, 1866,
SMP in **DOR**), but found at Osmington (78G, WBB) at
about this time. The leaves are always variegated with white
stripes here.

Lamium purpureum L.
RED DEAD-NETTLE, DUNCH-NETTLE
A common weed of bare, dry soil in arable fields or
gardens. 612. Rarely its flowers are white, as at Over
Compton (51Y), Coryates (68H), Lodmoor (68V),
Puddletown (79H) and Wareham (98, 1900, EFL), or pale
pink as at Witchampton (90Y).

189

Lamium hybridum Villars
CUT-LEAVED DEAD-NETTLE
An uncommon weed of arable, gardens, waste ground and tips, perhaps under-recorded. 35 + [7]. It smells like *Stachys sylvatica* when bruised.

Lamium amplexicaule L.
HENBIT DEAD-NETTLE
An occasional weed of arable, preferring dry calcareous soils but not restricted to them, and rare in the west. 89.

Galeopsis angustifolia Ehrh. ex Hoffm.
RED HEMP-NETTLE
Once common, now a rare weed of dry, arable fields, calcareous or sandy, now mainly in the north-east. 12 + [28]. JCMP found it common, and Good recorded it as frequent but found it in only 13 stands. (Old reports of *Galeopsis segetum* Necker from Upwey allotments (68S, 1926, MJA) and *G. ladanum* L., e.g. east of Weymouth (67, 1696, W Stonestreet) were probably errors for *G. angustifolia*.)

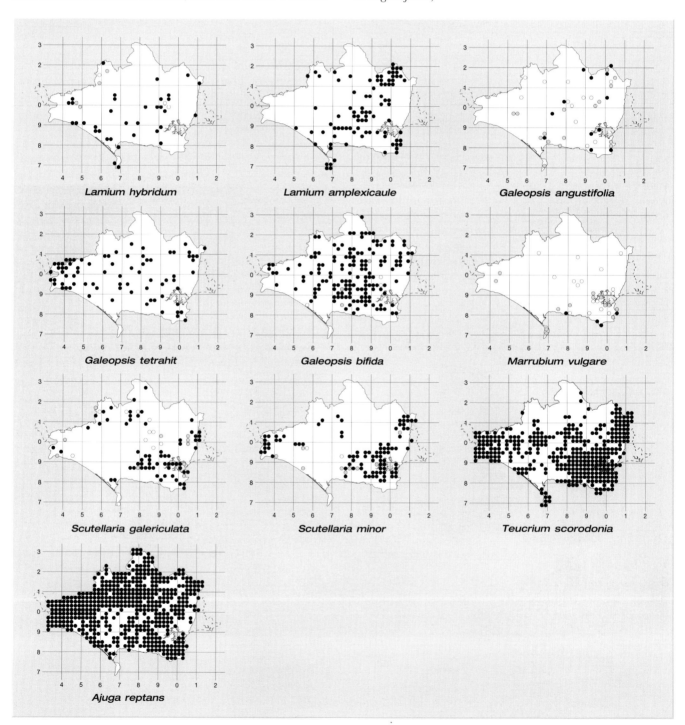

Lamium hybridum

Lamium amplexicaule

Galeopsis angustifolia

Galeopsis tetrahit

Galeopsis bifida

Marrubium vulgare

Scutellaria galericulata

Scutellaria minor

Teucrium scorodonia

Ajuga reptans

Galeopsis speciosa Miller
LARGE-FLOWERED HEMP-NETTLE
A rare casual from North Britain. [3]. Netherbury (49U, 1924, ML Moores in **BDK**); Hammiton Wood (59A, 1978, A Simon); Arne reserve (98T, 1970, B Pickess).

Galeopsis tetrahit agg.
Young or sterile plants could not be named. 194.

Galeopsis tetrahit L.
COMMON HEMP-NETTLE
An occasional arable weed, or among scrub. 77. White-flowered plants are often present.

Galeopsis bifida L.
BIFID HEMP-NETTLE
Frequent in wet or dry woods and hedgebanks, among bracken, or in calcareous or sandy arable. 150 + [6]. Least common along the coastal belt. White-flowered plants are uncommon.

Phlomis fruticosa L.
JERUSALEM SAGE
A Mediterranean shrub, occasionally planted as it has been for at least 200 years. 2 + [1]. [Morcombelake (39X, 1961–78, !)]; Radipole Ch (68Q, !); Sandbanks dunes (08N, !).

Melittis melissophyllum L.
BASTARD BALM
Rare, flowering sporadically, at the edge of old woods on acid clay soil. 3. Chetterwood (90U, 1874, B Mills in **NMW**, *et al.* in **DOR** and **LIV**, and 1995, BE and DAP); Boys Wood (00U, 1900, EFL, *et al.*, and 1983, AH, !); Birches Copse (01Q, 1920, AWG in **BDK**, *et al.*, and 1983, CE Ollivant). It does not occur further north in Britain.

Marrubium vulgare L.
WHITE HOREHOUND
A southern species of nutrient-rich waste places, rabbit warrens, cliff-tops and edges of arable fields. 4 + [31]. Rarely a casual in arable, or planted as a herb. It has declined from 'fairly frequent' in the 1890s through 'uncommon' in the 1930s to rare today. Post-1950 records: Stoke Abbott Mill (40K, 1976, AWG); Portland (67, 1954, JW Ash); Ringstead (78L, 1953, !); Bats Head (78V, 1960–93, !); St Aldhelm's Head (97 N and S, 1997, DAP, !); Ballard Head (08L, 1998, DAP); Parsons Barn (08L, 1964, RMB, !).

Scutellaria altissima L.
SOMERSET SKULLCAP
A rare denizen, Sherborne public park (61I, 1997, !). 1.

Scutellaria galericulata L.
SKULLCAP
Rather local in wet woods, marshes and brackish marshes, riverbanks and by ponds. 65 + [21]. While present in the larger river valleys, it may have declined in the central reaches of the Stour. It is absent from the dry parts of the Chalk and a large part of west central Dorset.

Scutellaria galericulata × *minor* = *S.* × *hybrida* Strail
Rare in wet, shaded sites, and possibly overlooked, as sterile colonies can be hard to name. 1 + [1]. Hurst Heath (78Z, !); Ridge (98N, 1914, CBG in **BMH**).

Scutellaria minor Hudson
LESSER SKULLCAP
Occasional in alder woods and sallow carr, along rides by acid streams or on wet, grazed heaths. 91 + [12]. Mostly in the Poole basin, with outliers in the west, also on the chalk near Bascombe Barn (78R, BE).

Teucrium scorodonia L.
WOOD SAGE
Frequent in many woodland rides, and mature, dry, sandy or calcareous grassland. 356. It is especially common along the coastal belt, where it tolerates salt-spray, but is scarce in deforested areas and also in much of the north, where it avoids poorly-drained soils.

[*Teucrium chamaedrys* L.
WALL GERMANDER
Rarely planted in dry places or on old walls, sometimes persistent. 2. Abbotsbury ruins (58M, 1961–77, ! in **RNG**); Milborne St Andrew (89D, 1860, JCMP in **DOR**).]

Ajuga reptans L.
BUGLE
Common in more or less damp woodland rides, in carr, damp grassland or lawns. 505. It prefers rich soils with impeded drainage, and is absent from a few deforested areas and from very acid heathland. A form with pink flowers occurs at Chideock (49F) and Lydlinch Common (71G), and one with white flowers at Bryanston (809).

Nepeta cataria L.
CAT-MINT
Uncommon, sporadic and in small numbers, in disturbed hedgebanks on calcareous soils. 5 + [23]. It has never been common, but its recent decline may be associated with mechanical mowing of verges, preventing seed production. Post-1960 records: Burton Freshwater (48U, 1928–70, AWG in **BDK**); Friar Waddon (68H, 1991–97, BE); Bedmill Farm (61C, 1956, JL Gilbert); Monkwood Hill Farm (70H, 1977, CA Hutchinson); Thornicombe (80R, 1989, !); Breach Common (82L, 1991, DAP); Boveridge (01M, 1993, GDF and JO). Attractive to cats of both sexes; it contains essential oils such as citronellal, geraniol and nepetalactones.

Nepeta × *faassenii* Bergmans
GARDEN CAT-MINT
A rare garden escape. 1. Two clumps on a verge at East Burton (88I, !). Unattractive to cats.

Glechoma hederacea L.
GROUND-IVY, ALEHOOF, HAY-MAIDEN
Common in woodlands, hedgebanks and sheltered grassland on calcareous or clay soils. 709. A form with pink flowers

191

occurs at Tyneham (88V, AH). The plant was once used to flavour ale, known locally as hay-maiden tea (Udal, 1889).

Prunella vulgaris L.
SELF-HEAL
Very common in woodland rides or short turf, in lawns, dune slacks and along paths. 711. It resists trampling and grows on most kinds of soil. Forms with white flowers have been seen at Winterbourne Steepleton (69B), Crossways (78P), North Wood (88H), New Swanage (08F) and Studland (08G), and with pink flowers at Guys Marsh (82F) and Studland dunes (08F). It contains pentacyclic triterpenes, but these have not been shown to heal anything.

Prunella vulgaris × *laciniata* = *P.* × *intermedia* Link
Probably present wherever *P. laciniata* grows, and contributing to its disappearance. 1 + [2]. Abbotsbury (58S, 1937, DM); Small Mouth (67T, 1983, AH); Easton (67V, 1932, JEL in **RNG**).

Prunella laciniata (L.) L.
CUT LEAVED SELF-HEAL
Rare in mature, short turf, usually calcareous. 2 + [2]. Abbotsbury (58S, 1937, DM); Small Mouth (67T, 1983, AH); Easton (67V, 1932, JEL); Herston (07D, 1939, JW Long and 1992, !). It is not known how this is spread, but it is usually transient, 'hybridised out of existence' by crossing with *P. vulgaris*. The flowers are usually cream-coloured.

Melissa officinalis L.
BALM
Occasional in small numbers as a garden escape in village lanes, on verges, pavements or tips. 89 + [12]. First record: Studland (08, 1911, CES), since when it has spread widely. Its essential oils include citral and citronellal, with pleasant scents.

Clinopodium ascendens (Jordan) Samp.
COMMON CALAMINT
Occasional, in small numbers, in dry, sunny hedgebanks, calcareous grassland or undercliffs. 40 + [12]. It prefers open sites on nutrient-poor soils with occasional disturbance. Rarely planted near houses. The robust form *baetica*, from Corfe Castle mound, was once given specific status (98R, 1866, SMP in **DOR**, *et al.*, 1937, HWP and NDS in **DOR** and 1999, RSRF & !).

Clinopodium vulgare L.
WILD BASIL
Frequent in dry woodland rides or hedgebanks, and in calcareous grassland. 308. Rare or absent in the Poole basin except on eutrophic verges. Forms with white flowers were seen at Birch Close (80W, !) and Gunville Down (91B, !).

Clinopodium acinos (L.) Kuntze
BASIL THYME
Once frequent, now rare in bare, calcareous arable or on dry banks. 7 + [34]. Good found it in 19 stands in the 1930s, since when it has much declined, especially on farmed sites. Post-1960 records: Bincombe tunnel (68Y, 1991, BE); Muckleford (69L, 1992, D Allen); Grimstone Down (69M, 1984, NCC); White Nothe (78Q, 1960, !); Marley Wood (88B, 1983, C Bass); Blandford Camp (90I, 1993, M Heath); Down Farm (01C, 1998, M Green); Badbury Rings (01M, 1960, BNSS); Sovell Down (91V, 1972, C Thomas); West Moors (00W, 1992, DAP); Blackbush Down (01M, 1963, KBR).

Origanum vulgare L.
WILD MARJORAM
Frequent in long or short calcareous turf, scrub or hedgebanks, verges and railway banks. 195. Absent from acid soils in the Poole basin except on a few eutrophic verges or tips, also from heavy clay soils except along the coastal belt. A distinct-looking form is grown in gardens, and was seen on a tip at Three Legged Cross (00X, 1982, ! in **RNG**). Forms with white flowers grew near New Barn (07E) and on Nine Barrow, Ballard and Ulwell Downs (08A and F, EAP).

Origanum majorana L.
POT MARJORAM
Casual on Crossways tip (78P, 1999, !). 1.

Thymus pulegioides L.
LARGE THYME
Occasional in dry, short turf. 28 + [9]. One ecotype grows in mature, calcareous turf, especially near the coast, while another prefers eutrophic banks or verges in heathland. White-flowered plants were seen at Church Knowle (98F, 1900, EFL) and Hartland Moor tramway (98M, 1960, RDG).

Thymus polytrichus A. Kerner ex Borbas
WILD THYME
Frequent in dry, short, calcareous turf, sometimes in fixed shingle or dunes. 246. Commonest on the Chalk and along the coastal belt, rare in eutrophiated heathland. Variable in habit and hairiness; Good gave seven named varieties. White-flowered plants were seen at West Hill, Corfe (98L, EAP) and Worth Matravers (97T, EAP).

Thymus vulgaris L.
CULINARY THYME
Naturalised on a wall at Upwey (68S, !), and planted in Colehill churchyard (01F, !). 2.

Lycopus europaeus L.
GYPSYWORT
Frequent in wet woods, fens, marshes, water-meadows or by streams and ponds. 174. Mostly in the north-west or in the Poole basin, scarce on the dry Chalk and the coastal belt.

Mentha arvensis L.
CORN MINT
Frequent both in damp woodland rides, and in damp arable nearby. 164. It seems indifferent to soil acidity, and its absence from the coastal belt may simply reflect the lack of woods there.

Mentha arvensis × *aquatica* = *M.* × *verticillata* L.
WHORLED MINT
Uncommon in colonies in woodland rides, grazed acid pastures, by streams and disturbed verges. 23 + [19]. Absent from the Chalk.

Mentha arvensis × *aquatica* × *spicata* = *M.* × *smithiana* R.A. Graham
TALL MINT
A rare and probably declining hybrid found by streams, ditches or damp verges. 4 + [19]. Post-1960 records: Higher Bere Chapel (30X, 1995, IPG and PRG); Parnham (40Q, 1976, AWG); Radipole (68Q, 1977, ! in **RNG**); Manns Cross (60F, 1987, SMB); Lower Whitechurch (89P, 1990, ! in **RNG**); Arne (98, 1968, DSR); Ulwell (08A, 1961, KG); New Barn (91Y, 1997, !). Older specimens are in **BDK** and **WRHM.**

(*Mentha arvensis* × *suavolens* = *M.* × *carinthiaca* Host
Reported as extinct in Dorset by Stace, but the report has not been traced.)

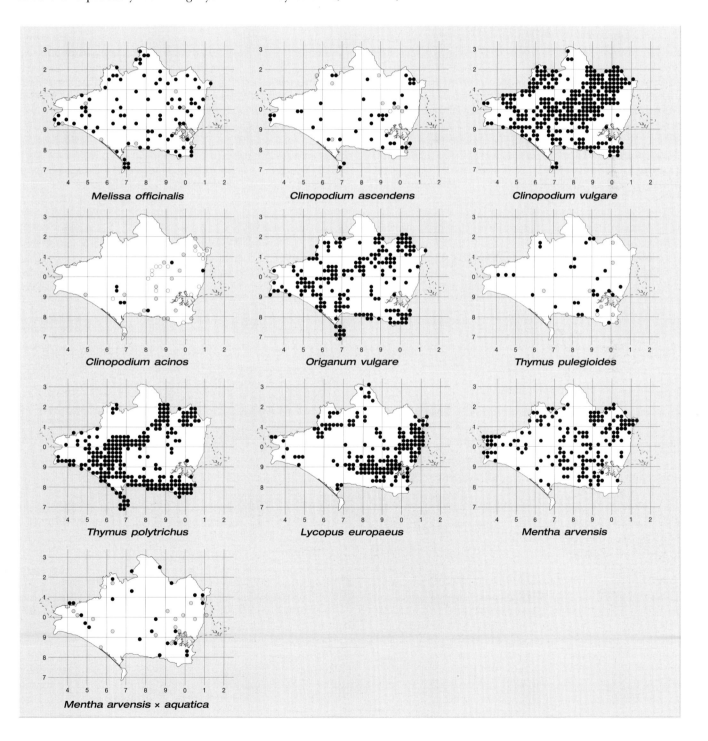

Melissa officinalis

Clinopodium ascendens

Clinopodium vulgare

Clinopodium acinos

Origanum vulgare

Thymus pulegioides

Thymus polytrichus

Lycopus europaeus

Mentha arvensis

Mentha arvensis × aquatica

Mentha arvensis × spicata = M. × gracilis Sole
BUSHY MINT
A rare hybrid of ditches, wet places or tips. 1. Post-1960 records: Charmouth (39R and W, 1965, AWG and 1980, DWT); Sherborne tip (61H, 1970, ACL); Shaftesbury (82R, 1960, !); Ulwell (08A, 1961, KG); Bear Cross (09M, 1965, MPY). Specimens are in **BDK** and **WRHM.**

Mentha aquatica L.
WATER MINT
Common in wet fields and marshes, by rivers, streams and ponds. 438. It survives winter flooding in ponds along the Winterborne valleys, and becomes locally dominant in autumn, when it is an important nectar source for insects.

Mentha aquatica × spicata = M. × piperita L.
PEPPERMINT
Occasional by streams and ditches, or as a garden throw-out near villages. 11 + [26]. Post-1960 records: (39M,1980, DWT); Over Compton (58Y, 1996, JAG); Radipole Lake (69Q, 1977, !); Flowers Farm lake (60H, 1982, JN); Sherborne and Oborne (61J and P, 1992, MJG); Crossways tip (78P, 1999, !);

tip at Scratchy Bottom (88A, 1999, !); Winfrith Heath (88D, 1963–81, !); East Stoke (88T, 1977, !); Winterborne Kingston (89N, 1994, !); Keysworth (98J, 1978, AH); Corfe Common (98K, 1969, DWT); East Morden (99D, 1964, !); Tarrant Gunville and Tarrant Hinton (91F and G, 1993, !).

Mentha spicata L.
SPEAR MINT
A garden escape or throw-out, forming small patches on verges, railway banks, waste ground or old tips, usually near houses and in drier habitats than other mints. 70 + [27]. Often reported as *M. longifolia* in the past. Increasing.

Mentha spicata × suavolens = M. × villosa Hudson
APPLE-MINT
A garden escape in similar habitats to *M. spicata*. 67 + [13] First record: Bloxworth (89X, 1905, WRL). Variable in hairiness. Increasing by invasive roots.

Mentha spicata × longifolia = M. × villosonervata Opiz
SHARP-TOOTHED MINT
One record, needing confirmation, from the Wriggle

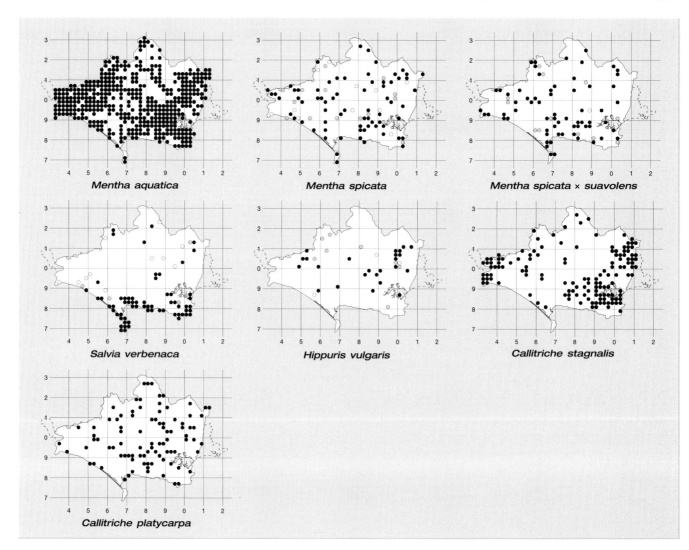

Mentha aquatica

Mentha spicata

Mentha spicata × suavolens

Salvia verbenaca

Hippuris vulgaris

Callitriche stagnalis

Callitriche platycarpa

River near Batcombe (60C, 1934, RDG in **WRHM** det. J. Fraser)

Mentha suavolens Ehrh.
ROUND-LEAVED MINT
A rare garden escape on verges or waste ground. 5 + [17]. Post-1960 records: Weymouth Sta (67U, 1995, !); Buckland Newton (60S, 1972, RDE); Yetcombe (60R, 1988, I and A McNeill); Plush (70B, 1992, !); Luckford Lake (88S, 1974, JFA); Canford Ch (09J, 1995, !); Edmondsham (01Q, 1993, GDF). Old reports of *M. rotundifolia* could be this or *M.* × *villosa*.

Mentha pulegium L.
PENNYROYAL
Rare, often transient, in wet woodland rides, pond margins, china-clay waste or a casual in arable or on tips with acid soil. 4 + [7]. Nottington (68R, 1926, MJA); [Holwell (60W, 1878, HH Wood)]; Rye Water Farm (61Q, 1993, SME); [Hurst Heath (78Z, 1926, MJA and 1962, RDG in **DOR**, !)]; Worgret Heath (98D, 1948, KG); frequent, Creech Heath (98G, 1996, RSRF & ! in **RNG**); Corfe Mullen (99Y, 1985, AH); Canford tip (09I, 1998, !); West Parley (09Y, 1932, FHH and 1937, DM); [Leigh Common (00F, 1799, RP and 1895, EMH)]; Holt Wood (00H, 1925, JHS and 1937, DM). It contains toxic essential oils including pulegone, which strongly stimulates the uterus and was once used as an abortifacient.

Mentha requienii Bentham
CORSICAN MINT
A garden escape on acid soil, Harmans Cross (98V, 1987, AH det. RM Harley). 1.

Lavandula × ***intermedia*** Lois.
GARDEN LAVENDER
Rarely planted on verges or waste ground. 4. Milton Abbey (70W); seedlings by the Customs House at Hamworthy (08E); Glebelands estate (08F); on wall, N. Haven (08I). In the past this has been cultivated for its essential oil.

Rosmarinus officinalis L.
ROSEMARY
Planted in churchyards, also in waste ground, and flourishing on Portland. Self-sown on walls at Charmouth (39R) and Wareham (98I). 14. It prefers calcareous soils; seen in 67W, 68B and Q, 69L, 60Z, 78CG, Q and S, 71Q, 81U, 08F and G.

(***Salvia pratensis*** L.
MEADOW CLARY
All seven reports of this from calcareous grassland are thought to be errors. It occurred as a casual opposite Park Lodge at Tollard Royal (91N, 1915, AWG), but in VC8.)

Salvia verbenaca L.
WILD CLARY
Rather local in dry grassland, fixed shingle or dunes. 63 + [15]. Most common along the coastal belt, scarce and often sporadic on sunny verges inland. It varies in the degree of leaf-dissection.

[***Salvia verticillata*** L.
WHORLED CLARY
This was once reported from Chickerell (68K, 1906, WBB).]

Salvia officinalis L
SAGE
Planted on a verge at Fishponds (39U, 1992, !). 1.

Salvia reflexa Hornem.
MINTWEED
A rare casual in arable. 2. Cutt Mill (71T, 1997, ! in **RNG**); above Little Wood, Clenston (80L, 1995, !).

[***Dracocephalum parviflorum*** Nutt.
AMERICAN DRAGON-HEAD
A casual on waste ground at Parkstone (09, 1937, NDS).]

HIPPURIDACEAE

Hippuris vulgaris L.
MARE'S TAIL
Locally abundant in large lakes, ponds and flooded pits, especially near the R. Allen, where it has been known for 200 years. 21 + [10]. Occasionally planted in garden or field ponds.

CALLITRICHACEAE

Note: The plastic morphology and infrequent fruiting of members of this family has resulted in 112 tetrad records of *Callitriche* agg.; very few records have been vetted by an expert.

Callitriche hermaphroditica L.
AUTUMNAL WATER-STARWORT
A northern species, with a single record from Chickerell pits (67P, 1997, CDP and DAP), where aquatic plants are sold. 1. Earlier reports from Wyke Regis and Little Sea were probably errors.

Callitriche truncata Guss.
SHORT-LEAVED WATER-STARWORT
Very local, but abundant in eutrophic water of the R. Axe and adjacent ponds. 2. Lower Holditch (30G, 1992, IPG and PRG); Forde Abbey lakes (30M, 1976, JGK and 1996, !, also CDP). An early record from Lynch (67N, 1928, ECW) has not been confirmed, and the site is now a caravan park.

Callitriche stagnalis Scop.
COMMON WATER-STARWORT
Common on more or less acidic mud in puddles, along woodland rides, and in acid ditches or ponds. 137. Frequent in the west, north-west and the Poole basin, absent from the Chalk.

Callitriche platycarpa Kutz
VARIOUS-LEAVED WATER-STARWORT
Frequent in nutrient-rich water, or on mud, in streams, ponds and ditches. 90.

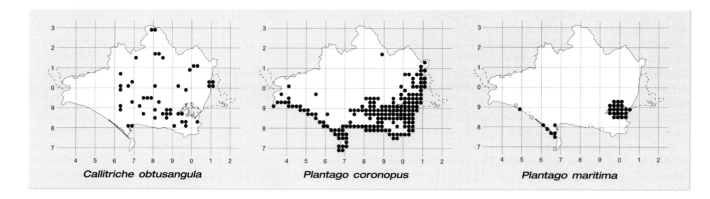

Callitriche obtusangula   Plantago coronopus   Plantago maritima

Callitriche obtusangula Le Gall
BLUNT-FRUITED WATER-STARWORT
Locally frequent in clear, fast-flowing water of chalk streams and rivers, less common in ditches and not seen on mud. 47.

Callitriche brutia Petagna
PEDUNCULATE WATER-STARWORT
Rare on wet, acid, peaty mud. 1 + [3]. Pool pond (88W, 1991, ! in **RNG**); Keysworth (98J, 1895, JCMP and 1978, !); High Moor, Talbot (09R, 1900, EFL). The specimen from Swyre (58J, 1895, JCMP in **DOR**) is *C. platycarpa*, and the report from Langton Herring (68A, 1905, WBB) is doubtful.

Callitriche hamulata Kutz ex Koch
INTERMEDIATE WATER-STARWORT
Scarce, usually in oligotrophic water. 3 + [12]. Newlands Batch (39W, 1980, DWT), needing confirmation; Black Hill, Bere (89M, 1961, !); Hinton meads (81C, 1994, ! det. CDP); Corfe Common (98Q, 1936, RDG and 1970, JB); Pamphill Common (90V, 1915, AWG in **BDK**); Little Sea (08H, 1935, RDG in **WRHM** det. WH Pearsall and 1995, BE and DAP); pond at Holt (00M, 1994, !). Early reports are unreliable, and may refer to *C. platycarpa* before that species was recognized.

PLANTAGINACEAE

Plantago coronopus L.
BUCK'S-HORN PLANTAIN
Locally frequent, either on cliff-tops and undercliffs on all types of soils, or along tracks or disturbances in dry heaths. 183. Common along the coastal strip, and in the Poole basin, currently extending its range to roadside verges and pavements inland, as it tolerates salt and trampling. It is the main constituent of lawns in sea-front gardens at West Bay (49Q). First record: Weymouth (67, 1775, T Yalden in **K**). Very variable in size and leaf-shape.

Plantago maritima L.
SEA PLANTAIN
Locally frequent on muddy shores and in salt-marshes,

along the shores of the Fleet, Radipole Lake and Poole Harbour. 29 + [7].

Plantago major L.
GREATER PLANTAIN
Common along verges and tracks, muddy pond margins, as a lawn weed and on tips. 733. Notably variable in size. Subsp. *intermedia* (Gilib.) Lange, a small form, occurs near the sea and at the upper edge of salt-marshes.

Plantago media L.
HOARY PLANTAIN
Frequent in short, dry, calcareous turf, where it is resistant to grazing, mowing or trampling. 321. Absent from areas with acid soil or heavy clays; sometimes restricted to churchyards in intensively farmed areas.

Plantago lanceolata L.
RIBWORT PLANTAIN
Very common in dry grassland, cliffs, verges and lawns. 732. It tolerates salt-spray, but dislikes acid soils. Variable in leaf-breadth; forms resembling *P. media* in this respect are not rare. A form with leaf rosettes replacing heads of flowers was found at Wareham (98I, !).

Plantago arenaria Waldst. & Kit.
BRANCHED PLANTAIN
A rare casual on new verges or tips. 1 + [1]. Lytchett Matravers (99M, 1985, AH in **RNG**); Parkstone (09, 1930–32, LBH in **RNG**, also HJW).

Littorella uniflora (L.) Asch.
SHOREWEED
Very local in shallow acid ponds or flooded pits. 7 +[13]. Restricted to the Poole basin apart from a dubious 19th century report from Small Mouth (67S). Post-1960 records: Moreton (78Z, 1978, RHD); Stoke Heath (88P, 1991–99, !); Povington (88Q, 1997, BE); lake near Coombe Keynes (88R, 1962, !); Stokeford Heath (88U, 1993, BE); Trigon (88Z, 1993, BE); Higher Hyde Heath (89K, 1995, BE); Creech (98B, 1989, DAP); Furzebrook (98G, 1961, DFW); Morden Decoy (99B, 1961, DFW); Little Sea (08H, 1895, JCMP, *et al.* in **RNG**, and 1995, BE and DAP); West Moors (00X, 1992, DAP and RMW).

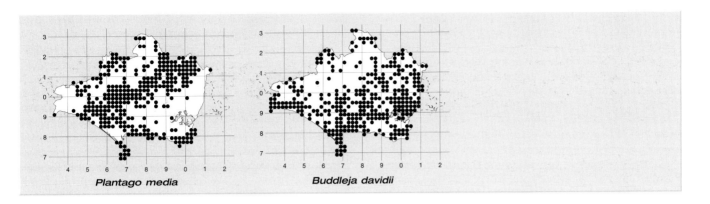

Plantago media   *Buddleja davidii*

BUDDLEJACEAE

Buddleja davidii Franchet
BUTTERFLY-BUSH
Frequent and well-established on undercliffs, limestone rocks and waste ground, and a garden escape on walls and pavements in and around towns and villages. 340. First recorded by Good in the 1930s, and spreading fast. Particularly abundant on Portland, where it perfumes the air in July. Forms with white flowers seen at Stourpaine tip (80U), and with deep purple flowers at Dorchester West Sta (69V); all forms are rich in nectar and attract insects.

Buddleja davidii* × *globosa* = *B.* × *weyeriana Weyer
WEYER'S BUTTERFLY-BUSH
First raised by Mr Van de Weyer at Smedmore Ho (97G) in 1914. 6. Rarely planted on verges, in 39U, 58Z, 68S, 69L, 70L and 89M. The flower-colour ranges from mauve to orange on the same bush.

Buddleja globosa Hope
ORANGE-BALL-TREE
Sometimes planted in parks, and self-sown on railway banks, waste ground and in old quarries. 11 + [1]. First record: Bryanston (80T, 1944, BNHS); also seen in 58S, 67W, 69V, 79X, 88F, 82D, 97P and Z, 99R, 07I and 09J. It grows fast to form large bushes, which can be cut back to the roots by heavy frosts.

OLEACEAE

Forsythia* × *intermedia hort. ex Zabel
Planted, since at least 1899, in parks, churchyards, village greens and rarely in field-hedges far from houses, but not self-sown. 24. Some records may be for other taxa in this genus.

Jasminum officinale L.
SUMMER JASMINE
A garden shrub, planted in churchyards or escaped. 2. West Milton (59D, 1993, LJM); Bincombe Ch (68X, !).

Jasminum nudiflorum Lindley
WINTER JASMINE
Planted at Broadmayne Ch (78I, !). 1.

Fraxinus excelsior L.
ASH
Locally dominant in woods on calcareous soils and a frequent hedgerow tree. 705. It is least common on very acid soils; on the chalk it rapidly colonises steep slopes, forming woods which do not appear on old maps, as at Okeford Hill (80E) and Coombe Bottom (81K). The largest measured ash is one 21 m × 1.3 m at Carey Ho (98E), but there is a pollard with a girth of 3.1 m at Melbury Park (50S) and a tree of 5.5 m girth near Dewlish House (79T). Some pollards in woods and stools in hedges may be centuries old. At Monmouth Close (00T) is the Monmouth Ash, a scion from the tree where the Duke of Monmouth was captured, ending his rebellion in the 17th century. The Cradle Tree at Turnworth (80D) was lost *c.* 1990, and Granny's Knitting Tree at Coombe Keynes (88M) was replaced in 1980. Some trees fail to fruit and are probably all-male. Var. *diversifolia* Ait., with entire leaves, has been seen in hedges near Badbury (90R, !) and Kingston Lacy (90W, RMW). Ash leaves are eaten by many moth and beetle larvae, and the flowers are taken by Woodpigeons. The tree is still coppiced for poles, and the timber is used for handles of axes, spades and hammers, as it has been since at least Roman times. Over 50 Dorset place names are based on 'ash'.

Fraxinus angustifolia Vahl
NARROW-LEAVED ASH
Occasionally planted in recent years in parks and on verges. 4. A tree at Melbury Park (50T) is 30 m × 2.1 m; also seen at Bere Regis (89M, ! in **RNG**), Upton CP (99W) and Canford School (09J). Easily overlooked as Common Ash, but the terminal buds are brown, not black.

Fraxinus ornus L.
MANNA ASH
Occasionally planted in parks or on verges. 7. Wheel House Lane (30S, IPG); Evershot and Melbury Park (50S, !); Bryanston (80T, !); Blandford St Mary (80X, !); Blandford by-pass, with many saplings (80Y, !); Northport (98E, !); Canford School 09J, !).

F. americana L., ***F. pennsylvanica*** Marsh and ***F. spaethiana***
Lingelsh. are planted at Canford School (09J, TH).

Syringa vulgaris L.
LILAC
Fairly common in hedges near villages or ruins, but sometimes far from houses. 256. It was on sale in Blandford in 1782, and has flowers which may be white or shades of mauve.

Ligustrum vulgare L.
PRIVET
Common at the edges of woods, in scrub and dry hedgebanks, mainly on calcareous soils. 597. A salt-resistant ecotype is dominant in thickets on calcareous undercliffs and occurs on the Chesil Bank. Least common in the Poole basin, and absent from heaths. Its scent attracts many insects when in flower, the leaves feed many moth larvae, and the berries are eaten by migrant thrushes, though toxic to men.

Ligustrum ovalifolium Hassk.
GARDEN PRIVET
Frequently planted in hedges near houses, occasional in plantations or on railway banks where it is presumably self-sown. 235. First report: planted at Abbotsbury (58S, 1899, MI). Neither JCMP nor Good mention it, so it must have spread rapidly.

Ligustrum lucidum Aiton f.
TREE PRIVET
Rarely planted as a street tree, or in parks. 4 + [1]. Milton Abbas (70W); Anderson (89Y); Wareham (98I); Charborough Park, (99J, 1914, D Morris); Canford School (09J).

Osmanthus × burkwoodii (Burkwood & Skipwith) P Green
A persistent garden shrub planted in parks and churchyards. 4. Abbotsbury (58S, 1981, !); Thornford (61B); Charborough Park (99I); Branksome Chine (08U).

Phillyrea latifolia L.
A garden shrub, planted in parks and churchyards. 3. Chideock Ch (49G, !); Beaminster cemetery (40V, BE & !); Abbotsbury ruins (58S, 1899, MI, also 1962, D McClintock in **RNG**, and 1998, !). The opposite leaves distinguish it from *Quercus ilex* when sterile.

Picconia excelsa (Aiton) DC.
A tree 18 m × 2.8 m in Abbotsbury gardens was planted in 1784.

SCROPHULARIACEAE

Verbascum blattaria L.
MOTH MULLEIN
An uncommon and sporadic casual on rail tracks, gardens or tips. 5 + [23]. Post-1960 records: Chickerell tip (68K, 1980, !); Dorchester West Sta (69V, 1992, !); Sherborne Park (61T, 1970, ACL); Knighton pits (78P, 1993, !); Bryanston (80T, 1969, GD Harthan); Blandford (80Y, 1996, !); Badbury

Rings (90R, 1962, HJB & 1980, NCC); Brownsea Is (08D and I, 1999, DCL, but perhaps confused with *V. phoeniceum*); Holt Heath tip (00M, 1993, DJG). Flowers rarely white.

Verbascum virgatum Stokes
TWIGGY MULLEIN
A rare casual, usually as single plants, on tips or in gardens. 2 + [4]. Beaminster (40V, 1937, AWG); Sherborne (61N, 1969, ACL); Crossways tip (78U, 1997, !); Carey (88Z, 1915, CBG in **BMH**); Canford Magna (09J, 1978, RHD); West Moors (00X, 1984, KJ Powrie).

Verbascum phoeniceum L.
PURPLE MULLEIN
A rare garden escape. 2. Winterborne Whitechurch allotments (80F, 1995, ! in **RNG**); Brownsea Is (08J, 1996, !). Flowers usually white.

Verbascum phlomoides L.
ORANGE MULLEIN
A garden escape, sometimes established for a few years by farmyards, verges, urban wasteland or tips. 15 + [4]. First record: Swanage Camp (08A, 1920, LFS Todd in **OXF**). White-flowered plants were seen near Red Bridge, Moreton (78U, !).

Verbascum densiflorum Bertol.
DENSE-FLOWERED MULLEIN
A rare casual on verges or tips. 2 + [1]. Charmouth (39W, 1988, RCP in **OXF** det IK Ferguson); Swanage Camp (08A, 1921, EW Nicholson in **OXF** det AW Thellung); Greenland Farm (08C, 1999, EAP, !).

Verbascum thapsus L.
GREAT MULLEIN
Frequent, but sporadic, in disturbed grassland, verges, quarries, rabbit warrens, fallow fields and tips. 390. Commonest on calcareous soils. Like all mulleins, the leaves, and sometimes the flowers, are eaten by larvae of the moth *Cucullia verbasci*.

Verbascum thapsus × nigrum = V. × semialbum Chaub.
Very rare in the north-east. [1]. Bottlebrush Down (01H, 1934, RDG in **WRHM**). It occurred with the parents by the A354 at Bokerley junction (01J, 1978–82, !) but in VC8.

[*Verbascum chaixii* Villars
NETTLE-LEAVED MULLEIN
Railway bank, Abbotsbury (58S, 1892, I Hawkins in **DOR**); Whatcombe (80F, 1897, JCMP in **DOR**). 2.]

Verbascum nigrum L.
DARK MULLEIN
Local in chalk turf, on verges and (in earlier years) on railway banks; noted in Bronze Age deposits. 27 + [16]. Frequent in the north-east, rare and sporadic or casual elsewhere. Forms with white flowers were seen at Fontmell Down (81T, 1994, J Beech).

Verbascum lychnitis L.
WHITE MULLEIN
A rare casual confused with other species. [4]. Stalbridge (71I, 1865, Mrs Allen in **DOR** as *V. pulverulentum*); garden, Little Monington (80T, 1970, A Lack); Badbury Rings (90R, 1978, J Dukes, needing confirmation); Wimborne St Giles (01F, 1904, EFL and 1919, H Wright).

Scrophularia nodosa L.
COMMON FIGWORT
Widespread, usually in small numbers, in old woods. 399. Absent from deforested areas, but surviving among the quarries and cliffs of north Portland. It is fertilised by wasps.

Scrophularia auriculata L.
WATER FIGWORT
Frequent in wet places by rivers, streams and ditches on rich muds, tolerant of shade. 527. It also grows on the floors of some quarries, as a pavement weed and on tips. Least common on the dry Chalk in the north-east. A form with pale yellow flowers occurs at Upwey (68S).

Scrophularia scorodonia L.
BALM-LEAVED FIGWORT
Very scarce along the coast; a Cornubian plant, it does not occur further east in Britain. 1 + [2]. Radipole Lake (67U, 1938, DM); Ower (98X, 1933–5, BNSS as *S. scopolii*); Goathorn (08D, 1930, B Van de Weyer and 1998, !); Furzey Is (08D, 1995, !); Brownsea Is (08D, 1995, FAW).

(Scrophularia scopolii Hoppe ex Pers.
All reports are thought to be errors for *S. scorodonia*.)

Scrophularia vernalis L.
YELLOW FIGWORT
Rare in plantations, scrub or disturbed verges. 3 + [6]. Lamberts Castle (39U, 1930s, AWG); Beaminster (40Q, 1993, RJS & O Davidson); Melbury Park (50, 1930s, RDG); Moreton Pit (78Z, 1937, W Van de Weyer); Clyffe (79W, 1960, Anon); East Stour (72W, 1972, RDE, *et al.*, and 1999, !); Affpuddle (89B and G, 1979, CA Hutchinson); Wyke Down (01C, 1996, GDF). It flowers in spring and produces seedlings in July.

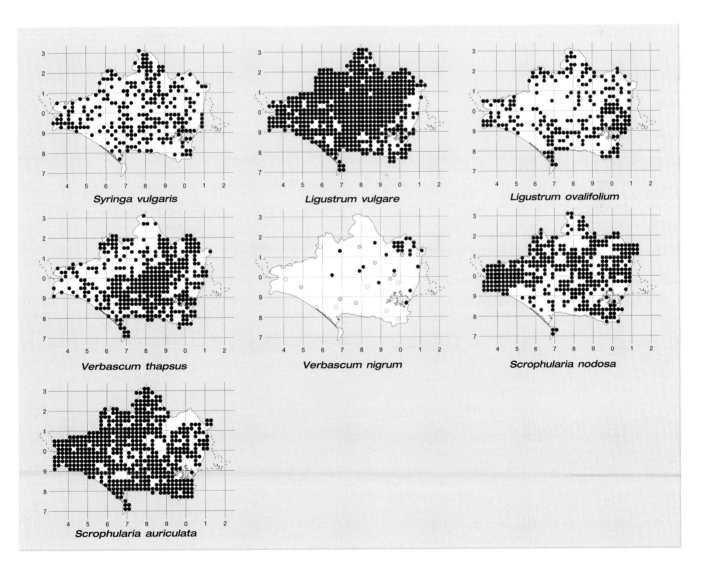

Syringa vulgaris

Ligustrum vulgare

Ligustrum ovalifolium

Verbascum thapsus

Verbascum nigrum

Scrophularia nodosa

Scrophularia auriculata

Scrophularia peregrina L.
NETTLE-LEAVED FIGWORT
Casual in a garden at Shapwick (90F, 1967–68, NM Hamilton in **RNG**). [1].

Mimulus moschatus Douglas ex Lindley
MUSK
Planted and later self-sown in wild gardens and plantations, or as a weed of moist pavements. 4 + [4]. Heywood Bottom (30L, 1968, JB); Forde Abbey (30M, !); Stoke Abbott (40K, 1927, AWG in **BDK**); Abbotsbury gardens (58S, 1899, MI and 1993, SG); Warmwell Ho (78N, 1938, RDG in **WRHM**); Broadstone (99X, 1921, JHS); Chettle Ho (91L, !); Gods Blessing Green (00G, !). It has no scent.

Mimulus guttatus DC.
MONKEYFLOWER
Locally frequent along clear streams, in cress beds and by ponds. 39 + [20]. First record: pre-1895. It may have decreased recently along the Stour and the Frome. Also seen as a casual on tips or pavements.

Mimulus guttatus × *luteus* = *M.* × *robertsii* Silverside
Only seen in cress beds along the Bere stream. 1. Hollow Oak (89L, 1922, AWG and 1989, BE, !).

Mimulus Cv. 'Malibu Series'
A colourful pavement weed seen at Dorchester (69V, 1998, !) and overwintering at Wareham (98I, 1998, !). 2. Flowers bright yellow or red.

(*Limosella aquatica* L. occurs on mud at Stanpit Wick (19L and R, 1991, RMW) in VC11.)

[*Calceolaria chelidonioides* Kunth
SLIPPERWORT
A casual garden escape. 3. Bradford Abbas (51S, 1867, JC Hudson in **BM**); Lewell Lodge (78J, 1928, ER Sykes in **BM**); Longthorns (89J, 1930s, RDG).]

Calceolaria integrifolia Murray was planted at Bincombe Ch (68X, 1993, !). 1.

Antirrhinum majus L.
SNAPDRAGON
Long established on cliffs, old walls, railway banks and in quarries, and casual in urban sites and tips. 83. Well naturalised on Portland. Flower-colour mostly pink, but variable.

Antirrhinum graniticum Rothm.
This may be the name of a procumbent pavement weed overwintering at Wareham (98I, 1998–99, !) sold as Cv. candelabra. 1.

Chaenorhinum minus (L.) Lange
SMALL TOADFLAX
Locally frequent in arable on calcareous soil, also on bare, disturbed verges, rail tracks and tips. 104. Seen on the

beach at Arish Mell (88K). Largely restricted to the Chalk or limestone.

Misopates orontium (L.) Raf.
WEASEL'S-SNOUT
Now scarce in arable, but locally frequent or sporadic in new or disturbed verges, allotments, gardens and tips. 45 + [25]. It prefers rich, freshly dug, moderately acid soil. JCMP thought it common, and Good found it in 21 of his stands. Probably decreasing.

Misopates calycinum (Vent.) Rothm.
PALE WEASEL'S-SNOUT
A rare casual, once seen at Binnegar tip (88Y, 1992, ! in **RNG**). 1. Flowers very pale pink.

Cymbalaria muralis P. Gaertner, Meyer & Scherb.
IVY-LEAVED TOADFLAX, MOTHER-OF-THOUSANDS
Frequent on village walls, in quarries, as a pavement weed and on tips. 341. First record: 1799, RP. Sparse in tetrads without villages, and in the east. White-flowered forms occur at Moreton (88E) and Culeaze (89L).

Cymbalaria pallida (Ten.) Wettst.
ITALIAN TOADFLAX
Escaped on a wall at Stapehill (00K, 1992, !) and likely to spread. 1. It also occured on a wall at Ebbesbourne Wake (92X, 1978, KEB), but in VC8.

Kickxia elatine (L.) Dumort.
SHARP-LEAVED FLUELLEN
Locally common in dry arable or fallow, on both calcareous and sandy soils, rare on clay. 108 + [23]. Occasional in mown, sandy firebreaks on the heaths, or at the backs of beaches. Most frequent in the north-east and the Poole basin, absent from the west. Good had it from 64 stands.

Kickxia spuria (L.) Dumort.
ROUND-LEAVED FLUELLEN
Locally frequent in dry arable, on both calcareous and sandy soils, often in stubble. 84 + [2]. Rare in tank tracks on the heaths, or at the backs of beaches. Mostly on the Chalk, and along the coastal belt between Bexington and Portland. A peloric form was found on Portland (67, 1882, SMP in **DOR**).

Linaria vulgaris Miller
COMMON TOADFLAX, BUTTER-AND-EGGS
Frequent in 'non-native' habitats, in colonies on field-edges, dry verges and railway banks. 301. Most frequent in calcareous soil, least common in the north-west.

Linaria vulgaris × *repens* = *L.* × *sepium* Allman
This hybrid was found on the rail track between Corfe and Swanage; the track was abandoned in the 1950s, but has recently been reinstated and the plant has reappeared. 1 + [1]. Langton Matravers (97Z, 1894, LVL in **DOR**, also 1912, AWG in **BDK** and 1915, CBG in **BMH**); New Barn (07E, 1999,!).

Linaria purpurea (L.) Miller
PURPLE TOADFLAX
A garden escape in dry places, sowing itself on walls, rail tracks, bare verges and tips. 209 + [2]. First record: Weymouth (67, 1870, SMP in **DOR**). Occasionally the flowers are pink, as at Crossways tip (78P) and Blandford (80Y); this is Cv. Canon Went.

Linaria purpurea × repens = *L.* × *dominii* Druce
On the rail track at New Barn (07E, 1999, !). 1.

Linaria maroccana Hook. f.
ANNUAL TOADFLAX
A casual pavement weed at Wareham (98I, 1998, !), Greenland Farm 08C, 1999, EAP) and on Canford Heath tip (09I, 1999, !). 3.

Linaria repens (L.) Miller
PALE TOADFLAX
A denizen, associated with rail tracks near Weymouth and Swanage, scattered elsewhere in artificial habitats.

12 + [6]. First record: Langton Matravers (97Z, 1895, LVL). Netherbury (49U, 1958, AWG in **BDK**); Portland Bill (66U, 1979, CE Richards); Castle Cove (67T, 1986, DAP); Rodwell (67U, 1991, JV Carrington, Portland (67W, 1989, !); Caundle Marsh (61W, 1997, JAG); Tyneham (88V, 1937, DM); Kimmeridge (97E, 1937, DM); between Worgret and Swanage (97Z, 98D, R and V, and 07J, 1912, AWG in **BDK**, *et al.*, and 1986, J Hicks); Harmans Cross (98Q and V, 1999, !); Upton Heath railway (99X, 1975, VJG); New Barn (07E, 1999, !); Ballard Down (08G, 1950, WAC); Cranborne (01M, 1993, GDF and JO).

[**Linaria supina** (L.) Chaz.
PROSTRATE TOADFLAX
A casual from ballast at Poole (09, 1866, JB Syme).]

(**Linaria arenaria** DC.
Reports of this from Bloxworth (89, 1934–38, BNSS) are taken to be errors. It is an alien which grows in sand dunes at Braunton Burrows in North Devon.)

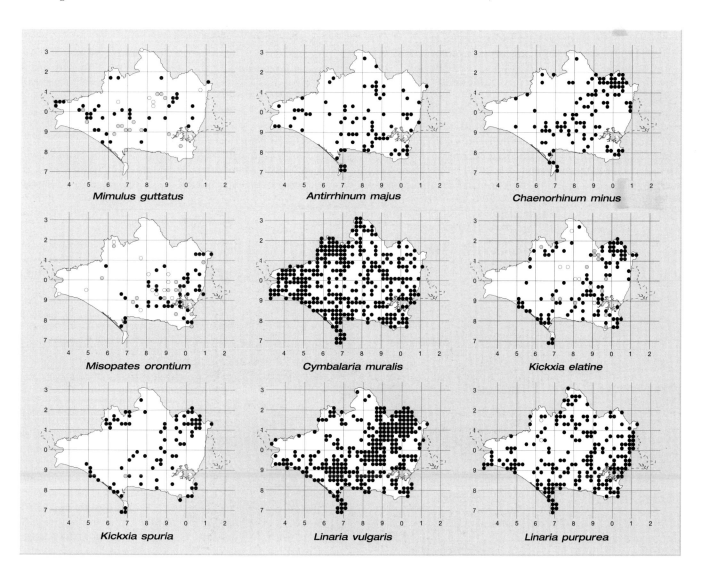

Mimulus guttatus

Antirrhinum majus

Chaenorhinum minus

Misopates orontium

Cymbalaria muralis

Kickxia elatine

Kickxia spuria

Linaria vulgaris

Linaria purpurea

Linaria triornithophora (L.) Willd.
This occurred spontaneously in a wild garden at Canford Cliffs (08P, 1996, JK Cooper in **RNG**). 1.

Digitalis purpurea L.
FOXGLOVE, DEADMAN'S BELLS
Locally abundant in cleared woodland, old hedgebanks and sheltered grassland, or a garden escape in waste ground. 465. It prefers well-drained, acid soils, including leached soils over Chalk. White-flowered plants are rare and probably garden escapes at Upton (99W and X); a peloric form was seen near Dorchester in 1999. Commonest in the west and in the Poole basin, and rare or absent from deforested areas, the coastal belt and much of the north. A representation was carved on corbels inside the church at Milton-on-Stour in the 19th century. The plant contains the toxic digitalin, used as a heart tonic.

Erinus alpinus L.
FAIRY FOXGLOVE
A rare alien established on a few old walls in or by gardens. 5 + [2]. Sherborne Abbey (61I, 1970, ACL); Kingston Maurward Ho (79A, !); Silton (72Z, RM Veall & !); Winfith Newburgh (88C, !); Spetisbury (90B, !); Merley (09E, 1979, MB); Stapehill (00K, !). Flowers pink or white.

Veronica serpyllifolia L.
THYME-LEAVED SPEEDWELL
Common in bare ground of woodland rides, heath tracks, verges, lawns, gardens and urban sites. 500. Variable in size; small forms accompany ephemerals on heath tracks, while a large form occurs in *Carex paniculata* swamps in Melbury Park (50S, !). Forms with white flowers have been seen at Blacknoll (88D), Hambledon Hill (81L) and north of Arne (98U).

Veronica austriaca subsp. *teucrium* (L.) D. Webb
LARGE SPEEDWELL
Once found as a garden escape on a verge at Cranborne (01L, 1982, GG Garbett). 1.

Veronica officinalis L.
HEATH SPEEDWELL
Frequent in dry places in woodland clearings, disturbed scrub or the edges of heaths, once seen on a wall. 236. Commonest on acid soils in the west and the Poole basin, also on leached soils over Chalk. Absent from deforested areas and Portland. The flower-colour varies from pale to mid-blue, as in abandoned bulb fields on Brownsea Is (08I), or white at Ballard Down (08F, EAP) and Poole Crematorium (09D).

Veronica chamaedrys L.
GERMANDER SPEEDWELL, BIRD'S-EYES
Common in woods, sheltered hedgebanks and mature grassland, forming colonies that are often galled. 707. A form with pale blue flowers occurred at East Down (80K), and forms with white flowers at Winfrith Heath (88D) and New Swanage (08F, EAP).

Veronica montana L.
WOOD SPEEDWELL
Frequent in old woods, both dry and wet. 354. Commonest in the north and west and absent from deforested areas, including Portland, and from very acid soils.

Veronica scutellata L.
MARSH SPEEDWELL
Scarce in marshes, either on nutrient-rich mud or in acid conditions. 28 + [23]. All recent records are from the Poole basin, or near Powerstock and Lewesdon. Good thought it frequent in the 1930s, but it has declined everywhere, either through drainage or from loss of ponds.

Veronica beccabunga L.
BROOKLIME
Frequent in wet, muddy places in woods, or by rivers, streams and ponds. 374. Absent from very acid pools and the drier parts of the Chalk.

Veronica anagallis-aquatica L.
BLUE WATER-SPEEDWELL
Occasional to frequent in clear streams, ditches and cress beds, sometimes in mud on tips. 136 + [8]. Found in all the main river basins, but scarce in the west. Var. *anagallidiformis* Bor., with a glandular raceme, has a few old records and one recent one from Ferncroft Farm (98D, 1930s, LBH and 1987, MHL).

Veronica anagallis-aquatica × *catenata*
= *V.* × *lackschewitzii* J. Keller
An uncommon hybrid, often very tall, found by streams and ditches, but not seen in the north or west. 10. First record: Ridge (98I, 1933, ECW in **RNG**); recently noted in 68P, 69R and V, 79Q, 71I, 88Y, 89L and Q, 80H, 81S, 90V and 01A.

Veronica catenata Pennell
PINK WATER-SPEEDWELL
Occasional by rivers, ditches and ponds or in marshes. 35 + [11]. Found in all the larger river basins, absent from the west.

Veronica acinifolia L.
FRENCH SPEEDWELL
A rare, alien weed in gardens. [1]. Beaminster (40V, 1937–76, AWG, !). It occurs in nursery gardens at Merriott, VC6 (41L, 1993, RSRF, !) from which it may spread.

Veronica arvensis L.
WALL SPEEDWELL
A common weed of bare, dry soil, in arable, gardens, quarries, wall-tops, pavements and waste ground. 579.

Veronica peregrina L.
AMERICAN SPEEDWELL
A rare, alien weed of gardens. 2 + [1]. Beaminster (40V, 1937–76, AWG in **BDK**, !); Frome St Quintin (50W, 1986, DAP); Cranborne garden centre (01L, 1994, GDF); likely to spread.

Veronica agrestis L.
GREEN FIELD-SPEEDWELL
Rare in arable, occasional in old gardens or building-sites and tips. 30 + [10]. JCMP found it 'not common', and Good had it from 26 stands, but lumped it with *V. polita*. The only records from arable are from 58E, 78Z, 79N, 80D and I.

Veronica polita Fries
GREY FIELD-SPEEDWELL
Uncommon in arable, much more frequent in old gardens

or disturbed verges on calcareous or sandy soils. 133 + [14]. JCMP found it common, Good did not distinguish it from *V. agrestis*.

Veronica persica Poiret
COMMON FIELD-SPEEDWELL
An introduced weed now abundant in arable, gardens and disturbed soil on all but the most acid soils. 675. First record: Weymouth (67, 1866, SMP in **DOR**). A form with white flowers occurred at Winterborne Kingston (89N, !).

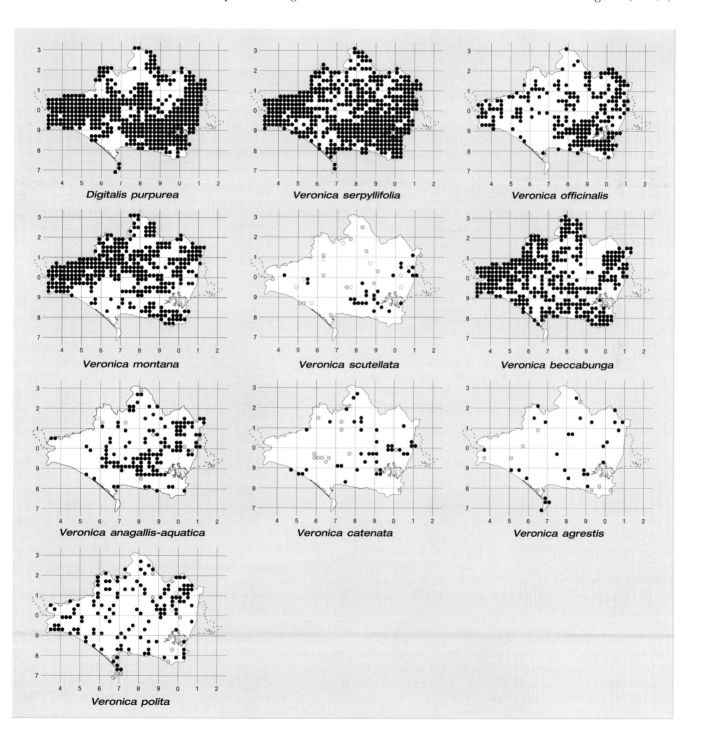

Veronica filiformis Smith
SLENDER SPEEDWELL
A locally abundant introduction in churchyard and other lawns. 203 + [13]. Rare in other habitats such as wood-pasture, riverbanks and old rail tracks. Unrecorded by both JCMP and RDG in their *Floras*; first record: Powerstock (59D, 1948, CD Chase), since when it has spread far and fast.

Veronica hederifolia L.
IVY-LEAVED SPEEDWELL
Common on bare soil in woods, old hedgebanks, arable, gardens and waste ground on all but the most acid soils; remains of seeds of neolithic age have been found. 531. It germinates in late winter and matures fast. Only 40% of recorders distinguished subsp. *hederifolia* (110 tetrads) from subsp. *lucorum* (Klett. & Richt.) Hartl. (101 tetrads); the distinction between them is not clear-cut, nor are there any obvious ecological differences.

Veronica longifolia L.
GARDEN SPEEDWELL
A transient escape on verges or railway banks. 2 + [5]. Batcombe (60G, 1984, M Collier); Black Head (78F, 1948, !); Kington Magna (72R, 1891, FWG); Winterborne Stickland (80H, 1999, !); Wareham (98I, 1953–55, !); Keysworth (98J, 1978, !); Beacon Hill (99S, 1949, WAC). On sale in Blandford in 1782; often reported as *V. spicata* or *V. spuria*.

Veronica gentianoides Vahl is planted on a wall at Buckhorn Weston Ch (72M, !). 1.

Hebe salicifolia (G. Forster f.) Pennell
KOROMIKO
Rarely planted in plantations, verges or churchyards. 8. Seedlings found in Abbotsbury gardens (58S, SG) and Dorchester pavements (69V, !); bushes reported from 49W, 78Q, 89E, 99Y, 08F and 09K.

Hebe brachysiphon Summerh.
Seedlings on a rail track at Winfrith (88D, 1992, DAP & ! in **RNG**), and on Wareham pavements (98I, !). 2.

Hebe dieffenbachii (Benth.) Cockayne & Allan
Self-sown on walls or pavements at Charlestown(67P, !), Wareham (98I, 1991, !) and Canford Heath (09C, !). 3.

Hebe barkeri (Cockayne) Wall.
Self-sown at Weymouth Sta (67U, 1995, ! in **RNG**). 1.

Hebe cupressoides Hooker f.
In woodland, Powerstock Common (59I, 1969, R Clarke in **RNG**). [1].

Hebe × franciscana (Eastw.) Souster
A somewhat frost-tender shrub planted as a hedge along the coast, rarely naturalised. 5. First record: Abbotsbury (58S, 1899, MI). Self-sown pavement weed, Lyme Regis (39F, 1999, !); Eype Ch, (49K, 1998, !); in plenty, limestone undercliff, Church Ope Cove (67V, 1961–98, !);

single bushes on cliffs at Weymouth (68V) and Swanage (08F). On sandy cliffs at Bournemouth (09V), but in VC11.

Sibthorpia europaea L.
CORNISH MONEYWORT
Very rare in a spruce plantation in the north-west. 1. A recently-found Cornubian plant which occurs no further east in Britain. Hewood Bottom (30R, 1992, BE, *et al.* in **CGE**). Planted in Abbotsbury gardens in 1899 (MI).

[*Melampyrum arvense* L.
FIELD COW-WHEAT
Reported from Bere field (89, 1799, RP); specimens found in 18th Century thatch at White Mill Barn (90K).]

Melampyrum pratense L.
COMMON COW-WHEAT
Locally frequent in old woods, mostly oak woods. 87 + [23]. Most frequent on acid soils in the west and the Poole basin, but also in woods on chalk, such as Garston Wood (00E), or clay, such as Clifton Wood (51R). Subsp. *commutatum* (Tausch ex A. Kerner) C. Britton is a broad-leaved form of calcareous woods, seen recently in Westley Wood (90F, 1900, EFL and 1994, !), with old records from 99C, 90S and Y and 07J. Any decline in this species could be due to the replacement of oaks by conifers.

Euphrasia anglica Pugsley
Uncommon in heathland or acid grassland, in the west and the Poole basin. 4 + [11]. Some colonies in the west may have hybridised with *E. nemorosa* (PFY). Thistle Hill (39M, 1935, RDG in **WRHM**); Cains Folly (39W, 1967, PFY); Beaminster (40V, 1993, LJM); Crook Hill (40Y, 1935, RDG in **WRHM**); Mapperton (59E, 1994, LJM); Askerswell Down (59J, 1958, B Welch); Kingcombe (59P, 1989, !); Corscombe (50C, 1935, RDG in **WRHM**); Toller Down (50G, 1935, RDG in **WRHM**); Rag Copse (50T, 1936, RDG in **WRHM**); Winfrith Heath (88D, 1936, RDG and 1953, !); Gore Heath (99F, 1942, NDS); Little Sea (08H, 1935, PMH); Sutton Holms (01K, 1951, PFY); Verwood Common (00Z, 1934, RDG in **WRHM**).

Euphrasia arctica Lange ex Rostrup subsp. *borealis* (F. Towns.) Yeo
A tall plant, local in old grassland, flushed heaths or eutrophic verges and cliff-tops. 12 + [9]. Restricted now to the south-east, but with two old records from the west. Specimens are in **BM, CGE, DOR** and **OXF.**

(*Euphrasia arctica × tetraquetra = E. × pratiuscula* F. Towns. was reported from Portland by RDG, but the specimen could not be named with confidence by PFY.)

Euphrasia arctica × nemorosa
Rare. 1 + [2]. Corfe Common (98Q, 1928, HWP, also 1980, AJS); probably this on Studland Heath (08H, 1999, !); Swanage (08F, 1928, HWP); Cranborne (01L, 1935, RDG; no specimen seen).

Euphrasia tetraquetra (Breb.) Arronde
Locally frequent in short, calcareous turf. 34 + [5]. Mainly in the coastal belt between Portland and Ballard Head, seen inland only at Eggardon Hill (59M, LJM and LMS), Hambledon Hill (81L, 1948, E Nelmes in **K**); Badbury Rings (90R, 1953, PFY and 1979–94, !) and Martin Down (01P, 1986, PET). Specimens are in **BDK, BM, K, OXF, RNG** and **WRHM**.

Euphrasia tetraquetra × *pseudokerneri*
Very rare. Badbury Rings (90R, 1953, PFY). [1].

Euphrasia nemorosa (Pers.) Wallr.
EYEBRIGHT
Much the commonest eyebright here, mostly in short calcareous turf, but sometimes in grazed heaths or eutrophic heathy verges. 234 + [8]. Frequent on the Chalk and in most of the Poole basin, absent from Portland, and scarce in the north and west.

Euphrasia nemorosa × *confusa*
Rare. 1 + [1]. Hunters Bridge (61Q, 1935, RDG det. PFY); Corfe Common (98L, 1980, AJS).

Euphrasia nemorosa × *micrantha*
Probably this on Bloxworth Heath (89V, !). It occurs in Avon forest park (10G, 1990, RP Bowman), but in VC11.

Euphrasia pseudokerneri Pugsley
Very local in short calcareous turf. 6 + [2]. The flowers are large though the plants are small. Rawlsbury Camp (70S, 1994, !); Hod Hill (81K, 1982, AJCB and 1989, !); Hambledon Hill (81L, 1999, RSRF); Melbury Hill (81U, 1989, !); Zigzag Hill (82V, 1969, !); Cranborne Chase (82?, 1923, JC Melville in **OXF**); Badbury Rings (90R, 1953, PFY); Round Down (07I, 1910, GCD in **OXF** and 1981, AJCB).

Euphrasia confusa Pugsley
Occasional in short, calcareous or heathy turf. 18 + [15]. A small and easily overlooked taxon. Specimens are in **BM, DOR, OXF, RNG** and **WRHM**.

(*Euphrasia ostenfeldii* (Pugsley) Yeo
Early reports (as *E. curta*) were all hybrids, det. PFY.)

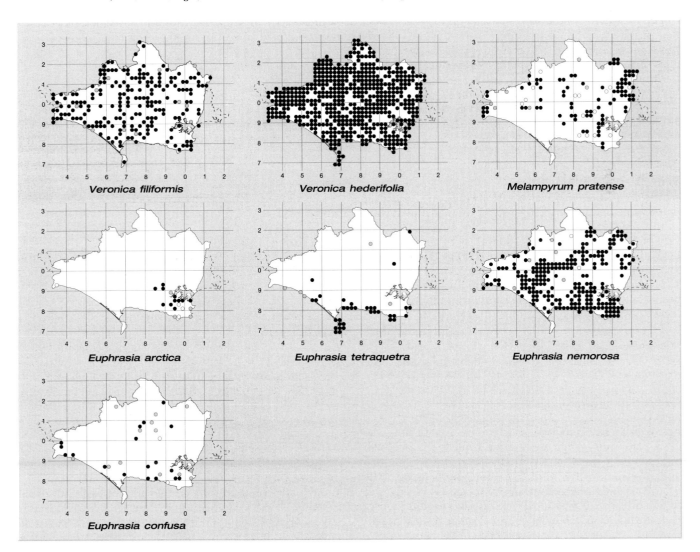

Veronica filiformis · Veronica hederifolia · Melampyrum pratense · Euphrasia arctica · Euphrasia tetraquetra · Euphrasia nemorosa · Euphrasia confusa

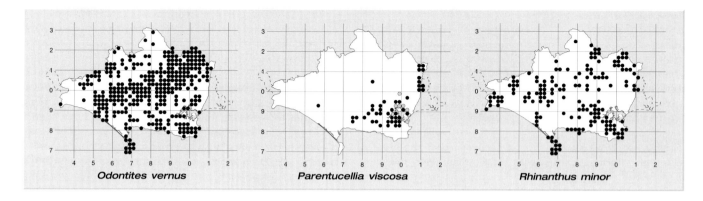

Odontites vernus   Parentucellia viscosa   Rhinanthus minor

Euphrasia micrantha Reichb.
A tall, spindly plant of grazed or trampled heaths. 3 + [4]. It is declining by hybridisation. Bloxworth Heath (89V, 1990–94, ! in **RNG**); Arne triangle (98N, 1913, CBG in **BMH**, *et al.*, and 1988, !); Wytch Heath (98S, 1910, CES); Slepe Heath (98T, 1913, CBG in **BMH** and 1960, !); Holton Heath (99K, 1979, ! in **RNG**); Gotham (01V, 1991, !).

Odontites vernus (Bellardi) Dumort
RED BARTSIA
Frequent in woodland rides, by grassy or heathy tracks, sometimes in arable. 325. Commonest on calcareous soils, local on sandy soils in the Poole basin and scarce on clay. Variable in stature and habit, occasionally becoming woody at base. Subsp. *vernus* and subsp. *serotinus* (Syme) Corbiere both occur, but were rarely recorded; they do not differ obviously in their ecology.

Parentucellia viscosa (L.) Caruel
YELLOW BARTSIA
Locally frequent in damp meadows and pastures, disturbed margins of heaths or verges. 46 + [7]. Once rare, now widespread in the Poole basin, with outliers at West Compton (59R, 1998, JAG; 1 plant) and Quarleston Ho (80M, !, on imported soil). Reports from Small Mouth and Portland need confirmation. First record: Poole (09, 1830, TBS). JCMP gave two sites in 1874 and four in 1895; the species is spreading from its original locality, and could have been introduced.

[*Rhinanthus angustifolius* C. Gmelin
GREATER YELLOW-RATTLE
Reported from Edmondsham (01Q, 1908, EFL).]

Rhinanthus minor L.
YELLOW-RATTLE
Locally frequent in mature grassland on sandy, alluvial or calcareous soils, either wet or dry. 166. It has declined because this type of grassland has mostly been ploughed and reseeded. Variable. Subsp. *stenophyllus* (Schur.) O. Schwarz, a late-flowering form from wet meadows, was once found at Ulwell (08F, 1908, EFL). Subsp. *calcareus* (Wilm.) E. Warb., with narrow leaves, is occasional in dry calcareous grassland on Portland and at Bere Regis (89M, 1897, JCMP);

Hambledon Hill (81L, !); Melbury Hill (81U, 1891, JCMP); Zigzag Hill (82V, !); Bokerley Dyke (01J, 1953–90, !).

Pedicularis palustris L.
MARSH LOUSEWORT
Very local, and in small numbers, in clearings in fens, marshy meadows or flushed bogs. 32 + [28]. Restricted to the west and the Poole basin, no longer common (JCMP, 1895) or frequent (RDG, 1930s). Its decline is due partly to drainage, and partly to lack of grazing, when the plants become smothered by coarse vegetation.

Pedicularis sylvatica L.
LOUSEWORT
Frequent on wet heaths or acid pastures, especially along tracks. 121 + [10]. Commonest in the west and the Poole basin. It occurs in sheltered chalk turf at South Field Down (69I, 1999, JHSC) and above Houghton (80B and C, 1968, TCEW and 1991, !).

Alonsoa cauliculata Ruiz Lopez & Pavon
A casual in a garden at Netherbury (49U, 1975, KN Davis in **RNG**). [1].

Orthocarpus pusillus Benth.
An American alien, on dredgings from a pond made in 1982. Talbot Heath (09R, 1994, Mrs Evelyn, also FAW, ! in **RNG**). 1.

Paulownia tomentosa (Thunb.) Steudel
FOXGLOVE-TREE
Rarely grown, and sometimes forming a tree outside gardens, as near a barn at Winterborne Clenston (80G, 1990–98, !; 9 m × 1 m). This tree blew down in 1998, but root suckers survive.

OROBANCHACEAE

Lathraea squamaria L.
TOOTHWORT
Local in old woods, mostly on the roots of hazel, rarely on other hosts such as blackthorn or plum. 41 + [15]. It prefers steep slopes on calcareous soils. Commonest in Cranborne Chase and the north-east, absent from deforested areas.

Lathraea clandestina L.
PURPLE TOOTHWORT

Planted and established in a few wild gardens on willow roots. 6. Forde Abbey (30M, 1992, S Kurzen); Melbury Park (50T, 1984, AH); Minterne Ho (60R and S, 1961, HE Beasley, *et al.* in **BDK**, and 1996, IPG, !); Lower Ansty (70R, 1986–97, A Stevens, !); Broadstone (09D, 1968, MPY, *et al.*, and 1985, A Grenfell).

Orobanche purpurea Jacq.
YARROW BROOMRAPE

Very rare and sporadic. 1. Small Mouth, east of the road (67S, 1937, DM and 1992, J Pyett, also 1997, DAP, !). It does not occur further north in Britain.

Orobanche rapum-genistae Thuill.
GREATER BROOMRAPE

A parasite on *Ulex* or *Cytisus*, now very rare. 1 + [22]. JCMP gave 12 localities, mostly in the south, which have nearly all been lost for no clear reason. Post-1929 records: Champernhayes (39N, 1995, MJG); Newlands Batch (39W, 1980, JN); Golden Cap (49B, 1874, E Fox, *et al.*, and 1981, JN, !); Upwey (68T, 1936, MJA); Ilsington (79L, 1929, CD Day in **DOR**); Studland (08A, 1895, AE Eaton, *et al.* in **BMH**, and 1948, !); Poole Civic Centre (09F, 1945, EN). Specimens are in **BDK, BMH, CGE, DOR, K** and **RNG.**

Orobanche elatior Sutton
KNAPWEED BROOMRAPE

Locally frequent as a parasite on *Centaurea scabiosa*, or rarely *Carduus nutans*, in tall, chalk grassland, often on ancient earthworks or verges. 27 + [11]. Commonest in the north-east. It does not occur further west in Britain.

Orobanche hederae Duby
IVY BROOMRAPE

Locally frequent on ivy, mostly in the coastal belt, and especially common on Portland. 16 + [3]. A yellow form occurred at The Verne (67W, BE).

(Orobanche artemisiae-campestris Vaucher ex Gaudin

A parasite of *Picris hieracioides*, reported from Portland (67, 1959, G Williams), but neither this nor any other record has been substantiated. A specimen from Bradford Abbas (51X, 1891, RDS Stephens in **DOR**) is probably *O. minor*. It occurs on the Is of Wight.)

Orobanche minor Smith
COMMON BROOMRAPE

An occasional, sporadic parasite of clover or *Lotus* in fields and verges, or on *Brachyglottis* 'Sunshine' in gardens. 66 + [3]. Var. *maritima* (Pugsley) Rumsey & Jury is locally abundant on *Daucus* on cliffs along the coast, and was once found on *Eryngium maritimum*. 5. FAW noted that the plant is eaten by rabbits. Specimens are in **BDK, BM, CGE, K, MANCH, NMW** and **RNG.**

BIGNONIACEAE

Catalpa bignonioides Walter
INDIAN-BEAN-TREE

An ornamental tree, rarely planted as single specimens in

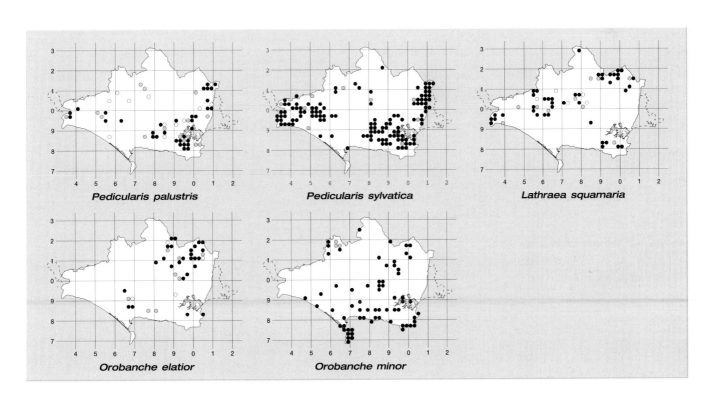

Pedicularis palustris

Pedicularis sylvatica

Lathraea squamaria

Orobanche elatior

Orobanche minor

parks. 2. Lower Lodge plantation (70V, !); Bryanston (80T, 1944, BNHS); Charborough Ho (99I, !).

ACANTHACEAE

Acanthus mollis L.
BEAR'S-BREECH
A persistent garden escape on cliffs, waste ground or verges near houses. 6 + [3]. West of Hardown Hill (49C, !); Eype Mouth (49K, !); Abbotsbury ruins (58M and S, 1960–78, !); Corscombe (50H, LJM); Fleet (68F, 1910, WBB); Swanage Cliffs (08F, EAP); Broadstone (09C, 1962–79, H Thomas); Wallis Down (09L, !); Cranborne (01R, GDF).

Acanthus spinosus L.
SPINY BEAR'S-BREECH
A persistent garden escape at and below Abbotsbury ruins (58M and S, 1960–98, !). 2.

LENTIBULARIACEAE

Pinguicula lusitanica L.
PALE BUTTERWORT
An inconspicuous plant of sheltered crannies in *Sphagnum* bogs. 47 + [13]. Rare in the west, mostly in the Poole basin. First record: 1762, W Hudson.

Pinguicula vulgaris L.
COMMON BUTTERWORT
Thought to be extinct in peat bogs until recently. 1 + [3]. Tadnoll (78Y, 1995–99, BE, !); between Arne and Stoborough (98N, 1865, JCMP in **DOR**); Poole (09, 1799, RP); Gotham (01V, 1909, EFL). A northern species at the edge of its range here; very rare in VC11.

(*Utricularia vulgaris* L.
GREATER BLADDERWORT
Because of confusion with *U. australis*, there is doubt whether this has ever been found in VC9, just as in VC11, but the plant occurs on the Somerset levels.)

Utricularia australis R. Br.
BLADDERWORT
Scarce in acid ponds, ditches and brackish reed-beds near the sea, flowering sporadically in response to eutrophication. 7 + [6]. Only in the Poole basin, where it has often been called *U. vulgaris*, from which it cannot be distinguished when sterile. Wool (88N, 1893, WRL in **LIV**); Wareham Common (98D, 1929, FHH and 1983, MHL); Stoborough (98I, 1895, JCMP); north of Bestwall Wood (98J, 1893, JCMP in **DOR**, *et al.*, and 1993, AH); Hartland Moor (98M and S, 1978, RB); The Moors (98N, 1937, RDG in **WRHM** and 1990, DAP); Morden Park Lake (99B, 1893, JCMP in **DOR** and 1936, RDG in **WRHM**); East Holton (99Q, 1937, RDG); Little Sea (08H, 1908, AWG in **BDK**, *et al.* in **RNG**, and 1991, AJB); South Haven (08I, 1978, SRD and 1992, !);

Brownsea Is (08J, 1997, KC); St Leonards Peats (10A, 1984, RMW), in VC11.

Utricularia intermedia Hayne
INTERMEDIATE BLADDERWORT
Very local in bog pools or streams, or among *Sphagnum* in old claypits. 4 + [9]. Only in the Poole basin, perhaps declining. Knighton Heath (88D, 1895, RPM); Furzebrook (98G, 1989–96, CDP, !); North of Wareham (98J, 1936, ECW in **RNG**); Hartland Moor (98M, 1934, HS Redgrave, *et al.*, and 1997, !); The Moors (98N, 1966, ! and 1981, AJB); Keysworth (98P, 1925, JHS); Corfe (98R, 1960, !); Slepe and Scotland Heath (98S, 1895, W Borrer, *et al.*, and 1980, AH); Bushey (98W, 1900, LVL); Morden Decoy (99A, 1874, JCMP and 1938, PMH); Lytchett Minster (99R, 1900, EFL); Little Sea (08H, 1955, !); Source of R. Bourne (09M, 1900, EFL).

Utricularia minor L.
LEAST BLADDERWORT
Local in bog pools, ditches and streams, flowering sporadically. 21 + [9]. Only in the Poole basin.

CAMPANULACEAE

[*Campanula patula* L.
SPREADING BELLFLOWER
Woods at Corfe Mullen (99, 1799, RP); Merley (09E, 1799, H Parker). No specimens seen; it appears sporadically in Hampshire.]

[*Campanula rapunculus* L.
RAMPION BELLFLOWER
Perhaps an escape. Corfe Mullen (99T, 1893, RPM); a report from Abbotsbury (58, 1937, G Cole) is doubtful.]

Campanula lactiflora M. Bieb.
MILKY BELLFLOWER
Planted, not naturalised, in the wild garden at Melbury Park (50T, 1971, !). [1].

Campanula persicifolia L.
PEACH-LEAVED BELLFLOWER
An occasional garden escape which seeds itself on calcareous or sandy soils, in waste places or on verges. 13 + [1]. Increasing. First record: Abbotsbury (58S, 1937, DM). Also seen in 39Z, 59K, 69A, 78Z, 79U, 88E (AJCB), 89M, 80T and Y, 82Q, 99F, 09I and 00T (GDF).

Campanula glomerata L.
CLUSTERED BELLFLOWER
Occasional in short, dry, calcareous turf. 109 + [7]. Widespread on the Chalk, scarce on limestone and absent from sandy or clay soils, except as a garden escape in 69V and 89N. It does not occur further west in Britain.

Campanula pyramidalis L.
CHIMNEY BELLFLOWER
Once seen as a casual garden escape on a wall at Wrackleford (69R, 1968, !). [1].

Campanula portenschlagiana

Campanula portenschlagiana Schultes
ADRIA BELLFLOWER

Frequent as a garden escape on brick or stone walls and pavements, but not seen in semi-natural habitats. 52. First noted 1964, but cultivated in 1899 (MI).

Campanula poscharskyana

Campanula poscharskyana Degen
TRAILING BELLFLOWER

Common as an escape in crevices of walls or pavements, less often on verges where it produces seedlings. 95. Rare away from houses, on banks or by sheltered streams, at Bowood (49P), Askerswell (59G), Bowden Hill (61J) and Broadstone Golf Course (09D). First record: 1957; spreading rapidly. Seen with white flowers at Upwey (68S).

Campanula latifolia L.
GIANT BELLFLOWER

A northern plant, established in a few woods, verges and wild gardens. 5 + [2]. Winyards Gap (40X, 1921, AWG in **BDK**, *et al.* in **WRHM**, and 1982, SPM Roberts); near Evershot (50R, 1886, WMR in **DOR**, and 50S, 1983, Mr Doyle); wild garden, Melbury Park (50T, 1937, DM and 1971–89, !); Bubb Down (50Y, 1936, RDG in **WRHM**); Abbotsbury gardens (58S, 1989, SG, !); Stubhampton Bottom (81X, 1937, DM and 1986, N Trebble); Durlston Bay (07I, 1925, W Whitwell).

Campanula trachelium L.
NETTLE-LEAVED BELLFLOWER

Local in old woods or nearby hedgebanks, rarely on sheltered calcareous banks in the open. 62 + [4]. Mainly on the Chalk. On the ironstone ridge at Abbotsbury (58S) it could be a garden escape, as it probably is at Dorchester (69V; with white flowers), Red Bridge pits (78Z) and White Lackington (79E; with white flowers). It does not occur further west in Britain.

Campanula rapunculoides L.
CREEPING BELLFLOWER

A persistent garden escape in shrubberies, verges and urban sites. 2 + [16]. Post-1960 records: Labour-in-vain Farm (58N, 1972, W Salmon in **RNG**); Abbotsbury (58S, 1978, RHD); Moreton (78Z, 1953, AK Harding); Sutton Waldron (81S, 1904, EFL and 1960, CS); Home Farm (91B, 1997, !); Wimborne (09E and 00A, 1981, C Tandy and 1983, MB); Poole (09A, 1977, RB); Cranborne (01L, 1967–79, WEA Evans).

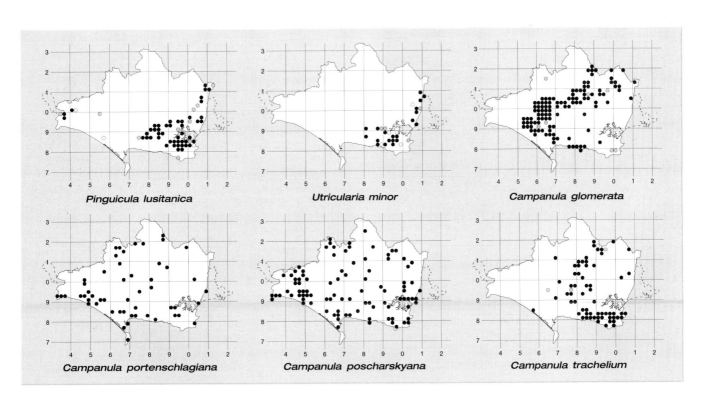

Pinguicula lusitanica Utricularia minor Campanula glomerata

Campanula portenschlagiana Campanula poscharskyana Campanula trachelium

Campanula rotundifolia L.
HAREBELL
Frequent in old, calcareous turf, less common in sandy turf with nutrient-poor soil. 206. Mainly on the Chalk, with scattered records in the Poole basin.

Campanula carpatica Jacq.
TUSSOCK BELLFLOWER
A rare pavement weed at Wareham (98I, 1998, !). 1.

Legousia hybrida (L.) Delarbre
VENUS'S-LOOKING-GLASS
Occasional in calcareous arable, rare in peaty arable or as a casual on waste ground. 36 + [30]. JCMP thought it frequent and gave 28 localities, Good also found it frequent and had it in 32 stands, so it may be declining.

Wahlenbergia hederacea (L.) Reichb.
IVY-LEAVED BELLFLOWER
Very local, decreasing and approaching extinction in open woodland or sheltered turf on acid soil. 2 + [7]. Mostly in the Poole basin. Powerstock Common (50, 1895, T Wainwright); Carey (88Z, 1949, H Bury); Bere Wood (89S, 1889, RPM in **DOR**); garden lawn, Broadstone (09C, 1976, RDG and 1991, AH); West Parley (09Y, 1867, JCMP in **DOR**); Mannington (00S, 1934, RDG in **DOR** and **WRHM**); West Moors (00R or S, 1908, EFL); Woolsbridge (00X + 10C, 1900, JCMP, *et al.*, and 1938, RDG in **DOR**); Crab Orchard (00Y, 1900, EFL, *et al.* in **DOR** and **WRHM**, and 1995, AJB and DAP).

Phyteuma spicatum L.
SPIKED RAMPION
Alien here, found in a garden at Houghton (80H, 1967–79, T Norman). [1]

Phyteuma orbiculare L.
ROUND-HEADED RAMPION
Very rare in calcareous turf. [1]. Gains Cross (81K, 1970, KEB; not refound); (Tinckley Down (91P, 1874, S Pitt), and Martin Down (01P, 1995, J Greaves) are both in VC8). Locally frequent in VC7, North Wiltshire.

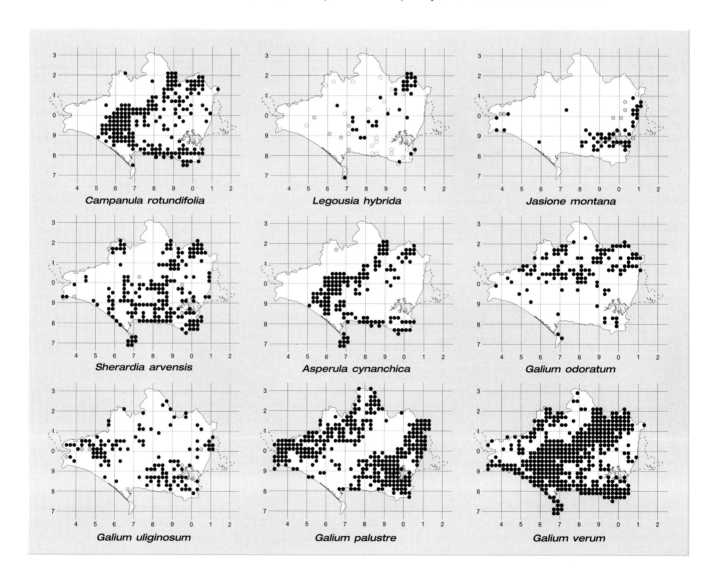

Campanula rotundifolia   Legousia hybrida   Jasione montana

Sherardia arvensis   Asperula cynanchica   Galium odoratum

Galium uliginosum   Galium palustre   Galium verum

Jasione montana L.
SHEEP'S-BIT
Local on dry, sunny banks, including railway banks, edges of heaths and on dunes. 49 + [10]. Almost restricted to acid soils of the west and the Poole basin, with outliers as at the Valley of Stones (58Y, 1980, D Moxon) and Plush (70G, 1991, DAP). Reports from Loscombe (59E) and Cowherd Shute Farm (82R) need confirmation. A form with white flowers was seen at Studland NNR (08H, EAP) and one with pink flowers at Upton CP (99W, !).

Lobelia urens L.
HEATH LOBELIA
This rare plant of damp, acid woodland rides has disappeared from three of its four Dorset sites but is locally frequent in the fourth. 1 + [3]. Hurst Heath (78Z, 1956, AK Harding, *et al.*, & 1991, !); Puddletown (79M, 1956, JG Lawn); Cockett Hill (99H, 1937, RDG in **WRHM** and 1942, BNHS); Lytchett Heath (99S, 1920, HC Hawley and 1937, RDG). It is somewhat sporadic, and could reappear after disturbance. There are old records from Axminster Heath (39, 1895, EG Baker in **DOR** and 1921, AWG) but in VC3. Dorset specimens are in **DOR, LANC, LIV** and **WRHM**. The plant contains the alkaloid lobeline which causes vomiting.

Lobelia erinus L.
GARDEN LOBELIA
Once seen as a casual in coppice at Milborne Wood (79Y, 1992, !); self-sown as a pavement weed in towns, or on tips. 30. Noted from 39F, L and R, 40Q, 58J, 67U, 68S, V and X, 69V, 79Y, 71X, 72Y, 88Z, 89D, 80Y, 82R, 98C, I and R, 99W, 07E and J, 08C, F and G, 09C, I, L and X. Flowers usually royal blue, less often white, pale blue or pink.

Pratia arenaria Hooker f.
A rare lawn weed, once established in the Avenue, Branksome (09K, 1955, D McClintock in **RNG**).

RUBIACEAE

Sherardia arvensis L.
FIELD MADDER
A frequent weed in dry arable, also in disturbed chalk turf, quarries, sometimes on sandy verges or in suburbia, known since the Iron Age. 191 + [6]. Mainly on calcareous soils and absent from heavy clays. Seen with white flowers at Ballard Down (08F, EAP).

Phuopsis stylosa (Trin.) Benth. & J.D. Hooker ex B.D. Jackson
CAUCASIAN CROSSWORT
An occasional, persistent garden escape with a foxy scent. 6 + [3]. Abbotsbury (58S, 1964, RW Butcher); Powerstock Common railway (59N, 1922, AWG in **BDK**) ; Corscombe (50C, 1993, LJM); Rampisham (50R, 1981, AJCB); Grove cliffs (67W, 1960, EH in **LANC**); Sherborne Castle (61N, 1972–80, !); Tolpuddle (79X, 1995, !); Melcombe Horsey (70L, 1994, !); garden verge, Durweston (80P, 1999, !).

Asperula cynanchica L.
SQUINANCYWORT
Frequent in old, short, calcareous turf. 155 + [1]. Specimens from Portland (66U, 67W, 77A, ! in **RNG**) and Ballard Down (08, 1938, F. Ballard in **NMW**) have orange roots and are subsp. *occidentalis* (Rouy) Stace, which is a taxon of dunes in west Britain, *fide* TCGR.

[*Asperula arvensis* L.
BLUE WOODRUFF
A rare casual once seen at Beaminster (40Q, 1934, AWG).]

[*Asperula tinctoria* L.
DYER'S WOODRUFF
Once planted in Abbotsbury gardens (58S, 1899, MI) and reported near here in 1964 by RW Butcher. Also reported as cultivated in Purbeck early this century; no specimens seen.]

Galium odoratum (L.) Scop.
WOODRUFF
Locally frequent in calcareous woods, and an occasional garden escape in village lanes. 123. Mainly in the north, but found in sheltered undercliffs on Portland and in north-facing woods in Purbeck.

(*Galium constrictum* Chaub. occurs in grazed turf by ponds in Hants. and Wilts. and should be looked for.)

Galium uliginosum L.
FEN BEDSTRAW
Frequent in wet woodland rides, water-meadows, marshes and flushed bogs. 122 + [7]. Mostly on alluvial soils or clays in the north-west and the Poole basin, absent from the chalk and scarce in the coastal belt. It favours moderately acid soils.

Galium palustre L.
COMMON MARSH-BEDSTRAW
Frequent by rivers, streams and ditches, in water meadows and marshes. 316. Absent from most of the Chalk, and the lowlands near Weymouth. Subsp. *elongatum* (C. Presl) Arcang. is a more robust polyploid in the same habitats. 63. It is under-recorded.

Galium verum L.
LADY'S BEDSTRAW
Common in dry grassland, mainly on calcareous soils but also on eutrophic verges in heathland. 420. Found on most of the Chalk and limestone, and along the coastal belt. Absent both from heavy clays in the north-west and from very acid soils in the Poole basin. Var. *maritimum* DC., a prostrate, depauperate form, occurs on fixed shingle at Small Mouth (67S, 1908, EFL and 1997, !).

Galium verum × *mollugo* = *G.* × *pomeranicum* Retz
Rare, usually as single plants near both parents, in calcareous soils or near the coast. 6 + [8]. West Bexington (58I, 1993, LJM); North Poorton (59E, 1997, IPG); Powerstock railway (59N, 1924, AWG);

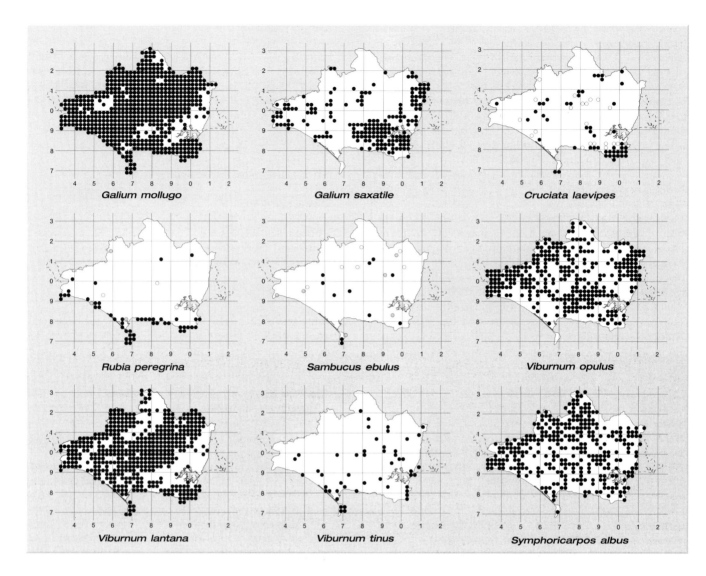

Galium mollugo   Galium saxatile   Cruciata laevipes

Rubia peregrina   Sambucus ebulus   Viburnum opulus

Viburnum lantana   Viburnum tinus   Symphoricarpos albus

Thornford (51W, 1962, SS); Ferrybridge (67T, 1930s, RDG and 1958, !); Grove Hill (68N, 1991, BE); Maiden Castle (68U, 1992, BE); Wyke Farm (61C, 1956, JL Gilbert); Westhill Lodge (61M, 1962, SS); Corfe (98R, 1912, CBG in **BMH** and 1953, CCT); Windmill Barrow (99I, 1890, RPM in **DOR**); Badbury Rings (90R, 1925, JHS and 1989, !); Durlston CP (07I, 1959, G Williams and 1990, H Murray); Ballard Down (08F, 1931–33, LBH).

Galium mollugo L.
HEDGE BEDSTRAW
Common at the edges of woods, in hedgebanks and grassland on calcareous or neutral soils. 593. Absent from very acid soils in the Poole basin. Subsp. *erectum* Syme, with a narrow panicle, is under-recorded but is probably uncommon in calcareous turf. 14.

Galium pumilum Murray
SLENDER BEDSTRAW
Scarce in short, calcareous turf. Much confused with *G.*

mollugo or *G. saxatile*. [3]. Specimens exist from Shipton Gorge (49V, 1954, WAC in **PVT**), Longthorns (89J, 1860–74, JCMP in **BM** and **DOR**) and Hod Hill (81K, 1916, AWG in **BDK**). Reported from central Dorset (60K, Q and R, 70K, 81Z, 82V and 91E, 1972, CA Jones in Moore, 1962), but not refound in any of her sites.

Galium saxatile L.
HEATH BEDSTRAW
Frequent in heaths; occasional in acid grassland, on Lower Greensand, and on anthills or leached soils overlying Chalk. 183. Widespread in the Poole basin, scattered elsewhere.

Galium aparine L.
CLEAVERS, CLADEN, GOOSEGRASS
An abundant weed of hedgebanks and nutrient-rich arable, frequent in disturbed soil and tips, less common in closed turf with nettles. 734. It germinates early in winter, and may have increased as a result of greater use of fertilizers.

[*Galium spurium* L.
FALSE CLEAVERS
A casual. 2. Wareham Sta (98E, 1902, EFL); Hamworthy junction (99V, 1900, EFL).]

[*Galium tricornutum* Dandy
CORN CLEAVERS
An annual weed of calcareous soils, rare in the 1890s and last seen in the 1970s. [21]. Recorded between 1900 and 1930 from 48Z, 40G and Q, 58I, 59E, 80S, 81Y, 97T, 01G and Q. Later records: Portland Bill (66Z, 1935, MJA); Studland (08, 1937, DM); Horton Farm (00Y, 1970s, RHD).]

Cruciata laevipes Opiz
CROSSWORT, BRIS, RICE
Local in woodland rides, calcareous scrub or grassland, rarely on eutrophic heathy verges. 49 + [21]. Its distribution pattern is unique, and includes the chalk scarp with the limestone of Portland and Purbeck. It has probably declined through destruction of habitat.

Rubia peregrina L.
WILD MADDER
Locally frequent along the coastal belt, among scrub on calcareous or clay undercliffs. 48 + [6]. Uncommon in woods or hedgebanks inland, e.g. Birdsmoor Gate (30V, 1997, IPG); North Poorton (59E, 1993, LJM and LMS); Yeovil junction (51S, 1956, IAR); Hod Wood (81K, 1640, J Parkinson, *et al.*, and 1993–99, !); Harley Wood (01B, 1984, J Edwards). In Somerset it is common inland, and its British distribution is largely Cornubian.

CAPRIFOLIACEAE

Sambucus racemosa L.
RED-BERRIED ELDER
Planted at Abbotsbury gardens (58S) and Canford School (09J, TH), but not naturalised.

Sambucus nigra L.
ELDER
Common in woodland clearings, hedges, old pits, undercliffs and waste ground. 731. Large trees seen at Ibberton (70Y; 1.65 m girth) and Challow Hill (98R). Var. *laciniata* L., with deeply-cut leaves, has been seen at Swanage quarries (07, 1911, CBG in **BMH**) and Gussage St Michael (91W, 1996, GDF). Cv. *albomarginata* is planted at Corscombe (50H, LJM). The tree is disliked by hedge-layers; its timber does not burn well, but the twigs are still used by children to make blowpipes. Both flowers and fruit are used by amateur winemakers, and the fruit is an important food for birds (Wilson, Arroyo & Clark, 1997); an infusion of the flowers causes mild sweating in man. The leaves contain cyanogenic glucosides. At isolated farms, a tree used to be planted to ward off witches.

Sambucus ebulus L.
DWARF ELDER
Widely scattered but scarce in hedges on calcareous soils. 11 + [12]. Nowhere does it grow in old woods, as it does in Europe, and it is probably a denizen. On sale in Blandford in 1782, and known from Portland since 1866 (SMP in **DOR**). At Shapwick (90G) it may have been introduced as a purgative by monks in the 15th century, and just survives today.

Viburnum opulus L.
GUELDER-ROSE
Frequent in wet woods and carr, occasional in hedges on dry sites on Chalk or Greensand. 323. It is least common along the coastal belt; inland it is increasingly planted in new hedges. The bark contains arbutin and scopuletin which act as muscle relaxants. Var. *sterile* DC., the Snowball-bush, is planted on Green Is (08D, !).

Viburnum lantana L.
WAYFARING-TREE
Common at the edges of woods, in scrub and old hedges and on undercliffs. 460. Mostly on calcareous soils, and absent from very acid soils of the Poole basin, also scarce in the north. Often planted in new hedges. Shoots after coppicing are said to have been used for the shafts of arrows. The form on undercliffs of east Portland (67V and 77A) has unusually narrow leaves.

Viburnum tinus L.
LAURUSTINUS
Self-sown on undercliffs and a few railway banks, often planted in plantations, hedges, parks and churchyards. 49. It flourishes on both calcareous and sandy soils. Well established at: Church Ope Cove (67V and 77A, 1966–97, !); Flaghead Chine to Canford Cliffs (08P and U, 1992–97, !); Spetisbury railway (90B, 1981, C Tandy and 1996, !); Durlston Head (07I, 1980, RJH Murray and 1997, !); Parkstone cutting (09K, 1995, !).

Viburnum rhytidophyllum Hemsley
WRINKLED VIBURNUM
Planted in parks since 1984, but not self-sown. 12. Noted in 60S, 61N, 79U, 71P, 89E, 80B and X, 90Q, 08D and J, and 09J and M.

Other *Viburnum* species and hybrids are planted on verges, in parks or churchyards, but none seed themselves. They include *V. × bodnantense* Stearn and *V. tomentosum* Thunb. at Canford School (09J, TH), *V. × burkwoodii* hort. at Maiden Newton Ch (59Y), and *V. × farreri* Stearn at Bere Regis (89M) and Tarrant Keyneston Ch (90H).

Symphoricarpos albus (L.) S.F. Blake
SNOWBERRY
Widely naturalised and sometimes subdominant in woods,

plantations and hedgebanks on all types of soil. 318. First recorded by Good in the 1930s, this is still spreading. It tolerates shade and is often found far from houses. The fruit is said to be toxic, though not to birds.

Symphoricarpos × chenaultii Rehder
HYBRID CORALBERRY
Occasionally planted on verges or in shade. 23. It is indifferent to soil, and is spreading fast.

Leycesteria formosa Wallich.
HIMALAYAN HONEYSUCKLE
Occasional by railways, on waste ground, walls and pavements, sown by birds from gardens. 42. First record: Sherborne (61I, 1970, ACL); spreading rapidly.

Lonicera pileata Oliver
BOX-LEAVED HONEYSUCKLE
Occasionally planted in plantations, villages or churchyards. 19. First record: Langton Long (90D, 1994, ! in **RNG**).

Lonicera nitida E. Wilson
WILSON'S HONEYSUCKLE
Commonly planted in village hedges, also on verges and railway banks, rarely in carr, on both calcareous and sandy soils. 276. Its fruits are inconspicuous, but seedlings have been seen at Lyme Regis (39F, LJM) and Plush (70B, !). First record: Stallen (61D, 1968, SS).

Lonicera involucrata (Richardson) Banks ex Sprengel
CALIFORNIAN HONEYSUCKLE
Planted at Upton CP (99W, 1992, !), not naturalised. 1.

Lonicera xylosteum L.
FLY HONEYSUCKLE
On sale at Blandford in 1782; in the 1930s reported from Shillingstone (81F, HJW) and Child Okeford (81G, HJW). One bush in a plantation at Walditch (49W, 1999, !). Rarely planted today, as at Portland Bird Observatory (66Z, RA) and Dorchester Public Park (69V, !). 1 + [2].

Lonicera tatarica L.
TARTARIAN HONEYSUCKLE
One bush in a hedge at Pimperne (90E, 1999, !), of garden origin.

Lonicera henryi Hemsley
HENRY'S HONEYSUCKLE
Rampant in waste ground near gardens at Easton (67W, 1996, ! det. EJ Clement). 1.

Lonicera japonica Thunb. ex Murray
JAPANESE HONEYSUCKLE
Much planted in gardens, and bird-sown on verges, railway banks, in village lanes, churchyards and tips. 47. First record: Stour Provost (82A, 1974, Duncliffe NHS). It covers a large area of an old railway embankment at Rodwell (67U, 1982–97, ! in **RNG**) and grows on cliffs at Stonebarrow (39W, MJG) and New Swanage (08F, EAP).

Lonicera periclymenum L.
HONEYSUCKLE, WOODBINE
Common in old woods, hedgebanks and among scrub on undercliffs on all but the most acid soils. 672. It twines round hazel rods, distorting them spirally, and such rods are used to make rustic walking sticks. An important foodplant for the White Admiral butterfly and many moths. A dwarf, prostrate form occurs on limestone undercliffs below St Aldhelm's Head (97S).

Lonicera caprifolium L.
PERFOLIATE HONEYSUCKLE
A rare garden escape, on sale at Blandford in 1782, but not grown today. [1]. Woolland (70T, 1891, WMR). The following later reports may refer to L. × italica: Silverlake Farm (61C, 1958, CT); Blandford (80Y, 1930s, RDG); Tarrant Gunville (91G, 1957, JAL).

Lonicera × italica Schmidt ex Tausch
GARDEN HONEYSUCKLE
A rare but persistent garden escape in hedges. 2. Whatcombe (80F, 1989, !); Smugglers Lane, Steepleton Iwerne (81Q, 1938, DM and 1996, !).

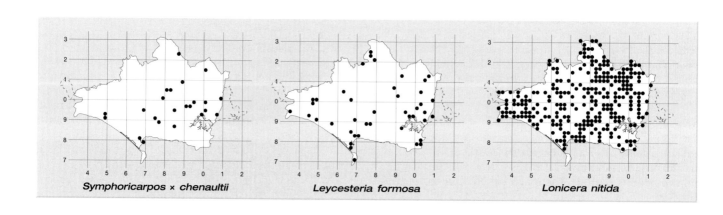

Symphoricarpos × chenaultii   Leycesteria formosa   Lonicera nitida

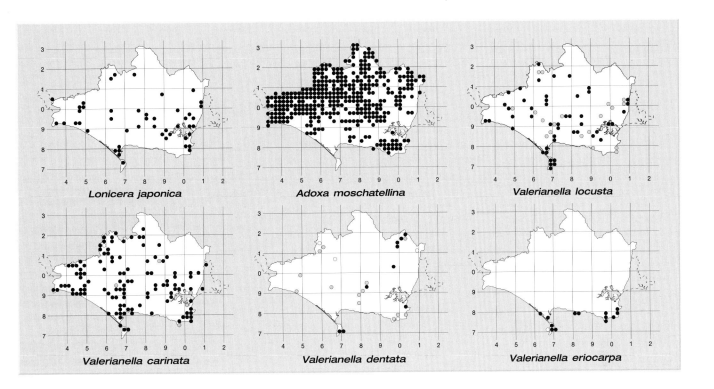

Lonicera japonica

Adoxa moschatellina

Valerianella locusta

Valerianella carinata

Valerianella dentata

Valerianella eriocarpa

Kolkwitzia amabilis Graebner
A relict bush in a derelict site at Milborne St Andrew (89D). 1.

Weigela florida (Bunge) DC.
Planted in parks or hedges. 3. West Stafford Ch (78J, !); Piddlehinton Camp (79I, 1992, !); Bryanston School (80T, 1945, BNHS).

Abelia chinensis R. Br. is planted at Colehill Ch (09F, !). 1.

ADOXACEAE

Adoxa moschatellina L.
MOSCHATEL
Locally abundant in old woods, both wet and dry, under alder, ash or oak. 360. Scarce or absent from deforested areas and from very acid soils near Poole Harbour.

VALERIANACEAE

Valerianella locusta (L.) Laterr.
COMMON CORNSALAD
Occasional in bare, dry places such as rail tracks, gardens, cliffs and old walls but seldom arable. 48 + [20]. Perhaps decreasing.

Valerianella carinata Lois.
KEEL-FRUITED CORNSALAD
Locally frequent in warm, bare, dry places; on new verges, old walls, in gardens and waste ground. 115 + [7]. It grows on most soils but dislikes heavy clays. Good gave only five localities, so it is increasing.

Valerianella rimosa Bast.
BROAD-FRUITED CORNSALAD
Rare or extinct as a weed of arable, on rail tracks or old walls. [28]. Post-1900 records: Burton Bradstock (48Z, 1928, AWG); Beaminster (40Q, 1924, AWG); Bere Hackett (51V, 1971, ACL); Portland (67, 1937, DM); Radipole (68Q, 1955, !); Preston (78B, 1925, CW Hewgill); East Holme (88Y, 1937, DM); Langton Matravers (97Z, 1916, CBG); Rempstone (98W, 1937, DM); Badbury (90R, 1918, JHS); Chilbridge Farm (90W, 1982, HJB); Gussage St Michael (91V, 1922, JHS); Swanage (07J, 1950, !); Poole (09, 1937, DM). JCMP had it in 10 sites, but Good did not find it in any of his stands. Specimens are in **BM, BMH, BRIST, CGE, DOR** and **LIV**.

Valerianella dentata (L.) Pollick
NARROW-FRUITED CORNSALAD
A rare and decreasing weed of arable and fallow, with remains dating back to the late Bronze Age. 9 + [17]. JCMP found it common, RDG frequent and in 19 stands. Now only on Portland and the north-east Chalk, also at Turners Puddle (89G, 1997, D Maxwell). Specimens are in **BDK, BM, BMH, DOR, RNG** and **WRHM**.

Valerianella eriocarpa Desv.
HAIRY-FRUITED CORNSALAD
A rare, native annual, long known on Portland, now found in bare, dry places on limestone or chalk cliff-edges or undercliffs along the coast. 13 + [1]. Specimens are in **BDK, BM, BMH, BRIST, CGE, DOR, LANC, RNG and SLBI**, so it is not necessary to collect more; Dorset is now the main stronghold of this plant in Britain.

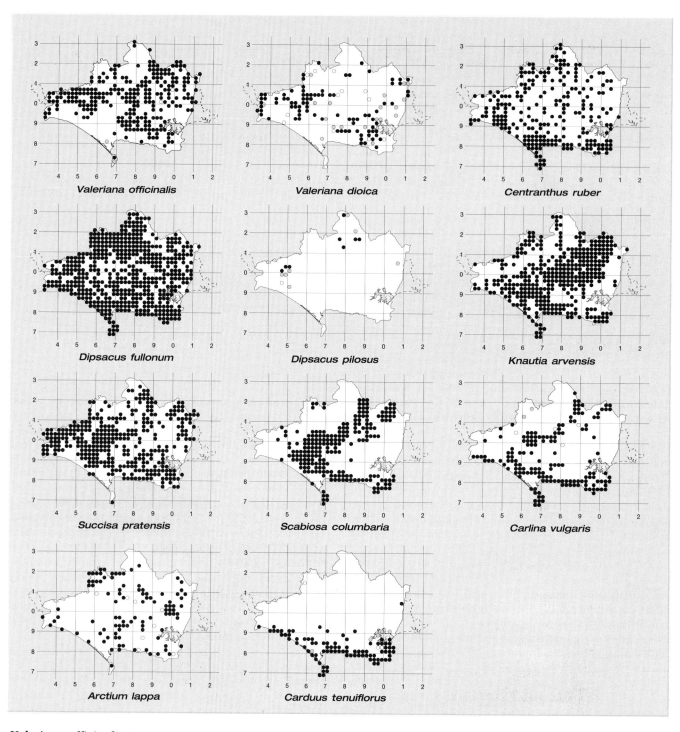

Valeriana officinalis L.
COMMON VALERIAN
Frequent in woodland rides, near rivers or in marshes; less common in hedgebanks and old quarries. 256 + [5]. Commonest in the west, on the Chalk, and in the main river valleys. Rare on Portland, and scarce or absent in deforested areas and along the coastal belt. Both var. *collina* Nyman, from calcareous woods, and var. *sambucifolia* Mikan ex Pohl, from marshes, are found. In the 18th century, roots were collected from Cranborne Chase for sale to apothecaries, for their sedative and pain-killing effects, though they contain alkaloids (valepotriates).

Valeriana dioica L.
MARSH VALERIAN
Local in alder woods, fens, marshes and flushed bogs. 89 + [32]. Commonest on Greensand in the west, and on alluvium in the Frome and Piddle valleys; absent from the Chalk, most calcareous clays and the coastal belt. Some sites have been destroyed by drainage or cultivation this century.

Centranthus ruber (L.) DC.
RED VALERIAN, CONVICT-GRASS
Now frequent on undercliffs and in old quarries, also a widespread escape in village lanes, on walls and waste ground. 244. Commonest along the coastal belt, especially on Portland. On sale at Blandford in 1782, recorded by Pulteney in 1799 and from Lulworth Cove in 1874. By 1895, JCMP knew eight localities, and it is still spreading. White-flowered forms have been seen in many villages and established at Uplyme, Portland Corfe, Swanage and Durlston CP; two distinct shades of red occur on Portland.

Dipsacus fullonum L.
WILD TEASEL, BRUSHES-AND-COMBS
Frequent, but sporadic, at the edges of woods, in rough grassland, on cliffs, verges or railway banks. 512. It often appears after disturbance.

Dipsacus sativus (L.) Honck.
FULLER'S TEASEL
A rare casual; Lodmoor tip (68V, 1955, !). [1].

Dipsacus pilosus L.
SMALL TEASEL
Very local and sporadic, in small numbers, usually in tall vegetation near water. 8 + [8]. Records are clustered around Beaminster in the west and Marnhull in the north. It germinates in the autumn, and will tolerate shade. Post-1960 records: Parnham (40Q, 1993, LJM and LMS); Meerhay (40W, 1997, IPG); Shatcombe Farm (50B, 1996, BE); Yardgrove Farm (71T, 1985, EJ Lenton); Sturminster Newton (71W, 1993, BE); Whistley Coppice (72Z, 1975, A Newton and 1994, RM Veall & !); Pen Mead Farm (81N, 1995, !); Fontmell Magna (81T, 1995, K Garrett, !).

Dipsacus laciniatus L.
CUT-LEAVED TEASEL
A rare garden escape in a village lane, Stoke Abbott (40K, 1995–96, IPG and D Maxwell, !). 1.

Cephalaria gigantea (Ledeb.) Bobrov.
GIANT SCABIOUS
A persistent garden escape on verges. 2. Piddlehinton (79E, 1992, !); Compton Abbas (81U, 1996, !).

Knautia arvensis (L.) Coulter
FIELD SCABIOUS, BLUE-BUTTONS, GIPSY-ROSE
Frequent in long grass, mostly on verges or railway banks. 366. It prefers calcareous soils. A white-flowered form was seen at Little Bredy (58Z, !).

Succisa pratensis Moench
DEVIL'S-BIT SCABIOUS
Locally abundant in woodland rides, nutrient-poor grassland, grazed heath and leached turf above Chalk, in both wet and dry sites. 339. Least common in deforested and prairie-farmed areas. A dwarf form occurs in chalk turf at Oakley Down (01D, ! in **RNG**). White-and pink-flowered forms occur at Corfe Common (98K) and Arne Triangle (98N).

Scabiosa columbaria L.
SMALL SCABIOUS
Frequent in short, calcareous grassland; rarely on eutrophic heathland verges or fixed shingle. 218. A pink-flowered form was seen at Badbury Rings (90R, FAW & !).

ASTERACEAE or COMPOSITAE

Echinops exaltatus Schrader
GLOBE-THISTLE
A garden escape on verges and tips. 1 + [2]. Stoke Abbott (40K, !); Lodmoor tip (68V, 1955, ! in **OXF**); Lewell (78J, 1960, RDG).

Echinops bannaticus Rochel ex Schrader
BLUE GLOBE-THISTLE
A well-established garden escape on verges and railway banks, in quarries and abandoned gardens. 10 + [5]. Little Bredy (58Z, !); Powerstock railway (59D, 1930s, AWG); Wraxall (50Q, 1972, RDE); Portland (97, 1966, BNSS); West Stafford (78J, !); Woodstreet (88M, !); Fontmell Magna (81T, 1995, K Garrett); Winspit quarry (97T, 1908–22, AWG and 1935, AH Carter in **RNG**); Dean Farm, Witchampton (90T, !); Swanage (07, 1916, AWG); Ferndown (00V, GDF); Woodlands (00U, GDF and JO); Wimborne St Giles (01G, JO); Cranborne (01R and S, GDF).

Carlina vulgaris L.
CARLINE THISTLE
Occasional in dry, calcareous grassland, sometimes in sandpits or on eutrophic verges in heathland. 159 + [5]. Mostly on scarp slopes of the Chalk or limestone, and especially common in the coastal belt.

Arctium lappa L.
GREAT BURDOCK
Occasional at the edges of woods, on river banks, in waste places and on cliff-tops, mostly in rich soils, once seen on Charmouth beach. 99 + [10]. Scattered, least common in the Poole basin. The root contains bitter arctiopicrin glycosides.

Arctium minus L.
LESSER BURDOCK
Common in woodland rides, hedgebanks, verges and waste ground on all but the most acid soils. 717. Subsp. *pubens* (Bab.) P. Fourn. and *nemorosum* (Lej.) Syme were not distinguished by most recorders and have very few known sites.

Carduus tenuiflorus Curtis
SLENDER THISTLE
Frequent on dry cliff-tops, undercliffs and rabbit warrens near the coast. 88 + [8]. Less common on dry, windswept ridges on the Chalk, or on disturbed heaths. It behaves as a Mediterranean annual, flowering in early summer, and rapidly drying up and disintegrating.

Carduus crispus L.
WELTED THISTLE
Frequent at the edges of woods, in rough grassland and hedgebanks, preferring clay soils. 313. It is least common on sands or alluvium in the Poole basin. A white-flowered form was seen at White Lackington (79E, !).

Carduus crispus* × *nutans* = *C.* × *stangii Buck
Uncommon in calcareous grassland, usually with both parents. 2 + [5]. Greenhill Barton (68X, 1993, !); East Chaldon (78W, 1993, S Philp); Hod Hill (81K, 1981, DWT); Sutton Waldron (81S, 1904, EFL); St Aldhelms Head (97S, 1967, CHS and R Jennings); Shapwick (90F, 1930s, RDG); Nine Barrow Down (08A, 1915, CBG in **BMH**).

Carduus nutans L.
MUSK THISTLE
Frequent but sporadic in dry calcareous grassland, sometimes abundant in fallow, also on cliff tops and waste ground. 298. Mostly on calcareous soils, rare on eutrophic verges in heathland, and absent from heavy clays in the north. A pale pink-flowered form was seen at Milborne Wood (79Y, !) and a white-flowered form at Puddletown (79M, !).

Cirsium eriophorum (L.) Scop.
WOOLLY THISTLE
Local and sporadic in dry, calcareous grassland, often in old quarries, rarely on tips. 40 + [8]. Commonest on limestone in the north-west and in Purbeck; rare on Portland, and extending west to Bexington. Thinly scattered on the Chalk, often casual. It does not occur further west in Britain.

Cirsium eriophorum* × *vulgare* = *C.* × *grandiflorum Kittel
Found on a verge between Abbotsbury and Burton Bradstock (58, 1932, JEL in **RNG**); possibly this in Durlston CP (07I, 1999, EAP). 1 + [1].

Cirsium vulgare (Savi) Ten.
SPEAR THISTLE
Common in rough grass, chalk turf, verges, cliffs, fallow fields, disturbed soil and the edges of heaths. 735. Tolerant of salt-spray. A fasciated plant with the florets replaced by spiny bracts occurred on Hambledon Hill (81L, 1991, ! in **RNG**). It germinates in late summer. Goldfinches eat the seeds of this and other thistles (Wilson, Arroyo & Clark, 1997). William Barnes devotes a poem to the thistle.

Cirsium dissectum (L.) Hill
MEADOW THISTLE
Occasional, but local, in fens, marshes, water-meadows and flushed bogs, rarely in mature chalk turf. 104 + [17]. Largely confined to the north-west and the Poole basin. Probably declining through drainage.

Cirsium dissectum* × *palustre* = *C.* × *forsteri (Smith) Loudon
Probably not uncommon wherever *C. dissectum* occurs, but overlooked. 3 + [4]. Birdsmoor Gate (30V, 1996, BE); Winfith Heath (88D, 1997, G Ellis, !); Throop Heath (89G, !); Creech Grange (98B, 1912, CBG in **BMH**); Slepe (98N, 1966, RDG in **DOR**); Ulwell (08A, 1909, CBG in **BMH**); Lower Row (00M, 1934, RDG in **WRHM**).

Cirsium tuberosum (L.) All.
TUBEROUS THISTLE
Very rare in long, chalk grassland. 2. Houghton Down (80C, 1982, RM, also 80B, 1991, ! in **RNG,** and 1992, A Halahan).

[***Cirsium oleraceum*** (L.) Scop.
CABBAGE THISTLE
A casual in the walled garden at Bryanston School (80T, 1944–45, BNHS).]

Cirsium acaule (L.) Scop.
DWARF THISTLE
Locally frequent in calcareous grassland. 315. It occurs on Chalk, Limestone, Forest Marble and Fullers Earth but is absent off these. In tall grass, it can develop a stem with up to four flowers, as at Gunville Down (91B), when it resembles *C. tuberosum*.

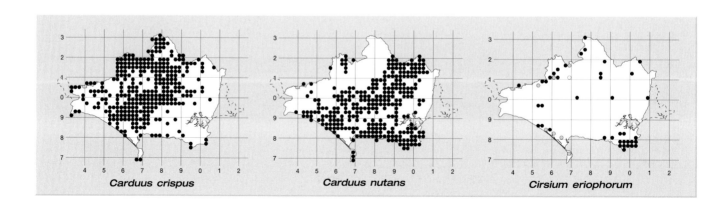

Carduus crispus   Carduus nutans   Cirsium eriophorum

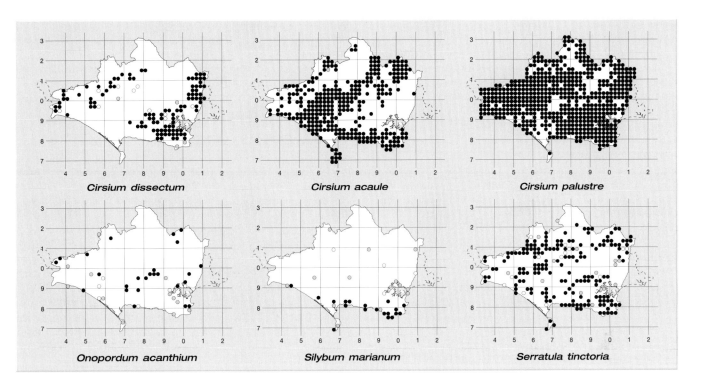

Cirsium dissectum · Cirsium acaule · Cirsium palustre

Onopordum acanthium · Silybum marianum · Serratula tinctoria

Cirsium acaule × palustre = C. × kirschlegeri
Schultz-Bip.
Rare in limestone grassland. [3]. Corfe (98, 1931–33, BNSS); Ashmore Down (91E, 1950, JF Hope-Simpson). Reports from Powerstock Common (59N, 1960) and Oakley Down (01D, 1977) need confirmation.

Cirsium palustre (L.) Scop.
MARSH THISTLE
Common in damp woodland rides, damp grassland, marshes, water-meadows and flushed bogs. 554. Sometimes in sheltered calcareous turf, and often accompanied by forms with white flowers. Least common in deforested areas.

Cirsium palustre × arvense
= C. × celakovskianum Knaf.
Rare. [2]. Blacknoll (88D, 1936, RDG in **WRHM**); by R. Piddle, Culeaze (89L, 1977, ! in **RNG**).

Cirsium arvense (L.) Scop.
CREEPING THISTLE
Abundant in old fields and verges, fairly constant in chalk turf, or a bad arable weed. 735. [Var. *setosum* (Willd.) Sledge, with scarcely prickly leaves, was found in the 1930s at Bindon Abbey (88N) and Chapmans Pool (97N)]. White-flowered forms are not uncommon. The main foodplant for the Painted Lady butterfly.

An immature specimen, labelled *Centaurea* sp., from Swanage Camp (08F, 1916, CBG) is in **BMH**. It could be a *Cirsium* sp. or *Picnomon acarna*.

Onopordum acanthium L.
COTTON THISTLE
Sporadic, and in small numbers, on chalk banks, field borders, verges and tips, 25 + [18]. Though grown in gardens, this has been known in the county since 1799, and behaves like a native plant. It is most often seen in the south and south-centre.

Cynara cardunculus L.
GLOBE ARTICHOKE
A garden relict in scrub above Lulworth Cove (88F, 1981, !). 1.

Silybum marianum (L.) Gaertner
MILK THISTLE
Regularly seen at the edges of sunny cliffs or by rabbit warrens along much of the coast, and as a rare casual inland. 19 + [17]. It behaves like a native plant at the edge of its range, germinating in late summer. The seeds contain silymarin, which protects against liver damage, even by *Amanita* sp.

Serratula tinctoria L.
SAW-WORT
Local, and sometimes abundant, in old grassland on chalk or clay, in acid grassland of woodland rides or heaths, and in water-meadows, flushed marshes and bogs. 172 + [29]. Often a relict in small numbers in hedgerows or on verges. Var. *reducta* Rouy, a dwarf form, was once seen on Wareham Heath (98, 1953, HW Woolhouse in **RNG**). White-flowered plants are rare.

Centaurea scabiosa L.
GREATER KNAPWEED
Frequent in tall, calcareous grassland. 348. Almost always on chalk or limestone, with a few outliers on eutrophic or limed verges in heathland. Not found on heavy clays in the north-west. A form with white flowers grew at Witchampton (90Y) and still grows at Badbury Rings (90L and R, !), and one with pale pink flowers at Puddletown (79M, !).

Centaurea montana L.
PERENNIAL CORNFLOWER
A persistent garden escape in village lanes, planted in churchyards, and on tips. 16 +[1]. First record: Lodmoor tip (68V, 1958, !). Seen recently in 39X, 68D and G, 62F, 78M, 72M, 80F and H, 81T, 97P, 91B, 00B, 01K and L.

Centaurea cyanus L.
CORNFLOWER
Once a rare weed of calcareous or sandy arable, now more often seen as a casual on disturbed verges, or on tips. 24

+ [27]. It is sometimes sown among 'wildflower' mixtures; forms with pink or white flowers are garden escapes. Seeds of medieval age have been dug up at Bere Regis. JCMP knew it from six sites and RDG saw it in only one stand. Specimens are in **BDK, DOR** and **RNG**.

[*Centaurea calcitrapa* L.
RED STAR-THISTLE
No specimens have been found, but this is reported from Bradford Abbas (51S, 1837, JB); Portland (67, 1930s, RDG); Arne (98, 1913, EBB); West Parley (09Y, 1900, RA Chudleigh).]

[*Centaurea solstitialis* L.
YELLOW STAR-THISTLE
This used to grow as a weed of lucerne fields, and on tips. 6. Recorded, without locality, by J Ray in 1695. Casual records: West Bay (49Q, 1921, AWG); Portland (67, 1872–83, HE Fox and Miss Isaac in **DOR**, and 1957–60, DM and MD Crosby); Weymouth Backwater (67U, 1820s, MF and 1935, MJA); Corfe (98, 1926, MJA *et al.*); Poole (09, 1919, F Spettigue); West Parley (09Y, 1900, RA Chudleigh).]

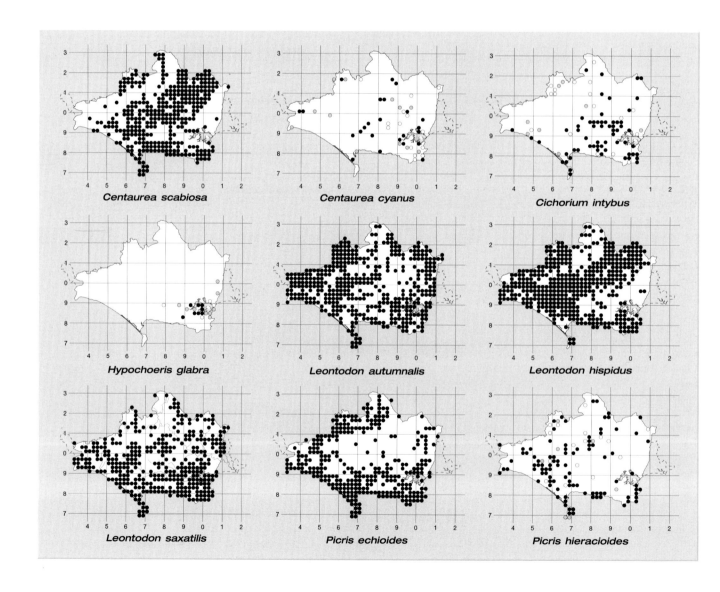

220

[*Centaurea melitensis* L.
MALTESE STAR-THISTLE
A rare casual of tips. 2. Weymouth (67, 1934, NDS); Poole (09, 1917, CBG).]

Centaurea diluta Aiton
LESSER STAR-THISTLE
A rare casual of tips. [4]. Beaminster (40Q, 1961, DMS Dupont); Sherborne (61I, 1970, ACL); Woodsford (78U, 1977, !); Poole (09A, 1963, !).

Centaurea nigra L.
COMMON KNAPWEED
Common in long grass, unfertilised pastures, verges and undercliffs. 649. Both radiate and the more common rayless forms occur, sometimes together, as at Red Bridge (78Z). A dwarf form occurs on exposed cliffs and hilltops at: Lynch (67P), Bincombe (68X), Osmington Mills (78F, ! in **RNG**), Poxwell (78L), Gad Cliff (87Z) and St Aldhelm's Head (97S). A rayless form with white flowers grows at Throop Heath (89F).

(*Centaurea nigra* × *jacea* = *C.* × *moncktonii* C. Britton has been reported near the coast at Burton Bradstock (48Z), East Creech (98G) and Durlston CP (07J), but no specimens have been seen, and all these may have been forms of *C. nigra* with narrow leaves or pale bracts.)

Centaurea ragusina L.
A relic of planting, surviving on waste ground at Weymouth Sta (67U, 1993–97, ! in **RNG** det. EJ Clement). 1.

Carthamus tinctorius L.
SAFFLOWER
A rare casual of tips. 2 + [4]. Beaminster (40Q, 1926–28, AWG); Radipole Lake (67U, 1926, MJA); Binnegar (88Y, 1992–94, !); Swanage (07J, 1929, LFS Todd); Parkstone (09F, 1899, HJG in **DOR**, as *Scolymus hispanicus*, and 1924–33, HJW); Canford Heath tip (09I, 1999, !).

Carthamus lanatus L.
DOWNY SAFFLOWER
Casual in arable at Stoke Abbott Mill (49I, 1976, AWG det JG Dony). [1].

Cichorium intybus L.
CHICORY
Occasional on verges and waste ground, rarely planted in fallow, but never in semi-natural habitats. 61 + [28]. It seems to have decreased in the north and west.

[*Arnoseris minima* (L.) Schweigger & Koerte
LAMB'S SUCCORY
Apart from some reports by RP in 1799 from the Chalk, this was once a rare weed of sandy arable in the Poole basin; suitable habitats survive. [11]. Pikes Wharf, Wareham (98I, 1862–75, JCMP in **DOR**); Slepe Farm (98S, 1916, CBG in **BMH**); Woodlands (00P, 1900, EFL); NE of West Moors Sta (00R, 1884, JCMP in **DOR**); Tricketts Cross (00V, 1937, VM Leather); Lions Bridge (00X, 1919, JHS); Verwood Lower Common (00Y, 1919, AWG and 1938, DM); Kings Farm, Woolsbridge (10D, 1938, RDG in **WRHM**).]

Lapsana communis L.
NIPPLEWORT
Common at the edges of woods, a persistent weed in hedgebanks and shaded gardens, on most types of soil except very acid or waterlogged ones; known since the Iron Age. 707.

Hypochoeris radicata L.
CAT'S-EAR
Common in short, dry turf, verges and lawns, and a colonist of bare, waste ground on both sandy and calcareous soils. 705. It resists trampling.

Hypochoeris glabra L.
SMOOTH CAT'S-EAR
Very local and sporadic in disturbed sandy arable, fallow or cliffs, rarely in dune slacks. 8 + [10]. Confined to the Poole basin. Post-1925 records: Buddens Farm (88U, 1937, RDG in **WRHM**); West Creech (88W, 1991, !); Worgret pits (98D, 1925, JHS); Ridge (98I, 1958, AWG); Sandford (98J, 1996, JHSC); Hartland Moor (98M, 1936, RDG in **WRHM** and 1977–97, !); Slepe (98N, 1927, FHH and 1976, JB); Scotland (98S, 1980–90, AJB); Arne (98U, 1977, VJG and 1999, !); Ower (98X, 1978, ! and 1991, BE); Fitzworth (98Y, 1991, BE); Gold Point (98Z, 1991, AJB); Goathorn (08C, 1935, FHH); Studland Heath and Little Sea (08G and H, 1950, AWG *et al.*); South Haven (08I, 1997 DAP); Sandbanks (08N, 1980, AJB); Talbot Heath (09S, 1965, MPY); Ferndown (00Q, 1927, FHH). Reports from Powerstock Common and Pamphill cannot be accepted without more evidence.

Leontodon autumnalis L.
AUTUMN HAWKBIT
Common in short, dry grassland on sandy or rarely calcareous soils. 425. Variable in size, and perhaps over-recorded in error for *Crepis capillaris*.

Leontodon hispidus L.
ROUGH HAWKBIT
Frequent in calcareous grassland, also on eutrophic verges in heathland. 441. First record: Lulworth Cove (88F, 1801, W Withering). Absent from acid soils in the Poole basin and from heavy clays in the north-west.

Leontodon saxatilis Lam.
LESSER HAWKBIT
Frequent in dry grassland on both sandy and calcareous soils. 366. Absent from heavy clays in the north-west.

Picris echioides L.
BRISTLY OXTONGUE
Common in disturbed soil, arable, verges, undercliffs, pavements and tips. 314. Commonest on heavy clays and along the coastal belt; scarce and sporadic on the Chalk or acid soils.

Picris hieracioides L.
HAWKWEED OXTONGUE
Frequent in tall, calcareous grassland. 95 + [23]. Most common along the coastal belt, on Portland, and on the

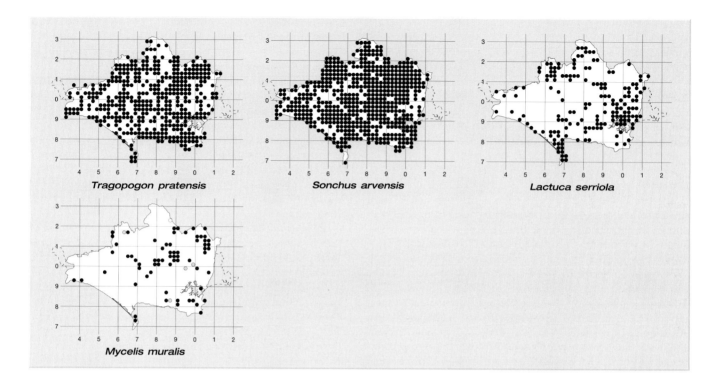

Tragopogon pratensis

Sonchus arvensis

Lactuca serriola

Mycelis muralis

north scarp of the Chalk in central Dorset. Probably introduced by the railway at Hamworthy (08E), and east of Gillingham (82D and I).

Scorzonera humilis L.
VIPER'S-GRASS
Locally abundant in one wet acid pasture, and in small numbers in two other sites in the Poole basin. 2. A rare native, with other British localities in Warwickshire and Glamorgan. Corfe Common (98K, 1997, S O'Connell); Ridge (98N, 1914, CIS, *et al.*, and 1990, RF, !); On verge, Arne Triangle (98N, 1978, JB and 1988, DAP, !); Poole (09A, 1927, LBH in **OXF**). There are specimens in **BM, BMH, DOR, LANC, LIV, OXF** and **RNG**.

Scorzonera hispanica L.
Surviving and 2 m tall in a vegetable plot at Symondsbury (49L, 1999, !); earlier found as a casual at Parkstone tip (09, 1929–33, HJW in **BM**). 1 +[1].

Tragopogon pratensis L.
GOAT'S-BEARD
Frequent in tall, calcareous grassland and dry, grassy verges. 422. Our plant is subsp. *minor* (Miller) Wahlenb. which is hardly distinct from subsp. *pratensis*.

Tragopogon porrifolius L.
SALSIFY
An uncommon casual, established for a few years on dry verges, railway banks or cliffs. 2 + [11]. Lyme Regis (39, 1938, DM); Salway Ash (49N, 1920, AWG); Weymouth railway (67, 1937, DM); Sherborne (61N, 1937, DM); East Weare (77B, 1988–90, AJB, !); Ringstead (78K, 1951, !);

Woodsford (79Q, 1820s, MF); Tolpuddle Ball (79M, 1997, PH Sterling, !); Stalbridge (71I, 1870s, Mrs Allen in **DOR**); Bourton (73Q, 1976, E Bange); Bere Regis (89M, 1872, JCMP in **DOR**); Sandbanks (08N, 1947, WAC); Poole railway (09, 1900–08, EFL). The root is edible.

(Sonchus palustris L.
MARSH SOW-THISTLE
A report from the R. Stour by RP in 1799 was probably an error for *S. arvensis*; suitable habitats exist round Poole Harbour, but the nearest native record is from Exbury in VC11.)

(**Sonchus palustris × arvensis** was reported from Winfrith by RDG, but no specimens have been seen.)

Sonchus arvensis L.
PERENNIAL SOW-THISTLE
Common and variable in arable, disturbed verges, undercliffs, marshes, and muddy beaches. 549. Subsp. *uliginosus* (M. Bieb.) Nyman, with a hairless involucre, is rarely recorded. Var. *angustifolius* Meyer, with narrow, almost entire leaves, has been seen at Stoborough Heath (98H, 1981, AH in **RNG**) and also at Dancing Ledge (07D, 1900, EFL).

Sonchus oleraceus L.
SMOOTH SOW-THISTLE, MILK-THISTLE
Very common in bare, disturbed soil, arable, gardens, cliffs, beaches, waste ground and tips. 717. Tolerant of salt-spray. Variable in leaf-shape. It germinates in summer.

Sonchus asper L.
PRICKLY SOW-THISTLE
Very common in bare, disturbed, nutrient-rich soil. 710.

In favourable conditions it can grow 2 m tall in a single season.

Lactuca serriola L.
PRICKLY LETTUCE
A frequent introduction in artificial habitats, disturbed verges, railway banks, quarries and tips, rarely on beaches. 166. First record: Swanage (07J, 1930s, RDG). Increasing fast, especially in the Poole basin and around Weymouth.

Lactuca virosa L.
GREAT LETTUCE
A scarce, native plant of tall calcareous grassland or on undercliffs. 5 + [9]. Confined to the east and decreasing; some early records may have been errors for *L. serriola*. Post-1900 records: Church Ope Cove (67V, 1908, WBB); West side of Portland (67W, 1860s, WBB); White Nothe (78Q, 1908, WBB and 1981, AH, !); Bats Head (78V, 1969, ! and 1992, BE); Gad Cliff (87Z, 1932, RDG in **DOR**); Durdle Door (88A, 1955, CCT, !); Kimmeridge (97E, 1971, !); Chapmans Pool (97N, 1925, JHS); St Aldhelm's Head (97S, 1986, AJB); Corfe Castle mound (98L, 1900, LVL, *et al.*, and 1955–99,!); Corfe cutting (98R, 1932, H Phillips in **BDK** and 1981, P Grey); Durlston CP (07I, 1985, AM).

(*Lactuca saligna* L.
LEAST LETTUCE
Reports from Cerne Abbas and Evershot (1799, RP) are unsupported by a specimen, while that from Corfe (98R, 1830, TBS) was probably an error for *L. virosa*.)

Cicerbita macrophylla (Willd.) Wallr.
COMMON BLUE-SOW-THISTLE
A persistent garden escape or throw-out forming patches on verges or tips. 10 + [3]. Morcombelake (49C, 1960, J Gardiner); Corscombe (50C and H, 1986, AH Aston and 1997, IPG); Trent (51Z, 1969, CA Howe and 1996, IPG); Winterbourne Steepleton (68J, 1979, J May); South Dorchester (68Z, 1979–81, S Renner); West Dorchester (69V, 1999, !); Winterbourne Abbas (69F, 1988, DAP and RF); Frampton Park (69H, 1993, LJM, !); Charminster (69R, !); Cerne (60R, 1972, RDE); Whatcombe (80F, !); Worth Matravers (97T, 1963, CHS and 1993, !).

Mycelis muralis (L.) Dumort.
WALL LETTUCE
Occasional in bare, dry, hard soil in woods and shaded gardens, or on old walls. 67 + [5]. Mostly on calcareous soils, and scarce in the west. JCMP only found this in eight sites, while Good thought it locally frequent, so it may have increased this century.

Taraxacum agg.
DANDELION, MALE, PISS-A-BED
Very common in wet or dry grassland, fallow and disturbed ground. 736. Too few collectors have studied the apomictic microspecies for what follows to be more than an interim account. The leaves are collected to feed rabbits, and the flowers to make a wine with diuretic properties and a high potassium content. It is a foodplant for many moths.

Section ***Erythrosperma*** (Lindb. f.) Dahlst.
Plants of dry grassland, especially short, calcareous turf; specimens in **BM and OXF**.
T. argutum Dahlst. Durlston CP (07I, 1977, GAM); a western taxon.
T. brachyglossum (Dahlst.) Dahlst. Dancing Ledge (97T, 1970, AJR); Durlston CP (07I, 1977, GAM).
T. fulviforme Dahlst. Lyme Regis (39, 1885, HNR); Swanage (07J, 1917, GCD).
T. fulvum Raunk Lulworth Cove (87J, 1970, AJR); Poole (09, 1883, Anon).
T. lacistophyllum (Dahlst.) Raunk Portland (67, PDS); Lulworth Cove (87J, AJR); St Aldhelm's Head (97S, !); Stoborough (98H, E Chicken); Badbury Rings (90R, 1892, EFL).
T. oxoniense Dahlst. Frequent
T. rubicundum (Dahlst.) Dahlst. Eggardon (59H, !); Badbury Rings (90R, 1892, EFL); Durlston CP (07I, GAM); (01, Anon).

Section ***Palustria*** (Lindeb.) Dahlst.
(Plants of wet meads, no certain records. Lewesdon Hill (40F, 1924, AWG in **BDK**); Lower Buckshaw and Sherborne (61, SS); Corfe Common (98Q, RP Bowman).)

Section ***Spectabilia*** (Dahlst.) Dahlst.
T. faeroense (Dahlst.) Dahlst. Wet meads. Corscombe (50D, LJM); Powerstock Common (50N, AJCB); Wareham (98J, DSR); Corfe Common (98Q, !).

Section ***Naevosa*** M. Christiansen
T. drucei Dahlst. Broadwindsor (40G, LJM); Corscombe (50D, LJM).

Section ***Celtica*** A. Richards
T. bracteatum Dahlst. Monkton Wyld (39I, LJM); Lulworth (88B, AJR); Ridge (98I, AJR).
T. britannicum Dahlst. Monkton Wyld (39I, LJM); Lulworth Cove (87J, AJR); West Morden (89X, AJR); Furzebrook (98H, AJR).
T. duplidentifrons Dahlst. Chideock (49G, LJM).
T. gelertii Raunk Stoborough and Higher Bushey (98H and R, Anon).
T. nordstedtii Dahlst. Marshwood and Wootton Hill (39N and Z, LJM); North Poorton (59E, LJM); Rampisham (50R, C Lister); Oakers Wood (89A, RJ Pankhurst); Whitefield (99D, !).
T. subbracteatum A. Richards Monkton Wyld (39I, LJM); Thorncombe (30R, LJM); Chideock (49G, LJM); Lulworth Cove (87J, AJR).

Section ***Hamata*** Oellgaard
T. atactum Sahlin & Soest Hod Hill (81K, !).
T. sahlinianum Dudman & A.J. Rich. Marshwood Ch (39Z, LJM).
T. boekmanii Borgv. Chideock (49G, LJM); West Bay (49Q, LJM).

T. hamatum Raunk. Lulworth Cove (87J, AM Boucher in **OXF**); (08, AM Boucher).

T. lamprophyllum M.Christiansen Melbury Osmond (50T, C. Lister); West Morden (89X, AJR); Corfe Common (98K, AJR).

(*T. lancidens* Hagend. *et al.* may occur, but its only record is from Holy City (20X, LJM) in VC3).

T. pseudohamatum Dahlst. Evershot (50S, !);Durlston CP (07I, GAM).

T. quadrans Oellgaard Morden (99D, AJR).

T. subhamatum M. Christiansen 08, AM Boucher.

Section ***Ruderalia*** Kirschner, Oellgaard & Stepanek
T. alatum H. Lindb. f. Stoborough (98I, AJR in **OXF**).

T. 'anceps' ined. Wootton Fitzpaine (39S, LJM); Bridport (49L, LJM).

T. ancistrilobum Dahlst. Portland (67Q, AJR in **OXF**).

T. cordatum Palmgren. Lulworth Cove (87J, AJR in **OXF**); Hod Hill (81K, !); Dancing Ledge (97Y, AJR).

T. croceiflorum Dahlst. Wootton Fitzpaine (39N, LJM); Lower Whatcombe (80K, E Chicken).

T. densilobum Dahlst. ? Studland beach (08G, !).

T. dilaceratum M. Christiansen Marshwood (39Z, LJM).

T. fasciatum Dahlst. Lyme Regis and Charmouth (39, LJM); Corfe Common (98K, AJR in **OXF**).

T. insigne Ekman Marshwood (39Z, LJM); Lulworth Cove (87J, AJR).

T. lepidum M. Christiansen Old meadow, Marshwood (39Z, LJM).

T. lingulatum Markl. Whitchurch Canonicorum (39X, LJM); Morden (99, AJR).

T. multicolorans Hagend., Soest & Zavenb. Bridport (49L, LJM).

T. obliquilobum Dahlst. Rampisham (50R, C Lister).

T. oblongatum Dahlst. Winterborne Kingston (89N, AJR in **OXF**); Durlston CP (07I, GAM in **BM**).

T. pannucicum Dahlst. Wootton Fitzpaine (39N, LJM); Lulworth Cove (87J, AJR); Peveril Point (07P, GAM); Studland (08, GAM in **BM**).

T. polyodon Dahlst. Wareham (98I, AJR in **OXF**).

T. pseudoretroflexum M. Christiansen Wootton Fitzpaine (39N, LJM).

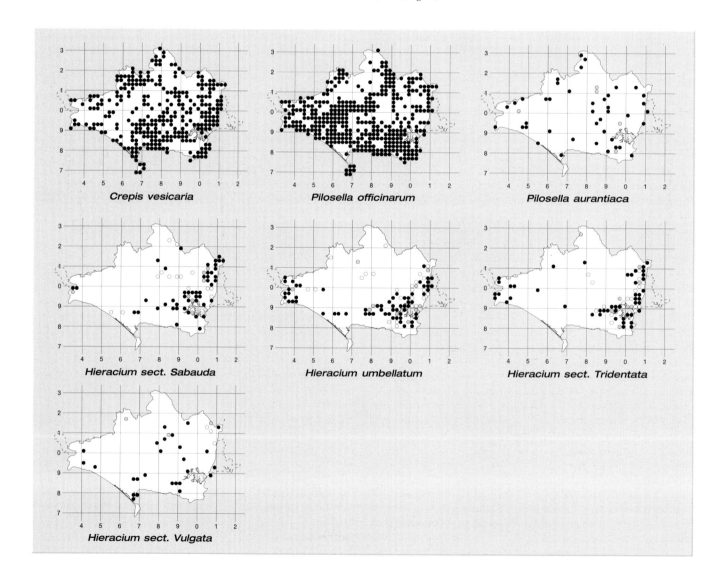

Crepis vesicaria

Pilosella officinarum

Pilosella aurantiaca

Hieracium sect. *Sabauda*

Hieracium umbellatum

Hieracium sect. *Tridentata*

Hieracium sect. *Vulgata*

T. pulchrifolium Markl. Chideock (49B, LJM).

T. rhamphodes G.E. Haglund Marshwood Ch (39Z, LJM).

T. sellandii Dahlst. Bridport (49L, LJM); Lulworth Cove (87J, AJR in **OXF**); Peveril Point (07P, GAM in **BM**).

T. stenacrum Dahlst. Winterborne Kingston (89T, AJR in **OXF**).

T. subcyanolepis M. Christiansen Portland (67, AJR in **OXF**).

T. subexpallidum Dahlst. Rampisham (50R, C Lister).

T. undulatiflorum M. Christiansen (08, AM Boucher).

Reports of *T. ekmannii, T. exacaulum, T. expallidiforme, T. hamatiforme, T. longisquameum, T. maculatum, T. pannulatiforme* and *T. porrectidens* are thought to be errors (Dudman & Richards, 1997).

Crepis biennis L.
ROUGH HAWK'S-BEARD
A casual of verges or disturbed calcareous grassland. 4 + [5]?. None of the 17 reports of this plant have been substantiated by a specimen. Reports before 1960 are suspect in that *C. vesicaria* was then thought to be rare, and many later reports by inexperienced botanists may be doubted. The plant is rare, and mostly transient, in Somerset and S. Hampshire. Maiden Newton chalkpit (59Z, 1988, RMW); Maiden Castle (68U, 1984, NCC); Dorchester (69V, 1980, RJH Murray); West Hill, Sherborne (61M, 1962, CT); Upton (78L, 1950, !); Wool (88N, 1938, RDG); Melbury Abbas (81Z, 1988, AM); The Moors (98N, 1965, DSR).

Crepis tectorum L.
NARROW-LEAVED HAWK'S-BEARD
Rare or overlooked, on a disturbed verge at Wyke Regis (67T, 1994, ! det. EJ Clement). 1.

Crepis capillaris (L.) Wallr.
SMOOTH HAWK'S-BEARD
Frequent to abundant in dry grassland and verges, on both calcareous and sandy soils. 638. Variable in size.

Crepis vesicaria L. subsp. *taraxacifolia* (Thuill.) Thell. ex Schinz & Keller
BEAKED HAWK'S-BEARD
Locally frequent in dry, disturbed habitats, verges, railway banks, village lanes and waste ground. 300. First record: Woolland (70T, 1880, WMR). Good found it uncommon in the 1930s, when DM only noted it in five sites. Since 1950 it has spread widely.

Pilosella peleteriana (Merat) F. Schultz & Schultz-Bip.
SHAGGY MOUSE-EAR-HAWKWEED
Local on sunny calcareous banks and cliffs along the coastal belt between White Nothe and Gad Cliff, also at West Weare on Portland. Found on both Chalk and limestone. 9. First record: Lulworth Cove (88A, 1953, CCT). Specimens are in **RNG** and **WRHM**. All Dorset material is subsp. *subpeleteriana* (Naeg. & Peter) Sell, det. PDS, unlike that from the Is of Wight. At Gad Cliff, some plants may be the hybrid **P. × longisquama** (Peter) Holub, but this needs confirmation.

Pilosella officinarum F. Schultz & Schultz-Bip.
MOUSE-EAR-HAWKWEED
Frequent in short, dry grassland, mostly calcareous, but also by heathy tracks. 383. Common on chalk and limestone, less common in the Poole basin, and scarce or absent on heavy clays. Once known as 'herb of nails' and used to avert injuries to horses feet.

Pilosella aurantiaca (L.) F. Schultz & Schultz-Bip. subsp. *carpathicola* (Naeg. & Peter) Sojah
FOX-AND-CUBS
An occasional garden escape on dry verges, railway banks, churchyards, quarries and walls. 41 + [5]. It grows on both calcareous and acid soils. First record: Church Knowle (98K, 1917, CBG in **BMH**). It has spread since the 1960s, probably due to gardeners.

Hieracium section *Sabauda* (Fries) Gremli
Occasional, in small quantity, in rides or clearings of old woods, also on heathy verges and railway banks. 47 + [21]. Rarely in calcareous grass, as at Bincombe tunnel (68Y, BE) and above Tyneham (88V, !). Mostly in the Poole basin or the extreme west, scattered over the chalk where it has gone from woods which have been replanted with conifers. Most of the specimens are *H. sabaudum* L. (*H. argutifolium* Pugsley), but *H. virgultorum* Jord. occurred at Talbot Heath (09R, 1900, EFL) and *H. vagum* Jord. at Slepe Heath (98T, 1953, CCT). Specimens are in **BM, BMH, GE, DOR, MAT, OXF** and **WRHM.**

Section *Hieracioides* Dumort
Hieracium umbellatum L.
Occasional in old woods, or verges, railway banks and tracks on acid soil. 62 + [36]. Mainly in the west and the Poole basin, with outliers on summit heaths. It has decreased in the north and in built-up regions of the east. Subsp. *bichlorophyllum* (Druce & Zahn) Sell & C. West, with ovate lower leaves, is probably frequent but has only been noted recently at Peters Gore (39U, LJM) and Redbridge (78Z, !). Specimens are in **BMH, CGE, DOR** and **MANCH.**

Section *Tridentata* (Fries) F. Williams
Occasional in woods, heathy tracks, verges, railway banks or sandy cliffs, always on acid soil. 47 + [27]. Mostly in the west and the Poole basin, with a few outliers in central Dorset, as at Piddles Wood (71W), where a large colony has grown for a century. About 25% of the specimens in **BM, BMH, CGE, DOR, K, OXF, RDG, RNG** and **WRHM** have been named by experts. Most material is *H. trichocaulon* (Dahlst.) Johans, but there are also records of: *H. calcaricola* (F.J. Hanb.) Roffey; Between Wareham and Bere Regis (89, 1900, JCMP in **DOR** det. PDS); Lytchett Matravers (99S, 1914, CBG in **BMH**) and *H. eboracense* Pugsl.; Holditch (30L, JAG); Pilsdon Pen (40A, JGK, !); Lewesdon Hill (40F, JGK, also LJM); Parkstone (09F, 1921, NDS in **BM** det. PDS).

Section *Vulgata* (Fries) F. Williams
Uncommon in small colonies by railways, on verges, near

churchyards or gardens, or on walls. 26 + [6]. Scattered, mostly on acid soils, or leached soils on chalk banks.

Hieracium acuminatum Jord.
Rare in the east. 3 + [1]. Holton Heath (99K, 1979, !); Meyrick Park (09W, !); Woolsbridge (00X, 1900, EFL det. HWP); Boulsbury Wood (01S, 1976, A Brewis, det. PDS).

Hieracium maculatum Smith
Scarce on railway banks, but confused with H. scotostictum. 5. Bridport (49R, 1982, WAC); Sherborne (61I, 1969, ACL in **DOR and WRHM**); West Stafford (78J, !); Stourpaine (80P, 1975, VJG); Spetisbury (90A and B, 1982, AH and 1987, Poole NHS); Alderholt (11B, 1982, RC Stern, !).

Hieracium scotostictum N. Hylander
An aggressive garden escape which is likely to spread. 7. Parnham (40Q, !);Weymouth (67T, U and Z, 1995, MC Sheahan); Blake's Farm (82A); Farnham (91M, ! in **RNG**); Knoll gardens (00K, 1992–96, !).

Hieracium exotericum Jordan ex Boreau
Most records in the *Vulgata* section have been called this, although few have been checked by experts. 11.

Section *Oreadea* (Fries) Zahn
Hieracium leyanum (Zahn) Roffey
Endemic to Portland, where it occurs in limestone turf and boulder scree or quarry spoil on both sides of the island. 2. West Weare (67W, 1885, WBB, *et al.,* and 1989, !); Verne plateau (67W, 1926, MJA); Shepherds Dinner (77A, 1879, WC Medlycott in **DOR**, also JCMP in **DOR**); East Weare (77B, 1910, GE Fulleylove in **CGE** and 1988, RF). The name is provisional.

Filago vulgaris Lam.
COMMON CUDWEED
Local in bare, dry places on heaths, old sandpits and sandy cliffs. 46 + [11]. Now restricted to the Poole basin, but there are a few old records from the west and centre. Those on the Chalk, e.g. Hod Hill (81K, 1968, BNHS), require confirmation.

[*Filago lutescens* Jordan
RED-TIPPED CUDWEED
It once occurred on dry, sandy soils. 2. By the railway between Broadstone and Parkstone (09, 1895, EFL in **DOR** and **LIV**); Mannington (00S, 1898, WRL in **LIV**), and between Mannington and the Cross Keys Inn (1919, JHS). It survived on Hengistbury Head (19Q) until 1927, but in VC11.]

[*Filago pyramidata* L.
BROAD-LEAVED CUDWEED
It used to occur in dry arable on the Chalk, or in sandy wastes. 6. Weymouth (67, 902, AS Montgomery in **LIV**); Luccombe Farm (80A, 1879, JCMP in **DOR**); Langton (80X, 1889, RPM in **BM**); Swyre (97J ?, 1930s, RDG – ambiguous, but the most likely of several Swyres); Godlingston (08A, 1867, JCMP in **DOR**, and 1874, W Borrer); between Studland and Old Harry (08L, 1848, J Woods

and 1867, JCMP in **DOR**). It was at Ashley Heath in VC11 (10H, MPY in **BM**) in 1965. Suitable habitat still exists in VC9.]

Filago minima (Sm.) Pers.
SMALL CUDWEED
Occasional in bare, dry, sandy places, arable, woodland or heathland tracks, or old sandpits. 61 + [18]. Mostly in the Poole basin with outliers in the west and near Sherborne.

Anaphalis margaritacea (L.) Bentham
PEARLY EVERLASTING
A garden escape on verges and tips. 3. Totnell (60J, 1980, F Goldsack); Buckhorn Weston (72M, 1995, !); Winterborne Zelston (89Y, 1987, !).

Gnaphalium sylvaticum L.
HEATH CUDWEED
Always sporadic, now rare if not extinct, in acid woodland rides or heathy tracks. [20]. JCMP had it from five sites in the 1890s. Post-1900 records: Ilsington Wood (79L, 1945, RDG in **DOR**); Bulbarrow (70S, 1915, AWG); Park Wood (88G, 1980, Anon); Stoke Heath (88P, 1977, !); East Stoke (88T, 1971, DWT); Clenston Wood (80G, 1937, AWG); Broadley Wood (80M, 1939, DM); East Holme (98C, 1937, DM); Morden (99C, 1920, AWG); Studland (08, 1937, DM); Wimborne (00, 1954, RA Graham); Verwood (00U, 1928, AWG); St Leonards (00V, 1948, KG). Reports from Oakers Wood (89A, 1983), West Moors Depot (00X, 1983) and Edmondsham (01L, 1979) are considered doubtful.

Gnaphalium uliginosum L.
MARSH CUDWEED
Frequent in bare, rutted tracks in woods, at the edge of ponds, and in damp arable. 356. Mostly in acid soils in the north and west, and in the Poole basin, less common on the Chalk.

Gnaphalium luteoalbum L.
JERSEY CUDWEED
Rare on cinder tips, and bare, disturbed, acid soil of tracks. 1. Holton Heath (99K, 1978, L Farrell and 1979–96, ! in **RNG**; also in a nearby rose bed).

Helichrysum bracteatum (Vent.) Andrews
STRAWFLOWER
A casual pavement weed. Wareham (98I, 1998, !). 1.

Inula helenium L.
ELECAMPANE
Very local, in small numbers, in hedgebanks, on verges and rarely in copses. 26 + [37]. Commonest on clay soils in the north, not seen recently in the west, and scarce on Wealden soils in Purbeck. A relic of ancient cultivation: the roots, a source of inulin, were used by herbalists for asthma and ailments of horses. It resembles *Telekia speciosa*, which has a different scent when bruised.

Inula conyzae (Griess.) Meikle
PLOUGHMAN'S SPIKENARD
Locally frequent in dry, calcareous grassland, including

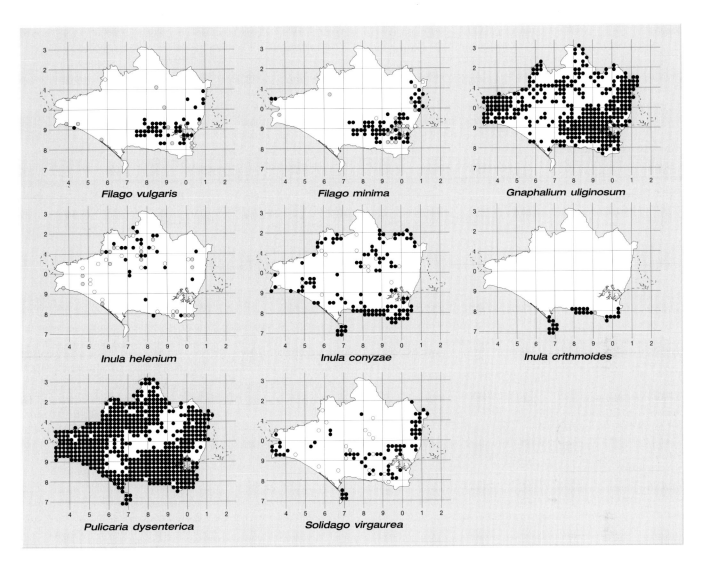

Filago vulgaris

Filago minima

Gnaphalium uliginosum

Inula helenium

Inula conyzae

Inula crithmoides

Pulicaria dysenterica

Solidago virgaurea

railway banks and cliffs. 118 + [13]. Commonest on the coastal belt and Portland, on the limestone in the north-west and the north-east Chalk. Scarce on dry, heathy banks, as at Dodpen Hill (39P), Trigon (88Z), Holton Heath (99K and Q) and Brownsea Is (08I).

Inula crithmoides L.
GOLDEN-SAMPHIRE
Locally frequent on calcareous cliffs or undercliffs, often very close to the sea, and on rocky beaches. 27 + [1]. Pulteney reported it from Poole in 1799, where the beaches are muddy. It now grows in many places along the coast between Portland and Swanage, but may be difficult to access.

Pulicaria dysenterica (L.) Benth
COMMON FLEABANE
Common in colonies in marshes and damp pastures, especially on clays and fairly acid soils. 545. It occurs close to the sea, but is absent from the drier parts of the Chalk and from built-up areas. The flowers attract many insects, despite its name.

[*Pulicaria vulgaris* Gaertner
SMALL FLEABANE
This once grew in heavily grazed pastures which flooded in winter. 5. Between Keysworth and Wareham (98J, 1875, JCMP in **DOR**); West Parley (09Y, 1894, EFL in **LIV**); Leigh Common (00F, 1889, JCMP in **DOR**); Gods Blessing Green (00G, 1932, FHH and 1937, DM); north of West Moors Sta (00R, 1896, EFL in **CGE and LIV,** also 1923, JHS). Suitable habitats in the Poole basin are sparse; the main British population now is in the Avon valley in VC11, with one site in 'new' Dorset.]

Telekia speciosa (Schreber) Baumg.
YELLOW OXEYE
A survivor in wild gardens, often in shade. 3. Forde Abbey (30M, !); Abbotsbury (58S, !); Spring Bottom (78K, 1953–96, !; probably planted here *c.*1900).

Solidago virgaurea L.
GOLDENROD
Locally frequent in rides or tracks in acid woods or heaths, and on acid railway banks. 72 + [18]. Mostly in the west and the Poole basin, with outliers on Greensand elsewhere.

An ecotype is frequent on limestone cliffs and quarries on Portland. Probably decreasing.

Solidago canadensis L.
CANADIAN GOLDENROD
A persistent garden escape, forming colonies on verges, railway banks and waste ground. 21 + [3]. It prefers acid soils, but occurs in a chalkpit at Godlingston (08A, WGT). First record: Worgret pits (98D, 1966, JB). Often confused with *S. gigantea*.

Solidago gigantea Aiton
EARLY GOLDENROD
In similar places to *S. canadensis*, but more frequent. 63 + [3] First record: Lodmoor (68V, 1955, !).

Solidago graminifolia (L.) Salisb.
GRASS-LEAVED GOLDENROD
An uncommon escape in forestry rides on acid soils, rarely grown in gardens. 2 + [3]. Hole Common (39H, 1978, J Fowles); Westford Farm (30H, 1928, AWG in **BDK** and 1996, JAG); Abbotsbury (58S, 1929, MAA); Bloxworth Heath (89R, 1989–97, ! in **RNG**); Swanage (08F, 1917, CBG). The flowers attract bees.

Aster novi-angliae L.
HAIRY MICHAELMAS-DAISY
Often grown in gardens, but only once seen on a tip at Sherborne (61H, 1970, ACL). 1.

Aster novi-belgii L. agg.
Few Michaelmas-daisies have been determined by experts in this difficult genus. All form colonies on verges, rail tracks, sandy cliffs, waste ground and old tips where they persist. 131. Some are naturalised on river banks or in brackish marshes in quantity.

Aster laevis L.
GLAUCOUS MICHAELMAS-DAISY
A critical plant, once reported from a verge near Sherborne tip (61H, 1970, SS). 1.

Aster pilosus Willd.
A large patch on a verge at New Barn, Sixpenny Handley (91Y, 1997, ! in **CGE** det. PFY). 1.

Aster novi-belgii Willd.
CONFUSED MICHAELMAS-DAISY
First record: Salt-marsh, Hamworthy (99, 1900, EFL). 74. Naturalised in more or less saline sites at Radipole Lake (67U), Lodmoor (68V,! det. PFY), R. Frome at Wareham (98I) and Keysworth Marshes (98P, RMW) but probably less common than its hybrids elsewhere.

Aster novi-belgii × *laevis* = *A.* × *versicolor* Willd.
LATE MICHAELMAS-DAISY
Probably the commonest, certainly the most colourful, taxon in this group. 30. First record: West Bay (49Q, 1993, LJM det ACL); well naturalised north-west of Radipole

Lake (68Q !), R. Frome at Frampton (69H, !) and Broadstone Golf course (09D, !). Ligules puple or, rarely, white.

Aster novi-belgii × *lanceolatus* = *A.* × *salignus* Willd.
COMMON MICHAELMAS-DAISY
Occasional. 16 + [4]. First record: Hampreston (09P, 1900, EFL). Other certain records: R. Brit at Beaminster (40Q, LJM det. ACL); Winterbourne Abbas (68E, LJM det. ACL); Swanage (08F, EAP); Studland (08G, EAP); Parley Common (09, 1938, RDG in **WRHM**).

Aster lanceolatus Willd.
NARROW-LEAVED MICHAELMAS-DAISY
An uncommon garden escape. 4. Chedington (40X, ! in **RNG**); Swanage (08F, EAP); Luscombe Valley (08P, ! det. PFY); Gussage All Saints (01A, !, det. ACL).

Aster tripolium L.
SEA ASTER
Frequent in salt-marshes and on muddy shores, occasional on rocky cliffs and undercliffs in Purbeck. 44 + [5]. All along the coast, commonest inside the Fleet and within Poole Harbour. A white-flowered form was seen on cliffs near Blackers Hole (07D, 1999, EAP).

(*Aster linosyris* (L.) Bernh.
A 19th century report from Portland has never been substantiated.)

Erigeron glaucus Ker Gawler
SEASIDE DAISY
Much planted on walls near the coast, and naturalised on sandy cliffs near Swanage and Canford Cliffs. 24. First record: Cliff west of Bournemouth (09Q, 1942, CIS). White-flowered forms occur further east along these cliffs, but in VC11.

Erigeron philadelphicus L.
ROBIN'S-PLANTAIN
A rare alien on verges and tips, usually transient. 2 + [2]. Abbotsbury Swannery (58S, !); Rodden (68B, 1959, !); Chickerell tip (68K, 1980, ! in **RNG**); Weymouth tip (68V, 1961, EH in **RNG**).

Erigeron karvinskianus DC.
MEXICAN FLEABANE
A garden escape, naturalised on walls or pavements in towns and villages, and in a chalk-pit above Sutton Poyntz (78C). 51. Planted at Abbotsbury in 1899 (MI); first record: Langton Matravers (97Z, 1952, !). Ligules dull pink or white.

Erigeron annuus (L.) Pers.
TALL FLEABANE
A garden escape, once reported from Portland (67, 1981, SM Luce). 1.

Erigeron acer L.
BLUE FLEABANE
Rather local in dry, nutrient-poor soils of heathy verges,

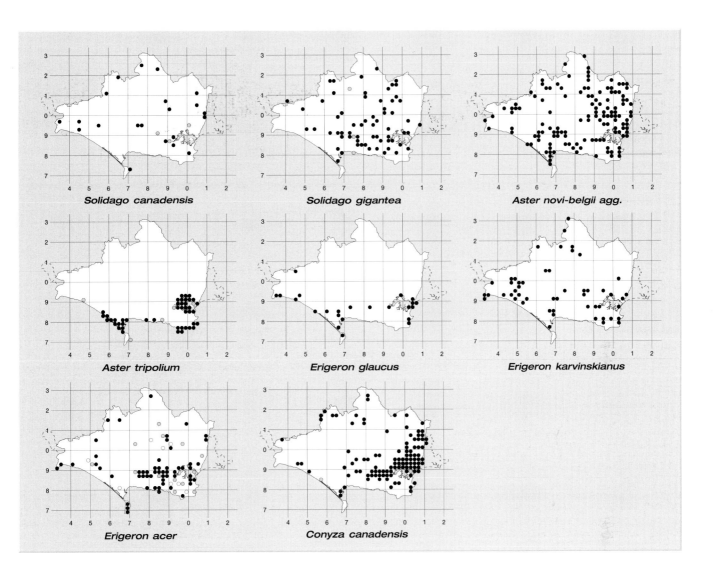

Solidago canadensis

Solidago gigantea

Aster novi-belgii agg.

Aster tripolium

Erigeron glaucus

Erigeron karvinskianus

Erigeron acer

Conyza canadensis

rail tracks and cliffs. 56 + [22]. Mostly in the Poole basin, but also on shallow limestone soils on Portland and along the coastal belt.

Erigeron acer × Conyza canadensis
= × *Conyzigeron huelsenii* (Vatke) Rauschert
Very rare, in a sand-pit with both parents at Binnegar (88Y, 1997, ! in **RNG**). 1.

Conyza canadensis (L.) Cronq.
CANADIAN FLEABANE
Locally frequent in bare, sandy fields, dry verges, rail tracks, nursery gardens, tips and urban sites. 107 + [3]. First Record: Swanage Camp (08F, 1917, CBG). Good found it scarce in the 1930s, but since 1950 it has spread widely and is now frequent in the Poole basin, and scattered elsewhere.

Conyza sumatrensis (Retz.) E. Walker
GUERNSEY FLEABANE
This occurs in the same artificial habitats as *C. canadensis*, and although it was not noticed until 1993 it is spreading fast. 14. Beaminster (40Q, PRG); Fortuneswell (67W, 1993, D Bevan); Sherborne (61I, IPG and PRG); Bovington Camp (88J, 1999, !); Blandford by-pass (80Y, EAP); Wareham Sta (98E, EAP); A35 from Lytchett Minster to Fleets Corner (99R and W and 09B, EAP, !); Greenland Farm (08C, !); South Haven (08I, EAP); Poole (09A, EAP); Canford tip (09I, !); Ferndown (00K, !). Plants from Poole (09A and B) are ***C. bilbaoana*** Remy, *fide* EJ Clement.

Callistephus chinensis (L.) Nees
CHINA ASTER
A casual of garden origin on tips. 2 + [2]. Lodmoor (68V, 1977, !); Sherborne (61H, 1970, ACL); Binnegar (88Y, !); Greenland Farm (08C, 1999, EAP).

Olearia spp. are increasingly planted. ***O. avicennifolia*** (Raoul) Hook. f. at Colehill Ch (01F, !); ***O. × haastii*** J.D.Hooker at Bryanston (80T, 1945, BNHS) and self-sown on the harbour wall at Weymouth (67U, 1979–82, ! in **RNG**, now gone); ***O. macrodonta*** Baker at Holworth Ho (78Q) and Green Is (08D).

Baccharis halimifolia L.
TREE GROUNDSEL
Self-sown at Hamworthy (99V, 1958, D McClintock in **RNG**). [1].

Bellis perennis L.
DAISY
Common in short turf, cliffs, verges, garden lawns and bare areas, less common on sandy soils. 728. Tolerant of salt-spray, mowing and trampling. William Barnes had a great affection for the 'Deaisy' in his poems.

Tanacetum parthenium (L.) Schultz-Bip.
FEVERFEW
A garden escape, known since the 18th century, now widespread as isolated plants on verges, village lanes, waste ground and tips. 277. It mostly occurs as the double-flowered form, and is still used as a herbal remedy for migraine, despite its powerful taste, due to alpha-pinene and parthenolide.

Tanacetum vulgare L.
TANSY
Occasional and perhaps native on riverbanks, especially by the Stour. 120 + [15]. Also in persistent patches on verges, waste places and tips, where it is probably an ancient escape from gardens. It was once used as an infusion to kill roundworms in children.

Tanacetum densum agg.
Planted on several old walls in Wimborne (00A, 1990–2000, ! det. EJ Clement). 1.

Seriphidium maritimum (L.) Polj.
SEA WORMWOOD
Very rare if not extinct in salt-marsh. [4]. Wyke Regis (67T, 1960, PJO Trist and 1983, AH); Patchins Point (98Z, 1888, JCMP in **DOR**, abundant in 1905, *et al.*, but only three plants in 1937, RDG); A single plant occurred on the verge of Arne village (98U, 1964–77, !). The species has also declined in south Hampshire, but is still found on the north Somerset coast.

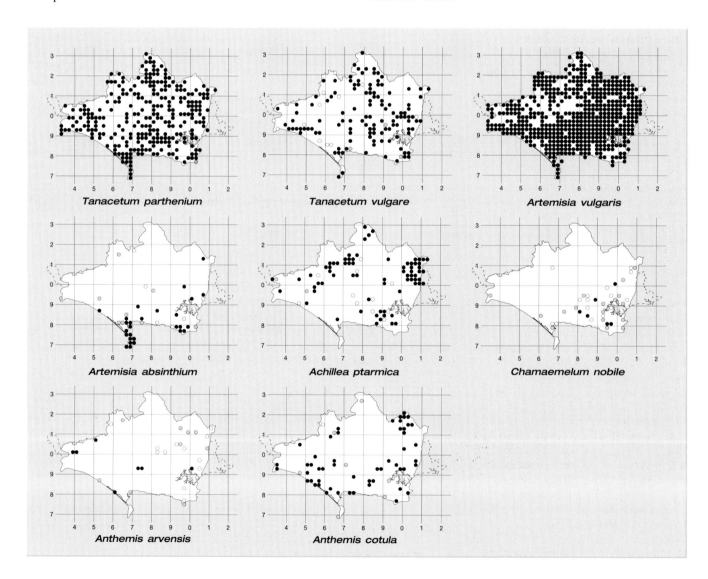

Artemisia vulgaris L.
MUGWORT
Common along verges, in urban wasteland and on tips.
526. It forms linear strips along many major roads, but
seldom spreads to semi-native habitats. Good found it
rather local and unevenly distributed in the 1930s, and it
is probably increasing.

Artemisia verlotiorum Lamotte
CHINESE MUGWORT
Rare, in waste places or cliff-tops. 1 + [1]. Herston (07E, 1973,
WGT); New Swanage (08F, 1981, AJCB and 1998, EAP).

Artemisia absinthium L.
WORMWOOD
Occasional in dry, sunny places on calcareous soils in
Portland and Purbeck, where it may be native. 28 + [15].
A rare casual inland. First record: Portland (67, 1832, TBS
in **QMC**). The oil from this contains anabsinthin, artabsin
and other substances and was once used as a vermifuge.

Artemisia abrotanum L.
SOUTHERNWOOD, OLD MAN
A rare garden escape. 2. Lodmoor tip (68V, !); Tincleton
(79Q, !). It does not flower here.

[*Artemisia biennis* Willd.
SLENDER MUGWORT
A rare casual. 2. Radipole Lake (67U, 1926, MJA); Swanage (08F,
1916, CBG in **BMH**).]

[*Artemisia annua* L.
ANNUAL MUGWORT
Reported without a locality in 1931 by L Abell.]

Santolina chamaecyparissus L.
LAVENDER-COTTON
Planted in churchyards and escaped on sandy cliffs. 3.
Lynch (67P, !); St Georges Ch, Portland (67W, !);
Flaghead Chine (08P, 1994, !).

[*Otanthus maritimus* (L.) Hoffsgg. & Link
COTTONWEED
It once occurred on sandy beaches. 3. East of Burton Bradstock
(48Z, 1760s, J Lightfoot in **BM**, also J Trehearne, RP and MF);
Brownsea Is (08, 1790s, Anon); Poole (09, 1650s, W. Hudson); It
survived longest at Mudeford (19V, 1879–91) but in VC11.]

Achillea ptarmica L.
SNEEZEWORT
Local in water-meadows or heathy commons. 70 + [14].
Mostly on the clays of the Poole basin or on the Greensand
in the north-west, absent from the Chalk. Garden escapes
on verges, as at Tarrant Gunville (91G, !), most have
double flowers. Perhaps declining through drainage.

Achillea millefolium L.
YARROW
Abundant in short and tall grassland, cliffs, verges, lawns

and waste ground. 726. It spreads by creeping roots, and
is resistant to trampling, mowing and salt-spray. Forms
with pink flowers seen in 79V, 80Y, 91I and S, but these
may be impermanent chimaeras. Its volatile oil is a source
of camphor and linalool.

Achillea filipendulina Lam.
FERN-LEAF YARROW
A garden escape, self-sown on a wall at Bridport (49R,
1972, KEB, also 1985, RCP in **OXF**, !). 1.

Achillea clavennae × *umbellata* = *A.* × *kolbiana* Sunderm.
Planted on a wall at Upwey (68S, !, det. EJ Clement); also
at Gaunt's Common (00H). 2.

Chamaemelum nobile (L.) All.
CHAMOMILE
Now very local on grazed heaths and commons. 6 + [33].
Both JCMP and RDG found it frequent in heathy districts,
and its marked decline is thought to be due to the cessation
of grazing. Reports from cornfields by WBB are probably
errors. Post-1950 records: Lyme Regis (39, 1974, J Fairley);
Blacknoll (88D, 1962–98, !); High Wood (88M, 1983, P
Grey); Snails Bridge (89W, 1995, JHSC & M Heath, also
1998, HR Winship & S Davani); Furzebrook (98G,
1959, !); abundant at Corfe Common (98K and Q, 1961–
99, !, *et al.*); Upton CP (99W, 1980, J Hadfield); Pamphill
cricket ground (90V, 1997, HR Winship, !); Broadstone
(09C, 1976, H Thomas). Specimens are in **DOR, MDH,
WAR and WRHM.**

Anthemis punctata Vahl subsp. *cupaniana* (Tod ex Nyman)
Ros. Fernandes
SICILIAN CHAMOMILE
A garden escape or throw-out. 6. Morcombelake (49B,
1993, MJG, !); on wall, Shipton Gorge (49V, 1999, !);
Quarr (72S, 1998, !); Durlston (07I, 1999, EAP); cliffs,
New Swanage (08F, 1999, EAP); Broadstone Golf Course
(09D, 1996, !).

Anthemis arvensis L.
CORN CHAMOMILE
Once a scarce weed in calcareous arable, now rare and
sporadic and mostly on new verges. 8 + [20]. JCMP gave
14 localities, and RDG had it from 15 stands, mostly in
the north-east. Post-1950 records: Bettiscombe (30V and
40A, 1998, IPG, introduced); West Bexington (58I, 1973,
C O'Shea); Halstock Leigh (50D, 1984, A Ferguson);
arable, Herbury (68A, 1994, !); Overcombe (68V, 1978, !);
Sherborne (61N, 1970, SS); Troytown by-pass (79G and
L, 1992, !); Chettle (91L, 1960, AWG in **BDK**); Swanage
(07J, 2000, DCL); Holes Bay (09B, 1993, !).

Anthemis cotula L.
STINKING CHAMOMILE, MADDERS
Once a frequent arable weed, now infrequent though
locally abundant. 53 + [7]. It prefers bare places on sandy
or slightly acid clay soils, and is transient on disturbed
verges or new sand-pits. Iron Age and medieval seeds have

been found. JCMP found it common, RDG had it in 137 stands; its decline is probably due to the use of weedkilling sprays.

Anthemis tinctoria L.
YELLOW CHAMOMILE
A rare garden escape. 3 + [1]. Beaminster (40V, 1998, MJG, also IPG); Cold Harbour, Sherborne (61N, 1969, ACL); Winfrith Newburgh (88C, 1999, !); New Swanage Cliffs (08F, 1998, EAP). Reports from 60J, 79K and 89V need confirmation.

Chrysanthemum segetum L.
CORN MARIGOLD, BOTHERUM
A locally abundant arable weed on well-drained soils, usually acid, but occasionally on chalk or limestone. 88 + [14]. Seeds occur in Roman deposits; JCMP found it abundant in the heathy districts, and Good had it from 57 stands. It is declining on farms where weedkillers are used, but appears on disturbed sites such as new verges and in broad-leaved crops such as kale.

Chrysanthemum coronarium L.
CROWN DAISY
Once found as a garden escape on a cliff-top tip at Grove (77B, 1959, !). [1].

Dendranthema grandiflora agg.
FLORIST'S CHRYSANTHEMUM
Once found as a garden throw-out on a tip at Three-legged Cross (00X, 1982, !). 1.

Leucanthemella serotina (L.) Tzelev
AUTUMN OXEYE
Once found as a garden throw-out at Lydlinch Common (71L, 1970, !). 1.

Leucanthemum vulgare Lam.
OXEYE DAISY, BUTTER DAISY, HORSE DAISY, MOON DAISY, SUN DAISY
Frequent in calcareous grassland, on cliffs, and sometimes abundant on bare, roadside banks. 571. Variable in size and in the number of heads per stem.

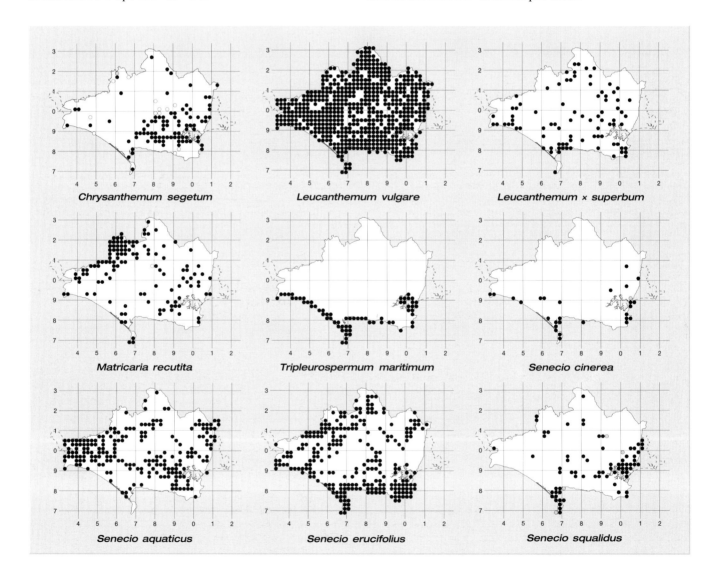

Chrysanthemum segetum

Leucanthemum vulgare

Leucanthemum × superbum

Matricaria recutita

Tripleurospermum maritimum

Senecio cinerea

Senecio aquaticus

Senecio erucifolius

Senecio squalidus

Leucanthemum × *superbum* (Bergmans ex J.Ingram) D.H. Kent
SHASTA DAISY
A frequent and persistent garden escape in village lanes, verges, railway banks, cliffs and tips. 85. It prefers calcareous soils. First record: Corfe (98, 1933, RDG).

Mauranthemum paludosum (Poir.) Vogt & Oberprieler
MARGUERITE
A self-sown, overwintering pavement weed at Wareham (98I, 1998–99, ! det. EJ Clement). 1.

Argyranthemum frutescens (L.) Schultz-Bip.
A self-sown seedling on Canford Heath tip (09I, 1999, !). 1.

Matricaria recutita L.
SCENTED MAYWEED
A locally common arable weed of well-drained soils. 122. Commonest on limestone and in the Poole basin. JCMP gives only three localities, and Good found it uncommon. Probably increasing. Its volatile oil contains many substances, and was once used to aid digestion.

Matricaria discoidea DC.
PINEAPPLE WEED
Common by farm tracks and as a pavement weed. 687. First recorded in 1908, and already well distributed in the 1930s.

Tripleurospermum maritimum (L.) Koch
SEA MAYWEED
Frequent on cliffs of clay, chalk or limestone, and on shingly or muddy beaches all along the coast. 66. Doubtfully distinct from *T. inodorum*, with which it hybridises. A double form was once seen on the Chesil Bank at Abbotsbury (58R, 1895, W Whitwell).

Tripleurospermum inodorum (L.) Schultz-Bip.
SCENTLESS MAYWEED
A very common arable weed on all types of soils, also in disturbed places, gardens and tips, known since the Bronze Age. 697. A double form occurred at Warren (89K, 1990, !).

(*Cotula coronopifolia* L.
BUTTONWEED
Reported from Stroud Bridge (89V, 1993, L Buckley), but perhaps erroneously. It usually grows on mud at the edge of ponds.)

(*Soliva pterosperma* (Juss.) Less. was found by FAW in a caravan park at Bournemouth, VC11, and may spread (Woodhead & Clement, 1997).)

Senecio cinerea DC.
SILVER RAGWORT
Locally abundant as a garden escape on sandy or limestone cliffs, rarely on verges or railway banks near houses elsewhere. 27 + [1]. First record: Swanage (07J, 1917, CBG); it still grows there.

Senecio cinerea × *jacobaea* = *S.* × *albescens* Burb. & Colgan
A transient hybrid, rarely seen on cliffs or waste ground near colonies of *S. cinerea*. 7 + [1]. Morcombelake (49B, !); South Perrott (40T, !); Bowleaze Cove (78A, !); Hod Hill (81K, TCGR); Wareham (98I, !); Swanage (07J, 1917, CBG); Bankes Arms, Studland (08G, !); Canford Heath tip (09I, !).

Senecio cinerea × *erucifolius* = *S.* × *thuretii* Briq. & Cavill.
Once seen in a garden at Ringstead (78K, 1977, !). 1.

Senecio inaequidens DC.
NARROW-LEAVED RAGWORT
An alien near Swanage (07J, 1999, DCL), which may spread. 1.

Senecio jacobaea L.
COMMON RAGWORT
Very common in dry, calcareous and neutral grassland, especially when overgrazed, and on verges, rarely on walls. 719. A notifiable weed, partly controlled on some verges by hand-pulling. It contains bitter alkaloids which are liver poisons that can kill horses, and is refused by rabbits. However it is eaten by larvae of the Cinnabar Moth. Hand-pulled or cut plants become sweet and so more palatable to stock.

Senecio jacobaea × *aquaticus* = *S.* × *ostenfeldii* Druce
This hybrid probably occurs wherever the parents meet, in wet woodland rides or pastures, but is under-recorded. 4. Birdsmoor Gate (30V, JE Hawksworth & GM Kay); Milborne St Andrew (89D, !); Corfe Common (98K, !); Gotham (01V, !).

(*Senecio jacobaea* × *erucifolius* was reported from Swanage (07J, 1930s, NDS), but is doubtful.)

Senecio aquaticus Hill
MARSH RAGWORT
Locally frequent in marshes, water-meadows and damp pastures. 204. Absent from the drier parts of the Chalk and limestone, including Portland.

Senecio erucifolius L.
HOARY RAGWORT
Frequent in mature, unimproved grassland, woodland rides, verges and railway banks. 250. Characteristic of more or less calcareous clay soils, and common along the coastal belt.

Senecio squalidus L.
OXFORD RAGWORT
Local, on disturbed, bare soil, cinderbeds and tips. 73 + [5]. It is particularly associated with rail tracks and stations, even when these are defunct; it also occurs as a pavement weed in Wareham, Weymouth, Swanage and Poole. In the latter site it extends to the beach. First record: Portland (67X, 1900, GCD), where it still persists on shingle near the defunct station.

Senecio squalidus × *vulgaris* = *S.* × *baxteri* Druce
A rare, casual hybrid of waste ground and tips. [3].
Beaminster (40Q, 1961, AWG in **BDK**); Woodsford (78U,
1977, !); Parkstone (09F, 1977, RDG).

Senecio vulgaris L.
GROUNDSEL
An abundant weed of dry arable, cliff-tops, beaches,
disturbed sites, gardens, urban wasteland and tips. 702.
The radiate form, var. *radiatus* Koch, is not uncommon.
(Var. *denticulatus* (O.F. Muell.) P.D. Sell has not been found,
but could occur on dunes).

Senecio vernalis Waldst. & Kit.
EASTERN GROUNDSEL
On a new verge, Puddletown by-pass (79M, 1999, ! det.
EJ Clement). 1.

Senecio sylvaticus L.
HEATH GROUNDSEL
Locally frequent in rides in acid woods, disturbed heaths
and rabbit warrens; rarely on clay cliffs near the sea. 160
+ [12]. Commonest in the Poole basin, and on Lower
Greensand soils.

Senecio viscosus L.
STICKY GROUNDSEL
Locally frequent and increasing in artificial habitats, rail
tracks, cinder beds, disturbed verges and tips. 64 + [7]. It
occurs on the beaches of Brownsea Is (08D, E and J) and
north of Arne (98U &Z). First record: Worgret (98D, 1922,
JHS).

Tephroseris integrifolia (L.) Holub
FIELD FLEAWORT
Very local, and in small numbers, in short, calcareous
turf. 5 + [10]. All but one of its localities are on the
north-east Chalk. It does not occur further west in Britain,
and is decreasing, due to ploughing of habitat
and subsequent failure of tiny populations. Post-
1900 records: Portland (67, 1949, WAC in **PVT** det. SME);
Hod Hill (81K and Q, 1767, RP, *et al.* in **DOR**, and 1983,
AH); Pimperne (90A, 1957, JAL, also F Partridge);
Badbury Rings (90R, 1895, JCMP, *et al.* in **DOR**, and
1989, AH); Gussage Hill (91W, 1887, LKW and 1989,
AH); Oakley Down (01D, 1937, DM, *et al.* in **DOR,** and
1995, BE); Bokerley Dyke and Blagdon Hill (01P,
1983–97, AH, !).

Delairea odorata Lemaire
GERMAN-IVY
A scarce garden escape at Sandbanks (08N, 1980,
AJCB). 1.

Brachyglottis compacta × *laxifolia* = *B.* **Cv. 'Sunshine'**
C. Jeffrey
SHRUB RAGWORT
Widely planted along the coastal belt. 13. Naturalised
on sandy cliffs or dunes at Swanage (07J, EAP), Sandbanks

(08N, RMW, !) and Flag Head Chine (08P, !), also
seen in 39F, 67X, 61N, 78B, F and K, 79Q, 88I, 89D
and 98V.

Ligularia dentata (A. Gray) H. Hara
LEOPARDPLANT
A garden throw-out at Merley Ho (09E, 1999, !).

Doronicum pardalianches L.
LEOPARD'S-BANE
A garden plant, long established in plantations and verges,
usually in shade and well inland. 14 + [3]. First record:
Moreton (88E, 1846, MF). Post-1970 records: Forde Abbey
(30M, !); Drimpton (40H, IPG); Halstock (50I, AJCB);
Melbury Bubb (50Y, AH); Oborne (61U, 1970, ACL);
Piddletrenthide (70A, !); Melcombe (70V, AH);
Winterborne Stickland (80H, !); frequent in Bryanston
School grounds (80T, 1942, BNHS and 1990, !); Fontmell
Parva (81H, !); Iwerne Courtney (81L, 1982, AH); New
Swanage (08F, EAP); Horton (00I, GDF); Edmondsham
(01Q, GDF); Cranborne (01R, GDF).

Doronicum pardalianches × *plantagineum* L.
= *D.* × *willldenowii* (Rouy) A.W. Hill
A rare garden escape in a copse at Broadwindsor (40L,
1994, JGK). 1.

Doronicum plantagineum L.
PLANTAIN-LEAVED LEOPARD'S-BANE
A rare garden escape in plantations, perhaps confused
with its hybrid. 1 + [3]. Forde Abbey (30S, AWG);
Stavordale Wood (58S, 1895, I Hawkins); Sherborne Park
(61T, 1876, JCMP in **DOR** and 1970, ACL); East Down
(80K, 1991, !).

Tussilago farfara L.
COLT'S-FOOT
Frequent on verges, in old quarries and disturbed
ground such as unstable clay undercliffs. 291. Its
extensive roots tolerate waterlogging and anoxic
conditions. Its distribution pattern is unusual,
and unrelated to underlying geology. The plant was
used by a Dorset well-sinker to indicate underground
water.

Petasites hybridus (L.) P. Gaertner, Meyer & Scherb.
BUTTERBUR
Occasional, forming colonies along streams, and
sometimes on verges. 143 + [2]. Male plants are common
in the west and north, but absent from acid soils and also
Purbeck. Probably a long established introduction by
farmers or herbalists.

Petasites japonicus (Siebold & Zucc.) Maxim.
GIANT BUTTERBUR
Introduced, and naturalised in wet places in a few wild
gardens. 6. Wootton Fitzpaine (39S, 1968, JB, *et al.*, and
1996, IPG); Abbotsbury gardens (58S, 1984–93, !); Litton
Cheney (59K, !); Melbury Bubb (50X, 1978, !); Spring

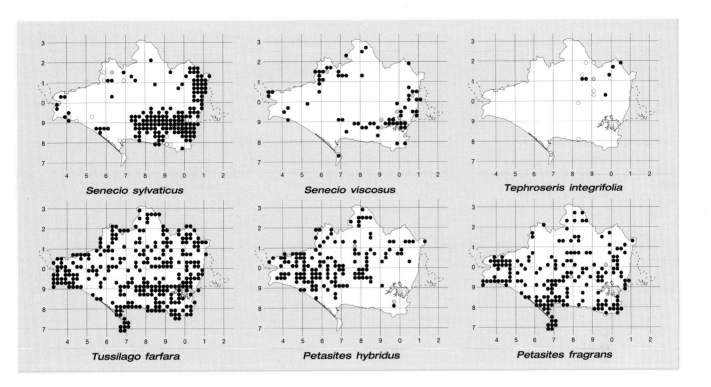

Senecio sylvaticus   Senecio viscosus   Tephroseris integrifolia

Tussilago farfara   Petasites hybridus   Petasites fragrans

Bottom (78K and L, 1948–93, ! in **RNG**); Shillingstone (81F, !).

Petasites fragrans (Villars) C. Presl
WINTER HELIOTROPE

A garden escape, now forming colonies in plantations, on verges and on sandy cliffs. 183 + [3]. It prefers light shade, which moderates the effects of winter frosts on its tender flowers. First record: Litton Cheney (59K, 1860, JCMP in **DOR**). Known in four sites in 1874, seven in 1895, becoming frequent by the 1930s and still spreading.

Calendula officinalis L.
POT MARIGOLD

Much grown in gardens, and escaping to waste ground and tips, often self-sown but transient. 34. It remains in flower in the coastal belt throughout a mild winter.

Osteospermum ecklonii (DC.) Norl.

A somewhat tender perennial planted on verges and outside gardens, as near Shaftesbury (82R).

Ambrosia artemisiifolia L.
RAGWEED

An uncommon casual found in allotments, near pheasant cages or on tips. 4 + [7]. Beaminster (40Q, 1963, AWG in **BDK**); Radipole Lake (68Q, 1976, DT Ireland); Hurst Heath (78Z, 1978, !); Binnegar tip (88Y, 1995, !); Bedchester (81S, 1996, !); Corfe (98R, 1915, CBG); Ower (98X, 1934, BNSS); Swanage (07J, 1999, DCL); Greenland Farm (08C, 1999, EAP);

Goathorn (08D, 1915, CBG in **BMH**):; Swanage Camp (08F, 1915, CBG in **BMH**); Studland (08G, 1917, CBG).

[*Ambrosia maritima* L. persisted for a few years at Middlebere Quay (98T, 1919–21, EJ Salisbury).]

[*Ambrosia psilostachya* DC. Swanage area (08, 1925, LBH in **K**).]

[*Ambrosia trifida* L. was a casual at Swanage Camp (08F, 1917, CBG in **BMH**).]

Xanthium strumarium L.
ROUGH COCKLEBUR

A rare casual, reported from Wareham in 1799 (RP). 2 + [1]. Recent records from Kimmeridge beach (97E, 1994, !) and Honeybrook Copse (00B, 1994, BE).

Xanthium spinosum L.
SPINY COCKLEBUR

A rare casual. 1 + [4]. Stoke Mill Farm (49I, 1976, AWG); Beaminster (40Q, 1950, AWG in **BDK**); Sherborne tip (61H, 1970, ACL); Radipole Lake (67U, 1926, MJA); West Moors (00W, 1995, GDF and JO).

Guizotia abyssinica (L.fil.) Cass.
NIGER

A casual from bird-seed, on fermenting rubbish in tips. 3 + [5]. Beaminster (40Q, 1963, AWG in **BDK**); The Nothe (67Z, 1977, !); Lodmoor tip (68V, 1997, !); Sherborne tip (61H, 1970, ACL); Binnegar tip (88Y, 1994–95, !); Fleets Corner (09B, 1977, !); Parkstone (09, 1930s, NDS); Canford Tip (09I, 1999, !).

Rudbeckia hirta L.
BLACK-EYED-SUSAN
A casual on a verge at Fiddleford (81B, 1992, ! in **RNG**); Greenland Farm (08C, 1999, EAP). 2.

Rudbeckia fulgida Aiton
A casual on Woodsford tip (78U, 1978, ! in **RNG**). [1].

Rudbeckia laciniata L.
CONEFLOWER
A garden escape. Sherborne tip (61, 1970, ACL); Dewlish (79U, 1991, !). 1 + [1].

Helianthus annuus L.
SUNFLOWER
Sometimes sown as a crop, or as a windbreak. 16. Noted as a relic in arable, allotments or tips, and as a bird-seed alien.

Helianthus tuberosus L.
JERUSALEM ARTICHOKE
Uncommon in corners of arable, probably as pheasant cover, and a casual on tips, mostly sterile. 8 + [2]. Rodwell (67U, 1960, !); Sherborne tip (61H, 1970, ACL); Lyddon Ho (70J, !); Binnegar tip (88Y, !); Wigmoor Coppice (81M, !); Almer (99E, !); Witchampton (90C, DAP); Greenland Farm (08C, !); Swanage (08F, EAP); Canford Tip (09I, !, in flower).

Helianthus tuberosus × rigidus = H. × laetiflorus Pers.
PERENNIAL SUNFLOWER
An occasional, persistent garden escape on verges and streambanks. 6 + [1]. Salway Ash (49N, R Maycock & A Woods); Kingcombe (59P, IPG); Yeovil (51T, 1956, IAR); Sherborne (61H, 1970, ACL and 61N, !); Hazelbury Bryan (70J, !); Marnhull (71U, CT & !); Canford Tip (09I, !).

Helianthus petiolaris Nutt.
LESSER SUNFLOWER
A casual in sandy arable, Broomhill Bridge (88D, 1990, ! in **RNG**). 1.

Galinsoga parviflora Cav.
GALLANT SOLDIER
A rare weed in allotments, gardens or urban sites, usually in well-drained sandy soil. 5 + [4]. Easily confused with sparsely hairy forms of G. quadriradiata, and early recorders did not realise that there were two alien species. Netherbury (49U, 1930, AWG); East Stoke (88T, DAP); Blandford (80Y, !); Wareham (98I, !); Knoll Hill (99T, !); Ballard Down (08G, 1988, WH Tucker); Parkstone (09F, 1930s, RDG); Bear Cross (09M, 1965, MPY); Talbot (09S, 1964, MPY).

Galinsoga quadriradiata Ruiz Lopez & Pavon.
SHAGGY SOLDIER
Now locally frequent as a weed of nursery gardens and disturbed verges on sandy soils, also in urban sites, pavements and tips. 32 + [3]. First record: Swanage (07J, 1949, RR Abell, and still there in 1994, WGT, !). Increasing fast, but flowering late and so overlooked.

Bidens cernua L.
NODDING BUR-MARIGOLD
Occasional, but sporadic, on bare mud by rivers, ditches and ponds. 21 + [23]. Mostly in the lower valleys of the Frome, Piddle and Stour, and scattered in the north-west. A ligulate form was seen by one of the lakes in Melbury Park (50S, 1981, !). Decreasing through loss of habitat.

Bidens tripartita L.
TRIFID BUR-MARIGOLD
Occasional on mud by rivers, ditches and ponds. 60 + [8]. Mostly along the larger river valleys in the east, and absent from the south-west. JCMP thought this was less frequent than B. cernua, but this is not so today.

Coreopsis tinctoria Nutt.
TICKSEED
A rare casual. Lodmoor tip (68V, 1957, !); East Stoke (88T, 1977, ! in **RNG**). [2].

Cosmos bipinnatus Cav.
MEXICAN ASTER
A garden escape on disturbed soil in waste ground and tips. 2 + [1]. Puddletown (79M, 1992, !); Fleets Corner (09B, 1977, !); Canford tip (09I, 1998, !). It is never more than casual.

[**Hemizonia pungens** Torrey & A. Gray was reported from Weymouth (67, 1930s, AWG).]

[**Madia sativa** Molina Casual in a sown ley, Beaminster (40R, 1942, AWG in **BDK** and **RNG**).]

Tagetes erecta L.
AFRICAN MARIGOLD
A garden escape. 1. Canford Tip (09I, !).

Tagetes minuta L.
SOUTHERN MARIGOLD
A casual, bird-seed alien. Netherbury (49U, 1923, AWG); Parkstone (09F, 1955, ER Banham). [2].

Tagetes patula L.
FRENCH MARIGOLD
A bedding-plant which escapes to tips and pavements. 4. Lodmoor tip (68V, !); Winterbourne Abbas (69F, !); East Stour (82B, !); Wimborne (00A, !).

Eupatorium cannabinum L.
HEMP-AGRIMONY
Common in woodland rides, wet fields, edges of streams, verges, railway banks and undercliffs; sometimes an urban weed. 506. Least common on clay soils in the north. Forms with white flowers have been seen at Black Head (78G), Moreton (78U), Arne triangle (98S), Agglestone Road (08G, EAP) and New Swanage (08F, EAP). Larvae of the Scarlet Tiger Moth feed on the leaves, and the flowers are a source of nectar for many insects.

MONOCOTYLEDONS

BUTOMACEAE

Butomus umbellatus L.
FLOWERING-RUSH
Frequent in shallow water at the edge of the River Stour from Five Bridges (72F) to Dudsbury (09U); less common in the Frome, Piddle and their ditches, and rarely in ponds, unless planted. 33 + [10]. Normally triploid, but a diploid form occurs at East Stoke (88T, CDP).

ALISMATACEAE

Sagittaria sagittifolia L.
ARROWHEAD
Frequent in the River Stour, also in the lower courses of the Frome, Piddle and Moors River, and south of Yeovil. 52 + [3]. Rarely planted in ornamental ponds, but sometimes in flooded pits.

Sagittaria latifolia L.
DUCK-POTATO
Introduced by two ponds west of Swanage (07E, 1999, EAP). 1.

Baldellia ranunculoides (L.) Parl.
LESSER WATER-PLANTAIN
Local, in and by muddy ponds and ditches on acid soils. 16 + [16]. Now mostly in the south of the Poole basin, with outliers near Bexington and Powerstock Common; gone from the north. Good only found it in four sites.

Alisma plantago-aquatica L.
WATER-PLANTAIN
Occasional on mud in or by the edge of rivers, lakes, ponds, new pits and ditches. 172. Mainly in the lowland river basins, and absent from most of the Chalk, with an outlier at Culverwell (66Z).

Alisma lanceolatum With.
NARROW-LEAVED WATER-PLANTAIN
Rare and scattered by muddy ponds or ditches. 5 + [3].

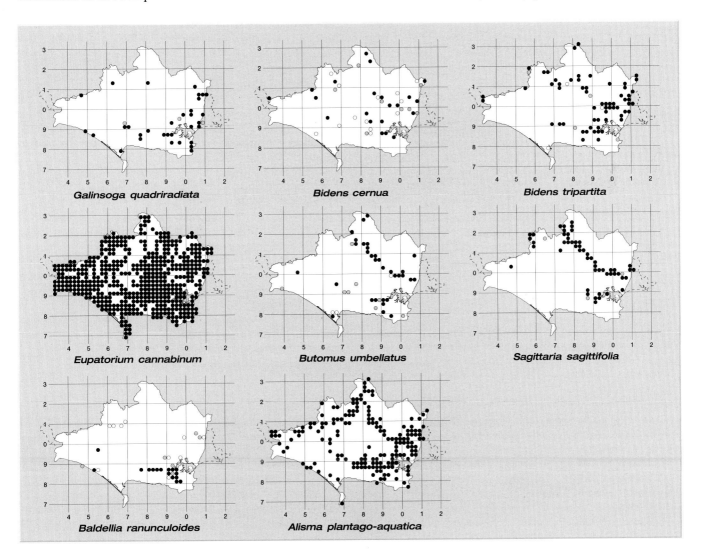

Galinsoga quadriradiata

Bidens cernua

Bidens tripartita

Eupatorium cannabinum

Butomus umbellatus

Sagittaria sagittifolia

Baldellia ranunculoides

Alisma plantago-aquatica

Post-1980 records: Burton Mere (58D, SME); West Bexington (58I, SME); Wareham Common (98D, CDP); Longham (09T, 1996, TCGR); Sutton Holms (01K, JHSC, !).

HYDROCHARITACEAE

Hydrocharis morsus-ranae L.
FROGBIT
Rare, floating in still water. 1 + [2]. Reported from East Stoke (88T, 1840, E. Fox) and Wareham (98, 1799, RP); A large colony in a flooded pit near Hyde (89Q, 1996, !), with native aquatics.

Stratiotes aloides L.
WATER-SOLDIER
Occasionally planted in garden ponds, escaped into ponds at Ilsington Wood (79L, !); Creech Heath (98G, NFS and BE, !), Creekmoor (99W, NFS) and Studland village (08G, EAP). 4. Abundant in Horseshoe Pond (09V, FAW) but in VC11.

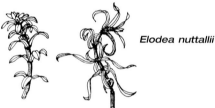

Elodea canadensis   *Elodea nuttallii*

Elodea canadensis Michaux
CANADIAN WATERWEED
Now frequent or dominant in some mature ponds, also in lakes, rivers and streams, often in shade. 57 + [12]. Most frequent in the valleys of the larger rivers. First record: Radipole (68, 1848, CC Babington). The inconspicuous flowers have not been reported.

Elodea nuttallii (Planchon) St John
NUTTALL'S WATERWEED
Fairly frequent in ponds and ditches in the Frome and Piddle basins, and scattered elsewhere. 28. First record: Moors River (10A, 1984, RMW), so it must be spreading.

Lagarosiphon major (Ridley) Moss
CURLY WATERWEED
An escape from aquarists in artificial ponds, old pits, and ponds in firebreaks. 33. First record: Guys Marsh (82K, 1983, AJCB).

APONOGETONACEAE

Aponogeton distachyos L. f.
CAPE-PONDWEED
A rare and usually transient escape from garden ponds. 7. It is well established in a stream near Swanage Sta (07J, 1997, MJG), and has also been seen in 58S, 78R, 89T, 97A, 99Z and 08P.

JUNCAGINACEAE

Triglochin palustre L.
MARSH ARROWGRASS
Local in marshes, fens and water-meadows, sometimes on undercliffs, as at Stair Hole (87J), Worbarrow Tout (87U) and Chapmans Pool (97N, AH). 30 + [14]. Commonest in the south-east.

Triglochin maritimum L.
SEA ARROWGRASS
Local in salt-marshes, and on muddy shores inside the Fleet and Poole Harbour. 29 + [4]. It has not been seen on Portland since 1881.

POTAMOGETONACEAE

Potamogeton natans L.
BROAD-LEAVED PONDWEED
Locally frequent to dominant in rivers, ditches, old pits and ponds, often in shade. 126 + [8]. It prefers some eutrophication, but is commonest in the Poole basin in waters which are not far from pH 7; absent from most of the Chalk, where ponds are scarce.

Potamogeton natans × *nodosus* = *P.* × *schreberi* G. Fischer
Rare in the upper reaches of the Rivers Stour and Cale. 3. First noted near Marnhull (71U, 1992, CT & !), later named by CDP as this hybrid, new to Britain (Preston, 1995), with additional sites in 71T and 72Q, in **CGE**. Probably seen by RDG (1968) and AJCB (1981) and recorded as *P. nodosus*.

Potamogeton natans × *lucens* = *P.* × *fluitans* Roth
Local, and only in the Moors River on the Hampshire border, where it is not rare. 3. Tricketts Cross and St Leonards bridge (00V and W, 1938, PMH in **BM**, *et al.*, and 1996, RSRF & !); Palmers Ford (10A, 1946, PMH in **BM**, and 1994, CDP in **CGE**).

Potamogeton polygonifolius Pourr.
BOG PONDWEED
Locally frequent in peaty streams and valley bogs, always in acid water, surviving in shade. 83. Very local in the west, and common in the heath districts of the Poole basin.

Potamogeton nodosus Poiret
LODDON PONDWEED
Only found in the middle reaches of the Stour, where it is locally frequent. 6. Durweston (80P and U, DAP and RF, !); Bryanston (80T and Y, 1965, HJB and 1988, DAP and RF, !); Shillingstone (81F, 1938, JE Dandy, *et al.*, and 1982–99, !); Haywards Bridge (81G, 1919, AWG in **BDK**, *et al.* in **BM, CGE, HDD, OXF and RNG**, and 1982–99, !).

Potamogeton lucens L.
SHINING PONDWEED
Local in the larger rivers. 16 + [5]. First found in the Stour

at Spetisbury (90B, 1867, JCMP in **CGE**). Specimens from the Stour are in **E**, from the Frome in **BM, BMH, LDS and LIV**, from the Piddle in **UEA** and from the Moors River in **BM and LTR**.

Potamogeton lucens × perfoliatus = P. × salicifolius Wolfg.
WILLOW-LEAVED PONDWEED
Rare in the lower Frome, with old records from the Stour (some as **P. lucens × alpinus**). 3 + [3]. Bindon Mill (88N, 1917, IM Roper, *et al.* in **BM, BMH, OXF, NMW and RNG**, and 1997, CDP and DAP); East Stoke (88T, 1987–97, CDP); Wareham (98I, 1893, EFL, *et al.* in **BM, BMH, CGE, LDS and RNG** and 1997, CDP and DAP); Sturminster Marshall (99P, 1891, JCMP); Wimborne (00A?, 1882–84, Anon in **BM**).

(**Potamogeton gramineus** L. was reported from the Stour in 1799 by RP; a report from Little Sea in 1976 was incorrect.)

Potamogeton alpinus Balbis
RED PONDWEED
Scarce in the lower basins of the Frome and Piddle,

perhaps lost from the Moors River. 4 + [5]. Wool Bridge (88N, 1916, CBG in **BMH**); Ford Heath (88U, NFS); Trigon (88Z, 1900, EFL); East Holme (98C and D, DAP and NFS); Wareham (98I and J, 1890, JCMP, *et al.*, and 1988, CDP); West Moors (00R, 1896, EFL in **LTR**); Tricketts Cross (00V, 1938, RDG in **BM and WRHM**). Specimens are in **CGE, K, LDS, LIV, NMW, OXF, SDN and UEA**.

Potamogeton perfoliatus L.
PERFOLIATE PONDWEED
Uncommon in the Frome, Piddle and lower Stour, with old records from the Brit and Crane and a recent one from the Bride. 20 + [5]. It prefers clear, fast-flowing calcareous streams, but also occurs in acid water at Little Sea.

[**Potamogeton friesii** Rupr.
FLAT-STALKED PONDWEED
There are good specimens of this from the R. Frome at Wareham (98I, pre-1914, in **BM, BMH, CGE and K**), and an unconfirmed report from Little Sea (08H, 1976, JO Mountford). 2.]

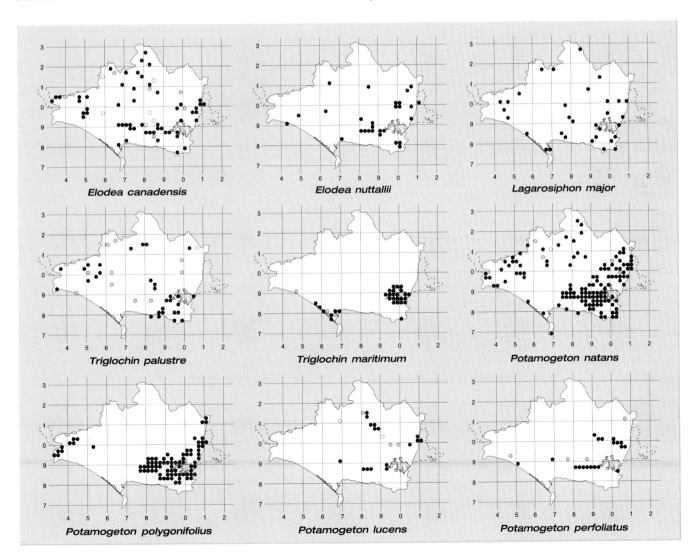

Elodea canadensis

Elodea nuttallii

Lagarosiphon major

Triglochin palustre

Triglochin maritimum

Potamogeton natans

Potamogeton polygonifolius

Potamogeton lucens

Potamogeton perfoliatus

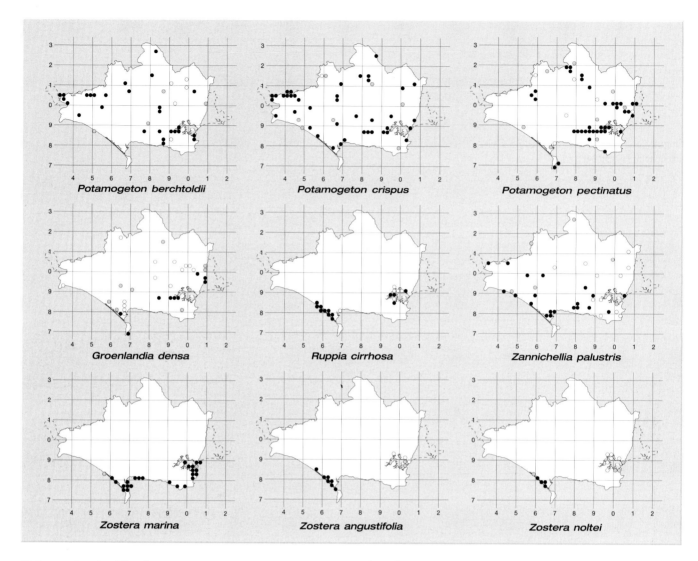

Potamogeton berchtoldii

Potamogeton crispus

Potamogeton pectinatus

Groenlandia densa

Ruppia cirrhosa

Zannichellia palustris

Zostera marina

Zostera angustifolia

Zostera noltei

Potamogeton pusillus L.
LESSER PONDWEED
Scattered in fairly mature lakes and streams in the south and south-east. 7 + [6]. Older reports confused this with *P. berchtoldii*. Seatown (49F, CDP); Chickerell (67P, CDP and DAP); Radipole Lake (68Q, 1963, ! in **BM**); Lodmoor (68V, 1999, CDP and DAP); Knighton pits (78N, BE and DAP); Encombe Lake (97P, 1937, NDS in **BM** and 1980–87, AH); Ridge (98I, 1915, IM Roper in **LDS** and 1962, ! in **BM**, also 98N, 1999, CDP and DAP); Little Sea (08H, 1882, HT Mennell in **BM**, *et al.* in **BMH**, and 1976, JO Mountford).

Potamogeton obtusifolius Mert. & Koch
BLUNT-LEAVED PONDWEED
Very local in lakes with acid water. 3. Coombe Keynes (88R, NFS); abundant in Little Sea (08H, 1973, M Loukes in **BM** and 1990–6, ACJ and DAP); Woodlands Park (00N, 1998, NFS).

Potamogeton berchtoldii Fieber
SMALL PONDWEED
Scarce in rivers, ditches, ponds and newly-flooded

pits. 27 + [9]. Scattered in the lowland river basins and in the north, also in Little Sea. Specimens are in **BDK, BM, BMH, CLEY, LIV, MANCH, SDN, UEA and WRHM.**

Potamogeton berchtoldii × acutifolius
= *P.* × *sudermanicus* Hagstr.
Very local in eutrophic ditches in both Frome and Piddle basins near Wareham (98I; Preston & Pearman, 1998). 1. Redcliff Farm and Stoborough meads (1920, JHS in **NMW**, *et al.* in **BDK, BM, NMW, OXF and RNG**, and 1997, CDP, !); Swineham (1996, B Pickess and DAP).

Potamogeton trichoides Cham & Schlecht.
HAIRLIKE PONDWEED
Local in still water of lakes or flooded pits. 7. Bowden Hill (61J, !); Chickerell (67P, CDP); Warmwell pit (78N,1999, !); Coombe Keynes (88R, 1962, ! in **BM** and 1990, NFS); Ford Heath (88U, NFS, !); Hyde House pits (89Q, !); Rockley Lake (99V, CDP and NFS). Tolerant of calcareous or acid water, but preferring oligotrophic conditions.

Potamogeton acutifolius Link
SHARP-LEAVED PONDWEED
Rare in eutrophic ditches north and south of Wareham (98I). Much less common than its hybrid, *P. × sudermanicus*. First found by JCMP in 1874, *et al.* in **BDK, BM, BMH, CGE, LIV** and **WRHM**. Also reported from The Moors (98N, 1968, DSR). 1 + [1].

Potamogeton crispus L.
CURLED PONDWEED
Occasional in the major rivers and adjacent ditches, more often seen in farm or garden ponds. 36. It prefers calcareous water.

Potamogeton pectinatus L.
FENNEL PONDWEED
Occasional, locally frequent in streams, ditches and ponds, and tolerant of polluted, rust-stained or brackish water. 41 + [9]. Commonest in the basins of the larger rivers, with colonies at Portland (66Z &77A), Melbury Park (50S), Encombe (97N) and in reed-swamp at South Haven (08I).

Groenlandia densa (L.) Fourr.
OPPOSITE-LEAVED PONDWEED
Now very local in ditches in water-meadows in the south and south-east. 9 + [24]. It was once much more widespread, but the reasons for its decline are unclear. Recent records: Portland Bird Observatory (66Z, 1982, RA); Chickerell pits (67P,1990, CDP and DAP); ditches east of Wool (88N, 1997, CDP and DAP); ditches near Wareham (98D and I, 1987–94, CDP, !); The Moors (98N, 1965, DSR in **UEA** and **WRHM**, and 1991, CDP); pond below Kingston Hill (98Q, 1976, RJ Pankhurst); by R. Stour, Canford Magna and Muscliff (09P, X and Y, 1991–97, FAW); Moors River (00V, 1978, NTH Holmes). Older specimens are in **BM, BMH** and **K**.

RUPPIACEAE

Ruppia maritima L.
BEAKED TASSELWEED
Scarce, but locally frequent in saline pools and ditches, and perhaps under-recorded. 7 + [8]. Its strongholds have always been inside the Fleet (recent records from 58S, 67J and 68A) and inside Poole Harbour (recently seen in 98I, N, T and 08H). Specimens are in **BDK, BMH, CGE and DOR**.

Ruppia cirrhosa (Petagna) Grande
SPIRAL TASSELWEED
Locally frequent to abundant in saline pools and ditches. 13 + [6]. Inside the Fleet it has some large colonies in 58R, S, V and W, 67J, N and P, 68A and F, and there are recent records from inside Poole Harbour from 98P, S and U and 09F. It has not been seen at Lodmoor (68V) since 1870, but could survive in the RSPB reserve.

Specimens are in **BDK, BM, BMH, CGE, DOR, K, LTR, NMW and UEA**.

ZANNICHELLIACEAE

Zannichellia palustris L.
HORNED PONDWEED
Rather local in streams, ponds and ditches, and tolerant of brackish water. 20 + [18]. Notably abundant in the lake at Bridehead (58Z). Easily overlooked, but it has probably declined in the north.

ZOSTERACEAE

Zostera marina L.
EELGRASS
Locally dominant below low-water mark on sandy or sheltered clay substrates, partly exposed at low spring tides, and often cast up on the beach by storms. 24 + [4]. Especially common inside Portland Harbour and in Studland Bay, absent on exposed coasts west of Abbotsbury. Eelgrass beds are an important habitat for small fish, molluscs and other invertebrates.

Zostera angustifolia (Hornem.) Reichb.
NARROW-LEAVED EELGRASS
Locally dominant in the Fleet between Abbotsbury and Small Mouth (Holmes, 1985). 7 + [5]. It is an important food for Swans, Brent Geese and ducks and a habitat for small invertebrates; dead material is abundant along the strandline. It probably still occurs in Poole Harbour, where it decreased markedly in the 1930s, but there are no recent records.

Zostera noltei Hornem.
DWARF EELGRASS
Locally abundant in the Fleet (67E, I, J and 68A). 4 + [10]. There are old records from nine tetrads inside Poole Harbour.

ARECACEAE

Phoenix dactylifera L.
DATE PALM
Seedlings seen at Sherborne tip (61I, 1970, ACL), Crossways tip (78P, 1999, !), Winterborne Kingston (89P, 1991–96, !) and Parkstone (09F, 1948, NDS); they never survive our winters.

Trachycarpus fortunei (Hook.) HA Wendl.
CHUSAN PALM
Quite large, isolated trees occur in a few plantations and parks, as at Little Wood (98B) and Brownsea Is (08J). 11. Seedlings are numerous at Abbotsbury gardens (58S), where some trees are a century old, and by a palm avenue in a garden at Rodwell (67Z). Other trees noted in 59K, 60S, 61N, 79E and W, 70W, 88V and 08U.

ARACEAE

Acorus calamus L.
SWEET FLAG
Long established by Sherborne Park Lake and the R. Yeo draining from it (61N), also by the Moors River (10A, 1984, RMW). 2 + [1]. There are old records from the mid-Stour (81K, 1840s, E. Fox). Its volatile oil contains asarone, and the plant may be carcinogenic.

Lysichiton americanus Hulten & St John
AMERICAN SKUNK-CABBAGE
In wild gardens on acid soil, and established by swamps in plantations. 11. Wootton Fitzpaine (39S, 1989, DAP and RF); Forde Abbey (30M, !); Beaminster (40V, LJM); Abbotsbury (58S, !); Melbury Park (50T, !); Cerne (60Q, 1981, AJCB and 1991, !); Minterne Parva (60R, 1984, !); Little Wood, Creech (98B, 1973–92, !); Brownsea Is (08I, !); Luscombe valley (08P, 1992, BE, !); St Leonards (10A, 1982–87, RMW).

Lysichiton camtschatcensis (L.) Schott
ASIAN SKUNK-CABBAGE
Planted by streams in wild gardens at Abbotsbury (58S) and Minterne Magna (60S). 2.

Zantedeschia aethiopica (L.) Sprengel
ALTAR-LILY
Planted in ponds and ditches and surviving outside gardens in the south. 7. Seen in 58S, 69J, 60Q, 78B and S, 98A and 99W.

Arum maculatum L.
LORDS-AND-LADIES, COWS-AND-CALVES, FROG'S-MEAT, SNAKE'S-FOOD
Common in woods, scrub, and hedgebanks on calcareous or neutral soils. 671. Scarce or absent from very acid soils in the Poole basin, as on the Arne peninsula, and not certainly recorded from Portland. The root is somewhat toxic.

Arum maculatum × italicum was once noted at Durlston CP (07I, 1964, AJS). [1].

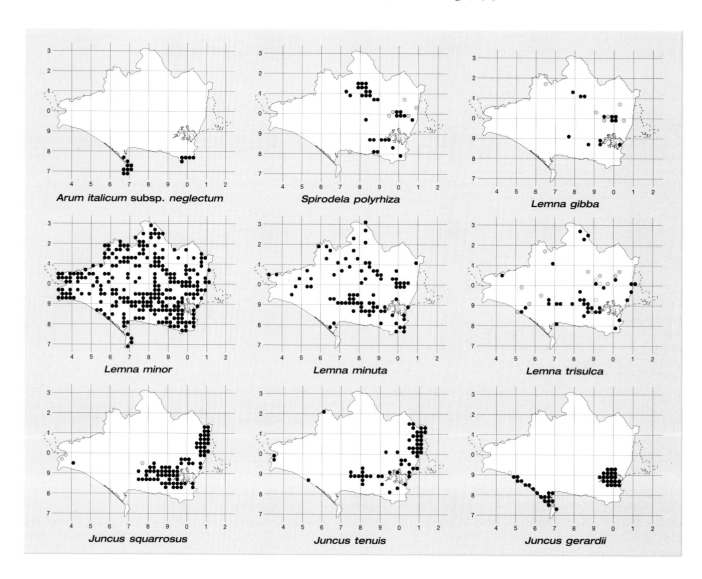

Arum italicum subsp. neglectum   Spirodela polyrhiza   Lemna gibba

Lemna minor   Lemna minuta   Lemna trisulca

Juncus squarrosus   Juncus tenuis   Juncus gerardii

Arum italicum Miller
ITALIAN LORDS-AND-LADIES
Subsp. **neglectum** (F. Towns.) Prime is local in limestone grassland and undercliffs on Portland and south-east Purbeck. 14 + [1]. Specimens are in **BDK** and **BM**. The roots were gathered to make Portland Sago or Arrowroot from the 16th to the 19th century, despite their toxicity; this was eaten in times of famine or sent to London for sale.
Subsp. *italicum* is rarely naturalised in churchyards, on verges or in hedgebanks near houses, sometimes as var. *pictum* with cream-veined leaves which appear in November. 17 + [5]. It was on sale in Blandford in 1782. Sizeable colonies occur at Coombe Keynes (88M, !) and Ridge (98I, JMB).

Dracunculus vulgaris Schott.
DRAGON ARUM
Appears sporadically on a verge in Wool (88N) as a garden escape. 1.

Arisarum proboscideum (L.) Savi
MOUSETAILPLANT
Once found in a hedge near Batcombe (60C, 1964, AE McR Pearse). [1]

LEMNACEAE

Spirodela polyrhiza (L.) Schleiden
GREATER DUCKWEED
Locally abundant in the Upper and Middle Stour, and local in a few eutrophic ponds, streams and ditches east of Dorchester. 32 + [6]. Not seen in flower.

Lemna gibba L.
FAT DUCKWEED
Occasional along the larger rivers, as well as the Yeo in the 1960s. 13 + [8]. Almost certainly under-recorded, as it does not fatten up until autumn.

Lemna minor L.
COMMON DUCKWEED
Widespread and often abundant in rivers, lakes and ponds or on mud. 269. Not seen in flower.

Lemna minor x 3  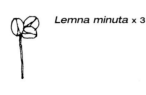 *Lemna minuta* x 3

Lemna trisulca L.
IVY-LEAVED DUCKWEED
Occasional in rivers, ditches and mature ponds, tolerant of shade and often deeply submerged. 34 + [14]. Most frequent in the Frome and Piddle basins, but probably under-recorded.

Lemna minuta L.
LEAST DUCKWEED
Although not noticed until 1988, this has spread fast to many rivers, streams and ponds, especially shaded ponds in woods or plantations, where it is sometimes dominant. 91. Most frequent in the Frome and Piddle basins.

(*Wolffia arrhiza* (L.) Horkel ex Wimm. is locally abundant in VCs 5 and 6, and easily overlooked. It occurred near the county boundary at Wincombe Park, Wilts (82X, 1946, JD Grose.)

COMMELINACEAE

Tradescantia virginiana L.
SPIDERWORT
Known as a garden plant since 1782, this escapes rarely. 2 + [1]. Buckhorn Weston (72M, !); Corfe Common (98Q, 1959, WB Alexander); Woodlands (00P, GDF). All records are transient.

Tradescantia fluminensis Vell. Conc.
WANDERING-JEW
A garden plant, established under a grating in Dorchester (69V, DAP, !). 1.

JUNCACEAE

Juncus squarrosus L.
HEATH RUSH
In wet or dry heaths. 76 + [3]. Frequent in the Poole basin, with very few records in the west: Champernhayes (39N, 1970, AWG); Lamberts Castle (39U, 1950, JMPB); Hardown Hill (49C, !). It forms rings on bare heath near Stroud Bridge (89V), owing to the plant dying in the centre.

Juncus tenuis Willd.
SLENDER RUSH
This introduction has spread along many tracks in heaths or acid woods, and near shallow pits at Crossways ((78U). 57. First record: Poole Park (09K, 1914, CBG in **BMH**). Locally frequent in the Poole basin, with outliers at Lamberts Castle (39U, 1927, AWG in **BDK** and 1938, DM); Dodpen Hill (39P, !); Limekiln Hill (58I, SME) and Sandford Orcas (62A, JAG).

(*Juncus compressus* L.
ROUND-FRUITED RUSH
This may have been collected at Crichel Pond (90Y, 1899–1905, WRL in **LIV**), but all other reports belong to *J. gerardii* or some other species. It may colonise the edge of old gravel-pits, as at Winkton Common (19M, 1994) in VC11.)

Juncus gerardii Lois.
SALTMARSH RUSH
Locally abundant in the upper parts of salt-marshes and on muddy shores between West Bay and Weymouth, and all round the inner shores of Poole Harbour and its islands. 39 + [2].

Juncus foliosus Desf.
LEAFY RUSH
Scarce or under-recorded, and mostly in the west. 7. It prefers damp places on acid or clay soils. Monkton Wyld Wood (39I, JGK); Pilsdon Pen (40A, JGK, also IPG); Hooke Park (59J and 50F, LJM and LMS); Chickerell (67P, CDP); Chapmans Pool (97N, BE); Slepe (98N, 1982, AJCB).

Juncus bufonius

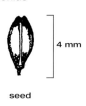

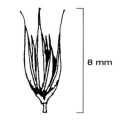

seed

flower with 6 tepals

Juncus bufonius L.
TOAD RUSH
Frequent in damp arable or tracks, at the edges of ponds and streams or on undercliffs. 405. Widespread, but scarce on the Chalk. Very variable in size and habit.

Juncus ambiguus

seed

flower

Juncus ambiguus Guss.
FROG RUSH
Rare or overlooked in wet, muddy places along the coast. 6 + [1]. Eype (49K, DAP and RF); Lodmoor (68V, 1999, CDP and DAP); Ringstead (78Q, DAP); Lulworth Cove (87J, AJ Showler); Chapman's Pool (97N, DAP); Wareham (98, 1916, GCD); Greenland Farm (08C, 1999, !).

Juncus subnodulosus Schrank.
BLUNT-FLOWERED RUSH
Local in fens and marshy fields, sometimes locally dominant, as at Whiteway and Povington (88Q and W, BE and DAP), The Moors (98N and P) and Corfe River (98R). 28 + [16]. Mainly in the southern Poole basin and the mid-west, and some sites have been lost through drainage.

Juncus articulatus L.
JOINTED RUSH
Frequent in marshy fields, wet woodland rides, wet heaths and by ponds on neutral to acid gleys. 326. Absent from most of the Chalk, and on Portland only at Culver Well (66Z).

Juncus articulatus × acutiflorus = J. × surrejanus Druce ex Stace
Rare or overlooked in wet places. 4. Low Down Farm (40B, IPG); Coppleridge (82N, L Biles); Ulwell (08A, EAP det. TA Cope); Little Sea and Studland Heath (08H, 1900, EFL, *et al.* in **DOR**, and 1992, B Radcliffe).

Juncus acutiflorus Ehrh. ex Hoffm.
SHARP-FLOWERED RUSH
Frequent and forming colonies on wet heaths and in marshes. 245. Absent from most of the Chalk.

Juncus bulbosus L.
BULBOUS RUSH
Frequent in bog pools and ditches, or by wet tracks in heaths and acid woods. 146. Local in the west, absent from the Chalk and limestone, and widespread in the Poole basin.

Juncus maritimus Lam.
SEA RUSH
Local in salt-marshes and on muddy shores, along the landward side of the Fleet, on an island in Radipole Lake, at Lodmoor and round the inside of Poole Harbour. 33 + [2]. Susceptible to trampling and human pressure, and lost from West Bay (49Q, 1949, AWG in **BDK**).

(**Juncus acutus** L.
All old reports of this were errors for *J. maritimus.*)

Juncus inflexus L.
HARD RUSH
Frequent in marshes and grassland on soils with impeded drainage. 493. More tolerant of lime than most rushes, but absent from most of the Chalk and scarce on Portland.

Juncus inflexus × effusus = J. × diffusus Hoppe
Usually as single clumps in marshes, but perhaps under-recorded. 5 + [15]. West Bourton (72U, RM Veall & !); Dungy Head (88A, !); Holes Bay (09A, !); Baiter (09F, !); Canford (09J, !). There are undated NCC records from Abbotsbury Castle (58T), Bere stream (89R) and Corfe Common (98Q).

Juncus effusus L.
SOFT-RUSH
The commonest rush in Dorset, sometimes locally dominant, in marshy fields, flushed wet heaths, and by ponds or lakes. 543. Absent or restricted to wooded valleys on the drier Chalk. The inflorescence may be compact or spreading.

Juncus conglomeratus L.
COMPACT RUSH
Frequent in wet grassland or heaths where there is impeded drainage. 254. Widespread in the north-west and the Poole basin, absent from the Chalk and limestone.

(*J. effusus × conglomeratus*, with a ridged stem and 30 vascular bundles, may well occur and have been overlooked.)

Luzula forsteri (Smith) DC.
SOUTHERN WOOD-RUSH
Local in old woods on calcareous soils. 19 + [9]. Most records from the north-east Chalk, with an outlier at Parnham (40Q, IPG). Specimens are in **DOR and WRHM**.

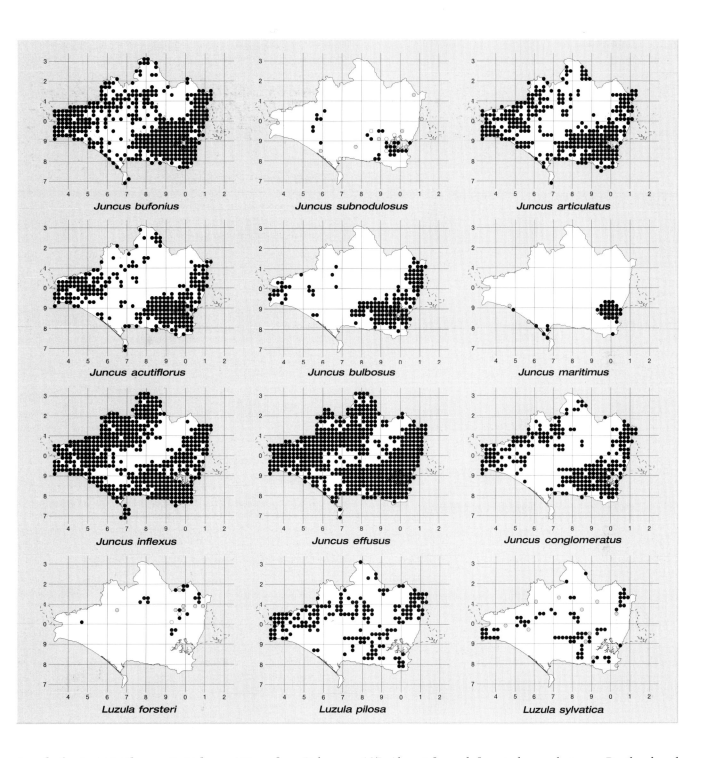

Juncus bufonius

Juncus subnodulosus

Juncus articulatus

Juncus acutiflorus

Juncus bulbosus

Juncus maritimus

Juncus inflexus

Juncus effusus

Juncus conglomeratus

Luzula forsteri

Luzula pilosa

Luzula sylvatica

Luzula forsteri × pilosa = L. × borreri Bromf. ex Bab.
Rare in calcareous woods in the north-east. 3 + [3]. Piddles
Wood (71W, 1920, AWG in **BDK**); Lytchett and
Charborough Park (99H and I, BE); High Wood, Badbury
(90R, BE); Woodcutts (91T, 1963, RDG in **DOR**);
Edmondsham (01Q, 1930s, RDG).

Luzula pilosa (L.) Willd.
HAIRY WOOD-RUSH
Occasional, on shaded banks or clearings in old woods.

187. Absent from deforested areas between Portland and
Dorchester, the coastal strip, and urbanised Poole.

Luzula sylvatica (Hudson) Gaudin
GREAT WOOD-RUSH
Local, but often forming large colonies on banks in old
woods, surviving when the woods have been replaced by
plantations. 73 + [12]. Good found it widespread but
uncommon in the 1930s, as it is today. It is absent from
deforested areas and very acid soils.

Luzula luzuloides (Lam.) Dandy & Wilm.
WHITE WOOD-RUSH
Introduced, as a single clump in Ensbury Wood (09Y, 1991, FAW). 1.

Luzula campestris (L.) DC.
FIELD WOOD-RUSH
Frequent in dry, unimproved grassland and lawns, also in grassy heaths, but easily overlooked in grazed turf. 432. Least common in intensively-farmed or built-up areas.

Luzula multiflora (Ehrh.) Lej.
HEATH WOOD-RUSH
Frequent in dry and wet heaths, acid woodland rides or verges. 172. Scattered in the north and west, absent from the Chalk and frequent throughout the Poole basin. Most Dorset material is subsp. *congesta* (Thuill.) Arcang., with compact heads.

CYPERACEAE
(Pearman, 1994)

Eriophorum angustifolium Honck.
COMMON COTTONGRASS
Locally frequent in acid bogs. 105 + [3]. In most parts of the Poole basin, rare on Greensand in the west, and even rarer in calcareous mires at Eype (49F), Seacombe (97Y, 1991, J Graham) and Culver Well on Portland (66Z). In the latter site it has survived since before 1856.

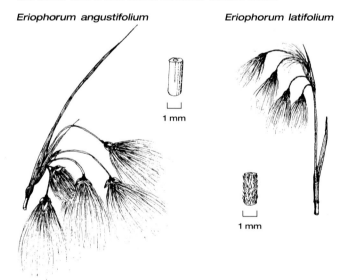

Eriophorum angustifolium   *Eriophorum latifolium*

1 mm   1 mm

Eriophorum latifolium Hoppe
BROAD-LEAVED COTTONGRASS
Rare in calcareous fens or flushes. 2 + [8]. Aunt Mary's Bottom (50L, 1986–98, DAP); Winfrith Heath (88D, 1968, !, probably gone); Hartland Fen (98M, 1970, SBC); The Moors (98N, 1965, DSR, !); Corfe River (98R, 1997, RMW, !). Old records include Powerstock Common (59I, 1933, AWG in **BDK**); East Morden (99C, 1895, JCMP); Cliff springs between Dancing Ledge and Tilly Whim (97Y and 07I, 1914, CBG in **BMH**, *et al.*, and 1920, JHS); Ulwell

and Punfield (08A and F, 1895, JCMP and 1911, CBG in **BMH**); Tricketts Cross (00V, 1938, RDG in **DOR** and **WRHM**). It has undoubtedly decreased, but it has not been refound in 25 of Good's sites, and some or all of these may have been *E. angustifolium*.

Eriophorum gracile Koch ex Roth
SLENDER COTTONGRASS
Once rare, now probably extinct in bogs. [5]. Three Barrow Heath and New Mills Heath (98H and M, 1930, FHH and 1940, JW Haines in **BM**); Godlingston Heath (08B, 1899–1909, EFL); South of Little Sea (08G and H, 1893, EFL in **LIV**, *et al.*, and 1955, !; the area where it grew, Pipley swamp, is now dense carr).

Eriophorum vaginatum L.
HARE'S-TAIL COTTONGRASS
Rare in wet heaths and bogs. 7 + [7]. Very local in the central Poole basin, and slightly less so near the Moors River on the Hampshire border. Recent records: Winfrith Heath (88D, 1986, SBC); Woolbridge and South Heaths (88P, 1981, JN and 1991, BE); East Holme (88Y, 1978, BSBI, also VJG); Turners Puddle (89F, 1996, BE); Hyde Heath (89K, 1984, RM); SE of Creech (98G, 1991, JRW); Morden Bog (99A, 1973, SBC); Parley Common (09Z, 1938, DM and 1974–81, HCP); Tricketts Cross (00V, 1995, GDF); Cranborne Common (01V, 1995, GDF); Lions Hill (10B, 1981, HCP).

Trichophorum cespitosum (L.) Hartman
DEERGRASS
Frequent, but never dominant, on dry to moist heaths, where it is one of the earliest plants to flower. 96 + [2]. Mostly in the Poole basin, with an outlier at Fishpond Bottom in the west (39U, 1977, JB). Grazed by deer.

Eleocharis palustris (L.) Roemer & Schultes
COMMON SPIKE-RUSH
Widespread but local in marshes and fens, and sometimes dominant in mature ponds. 156. Mostly in the north-west and in the Poole basin, where it prefers moderately eutrophic sites and avoids very acid places. There is an outlier on limestone at Culver Well (66Z). Almost all our plants are subsp. *vulgaris* Walters, but subsp. *palustris* has been found at Beaminster (40, 1921, AWG in **BDK**), by an angler's pond at Zelston (89Y, 1999, !) and among reeds at South Haven (08I, 1993, RSRF & ! in **RNG**).

Eleocharis uniglumis (Link) Schultes
SLENDER SPIKE-RUSH
Rare, at the edges of brackish or coastal marshes. 14 + [1]. Mainly around the shores of Poole Harbour, also at Ridge Cliff (39W, AWJ); Abbotsbury (58S, DAP); Culver Well (66Z, 1868, J Woods, *et al.*, and 1986, DAP); Radipole Lake (67U, 1960, RGB Roe) and Lodmoor (68V, 1989, RSPB). Specimens are in **BDK, BMH** and **RNG**.

Eleocharis multicaulis (Smith) Desv.
MANY-STALKED SPIKE-RUSH
Occasional in wet heaths and bogs, in wetter sites than

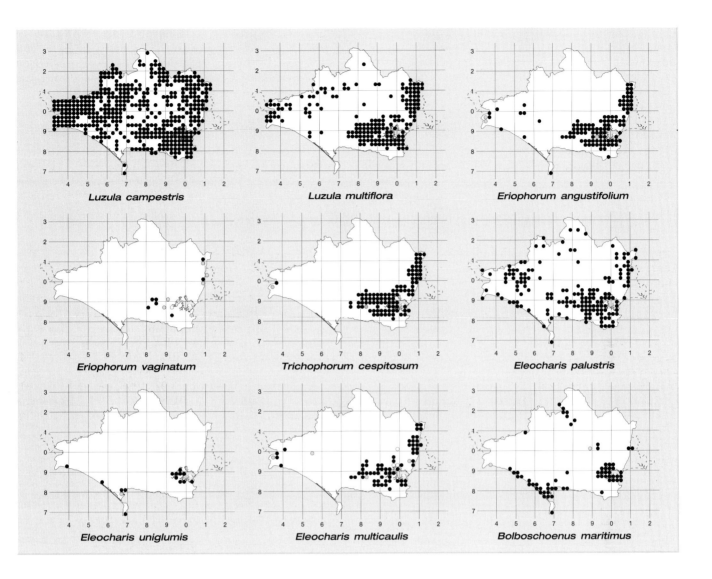

Luzula campestris

Luzula multiflora

Eriophorum angustifolium

Eriophorum vaginatum

Trichophorum cespitosum

Eleocharis palustris

Eleocharis uniglumis

Eleocharis multicaulis

Bolboschoenus maritimus

Trichophorum cespitosum, which it resembles. 65 + [7]. Mostly in the Poole basin, with a few sites in the west.

Eleocharis quinqueflora (F. Hartmann) Schwarz
FEW-FLOWERED SPIKE-RUSH
Scarce in fens or flushed bogs, in short vegetation. 6 + [9]. Champernhayes (39N, 1931, AWG in **BDK**); Rampisham (50L, DAP); Culver Well (66Z, 1896, JWW); two sites near Blacknoll (88D, AJB, also DAP); Stoborough (98H, 1986, AH & ! in **RNG**); Corfe Common (98K and Q, DAP in **RNG**, also FAW); Hartland Moor (98M, 1953–68, PJ Newbould); Slepe (98N, DAP); Morden Decoy (99A, 1895, JCMP in **DOR**); Verwood Common (00Y, 1930, AWG); Little Sea (08H, 1912, HJR in **RNG** and 1955, !); South Haven (08I, 1955, !).

Eleocharis acicularis (L.) Roemer & Schultes
NEEDLE SPIKE-RUSH
Very local and sporadic on mud in or by lakes and rivers. 3 + [3]. Coombe Keynes lake (88L and R, DAP, !); Holme Bridge (88Y, 1915, CBG in **BMH**, also JHS in **NMW**);

Gallows Hill (89K, !); Creech Pond (98B, 1961, DFW); Little Canford (09P, 1961, DFW).

Eleocharis parvula (Roemer & Schultes) Link ex Bluff, Nees & Schauer
DWARF SPIKE-RUSH
Rare or extinct at or below LWM on estuarine mud. Suitable habitat exists in the south-east corner of Poole Harbour, where the plant may still survive. [2]. Little Sea (08H, 1870, JCMP in **DOR**, *et al.* in **BDK, K, LIV** and **RNG**, last seen *c*.1976); South of Redhorn Quay (08H, 1935, RDG and C Day in **WRHM**); Ower (98X, 1933–35, BNSS). It still flourishes in Stanpit Marsh in VC11, where FAW has commented on how hard it is to find.

Bolboschoenus maritimus (L.) Palla
SEA CLUB-RUSH
Locally frequent in salt-marshes, brackish ditches and mouths of streams near the sea. 62 + [3]. Occasional along the coast from Lyme Regis to Ringstead Bay, especially inside the Fleet, and frequent on the shores of Poole Harbour.

It also occurs inland along the Stour between Higher Nyland and Sturminster Newton, and between Durweston and Spetisbury, also in the Moors River near Tricketts Cross (00V and 10A). Outliers include Sutton Bingham Reservoir (50P), Culver Well (66Z), a chalk spring at Sutton Poyntz (78C), a ditch at Crossways (78P) and Encombe Lake (97P).

Scirpus sylvaticus L.
WOOD CLUB-RUSH
Very locally dominant in marshy places in sun or shade.

26 + [12]. Scattered in the north and west, with outliers at Corfe claypits (98L and R, BE, DAP and NFS) and West Moors (00W, 1928, FHH and 1995, GDF and JO). Decreasing, and not refound in some old localities.

Scirpoides holoschoenus (L.) Sojak
ROUND-HEADED CLUB-RUSH
Very rare as a denizen. 1 + [1]. On made ground, Holes Bay (09B, 1991–97, BE, !); Parkstone (09F, 1918, E Rothwell in **ALT**).

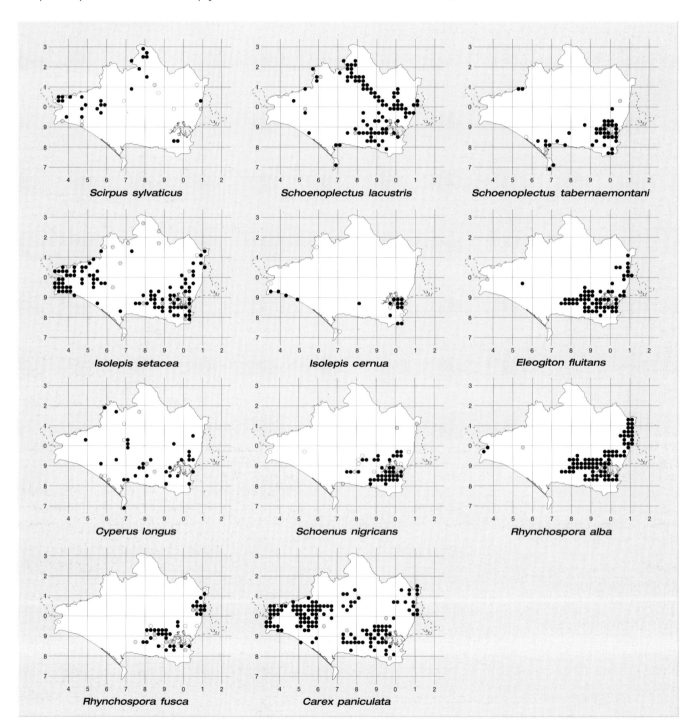

Schoenoplectus lacustris (L.) Palla
COMMON CLUB-RUSH
Frequent in the larger rivers, and locally dominant in some large ponds or flooded pits. 92 + [4]. Common in the Stour, Allen, lower Frome and Moors Rivers, with outliers in the north and west, and on Portland (67V).

Schoenoplectus tabernaemontani (C. Gmelin) Palla
GREY CLUB-RUSH
Occasional in more or less brackish ponds and ditches near the coast. 40 + [5]. Most frequent around Poole Harbour, and at Slepe (98S) it occurs in a bog with *Eriophorum*. Outliers include some inland sites, such as Sutton Bingham Reservoir (50J and P), Portland (66Z and 77A) and near Wimborne (09E). Much confused with *S. lacustris*.

Isolepis setacea (L.) R. Br.
BRISTLE CLUB-RUSH
Local and sporadic in bare, wet, acid soil of pond margins, tracks and firebreaks. 97 + [16]. Widespread in the west and the Poole basin, absent from the Chalk, and lost from old sites in the north and on Portland.

Isolepis cernua (Vahl) Roemer & Schultes
SLENDER CLUB-RUSH
Rare and in small numbers in bare, wet, acid soil, almost always near the coast. 16 + [10]. Inland records are from Winfrith Heath (88D, 1963–78, !) and Corfe Common (98Q, 1993, RMW and 1999, DAP, !).

Eleogiton fluitans (L.) Link
FLOATING CLUB-RUSH
Locally frequent to dominant in ponds, flooded pits or ditches with acid water. 65. Mostly in heathland of the Poole basin, with outliers at Fishpond Bottom (39U) and Powerstock Common (59N) in the west.

Blysmus compressus (L.) Panzer ex Link
FLAT-SEDGE
Very local, and decreasing, in grazed, damp pastures. 2 + [7]. Wyke Regis (67T, 1914, Mr Chapman in **WRHM**); Empool and Warmwell Mill (78I and N, 1878, WBB); Roke Pond (89J, 1874, JCMP, *et al.*, and 1991, DAP, !); Cowgrove (99Z, 1900, EFL, *et al.* in **BDK** and **BMH**, and 1948, KG); Spetisbury (90B, 1930s, RDG); Leigh Common (00F, 1930s, AWG); Wimborne St Giles (01F, 1963, KBR and 1990, DAP: abundant here).

Cyperus longus L.
GALINGALE
Very local as a native plant in marshes and by rivers; also a garden escape or planted by ponds or flooded pits. 12 + 18p + [8]. A record from 1688 by J Newton from a Chapel in Purbeck is puzzling. The colony at Ulwell (08F, 1847, E Hussey, *et al.*, and 1997, BE) has a long history, but is now reduced to a few stems. The largest native population is at Turners Piddle (89G, AH). In some sites, such as Culver Well (66Z), it is a matter of opinion how the colony originated.

Cyperus eragrostis Lam.
PALE GALINGALE
A garden escape, becoming naturalised. 4. Littlewindsor (40M, DAP); Sterte (09A, RMW); Canford Pits (09I, !); Alderholt (11G, GDF).

[***Cyperus fuscus*** L.
BROWN GALINGALE
Inconspicuous on eutrophic mud. [3]. Bere Regis (89M, 1881, JCMP in **DOR** and 1893–1909, EFL in **LIV**); Cowgrove (90V, 1929, HH Haines); Hartland Moor (98M, *c*.1910, IM Roper in her copy of Linton's *Flora of Bournemouth*).]

Cyperus involucratus Rottb.
Planted by a stream at Sutton Poyntz (77B, 1994–99, !), but unlikely to survive a severe winter. 1.

Schoenus nigricans L.
BLACK BOG-RUSH
Locally abundant in fens, flushed areas in bogs, and sometimes in brackish sites or on wet undercliffs. 38 + [12]. Now restricted to the Poole basin, but there are old records from Monkton Wyld (39N, 1895, FT Richards) and Powerstock Common (59I, 1840s, E Fox).

Rhynchospora alba (L.) Vahl
WHITE BEAK-SEDGE
Locally common in wet heaths and bogs, on bare or nearly bare peat. 83 + [2]. Mainly in the Poole basin, with a few outliers in the west.

Rhynchospora fusca (L.) W.T. Aiton
BROWN BEAK-SEDGE
Local in bogs and wet heaths, usually in bare or grazed sites. 34 + [20]. Now restricted to the Poole basin, with an outlier at Champernhayes (39N, AWG, 1924 and 1955, !) which may have gone. The region around Morden Bog (99) now holds the bulk of the British population of this plant.

Cladium mariscus (L.) Pohl
GREAT FEN-SEDGE
Very locally dominant in fens. 5 + [3]. [Reported from the Fleet at Weymouth (67, 1799, RP) and Abbotsbury Swannery (58S, 1874, JCMP)]; Coombe Keynes Lake (88L and R, 1895, JCMP, *et al.*, and 1990, !); Keysworth (98J, 1900, EFL, *et al.* in **BMH** and **WRHM**, and 1986, DAP); The Moors (98N, 1966, DSR, *et al.* in **DOR**, and 1998, DAP, !); Morden Park Lake (99B, 1913, CBG in **BMH**); Whitefield or Morden Millpond (99C, 1895, JCMP and 1990, BE, !).

Carex paniculata L.
GREATER TUSSOCK-SEDGE
Locally frequent to dominant in alder woods, carr, by streams and flushes on acid soils, where it forms huge tussocks if undisturbed. 134 + [3]. Common on Greensand in the north-west, absent from the Chalk, and local in the Poole basin.

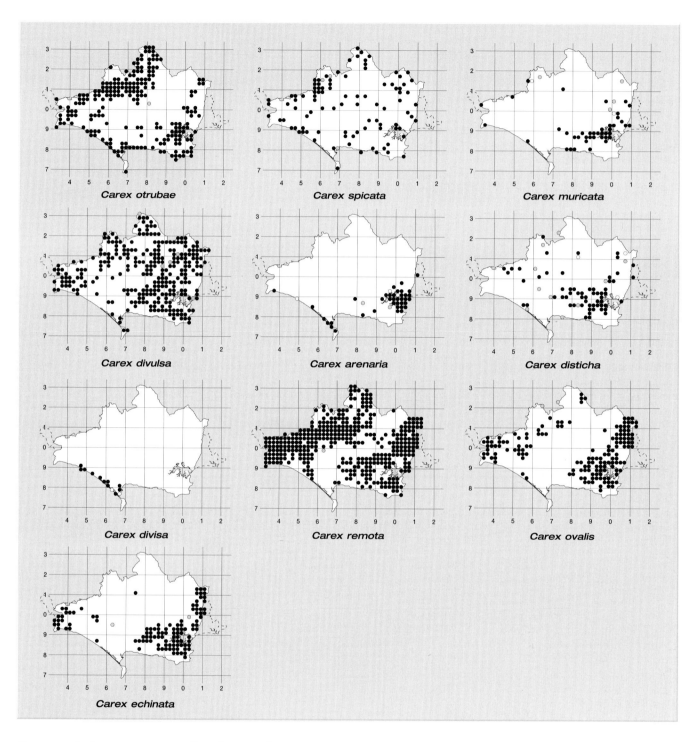

Carex otrubae

Carex spicata

Carex muricata

Carex divulsa

Carex arenaria

Carex disticha

Carex divisa

Carex remota

Carex ovalis

Carex echinata

Carex paniculata × remota = C. × boenninghausiana Weihe
Rare in carr. 5. Hooke Park (59J, DAP); Redholm Coppice (59P, DAP); Rooksmoor (71K, DAP and D Elton); Bere Stream (89L, DAP); Moors River (10A, 1987, RWD).

Carex diandra Schrank
LESSER TUSSOCK-SEDGE
Very rare in peaty ditches. 1 + [2]. The Moors (98N, 1965, DSR and 1998, AH and DAP, only four plants). [East

Morden (99C, 1888, JCMP in **DOR**); Seldown (09, 1830, TBS)].

(**Carex vulpina** L. all reports have been found to be *C. otrubae*, including plants from The Moors and from iron-rich ditches near Holton Heath Sta (99K), which are atypical.)

Carex otrubae Podp.
FALSE FOX-SEDGE
In fairly small numbers in ditches and wet places on clay

soils. 216. Most frequent in the north and west, also on wet undercliffs along the coast; scarce in iron-rich, acid ditches in the Poole basin and absent from the Chalk.

Carex otrubae × remota = C. × pseudoaxillaris K. Richter

A rare hybrid in wet, acid sites, often in shade. 3 + [5]. Clanden Hill (40L, 1921, AWG in **BM** and 1998, PRG); Horn Hill (40R, 1923–27, AWG in **BDK**); Mount Pleasant lane, Kingcombe (59P, DAP); Leigh (60J, 1895, WMR and 1987, DAP); between Ulwell and Swanage Sta (07J, 1871, JCMP in **DOR**, *et al.*, and 1914, CBG in **BMH**). There are pre-1900 reports from Lyme Regis (39) and Langton Matravers (97); reports from Affpuddle Heath (89B) and Castle Hill Wood (01R) need confirmation.

Carex spicata Hudson

SPIKED SEDGE
Widespread, though not in quantity, in dry hedgebanks on chalk or clay soils. 95 + [4].

Carex muricata L. subsp. lamprocarpa Celak

PRICKLY SEDGE
Occasional in grassy heaths, dry verges or railway banks on acid soils. 41 + [7]. This has been confused with *C. spicata*, but certainly occurs in many places in the Poole basin as well as near Holditch (30G, IPG) and Yeovil junction (51S, IPG) in the north-west.

Carex divulsa Stokes

GREY SEDGE
Occasional in dry hedgebanks and verges, woodland rides or scrub, always in small numbers. 276. It prefers calcareous soils. Subsp. *leersii* (Kneucker) W. Koch grows in similar habitats and in churchyards. 18. It is restricted to the Chalk, mostly in Purbeck and the north-east.

Carex arenaria L.

SAND SEDGE
Locally abundant on sandy cliffs and in dunes between Studland and Branksome Chine, and round the inner shores of Poole Harbour. 38 + [5]. A surprising outlier is on Portland (67W, BE), but a report from Lulworth Cove (88F) cannot be accepted without confirmation. It occurs in a few dry, sandy places inland, as at Red Bridge (78Z, BE), East Burton Heath (88I, 1960, RDG in **DOR**), Povington (88W, BE), Broadstone suburbs (99X, !), Oakdale cemetery (09G, FAW), Canford Heath (09I, RMW), Talbot Heath (09R, 1983, RM) and St Leonard's Hospital (10A, 1999, BE).

Carex disticha Hudson

BROWN SEDGE
Local in eutrophic water-meadows, marshes and fens. 60 + [15]. Commonest in the lower basins of the Frome and Piddle, but not along the Stour, and scattered elsewhere. Probably declining through ploughing and draining of its habitats.

Carex divisa Hudson

DIVIDED SEDGE
Very local in wet brackish marshes. 11 + [2]. Only

between West Bay and Weymouth, especially along the landward side of the Fleet, also reported from Culver Well (66Z, 1988, RF). Curiously, absent round Poole Harbour.

Carex remota L.

REMOTE SEDGE
Common in alder and other woods, often by shaded streamlets. 374. It dislikes competition, but can grow in deep shade. Absent from the drier parts of the Chalk, Portland, and very acid soils in the Poole basin.

Carex ovalis Gooden.

OVAL SEDGE
Occasional in dry, grassy heaths, rough commons and woodland rides, sometimes in water-meadows, but always on acid soils. 151 + [2]. It tolerates light grazing and slightly brackish sites.

Carex echinata Murray

STAR SEDGE
Locally common in peat bogs, sometimes in grazed marshes on acid soils, or in very wet woods. 110 + [2]. Occasional in the west, absent from the Chalk and Portland, and frequent in the Poole basin. There are outliers at Abbotsbury Castle (58N, NCC), Langford (69H, 1969, DA Wells; perhaps lost), Bascombe Barn (78R, BE), and Deadmoor Common (71K, DWT).

Carex dioica L.

DIOECIOUS SEDGE
Rare and decreasing in mildly eutrophic bogs. 4. Extant sites are: South-east of Blue Pool (98G, DAP); Halfway Bog (98H, 1896, JWW, *et al.*, and 1988, AJB and DAP); Hartland Moor (98M, 1915, CBG in **BMH**, *et al.*, and 1994, SBC in **WRHM**); Slepe bog (98S, 1992, DAP); Agglestone bog (08, 1858, JCMP, *et al.*, 08B, 1988, ! and 08G, 1988, DAP). There are older records from 59I, 50L, 78N and Y, 98L, N and W.

[Carex elongata L.

GINGERBREAD SEDGE
Rare, but forming large colonies in swampy carr near the Moors River on the Hampshire border. [1]. Tricketts Cross (00V, 1938, VM Leather, *et al.* in **K** and **RNG**); Moors River (10A, 1982–85, RMW; probably all now in VC11).]

Carex curta Gooden.

WHITE SEDGE
Very local in carr or iron-rich ditches. 5 + [1]. Only in the east near the Hampshire border. Slop bog (00Q, 1995, DAP); between Mannington and West Moors (00R?, 1891, JCMP in **DOR**); Tricketts Cross (00V, RMW); West Moors Depot (00W, DAP); Cranborne Common (01V, 1963, KBR, *et al.*, and 1992, DAP, also 11A, 1982, RM Veall). A report from Arne (98U, 1981, MRH) was dubious, and records from South Parley Common (09Y, RP Bowman), Moors River (10A, RMW) and St Leonards (10B, AJB) are in VC11.

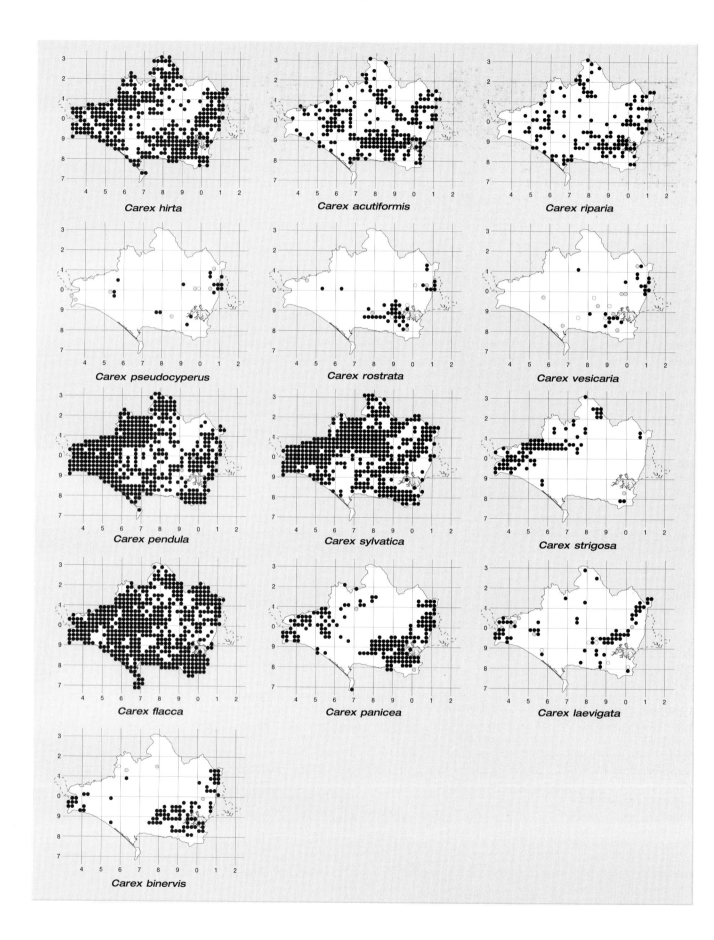

Carex hirta L.
HAIRY SEDGE
Common in damp, unimproved grassland, often on clay or by winterbornes, and able to push through tarmac in new suburbs. 393. Least common on the dry Chalk. It is resistant to trampling and grazing. Var. *subhirtaeformis* Pers., with hairless leaves, is uncommon.

Carex lasiocarpa Ehrh.
SLENDER SEDGE
Rare and local in boggy ditches and reed-swamps. 3 + [1]. Tadnoll (78Y, 1993, DAP and J Topp); Keysworth (98J, 1900, EFL, *et al.* in **BDK, BMH** and **RNG**, and 1994, DAP); Morden Bog (99A, 1895, WRL in **LIV**, *et al.* in **WRHM**, and 1990, AJB and DAP, !); Turlin Moor (99Q, 1915, CBG in **BMH**).

Carex acutiformis Ehrh.
LESSER POND-SEDGE
Locally common along rivers and by ponds; also in dense colonies in wet woods, where it rarely flowers. 228. Commonest in the larger river basins, rare in the west and on the Chalk, and absent from Portland.

(**Carex acutiformis** × **vesicaria** = **C. × ducellieri** Beauverd
This occurs in acid carr at Ebblake (10D, 1997, RP Bowman), but in VC11.)

Carex riparia Curtis
GREATER POND-SEDGE
Locally dominant by rivers and ditches in eutrophic water-meadows. 143. Mostly in the valleys of the larger rivers, and rare on the Chalk.

Carex pseudocyperus L.
CYPERUS SEDGE
Local and scattered in isolated ponds, flooded pits and ditches with fairly eutrophic water, sometimes in shade. 17 + [9]. Most frequent in the east, absent from the north and from sites near the coast.

Carex rostrata Stokes
BOTTLE SEDGE
Local in ditches in water-meadows and fens, less often in pools in somewhat eutrophic bogs. 38 + [7]. Widespread in the Poole basin, with outliers in atypical habitats at Mosterton (40M, 1930s, AWG), Park Pond (50F, BE) and Marr's Cross (60F, 1974, DA Wells and 1993, BE).

Carex rostrata × *vesicaria* = *C.* × *involuta* (Bab.) Syme
A rare and critical hybrid, found by the R Allen south of Witchampton (90X, 1990, D Exton & D Palmer). 1 + [3?]. Old records, needing confirmation, from East Holme (98D, NCC); Wareham (98I, 1915, CBG in **BMH**); and Norden (98L, 1915, CBG in **BMH**).

Carex vesicaria L.
BLADDER SEDGE
Uncommon, and in small numbers, by ponds and ditches with eutrophic, acid, peaty soils. 20 + [12]. Good's reports

of this from the Stour basin may have been errors. At present it is scattered in the Poole basin with an outlier at Rooksmoor (71K, 1990, DAP and D Elton).

Carex pendula Hudson
DROOPING SEDGE
Often dominant in oak and wet alder woods on clay soils in the north and west. 452 + [2]. Scattered in woods elsewhere, and it may be a garden escape in most habitats in the east, where Good had very few records. These habitats include pond margins, chalk scrub, limestone quarry-waste on Portland and tips, where it is invasive but not in great quantity. It is probably increasing.

Carex sylvatica Hudson
WOOD-SEDGE
Frequent in the drier parts of old woods, especially along rides, and sometimes abundant, as at Ashley Chase (58N). 438. Absent from deforested areas such as Portland, and between Weymouth and Dorchester, also from urbanised and suburbanised Poole.

Carex strigosa Hudson
THIN-SPIKED WOOD-SEDGE
Usually in small numbers in wet places in old woods. 89 + [1]. It occurs in most of the fragments of old woodland in the north and west, with outliers at Langton Wood (97Z) and near Edmondsham (00P, 01Q and R). Overlooked by earlier botanists, but not by DAP.

Carex flacca Schreber
GLAUCOUS SEDGE
The commonest Dorset sedge, which occurs in dry or wet calcareous grassland, grassy woodland rides, and sometimes in heaths. 520. It tolerates grazing, but not shade.

Carex panicea L.
CARNATION-SEDGE
Frequent on wet, acid soils, on the Greensand in the north and west and on heaths in the Poole basin. 174. Absent from chalk and limestone, apart from two Portland records which need confirmation: Culver Well (66Z) and Tout Quarry (67W, RJS).

[*Carex depauperata* Curtis ex With.
STARVED WOOD-SEDGE
Only one certain record, from a hedgebank between Cranborne and Damerham (01S?, 1925, HHHaines in **LTR**). [1]. It seeds itself in my garden at Winterborne Kingston.]

Carex laevigata Smith
SMOOTH-STALKED SEDGE
Local, by streams or flushes in alder or oak woods, or in damp woodland rides on acid soils. 67 + [14]. Scattered in the north and west, and in an arc around the outer edge of the Poole basin.

Carex binervis Smith
GREEN-RIBBED SEDGE
Fairly common in dry or moist heaths. 88 + [4]. Found in

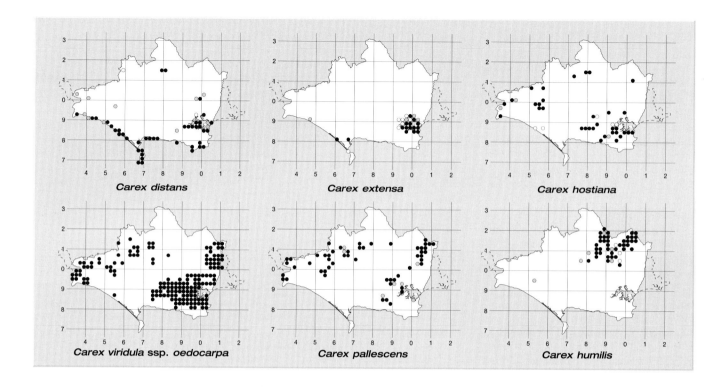

the extreme west and the Poole basin, with outliers at Abbotsbury (58N), Mount Pleasant (59P) and White Horse Common (60J): absent from the Chalk.

Carex distans L.
DISTANT SEDGE

Occasional in marshes and wet undercliffs along the coast, where it tolerates salt. 41 + [13]. The only recent inland records are from fenny meadows near Hinton St Mary (71X and 80C, DAP & !) and Tadden (90V, 1999, BE), but there are several reports from the 1930s which have not been refound.

Carex punctata Gaudin
DOTTED SEDGE

Scarce, sporadic and decreasing in damp sandy ground near Poole Harbour. 5 + [6]. Keysworth (98P, 1920, JHS); Arne Big Wood (98U, 1991, AJB); Holton Heath (99K, 1922, JHS); Hamworthy (99R and V, 1890, EFL in **CGE**, et al., and 1956, !); Goathorn (08D, 1917, CBG in **BMH**); near Knoll Ho, Studland (08G, 1983, RWD and 1999, EAP); Studland Heath, South Haven and Little Sea (08H, 1895, JCMP, et al. in **BMH** and **LANC**, and 1995, DAP); North Haven (08I, 1895, W Fisher); Brownsea Is (08J, 1971, WA Stewart and 1995, DAP); Loscombe Valley and Lilliput (08P, 1882, JCMP, et al., and 1998, DAP); Longfleet (09A, 1900, EFL and 1928, LBH in **DOR**).

Carex extensa Gooden.
LONG-BRACTED SEDGE

Scarce in salt-marshes, especially around the inside of Poole Harbour. 15 + [1]. It survives as outliers at Herbury

(68A, 1995, IPG and PRG), Radipole Lake (68Q, 1895, JCMP, et al., and 1977, !) and on cliffs at Blackers Hole (07D, 1922, JHS and 1998, BE). Good's inland records are now thought to be errors.

Carex hostiana DC.
TAWNY SEDGE

Uncommon in eutrophic water-meadows, fens and rushy pastures, or in firebreaks on heaths. 33 + [9]. Scattered in the west and in the Poole basin, with outliers at Lydlinch Common (71G), Hinton meads (71X and 80C) and Wimborne St Giles (01F).

Carex hostiana × viridula = C. × fulva Gooden.

A rare hybrid. 4 + [5]. Champernhayes and Coneys Castle (39N and T, 1932–35, AWG in **BDK**); Pilsdon Pen (40A, IPG); Brackets Coppice (50D, JGK); Cat Range Head (88R, P Selby); Corfe Common (98K, 1897, EFL in **BM** and 1993, RMW); between Godlingston Farm and Herston (07E, 1929, CES); Verwood (00Z, 1909, EFL).

Carex viridula Michaux

The three subsp. have distinct habitats; old records are often muddled.

Subsp. **brachyrrhyncha** (Celak) B. Schmid
LONG-STALKED YELLOW-SEDGE

Rare, in relict fens, and once in flushed springs by the sea. 4 + [8]. Bedmill Farm (61C, 1956, JL Gilbert); Thornford (61G, 1971, SS); Povington Heath (88W, AJB); Turners Puddle (89G, DAP and RMW); Whitefield (89W, 1936, RDG in **WRHM**); above Dancing Ledge (97Y, 1914, CBG in **BMH**); Hartland Moor (98M, 1955, PJ Newbould); The Moors (98N, 1965, DSR and 1998, DAP); Wimborne

St Giles (01G, DAP). Reports from Pilsdon Pen (40A), Hilfield (60H) and Kingswood Farm (08B) need confirmation.

Subsp. *oedocarpa* (Andersson) B. Schmid
COMMON YELLOW-SEDGE
Frequent along wet tracks in heaths, in acid woodland rides, or on bare peat in bogs; rarely in wet neutral grassland. 156 + [4]. Scattered in the west and north, and frequent in the Poole basin.

Subsp. *viridula*
SMALL-FRUITED YELLOW-SEDGE
Rare in grazed fens or sandy dune slacks with short vegetation. 3 + [5]. Knights in the Bottom (59J, LJM); Langford meadow (69H, 1970, DA Wells); Winfrith Heath (88D, 1978, !); Hartland Moor (98M, 1960–70, SBC); The Moors (98N, 1965, DSR); Little Sea and north of Studland (08G and H, 1895, WMR, *et al.* in **BDK, BMH** and **DOR**, and 1995, BE and DAP, !).

Carex pallescens L.
PALE SEDGE
Uncommon and in small numbers, in grassy woodland rides, scrub or damp grassland, either neutral or mildly acid. 53 + [8]. Scattered in the north and west, absent from the Chalk, and found in an arc around the outer edge of the Poole basin.

[Carex digitata L.
FINGERED SEDGE
Once collected from a wood south of Wool (88, 1912, RVS in **BMH**); it usually grows in old woods on calcareous soil.]

Carex humilis Leysser
DWARF SEDGE
Locally abundant in fragments of short, dry, calcareous turf in the north-east, often on slopes facing west or south-west. 39 + [9]. It may once have grown further west on the Chalk, as there are old reports from Eggardon Hill (59H, 1955, AS Thomas) and Rawlsbury Camp (70S). There are still large populations at Enford Bottom (80P), Hambledon Hill (81L), Blandford Camp (90I and J), The Cliff (90N and T), Tenantry Down (01B) and elsewhere, but the main stronghold of this sedge is in Wiltshire. It resists grazing, and also close mowing on army land.

Carex caryophyllea Latour.
SPRING SEDGE
Frequent in short, dry turf in calcareous or, less often, in neutral grassland; scarce in damp sites. 229. Commonest along the northern scarp of the Chalk in mid-Dorset, and along the chalk ridge between Preston and Ballard Head.

Carex montana L.
SOFT-LEAVED SEDGE
Rare in acid woodland rides. 1. Castle Hill Wood (01R, 1909, EFL in **BIRA** and **CGE**, *et al.* in **BMH** and **RNG**, and 1992, !).

Carex pilulifera L.
PILL SEDGE
Frequent in sandy woodland rides, dry heaths and leached areas on chalk summits. 171 + [3]. It is much grazed by deer. Var. *longibracteata* Lange was found on Studland Heath (08H, 1917, CBG).

Carex limosa L.
BOG-SEDGE
Rare in wet bogs and bog-pools, with *Sphagnum*. 3 + [6]. Moreton (88E?, 1937, DM); East Burton Heath (88I, 1932, RDG); Morden (98E, 1990, AJB and DAP); Halfway Bog (98H, 1929, AWG); Hartland Moor (98M, 1916, CBG in **BMH**, *et al.* in **WRHM**, and 1994, DAP); The Moors (98N, 1966, !); Morden Decoy (99A, 1895, JCMP, *et al.*, and 1955, !); Waterloo (09A, 1900, LVL); Slop Bog (00V, 1936, RDG, *et al.*, and 1995, GDF).

Carex acuta L.
SLENDER TUFTED-SEDGE
Uncommon, or under-recorded, by rivers, ditches and lakes, in carr, or in marshes with *C. acutiformis*. 26 + [6]. Mostly in the upper Stour basin, and the lower basins of the Frome, Piddle and Moors Rivers. Some of the 1930s reports may be errors. It does not occur further west in Britain.

Carex acuta × nigra = C. × elytroides Fries
A rare and critical hybrid. 1 + [1]. Poole (09A, 1904, EFL); meadows near Wareham (98D &/or I, 1900, EFL, also 1936, JEL in **RNG** and 1986, DAP; this may have been *C. acuta × acutiformis*, but the site is now a by-pass).

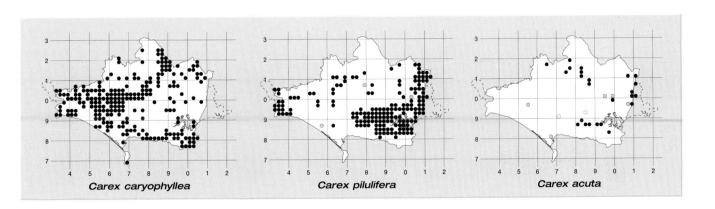

Carex caryophyllea   Carex pilulifera   Carex acuta

Carex nigra (L.) Reichard
COMMON SEDGE

Occasional in wet, acid grassland or heath, sometimes in fens, often reported in error. 120 + [2]. Mainly on the Greensand in the north and west and in the Poole basin, with an outlier at Culverwell (66Z). Variable. Var. *chlorostachya* Reichb., with pale green fruits, was found at Lytchett Matravers (99M, 1930s, NDS), and a markedly tufted form is not rare.

(*Carex elata* All. was reported from Rempstone in 1830, but never confirmed.)

Carex pulicaris L.
FLEA SEDGE

Uncommon in fens and flushed areas in bogs, but not in very acid sites. 50 + [9]. Found in the north and west, and in the Poole basin, absent from the Chalk except near Flowers Barrow (88Q, 1975, TCEW and 1998, BE, growing with *Brachypodium pinnatum*). Declining through drainage.

POACEAE or GRAMINEAE

Semiarundinaria fastuosa (Lat.-Marl. ex Mitford) Makino ex Nakai

Planted as a garden hedge near Red Bridge, Moreton (78Z, !). 1.

Yushania anceps (Mitf.) Lin
INDIAN FOUNTAIN-BAMBOO

Naturalised in school grounds, Knoll Hill (99T, !) and planted at Kingston Lacy (90V). 2.

Fargesia murielae (Gamble) T.P. Yi
CHINESE FOUNTAIN-BAMBOO

Planted in a wild garden at Studland Bay Ho (08G, !) and Ferndown (00Q, !), in flower. 2.

Pleioblastus pygmaeus (Miq.) Nakai
DWARF BAMBOO

Probably this at Stratton Ch (69L, !). 1.

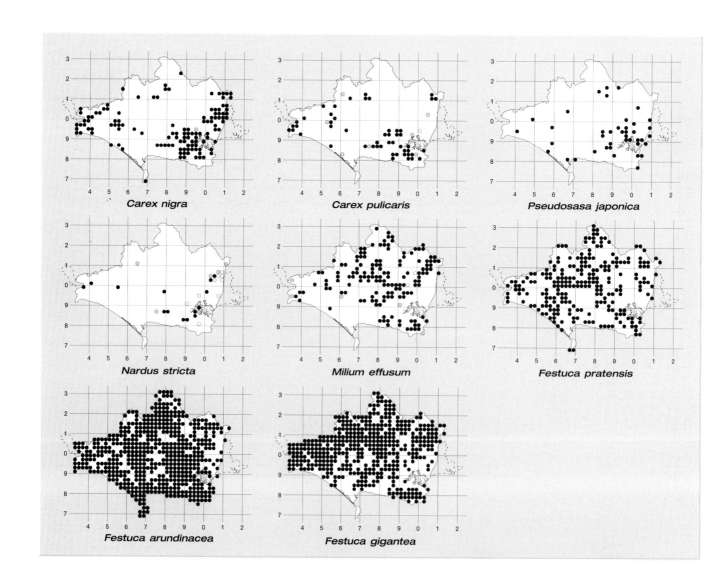

Carex nigra

Carex pulicaris

Pseudosasa japonica

Nardus stricta

Milium effusum

Festuca pratensis

Festuca arundinacea

Festuca gigantea

Pleioblastus simonii (Carriere) Nakai
SIMON'S BAMBOO
In a wood at Knowl Hill, Studland (08G, !) in flower; sterile near Merley Ho (09E, !). 2.

Sasa palmata (Burb.) Camus
BROAD-LEAVED BAMBOO
Forming thickets in wild gardens or plantations. 17. Large colonies at Bryanston School (80T, !) and All Saints Ch, Branksome (09Q, !).

Sasa veitchii (Carriere) Rehder
VEITCH'S BAMBOO
Rare in plantations. 3. Wootton Fitzpaine (39S, 1983, TBR); Parnham (40Q, !); Post Green (99L, !).

Sasaella ramosa (Makino) Makino
HAIRY BAMBOO
With other bamboos in a plantation near Merley Ho (09E, 1999, !). 1.

Pseudosasa japonica (Siebold & Zucc. ex Steudel) Makino ex Nakai
ARROW BAMBOO
Naturalised in plantations, churchyards and grounds of big houses. 42. Seen in flower at Corfe Castle (98R, 1990, ! in **RNG**); after flowering, clumps become moribund.

Phyllostachys aurea (Carriere) Riv. & C. Riv.
Planted at Abbotsbury Swannery (58S), Godlingston (08A, EAP) and Alderholt Ch (11B). 3.

Phyllostachys nigra (G. Lodd.) Monro
Planted by a pond at Upton CP (99W, !). 1.

Leersia oryzoides (L.) Sw.
CUT-GRASS
Rare or extinct by a eutrophic ditch north of Wareham (98J, 1897, EFL in **RNG** and **LIV**, *et al.* in **BDK, BMH** and **LIV**, last seen 1965, DSR, !). [1]. The site has been heavily grazed in recent years.

Nardus stricta L.
MAT-GRASS
Local and in small numbers in dry heaths. 14 + [18]. Most records are from the Poole basin, with outliers in the west: Lamberts Castle (39U, 1988, NCC); Hardown Hill (49C, 1953, EN); Pilsdon Pen (40A, 1928, AWG and 1998, IPG); Kingcombe (59P,1999, BE); Seven Ash Common (61K, 1976, C Burchardt). It was said to be locally abundant in 1897, but Good only found it in 12 sites; probably decreasing, like other northern plants in the county.

Anemanthele lessoniana (Steud.) Veldkamp
An ornamental grass planted in Abbotsbury gardens and sowing itself there (58S, SG); pavement weed, Walditch (49W, 1999, !); Parnham (40Q, 1999, !). 3. Likely to spread.

Milium effusum L.
WOOD MILLET
Frequent in many old woods, usually on calcareous soils, but not in new plantations. 146 + [7]. Scarce in the west, and absent from Portland and deforested areas.

Festuca pratensis Hudson
MEADOW FESCUE
Widespread and locally abundant in unimproved neutral grassland on clay soils, surviving on verges in intensively farmed regions. 240. Absent from heathland and from chalk turf. The hybrid with *F. arundinacea* should be looked for.

Festuca arundinacea Schreber
TALL FESCUE
Abundant in neutral grassland and verges, especially on clay soils. 540. Dwarf forms are frequent along the coast, mostly on undercliffs, but also on shingle behind the Chesil bank.

Festuca gigantea (L.) Villars
GIANT FESCUE
Frequent in woods and plantations, and occasional in hedgebanks. 437. Absent from Portland, also from very acid soils, and scarce in deforested areas.

Festuca gigantea × *rubra* probably occurred in dry valley pasture near scrub at Cogden Farm (58E, 1994, ! in **LTR**). The inflorescence was like *F. gigantea*, but sterile, and the lower leaves were narrow like those of *F. rubra*.

Festuca heterophylla Lam.
VARIOUS-LEAVED FESCUE
Very rare on shaded verges. 2. Stoke Common (88S, 1992, !); Briants Puddle (89B, 1991, !).

Festuca arenaria Osbeck
RUSH-LEAVED FESCUE
Locally abundant in dunes, especially on the outer dune with *Ammophila*. 5 + [1]. Studland, South Haven and Sandbanks (08G, H, I, M and N); Poole (09A, 1925, Anon in K and 1964, !); probably on inaccessible sandy cliffs between Flag Head and Branksome, since it grows on similar cliffs near Bournemouth in VC11. Specimens are in **BDK, CGE** and **K**.

Festuca rubra L.
RED FESCUE
Abundant in dry grassland everywhere, also on undercliffs and shingle beaches. 739. Intolerant of shade and of very acid soils. Very variable. Most inland records are subsp. *rubra*, while on cliffs and beaches the dwarfer, often pruinose, subsp. *juncea* (Hackel) K. Richter is frequent. Subsp. *commutata* Gaudin, tufted and lacking rhizomes, is rarely reported but probably frequent in lawns and verges. Subsp. *megastachys* Gaudin, tall and with upper leaves flat, is rare in shaded places: Chedington Wood (40Y, IPG); Moreton (88E, !); Clouds Hill (89F, !). The densely-matted Subsp. *litoralis* (G. Mey) Auquier probably occurs in the upper parts of salt-marshes at Poole (08H and I,

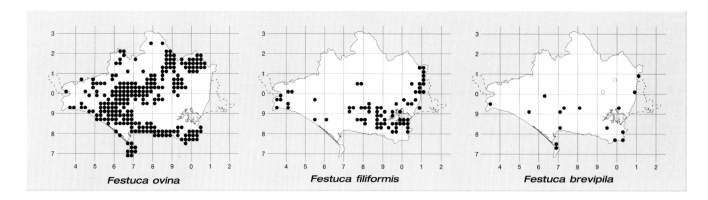

Festuca ovina   Festuca filiformis   Festuca brevipila

1999, EAP), but subsp. *rubra* also occurs there and the two are hard to distinguish.

Festuca ovina L.
SHEEP'S FESCUE
Abundant or locally dominant in short, calcareous turf, colonising freshly bared soil. 254. Its distribution-pattern follows that of chalk and limestone fairly closely, where the plants are probably subsp. *hirtula* (Hackel ex Travis) M. Wilkinson. Old reports from heathy grassland could have been subsp. *ovina*, but may have been confused with *F. filiformis*.

Festuca filiformis Pourret
FINE-LEAVED SHEEP'S-FESCUE
Fairly frequent on dry heaths, often in disturbed sites. 74. Found on summits in the west, and in the Poole basin, where it has been confused, or lumped, with *F. ovina*. Reports from calcareous turf at Dancing Hill (61H) and chalk scree at Ballard Cliff (08F) are unlikely to be correct.

Festuca brevipila R. Tracey
HARD FESCUE
An early-flowering grass, much planted on new verges. 15 + [2]. It has been reported under several names, but is under-recorded. A specimen from the A35 at Askerswell (59K, !) is in **RNG**.

Festuca pratensis × Lolium perenne =
× Festulolium loliaceum (Hudson) P. Fourn.
Occasional, as single clumps, in neutral or calcareous grassland or water-meadows. 18 + [8]. Not reported from acid soils, nor from 'improved' grassland.

Festuca rubra × Vulpia fasciculata =
× Festulpia hubbardii R. Cotton & Stace
A hybrid found at rare intervals at Small Mouth (67S, 1790s, S Goodenough in **K**, 1918, J Tulleylove in **K** and 1933, AL Still in **RNG**). [1].

Lolium perenne L.
PERENNIAL RYE-GRASS, EVER-GRASS
Abundant in sown leys, verges, parks, lawns, sports grounds and footpaths. 734. Resistant to trampling and grazing. Variable in size, the cultivated forms being the largest; forms with a branched inflorescence are rare.

Lolium perenne × multiflorum = L. × boucheanum Kunth.
Scattered, usually as single plants, in verges, disturbed ground and tips. 12. First record: Weymouth tip (68V, 1960, EH in **LANC**). Also in 39L, 59H, 69Z, 61I, 72V, 88Y, 89X, 98D and 01A.

(*Lolium perenne × rigidum* may be the correct name of a specimen from arable at Warren (89K, 1990, ! in **RNG**).)

Lolium multiflorum Lam.
ITALIAN RYE-GRASS
Frequently planted as a fodder crop, and a relict on verges, by farms and tips. 302.

Lolium rigidum Gaudin
MEDITERRANEAN RYE-GRASS
Once found as a casual at Woodsford tip (78U, 1977, ! in **RNG**). [1].

Lolium temulentum L.
DARNEL
A rare casual on tips, e.g. Parkstone (09F, 1830, TBS and 1937, JFGC in **OXF**). [8]. It was last seen at Lodmoor tip (68V, 1963, !) and Sherborne tip (61I, 1970, ACL).

Vulpia fasciculata (Forsskal) Fritsch
DUNE FESCUE
Rare, but persistent, on dunes, fixed shingle, or sandy banks inland. 7 + [1]. Wyke Regis (67N, 1984, RA); Small Mouth (67S, 1788, J Lightfoot, *et al.* in **BDK, BM** and **DOR**, and 1992, DAP & !); Wareham Common (98D, 1989, RD Porley); Studland dunes (08G, 1895, JCMP and 1985, JC Davis, also 08H, 1990, C Flynn); South Haven (08I, 1895, JCMP, *et al.*, and 1992, C Flynn); Sandbanks (08N, 1980, AJB).

Vulpia bromoides (L.) Gray
SQUIRRELTAIL FESCUE
Occasional in dry, disturbed, sandy grassland, by rabbit warrens, heath tracks, sandy verges and rail tracks. 156. It dislikes competition, and is commonest in the Poole basin, notably on Brownsea Is, and scattered elsewhere.

Vulpia myuros (L.) C.C. Gmelin
RAT'S-TAIL FESCUE
Occasional in bare, dry places on verges, rail tracks, village

paths, pavements and wall tops. 65 + [11]. Commonest in the Poole basin, but widespread and perhaps increasing.

Vulpia ciliata Dumort subsp. *ambigua* (Le Gall) Stace & Auq.
BEARDED FESCUE
Once restricted to dunes, this has recently colonised bare sandy banks or verges in heathland. 8 + [5]. Small Mouth (67S, 1895, JCMP and 1991, !); Winfrith Heath (88D, 1995, BE); Higher Hyde Heath (88P, 1993, BE); Gallows Hill (89K, 1990, !); Langton Matravers (97Z?, 1921, CES in **K**, and 1933, AL Still in **RNG**); Hartland Moor and Stoborough Heath (98H, 1957, JH Hemsley in **WRHM** and 1998, !); Rockley Sands (99V, 1991, A Leftwick); Studland dunes (08G, 1895, JCMP, *et al.*, and 1994, !); War Hill (08H, 1992, C Flynn): South Haven (08I, 1879, H Trimen, *et al.*, and 1991, DAP); North Haven (08N, 1895, JCMP and 1969, !); Parkstone (09F, 1936, HNR).

(*Vulpia unilateralis* (L.) Stace
A report from Five Barrow Hill (88R, !) was an error.)

Cynosurus cristatus L.
CRESTED DOG'S-TAIL
Frequent in unimproved, dry pastures on calcareous or neutral soils. 585. Taller forms are rare in marshes, as at Lytchett Bay (99R, !), and forms lacking anthocyanin are occasional. In Wiltshire, the stems were used for weaving bonnets.

Cynosurus echinatus L.
ROUGH DOG'S-TAIL
A rare casual in dry, bare soils, perhaps adventive from Guernsey. 1 + [4]. Manor Farm, Toller (59U, 1999, N Spring); Weymouth (67 or 68, 1926, MJA and 1935, NDS); Small Mouth (67S, 1960–61, RDG in **DOR**, !); Corfe (98R, 1933, BNSS); Parkstone (09F, 1937, DM); abundant and established on sandy cliffs at Bournemouth in VC11.

[*Lamarckia aurea* (L.) Moench
GOLDEN DOG'S-TAIL
Once found in an old garden at Upton (99W, 1906, HJG).]

Puccinellia maritima (Hudson) Parl.
COMMON SALTMARSH-GRASS
Locally abundant on muddy shores and at the lower edge of salt-marshes. 30 + [5]. Found on the landward side of the Fleet and all round the inside of Poole Harbour, with outliers at West Bay and east Portland (77B); old records from Portland Bill (66), Lulworth Cove (87J) and Kimmeridge (97E, 1937, DM) have not been confirmed recently. Variable in habit and size of panicle.

Puccinellia distans (L.) Parl.
REFLEXED SALTMARSH-GRASS
Uncommon, sporadic and in small numbers on hard, bare ground near the sea. 8 + [7]. Post-1950 records: West Bay (49K, 1895, JCMP, *et al.* in **BDK,** and 1972–92, ! in **RNG**); Langton Herring (68A, 1994, SME); Neck of Portland (67S, 1992, !); Radipole Lake (67U, 1955, !); Bowleaze Cove (78A, 1993, !); Keysworth (98P, 1989, RMW); Fitzworth (98Y, 1991, BE and DAP); Rockley Sands (99Q, 1999, !); Studland (08G, 1956, CHS); Poole (09A and B, 1830, TBS, *et al.*, and 09F, 1989–99, !). Decreasing, and not yet noted along major roads.

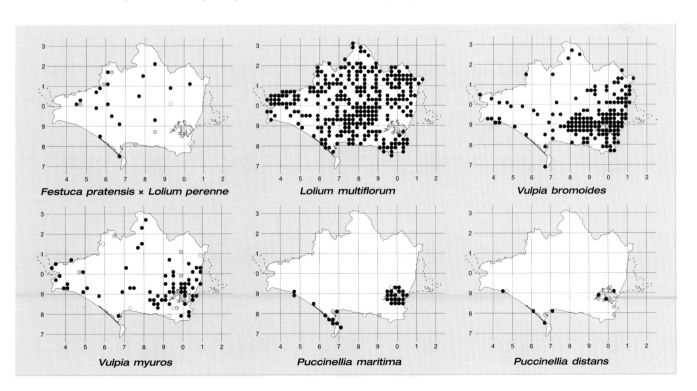

Festuca pratensis × Lolium perenne

Lolium multiflorum

Vulpia bromoides

Vulpia myuros

Puccinellia maritima

Puccinellia distans

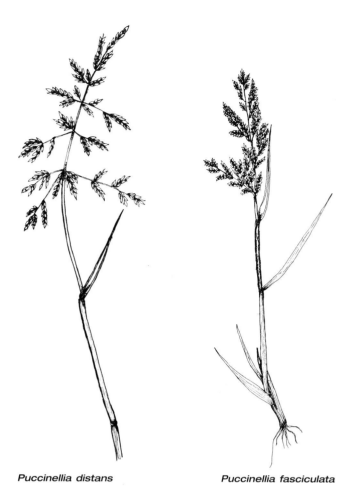

Puccinellia distans   Puccinellia fasciculata

Puccinellia fasciculata (Torrey) E. Bick.
BORRER'S SALTMARSH-GRASS
Rare on hard ground near the sea. 3 + [4]. West Bay (49K, 1936, MJA and 1993, !); The Nothe (67Z, 1977–90, !); Lodmoor (68V, 1958, !, *et al.*, and 1964, J Mason); Hamworthy (99V, 1874, JCMP and 1956, !); North Haven (08I, 1993, ! in **RNG**).

Puccinellia rupestris (With.) Fern. & Weath.
STIFF SALTMARSH-GRASS
Uncommon, sporadic and in small numbers on hard

ground near the sea. 10 + [16]. Most recent records are between West Bay and Bowleaze Cove, with a few round Poole Harbour and an outlier at Peveril Point (07P, 1895, JCMP, *et al.*, and 1952, DE Coombe).

Briza media L.
QUAKING-GRASS
Locally frequent in short calcareous turf. 331. Much less common by nutrient-enriched heath tracks, or in damp meadows and fens with tall vegetation. Least common in the north and in the Poole basin. Forms lacking the normal purple anthocyanin are occasional.

Briza minor L.
LESSER QUAKING-GRASS
Scarce and sporadic in sandy arable. 12 + [21]. Mainly in the Poole basin, with a few old records from south-east Purbeck and one from Stoke Abbott (40K, 1930, AWG). It has decreased during this century, but Good only found it in 10 stands in the 1930s.

Briza maxima L.
GREATER QUAKING-GRASS
Often grown in gardens, and escaping as a pavement weed for a few seasons; perhaps naturalised on Canford Cliffs (08P and U, 1961, HJB and 1996, RMW). 24 + [5]. It prefers well-drained, sandy soils, and is nowhere abundant as it is on Bournemouth cliffs in VC11; two plants on Hod Hill (81K, 1996, AH).

(**Poa infirma** Kunth
Atypical plants on Bournemouth Cliffs may be this, but only in VC11.)

Poa annua L.
ANNUAL MEADOW-GRASS
Abundant in arable, gardens, lawns, tracks, pavements, shingle and sandy beaches. 736. Very variable in size and luxuriance. Tolerant of trampling and salt-spray. The seeds are eaten by partridges and small birds (Wilson, Arroyo & Clark, 1997).

Poa trivialis L.
ROUGH MEADOW-GRASS
Abundant in damp grassland and fens. 725. Also frequent

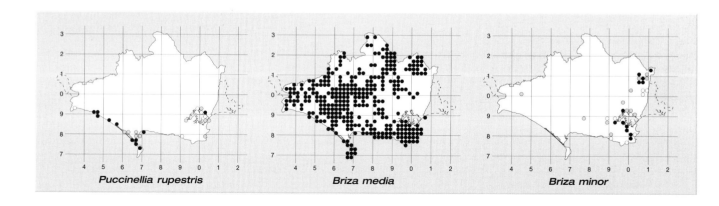

Puccinellia rupestris   Briza media   Briza minor

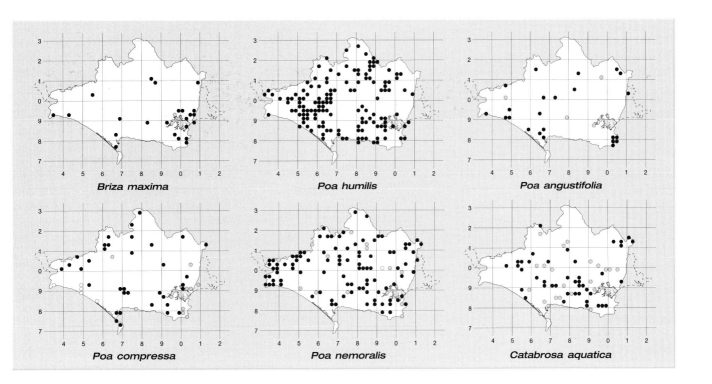

Briza maxima   Poa humilis   Poa angustifolia

Poa compressa   Poa nemoralis   Catabrosa aquatica

as an arable weed, and as a depauperate form in woodland rides, as it tolerates some shade.

Poa humilis Ehrh. ex Hoffm.
SPREADING MEADOW-GRASS
This ill-defined taxon occurs in dry chalk turf, wet pastures and fens, enriched heathy verges and at the back of shingle beaches. 140. It is probably under-recorded or confused with *P. pratensis*.

Poa pratensis L.
SMOOTH MEADOW-GRASS
Abundant in dry grassland, lawns, verges, dunes, wall tops and pavements, but not in shade. 729. Very variable in size.

Poa angustifolia L.
NARROW-LEAVED MEADOW-GRASS
Uncommon in dry grassland and on tops of old limestone walls. 23 + [5]. Probably under-recorded.

Poa compressa L.
FLATTENED MEADOW-GRASS
Local in disturbed calcareous turf, often in new verges, arable, pavements, sandpits, wall tops and tips. 35 + [15].

(**Poa palustris** L. was reported from Holnest (60N, 1979, AJCB) but needs the support of specimens. It is usually a plant of shallow lakes or river valleys.)

Poa nemoralis L.
WOOD MEADOW-GRASS
Locally frequent in acid or calcareous woods, in beech

plantations and shrubberies in large estates, rarely on walls. 99 + [13]. It has an unusual distribution.

Poa bulbosa L.
BULBOUS MEADOW-GRASS
Rare in bare, dry, sunny places on sandy or calcareous soils. 3 + [1]. Sherborne Old Castle (61T, 1970, ACL); East Portland (77B, 1988, RF, !); South Haven verge (08I, 1990–94, !); Sandbanks (08N, 1996, RMW). A Mediterranean species which may be increasing.

Dactylis glomerata L.
COCK'S-FOOT
Abundant in coarse grassland of pastures, verges, undercliffs and lawns, but not in acid soils on heaths. 736. Plants resembling subsp. *hispanica* (Roth) Nyman occur low down on calcareous cliffs at Portland (67V), White Nothe (78Q), Dungy Head (87E), Mupe Bay (88K) and probably elsewhere.

Dactylis polygama Horv.
SLENDER COCK'S-FOOT
Once seen in a plantation at Sherborne Park (61N, 1970, ACL). [1].

Catabrosa aquatica (L.) P. Beauv.
WHORL-GRASS
Although JCMP found this 'generally distributed' in the 1890s, it is not so today. 42 + [22]. Local in clear, shallow streams, ditches and cress-beds, especially in the Frome and Piddle valleys, but gone from most of the Stour basin. Once found by a stream on the beach at Kimmeridge (97E, 1933, AL Still in **RNG**), as it often is in north Britain.

Catapodium rigidum (L.) C.E. Hubb.
FERN-GRASS
Local, and in small numbers, in bare places in calcareous turf, quarries, old walls, ashbeds and as a pavement weed. 148 + [7]. Commonest along the coastal strip and rarely in disturbed sandy heath, as at Stokeford Heath (88U), Gallows Hill (89K) and Cranborne Common (11A).

Catapodium marinum (L.) C.E. Hubb.
SEA FERN-GRASS
Occasional in bare places near the sea, on shingle, cliffs and old walls. 53 + [3]. It occurs locally all along the coast, and has recently colonised the A354 at Ridgeway Hill (68S, 1999, !), the A352 near Shitterton (89H, 1999, !) and the central reservation of the A35 near Poole (99W, 1996, DAP).

Parapholis strigosa (Dumort.) C.E. Hubb.
HARD-GRASS
Occasional in the upper parts of salt-marshes, on muddy shores, clay cliffs or shingle beaches, extending to garden lawns at West Bay (49K). 23 + [13]. Commonest between West Bay and Bowleaze Cove, and on Portland, and perhaps declining round Poole Harbour.

Parapholis incurva (L.) C.E. Hubb.
CURVED HARD-GRASS
Rare on hard ground near the sea, on clay or limestone cliffs. 16 + [4]. Most frequent on Portland and in south-east Purbeck.

Hainardia cylindrica (Willd.) Greuter
ONE-GLUMED HARD-GRASS
A rare, but easily overlooked, casual once found at Eastbury tip (61I, 1970, ACL). [1].

Glyceria maxima (Hartman) O. Holmb.
REED SWEET-GRASS
Locally frequent by eutrophic rivers, lakes and ditches. 94 + [5]. Mainly in the basins of the lower Frome and Piddle, the upper Stour and the Allen, with outliers in the north and west. The form *picta*, with leaves striped with white and pink, has escaped into ponds near Allan's Farm (81U) and Cowherd Shute Farm (82S).

Glyceria fluitans (L.) R. Br.
FLOATING SWEET-GRASS
Widespread in muddy woodland rides, ponds, streams, ditches and temporary pools. 380. Rare on the Chalk and in suburban Poole.

Glyceria fluitans × *notata* = *G.* × *pedicellata* Townsend
Scarce in marshes and water-meadows, mostly in the north. 5 + [14]. Recent records: Frome St Quintin (50W, 1974, DFW); Purse Caundle and Poyntington (61Y and 62K, 1996, JAG); Devil's Brook (79T, 1961, DFW); Stock Gaylard Park (71G, 1997, !); Hinton St Mary (81C, 1976, DWT); Redcliff (98J, 1981, AJCB); Swanage (07J, 1999, EAP). Either decreasing or overlooked recently.

Glyceria declinata Breb.
SMALL SWEET-GRASS
Occasional, but widespread, on mud by streams, ditches and ponds. 63 + [8].

Glyceria notata Chevall.
PLICATE SWEET-GRASS
On mud in similar places to the other species. 122. Less common than *G. fluitans* except in the north-west.

Melica uniflora Retz.
WOOD MELICK
Frequent on dry banks in woods or old hedges. 186. Commonest in the north-east, and absent from deforested areas in south-central Dorset and on Portland. Good found it uncommon in the 1930s.

Helictotrichon pubescens (Hudson) Pilger
DOWNY OAT-GRASS
Occasional to frequent in tall, calcareous grassland. 230. Well distributed on the Chalk and limestone, absent from heavy clays and the Poole basin heaths, except on a few enriched verges.

Helictotrichon pratense (L.) Besser
MEADOW OAT-GRASS
Occasional to frequent in tall, calcareous grassland, often with *H. pubescens*, but easier to detect by its glaucous leaves when not in flower. 208.

Arrhenatherum elatius (L.) P. Beauv. ex J. & C. Presl
FALSE OAT-GRASS
Abundant in rough grassland, especially on clay, and on chalk where downwash from steep slopes has deepened the soil. 731. Var. *bulbosum* (Willd.) St Amans, with a swollen stem-base, is adapted to dry conditions on cliffs, quarries and verges; it is under-recorded, but noted in 67W, 79G, 89G, 90S, 91Y, 01H and 11H.

Avena barbata Pott ex Link
SLENDER OAT
Casual at Sandbanks (08N, 2000, !).

[*Avena strigosa* Schreber
BRISTLE OAT
A rare casual, last reported from Poole (09F, 1930, LBH); a report from Durlston CP in 1977 is not accepted.]

Avena fatua L.
WILD-OAT, CHEAT
A frequent arable weed on most soils; also found on tips, rarely on beaches. 365. Good only found it in two of his arable stands, so it must have increased recently, perhaps because it resists many herbicidal sprays.

Avena sterilis L. subsp. *ludoviciana* (Durieu) Nyman
WINTER WILD-OAT
Once seen in arable on Bere Down (89N, 1989, !) 1.

Avena sativa L.
OAT
Sometimes grown as a crop, and a common relict on waste ground near barns or on tips, but only casual. 29. Introduced on disturbed verges from straw packing of pipes.

Gaudinia fragilis (L.) P. Beauv.
FRENCH OAT-GRASS
Rare and inconspicuous, but apparently native, in fairly damp, neutral grassland on clay. 9. Nash Farm (39U, 1999, DAP); Mutton Street, frequent (39Z and 49E, 1989, AH, also ! in **RNG** and 1999, IPG); Crabbs Bluntshay Farm (49D, 1999, FG); between Chickerell and Putton (68K, 1980, !); east of Redcliff Point (78A, 1999, !); near Riviera Hotel (78B, 1999, DAP); Westbrook Farm (72X, 1994, DAP and DG); Pinnocks Moor (01Q, 1998, JHSC). Good may have overlooked this June-flowering grass because it is unrecognisable by University vacation time, or it may have arrived recently.

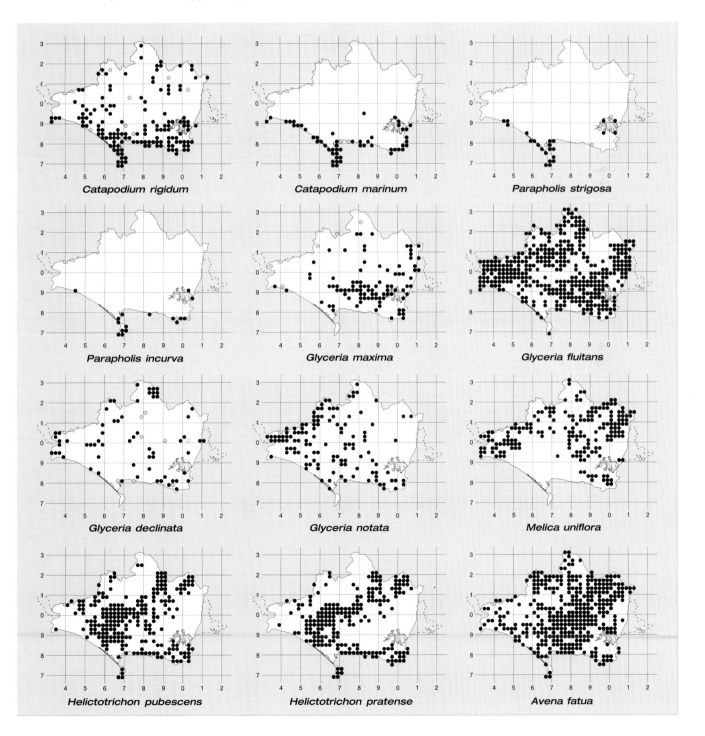

Catapodium rigidum

Catapodium marinum

Parapholis strigosa

Parapholis incurva

Glyceria maxima

Glyceria fluitans

Glyceria declinata

Glyceria notata

Melica uniflora

Helictotrichon pubescens

Helictotrichon pratense

Avena fatua

Trisetum flavescens (L.) P. Beauv.
YELLOW OAT-GRASS
Common in calcareous grassland, and frequent in neutral grassland on calcareous clay. 381. Absent from acid soils and heaths.

Koeleria macrantha (Ledeb.) Schultes
CRESTED HAIR-GRASS
Occasional in short, calcareous turf. 180. Rarely in disturbed sandy verges, as at Gallows Hill (89K). A distinct-looking

ecotype, once known as *K. glauca*, has grown on stabilised shingle at Small Mouth for many years (67S, 1869, SMP in **DOR**, *et al.* in **BMH, DOR** and **RNG**, and 1997, !).

Deschampsia cespitosa (L.) P. Beauv.
TUFTED HAIR-GRASS
Frequent to abundant in old grassland, especially on clay, on verges, and able to survive shade in a sterile state. 534. There is an unexplained gap in its distribution around Dorchester. Subsp. *parviflora* (Thuill.) Dumort, with small

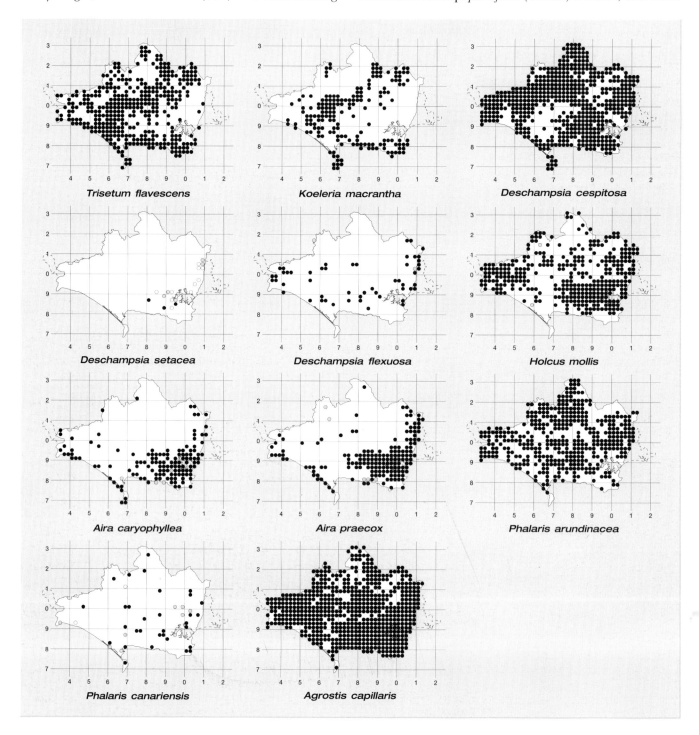

Trisetum flavescens

Koeleria macrantha

Deschampsia cespitosa

Deschampsia setacea

Deschampsia flexuosa

Holcus mollis

Aira caryophyllea

Aira praecox

Phalaris arundinacea

Phalaris canariensis

Agrostis capillaris

spikelets, occurs in shady places in woods, as at Corscombe (50 D and H), Dancing Hill (61I), Duddle Heath (79F) and Kingsettle Wood (82S).

Deschampsia setacea (Hudson) Hackel
BOG HAIR-GRASS
Rare and decreasing in shallow bogs and temporary heathland pools. 3 + [15]. All sites are in the Poole basin, but it has gone from many of them. Post-1930 records: two sites, Winfrith Heath (88D, 1997, DAP); Povington (88W, 1998, BE); Cold Harbour (88Z, 1953, !); Hartland Moor (98M, 1991, AJB); The Moors (98N, 1937, RDG in **WRHM**); Slepe Heath (98N?, 1934, HS Redgrave in **K**); Morden Decoy (99A, 1953, !); Holt Heath (00S, 1918, JHS); Verwood Lower Common and Gundry's enclosure (00X, 1938, RDG in **DOR**, also PMH in **K**); Horton Lower Common (00Y, 1935, RDG). A northern plant in Britain, perhaps declining for climatic reasons.

Deschampsia flexuosa (L.) Trin
WAVY HAIR-GRASS
Locally frequent on dry, acid soils on heaths, often under pines, and in small numbers on leached soil of chalk summits. 52 + [2]. It is found mainly in the far west and the Poole basin, where it is markedly less common than it is in VC11.

Holcus lanatus L.
YORKSHIRE-FOG
Abundant in most types of grassland, woodland rides, verges, undercliffs and lawns. 735. Absent from freshly sown leys, as it prefers leached or infertile soils. Forms lacking purple anthocyanin are not rare.

Holcus mollis L.
CREEPING SOFT-GRASS
Locally frequent to co-dominant in acid woodland rides, often with bracken and bluebells. 338 + [2]. It also occurs in dry, acid hedgebanks, but is scarce on the Chalk and limestone and absent from deforested areas.

(*Corynephorus canescens* (L.) P. Beauv. was reported from the north shore of Poole Harbour (09, 1799, RP), but no specimen has been found; it is now restricted to East Anglia in Britain.)

Aira caryophyllea L.
SILVER HAIR-GRASS
Locally frequent in bare, dry, sandy soil, heathy verges and rail tracks. 132 + [4]. It is also found on stabilised shingle beaches, and on bare, shallow calcareous soil on cliff-tops. Its range includes the west, the coastal belt including Portland, and the Poole basin heaths.

Aira praecox L.
EARLY HAIR-GRASS
Common in bare, dry, sandy soil of heath tracks, on rail tracks, fixed shingle and wall tops. 166 + [5]. It also occurs in dry, leached soil on the summits of chalk and limestone hills in outlying sites. Occasional in the west, along the

landward side of the Chesil bank, and frequent in the Poole basin. Not known on Portland.

Anthoxanthum odoratum L.
SWEET VERNAL-GRASS
Frequent to abundant in dry neutral to acid grassland, in some woodland rides and a frequent lawn weed. 617. Tolerant of grazing and least common on the Chalk. Its coumarin content imparts a pleasant scent to hay and makes the stalks chewable.

Anthoxanthum aristatum Boiss.
ANNUAL VERNAL-GRASS
A rare casual in sown leys. 1? + [6]. Beaminster (40Q, 1922, AWG in **BDK**); Wareham (98I, 1915, CBG in **BMH**); Corfe (98, 1911, CBG in **BMH**); Studland (08, 1920s, Lady Davy in **RNG**); Poole (09A, 1917, CBG); Verwood (00Z, 1930s, RDG); St Leonards (10A, 1982–87, RMW, probably in VC11).

Phalaris arundinacea L.
REED CANARY-GRASS
Locally common by streams, ponds and ditches and sometimes forming patches in drier disturbed ground such as verges. 409. Var. *picta* L. (Gardener's Garters) also has invasive roots, and is occasionally seen as a throw-out: Owermoigne (78U); Bere Heath (89R); Brickfields Business Park, Gillingham (82C); Green Is (08D); Dewlands Common (00U).

Phalaris aquatica L.
BULBOUS CANARY-GRASS
Rarely planted as a crop in arable fields, and surviving as a relict. 3. Langton Herring (68A, 1994, ! in **RNG**); Caesar's Camp (91H, 1997, !); Horton (00I, 1999, !).

Phalaris canariensis L.
CANARY-GRASS
Sporadic as a bird-seed alien in gardens or on pavements, and apparently persisting on tips, where it may be introduced each year. 31 + [11].

Phalaris minor L.
LESSER CANARY-GRASS
A rare casual in gardens and tips, perhaps introduced from Guernsey. [3]. Beaminster (40Q, 1922, AWG in **BDK**); Corscombe (50C, 1920s, AWG); Weymouth tip (67U, 1951, !).

Phalaris paradoxa L.
AWNED CANARY-GRASS
Rarely undersown for pheasant food in arable crops, or a more or less persistent weed. 3 + [3]. Frankham Farm (51Q, 1994, JAG); Weymouth (67, 1930s, RDG); Langton Herring (68A, 1984, Anon in **RNG** and 1996, SME); Eastbury (61I, 1970, ACL); Punfield (08F, 1847–72, J Hursey); Cranborne Castle (01L, 1984, ! in **RNG**).

Agrostis capillaris L.
COMMON BENT, BENNETS
Abundant in dry grassland with some preference for slightly acid soils. 603. It is commonly planted in garden lawns.

Agrostis capillaris × *stolonifera* = *A.* × *murbeckii* Fouill. ex P. Fourn.

Perhaps frequent in old grassland and churchyards, but only noted as probable at Winfrith Newburgh Ch (80C) and Shapwick (90G). 2.

Agrostis gigantea Roth

BLACK BENT

Locally abundant as a weed in cereal crops on dry, light soils. 141. Probably under-recorded, but uncommon on heavy clays and also on very acid soils, neither of which are much cultivated.

Agrostis stolonifera L.

CREEPING BENT

Abundant in more or less damp grassland, verges, undercliffs, overgrown ponds and lawns. 722. Very variable. Var. *stolonifera* (L.) Koch, with rigid, dense-flowered panicles, occurs on clay undercliffs at Eype Mouth (49K, AH, !), Osmington Mills (78F, 1930s, LBH in **BM**) and Swanage (08F, 1900, WMR).

Agrostis curtisii Kergeulen

BRISTLE BENT

Locally abundant in heathland, often increasing after disturbance or fires; rare in leached soils overlying chalk summits. 154. Mostly in the west and the Poole basin, absent from the north and north-west; its British distribution is mainly Cornubian.

Agrostis canina L.

VELVET BENT

Local in damp, acid grassland or damp heaths, rarely in marshes on clay soils. 86. Under-recorded, as old records are lumped with those of *A. vinealis*. Widespread in the west and the Poole basin, absent from the Chalk and limestone.

Agrostis vinealis Schreber

BROWN BENT

Locally frequent in dry, acid woodland rides, heaths and dry, acid grassland, and rarely in leached soil on chalk summits. 77. Mostly in the west or the Poole basin. Old

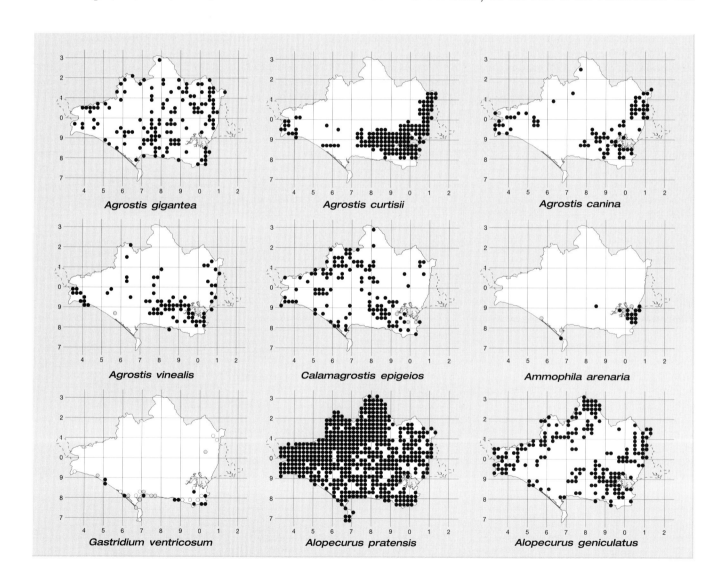

records are confused with *A. canina*, with which I have never seen it grow, though its range is much the same.

[***Agrostis stolonifera* × *Polypogon monspeliensis* =** **× *Agropogon littoralis* (Smith) C.E. Hubb.**
It once occurred in dune slacks east of Little Sea (08H, 1895, JCMP, also 1912, CBG in **BMH**, 1920, JHS and 1922, AWG in **BDK**). 1. Both topography and habitat here have changed, and Good failed to find it in the 1930s.]

***Calamagrostis epigeios* (L.) Roth**
WOOD SMALL-REED
Locally frequent, forming patches in woodland rides, old grassland or damp verges on clay. 121 + [5]. Rarely in heathland, or leached calcareous soil, and tolerant of brackish conditions near the sea. First record: 1571, M L'Obel. Its distribution-pattern is unusual, mostly on clay in the north-west and south-east, but absent from deforested areas, Portland and the north-east Chalk.

***Calamagrostis canescens* (Wigg.) Roth**
PURPLE SMALL-REED
Easily overlooked when growing in wet woods in a starved state. [2]. There are authenticated records from Luckford Copse (88S, 1893, JCMP in **BM**) and Tricketts Cross (00V, 1927, HH Haines in **LTR**), and reports from West Moors (00R, 1930, FHH and 1937, DM); other specimens named as this have proved to be *C. epigeios*. EJ Clement has pointed out that this is one of the few British grasses whose stems are branched.

***Ammophila arenaria* (L.) Link**
MARRAM
Locally abundant on foredunes and sandy cliffs between Studland and Canford Cliffs, less common along the inner shores of Poole Harbour. 15 + [5]. Outliers, where the plant occurs in small numbers, are at Abbotsbury (58S, 1937, DM); Small Mouth (67S, 1967–91, !); Radipole Lake (68Q, 1924, RDG and C Day); inland on army land at Gallows Hill (89K, 1998, BE and DAP); Swanage (07J, 1895, JCMP). An important dune-forming plant receiving some protection from human pressure by fencing at Studland; it dies off as the inner dunes become stabilised.

***Gastridium ventricosum* (Gouan) Schinz. & Thell.**
NIT-GRASS
Once a rare cornfield weed, now sporadic and decreasing, and mostly on bare clay on undercliffs or in clearings of gorse scrub along the coastal belt. 11 + [21]. First record: in arable between Weymouth and Radipole (67, 1775, T Yalden). Good only found it in one arable field in the 1930s. Populations today are small, except at East Bexington (58D and E, 1991, CDP and DAP) and Durlston CP (07I, 1982, RJH Murray and 1998, DAP).

***Lagurus ovatus* L.**
HARE'S-TAIL
A rare casual along the coast, or a garden escape inland. 3 + [2]. Weymouth Sta (67U, 1950, NDS); Thornford (61B,

1993, IPG and PRG); Winfrith Newburgh (88C, 1998, !); Knoll Ho foredunes (08G, 1998, !); Broadstone (09C, 1895, JCMP). Well established on sandy cliffs at Southbourne in VC11.

***Apera spica-venti* (L.) P. Beauv.**
LOOSE SILKY-BENT
A rare casual of waste ground and tips, not found in arable as elsewhere in Britain. 2 + [5]. West Bay (49Q, 1924, AWG); Beaminster (40Q, 1932, AWG in **BDK**); Sherborne tip (61H, 1981, AJCB); Wolfeton Ho (69R, 1991, RCP in **OXF**); Poole Park (09F, 1908, HJG); West Moors Sta (00R, 1915, EFL).

***Mibora minima* (L.) Desv.**
EARLY SAND-GRASS
Very rare in bare, sandy soil or dunes. 1 + [2]. Garden of Peake Ho, Corfe (98R, 1964–67, JB); dunes, Studland (08G, 1993, FAW and 1999, EAP, !); Stewart's Nursery, Ferndown (00Q, 1928, FHH, *et al.* in **DOR, LANC, LIV** and **WRHM**, last seen 1968, V Follett).

***Polypogon monspeliensis* (L.) Desf.**
ANNUAL BEARD-GRASS
Rare on muddy shores, on the damp floor of a limestone quarry, and as a bird-seed alien in tips and gardens inland. 8 + [6]. Dorchester (69V, 2000, !); Yeolands Quarry, Portland (77A, 1996, RMW and C Steele); Winspit (97T, 1915, EBB); Slepe Moor (98T, 1937, RDG in **WRHM** and 1996, B Pickess and DAP); Gold Point (98U, 1937, DM and 1958, NW Moore in **WRHM**); in a garden, Lytchett Matravers (99M, 1992, !); Swanage (08F, 1998, EAP); [Studland and Little Sea, 1874, JCMP, *et al.* in **BDK, BMH, HDD** and **K,** last seen 1929, AWG)]; Greenland Farm (08C, 1999, EAP and DCL); Old Harry (08L, 1978, RHD); Lilliput (08P, 1919, EFL); Poole and Parkstone (09, 1929, LBH); Hatch Pond (09B, 1996, Anon); Canford Heath tip (09I, 1999, !).

***Polypogon viridis* (Gouan) Breistr.**
WATER BENT
Rare on waste ground or pavements near the coast. 5. Broadwey (68R, 1999, !); Hamworthy (09A, 1999, !); Darbys Corner (09C, 1995, !); Church Road, Parkstone (09F, 1992, C Flynn); Canford tip (09I, 1998, ! in **RNG**).

***Alopecurus pratensis* L.**
MEADOW FOXTAIL
Frequent to abundant in unimproved neutral or calcareous grassland. 543. Intensive farming in central Dorset has often reduced this to the status of a relict on verges.

***Alopecurus geniculatus* L.**
MARSH FOXTAIL
Occasional, rarely dominant, in wet grassland or by ponds. 200. Absent from the drier parts of the Chalk and limestone, and probably under-recorded.

***Alopecurus geniculatus* × *bulbosus* = *A.* × *plettkei* Mattf.**
Locally frequent in wet, somewhat brackish pastures near the sea where the parents meet. 10. West Bay (49Q, DAP);

Burton Mere (50D and E, DAP); Abbotsbury (58S, 1932, JEL in **RNG** and 1989, DAP); Lodmoor (68V, BE); Swineham (98I and N, DAP); Kesworth (98P, !); Fitzworth (98Y, DAP); Lytchett Bay (99R, DAP and DG).

Alopecurus bulbosus Gouan
BULBOUS FOXTAIL
Locally abundant in wet meadows behind estuaries, and in the upper parts of salt-marshes. 10 + [4]. On the coast between West Bay and Lodmoor, and along the inner shores of Poole Harbour.

Alopecurus aequalis Sobol.
ORANGE FOXTAIL
Rare, but easily overlooked, in ponds or flooded claypits. 2 + [1]. Chickerell (67P, 1988, DAP); Ridge (98I, 1970, JB); Pamphill (90V, 1988, RF in **RNG**, also DAP).

Alopecurus myosuroides Hudson
BLACK-GRASS
Locally abundant as an arable weed in dry soils. 76. Mostly in the north, also around Langton Herring (68) and in south Purbeck. First record: Portland (67, 1865, SMP in **DOR**), but it has not been seen there since. JCMP found it common, while Good only had it in 13 of his stands. Its unusual distribution-pattern now may reflect individual farmer's use of selective weedkillers.

Phleum pratense L.
TIMOTHY
Frequent in unimproved grassland, and sometimes an arable weed. 604. It was once, and perhaps still is, a constituent of seeds mixtures for ley grassland. Variable in size.

Phleum bertolonii DC.
SMALLER CAT'S-TAIL
Frequent in short calcareous turf, and rarely in short turf on sandy or clay soils. 362. Variable.

Phleum arenarium L.
SAND CAT'S-TAIL
Very local on fixed shingle or bare soil on cliff-tops. 2 + [4]. Burton Freshwater (48U, 1930, AWG and 1960, !); West Bay cliffs (49Q, 1932, AWG and 1994, DAP); Small Mouth (67S, 1895, JCMP and 1960–97, !); Chesil bank, 67J and N, 1960, !); Weymouth (67, 1799, RP); North shore, Poole (09, 1799, RP). Reports from Black Ven (39L), Bindon Hill (88) and Purbeck (97) are not accepted without confirmation.

(*Phleum graecum* Boiss. & Heldr. was reported from army camps at Corfe and Ulwell by CBG in 1915–16, but his specimens in **BMH** are perennial *P. bertolonii*.)

[*Bromus arvensis* L.
FIELD BROME
Casual found in cornfields east of Almer (99E, 1896, EFL). 1. Not found by Good in any of his stands, and his report of this from

Egliston (88V, 1933) lacks a specimen. WFS diarists once used this name for *B. hordeaceus*.]

Bromus commutatus Schrader
MEADOW BROME
Locally frequent in old, unimproved meadows on chalk, clay or alluvium. 70 + [2]. Scattered, but absent from acid soils in the Poole basin and most of the intensely farmed Chalk, where tall forms sometimes appear as weeds in cereal crops. Perhaps under-recorded or confused with *B. hordeaceus*.

Bromus racemosus L.
SMOOTH BROME
Local in unimproved meadows and water-meadows on calcareous or clay soils. 27 + [11].

Bromus hordeaceus L.
SOFT-BROME, BOB-GRASS
Abundant in dry grassland, on verges, and sometimes as an arable weed. 690. Variable in size. Subsp. *ferronii* (Mabille) P.M. Smith occurs all along the coast, at cliff-edges or on fixed shingle. 45. It is adapted to dry soils, high winds and nitrification from bird-droppings. Subsp. *thominei* (Hardouin) Braun-Blanquet is much less common along tracks and bare places on cliffs. 4. West Weare (67W, ! in **RNG**); Worbarrow Bay (87U, !); Hambury Tout (88A, !); Durlston CP (07I, 1912, CBG in **BMH** and 1990, ! in **RNG**). A tall form with long anthers, provisionally called '*longipedicillatus*' has been found at Stoborough (98C), Wareham walls (98I, ! det. LMS) and Marsh Br (99E).

Bromus hordeaceus × *lepidus* = *B.* × *pseudothominei*
P.M. Smith
Scarce in dry, meadows and verges. 4 + [5]. Small Mouth (67S, 1960, PJO Trist); Fortuneswell (67W, 1960, J Rogerson); Rye Water Farm (61Q, 1990, DWT); under Creech Wood (88W and 98B, 1997, BE); Blandford (80Y, 1930s, NDS); Wareham (98I, 1930s, NDS); Blandford Camp (90I, 1999, JHSC): Talbot (09S, 1964, MPY).

Bromus lepidus O. Holmb.
SLENDER SOFT-BROME
Uncommon and sporadic in dry, disturbed verges and tracks, on both calcareous and sandy soils. 16 + [16].

[*Bromus interruptus* (Hackel) Druce
INTERRUPTED BROME
A casual, once found in a sainfoin field between Verwood Sta and Edmondsham (01, 1902, EFL in **LIV**). 1.]

[*Bromus secalinus* L.
RYE BROME
A cornfield weed since Roman times, now a rare casual. JCMP gives 14 localities between 1830 and 1895, but Good only found it once in the 1930s. Post-1920 records: Powerstock (59, 1930s, RDG); Portland Bill (66Z, 1932, JEL in **RNG**); Badbury Rings (90R, 1920s, JHS); Sandbanks (08N, 1964, JB).]

[**Bromus japonicus** Thunb. ex Murray
THUNBERG'S BROME
A casual once found at Parkstone (09F, 1930s, LBH).]

Bromopsis ramosa (Hudson) Holub
HAIRY-BROME
Frequent, but rarely in quantity, in woods and shaded hedgebanks. 415. Rare in deforested areas and absent from very acid soils.

Bromopsis benekenii (Lange) Holub
LESSER HAIRY-BROME
Easily overlooked or mistaken for *B. ramosa*, in similar habitats. [1]. Near the outflow of Sherborne Park Lake (61N, 1970, ACL in **K**).

Bromopsis erecta (Hudson) Fourr.
UPRIGHT BROME
Locally abundant in tall, calcareous grassland and verges, and sometimes on limed or enriched verges and tracks in heathland. 131. Gaps in its distribution on the Chalk are probably due to its destruction by cultivation.

Bromopsis inermis (Leysser) Holub
HUNGARIAN BROME
A rare introduction. 1 + [1]. Sherborne School (61I, 1969, ACL); Baiter (09F, 1989, ! in **RNG**).

Anisantha diandra (Roth) Tutin ex Tzvelev.
GREAT BROME
Rarely established in warm, dry places on cliffs. 5. East Portland (77A and B, 1984, RA, also 66U, !); sandy cliffs, Poole (08N, P and U, 1992, RMW); well established further east in VC11.

Anisantha rigida (Roth) N. Hylander
RIPGUT BROME
An increasing casual in bare, dry places. 4 + [1]. Easton tip (67V, 1990, ! in **RNG**); Frome Hill (78E, 1997, !); Lulworth Cove shore (88F, 1960, !); disturbed dunes, Sandbanks (08N, 1991, !); Holes Bay (09B, 1991, !).

Anisantha sterilis (L.) Nevski
BARREN BROME
Abundant in dry hedgebanks, verges, village lanes, and often as a weed in arable, gardens and pavements; seeds

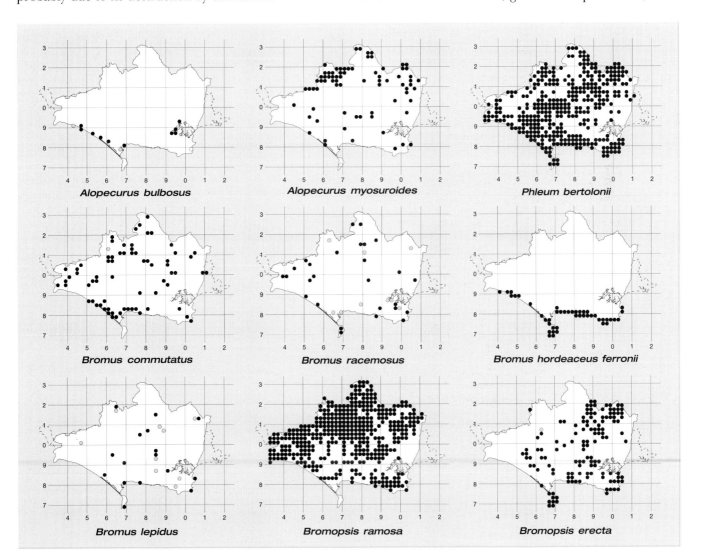

Alopecurus bulbosus

Alopecurus myosuroides

Phleum bertolonii

Bromus commutatus

Bromus racemosus

Bromus hordeaceus ferronii

Bromus lepidus

Bromopsis ramosa

Bromopsis erecta

are known from the Iron Age. 710. Most common in the coastal belt. It germinates in autumn.

Anisantha tectorum (L.) Nevski
DROOPING BROME

An uncommon casual on new verges or waste places, usually near houses. 2 + [6]. Beaminster (40Q, 1965, AWG in **RNG**); Weymouth (67, 1930s, RDG); Sherborne (61I, 1970, ACL); Durweston (80P, 1891, EFL in **LIV**); Wareham by-pass (98D, 1989, ! in **RNG**); Norden Farm (98L, 199, PD Stanley); Corfe village (98R, 1966, JB); Parkstone (09F, 1960, MPY).

Anisantha madritensis (L.) Nevski
COMPACT BROME

Once confined to a very few old walls, now increasing in the coastal belt, in dry waste places. 6 + [2]. West Bay (49Q, 1921, AWG); Abbotsbury Ch and walls (58S, 1960–99, !); Small Mouth and Ferry Bridge (67S and T, 1995, !); wall, Broadwey (68R, 1995, !); urban footpath, North Haven (08I, 1999, !); Sandbanks (08N, 2000, !); Branksome Park (09K, 1945–51, KG).

Ceratochloa carinata (Hook. & Arn.) Tutin
CALIFORNIA BROME

Sporadic on verges and by farm tracks, mostly casual but increasing. 13 + [2]. First record: Swanage (08J, 1915, CBG in **BMH**, as *C. cathartica*, with which it is often confused); still there on allotments, 1998, DCL.

Ceratochloa cathartica (Vahl) Herter
RESCUE BROME

A scarce casual in gardens or allotments. 1 + [2]. Beaminster (40Q, 1920–28, AWG); Knitson Farm (08A, 1998, EAP); Upton (09A, 1906, HJG).

Brachypodium pinnatum (L.) P. Beauv.
TOR-GRASS

Locally dominant in calcareous grassland and cliffs on calcareous clay, and in small patches on dry, heathy verges where these have been enriched in some way. 188 + [5]. Irregularly scattered over the Chalk, and along the coastal belt between Swyre (58I) and Studland 08), also along the railway east of Gillingham (82I). Forms with branched

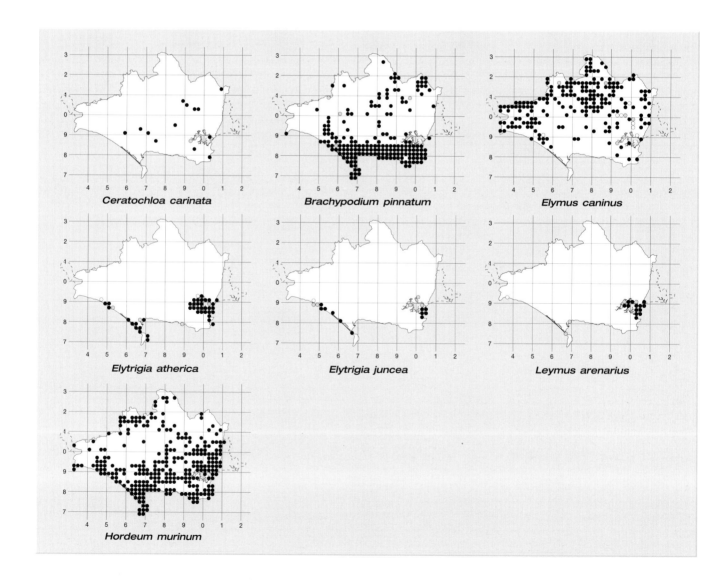

Ceratochloa carinata   Brachypodium pinnatum   Elymus caninus

Elytrigia atherica   Elytrigia juncea   Leymus arenarius

Hordeum murinum

inflorescences are not rare. Disliked by grazing animals, and also by farmers, who try (unsuccessfully) to destroy it by fire. Both the type and ssp. *rupestre* (Hort.) Schiebler occur.

Brachypodium sylvaticum (Huds.) P. Beauv.
FALSE BROME
Frequent everywhere in woods and hedgebanks. 689.

Elymus caninus (L.) L.
BEARDED COUCH
Occasional, rarely in quantity, at the edge of woods, in sheltered hedgebanks, river banks and village lanes. 191 + [11]. Much commoner in the north than in the south, scarce on acid soils in the Poole basin, and absent from deforested areas.

Elytrigia repens (L.) Desv. ex Nevski
COMMON COUCH, SQUITCH
Abundant in disturbed grassland, hedgebanks, arable and gardens with both open and closed vegetation. 728. Variable in size and awn-length; a form with a spike of spikes occurred at Greenhill Barton (68X, !). Subsp. *arenosa* (Spenner) A. Love, whose leaf-sheaths have hairless margins, is locally dominant at the top of salt-marshes around the inner shores of Poole Harbour, and is confused with *E. atherica,* with which it sometimes grows.

Elytrigia repens* × *atherica* = *E.* × *oliveri (Druce) Kergeulen
Perhaps overlooked, with only one record from East Fleet (67J, 1988, DAP and ARG Mundell in **RNG**).

Elytrigia repens* × *juncea* = *E.* × *laxa (Fries) Kergeulen
A critical hybrid, found near the beach. 1 + [4]. Specimens collected by CBG in **BMH** are wrongly named. Burton Bradstock (48Z, 1912, HJR); Neck of Portland (67S, 1911, WBB); Sandsfoot (67T, 1908, GCD); Lodmoor (68V, 1908, GCD); south of Redend Point (08L, 1991, !).

Elytrigia atherica (Link) Kergeulen ex Carreras Martinez
SEA COUCH
Frequent in the upper parts of salt-marshes, less common on shingle beaches and low clay cliffs. 39 + [6]. It occurs along the coast between Burton Bradstock and Radipole Lake, including East Portland, and was once at Gad Cliff (87Z, 1932, RDG). Some records from inside Poole Harbour may be *E. repens* subsp. *arenosa,* but it certainly occurs north of Arne (98U and Z), at Holes Bay (99V) and near Greenland Farm (08C).

Elytrigia atherica* × *juncea* = *E.* × *obtusiuscula (Lange) N. Hylander
A scarce and critical hybrid of the upper shore. 3 + [7?]. Burton Mere (58D, 1993, LJM); Small Mouth (67S, 1954, DO Jones and 1987, ! in **RNG**); South Haven (08I, 1911, CBG in **BMH** and 1982, AJB). Earlier reports by JCMP and CBG in **BMH** are doubtful.

Elytrigia juncea (L.) Nevski
SAND COUCH
Local on undisturbed, sandy foreshores or shingle beaches. 11 + [6]. At Studland Bay it forms small dune ridges before *Ammophila* takes over. Rare between West Bay and Small Mouth, and more frequent from Studland to North Haven.

Leymus arenarius (L.) Hochst
LYME-GRASS
Rare on beaches between Charmouth and Lodmoor, and locally frequent on outer dunes and sandy cliffs between Studland and Canford Cliffs. 13 + [5]. There are a few records from inside Poole Harbour, including Hamworthy, Shipstal point, Gold Point and Brownsea Is, and it has been planted on an island in the Stour at Ensbury (09X, FAW).

Hordeum vulgare L.
SIX-ROWED BARLEY
Rarely planted as a crop today, as at Winterborne Kingston (89N, 1996, !), and a scarce casual, e.g. Goathill (61T, 1996, FAW); Stalbridge Common (71N, 1996, !). 7.

Hordeum distichon L.
TWO-ROWED BARLEY
A major crop-plant and a frequent, but under-recorded, relict in later crops, by verges and farm tracks. 23+. The main cultivars grown today are winter-sown Fanfare, Intro and Regina or spring-sown Chariot and Optic, but these are replaced every few years. Used for cattle food and for brewing beer. Barley seeds are found in archaeological excavations as far back as the Neolithic period.

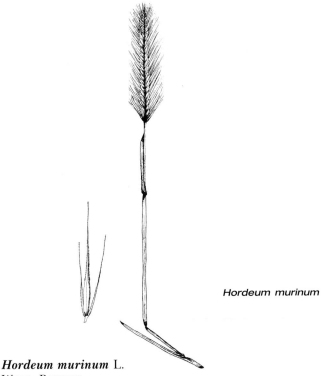

Hordeum murinum

Hordeum murinum L.
WALL BARLEY
Frequent in the coastal belt in grassland, cliff-tops and

stabilised shingle beaches. 276. Less frequent inland in unnatural habitats such as verges, village lanes and pavements, least common in the north and west. It resists trampling. Subsp. *leporinum* (Link) Arcang., with a short central floret, is a rare casual once seen at Wareham by-pass (98I, 1989, ! in **RNG**).

Hordeum jubatum L.
FOXTAIL BARLEY

A rare casual of verges and waste places. 1 + [4]. Weymouth (67, 1930s, RDG); south of Small Mouth (67S, 1988, DAP); Corfe Army Camp (98R, 1915, CBG in **BMH**); Hamworthy (99V, 1967, AWG in **BDK**); Swanage (07, 1930s, RDG).

Hordeum secalinum L.
MEADOW BARLEY

Locally abundant in unimproved, neutral pasture on clay along the coastal belt, including Portland, decreased and often as a relic in the north-west. 134 + [3]. It is absent from almost all the Chalk, and from acid soils in the Poole basin.

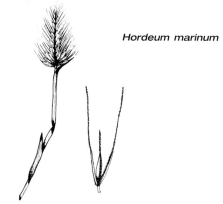

Hordeum marinum

Hordeum marinum L.
SEA BARLEY

Rare and decreasing in hard ground near the sea, where it could be overlooked if heavily grazed. 1 + [6]. West Bay (49Q, 1928, AWG); north-west of Burton Mere (58D, 1993, LJM and LMS); Wyke and Small Mouth (67S and T, 1836, R. Blunt in **WAR**, *et al.*, and 1967, JB); Weymouth Backwater (68Q, 1861, RF Thompson in **DOR**, and 1930s, DM); Lulworth Cove (87J, 1895, WMR); Parkstone (09F, 1930s, RDG).

Secale cereale L.
RYE

Occasionally planted as a crop since the early Iron Age, and found as a casual in arable, fallow, by verges and farm tracks, but not persisting. 24.

Triticum aestivum L.
BREAD WHEAT

A major crop-plant since the late Bronze Age, and a frequent casual in waste places. 34+. An awned form is sometimes planted, but the main cultivars today are spring-sown Chablis or winter-sown Consort, Equinox or Hereward. Despite intensive spraying, wheat crops are still attacked by mildews and rusts.

Triticum turgidum L.
RIVET WHEAT

Casual at Sherborne tip (61H, 1966, ACL).

[*Triticum dicoccum* Schrank and *T. spelta* L. seeds are found in archaeological excavations from the Neolithic to the Iron Age, but are neither grown nor seen today.]

Danthonia decumbens (L.) DC.
HEATH-GRASS

Occasional, rarely in quantity, in old grassland and heaths, rarely in fens or bogs. 243 + [3]. It tolerates a wide range of pH and moisture conditions on all types of soil, and resists trampling. Its absence from much of the Chalk today is probably due to intensive agriculture. There is a relict site at Portland Bill (66U).

Cortaderia selloana (Schultes & Schultes f.) Asch. & Graebner
PAMPAS-GRASS

Extensively planted for about a century near Abbotsbury gardens and Swannery (58S, SG), but while it produces seeds, no seedlings have been reported. 28 + [1]. There is a large colony on made-ground at Lodmoor which must have been self-sown (68V, ! in **K**), and the same is true for clumps of a dwarf form at Darbys Corner (09C, 1995, !) and on cliffs west of Branksome Chine (08U).

Cortaderia richardii (Endl.) Zotov
EARLY PAMPAS-GRASS

In similar places to *C. selloana*, perhaps under-recorded. 5. Lyme Regis cliffs (39F, 1980s, TBR and 1999, !); self-sown at Lodmoor old tip (68V, 1997, ! in **K**); planted at West Stafford Ch (78J) and by a pit west of Huish (99D); several clumps at Canford Tip (09I, 1999, !).

Molinia caerulea (L.) Moench.
PURPLE MOOR-GRASS

Abundant or dominant in wet heaths and conifer plantations, bogs, and western summit heaths. Also occasional in acid woods, grazed pastures, fens, swamps and marshes on acid soils. 196. First record: Canford Heath (99, 1551, W Turner). The tall form in fens and wet, neutral grassland is usually subsp. *arundinacea* (Schrank) K Richter, as at Dodpen Hill (39P, LJM), Hooke Park (59J, LJM), Lydlinch Common (71G, !), Deadmoor Common (71K, R Sherlock) and Whitefield (99B and C, !). Grazed by deer, but disliked by farm stock.

Phragmites australis (Cav.) Trin. ex Steudel
COMMON REED

Locally dominant in shallow water, in brackish estuaries, and at the edges of lakes and larger rivers. 257. Inland, *Phragmites* colonises flooded pits, is present in most fens, and occasionally grows in hedges when the nearby wetland has been drained. It is absent from the drier parts of the Chalk and most of Portland. Var. *stolonifera* G.F.Mey., with long aerial roots, is probably frequent on wet clay cliffs, as at Lyme Regis (39L), Sandsfoot (67T), Osmington Mills

(78K) and Swanage (07J). Extensive reedbeds occur along the coast at Burton Mere (58D), by the Fleet at Abbotsbury Swannery (58S), at Radipole (= Reedy Pool) Lake RSPB reserve (68Q), South Haven (08I), and especially around the inner shores of Poole Harbour. They are an important refuge for deer and many birds, including Bearded Tits which feed off the seeds, and for nests of Reed Buntings and Reed Warblers. The leaves are food for certain Wainscot and other moths. Reeds are harvested at Abbotsbury and Radipole, partly for conservation but also to make long-lasting thatched roofs.

Cynodon dactylon (L.) Pers.
BERMUDA-GRASS

Rare on sandy cliffs and beaches, and surviving in verges near the sea, but only in the south-east. 6 + [2]. Very resistant to trampling. Brownsea Is (08E, 1961–77, HJB); Studland Bay (08H and I, 1846, WD Bromfield in **MDH**, *et al.* in **BDK, DOR, LIV, OXF** and **RNG**, last seen 1935; Haven Hotel pavements (08I, 1999, DCL); Sandbanks (08N, 1934, RDG, *et al.*, and 1992, !); Lilliput (08P, 1895, JCMP, *et al.* in **BDK, DOR, OXF** and **RNG**, and 1980, AJB); Canford Cliffs (08U, 1992, RMW); Sterte (09A, 1945, FR, *et al.*, and 1980, AJB); [Parkstone (09F, 1790s, T Velley in **LIV** and 1896, EFL in **LIV.**]; Baiter (09F, 1989, !). It occurs at the base of sandy chines further east in VC11.

Spartina maritima (Curtis) Fern.
SMALL CORD-GRASS

Rare in salt-marshes inside Poole Harbour. 2 + [1]. Keysworth (98P, 1965, CE Hubbard, also DSR); Brands Bay (08C, 1999, EAP and 08H, 1980, AJCB); Poole Harbour (08, 1951, HJ Killick).

Spartina maritima × *alterniflora* = *S.* × *townsendii* Groves & J. Groves
TOWNSEND'S CORD-GRASS

First recorded from salt-marsh at Ower (98, 1899, JCMP in **K**), this spread very fast in Poole Harbour, but is now largely replaced by its amphidiploid progeny, with which it has been confused. 3 + [8]. Still at Lodmoor (68V, 1955–99, !) and Brands Bay (08C and H, 1999, EAP). In 1965, CE Hubbard found it at Keysworth (98P), Arne Bay, Fitzworth Point and Ower (98Y and Z), Holes Bay (99V and W, 09A and B), Brands Bay, Furzey and Green Is (08C and D).

Spartina anglica C.E. Hubb.
COMMON CORD-GRASS

Locally dominant in salt-marshes fringing Poole Harbour, extending to the lagoon on Brownsea Is. 23. Elsewhere reported from Small Mouth (67T, 1927, MJA), Radipole Lake (68Q, 1937, MJA) and Lodmoor (68V, 1955–97, !). In Poole Harbour, it increased greatly to a maximum in 1917–24, when it trapped about 4% of the sediment brought down by the Frome, and was exported to North Europe and Australia as a mud-binder. It is now dying back, either through erosion, failure to grow on anoxic mud or some unknown cause such as poisoning by antifouling agents containing tributyltin. Since 1961, about three quarters of the seed heads have been infested with ergot (Raybould, 1997).

Spartina alterniflora Lois.
SMOOTH CORD-GRASS

Planted in a salt-marsh at Keysworth (98P, 1963, CE Hubbard). [1].

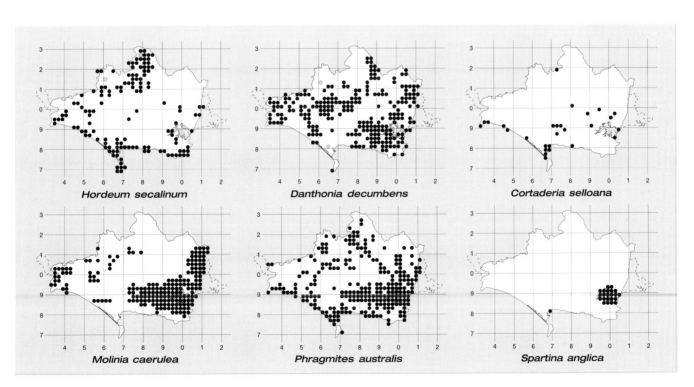

Hordeum secalinum

Danthonia decumbens

Cortaderia selloana

Molinia caerulea

Phragmites australis

Spartina anglica

273

Panicum capillare L.
WITCH-GRASS
Casual, preferring warmer countries than Britain. 3 + [2]. In maize, Forde Abbey (30S, 1997, IPG); among carrots, Beaminster (40Q, 1952, AWG in **BDK**); bird-seed alien, Winterborne Kingston (89N, 1998, !); Greenland Farm (08C, 1999, EAP det. EJ Clement, !); Swanage Camp (08F, 1917, CBG in **BMH**).

Panicum miliaceum L.
COMMON MILLET
Occasionally strip-planted in arable for pheasants, also casual on tips and a pavement weed from bird-seed. 25 + [13].

Echinochloa crus-galli (L.) P. Beauv.
COCKSPUR
Once a mere casual in gardens and on tips, now becoming an arable weed in maize and in fruit-farms. 23 + [12]. Often a bird-seed alien, and on verges.

Echinochloa esculenta (A. Braun) H. Scholz
JAPANESE MILLET
Strip planted in arable at Cutt Mill (71T, 1997, ! and 71Y, 1999, !), and a casual at Crossways tip (78P, 1999, !) and Binnegar tip (88Y, 1992, ! in **RNG**). 4.

Echinochloa colona (L.) Link
SHAMA MILLET
A rare casual. 1. Parkstone (09F, 1908, HJG); Kinson (09T, 1995, FAW).

Echinochloa frumentacea Link
WHITE MILLET
A rare, casual, bird-seed alien of waste ground and tips. 3 + [1]. Conygar Hill (49R, 1989, ! in **RNG**); Abbotsbury (58S, 1988–92, ! in **RNG**); Sherborne tip (61I, 1970, ACL); Crossways tip (78P, 1999, !).

Setaria pumila (Poiret) Roemer & Schultes
YELLOW BRISTLE-GRASS
Seen in arable at Lewell Mill (78J, 1994, DAP and RMW) and Cutt Mill (71Y, 1999, !), more often a casual bird-seed alien of tips and pavements. 14 + [6]. There is a specimen in **BMH**.

Setaria verticillata (L.) P. Beauv.
ROUGH BRISTLE-GRASS
A rare casual of nursery gardens and waste places. 4 + [5]. The only recent records are from Hurst Farm (79V, 1994, M Chandler), Swanage (07J, 1999, DCL), New Swanage (08F, 1998, EAP) and Canford tip (09I, 1999, !). The first and last of these were var. *ambigua* (Guss.) Parl. Specimens are in **BDK, CGE, E** and **K.**

Setaria viridis (L.) P. Beauv.
GREEN BRISTLE-GRASS
An occasional casual in gardens, building-sites or tips. 10 + [17].

Setaria italica (L.) P. Beauv.
FOXTAIL BRISTLE-GRASS
A casual on tips, whose panicle varies greatly in length and breadth; small forms may lack bristles. 7. Sherborne (61H, 1970, ACL and 1989,!); Crossways (78P, 1999, !); Binnegar (88Y, 1957 and 1993–94, !); Arne (98U, 1993, !); Canford Tip (09I, 1999, !); Three-legged Cross (00X, 1982, !).

[***Digitaria ischaemum*** (Schreber ex Schweigger) Muhlenb.
SMOOTH FINGER-GRASS
A casual once reported from Parkstone (09F, 1905, HJG).]

Digitaria sanguinalis (L.) Scop.
HAIRY FINGER-GRASS
A rare casual, often as a pavement weed. 2 + [4]. Post-1960 records: Wareham Sta (98J, 1970, NR Webb in **WRHM** and 1979, MB); Hamworthy (09A, 1977, ! in **RNG**); Poole Park (09F, 1992, C Flynn); Wimborne (00A, 1994, !).

Sorghum halepense (L.) Pers.
JOHNSON-GRASS
An occasional casual on waste ground or tips. 2 + [3]. Post-1960 records: in arable, Stoke Abbott Mill (49I, 1976, AWG); Lodmoor tip (68V, 1997–98, !); Sherborne tip (61I, 1970, ACL); Wareham (98I, 1991–92, !); Hamworthy Docks (09A, 1977, !).

Zea mays L.
MAIZE
During the last 20 years, this has been planted as a crop

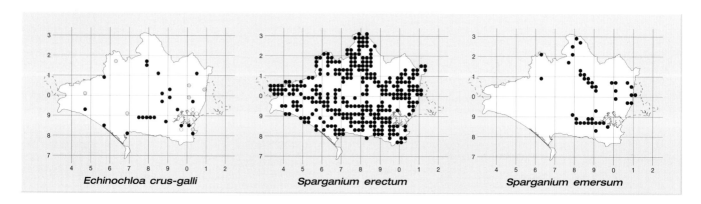

Echinochloa crus-galli   Sparganium erectum   Sparganium emersum

for cattle fodder in the south, and it occurs as a relict in subsequent crops or on tips. 6. Today's favoured cultivars are Lincoln and Reynard.

SPARGANIACEAE

Sparganium erectum L.
BRANCHED BUR-REED
Frequent in streams, rivers and their ditches. 313. The four subspecies have seldom been distinguished by recorders, but subsp. *erectum* is frequent and subsp. *neglectum* (Beeby) K. Richter is widespread. Subsp. *microcarpum* (Neuman) Domin. occurs in the Bere Stream (89M, 1900, EFL) and both north and south of Wareham (98D), while subsp. *oocarpum* (Celak) Domin. is recorded from the Moors River (00V, 1976, A Brewis).

Sparganium emersum Rehmann
UNBRANCHED BUR-REED
Occasional in flowing water of the Stour, Frome, Piddle and Moors Rivers, with outliers in the Yeo (61C) and at Sandford Orcas (62F). 42. Absent from the west.

(*Sparganium angustifolium* Michaux was reported from Corfe Heath (98, 1799, RP) but never confirmed.)

Sparganium natans L.
LEAST BUR-REED
A northern species, surviving in fen ditches and a pond at The Moors (98N, 1937, RDG in **DOR** and **WRHM**, *et al.* in **RNG**, and 1992, DAP, !). 1.

TYPHACEAE

Typha latifolia L.
BULRUSH
This rapidly colonises nutrient-rich ponds, flooded pits and ditches, and becomes temporarily dominant. 254. It is absent from drier parts of the county, including most of the Chalk.

Typha latifolia × *angustifolia* = *T.* × *glauca* Godron
An easily overlooked hybrid, but rare. 1 + [3]. Beaminster

(40Q, 1925, AWG); Priors Down lake (71P, 1996, !); Ridge (98I, 1917, CBG); Little Sea (08H, 1977, JO Mountford).

Typha angustifolia L.
LESSER BULRUSH
Uncommon in lakes, ponds and by rivers. 22 + [12]. Planted in some places, such as Abbotsbury Swannery (58S) and Heedless William's Pond (79F), but native in at least half its sites, as near Keysworth Farm (98P, JHSC). Good found it in 13 stands in the 1930s.

PONTEDERIACEAE

Pontederia cordata L.
PICKERELWEED
Occasionally planted in ornamental ponds, or an escape, but rarely surviving for long. 4 + [1]. Warmwell Down (78M, 1993, !); [Worth Matravers (97T, 1991, !)]; Wimborne (09, 1949, WM Reynolds in **BM**); Stapehill (00K, 1992, !); Dewlands Common (00U, 1991, DAP).

LILIACEAE

Narthecium ossifragum (L.) Hudson
BOG ASPHODEL
Locally frequent to abundant in wet heaths and bogs. 91 + [4]. Scarce in the far west, frequent in the Poole basin, with outliers at Gorwell (58T, 1895, JCMP) and Lower Kingcombe (59P, 1989, !).

Asphodelus fistulosus L.
HOLLOW-LEAVED ASPHODEL
A rare casual, but not from gardens. 1 + [1]. Corfe Common (98Q, 1982, RM); Holton Heath (99K, 1917, F Henley).

[*Simethis planifolia* (L.) Gren.
KERRY LILY, BOURNEMOUTH LILY
This survived in Branksome for at least 67 years, perhaps introduced with Bournemouth Pines. The area where it grew is now a public park or built over, though some of its habitat, sandy pine wood, remains. Branksome Park (09K, 1847, C. Wilson, *et al.* in **ALT, BMH** and **DOR**, last seen 1914, CBG in **BMH**). WBB saw

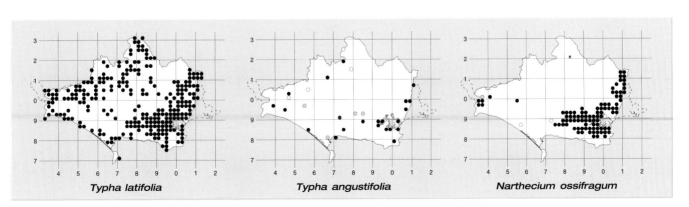

Typha latifolia   Typha angustifolia   Narthecium ossifragum it in 1908 in Mr Packer's cemetery, which has not been traced. Some bulbs were transferred from the wild to Bournemouth Public Gardens in 1909.]

Hosta sieboldiana (Hook.) Engl. was planted at Cowherd Shute Farm (82S, 1995, !).

Hemerocallis fulva (L.) L.
ORANGE DAY-LILY
An occasional, persistent, garden escape or throw-out in woods, verges, sandy cliffs and tips. 26.

Kniphofia uvaria (L.) Oken
RED-HOT-POKER
A persistent garden escape in churchyards, on verges or old tips. 11. **K. × praecox** Baker has been planted at Piddlehinton Camp (79I, 1950–99, !) and on dunes at Sandbanks (08N, 1991, !) and some other records may be this.

Colchicum autumnale L.
MEADOW SAFFRON, NAKED MAIDENS
Very local in old woods, and once also in old meadows from which there are no recent records. 14 + 2p + [19]. Decreasing, but easily overlooked out of season. All recent records are in west-central Dorset, between Halstock and Milton Abbey. The outlier at Creech Grange (98B, 1900, EFL and 1914, CBG in **BMH**) has not been refound, and plants from Powerstock Ch (59D, !) and Stockford Ho (88T, !) were certainly introduced. A source of the toxic alkaloid colchicine, which inhibits cell processes in microtubules, and is used artificially to double chromosome numbers.

Colchicum bivonae Guss. was planted in grass at Bride Head (58Z, 1982, !).

Gagea lutea (L.) Ker Gawler
YELLOW STAR-OF-BETHLEHEM
Very local in old woods, hedgebanks or stream-banks in the north and west. 4 + [4]. Sleech Wood (39G, 1950, JPMB); Hilton (70R, 1859, JCMP in **DOR**, *et al.*, and 1930–54, AWG); Stoke Wake (70N, 1934, VM Leather, *et al.*, and 1991, AH & !, thousands of sterile plants); Coombe Bottom, Ibberton (70Y and 80D, 1977, RJ Pankhurst and 1996, IPG and PRG); Kings Stag (71F, 1978, FR Green and 1984, !, flowering well); Hinton St Mary (71Y, 1904, E Acton). Some DWT records from SY59 and ST50 are not accepted.

Tulipa sylvestris L.
WILD TULIP
Very locally established in hedgebanks, parks and old gardens, probably introduced. 4 + [13]. Post-1950 records: Winterbourne Monkton (68T, 1985, AJ Norman); Leigh (60E, 1895, I Thompson, *et al.* in **WRHM** and 1991, AH & !); Rew (70C, 1974, J Walton); Mappowder Old Rectory (70I, 1976, J Walton and 1999, WGG Woodhouse); Child Okeford (81G, 1978, GD Harthan); Kimmeridge (97E, 1978, RB); Tarrant Gunville (91G, 1956, HJ Moore). It may survive in a sterile state at Tarrant Crawford (90B, 1921, AWG in **BDK**) and Studland (08G, 1915, CBG in **BMH**).

Tulipa gesneriana L.
GARDEN TULIP
Perhaps established on sandy cliffs near Poole (08P, 1992,

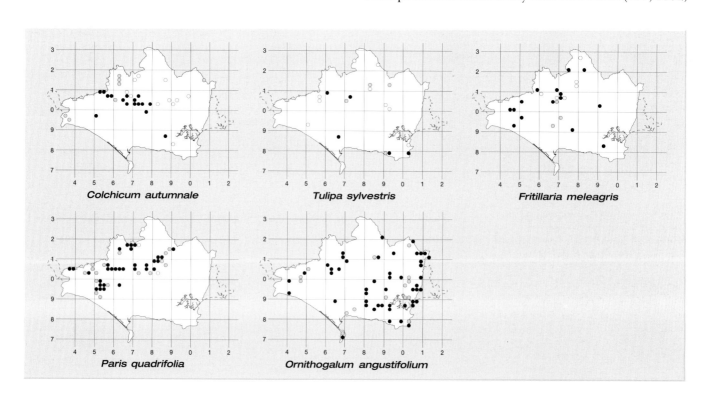

Colchicum autumnale

Tulipa sylvestris

Fritillaria meleagris

Paris quadrifolia

Ornithogalum angustifolium

RMW), otherwise seen planted in a few churchyards, or as a garden escape on verges. 11. Noted in 40K, 68C, 78Z, 81A, 98J, 08G, 07J, 01H and K and 11G.

Fritillaria meleagris L.
FRITILLARY, DROOPING BELL-OF-SODOM
Recorded from wet pastures in the north for 200 years, and perhaps native. 1 + 14p + [9]. Also planted in churchyards, orchards and gardens as shown on the map. Possibly native sites: Buckland Newton (60E, 1877, HH Wood in **DOR** and 1978, RHD); Pulham (70D and E, 1879, JCMP, *et al.*, and in the churchyard, 1997, !; said to have been transplanted here when the original site was ploughed); Piddles Wood (71W, 1799, RP and 1937, DM).

Lilium martagon L.
MARTAGON LILY
A long-established garden relic in plantations on large estates. 5 + [4]. Forde Abbey (30M, !); Vearse Farm, Bridport (49L, 1988, J Neal); Stoke Abbott (40K, 1936, AWG and 1995, BE); Bingham's Melcombe (70R, 1915, AWG and 1992, !); Bryanston School (80T, 1956, GD Harthan); Smedmore Grove (97J, 1865, JCMP in **DOR**); Langton Long (90D, 1895, S Lenton in **DOR**); Kingston Lacy (90R, 1931, FHH, *et al.*, and 1982, KA Hearn).

Lilium pyrenaicum Gouan
PYRENEAN LILY
Planted in Forde Abbey grounds (30M, 1992, !). 1.

Lilium candidum L.
MADONNA LILY
A rare garden throw-out. 2. Corscombe (50H, 1990, !); Gussage St Michael (91W, 1996, GDF).

Convallaria majalis L.
LILY-OF-THE-VALLEY
Possibly native in Martin Wood (00Q, 1947, J Parke and 1976, W Ingram); otherwise a long-established garden escape in plantations, usually on sandy soils, or obviously planted. 9 + [5]. Established as follows: Bridport railway bank (49R, 1950, J Warrington); The Cliff, Bryanston (80T, 1840, FWG and 1938, DM); Creech Grange Woods (98B, 1799, RP, also 1914, CBG in **BMH** and 1973, !); Carey (98D, 1990, !); East End (99Z, !).

Polygonatum multiflorum (L.) All.
SOLOMON'S-SEAL
Locally common in a few woods in the north and north-east, usually on chalk. 3 + [2]. Much confused with its hybrid, and found by neither JCMP nor RDG. Colmer Coppice (30V, 1998, IPG); Dutnoll Coppice (79E, 1992, !); Sutton Down (81S, 1936, J Parke); Ball's Copse (91, 1920s, JHS) and Farnham Chase (91, 1870s, JC Dale) perhaps both in VC8; Woodcutts (91U, 1973, DJG); Garston Wood (01E, 1937, DM, *et al.*, and 1992–98, !). It used to be frequent in woods north of the county boundary, in VC8.

Polygonatum multiflorum × *odoratum* = *P.* × *hybridum* Bruegger
GARDEN SOLOMON'S-SEAL
A denizen or garden escape, usually in very small numbers in plantations. 14 + [8]. Often confused with *P. multiflorum*. First record: Marshwood (39Z, 1927, AWG). Noted in 39R and W, 30L and Q, 50T, 61N, 68J, L and X, 70V, 89T, 80Y, 98V, 99E, J and W, 90V, 91V, 07J, 00Z and 01G. Probably this at Zigzag Hill plantation (82V, 1973, P Bodman and KM Godfrey, recorded as *P. multiflorum*).

(*Polygonatum odoratum* (Mill.) Druce
All reports are errors.)

Paris quadrifolia L.
HERB PARIS
Scarce in old woods, less often in plantations, in the north and west, usually in small colonies. 31 + [19]. Absent from the south and east, where the rainfall is perhaps too low for it.

Camassia quamash (Pursh) Greene
QUAMASH
Six plants on a verge outside a house above Milton Abbas (80B, 1996–99, !). It spreads in gardens.

Ornithogalum pyrenaicum L.
SPIKED STAR-OF-BETHLEHEM
Very rare and unlikely to be native. Durlston CP (07I, 1986, AM); Studland (08, 1930, DM and VM Leather).

Ornithogalum angustifolium Boreau
STAR-OF-BETHLEHEM
In small woods, grassland, village lanes, churchyards and disturbed dunes, but nowhere looking native. 48 + [16]. Mostly in small numbers, but able to spread by seed.

Ornithogalum nutans L.
DROOPING STAR-OF-BETHLEHEM
A rare garden outcast in lanes or hedgerows, in one case planted in dunes. 5 + [1]. First record: Claysmore School (81S, 1957, JAL). Muston (89T, 1991, !); Sandbanks (08N, 1990, !); Bowldish Pond (01H, 1994, GDF); Edmondsham (01K, 1994, GDF); Alderholt (11G, 1994, GDF).

Scilla siberica Haw
SIBERIAN SQUILL
A garden escape. 4. Naturalised in Forde Abbey grounds (30M, !); Cerne Abbas (60Q, !); Tarrant Crawford (90Y, !) and Glebelands estate (08F, !).

(*Scilla verna* Hudson has been reported in error from Portland.)

Scilla peruviana L.
PORTUGUESE SQUILL
A few clumps are established at St Georges Ch, Portland (97V, !). 1.

Hyacinthoides italica (L.) Rothm.
ITALIAN BLUEBELL
Planted at Abbotsbury Gardens (58S) and established in small numbers in two plantations at Ringstead (78K, 1961–84, !). 1.

Hyacinthoides non-scripta (L.) Chouard ex Rothm.
BLUEBELL, GREYGLE or GRIGGLE
Locally abundant or co-dominant in most old woods, also in old hedgebanks, under bracken and in the open on brackeny slopes and undercliffs, especially in the west. 621. White-flowered forms are uncommon, and pink-flowered plants have only been seen on Green Is (08D). Var. *bracteata* Druce, with extra-long pigmented bracts, has been seen in 60G, 89T, 81M and 99C (! in **RNG**). Its sap is irritant and has poisoned horses.

Hyacinthoides non-scripta × *hispanica*
GARDEN BLUEBELL
Commonly naturalised near houses, in verges, village lanes and churchyards, but mostly in artificial habitats. 218. White-flowered forms are not uncommon, and pink-flowered forms have been seen at Worgret (88Y) and Gussage All Saints (91V).

Hyacinthoides hispanica (Miller) Rothm.
SPANISH BLUEBELL
Uncommon, and hard to tell from its hybrid. 2. Along low cliffs, Burton Bradstock (48Z, !); plantation above Durlston Bay (07I, !).

Hyacinthus orientalis L.
HYACINTH
Planted in churchyards, verges or roundabouts, and as a garden throw-out on waste ground or tips. 8. Bulbs can survive for some years in such sites. Noted in 58Z, 68Y, 70D, 72W, 89M, 98E, and 08F and U.

Chionodoxa forbesii Baker
GLORY-OF-THE-SNOW
Naturalised in Forde Abbey grounds (30S), planted in Dorchester (69V) and a garden throw-out at Middle Lodge, Bryanston (80T) and Glebelands estate (08F). 4.

Muscari neglectum Guss. ex Ten.
GRAPE-HYACINTH
A rare garden escape in village lanes or churchyards. 2 + [1]. Sherborne tip (61I, 1970, ACL); Iwerne Courtney Ch (81S, !); Sixpenny Handley (01D, GDF). Many records confused with *M. armeniacum*, but an old one from Portland (66U, 1878, JWW and 1891, JCMP in **DOR**, as *M. racemosum*) might have been native.

Muscari armeniacum Leichtlin ex Baker
GARDEN GRAPE-HYACINTH
Frequently escaped or planted on verges, in village lanes and churchyards. 78. It increases rapidly by seed and by offset bulbs.

Muscari comosum (L.) Miller
TASSEL HYACINTH
A rare garden escape. 2 + [1]. In pasture, Langton Herring (68B, 1990, UH Bowen); Langton Matravers (97Z, 1929, G Baring); Durlston CP (07I, 1990, H Murray).

(*Allium schoenoprasum* L.
CHIVES
Widely grown in gardens, and reported from Girdlers Coppice (71W) and Talbot Heath (09R) on insufficient evidence.)

Allium cepa L.
ONION
Much grown in gardens since at least 1782, when eight cultivars were on sale in Blandford, and seen on tips.

Allium roseum L.
ROSY GARLIC
Scarce, but slowly increasing, in patches on verges, waste ground, churchyards and on a shingle beach. 12. Lyme Regis Park (39L, 1991, !); planted in Abbotsbury gardens in 1899 (MI), it has escaped to several nearby sites including the back of the Chesil bank (58M and S, 1960, KG, *et al.* in **LANC**, and 1997, !); Portland Bill (66Z, 1994, !); east Weymouth (67Z, 1987, !); north of Fleet (68F, 1998, !); Overcombe (68W, 1994, !); Charminster (69R, 1972, RDE and 1997, !); Leigh (60E, 1996, IPG and PRG); Ringstead (78L, 1997, FAW); Galton (78S, 1999, !); Herston and Leeson Ho (07E, 1977-94, WGT).

Allium neapolitanum Cirillo
NEAPOLITAN GARLIC
A rare garden escape. 2 + [1]. Bridport (49R, 1998, JAG); Litton Cheney (59K, 1992–95, !); Abbotsbury (58M, 1961, !, also D McClintock in **RNG**).

Allium subhirsutum L.
HAIRY GARLIC
A rarely grown casual garden escape. 1. Pimperne (90E, 1994, !).

Allium moly L.
YELLOW GARLIC
Often grown in gardens, casual outside them. 1 + [1]. Lodmoor tip (68V, 1960, !); Durlston CP (07I, 1988).

Allium triquetrum L.
THREE-CORNERED GARLIC
Locally naturalised as a garden escape in plantations, churchyards and verges. 36. Mostly in the coastal belt, where it is not casual as Good implies, except on tips. First record: Weymouth (67, 1930, AWG).

Allium paradoxum (M. Bieb.) G. Don
FEW-FLOWERED GARLIC
An easily overlooked garden escape, very locally established in the north-east. 3. Stubhampton Bottom (91C, 1986, AH); Well Bottom (91D, 1981–86, R Legge); Ashmore (91D, !); Melbury Down (91E, 1997, P Amies).

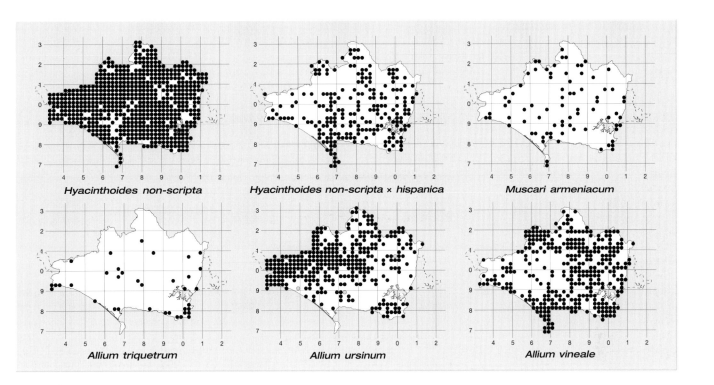

Hyacinthoides non-scripta | Hyacinthoides non-scripta × hispanica | Muscari armeniacum

Allium triquetrum | Allium ursinum | Allium vineale

Allium ursinum L.
RAMSONS, SIVES
Locally dominant in old woods, often in wet ravines, also in plantations, shaded hedgebanks and churchyards, as at Wimborne Minster (09E, !). 331 + [3]. Commonest in the north and west, local on the Chalk and in Purbeck, and absent from deforested areas, Portland and most of the Poole basin. (Reports from east Portland are probably errors for *A. triquetrum*). Colonies occur near bluebells, but rarely interpenetrate. Its distribution suggests a need for high rainfall; I have seen colonies browsed by straying sheep.

Allium oleraceum L.
FIELD GARLIC
Scarce in tall, calcareous grassland, mostly near the coast. 7 + [7]. Abbotsbury (58, 1961, J Tinegate); Fleet (68F, 1955, DM and RDG); between Nottington and Radipole (68R, 1895, JWW); Sherborne (61H, 1969, ACL and 61I, 1987, DAP); White Nothe (78Q and V, 1969–93, !, also MJG); Gad Cliff (87Z, 1870, JCMP in **DOR**, *et al.* in **WRHM** and 1993, !); Lulworth Cliffs (88F, 1993, DAP, !); Tyneham Cap (88V, 1870, JCMP in **DOR** and 1928, AWG); Nutford Farm (80U, 1934, RDG in **WRHM**); Compton Abbas (81U, 1904, EFL); St Aldhelm's Head (97S, 1997, MJG); Melbury Down (91E, 1991, AJB).

Allium carinatum L.
KEELED GARLIC
A rare alien established on verges. 2 + [3]. Maiden Newton (59Y, 1935, RDG in **WRHM** and 1977, E Busfield); Langton Cross (68G, 1995, !); Ansty (70R, 1890, Miss Windhams in **DOR**); Hilton (70W, 1977, G Soane and 1983, A Higgs); Winterborne Houghton (80H, 1978, T Norman).

Allium ampeloprasum L.
WILD LEEK
Very local, and perhaps an ancient introduction, on low cliffs or behind the Chesil bank in the south-west. 3 + [1]. Eype (49K, 1985, AH); West Bexington (58I, 1865, JCMP in **DOR** and 1981, AH in **RNG**); East Bexington to Abbotsbury (58M, 1953, DM, *et al.*, and 1991, SME, !); an old record from Little Bredy (59, 1869, TBS) probably refers to a coastal site. Its British distribution is Cornubian (Fitzgerald, 1988).

Allium porrum L.
LEEK
A garden throw-out at East End (99Z, !).

Allium vineale L.
WILD ONION
Frequent on verges, rough grassland, village lanes, churchyards and sometimes in arable. 337. Widespread, most frequent along the coastal belt.

Allium nigrum L.
BROAD-LEAVED LEEK
A rare garden escape which can become a pest in gardens on both acid and calcareous soils. 5. Winterborne Kingston (89P, 1990, !); Bryanston grounds (80T, 1996, !); Paradise Farm (81W, 1966, ! in **RNG**); Fernhill Ho, Witchampton (90Y, 1996, !); Studland Bay Ho (08G, 1996, !).

Nectaroscordum siculum (Ucria) Lindl.
HONEY GARLIC
A rare but persistent garden escape. 5. Horn Hill (40Q and R, 1960, D Griffiths, *et al.*, and 1997, IPG); Abbotsbury

ruin (58M, 1899, MI, *et al.* in **RNG**, and 1960–98, !); Little Bredy (58Z, 1993, !); Folke Ch (61L, 1998, !). Beloved by dried-flower arrangers, but not by those with sensitive noses.

Tristagma uniflorum (Lindl.) Traub
SPRING STARFLOWER
A rare, garlic-scented, garden escape near the coast. 3. Portland Bill (66U, 1991, A Daly); Swanage (07J, 1960, WB Alexander, !); Sandbanks (08N, 1992, RMW, !); Branksome Chine cliffs (08U, 1997, O Linford).

Crinum × powellii Baker
Planted at Loders Ch (49V, 1999) and Brownsea Is Ch (08I, 1999).

Leucojum aestivum L.
SUMMER SNOWFLAKE
Subsp. *aestivum* is very local in wet woods in the basins of the Piddle, Stour and Allen. 10 + [3]. Waddock (79V, 1986, DM Exton & QG Palmer); Turners Piddle (89G, 1977, G Soane, *et al.*, and 1981–97, !; it looks native here but may be an ancient introduction); Culeaze (89L, 1994, AH); Is in Stour, Langton Long (80X, 1978, RHD and 1989, DAP and RF); Steepleton Ho (81Q, 1994, !); Corfe Mullen (99U, 1920s, JHS); Is in Stour, Spetisbury (90B, 1895, EK Chambers and 1994, !); Is in Stour, Charlton Marshall (90C, 1978, L Farrell and 1989, DAP and RF); Shapwick (90G, 1891, EFL in **LIV**, *et al.*, and 1989, DAP and RF); Tadden (90V, 1846, JC Dale, *et al.* in **BDK** and **RNG**, and 1989, DAP and RF); High Hall Decoy (00B, 1990, DAP and RMW). Subsp. *pulchellum* (Salisb.) Briq. is an ill-defined taxon much planted in gardens and churchyards, and escaped in plantations or by streams. 22 + [3]. It is sometimes reported as *L. vernum*.

Leucojum vernum L.
SPRING SNOWFLAKE
Very local, long established by one shaded stream where it might be native, and an established garden throw-out elsewhere. 7. Wootton Fitzpaine (39S, 1866, JCMP in **DOR**, *et al.* in **LIV, RNG** etc., and 1996,

FAW, *c.*5,000 plants); Loscombe (59D, 1982, HG Darby); Minterne Parva (60R, 1984–97, !); Creech Grange (98B, 1917, CBG); Norden Wood (98L, 1979, AH and AM Ridge); Cole Hill (00A, 1978, L Farrell); Knowle Hill, Woodlands (00J, 1990, H Hunt and 1998, RMW).

Galanthus nivalis L.
SNOWDROP
Well established by shaded streams such as the Lydden and the Allen, in grounds of big houses and many churchyards. 334. Also widely planted or escaped in village lanes, rarely as a double form as at Lulworth (88A), but avoiding very acid soils in the Poole basin. First record: 1830. Large populations occur at Herringston (68Z), Kingston (97N) and Kingston Lacy Park (90V); at the latter site it is a tourist attraction in February.

Galanthus plicatus M. Bieb.
PLEATED SNOWDROP
A rare garden escape seen at Long Burton (91M, 1998, JAG). 1.

Galanthus elwesii Hook. f.
GREATER SNOWDROP
Occasionally planted in churchyards, less common in plantations or on verges. 17. Oathill (40C, PRG); East Fleet (68F); Alweston (61S, JAG); Holwell Ch (61V &W); Osmington Ch (78G, ! in **RNG**); Warmwell Ch (78M); Tolpuddle (79X); Stalbridge (71J); East Holme Ch (88Y); Affpuddle Ch (89B); Culeaze Ho (89L); Anderson Manor (89Y); Hambledon Hill (81L); Sutton Waldron Ch (81S); Studland (08G); Brownsea Is (08I).

Galanthus alpinus Sosn.
CAUCASIAN SNOWDROP
Planted in grassland in Forde Abbey grounds (30M, 1996, !). 1.

Galanthus ikariae Baker
GREEN SNOWDROP
Planted in grassland at Swanage park (07J, 1993, ! in **RNG**). 1.

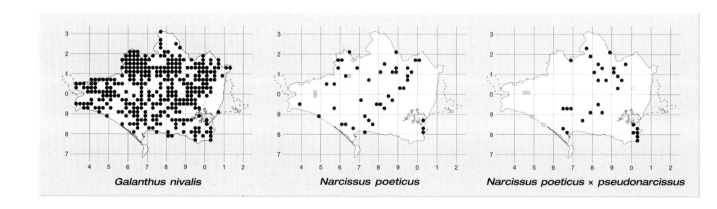

Galanthus nivalis   *Narcissus poeticus*   *Narcissus poeticus* × *pseudonarcissus*

Narcissus tazetta L.
BUNCH-FLOWERED DAFFODIL
A rarely established alien. 4. Subsp. *italicus* (Ker-Gawler) Baker is naturalised near Portland Bill (66U) and at St Georges Ch (67V and W, 1990–97, ! in **RNG**); Subsp. *aureus* (Loisel.) Baker occurred as a garden throw-out at Baiter (09F, 1989, !).

Narcissus tazetta × poeticus = N. × medioluteus Miller
PRIMROSE PEERLESS
Rarely planted in the grounds of big houses or on verges. 13 + [7]. First record: High Hall (00B, 1866, JCMP in **DOR**). Recently seen at Coryates (68H), Buckland Ripers (68L), Stratton (69R), Culeaze (89L), West Orchard (81I), Tarrant Crawford (90G), Pimperne (91A), Gussage St Andrew (91S), Durlston CP (07I), New Swanage (08F), Holes Bay (09B), Wimborne (09E) and Garston Wood (01E).

Narcissus poeticus L.
PHEASANT'S-EYE DAFFODIL
Planted and persisting on verges in villages, sometimes in churchyards or old meadows. 42 + [5]. Some survive as relics of the old bulbfields on Brownsea Is (08I).

Narcissus poeticus × pseudonarcissus = N. × incomparabilis Miller
NONESUCH DAFFODIL
Occasionally planted on verges and in churchyards. 30. Modern cultivars have a wide range of flower-colours.

Narcissus pseudonarcissus L.
DAFFODIL, BELLFLOWER, LENT-LILY
Subsp. *pseudonarcissus*. Locally frequent in old woods on neutral to slightly acid soils, and persisting in a few old meadows and hedgebanks. 81 + [22]. Sometimes planted in churchyards, and in the old bulbfields on Brownsea Is. Commonest in the west and north-west, and in an arc outside the more acid soils of the Poole basin, but absent from deforested and urban areas. Large populations occur near Bettiscombe (30V) and Charborough Park (99I).

Subsp. major (Curtis) Baker
SPANISH DAFFODIL
Frequently planted in many forms and flourishing in plantations, on verges, round village name-signs and in churchyards. 250. Cultivated as a crop at Henbury (99P) and Lamb's Green (99Z), and surviving in quantity on Brownsea Is.

Narcissus pseudonarcissus × cyclamineus = N. Cv. 'FEBRUARY GOLD' and 'TETE-A-TETE'
Increasingly planted on verges and in churchyards. 8. Noted from Dorchester (79A), Woodsford (79Q), Whitechurch (89J), Lytchett Minster (99L), Spetisbury (90B), Kingston Lacy (90Q), Studland (08G) and Alderholt (11B).

Narcissus minor L.
Planted near Durlston CP Lighthouse (07I, 1999, EAP). 1.

Narcissus jonquilla L.
JONQUIL
Once seen planted at Woodsford Ch (79Q, 1997, !). 1.

Asparagus officinalis L. subsp. officinalis
GARDEN ASPARAGUS
Single clumps or seedlings occur as escapes or relics of cultivation in waste places, beaches, railway banks or on walls. 14 + [5]. They do not persist long.

Subsp. prostratus (Dumort.) Corb.
WILD ASPARAGUS
A very rare maritime plant with a long history at Portland, also near Poole where it is now extinct. 1 + [5]. Chesil bank south of Moonfleet (68A, 1970s, JKH); Portland Bill (66Z, 1551, W Turner, also north of Cave Hole, 1876, WBB, and JCMP); Small Mouth (67S, 1551, W Turner, *et al.* in **BDK, BIRM, CGE, DOR, LIV** and **NMW** and 1997, PH Sterling and RJS, !); North of Ferry Bridge (67T, 1883, JCMP in **DOR** and 1907, WBB); [Sandbanks and Lilliput (08, 1799, RP, also 1902, KG Firbank and 1920s, JHS);

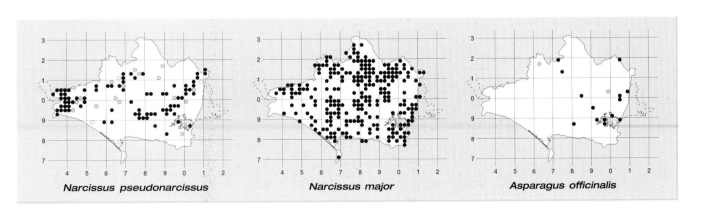

Narcissus pseudonarcissus   Narcissus major   Asparagus officinalis

Ruscus aculeatus L.
BUTCHER'S-BROOM

Local, in small numbers, in old woods and hedgebanks, sometimes in plantations. 169 + [7]. It is rarely planted for hedging or in churchyards. It grows on both acid and calcareous soils and extends to cliff-top scrub at Weymouth (67), Durdle Door (88A), and Old Harry (08L); most frequent in the east and not known on Portland.

IRIDACEAE

Libertia formosa Graham
CHILEAN-IRIS

Naturalised in the gardens and the Swannery at Abbotsbury (58S, 1961, D McClintock in **RNG** and 1984–96, !, where it was planted before 1899 (MI)); also at Studland (08G, 1998, EAP). 2.

Sisyrinchium montanum E. Greene
AMERICAN BLUE-EYED-GRASS

A garden escape in waste places, which does not persist.

[5]. Grimstone (69M, 1930s, RDG); Bere Heath (89, 1926, WA Ffooks); Hamworthy (99V, 1920, HJG); Wimborne (09, 1961, RDG); Turbary Common (09S, 1975–79, AK Hunt).

Sisyrinchium striatum Smith
PALE YELLOW-EYED-GRASS

A increasing garden escape, self-sown in waste places. 9. Moreton (88E, 1998, !); Bovington Camp (88J, 1996, !); Cowherd Shute Farm (82S, 1994, !); East Walls, Wareham (98I); Newton Farm (99G); Lytchett Matravers (99M, 1999, !); New Swanage cliffs (08F, 1998, EAP); Holes Bay (09B, 1993, !); Canford Heath tip (09I, 1999,!).

Hermodactylus tuberosus (L.) Mill.
SNAKE'S-HEAD IRIS

More or less naturalised near Studland. 2. Studland (08, 1867, Mr Smithies in **DOR**); Glebelands estate (08F, 1994, FAW, !); Studland Bay Ho (08G, 1996, !).

Iris germanica L.
BEARDED IRIS

An occasional garden escape, surviving on waste ground

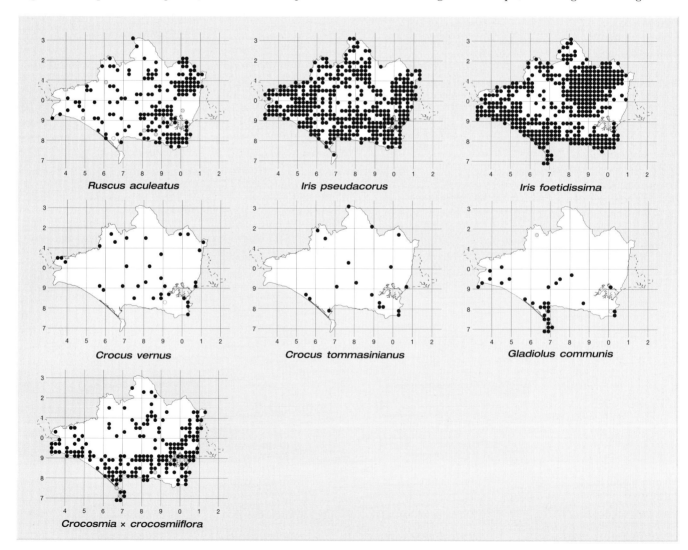

or casual on tips. 14. Noted in 39R, 49Q, 67V, 68K and V, 78P and X, 71H, 72M, 88C and P, 08F and U, 09I.

Iris sibirica L.
SIBERIAN IRIS

A rare, but established, garden escape on verges at Tadnoll (78Y, 1989–99, ! in **RNG**) and Bookham (70C, 1994, !); in a shrubbery at Parnham Ho (40Q, !). 3.

Iris pseudacorus L.
YELLOW IRIS, LAVERS, SWORD-FLOWER

Locally frequent by rivers, streams, ditches and lakes, tolerating shade but then not flowering. 404. Found in the basins of all the larger rivers, but absent from the drier Chalk and limestone. A pale-yellow flowered form grows at Winterbourne Abbas (69B).

[*Iris spuria* L.
BLUE IRIS

Known from a damp meadow at Chickerell for *c*.100 years, and deliberately ploughed out in 1972. The original site was at 643814 (1942, CRS Bradley and 1971, JEL in **RNG**); the only other British site is in Lincolnshire.]

Iris orientalis Miller
TURKISH IRIS

A garden escape, occasionally established on waste ground. 6. Abbotsbury (58M, 1960–98, ! in **RNG**); between Small Mouth and Wyke (67T, 1992, !); Grove, and near St Georges Ch (67Y, 1990, !); verge, Lodmoor (68V, 1992–99, !); Blacknoll (88D, 1991–99, !); Wareham (98I, 1978, !).

Iris foetidissima L.
STINKING IRIS, BLOODY BONES

Locally frequent at the edges of woods, in scrub and dry, rough grassland. 419. Commonest along the coastal belt and on the Chalk, but absent from very acid soils in the Poole basin. Ungrazed by rabbits, and seen as an epiphyte on maple at Chetterwood (90N). Var. *citrina* Bromf., with dull, pale yellow petals, has eight old records from the south-east, but the only recent sightings have been in gardens.

Crocus vernus (L.) Hill
SPRING CROCUS

Commonly planted in churchyards, also on verges near houses. 34 + [2]. Very large populations occur at Forde Abbey (30M and S, 1895, Mrs Roper and 1995, !); near Studland Ch (08G, 1913, RVS, *et al.* in **BDK** and **BMH**, and 1964–96, !); and Alderholt Ch (11B, 1998, !). Variable in size and colour of flowers, which may be dark or pale purple, or white.

Crocus tommasinianus Herbert
EARLY CROCUS

Frequently planted in churchyards, and escaped on verges and cliff-tops, where it spreads rapidly by seed. 19. Flowers usually pale, rarely dark purple. First record: Poyntington (62K, 1970, ACL).

Crocus chrysanthus (Herbert) Herbert
GOLDEN CROCUS

Occasionally planted in churchyards. 3. Abbotsbury (58S, !); Weymouth (67, 1993, M Burnhill); Piddletrenthide (70A, !).

Crocus flavus × *angustifolius* = *C.* × *stellaris* Haw
YELLOW CROCUS

Frequently planted in churchyards and village lanes. 23.

Crocus speciosus M. Bieb.
BIEBERSTEIN'S CROCUS

An autumn-flowering plant, planted in the grounds of Birkin Ho (79A, 1996, !).

[*Gladiolus illyricus* Koch
WILD GLADIOLUS

Once reported from a brackeny hedgebank at Ensbury (09T, 1874, JH Austen); it survives in the New Forest in VC11.]

Gladiolus communis L. subsp. *byzantinus* (Miller) A.P. Ham.
EASTERN GLADIOLUS

Sometimes grown in gardens, and a well established escape along the coastal belt. 31 + [2]. Often confused with other *Gladiolus* species. First record: Langton Matravers (07E, 1914, CBG in **BMH**). Specimens are in **DOR** and **LANC.**

Crocosmia paniculata (Klatt) Goldblatt
AUNT-ELIZA

An occasional garden relict or escape. 9. First record: Spring Bottom (78K, 1982, ! in **RNG**). Also seen at Lyme Regis (39L), Parnham (40Q), Beaminster (40W), Stokeford Heath (88U), Bere Heath (89W), Blandford St Mary Ch (80X), West Moors (00W) and Three-legged Cross (00X).

Crocosmia × *crocosmiiflora* (Lemoine) N.E. Br.
MONTBRETIA

More or less naturalised in grounds of large houses, on verges and cliffs, also a garden throw-out on tips. 195 + [4]. Commonest along the coastal belt and on acid soils, but neither in quantity nor in native habitats.

AGAVACEAE

Yucca filamentosa L.
ADAM'S-NEEDLE

Planted at Iwerne Courtney Ch (81R, 1995, !).

Yucca gloriosa L.
SPANISH-DAGGER

Planted at Yetminster Ch (51V, 1996, !), Bagwell Farm (68F, !) and Dorchester Industrial estates (69V, !); two flowering plants on a dry bank near Herrison, away from houses (69X, 1998, !); Sandbanks W. beach (08N, !). 5.

Cordyline australis (G. Forster) Endl.
CABBAGE-PALM

Planted and flowering on verges and roundabouts at Weymouth and Poole. It survives at Abbotsbury ruins (58M,

1961–78, !), in a hedge at Westham allotments (67U, !), and is naturalised on Bournemouth cliffs in VC11.

Phormium tenax Forster & Forster f.
NEW ZEALAND FLAX
Planted in wild gardens at Abbotsbury, north Portland, Weymouth, Sherborne and Studland. 3. More or less naturalised on cliffs at New Swanage (08F, 1998, EAP, !) and on dunes at Sandbanks (08N, !); planted by a pit west of Huish (99D). It produces seeds, but no seedlings have been reported.

DIOSCOREACEAE

Tamus communis L.
BLACK BRYONY
Frequent at the edges of woods and in hedges. 593. Widely distributed, but least common on acid soils. The leaves sometimes turn a dark purplish-black colour, and the roots were once used to rub onto bruised or rheumatic tissues, but the sap can cause dermatitis.

ORCHIDACEAE
(see Jenkinson, 1991)

Cephalanthera damasonium (Miller) Druce
WHITE HELLEBORINE
Occasional, in small numbers, in woods and plantations, especially on bare ground under beech. 20 + [18]. Grazed by both deer and rabbits. Recent records are mostly from the north-eastern Chalk, with two outliers in the south-east. Old records extend this range further west on the Chalk, so the plant may be decreasing, or the habitat it prefers may be disappearing.

[**Cephalanthera longifolia** (L.) Fritsch
NARROW-LEAVED HELLEBORINE
Found in woods under beech, but not recently, and perhaps misidentified. Gussage St Michael (91, 1887, LJW); Creech Hill, Cranborne (01L, 1891, TW Smart).]

Epipactis palustris (L.) Crantz
MARSH HELLEBORINE
Very local in open fens, especially on undercliffs, and in

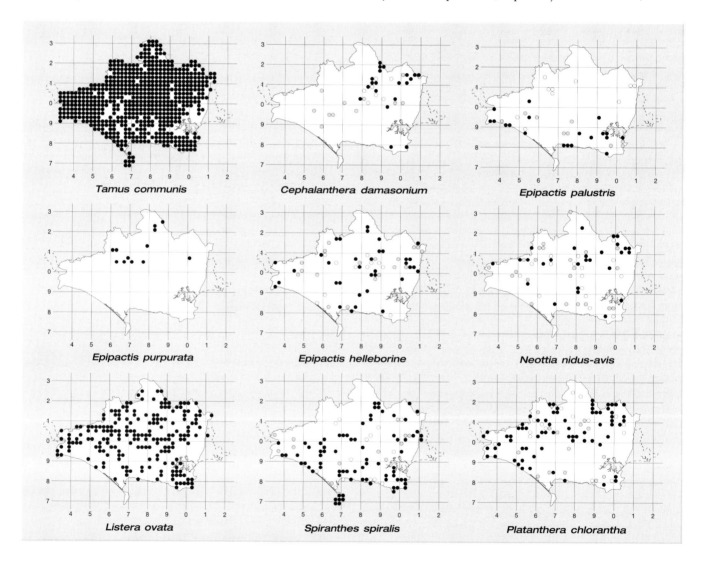

flushed bogs. 19 + [19]. Good found it mainly in the coastal belt where it still flourishes, but in the 19th century it occurred in the north, where sites have been lost through drainage. It cannot compete with dense reeds.

Epipactis purpurata Smith
VIOLET HELLEBORINE

Rare and in small numbers in old woods, mainly in old hazel copses. 12. Absent from deforested areas and the south. Twisting Alders Coppice (60H, DRS, N Spring & O Rackham); Buckland Newton (60S, 1908, CW Dale and 1991, MNJ); Dungeon Hill (60Y, 1905, CW Dale and 1989, MJG); Knighton Common (61A and F, 1992, CT & !); Horse Close Copse (70C, 1991–96, AH); Brooks and Moots Copses (70S, 1993, BE); Girdlers Coppice (71W, 1988, DAP, also AH); Duncliffe Wood (82F and G, 1989, N Trebble, also MNJ); Kingsettle Wood (82S, 1990, DNHS); Hinton Martell (00D, 1995, GDF). Often browsed by deer. Either increasing, or the north is now better known botanically than of old.

Epipactis helleborine (L.) Crantz
BROAD-LEAVED HELLEBORINE

Occasional, in small numbers, in woods and plantations, often in deep shade. 35 + [31]. Widespread, but probably transient, as it has not been refound in many old sites, where some old records might refer to other species in this genus. Fertilized by wasps.

Epipactis leptochila (Godfery) Godfery
NARROW-LIPPED HELLEBORINE

Rare, and in small numbers in calcareous woods, often under hazel or rarely beech. 4. Grazed by deer and often sterile. Poxwell Big Wood (78L, 1993, !); Parnholt Woods (71, 1980, J Fryer); Great Wood, Creech (98A, 1976, RB and 1983–85, AH and MNJ); Hill Coppice and New Town (91Z, 1936, RDG, *et al.*, and 1996, GDF); Garston Wood (01E, 1990–93, B Last).

Epipactis phyllanthes G.E. Smith
GREEN-FLOWERED HELLEBORINE

Rare in old woods on acid soils. 3. Hurst Heath (78Z, 1984, MNJ and 1993, BE); *c.*100 plants, Oakers Wood (89A, 1982, JKH, *et al.*, and 1991, MNJ, !); West Moors (00R, 1991, MNJ).

Neottia nidus-avis (L.) Rich.
BIRD'S-NEST ORCHID

Very local and in small numbers in woods, often in deep shade. 27 + [38]. A saprophyte, it requires deep humus, usually under oak or hazel but sometimes under mature beech, pine or yew; the underlying soil may be calcareous or acid. Sporadic or transient, and not refound in many old sites.

Listera ovata (L.) R. Br.
COMMON TWAYBLADE

Occasional in woods, plantations, fens and shaded hedgebanks, well distributed except in deforested areas. 202. Usually in small numbers, but abundant in calcareous plantations at Kingston (97P) and Jubilee Wood (90M), where it grows in bare places with little competition. Scarce in open chalk grassland at Fontmell Down (81Y and Z) and Zigzag Hill (82V).

[*Listera cordata* (L.) R. Br.
LESSER TWAYBLADE

Always hard to find, this once grew on acid soils under heather. 2. Summit of Golden Cap (49B, 1892, HNR and WF); Branksome Chine (09K, 1895, E Armitage, also EFL); near Bournemouth (09, 1856, JH Austen in **K**, perhaps in VC11, as was the specimen from near Boscombe Chine (19A, 1853, Rev Robarts in **K**). It still grows on Exmoor in Somerset.]

Spiranthes spiralis (L.) Chevall.
AUTUMN LADY'S-TRESSES

Local in short, calcareous turf, rare on heathy verges, and sometimes in garden lawns. 78 + [27]. Widely distributed on the Chalk and limestone, especially in the coastal belt and on Portland, but sporadic or transient in some sites. Large populations seen at Bincombe Hill (68X), Melbury Hill (81U), Fontmell Down (81Y and Z), Badbury Rings (90R) and Cowgrove (90V).

Hammarbya paludosa (L.) Kuntze
BOG ORCHID

Rare and sporadic on *Sphagnum* mats in valley bogs. 8 + [7]. Now only in the southern Poole basin, with an extinct outlier in the west. Its flowering time is variable, and it is easily overlooked. [Champernhayes (39N, 1895, GL)]; Winfrith Heath (88D, 1961–78, ! and 1995, HGA Bates); Stoke Heath (88P, 1971, C Pepin and 1989, BE); Oakers bog (89A, 1995, M Forster); West Holme Heath (98B, 2000, BE); Worgret Heath (98D, 1976, B Pickess and 1991, MNJ); Stoborough Heath and Halfway Inn (98H, 1895, JCMP, *et al.*, and 1999, DCL); Ridge (98I, 1917, CBG in **BMH**); Hartland Moor (98M, 1938, DM, *et al.*, and 1995, T Brodie-Jones); Slepe Heath (98N, 1953 AWW in **HIWNT**, *et al.*, and 1995, DAP); Middlebere crossing (98S, 1961–70, SBC); Morden Decoy (99A, 1895, JCMP); Studland Heath (08B and H, 1895, JCMP and 1973, SBC); [Source of the Bourne, West Howe (09M, 1895, WMR)].

Herminium monorchis (L.) R. Br.
MUSK ORCHID

Very rare in short, chalk turf. 1. Batcombe Down (60H, 1874, T Marshall, *et al.* in **DOR**, and 1999, M. Greenhill); one plant was found at Hillfield Hill in the same tetrad in 1980. Reports from Badbury Rings (90R) and Loders (1846, J. Jones) are errors; old records from Rushmore Down (91J) and Ebbesborne Wake (92V) are correct, but in VC8.

Platanthera chlorantha (Custer) Reichb.
GREATER BUTTERFLY-ORCHID

Occasional, in small numbers, in old woods, plantations and scrub on calcareous soils. 71 + [30]. Widespread, but absent from deforested areas and most of the Poole basin; in some years it is locally abundant in chalk grassland. Var. *tricalcarata* Hemsl., with three spurs per flower, was found at The Holts (61X, 1907). According to JS Udal (Udal, 1889), an extract was used to make a bright green ointment to treat ulcers, but it would be hard to find sufficient material for this today.

Platanthera bifolia (L.) Rich.
LESSER BUTTERFLY-ORCHID

Rare and in small numbers in flushed, wet heaths or sheltered chalk turf. 10 + [26]. It has declined in recent years. Post-1970 records: Fishpond bog (39U, 1985, HR Cook); Batcombe (60B and C, 1962–84, MH Collier); Hurst Heath (78Z, 1938, RDG in **WRHM** and 1972, L Diaper); Winfrith Heath (88D, 1960–80, !); Bere Wood (89S, 1990, BE); Bloxworth Heath (89W, 1936, RDG in **WRHM** and 1991, MNJ); Hartland Moor (98M, 1980, D Simcock); Upton Heath (99X, 1982, G Dutson); Sovell Down (91V, 1968, HCP); Kingswood bog (08B, 1981, HCP); Spur bog (08H, 1895, JCMP, *et al.*, and 1993, MNJ); Ackling Dyke (01I, 1968, HCP).

Anacamptis pyramidalis (L.) Rich.
PYRAMIDAL ORCHID

Local in dry, calcareous grassland, undercliffs, old quarries, verges and railway banks. 127 + [8]. Scattered on the Chalk and limestone, especially at Portland, south-east Purbeck and the north-east, with only one large population, at Fontmell Down (81Y and Z). Once seen on fixed shingle at Small Mouth (67S), and on heathy verges at Studland (08G, 1999, Rees Cox) and Canford Heath (09D, 1999, !).

Anacamptis (Orchis) morio (L.) R. M. Bateman, Pridgeon & M. W. Chase
GREEN-WINGED ORCHID

Local in short, calcareous turf, old pastures, rarely on sandy heaths or verges, and once seen in fen at Wareham Common (98D). 67 + [36]. Absent from the intensively farmed parts of the Chalk. Forms with white flowers were seen at Powerstock Common (59, 1980, D Fowler) and Corfe Mullen (99T,1991, MNJ). Large populations occur at West Bexington (58I), Bindon Hill (88F and K), East Morden (99C), Corfe Mullen (99T), West Howe (09M) and Northleigh Ho (00F), and MNJ has noted a wide range of colours in some of these.

Gymnadenia conopsea (L.) R. Br.
FRAGRANT ORCHID

Occasional in dry, calcareous turf. 53 + [16]. Good found it rare in the 1930s. Forms with white flowers occur among the larger populations in the north-east Chalk, in 81K and T, 91W and 01J.

Gymnadenia densiflora (Wahlenb.) Dietrich
(*Gymnadenia conopsea* subsp. *densiflora* (Wahlenb.) Camus, Bergon & A. Camus)

Rare in fens. 2. Woolcombe (59M, 1986, NCC); Whitmore Bottom (99C, 1989, MNJ). Plants from a wet meadow at Winterhays (50Z, 1980, DCL) may be this, but need critical examination.

Gymnadenia borealis (Druce) R. M. Bateman, Pridgeon & M. W. Chase
(*Gymnadenia conopsea* subsp. *borealis* (Druce) F. Rose)

Scarce in heaths or flushed bogs. 1. Middlebere Heath (98H, 1937, DM, also 1954, M Richards & 1993, MNJ). Probably this at Winfrith Heath (88D, 1986, AH, !) [& Little Sea (08G, 1895, JCMP)].

Gymnadenia conopsea × *Dactylorhiza viridis* = × *Dactylodenia jacksonii* (Quirk)
Reported from calcareous turf at Hod Hill (81K, 1970s, RDG) and Badbury Rings (90R, 1950s, V Follett). [2].

Gymnadenia conopsea × *Dactylorhiza fuchsii* = × *Dactylodenia st-quintinii* (Godfery) J. Duvign.
Very rare and transient in calcareous turf. 2 + [2]. Eggardon Hill (59L, 1965, AG Bodman); Kingcombe (59P, 1990, AH & T Norman); Wool (88N, 1917, GCD); Fontmell Down (81Y, 1981, MNJ).

Gymnadenia borealis × *Dactylorhiza maculata* = × *Dactylodenia legrandiana* (Camus) Peitz
Once reported from Middle bog, Hartland Moor (98H, 1954, M Richards).

Dactylorhiza (Coeloglossum) viridis (L.) R. M. Bateman, Pridgeon & M. W. Chase
FROG ORCHID

Rare and decreasing in short, calcareous turf. 24 + [20]. Almost confined to the Chalk, with outliers on a calcareous railway bank at Powerstock Common (59N, 1895, E Fox and 1992, JGK) [& Langton Matravers (97Z, 1895, LVL)]. There are no large populations, but the largest are at Fontmell Down (81Y) and Badbury Rings (90R). MNJ has observed this being pollinated by soldier beetles, *Cantharis* sp.

Dactylorhiza fuchsii (Druce) Soo
COMMON SPOTTED-ORCHID

Locally frequent in woodland rides, dry, calcareous grassland, clay cliffs, marshes, verges and flushed heaths. 295. First record: Up Cerne (60L, 1696, W Stonestreet). Absent from deforested areas, not reported from Portland, and often transient elsewhere. White-flowered plants seen by MNJ at Powerstock Common (59N), Fontmell Down (81Y) and Badbury Rings (90R).

Dactylorhiza fuchsii × *maculata* = *D.* × *transiens* (Druce) Soo
Rare, and very hard to determine with confidence (Jenkinson, 1992). 2 + [5]. Bettiscombe (30V, 1998, IPG); Culeaze (89L, 1927, GCD); Wareham (98I, 1917, GCD); Corfe Common and Arne (98Q and U, 1943, AJA Dunston in **BM**); Edmondsham (01R, 1980, JN); St Leonards (10B, 1992, MNJ).

Dactylorhiza fuchsii × *incarnata* = *D.* × *kerneriorum* (Soo) Soo
A rare triploid hybrid in wet meadows and bogs. 1? + [3]. Barrowland Farm (59N, 1938, AWG in **WRHM**); Hyde Ho (89Q, 1937, RDG in **WRHM**); Corfe Common (98Q, 1981, MNJ); Spur bog (08H, 1996, D Tennant; this was probably *D.* × *carnea, fide* EAP).

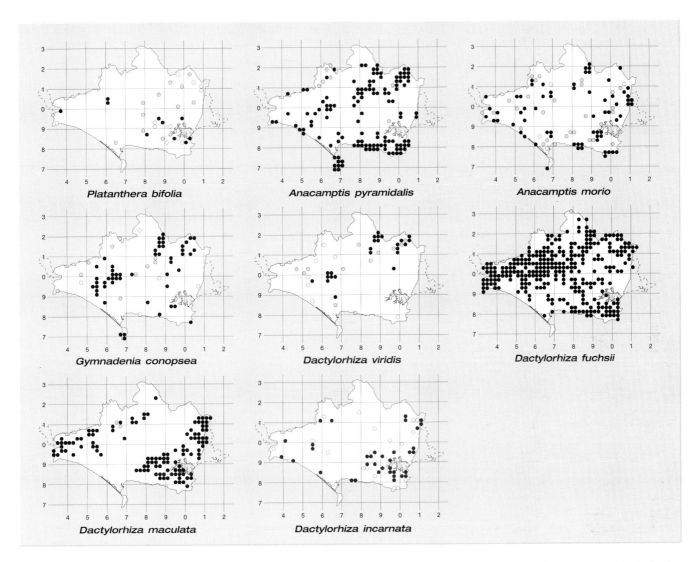

Platanthera bifolia

Anacamptis pyramidalis

Anacamptis morio

Gymnadenia conopsea

Dactylorhiza viridis

Dactylorhiza fuchsii

Dactylorhiza maculata

Dactylorhiza incarnata

Dactylorhiza fuchsii × praetermissa = D. × grandis
(Druce) P. Hunt
A fairly common triploid hybrid, occurring as single plants or clumps in fens, marshes and old claypits. 18 + [10]. Post-1980 records from 40G, 50S, 60F, 78P, 88P, 81U, 97U, 98N and Q, 99C, 90V, 08H, 00K and S, 01G, P and Q, 10B. Good's specimens are in **WRHM**, and others in **BM**. Old recorders confused this with *D. praetermissa* var. *junialis*.

Dactylorhiza maculata (L.) Soo
HEATH SPOTTED-ORCHID
Locally frequent in damp, acid pastures, wet heaths and bogs. 116 + [2]. Commonest in the Poole basin, occasional in the north and west, and absent from the Chalk, apart from a wet meadow at Hinton St Mary (71X and 81C, 1994, MNJ, !). Old records confuse or lump this with *D. fuchsii*. Forms with white flowers occurred at Coombe Heath (88S, !) and Verwood (01V, MNJ).

Dactylorhiza maculata × incarnata = D. × carnea
(Camus) Soo
A triploid hybrid, with single plants recorded from a few

water-meadows, flushed or grazed bogs in the Poole basin. 4 + [2]. Hartland Moor (98M, 1953–58, DSR and PJ Newbould); Spur bog (08H, 1999, EAP);Wimborne St Giles (01G, 1982, MNJ); Edmondsham (01Q, 1909, EFL); Black Moor (10E, 1980, HCP); Cranborne Common (11A, 1982, !).

Dactylorhiza maculata × praetermissa = D. × hallii
(Druce) Soo
A triploid hybrid, rarely found in wet places and flushed bogs. 6 + [9]. Post-1980 records are all from the Poole basin. Winfrith Heath (88D, 1980, L Diaper); Middlebere Heath (98N and S, 1980, L Diaper and 1995, P Oswald); Corfe Common (98Q, 1986, MNJ); Woodlands and by R. Crane, Edmondsham (00U and 01R, 1980, JN); St Leonards (10B, 1992, MNJ). Old specimens are in **BM** and **WRHM**.

Dactylorhiza incarnata (L.) Soo
EARLY MARSH-ORCHID
Local in marshes, water-meadows, wet undercliffs and flushed bogs, rarely in dry grassland. 36 + [23]. Mainly in the Poole basin, and scattered elsewhere off the Chalk. Subsp. *incarnata*, with flesh pink flowers, is the less common

form, while subsp. *pulchella* (Druce) Soo, with deep pink flowers which appear before those of *incarnata* is more often in flushed bogs. Specimens are in **BM, BMH** and **WRHM.**

Dactylorhiza incarnata × praetermissa = D. × wintoni (Camus) P. Hunt
Rarely reported from wet places. 2 + [1]. Culeaze (89L, 1927, GCD); Slepe ford (98S, 1958, DSR and 1985, MNJ); Wimborne St Giles (01G, 1982, MNJ).

Dactylorhiza praetermissa (Druce) Soo
SOUTHERN MARSH-ORCHID
Locally frequent in fens, wet undercliffs, marshy fields, flushed bogs and rarely in sheltered north-facing chalk turf. 137 + [16]. Mostly in west to central Dorset and the Poole basin, least common on the Chalk. Variable. Var. *junialis* (Verm.) Senghas, with ring-spots on the leaves, is occasional; a form resembling *D. purpurella* occurs at Waterston (79H, 1982, DFW), and forms resembling *D. traunsteineri* occur at Hinton St Mary (71X and 81C) and Studland Heath (08H).

Dactylorhiza traunsteineri (Sauter ex Reichb.) Soo
NARROW-LEAVED MARSH-ORCHID
This controversial species (Jenkinson, 1992) occurs in a rich water-meadow at Wimborne St Giles (01G, 1997, MNJ), and one plant was reported from Studland Heath (08H, 1993, MNJ).

Neotinea (Orchis) ustulata (L.) R. M. Bateman, Pridgeon & M. W. Chase
BURNT ORCHID
A rare and decreasing species of short, dry, chalk turf. 4 +

[13]. Now found only on the north-east chalk, though once more widespread. Post-1910 records: Higher Houghton (80B, 1968, T Norman); Hod Hill (81K, 1916, AWG); Fontmell Down (81Y, 1983, MNJ); The Cliff, Tarrant Rushton (90K, 1915, AWG); The Foreland (08L, 1912, CBG in **BMH**); Bokerley Dyke, Pentridge Down and Martin Down (01I, J, N and P, 1988, MNJ, *et al.*, and 1994, FR). The record from Nutford Field (80Y, 1799, RP) was probably the tall form which grows in water-meadows.

Orchis mascula (L.) L.
EARLY-PURPLE ORCHID, GIDDY-GANDER, GOOSEY-GANDER, SINGLE CASTLE, SOLDIERS
Occasional, often sporadic, in old woods, hedgebanks, calcareous grassland, clay cliffs and verges, and once in a wet wood among *Carex paniculata* at Chester Hill (58W). 264 + [6]. Rarely in large numbers, except at The Rings, Corfe (98K). Widespread, but absent from very acid soils in the Poole basin. Forms with white flowers were seen at Woolland ((70T), Bere Wood (89S, 1989–93), and Duncliffe Wood (82F; all MNJ), and forms with unspotted leaves are occasional.

(Orchis (Aceras) anthropophora (L.) All.
MAN ORCHID
A few plants occur at Martin Down in VC8 (01P, 1994, FR), but there are no records from VC9.)

Himantoglossum hircinum (L.) Sprengel
LIZARD ORCHID
Very rare and sporadic in tall grassland. 1 + [2]. Langton Herring (623824, 1933, RDG in **DOR**, *et al.*, and 1955, JKH, later dug up); Piddlehinton (79D, 1922, Miss Lovelace in

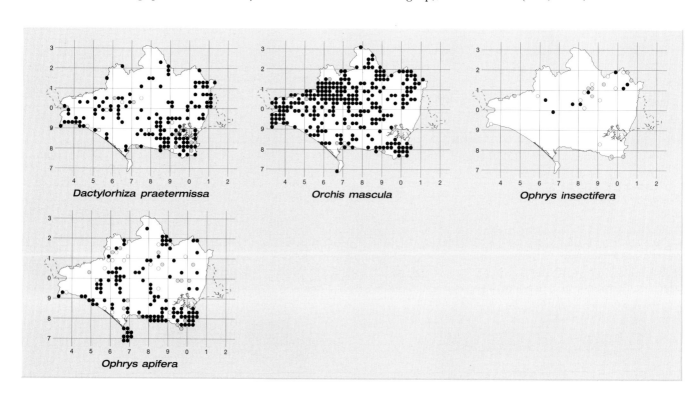

Dactylorhiza praetermissa

Orchis mascula

Ophrys insectifera

Ophrys apifera

LIV); Slepe (99G, 1987, MG Morris, *et al.*, and 1999, !, *c.*23 plants). Good thought this an indicator of climatic warming.

Ophrys insectifera L.
FLY ORCHID

Local, sporadic and in small numbers in old woods or beech plantations on calcareous soil, often in deep shade. 8 + [20]. Most recent records are from the north-east Chalk, but it used to occur in Purbeck. Lyme Regis (39, 1895, JCMP); Seaborough Hill (40I, 1960s, AWG); Bubb Down (50Y, 1895, MI); Came Wood (68X, 1950, JKH); Bramble Bottom (69U, 1983, RM); Gore Hill and Hendover Copse (60H, 1962, MH Collier, *et al.*, and 1996, AH); Binghams Melcombe (70R, 1846, CW Bingham and 1991, BE); Milton Abbey Wood (80B, 1895, JCMP and 1992, BE); Enford Bottom (80P, 1952, GD Harthan and 1998, !); Bryanston (80T, 1956, BNHS); Hod Wood (81K, 1916–32, AWG); Steepleton (81Q, 1943, BNHS); East Man (97T, 1961, CHS); Creech Grange (98B, 1912, CBG in **BMH**); Tarrant Gunville (91G, 1957, JAL); Handley Common (91Z, 1976–83, RB); Durlston (07I, 1930, RDG); Wimborne St Giles (01F, 1932, FHH and 1980, AH); Creech Hill Wood (01L, 1895, TW Smart and 1982, !); Martin Down (01P, 1973, L Diaper); Perry Copse (01S, 1960, KBR and HJB).

Ophrys sphegodes Miller
EARLY SPIDER-ORCHID

Locally frequent in short, limestone turf. 7 + [1]. Only in south Purbeck; populations and heights of plants vary from year to year (Burt, 1987; Hutchins, 1987). Winspit (97S, 1934, W Harrison in **LIV**, and 1983, RS Cropper, !); south of Worth Matravers (97T, 1874, JCMP, *et al.*, and 1991, J Meiklejohn); above Dancing Ledge (97Y, 1874, RPM, *et al.*, and 1991, !); Corfe Castle (98R, 1880, HNR); Langton Matravers (07D, 1926, L Anstis in **LIV**, *et al.*, and 1992, !); in a garden, Priest's Way (07E, 1991–94, WGT); Durlston CP (07I, 1889, EFL in **LIV**, *et al.*, and 1997, !); Swanage (07J, 1920, AWG in **BDK**, *et al.*, and 1991, WGT). There are specimens in **BDK, LANC, LIV, RNG, WAR** and many other herbaria; there is no need to collect more.

Ophrys apifera Hudson
BEE ORCHID

Occasional, in small numbers, in calcareous turf, especially on old earthworks, cliffs and quarries. 117 + [35]. Rarely on limed, heathy verges or in gardens; often transient. Commonest on the Chalk and along the coastal belt. The novelist J. M. Faulkner picked basketfuls on Nine Barrow Down in the 1870s. In dry summers the plants shrivel and turn black without flowering, but the tubers survive. The lip-colour is variable and may fade with age. Var. *bicolor* occurred on Fontmell Down (81Y, 1993–96, NJ Heywood det. MNJ), var. *chlorantha* occurred at Bindon Hill (88F, 1968, MJ Halsall), Ulwell (08A, 1993, MNJ) and Cranborne (01L, 1994–96, MNJ), while var. *trollii* (Heg.) Reichb.f. has been noted in 49Q, 67V, 77A, 80T, 81Y and Z, 82V, 97Y and 07E.

(*Ophrys fuciflora* (Crantz) Moench was reported in error from Bere Regis in 1891.)

[*Ophrys bertolonii* Moretti
MIRROR ORCHID

This was planted by a gardener on Round Down (07I, 1976–78, R Webster) but is gone now.]

CHAPTER 7

DORSET BRYOPHYTES

Bryophytes include about 35,000 haploid species which produce diploid capsules containing haploid spores. Most species bear leaves which are only one cell thick.

There is little published work on Bryophytes in Dorset. Over 100 years ago, Dale produced a list of 111 mosses from Glanvilles Wootton and nearby (Dale, 1883), since when no systematic list has appeared other than those in the Census Catalogues by the British Bryological Society (BBS). Even earlier, in 1879, HH Wood was advertising 160 species of dried bryophytes for sale, mostly from near his home in Holwell; some are in **BM**. The next publication was in 1943, on Sphagna from Purbeck (Dunston, 1943), which has been superseded by an improved account (Edwards, 1997). Miscellaneous short papers since then (Pickess and Giavarini,1981; Prentice, 1983; Stern, 1982; Read, 1950–1951), and particularly accounts of BBS visits to the county (Hill, 1970, 1977 and 1995) have added new county records or localities. The paucity of publications does not reflect the amount of bryological work carried out. The absence of a resident bryologist for most of the last two centuries has hindered systematic recording, but the following collectors are among those worthy of mention:

WPC Medlycott (1879); 25 common species in **DOR** from near Weymouth, + 1 rarity in **BM**.

Miss G Lister (1890–1904); 160 species in **DOR** from near Lyme Regis.

EM Holmes (1906–08); 62 species in **DOR** from Swanage and Studland, and more in **CGE and NMW**.

AW Graveson (1915); a few common species from near Blandford.

HW Monckton (1923–24); three species in **RNG**.

HH Knight and WE Nicholson (1920s); a number of rarities from Portland in **CGE**.

EW Jones (1933–37); 47 species in **WRHM** from near Studland, + rarities elsewhere.

AEA Dunston (1942–43); 13 Sphagna in **DOR** from south Purbeck.

HJM Bowen (1955–); 90 species in **BMH**, **DOR**, **PVT** and **RNG**.

F Rose (1960–); many rarities in **PVT**.

MO Hill, with Mrs M Milnes-Smith and RC Stern (1964–77); collected a large amount of data for a proposed Bryophyte Flora of Dorset, which I have been privileged to see, + **PVT**.

VJ Giavarini (1977–); 101 species in **DOR**.

CJ Hadley (1985–93); several bryophyte surveys for DWT near Swanage.

Miss HC Prentice (1988–); an unpublished thesis including detailed distribution maps of 30 heathland species in the north Poole basin.

BJ Edwards (1990–); numerous records, and a scholarly report (Edwards, 1997).

Other distinguished bryologists who have visited the county include MFV Corley, AC Crundwell, Mrs JA Paton, R Porley, EC Wallace and many more.

The county is not particularly rich in bryophytes. The main reason is a climatic one, as the rainfall is much lower than in Devon and Cornwall to the west and, except perhaps on Portland, our winter temperatures are lower. As regards edaphic factors, Dorset has a wide variety of habitats, but most are of limited area. Old woodlands are small, and less rich than the New Forest to the east, but have a unique plant in *Plagiochila norvegica*. Heaths and bogs are widespread, so that 16 species of *Sphagnum* are known, with associated hepatics. Limestone cliffs on Portland and in Purbeck have many rarities such as *Acaulon triquetrum*, *Eurhynchium meridionale*, *Pottia wilsonii*, *Southbya nigrella* and the recently discovered *Bartleya ohioensis*. Montane species, such as *Grimmia trichophylla*, *Hedwigia ciliata* and *Racomitrium* spp. are rare. Alien species are few in number, but *Lophocolea bispinosa* and *Telaranea murphyae* are notable recent arrivals.

The only commercial use of moss was in the late 19th century, when mixed moss was raked from north-facing grassland on Okeford Hill and sent by rail, in sack-loads, to Covent Garden (Dorset Fed. Women's Institutes, 1990). Small amounts, and also *Sphagnum*, are probably still collected for hanging baskets, hyacinth bowls and orchid culture. Thomas Hardy liked to scrape moss from

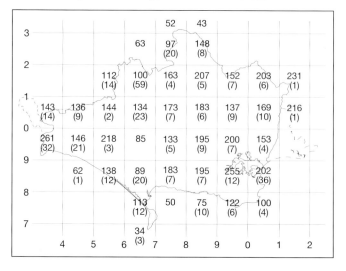

Figure 16. Number of bryophyte species (+ old records) per hectad in Dorset.

gravestones in Stinsford churchyard with a wooden scraper that he himself made.

The ensuing list has been pruned to save space, in the hopes of a fuller Bryophyte Flora to come. The nomenclature and order of taxa is that of Smith (1978 and 1990), despite current changes (Blockeel & Long, 1998). Habitats are summarised, and the number given is the total number of 10 km × 10 km grid squares from which the species has been recorded: species maps on this scale are available (Hill, Preston & Smith, 1991–1994). This number gives an objective measure of how widespread a species is. Localities are only given for the scarcer species, and herbarium specimens are cited where known; many herbaria have yet to be checked for these, notably **CGE** and **NMW,** but with so many critical species it seems wise to list where vouchers may be found. The net total of species reliably reported from Dorset, is 425, of which 324 are mosses and 101 liverworts; only five moss species have not been seen for a century and are probably extinct.

HEPATICAE
LIVERWORTS

Trichocolea tomentella (Furh.) Dum. Rare, by streams in alder woods in the west. 4. Wyld Warren, MOH; Monkton Wyld Wood, BE; Redholm Coppice, !, also BBS; Chicks Copse, !; Aunt Mary's Bottom, JRW; Hill Wood, Melcombe, EWJ. **DOR.**

Telaranea murphyae Paton. Very rare under *Rhododendron*. 1. Branksome Chine, N. Woods det. M Newton.

Kurzia pauciflora (Dicks.) Grolle. Uncommon in wet, acid habitats in the Poole basin and the west, among other bryophytes. 12. **WRHM.**

Kurzia sylvatica (Evans) Grolle. Very rare, on a heathy bank. 1. Wyld Warren, MOH.

Lepidozia reptans (L.) Dum. Occasional on rotting logs or sheltered acid banks. 17. **DOR.**

Calypogeia fissa (L.) Raddi. Frequent in wet, acid places. 28. **DOR, WRHM.**

Calypogeia muelleriana (Schiffn.) K. Muell. Less common than *C. fissa,* in similar places. 14. **DOR, WRHM.**

Calypogeia sphagnicola (H. Arn. & J. Perss.) Warnst. & Loeske. Uncommon among *Sphagnum* in carr or bogs in the Poole basin. 4. Moreton Heath, FR, also MOH; Bryants Puddle Heath, JAP; Holt Heath, MOH. **BBSUK.**

Calypogeia arguta Nees & Mont. Occasional in acid places. 19. **DOR, WRHM.**

(*Calypogeia neesiana* (Mass. & Carest.) K. Muell. is known from SZ19 in VC 11.)

Cephalozia bicuspidata (L.) Dum. Frequent in acid places. 24. **DOR, WRHM.** Var. *lammersiana* (Hueb.) Breidl is rare. 2.

Cephalozia macrostachya Kaal. is scarce among *Sphagnum* or *Molinia* in the west and the Poole basin. 8+. **BM, CGE, NMW.** Some records need confirmation.

Cephalozia lunulifolia (Dum.) Dum. Occasional in wet heaths or bogs. 12.

Cephalozia connivens (Dicks.) Lindb. Occasional in wet heaths and bogs in the west and the Poole basin. 14. **DOR, WRHM.**

Nowellia curvifolia (Dicks.) Mitt. Very local on stumps or logs, often in fruit. 4. Powerstock Common, JA; Melbury Park, BBS; Bracket's Coppice, BE; Brownsea Is, FR; Delph Wood, VJG.

Cladopodiella fluitans (Nees) Buch. Uncommon among *Sphagnum,* or on moist, acid soils in the Poole basin. 8. **BBSUK, CGE, DOR.**

Cladopodiella francisci (Hook.) Joerg. In similar habitats to *C. fluitans.* 9. Winfrith Heath and Clouds Hill, MOH; Coombe Heath, BE; Morden Heath, J Milsom; Slepe Heath, VJG; Studland Heath, WEN; Canford Heath and Cripplestyle, MOH; Holt Heath and Cranborne Common, BE. **BBSUK, CGE.**

Odontoschisma sphagni (Dicks.) Dum. Local among *Sphagnum* in the west and the Poole basin.12. **CGE, DOR, LDS, WRHM.**

Odontoschisma denudatum (Mart.) Dum. Scarce and local along rills at the upper edge of *Erica tetralix* heath, also on peaty banks and rotten logs. 8. **BRIST.**

Cephaloziella rubella (Nees) Warnst. Rare on wet heaths in the Poole basin. 3. Telegraph Hill, RCS.

Cephaloziella hampeana (Nees) Schiffn. Rare on wet heaths in the west and the Poole basin. 3. St Oswald's Bay, BE, !; Morden Heath, JAP; Studland Heath, EWJ. **WRHM.**

Cephaloziella baumgartneri Schiffn. Very local on sheltered limestone on Portland. 2. First record: HH Knight and WEN (1922 in **CGE**), and later seen by many at Church Ope Cove, King Barrow quarry, Verne, West Weare and Yeolands quarry. **BBSUK, BM, CGE, NMW, RNG.**

Cephaloziella divaricata (Sm.) Schiffn. Local among other bryophytes on wet heaths in the west and the Poole basin. 11. **DOR.**

(*Barbilophozia attenuata* (Mart.) Loeske has been found on sheltered peat in SZ19, VC 11.)

Lophozia ventricosa (Dicks.) Dum. Scarce on wet, heathy soil, among *Sphagnum* or on rotten logs, in the Poole basin, also at Valley of Stones. 8. Var. *ventricosa* at Knowle Hill, BBS.

Lophozia excisa (Dicks.) Dum. Rare on wet heaths. 2. Stokeford Heath, BBS; Gallows Hill, Wool Heath and West Moors, BE.

Lophozia capitata (Hook.) Macoun. On a wet, acid track, Stroud Bridge, BE det. MOH.

Lophozia incisa (Schrad.) Dum. Only recorded from Studland Heath (1933, EWJ in **WRHM**).

Lophozia bicrenata (Schmid. ex Hoffm.) Dum. Rare in wet, acid habitats. 6. Black Down, MOH; north side, Sutton Waldron Ch, ! det. MOH; Knighton Pits and Higher Hyde, BE; Morden Heath, E Armitage; Newton Heath, C Bloom.

Leiocolea badensis (Gott.) Joerg. Rare, on sheltered limestone on Portland or on the Chalk. 4. West Weare, BBS; Yeolands quarry, !; Okeford Hill, JA; Stubhampton Bottom, MM.

Leiocolea turbinata (Raddi) Buch. Occasional on sheltered, calcareous rock or soil. 24. **DOR, LDS.**

Gymnocolea inflata (Huds.) Dum. Locally frequent with *Erica tetralix* on peat which is wet in winter, even if dry in summer. 11. **DOR, WRHM.**

(*Ancistrophyllum minutum* (Schreb.) Schust. has been found near Christchurch in VC 11.)

Tritomaria exsectiformis (Breidl.) Loeske. Scarce on wet heaths in the Poole basin. 4. Creech Heath, BBS; Studland Heath, EWJ in **WRHM**; Newton Heath, B Ottley; Boveridge and Telegraph Hill, RCS.

(*Jamesoniella autumnalis* (DC.) Steph. has a doubtful report from SU10 in VC 11.)

Mylia anomala (Hook.) S.F. Gray. Occasional among *Sphagnum* or in wet, peaty rills with *Erica tetralix*, in the Poole basin. 11.

Jungermannia atrovirens Dum. Rare on wet, shaded rock in the west. 1. Hooke Park, JA; Loscombe, BBS.

Jungermannia gracillima Sm. Occasional on wet, acid soil or china clay spoil. 15. **DOR, WRHM**.

Jungermannia hyalina Lyell. Rare in wet woodland rides. 1. Honeycomb Wood, EWJ.

Nardia scalaris S.F. Gray. Occasional on acid soils, in cliff ravines in the west or on peat in the Poole basin. 13. **DOR, WRHM**.

(*Nardia geoscyphus* (De Not.) Lindb. has been found near Christchurch in VC 11 by JAP.)

Marsupella funckii (Web. & Mohr.) Dum. Once seen at Houghton Wood, EWJ.

Diplophyllum albicans (L.) Dum. Locally frequent on acid banks in woods. 23. **DOR**.

Diplophyllum obtusifolium (Hook.) Dum. Recently found on acid banks in the Poole basin. 2. Stokeford Heath and Clouds Hill, BBS; Higher Hyde Heath, BE.

Scapania compacta (A. Roth) Dum. Scarce on sheltered banks. 3. Moreton Station, MOH; Charity Wood, MM; Creech Heath, BBS.

Scapania irrigua (Nees) Nees. Scarce on damp, acid soil, usually in shade. 5. Wootton Ho, MOH; Piddles Wood, JA, also MM; Houghton Wood, EWJ; Farnham Woods and Scrubbity Barrows, MOH and MM; Studland, EWJ. **WRHM**.

Scapania undulata (L.) Dum. Scarce on rocks or tree roots by streams. 7. Wyld Warren, MOH; Dodpen Hill, BE; Redholm Coppice, BBS; Stoborough, WR Sherrin; Morden Park, BE; Delph Wood, VJG; Parkstone Golf Course, RCS; Cranborne Common, RCS.

Scapania nemorea (L.) Grolle. Scarce on rotten logs or wet banks. 4. Piddles Wood, BBS; Oakers Wood, MOH; Bere Wood, EWJ; Woodlands and Deer Park Wood, EWJ.

Scapania aspera M. & H. Bern. Scarce in sheltered, chalk turf. 3. Eastcombe Wood, MM; Corfe Castle, JA; Knowle Hill, FR; Godlingston Hill, EMH, also BE.

Scapania gracilis Lindb. Rare in acid habitats. 3. Rampisham, JA; Studland, EWJ in **WRHM**; Ashley Heath, RCS.

Lophocolea bidentata (L.) Dum. Common on wet banks or logs, but not seen on Portland. 36. **DOR**.

Lophocolea bispinosa (Hook. f. & Tayl.) Gott. *et al.* An antipodean plant, first found in in wet, sandy places at Clouds Hill (!, det. JAP) and now locally abundant. 4. Knighton pits, !; Warmwell, Buddens Farm and Red Bridge pits, BE; Higher Hyde and Winfrith Heaths, BE; West Moors, BE. **DOR**. At Knighton it is associated with *Epilobium brunnescens*, also from New Zealand.

Lophocolea heterophylla (Schrad.) Dum. Very common on old logs. 37. **DOR**.

Lophocolea fragrans (Moris & De Not.) Gott. *et al.* Rare on bark or stone in sheltered ravines, mainly in the west. 3. Lyme Regis, WEN; Sleech Wood, BE; Monkton Wyld, MOH; Up Cerne Wood, BE; Ochill Wood, Houghton, BE.

Chiloscyphus polyanthos (L.) Corda. A variable species, occasional on logs or wet soil in woods, also by the R. Axe and R. Stour. 16. **DOR, WRHM**.

Plagiochila porelloides (Torrey ex Nees) Linden. Uncommon on sheltered limestone or banks in calcareous woods. 10. **DOR**. Some records need confirmation.

Plagiochila asplenioides (L. emend. Tayl.) Dum. Locally frequent on sheltered banks in woods in the north and west. 29. **DOR**.

Plagiochila norvegica Blom & Holten. Very rare on stone in Hendover Coppice, B Collins det. JAP, also BE, MOH and !. The only British locality, otherwise in Scandinavia.

Southbya nigrella (De Not.) Henriques. A nationally rare species confined to Portland and the Is of Wight. Local and evanescent in shallow calcareous soil in spring. 2. First found by WEN in 1921, and later seen by many others at Bowers quarry, Church Ope Cove, Portland Bill, Verne and West Weare. BE has found it in at least 12 sites. **BBSUK, BM, CGE, NMW**.

Radula complanata (L.) Dum. Frequent on eutrophic bark of elder, sycamore and other trees. 30. **DOR, WRHM**.

(*Ptilidium ciliare* (L.) Hampe occurs in grazed wet heath at Cowards Marsh in VC 11, HCP.)

Ptilidium pulcherrimum (G. Webb) Vainio is rare on branches of trees, especially sallow. 3. Stonebarrow Hill, MOH; Arne, JRW; Morden bog, DSR, also MOH.

Porella arboris-vitae (With.) Grolle. Rare on sheltered limestone rocks, in chalk turf or on bark. 4. On ash, Kingcombe, BE; Portland, !; Corfe railway cutting, !; Knowle Hill, FR, also BBS, BE; Great Wood, Creech, EWJ; Godlingston Hill, BE; Ballard Down, EWJ. Tastes bitter.

Porella obtusata (Tayl.) Trev. Scarce on sheltered rock or in chalk turf. 4. Valley of Stones, BBS; Church Ope Cove, JA, also BE; Knowle Hill and West Hill, FR; Stonehill Down, OE Balme in **WRHM,** also VJG.

Porella platyphylla (L.) Pfeiff. Abundant on old limestone walls, less common at bases of trees. 33. **DOR**.

Frullania tamarisci (L.) Dum. Widespread but not common on bark, sometimes on cliffs or sarsens and rarely in sheltered chalk turf. 30. **BBSUK, DOR, LDS, WRHM**.

Frullania fragilifolia (Tayl.) Gott. *et al.* Very rare on sarsen stones or chert. 2. Valley of Stones, !; below Grove, BE. **DOR**.

Frullania dilatata (L.) Dum. Very common on bark, less so on sheltered rock. 36. **DOR**. Variable in colour and luxuriance; some *Frullania* spp. are said to cause allergies.

Lejeunea cavifolia (Ehrh.) Lindb. Occasional on tree bases in sheltered ravines, usually on calcareous soil, and on limestone boulders on undercliffs. 10. **DOR**.

Lejeunea lamacerina (Steph.) Schiffn. Scarce in sheltered ravines, mostly in the west. 5. Monkton Wyld, Prime Coppices and Wootton Fitzpaine, MOH; Birdsmoor Gate, MOH; Shave Cross, MOH; Abbotsbury Castle, MOH; Great Wood, Creech, MOH, also BBS.

Lejeunea ulicina (Tayl.) Gott. *et al.* Frequent on sheltered tree trunks. 34. **DOR, WRHM.**

Cololejeunea rossettiana (Mass.) Schiffn. Local, mostly among sheltered limestone rocks of Portland and Purbeck. 4. Melbury Park, RD Porley; Church Ope Cove and East Weare, BBS, also BE; Emmetts Hill, BE; St Aldhelm's Head, BE and !. **DOR.**

Cololejeunea minutissima (Sm.) Schiffn. Occasional on sheltered bark, mostly elder and near the coast. 17. **DOR, WRHM.**

Marchesinia mackaii (Hook.) S.F. Gray. Local in crevices of limestone boulder scree. 3. Portland, HH Knight and WEN, *et al.* at Church Ope Cove, Durdle Pier etc, !; Emmetts Hill, BE; St Aldhelm's Head, BE and !. **CGE, DOR, RNG.**

Fossombronia foveolata Lindb. Rare on wet, acid clay. 3. Stokeford Heath, NAS, also BE, !; Oak Hill, BE; Blue Pool, AC Crundwell.

Fossombronia husnotii Corb. Rutted acid track, Bere Wood, BE.

Fossombronia pusilla (L.) Nees. Occasional in woodland tracks, cart ruts, by streams or in damp arable. 17.

Fossombronia wondraczekii (Corda) Dum. Occasional in rutted tracks or damp, disturbed, sandy soil. 6. Piddles Wood, MM; South Heath, !; Studland, EWJ; Creech and Wytch Heaths, !; Upton Heath, !; Threshers Heath, BE; Gore Heath, MOH; Farnham Woods and Deer Park Wood, MOH. **DOR, WRHM.**

Fossombronia incurva Lindb. Rare on damp, gravelly soil. 1. Clouds Hill, D Long.

Pellia epiphylla (L.) Corda. Very common by streams and ditches on acid soils, but not seen on Portland or much of the Chalk. 29. **DOR.**

Pellia neesiana (Gott.) Limpr. Scarce in marshy places. 4. Cains Folly and Stonebarrow Hill, MOH; Piddles Wood, JA; Clouds Hill, BBS; Farnham Woods, MOH.

Pellia endiviifolia (Dicks.) Dum. Common by streams and ditches on calcareous soils. 33. **DOR.**

(*Pallavicinia lyellii* (Hook.) Carruth. has been reported from bogs in SU10 in VC 11.)

Blasia pusilla L. Uncommon on bare, moist clay. 5. Cains Folly, MOH; Eype, BBS; Clouds Hill, BE; Studland Heath, EWJ, also !; Boveridge, RCS.

Aneura pinguis (L.) Dum. Locally frequent on wet ground by springs, streams in woods, or on grazed heaths. 23. **DOR, WRHM.**

Cryptothallus mirabilis Malmb. Discovered in 1948, this parasite of mycorrhiza grows under *Molinia* tussocks or buried in *Sphagnum* and so is under-recorded. 5. Wytherston Marsh, BE; Oakers bog, D Long; Stony Down, RCS, also BE; Studland, MOH; Godlingston Heath, PH Pitkin; Uddens Plantation, BE.

Riccardia multifida (L.) S.F. Gray. Uncommon in wet, acid sites in the west and the Poole basin. 8. **DOR.**

Riccardia chamaedryfolia (With.) Grolle. Occasional in bogs and fens, or on damp basic soil. 16. **DOR, WRHM.**

Riccardia incurvata Lindb. Rare on damp gravelly soil. 1. Clouds Hill, D Long.

Riccardia palmata (Hedw.) Carruth. Rare on logs. 1. Rampisham, JA, also MOH.

Riccardia latifrons (Lindb.) Lindb. Occasional in *Sphagnum* bogs or peat in the Poole basin. 8. Moreton and Winfrith Heaths, MOH; Oakers bog, BBS; Hartland Moor, JAP and PF Hunt; Canford Heath, MOH; Holt Heath, MOH.

Metzgeria furcata (L.) Dum. Common in bark crevices of trees, less common on sheltered rock; rare on bryophytes in lichen heath, Bindon Hill, BE. 36. **DOR.**

Metzgeria temperata Kuwah. Occasional on bark, especially elder and willow. 16. **DOR.**

Metzgeria fruticulosa (Dicks.) Evans. Occasional on bark, especially elder. 13. **DOR.**

(*Sphaerocarpus michelii* Bellardi. This rare casual of damp, acid arable has records from ST51and 61, probably in Somerset.)

Sphaerocarpus texanus Aust. A rare casual of damp, acid arable, twice seen near Ferndown, ECW, 1937–39, in **NMW.**

Lunularia cruciata (L.) Dum. Common on damp, sheltered paths, or in glasshouses. 33. **DOR.**

Conocephalum conicum (L.) Underw. Frequent on vertical stone or soil by sheltered streams, occasionally fruiting; not seen on Portland. 31. **DOR.**

Reboulia hemisphaerica (L.) Raddi. Uncommon on sheltered soil. 4. Loscombe, BBS; Binnegar, MOH; Corfe Castle, JA; West Hill, PJ Wanstall; Canford Heath, PG Biddle.

Marchantia polymorpha L. Occasional on soil, especially near buildings, often in fruit. 13. **DOR.** Resistant to weedkillers, and often an abundant weed in nursery gardens and greenhouses.

Ricciocarpos natans (L.) Corda. Rare. 1. In a pond at Motcombe, E Biles.

Riccia cavernosa Hoffm. emend Raddi. Uncommon on rich mud. 3. Sutton Bingham Reservoir, JA, also !; Winfrith Heath, BE; seasonal pond, Winterborne Kingston, ! det. MOH.

Riccia huebneriana Lindenb. Rare on mud. 1. Woodlands Park, EWJ.

Riccia fluitans L. emend. Lorbeer. Locally common in fen ditches. 2. Northport, !; Green Pool, Furzebrook, DG Hewett; Ferndown, WJ Read (1950). **DOR, WRHM.**

Riccia sorocarpa Bisch. Occasional in damp, sandy arable. 11. **DOR.**

Riccia glauca L. Occasional in damp, sandy arable, many records are old ones. 12. **DOR.**

ANTHOCEROTAE

Anthoceros punctatus L. Uncommon on damp clay or acid soil, on undercliffs or by ditches. 4. Cains Folly, MOH; Eype, JA, also BBS; Loscombe, BBS; Tadnoll, ! det. EV Watson. **OXF.**

Anthoceros agrestis Paton. Rare in wet places. 2. Wareham, AC Crundwell; Gears Mill, MM.

Phaeoceros laevis (L.) Prosk. Locally abundant on clay undercliffs, or on streambanks. 4. Thorncombe Beacon, BBS, also !; Abbotsbury gardens, JAP; Aller, !; Corfe Castle, BBS.

MUSCI
MOSSES

SPHAGNOPSIDA
(see Edwards, 1997; many specimens are in **WRHM**)

Sphagnum papillosum Lindb.
Frequent in bogs in the west and the Poole basin. 13. **CGE, DOR, OXF.** Remains have been found in material of Roman age at Brownsea Is.

Sphagnum palustre L. Local in carr, and sometimes in grazed bogs. 18. **DOR.**

Sphagnum magellanicum Brid. Local in bogs, hanging over runnels, in the west and the Poole basin. 12. **BM, CGE, DOR.**

Sphagnum squarrosum Crome. Occasional in eutrophic carr in the west and the Poole basin, locally abundant near Little Sea. 8. **DOR.**

Sphagnum teres (Schimp.) Angstr. Rare in carr. 3. Kingcombe, SME det. MOH; Oak Hill, ! det. RE Daniels; Corfe Common, AEA Dunston. **DOR.**

Sphagnum fimbriatum Wils. Occasional in carr. 11. **DOR.**

Sphagnum capillifolium (Ehrh.) Hedw. Frequent in wet heaths and bogs in the west and the Poole basin. 14. **BM, CGE, DOR.**

Sphagnum subnitens Russ. & Warnst. Occasional in carr, fen or grazed bogs in the west and the Poole basin. 14. **DOR.**

Sphagnum molle Sull. Uncommon or overlooked in wet heaths in the Poole basin. 8. **DOR.**

Sphagnum compactum DC. Frequent on peat in open heaths in the west and the Poole basin. 13. **BM, CGE, DOR, OXF.**

Sphagnum auriculatum Schimp. Frequent, submerged in bog pools. 18. Var. *inundatum* (Russ.) MO Hill is less common and not usually submerged; it occurs in fens, marshes and bogs. 9. **DOR**

(*Sphagnum subsecundum* Nees may once have occurred in SY39, but most reports are errors.)

Sphagnum contortum Schultz. Rare in base-rich flushes in the west. 1. Champernhayes, FR; Wyld Warren, MOH.

Sphagnum cuspidatum Hoffm. Occasional in wet bogs and acid pools in the Poole basin. 12. **CGE, DOR, OXF.**

Sphagnum tenellum (Brid.) Brid. Usually as scattered stems among other Sphagna in bogs. 12. **BM, CGE, DOR, NY.**

Sphagnum pulchrum (Braithw.) Warnst. Locally abundant in shallow valley bogs in the Poole basin. 9. **BM, DOR, NMW.**

Sphagnum recurvum P. Beauv. Occasional in bogs and carr in the west and the Poole basin. 15. **BBSUK, DOR.** Var. *tenue* Klinggr. is rare, at Redholm Coppice, BBS. Var. *amblyphyllum* (Russ.) Warnst. is occasional in *Betula/ Molinia* woodland, BE.

(*Sphagnum quinquefarium* (Braithw.) Warnst. was reported in error by AEA Dunston.)

BRYOPSIDA

Tetraphis pellucida Hedw. Occasional on sheltered, rotting logs, peaty banks and *Molinia* tussocks. 14. **DOR.**

Polytrichum longisetum Sw. ex Brid. Rare on acid soil in woods. 3. Uplyme, GL; Piddles Wood, MFV Corley; Studland, L Crowson. **DOR.**

Polytrichum formosum Hedw. Common on acid or leached soil in woods. 32. **DOR.**

Polytrichum commune Hedw. Locally common in flushed bogs, or by ponds in woods. 22. **DOR.** Var. *perigoniale* (Michx.) Hampe is scarce. Stokeford Heath, !.

Polytrichum piliferum Hedw. Locally common on dry, acid soil, on banks in heaths and on shingle on the Chesil bank. 22. **DOR, WRHM.**

Polytrichum juniperinum Hedw. Locally abundant on dry heaths and old sandpits, occasional on leached soil elsewhere but not seen on Portland. 35. **DOR.**

Pogonatum nanum (Hedw.) P. Beauv. Uncommon, but widespread on dry, sandy banks. 7. **DOR.**

Pogonatum aloides (Hedw.) P. Beauv. Occasional on dry acid banks. 19. **DOR.**

Pogonatum urnigerum (Hedw.) P. Beauv. Uncommon, but widespread on dry, acid banks and cliffs. 6.

Atrichum undulatum (Hedw.) P. Beauv. Locally abundant on the ground in old woods or on shaded banks, and regularly fruiting. 36. **DOR.**

(*Atrichum angustatum* (Brid.) Br. Eur. Found at Uplyme, GL in **DOR**, but probably in VC 3.)

Archidium alternifolium (Hedw.) Schimp. Uncommon on more or less disturbed heaths or sandy banks. 9. **DOR.**

Pleuridium acuminatum Lindb. Local in disturbed sandy soil, woodland rides, heath tracks or arable. 24. **DOR.**

Pleuridium subulatum (Hedw.) Lindb. In similar habitats to *P. acuminatum* but less common. 12. **DOR, WRHM.**

Pseudephemerum nitidum (Hedw.) Reim. Occasional, in small quantity, in rutted woodland tracks or damp arable. 18.

Ditrichum cylindricum (Hedw.) Grout. Occasional or overlooked on acid soil in woods, damp banks or arable. 18.

Ditrichum flexicaule (Schimp.) Hampe. Occasional in calcareous turf, on the Chalk and on Portland. 11. **BM, DOR, NMW.**

Ditrichum heteromallum (Hedw.) Britt. Uncommon in acid woodland rides. 5. Wootton Hill, MOH; Clouds Hill, !; Morden Decoy, !; Three-legged Cross, RCS.

Blindia acuta (Hedw.) Br. Eur. A montane plant, reported from wet stone by the R. Stour at Longham, Mr Southey.

Seligeria pusilla (Hedw.) Br. Eur. Rare and inconspicuous on shaded calcareous rock. 1. Hooke Park and Eggardon Hill, JA.

Seligeria acutifolia Lindb. Rare on chalk. 1. Knowle Hill, M Yeo in **BBSUK.**

Seligeria paucifolia (Dicks.) Carruth. Scarce on shaded chalk pebbles, mostly in the north-east. 11.

Seligeria calcarea (Hedw.) Br. Eur. Scarce on shaded calcareous rock from Lyme undercliff to the north-east. 8. **DOR.**

Ceratodon purpureus (Hedw.) Br. Locally abundant on dry, more or less acid soils, especially on heaths; in calcareous regions it colonises ash-beds and gravel paths. All grid squares. **DOR.**

Dichodontium pellucidum (Hedw.) Schimp. Scarce on

damp stone or soil in or by streams in the north and west. 5. Mintern's Hill, JA; Bracket's Coppice, BBS; Sydney Coppice, BE; Hooke Park, JA; Melbury Park, BBS; Shaftesbury, MM.

(*Dicranella palustris* (Dicks.) Crundw. ex Warb. was reported from Creech (1965, CHS) but needs confirmation.)

Dicranella schreberana (Hedw.) Dix. Frequent on calcareous clay arable. 20. **DOR**.

Dicranella rufescens (With.) Schimp. Scarce on damp, acid soil. 6. Empool Heath, JA; Bovington Heath, !; Melbury Abbas, MM; Peveril Point, VJG; St Leonards, HCP. **DOR**.

Dicranella varia (Hedw.) Schimp. Frequent on damp banks and paths, and forming tufa on wet undercliffs. 35. **DOR**.

Dicranella staphylina Whitehouse. Locally frequent in arable fields. 21. **DOR**.

Dicranella cerviculata (Hedw.) Schimp. Scarce on damp, peaty banks in the Poole basin. 3. Studland Heath, CH Binstead; Brownsea Is, JA; Cranborne Common, MOH.

Dicranella heteromalla (Hedw.) Schimp. Abundant on hard acid banks in woods. 36. **DOR**.

Bartleya ohioensis Robins. Found on a dripping cliff below Dungy Head, ! det. J-P Frahm. If correct, new to Britain. It was sterile, and could be a *Dicranella* or a *Campylopus*.

Dicranoweisia cirrata (Hedw.) Lindb. Common on eutrophic bark, especially elder, also on old wooden fences or thatch, and sometimes on stone. Not reported from Portland. 31. **DOR**.

(*Dicranum polysetum* Sw. has been recorded from acid soil at Ashley Heath, ECW, in VC 11.)

Dicranum bonjeanii De Not. Occasional in flushed *Erica tetralix* heath, or in old chalk turf. 14. **DOR, WRHM.**

Dicranum scoparium Hedw. Common in acid woodland, on heaths, and in leached grassland on chalk summits; occasionally seen as an epiphyte. 37. **DOR**.

Dicranum majus Sm. Occasional on acid soil in old woods in the north and west. 16. **DOR**.

Dicranum spurium Hedw. Occasional on heaths. 7. Owermoigne Heath and Moreton, ECW; Povington Heath, ECW; Arne Heath, DSR; Canford Heath, MOH; Horton Common and Barnsfield Wood, HCP. **BBSUK.**

(*Dicranum montanum* Hedw. is spreading in VC 11, and has reached SU10; it may occur on oligotrophic bark in the east.)

Dicranum tauricum Sapehin. A colonist, once found on a stump at Rolfs Wood, MM, and likely to spread.

(*Dicranodontium denudatum* (Brid.) Broth. was reported from Parkstone in 1940 by DS Wyard, probably in error.)

(*Campylopus subulatus* Schimp. was found on a gravel path in the New Forest, SU10, in VC 11 by JAP.)

Campylopus fragilis (Brid.) Br. Eur. Very local on mature or degenerate dry heaths, in the west and the Poole basin; at base of *Ammophila* in dunes, Studland, BE. 8. **DOR, OXF, WRHM.**

Campylopus pyriformis (Schultz) Brid. Frequent on heaths, especially after a fire, and sometimes on dead wood. 28. **DOR, WRHM.**

Campylopus paradoxus Wils. Locally frequent on peat in dry heaths, rarely on old roofs. 24. **DOR**.

(*Campylopus atrovirens* De Not. Reports from wet heaths are errors; it occurs in Devon.)

Campylopus introflexus (Hedw.) Brid. An alien moss which has become abundant on peaty heaths as an early colonist, and on rotting wood far from heathland. 27. **DOR**. EW Jones did not find it when he surveyed Studland Heath in 1933; I saw it at Champernhayes in 1973.

Campylopus brevipilus Br. Eur. Local on heaths in the west and the Poole basin. 12. **CGE, DOR, NMW, OXF, WRHM.**

Leucobryum glaucum (Hedw.) Angstr. Locally abundant in woods on acid soils, or on mature heaths. 16. **BM, DOR, WRHM**.

Leucobryum juniperoideum (Brid.) C. Muell. Only one record, from Piddles Wood, F Foster.

Fissidens viridulus (Sw.) Wahlenb. The variable aggregate species is frequent on shaded banks, or on rocks in streams. 35. **DOR**. Var. *viridulus* and *F. pusillus* Wils. are occasional. Var. *tenuifolius* (Boul.) AJE Smith occurs on shaded chalk pebbles.

Fissidens incurvus Starke ex Roehl. Occasional on sheltered calcareous soil. 22. **DOR**.

Fissidens bryoides Hedw. Common on clay banks, often in woods. All grid squares. **DOR**.

(*Fissidens curnovii* Mitt. was reported from Lyme Regis by W Mitten, perhaps from VC 3; it likes deep shade.)

Fissidens rivularis (Spruce) Br. Eur. Very rare, on wet rock at Loscombe in the west, MOH.

Fissidens crassipes Wils. ex Br. Eur. Occasional on calcareous rock near streams. 19.

Fissidens exilis Hedw. Occasional on clay banks. 10.

Fissidens taxifolius Hedw. Very common on soil in woods or hedgebanks. All grid squares. **DOR**.

Fissidens cristatus Wils. ex Mitt. Occasional in calcareous turf. 23. **DOR**.

Fissidens adianthoides Hedw. Fairly frequent in calcareous turf. 23. **DOR**.

Encalypta vulgaris Hedw. Uncommon on calcareous rocks and old walls, most frequent on Portland; in calcareous turf above cliffs, St Aldhelm's Head, BE. 9. **BBSUK, BM, CGE, DOR, NMW, OXF**.

Encalypta streptocarpa Hedw. Occasional on calcareous rocks, old walls, and shallow turf. 21. **BM, DOR.**

Tortula ruralis (Hedw.) Gaertn., Meyer & Scherb. Frequent on dry, sunny calcareous rocks, walls or roofs. 28. **DOR**. Subsp. *ruraliformis* (Besch.) Dix. is locally frequent on sand dunes, and in inland artificial habitats on sandy tracks or old asphalt, where it is eaten by slugs.

Tortula intermedia (Brid.) De Not. Common on dry calcareous rocks and walls. 38. **DOR**.

Tortula laevipila (Brid.) Schwaegr. Frequent on the bark of old trees, often as its scarcely distinct variety *laevipiliformis* (De Not.) Limpr. 36. **DOR**.

Tortula muralis Hedw. Very common on limestone, old bricks and man-made stonework. All grid squares. **DOR**.

Tortula marginata (Br. Eur.) Frequent on damp, shaded walls or limestone boulders. 22. **BM, DOR**.

Tortula subulata Hedw. Occasional on sheltered banks and wall-tops. 16. **BM, DOR, OXF**.

Tortula papillosa Wils. ex Spruce. Occasional on bark, mostly on elder. 16. **DOR**.

Tortula latifolia Bruch ex Hartm. Occasional on bark of trees by rivers, especially along the Stour. 21.

(*Tortula vahliana* (Schultz) Mont. A critical plant, whose reported occurrences at Houns Tout and Newton Heath need confirmation before they can be accepted.)

Aloina aloides (Schultz) Kindb. Occasional on shallow, calcareous soils, on cliffs and old walls. 20. **BM, DOR, NMW, OXF.** Var. *ambigua* (Br. Eur.) Craig has only been seen in the west, at Charmouth and Eype.

Aloina brevirostris (Hook. & Grev.) Kindb. Once found by the old railway at Spetisbury, MM.

Desmatodon convolutus (Brid.) Grout. Rare on cliffs. 4. Eype, BBS; Seatown, MOH; West Weare, BBS; Durdle Door, MOH; Worborrow Bay, BE; Swanage, EMH. **NMW.**

Pterygoneurum ovatum (Hedw.) Dix. Rare on shallow, calareous soil. 2. Winspit, AR Finch; Durlston, MFVC.

Pottia caespitosa (Bruch ex Brid.) C. Muell. Rare on chalky soil. 3. White Nothe undercliff, BBS; Badbury Rings, MCF Proctor, also BE; Gussage Hill, MM.

Pottia starkeana (Hedw.) C. Muell. Frequent in calcareous arable or bare soil. 27. Subsp. *minutula* (Schleich. ex Schwaegr.) Chamberlain is the common form. Subsp. *starkeana* and var. *brachyodus* C. Muell. are found mostly along the coast. Subsp. *conica* (Schleich. ex Schwaegr.) Chamberlain, is uncommon. **CYN, DOR, OXF, NMW.**

Pottia commutata Limpr. Rare in shallow, calcareous soils of Portland and Purbeck. 5. Portland Bill, JA; West Weare, !; Church Ope Cove, APC; Cheynes Weare and below Gad Cliff, BE; Emmetts Hill, ECW, also FR.

Pottia wilsonii (Hook.) Br. Eur. Very rare on shallow, calcareous soil at Durlston, MFVC.

Pottia crinita Wils. ex Br. Eur. Scarce on shallow, calcareous soils near the coast. 7. Portland, EMH; Durdle Door, JA; Lulworth Cove, BBS; Chapmans Pool and Durlston Head, MFVC; Studland, EMH. **BRIST, DOR, NMW.**

Pottia lanceolata (Hedw.) C. Muell. Occasional in sparse calcareous turf. 15. **DOR.**

Pottia intermedia (Turn.) Fuernr. Occasional in arable. 10.

Pottia truncata (Hedw.) Fuernr. Abundant in calcareous or sandy arable. 34. **DOR, RNG, WRHM.**

Pottia heimii (Hedw.) Fuernr. Scarce on shallow or saline calcareous soils near the coast. 8. Abbotsbury, MOH; Portland Bill, JA, also MOH; Ferry Bridge, BE; Lulworth Cove, BBS; Chapmans Pool, EWJ; Durlston Head, MFVC; Peveril Point, EWJ. **BM, NMW.**

Pottia bryoides (Dicks.) Mitt. Rare on shallow, calcareous soils. 3. Charmouth and Seatown, MOH; Weymouth, APC; Bincombe Ch, ! det. MOH.

Pottia recta (With.) Mitt. Occasional in calcareous arable or bare places in chalk turf. 18. **DOR.**

Phascum cuspidatum Hedw. Frequent in arable or bare soil. 37. **DOR.** Var. *piliferum* (Hedw.) Hook. and Tayl. only noted from cliffs near Portland Bill, JAP, also BE; Winspit, BE.

Phascum curvicolle Hedw. Scarce in short, calcareous turf. 9. Uplyme, GL; Eggardon Hill, BBS; Church Ope Cove, APC; Melbury and Gussage Hills, MM; Bindon Hill, St Aldhelm's Head, Badbury Rings and Godlingston Hill, BE; Blagdon Hill, MM. **DOR.**

Phascum floerkeanum Web. & Mohr. Rare in old chalkpits or short turf. 3. Belchalwell Street, JA; Hambledon Hill, BE; Melbury Hill, MCF Proctor.

Acaulon muticum (Brid.) C. Muell. Scarce on cliff-tops or in arable. 3. Uplyme, GL; Stonebarrow Hill, MOH; Thorncombe Beacon, MOH; Ashley Wood, MM. **DOR.**

Acaulon triquetrum (Spruce) C. Muell. A very local, winter ephemeral of bare, calcareous soil near the coast. 4. Portland, EMH, also W Mitten; Durdle pier, BE; below Hambury Tout, BE, !; Emmetts Hill, ECW, also BE. **NMW, NY.** It does not occur further north in Britain.

Barbula convoluta Hedw. Abundant on hard-packed soil. All grid squares. **DOR.**

Barbula unguiculata Hedw. Common on calcareous soils, undercliffs, arable and old walls. All grid squares. **DOR.**

Barbula hornschuchiana Schultz. Frequent on calcareous soil and arable. 27. **DOR.**

Barbula revoluta Brid. Common on calcareous soils, undercliffs and old mortared walls. 31. **DOR.**

Barbula acuta (Brid.) Brid. Rare in chalk turf. 2. Zigzag Hill, MM; Ballard Down, EMH; on rubble, South Haven peninsula, EWJ. **NMW.**

Barbula fallax Hedw. Abundant on calcareous soil, paths and arable. All grid squares. **DOR.**

(*Barbula spadicea* (Mitt.) Braithw. was reported in a Census Catalogue, but never confirmed.)

Barbula rigidula (Hedw.) Mitt. Frequent on calcareous rocks and walls. 30. **DOR.**

Barbula nicholsonii Culm. Rare on seasonally inundated stone. 1. Hayward Bridge, ECW and RCS.

Barbula trifaria (Hedw.) Mitt. Frequent in small amounts on rocks and walls. 34. **DOR.**

Barbula tophacea (Brid.) Mitt. Locally frequent on wet, calcareous clay or stone, especially on cliffs, forming tufa. 23. **DOR.**

Barbula vinealis Brid. Abundant on calcareous rocks and walls. All grid squares. **DOR.**

Barbula cylindrica (Tayl.) Schimp. Common on damp, sheltered walls. 35. **DOR.**

Barbula recurvirostra (Hedw.) Dix. Common on calcareous rocks and walls, and sometimes on soil, 35. **DOR.**

(*Gymnostomum calcareum* Nees & Hornsch. Very rare on sheltered limestone. 1. Near Lyme Regis, perhaps in Devon.)

(*Gymnostomum aeruginosum* Sm. and *G. recurvirostrum* Hedw. were reported from calcareous cliffs at Durlston by CJ Hadley, but were probably some other species in this difficult genus.)

Gymnostomum viridulum Brid. Locally common on very shallow, calcareous soil in quarries and undercliffs on Portland. 2. First found here by WEN, and seen later by BBS, BE, MFVC, MV Fletcher, MOH and !.

Gyroweisia tenuis (Hedw.) Schimp. Occasional on damp, shaded, calcareous stone. 15. **BM, OXF.**

Leptobarbula berica (De Not.) Schimp. Rare, perhaps recently introduced, on shaded and sheltered church walls. 2. Burton Bradstock Ch and Melbury Osmond Ch, HLK Whitehouse.

Eucladium verticillatum (Brid.) Br. Eur. Local on damp calcareous stonework, mostly in the west, also on Southwell cliffs, ! and in dripping caves at Winspit, BE. 14. **BM.**

Weissia controversa Hedw. Occasional on sandy banks and cliffs. 24. **DOR**. Var. *crispata* (Nees & Hornsch.) Nyholm is less common on calcareous soils. 5.

Weissia rutilans (Hedw.) Limpr. Scarce on sandy banks. 4. Wootton Ho, MOH; Holwell, HHW; Creech Heath, BBS; Studland, Mr Parsons. **BM, CYN**.

Weissia tortilis (Schwaegr.) C. Muell. Rare in old, calcareous turf. 3. West Weare, ECW; Knowle Hill and Stonehill Down, FR, also BBS; Ballard Cliff, BE. **BBSUK, WRHM**.

Weissia microstoma (Hedw.) C. Muell. Occasional on basic soil in grassland. 20. **DOR, WRHM**. Var. *brachycarpa* (Nees & Hornsch.) C. Muell. is scarce on wetter soils. 4.

Weissia squarrosa (Nees & Hornsch.) C. Muell. Rare, on damp soil. 1. Uplyme, GL in **DOR**, perhaps in VC 3.

Weissia rostellata (Brid.) Landb. Very rare, in damp pasture. 1. Mappowder, JA.

Weissia sterilis Nicholson. Rare in chalk turf. 4. Maiden Castle, ECW; Ridgeway Hill, BE; Badbury Rings, BE; Swanage, CH Binstead. **NMW**.

Weissia longifolia Mitt. var. *angustifolia* (Baumg.) Crundw. & Nyh. Occasional in short calcareous turf. Hybrids with *W. controversa* were seen at Knowle Hill, BBS.

Oxystegus sinuosus (Mitt.) Hilp. Common on damp, sheltered rocks and walls. 37. **DOR**.

Trichostomum crispulum Bruch. Frequent on basic soils, rocks or mortared walls. 26. **DOR, NMW**.

Trichostomum brachydontium Bruch. Occasional on basic soils, rocks and cliffs. 16. **DOR, WRHM**.

Tortella tortuosa (Hedw.) Limpr. Uncommon in old, calcareous turf or among limestone rocks. 12. **BM, DOR**.

Tortella flavovirens (Bruch) Broth. Locally frequent on blown sand at Small Mouth, !, and on cliffs at Portland and along the coast to Studland. 8. **CYN, DOR, NMW, WRHM**.

Tortella inflexa (Bruch) Broth. Scarce in sheltered, chalk turf or on chalk pebbles. 5. Giant Hill, Enford Bottom, Clubmen's Down and Fontmell Down, BE; Hambledon Hill, MM: Great Wood, Creech, FR; Knowle Hill, BBS.

Tortella nitida (Lindb.) Broth. Occasional on basic rocks and walls near the coast, and inland at Melcombe Horsey Ch, !. 9. **DOR, NMW, OXF, RNG**.

Pleurochaete squarrosa (Brid.) Lindb. Scarce on open calcareous soils in the north-east, and along the coast east of Wyke Regis; on stabilised shingle at Ferry Bridge, BE., **BM, RNG**.

Trichostomopsis umbrosa (C. Muell.) H. Robinson. Rare or overlooked on sheltered and shaded brick walls. 1. Evershot Ch, HLK Whitehouse, also !.

(*Leptodontium flexifolium* (With.) Hampe has not been found closer than Axminster, VC 3, GL in **DOR**.)

Leptodontium gemmascens (Mitt. ex Hunt) Braithw. A rare or overlooked spring ephemeral on wet thatch. 2. Higher Bockhampton, BBS, also !; Briantspuddle, BBS.

Cinclidotus fontinaloides (Hedw.) P. Beauv. Occasional on seasonally inundated stone, especially by the Stour and lower Frome. 14. **DOR, WRHM**.

Cinclidotus mucronatus (Brid.) Mach. Occasional on rock and bark seasonally inundated by streams. 27. **DOR**.

Schistidium apocarpum (Hedw.) Br. Eur. Common on calcareous rock and old walls, often in shelter and in churchyards. All grid squares. **DOR**.

Grimmia pulvinata (Hedw.) Sm. Abundant on calcareous rocks and walls. All grid squares. **DOR**.

Grimmia orbicularis Bruch ex Wils. Rare on calcareous rocks in the south and west. 4. Lyme Regis, HH Knight and WEN; Black Down, ECW; Portland quarries, !; West Weare, RCS; Winterbourne Monkton, HH Knight and WEN; Preston, ECW. **CGE, NMW**.

Grimmia trichophylla Grev. Rare on sarsens, chert or sandstone. 5. Grey Mare and her Colts, !; Valley of Stones, !, also BBS; East Weare, BE; Rye Water Farm, BE, !; Studland, EMH. **DOR, NMW, RNG**.

Racomitrium heterostichum (Hedw.) Brid. Scarce on acid rocks or tiled roofs. 5. Fifehead Magdalen, MOH; Lytchett Minster Ch, MM; Studland Heath, Mr Parsons; Colehill Ch and Alderholt Ch, !. **CYN**.

Racomitrium lanuginosum (Hedw.) Brid. Scarce on acid heaths, and once on a leached patch of *Calluna* over chalk, very rare in chalk turf. 9. Champernhayes, GL; Powerstock Common, BBS; East Burton Heath, ECW; Winfrith Heath, FR, also MOH and !; Lower Hyde Heath, BE; Melbury Down, JH Tallis; Creech Heath, BBS; Knowle Hill, FR; Stoborough Heath, BE; Decoy Heath, JH Tallis; Lytchett Minster, MOH; Canford Ch roof, !; Godlingston Heath, T Laflin; Studland Heath, EWJ; Verwood, HHW; West Moors, BE. **BM, DOR, NMW**.

Racomitrium ericoides (Brid.) Brid. Rare on gravelly heaths. 2. Houghton Wood, EWJ; Studland Heath, EWJ. **DOR, WRHM**.

[*Racomitrium elongatum* Frisvoll. Very rare or extinct. Verwood, HHW in **BM**.]

Ptychomitrium polyphyllum (Sw.) Br. Eur. Rare on acid stone. 2. Powerstock Common railway, JC Duckett; on sarsen, Coombe valley, Chalbury, A Morrison det. MOH.

Funaria hygrometrica Hedw. Locally abundant on old bonfire-sites, rarely on charcoal, also on rich soil. All grid squares. **DOR**.

Funaria muhlenbergii Turn. Very rare in chalk turf. 1. West Hill, Corfe, PJ Wanstall, also FR *et al.* **WRHM**.

Funaria pulchella Philib. Rare in limestone soil on Portland. 1. West Weare, EWJ; Verne, BJ O'Shea. **BBSUK**.

Funaria fascicularis (Hedw.) Lindb. Scarce in marshes or damp arable. 6.

Funaria obtusa (Hedw.) Lindb. Rare in damp, sandy or peaty soil. 5. Uplyme, GL; Glanvilles Wootton, CW Dale; Studland Heath, EWJ. **DOR, WRHM**.

Physcomitrium pyriforme (Hedw.) Brid. Local on bare mud. 26. **DOR**.

Physcomitrella patens (Hedw.) Br. Eur. Uncommon on bare mud, absent near the coast. 12.

Ephemerum recurvifolium (Dicks.) Boul. Uncommon on bare, calcareous soil, on undercliffs and in arable. 7. Eggardon Hill, MFV Corley; West Weare and Southwell, BE; Doles Ash Farm, MM; Fossil Forest, Gad Cliff, Hill Bottom, Compton Down and Hambledon Hill, BE; Ashmore Down, MOH.

Ephemerum serratum (Hedw.) Hampe. Occasional in rutted woodland rides or damp, acid, arable. 16. **DOR**.

Splachnum ampullaceum Hedw. Rare, sporadic and declining on cow dung in grazed bogs. 4. Champernhayes, GL; Moreton Heath, WBB; Hartland Moor, SBC; Uddens Moor, HW Monckton. **BM, DOR, RNG**.

Schistostega pennata (Hedw.) Web. & Mohr. Very rare inside caves or rabbit burrows. 1. North Chideock, MOH; Chideock, SME.

Orthodontium lineare Schwaegr. An alien, first recorded *c.*1920, and now frequent on stumps or sometimes on peaty soil in woods. 18. **DOR**.

Leptobryum pyriforme (Hedw.) Wils. Local on bonfire-sites or on soil in glasshouses. 12. **WRHM**.

Pohlia nutans (Hedw.) Lindb. Frequent on acid soil. 24. **DOR**.

(*Pohlia drummondii* (C. Muell.) Andrews is rare in the New Forest, and has been seen in SU10, JA, but in VC 11.)

Pohlia proligera (Kindb.) Lindb. Occasional on disturbed acid soil. 16.

Pohlia lutescens (Limpr.) Lindb. f. Uncommon on clay soils. 9.

Pohlia carnea (Schimp.) Lindb. Common on damp, clay soil. All grid squares. **DOR, WRHM**.

Pohlia wahlenbergii (Web. & Mohr) Andrews. Frequent on damp soils in woodland. 21. **DOR**. Var. *calcarea* (Warnst) Warburg is scarce in calcareous grassland.

Epipterygium tozeri (Grev.) Lindb. Occasional in damp, shaded ± acid clay banks. 13. **BM, DOR**.

Bryum pallens Sw. Occasional on damp soils, or on old walls. 18. **DOR.**

Bryum algovicum Sendtn. ex C. Muell. Rare on sand dunes, or calcareous soil. 4. Small Mouth; Studland dunes.

Bryum inclinatum (Brid.) Bland. Uncommon on sandy or calcareous soils. 7.

[*Bryum knowltonii* Barnes. Once found in dune slacks at Studland, W Mitten.]

Bryum intermedium (Brid.) Bland. Rare by streams. 1. Uplyme, GL; The Spittles undercliff, MOH. **DOR**.

Bryum donianum Grev. Uncommon on sandy soils. 6. **DOR**.

Bryum capillare Hedw. Abundant on rocks and walls, also on bark, especially elder. All grid squares. **DOR**.

Bryum subelegans Kindb. Older workers lumped this with *B. capillare*. It may be frequent on eutrophic bark. 5. Nettlecombe Tout and Whiteway Hill, MOH; Bonsley Common, MM; Durweston, BBS; Child Okeford, J Gardiner; Zigzag Hill, MM.

Bryum torquescens Bruch ex De Not. Occasional on basic soils, but lumped with *B. capillare* until 1973. 6. Lyme Regis, GL; Charmouth, WR Sherrin; Grove, Verne and West Weare, MOH; Child Okeford, MM; Hambledon Hill, S Greene; Godlingston Hill, MOH; Ulwell, MM. **BM, DOR**.

Bryum canariense Brid. Rare in limestone crevices or old chalk turf. 3. Church Ope Cove, BBS; below Grove, MOH; Knowle Hill, BBS.

Bryum creberrimum Tayl. Rare in crevices of walls. 2. Uplyme, GL; Lytchett Minster, MOH. **BBSUK, DOR.**

Bryum pallescens Schleich. ex Schwaegr. Overlooked, perhaps increasing. 1. Under electricity pylon, Briantspuddle, BBS; possibly this on the undercliff at Thorncombe Beacon, !.

Bryum pseudotriquetrum (Hedw.) Schwaegr. Occasional in wet flushes on heaths or in marshes. 18. **DOR, WRHM**.

Bryum caespiticium Hedw. Frequent on waste ground or walls. 16. **DOR**.

Bryum alpinum Huds. ex With. Scarce on wet, heathy tracks. 4. Champernhayes, GL, also MOH; Moreton Station, ECW, also MOH; Cranborne Common, MOH. **DOR, NMW**.

(*Bryum mildeanum* Jur. All reports are erroneous.)

Bryum bicolor Dicks. Abundant in arable and waste ground. All grid squares. **DOR**.

Bryum gemmiferum Wilcz. & Dem. Probably frequent. 5.

Bryum gemmilucens Wilcz. & Dem. Rare in woodland rides. 1. Farnham Woods and Scrubbity Barrows, MOH.

Bryum dunense Smith & Whitehouse. Rare in sandy soils or undercliffs. 4. Weymouth, Mr Wigginton; White Nothe, BBS; Morden Heath, MOH; Ballard Down, BBS; Brownsea Is, CCT. **RNG**.

Bryum argenteum Hedw. Abundant in arable and waste ground, or a pavement weed. All grid squares. **DOR**.

Bryum erythrocarpum agg. A difficult group revised in 1964, so lacking old records; mostly weedy.

Bryum radiculosum Brid. Common in calcareous arable or on old walls. 32. **DOR**.

Bryum ruderale Crundw. & Nyh. Frequent in arable or bare soil. 24.

Bryum violaceum Crundw. & Nyh. Uncommon in arable. 5. Recorded from 39, 89, 99, 90 and 91. **DOR**.

Bryum klinggraeffi Schimp. Occasional in arable. 14.

Bryum sauteri Br. Eur. Scarce on waste ground in the west. 3. Catherston Leweston and Wootton Fitzpaine, MOH; Birdsmoor Gate, MOH; Symondsbury, MOH.

Bryum tenuisetum Limpr. Rare on acid soil. 1. Studland Heath, EWJ.

Bryum microerythrocarpum C. Muell. & Kindb. Occasional in arable or disturbed soil. 9.

Bryum bornholmense Winkelm. & Ruthe. A non-weedy species of sandy soils, often on bonfire-sites. 9.

Bryum rubens Mitt. Abundant in arable and disturbed soils. 35. **DOR**.

Rhodobryum roseum (Hedw.) Limpr. Scarce on anthills in old calcareous turf, not seen on Portland. 7. Uplyme, GL; Valley of Stones, MOH; Eggardon Hill, MOH; Longcombe Bottom, MM; Corfe, JA; Fontmell Down, !; Ashmore Down, JRW. **DOR**.

Mnium hornum Hedw. Abundant on leached soil in old woods and shaded banks, also on stumps, and heaths recovering from 'improvement'; not on Portland. 37. **DOR**.

Mnium stellare Hedw. Uncommon on sheltered, acid soil in woods, or on logs, mainly in the west. 7. Loscombe and Beaminster, MOH; Eggardon Hill, JA; Hooke Park and Bracket's Coppice, BBS; Corscombe, MOH; Sherborne Golf Course, MOH; Oakers Wood, BBS; Eastcombe Wood, MM.

Rhizomnium punctatum (Hedw.) Kop. Locally frequent on mud, tree bases and logs in sheltered carr. 31. **BM, DOR**.

Plagiomnium cuspidatum (Hedw.) Kop. Scarce on soil or branches in wet, shaded places. 3. Black Down, ECW; Cornford Bridge, MOH. **NMW.**

Plagiomnium affine (Funck) Kop. Occasional in woodland rides or sheltered lawns. 20. **DOR.**

Plagiomnium elatum (Br. Eur.) Kop. Rare in carr. 4. Eggardon Hill, MFVC; Sydling Water, C Nicholls; Bainly Bottom, MM; Corfe Common, JA.

Plagiomnium undulatum (Hedw.) Kop. Common in sheltered woods and banks, and in shaded lawns, especially in churchyards. 35. **DOR.**

Plagiomnium rostratum (Schrad.) Kop. In small amounts in sheltered woods or banks. 20. **WRHM.**

Aulacomnium palustre (Hedw.) Schwaegr. Frequent in bogs, mainly in the Poole basin or the west. 18. **DOR, NMW.**

Aulacomnium androgynum (Hedw.) Schwaegr. Mostly on logs, also on damp peat, avoiding the Chalk. 16.

Bartramia pomiformis Hedw. Local, perhaps decreasing, on shaded acid banks. 14. **BM, DOR, NMW, RNG, WRHM.**

Philonotis arnellii Husn. Rare or extinct on soil banks. 3. Moreton Heath, HHW; Morden, BBS; Tollard Royal, EWJ (perhaps in VC 8). **BBSUK, BM.**

Philonotis caespitosa Wils. ex Milde Very rare in wet, acid soil. 1. Abbotsbury Castle, MOH.

Philonotis fontana (Hedw.) Brid. Very local in flushes and marshes. 14. **DOR, NMW.**

Philonotis calcarea (Br. Eur.) Schimp. Very rare in fens or flushes. 2. Uplyme, GL (perhaps in VC 3); Corfe Common, ECW, also BE and VJG. **DOR, NMW.**

[*Breutelia chrysocoma* (Hedw.) Lindb. A montane plant on wet heaths, once found at Champernhayes, GL in **DOR.**]

Zygodon viridissimus (Dicks.) R. Br. Very common on bark, especially in crevices, most often on elder. 38. **DOR.** Var. *stirtoni* (Schimp. ex Stirt.) Hagen. is uncommon, on stone. 4. **CGE, NMW.**

Zygodon baumgartneri Malta. Occasional on bark of large trees in old woods and parks. 13. **DOR.**

Zygodon conoideus (Dicks.) Hook. & Tayl. Occasional on bark, especially on elder. 11.

Orthotrichum striatum Hedw. Rare and decreasing on bark. 8. Lyme Regis, GL; Stanton St Gabriel, MOH; Hooke Park, JA; Holwell, HHW; Sherborne, H Boswell; Oakers Wood, BBS; Spetisbury, MM. **BM, DOR, OXF.**

Orthotrichum lyellii Hook. & Tayl. Widespread on bark of large hedgerow trees, especially ash. 33. **DOR.**

Orthotrichum affine Brid. Common on bark of elder, sallow and other trees. 35. **DOR.**

[*Orthotrichum rivulare* Turn. Once found on seasonally inundated bark near Holwell, HHW in **BM, OXF.**]

Orthotrichum sprucei Mont. Scarce on silty bark of riverside trees, especially by the Stour. 9. **BM, NMW, OXF.**

Orthotrichum anomalum Hedw. Frequent on calcareous rocks, walls and gravestones. 32. **DOR.**

Orthotrichum cupulatum Brid. Rare on limestone. 3. Glanvilles Wootton, CW Dale; R. Frome walls, Wareham, FR in **WRHM.**

Orthotrichum tenellum Bruch ex Brid. Occasional on bark of elder or other trees in the coastal belt. 17. **DOR, NMW, OXF.**

Orthotrichum diaphanum Brid. Common on bark of old trees, sometimes on wood or stone. All grid squares. **DOR.**

Orthotrichum pulchellum Brunton. Rare on sheltered bark, especially elder and willow, also on granite. 6. Uplyme, GL; Wyld Warren, MOH; Netherbury Ch, BE and !; Valley of Stones, BBS; Powerstock Common, BE; Askerswell Down, JA; Eggardon Hill, MOH; Redholm Coppice, BBS; Tadnoll, BBS; Creech Heath, BBS; Sherford, BE. **DOR.**

Ulota crispa (Hedw.) Brid. Common on branches of trees in carr or wet woods. 32. **DOR.** Var. *norvegica* (Groenvall) Smith and Hill is less often recorded in the same habitat. 10.

Ulota phyllantha Brid. Occasional on the boles of large trees in the west and along the coast. 17. **DOR, NMW.**

Hedwigia ciliata (Hedw.) P. Beauv. (Probably all the segregate *H. stellata*). Rare and local on sarsen and conglomerate stones, or on old tiled roofs. 4. Valley of Stones, !, also VJG and BBS; Kingston Russell stone circle, T Laflin; Chalbury, A Morrison; Agglestone, EMH; Canford Ch, !. **DOR.**

Fontinalis antipyretica Hedw. Locally abundant in clear streams, ponds and flooded pits, often in shade. 31. **DOR, WRHM.** Resistant to weedkillers used to kill vascular plants in angler's ponds.

Climacium dendroides (Hedw.) Web. & Mohr. Scarce and local in grassy marshes. 8. Mary's Well, JRW; Aunt Mary's Bottom, BE and JRW; Langford Mead, JRW; Birts Hill, JA; Thorncombe, BE; Tadnoll, BBS; Clouds Hill, !; Wareham Common, JRW. **BMH, DOR.**

Cryphaea heteromalla (Hedw.) Mohr. Common on elder and other bark in shelter, usually in fruit. 37. **DOR, WRHM.**

Leucodon sciuroides (Hedw.) Schwaegr. Local but widespread on ash and other large trees, less common on limestone. 34. **CGE, DOR, WRHM.** Var. *morensis* (Limpr.) De Not. is much larger and rarer. 2. Eype, R.Parker; on rocks below Grove, BE.

(*Antitrichia curtipendula* (Hedw.) Brid. might have occurred in the last century, and is listed in one Census Catalogue, but there are no certain records.)

Pterogonium gracile (Hedw.) Sm. Rare on bark of old trees or on sheltered limestone. 9. Valley of Stones, BBS; Melbury Park, FR, also BE, !; Burl Moor, BE; Portland, EMH; Grove, BE; Green Hill, Hilton, BE; Lydlinch, HHW, also BE; Swanage, HW Monckton. **DOR.**

Leptodon smithii (Hedw.) Web. & Mohr. Widespread on boles of ash and other large trees, sometimes on tiled roofs and walls; there is an early 19th century record from near Weymouth, M Goult. 33. **BM, DOR, NMW, OXF, WRHM.**

Neckera crispa Hedw. Local in sheltered calcareous grassland or in old quarries. 19. **BM, DOR, SLBI.**

Neckera pumila Hedw. Frequent on bark in woods or on trees in sheltered valleys. 32. **DOR, WRHM.**

Neckera complanata (Hedw.) Hueb. Common on bark in woods, or on soil of shaded banks, also in old calcareous turf. 38. **DOR, RNG.**

Homalia trichomanoides (Hedw.) Br. Eur. Occasional on sheltered roots and banks, less common than *Neckera* spp. and mainly in the north. 25. **DOR.**

Thamnobryum alopecurum (Hedw.) Nieuwl. Locally

dominant on calcareous soil in woods and plantations, and in sheltered churchyards. 35. **DOR**.

Hookeria lucens (Hedw.) Sm. Local, beside streams in wet woods, mostly in the west with a few sites in the Poole basin. 14. **DOR, RNG**.

Myrinia pulvinata (Wahlenb.) Schimp. Rare on seasonally inundated bark or wood in the valley of the upper Stour. 3. Holwell, HHW, also JA and MOH; Cutt Mill, MM; Child Okeford, MM. **BM**.

Habrodon pusillus (De Not.) Lindb. Very rare on ash bark. 1. Hill Bottom, TL Blockeel in **BBSUK**, also BE.

Leskea polycarpa Hedw. Occasional on seasonally inundated bark or wood. 29. **DOR**.

Heterocladium heteropterum (Bruch ex Schwaegr.) Br. Eur. var. *flaccidum* Br. Eur. Scarce on shaded flints in woods, or by streams. 7. Monkton Wyld, MOH; Wynford Eagle, EWJ; Crook Hill, MOH; Kingcombe, BE; Hendover Copse, BE and !; Woolland Hill, BBS; Milton Park Wood, !; Bulbarrow, JA. **DOR**.

Anomodon viticulosus (Hedw.) Hook. & Tayl. Locally frequent in shelter, in calcareous turf or by rocks, around old trees and in hedgebanks. 31. **DOR**.

Thuidium abietinum (Hedw.) Br. Eur. subsp. *hystricosum* (Mitt.) Kindb. Scarce in old chalk grassland. 3. Flowers Barrow, !; Compton Abbas, MM; Zigzag Hill, MM, also BE and NAS. **DOR**.

Thuidium tamariscinum (Hedw.) Br. Eur. Abundant on humus in woods and plantations and occasional in sheltered turf. 35. **DOR**.

Thuidium delicatulum (Hedw.) Mitt. Very rare in calcareous turf. 2. Kingcombe, SME, det. MOH; Knowle Hill, G Bloom.

Thuidium philibertii Limpr. Rare in chalk turf. 3. Maiden Castle, JA and EF Warburg; Knowle Hill, FR. **DOR**.

Cratoneuron filicinum (Hedw.) Spruce Frequent in marshes, by calcareous springs and on wet undercliffs. 35. **DOR, OXF**.

Cratoneuron commutatum (Hedw.) Roth. Occasional in fens and marshes, sometimes forming tufa in wet woods. 12. **DOR**. Var. *falcatum* (Brid.) Monk. is scarce. 4. Champernhayes, GL; Winfrith Heath, FR; Corfe Common, MOH. **DOR**.

Campylium stellatum (Hedw.) J. Lange & C. Jens. Occasional in fens and marshes. 16. Var. *protensum* (Brid.) Bryhn is occasional in sheltered chalk grassland. 15. **BM, DOR, NY, RNG**.

Campylium chrysophyllum (Brid.) J. Lange. Occasional in sheltered chalk grassland, also in dune slacks. 19. **BM, OXF**.

Campylium polygamum (Br. Eur.) J. Lange & C. Jens. Rare in wet grassland, flushes or dune slacks. 3. Lulworth Cove, BBS; Worbarrow Bay, BE det. MOH; Studland, WR Sherrin, also EWJ in **WRHM**.

Campylium elodes (Lindb.) Kindb. Rare in marshes or flushed bogs. 2. Holwell, HHW; Winfrith Heath, MOH. **BM**.

(*Campylium calcareum* Crundw. & Nyh. Reports of this calcicolous species from Portland and Hambledon Hill need confirmation. It is quite likely to occur.)

Amblystegium serpens (Hedw.) Br. Eur. Abundant in woods, hedgebanks, sheltered grassland, on logs and under walls. All grid squares. **DOR**.

Amblystegium fluviatile (Hedw.) Br. Eur. Rare on rock or wood by rivers. Reported at Trill Bridge, Mr Southey, and Moors River, NT Holmes.

Amblystegium tenax (Hedw.) C. Jens. Uncommon on seasonally immersed rocks and tree bases by rivers. 11. **DOR**.

Amblystegium varium (Hedw.) Lindb. Uncommon on bark by streams. 6. Mappowder, JA; Shillingstone, JA, also MM.

Amblystegium riparium (Hedw.) Br. Eur. Frequent in fens, marshes or by rivers; often submerged. 34. **DOR**.

Drepanocladus aduncus (Hedw.) Warnst. Local in marshes, shallow ponds and dune slacks. 14. **DOR**.

Drepanocladus fluitans (Hedw.) Warnst. Occasional in shallow heath pools in the Poole basin. 8. **DOR, WRHM**.

Drepanocladus exannulatus (Br. Eur.) Warnst. Uncommon in marshes and bogs in the Poole basin, with outliers at Kingcombe, BBS, and Child Okeford, MM. 7. **DOR**.

Drepanocladus revolvens (Sw.) Warnst. Scarce in flushes in bogs. 5. Champernhayes, GL; Winfrith Heath, !, also FR, MOH; Corfe Common, JA, also VJG; Hartland Moor, PJ Newbould; Blue Pool, BE; Studland Heath, BE. **DOR**.

Drepanocladus cossonii (Schimp.) Loeske. Rare in a base-rich flush. 1. Corfe Common, BE det. MOH.

Sanionia uncinata (Hedw.) Loeske. Rare in marshes and bogs. 3. Glanvilles Wootton, HHW; Creech Heath, BBS; Corfe, EMH; Brownsea Is, FR. **DOR, OXF**.

Hygrohypnum luridum (Hedw.) Jenn. Scarce in or by flowing water, or in church gutters. 4. Bracketts Coppice, BBS; Hooke Park, J Gardiner; Glanvilles Wootton, CW Dale; Milton Abbas, RCS; Bryanston, Mr Southey.

Scorpidium scorpidioides (Hedw.) Limpr. Rare in flushed bogs or old claypits. 3. Champernhayes, GL; Winfrith Heath, BE; Blue Pool and Furzebrook, BE; Stoborough, VJG; Hartland Moor, SBC. **DOR**.

Calliergon stramineum (Brid.) Kindb. Scarce in marshes and flushed bogs. 7. Champernhayes, GL; Redholm coppice, BBS; Oakers bog, BBS; Hartland Moor, MOH; Holt Heath, HCP; Cranborne Common, MOH; Gotham and Alderholt Common, HCP. **DOR**.

Calliergon cordifolium (Hedw.) Kindb. Scarce in marshes, extending to carr. 10. **BBSUK, DOR, WRHM**.

Calliergon giganteum (Schimp.) Kindb. Rare in flushed marshes or carr in the Poole basin. 4. Moreton Heath, ECW; Corfe Common, JA; Holton Heath, VJG; Studland Heath, JAP. **DOR**.

Calliergon cuspidatum (Hedw.) Kindb. Abundant in marshes, wet grassland and sheltered calcareous turf. All grid squares. **DOR**.

Isothecium myurum Brid. Frequent at the bases of large trees, mostly oaks, in woods. 33. **DOR**.

Isothecium myosuroides Brid. Frequent to locally abundant in old woods. 35. **DOR**.

Isothecium striatulum (Spruce) Kindb. Local and rare on sheltered and shaded limestone or tufa in south Purbeck. 2. Langton West Wood and Talbots Wood, BE; Blashenwell, EMH, also ECW. **DOR, NMW**.

Scorpiurium circinatum (Brid.) Fleisch. & Loeske. Frequent among calcareous rocks near the coast in the south and west, especially on Portland. 22. **BM, DOR, NMW**.

Homalothecium sericeum (Hedw.) Br. Eur. Very common on bark, also on limestone or concrete walls, forming rings which decay in the centre. 38. **DOR**. A favourite substrate for the lichen *Bacidia sabuletorum*.

Homalothecium lutescens (Hedw.) Robins. Abundant in dry calcareous grassland or dunes, and on eutrophicated verges in heathland. 32. **BM, DOR, OXF**.

Brachythecium albicans (Hedw.) Br. Eur. Locally frequent to abundant in dry, sandy turf. 27. **DOR**.

Brachythecium glareosum (Spruce) Br. Eur. Occasional in calcareous soils, often on undercliffs. 9.

[*Brachythecium salebrosum* (Web. & Mohr) Br. Eur. Rare, overlooked or extinct in woods. 1. Bloxworth, HHW in **BM**.]

Brachythecium mildeanum (Schimp.) Milde. Scarce in marshes or dune slacks. 4. Kingcombe, BBS; Tadnoll, MOH; Stokeford Heath, BBS; Corfe, CH Binstead; Studland Heath, EWJ. **BBSUK**.

Brachythecium rutabulum (Hedw.) Br. Eur. Abundant on rich soil, stone and logs in woods, hedgebanks and grassland; variable. All grid squares. **DOR**.

Brachythecium rivulare Br. Eur. Widespread but local on rocks and trees by streams or in marshes. 26. **DOR, WRHM**.

Brachythecium velutinum (Hedw.) Br. Eur. Frequent on sheltered stones in woods, hedgebanks or gardens. 30. **DOR**.

Brachythecium populeum (Hedw.) Br. Eur. Uncommon on stone or bark in shelter of woodland, in nutrient-rich sites. 9. Wootton Fitzpaine and Monkton Wyld, MOH; Minterns Hill, MOH; Valley of Stones, BE; Eggardon Hill, BBS; Cerne Abbas, MOH; Piddles Wood, JA; Ringstead, BBS; Woolland Hill, MOH. **DOR**.

(*Brachythecium appleyardiae* McAdam & A.J.E. Smith is a very rare plant from a wall in ST 82 in VC 8.)

Pseudoscleropodium purum (Hedw.) Fleisch. Abundant in calcareous turf, also on heaths and other habitats. 37. **DOR**.

Scleropodium cespitans (C. Muell.) L. Koch. Occasional on the ground in woods and by streams, commonest in the north. 23. **DOR**.

Scleropodium tourettii (Brid.) L. Koch. Scarce in short turf on cliffs and dry banks. 6. Stonebarrow Hill, GL; Portland, BBS; West Weare, MOH; Grove and East Weare, BE; Black Down, ECW; Lulworth Cove, ECW, also BBS; Redhorn Quay, EWJ. **DOR, NMW, WRHM**.

Cirriphyllum piliferum (Hedw.) Grout. Widespread but local in calcareous woods and hedgebanks. 35. **DOR**.

Cirriphyllum crassinervium (Tayl.) Loeske & Fleisch. Widespread but local in woods or hedgebanks, rarely on roofs. 24. **DOR**.

Rhynchostegium riparioides (Hedw.) C. Jens. Locally dominant on stone or concrete of small waterfalls in calcareous streams. 37. **DOR**.

Rhynchostegium murale (Hedw.) Br. Eur. Occasional on sheltered calcareous rocks and walls. 20. **BM, DOR**.

Rhynchostegium confertum (Dicks.) Br. Eur. Common on sheltered rocks and walls, in woods and on eutrophic bark, especially elder. 35. **DOR**.

Rhynchostegium megapolitanum (Web. & Mohr) Br. Eur.

Scarce in dry soil, mostly near the coast. 5. Chesil Beach, ECW; Grove cliffs, MOH; Lulworth Cove, T Laflin; Hammoon, MM; Chapmans Pool, MFVC.

Eurhynchium striatum (Hedw.) Schimp. Frequent on the ground in woods and sheltered hedgebanks. 35. **DOR**.

Eurhynchium meridionale (Br. Eur.) De Not. Local on limestone rocks on Portland, for years the only British site for this Mediterranean moss. 4. First found by WPC Medlycott (1881 in **BM**) and later by many others. Portland Bill, HH Knight; West Weare, BBS; Tout Quarry, CJ Hadley; Church Ope Cove BBS, also BE; also on Bindon Hill, BE. **BBSUK, BM, CGE, E, NMW, RNG**.

Eurhynchium pumilum (Wils.) Schimp. Locally frequent on sheltered banks in woods or hedges on calcareous soil. 32. **DOR**.

Eurhynchium praelongum (Hedw.) Br. Eur. Abundant in sheltered, nutrient-rich sites; woods, hedgebanks, thin turf or on logs. All grid squares. **DOR**.

Eurhynchium swartzii (Turn.) Curn. Common in calcareous turf, hedgebanks and fallow, less tolerant of shade than *E. praelongum*. All grid squares. **DOR**.

Eurhynchium schleicheri (Hedw. f.) Milde. Scarce on leached soil of banks in woods or road cuttings. 5. Lyme Regis, AC Crundwell; Hooke Park, BBS; Trent, JA; Child Okeford, MM; Great Wood, Creech, ECW, also MOH.

Eurhynchium speciosum (Brid.) Jur. Rare or overlooked on wet soil near streams. 2. Eype undercliff, MOH; R. Stour near Hod Wood, MM.

Rhynchostegiella tenella (Dicks.) Limpr. Common by sheltered, calcareous rocks or walls, as in churchyards. 38. **DOR**.

Rhynchostegiella curviseta (Brid.) Limpr. Occasional on stone or bases of trees near streams. 19.

Rhynchostegiella teesdalei (Br. Eur.) Limpr. Occasional on shaded and sheltered stone near streams. 13.

Entodon concinnus (De Not.) Paris. Local in old chalk turf, mostly in the north-east. 10. Dorsetshire Gap, MM; West Creech Hill and Stonehill Down, BE; north of Blandford, FR; Shillingstone Hill, Hambledon Hill and Compton Abbas, MM; Clubmen's Down, Enford Bottom, Fontmell Down and Preston Hill, BE; West Hill, ECW; Knowle Hill, FR; Badbury Rings, T Laflin; Gussage Hill, MM; Sovell Down, RCS; Ackling Dyke and Oakley Down, BE; Bokerley Dyke, FR.

Plagiothecium latebricola Br. Eur. Occasional on decaying tussocks of *Carex paniculata*, or on logs in wet woods. 10.

Plagiothecium denticulatum (Hedw.) Br. Eur. Frequent on acid banks and on logs and stumps in shade and shelter. 26. **DOR**.

Plagiothecium curvifolium Schlieph. in Limpr. Scarce on logs and stumps in woods in the north-east. 8.

(*Plagiothecium laetum* Br. Eur. A plant with fusiform gemmae from Bere Wood (! det. CJ Hadley) may have been this, or perhaps the closely similar *P. curvifolium*.)

Plagiothecium succulentum (Wils.) Lindb. Occasional to frequent on sheltered, acid banks in woods, but absent from the south-west. 23. **DOR**.

Plagiothecium nemorale (Mitt.) Jaeg. Frequent on sheltered, acid banks in woods. 32. **DOR**.

Plagiothecium undulatum (Hedw.) Br. Eur. Locally frequent on sheltered, acid soil in woods. 22. **DOR**.

(*Isopterygium pulchellum* (Hedw.) Jaeg. was reported from Townsend Reserve, Swanage by CJ Hadley, probably in error.)

Isopterygium elegans (Brid.) Limpr. Locally frequent on hard, acid banks in shade and shelter. 29. **DOR, WRHM**.

Taxiphyllum wissgrillii (Garov.) Wijk. & Marg.Occasional on shaded, calcareous stones or banks in the shelter of woods, or at the base of trees. 14.

Hypnum cupressiforme Hedw. Abundant on rocks, walls, soil, bark and logs. All grid squares. **DOR**. Very variable. Var. *resupinatum* (Tayl.) Schimp. is not uncommon on bark, logs and walls. 16. **DOR**. Var. *lacunosum* Brid. is frequent on exposed rocks and walls, and in calcareous turf. 22. **DOR**.

Hypnum imponens Hedw. Rare on wet heaths. 2. Spur bog, Studland Heath, EWJ; West Moors depot, NAS. **WRHM**.

Hypnum mammillatum (Brid.) Loeske. Locally dominant on the boles of large trees in woods. 28. **DOR**.

Hypnum jutlandicum Holmen & Warncke.Locally abundant in dry heaths and open plantations on acid soils. 23. **DOR**.

Hypnum lindbergii Mitt. Uncommon by sandy tracks or ditches in woods. 9. **DOR**.

Ctenidium molluscum (Hedw.) Mitt. Locally abundant in calcareous grassland, and sometimes in woods or fens. 32. **BM, DOR**. Var. *sylvaticum* F. Rose is scarce on acid soil in conifer plantations.

Rhytidiadelphus triquetrus (Hedw.) Warnst. Locally frequent in sheltered calcareous turf, open woods or flinty slopes; least common in the coastal belt. 32. **DOR**.

Rhytidiadelphus squarrosus (Hedw.) Warnst. Locally abundant in mown turf, as in churchyards, also in woodland rides and heaths. 35. **DOR**.

Rhytidiadelphus loreus (Hedw.) Warnst. Local, and perhaps decreasing, on the ground in old woods, rarely in heathland as at Sandford Heath, BE. 19. **DOR**.

Pleurozium schreberi (Brid.) Mitt. Locally abundant in heaths and open conifer plantations on acid soils. 25. **DOR, OXF, WRHM**.

Hylocomium brevirostre (Brid.) Br. Eur. Uncommon in sheltered, calcareous woodland or grassland in Purbeck and the north. 14. **BM**.

Hylocomium splendens (Hedw.) Br. Eur. Local in heathland, also in sheltered, calcareous turf. 26. **DOR**.

CHAPTER 8

DORSET LICHENS

Lichens include about 13,500 species of fungi which form self-supporting associations with photosynthetic algae.

There are two previous lichen Floras of Dorset, one by myself in 1976 and an unpublished one by Holmes in 1906. Background data (given in Bowen, 1976) will not be repeated here, other than to confirm that Dorset suffers relatively little from air pollution by sulphur dioxide, but that intensely-farmed regions have a eutrophic flora which implies local pollution by ammonia. Since 1976 there has been an upsurge in recording activity, particularly by B Edwards, VJ Giavarini and PW James, as well as four new developments:

1. The publication of a new *Lichen Flora of Great Britain* (Purvis, Coppins, Hawksworth, James & Moore, 1992) has improved the standard of determining species, elevated some lower taxa and sunk others, as well as changing many names. The main source of nomenclature here is the Checklist by some of the authors of the Flora (Purvis, Coppins & James, 1994).
2. More intensive studies of old woods and parks has added new records, notably from the rich area of Melbury Park (ST50), and discovered previously unknown fragments of ancient woodland on the army ranges at Povington (SY88) and nearby at Creech Grange (SY98). The loss of virtually all large elms has caused a decline of many species, and the probable extinction of *Caloplaca luteoalba*.
3. Research on some of the calcareous grassland by OL Gilbert has added new taxa and improved our knowledge of this habitat (Gilbert, 1993).

4. Unpublished investigations of 182 of our ancient churchyards, at the instigation of TW Chester, has uncovered 13 churchyards with more than 100 species, including one with 178 species at Iwerne Courtney (ST81), which is of national importance. The net result has been to increase the number of taxa known from the county from 497 in 1976 to 652 today; only 16 of the latter have not been seen recently. No doubt more species will be discovered, and further records may be found in herbaria which have not been scanned for Dorset material. The county is therefore one of the richest in lowland Britain. It has more lichens than Hampshire, which has 590, mainly because it has more coastal habitats. It is certainly not as rich as Devon, which is much larger and has a wider range of altitude and rainfall.

In the ensuing list, taxa are given in alphabetical order by genus, grid references and dates are mostly omitted, and finders are distinguished by their initials; note that EMH and GL were working about a hundred years ago, and the Census Catalogue by WW was published in 1953. Habitats are summarised, and the number given under each entry is the total number of 10 km × 10 km Grid Squares from which a plant has been recorded. The maximum number of such squares is 39, if a few fragments are neglected, and localities are not usually given for plants recorded from more than eight squares. In view of the frequent taxonomic revisions, herbarium specimens are cited where known, but specimens of many rarities are in private collections.

LIST OF DORSET LICHEN SPECIES

Absconditella lignicola Vezda & Pisut. Rare or overlooked on *Calluna* stems, also on dead larch twigs. 3. Oak Hill, VJG in **E** det. BJC; Studland Heath, VJG det. BJC; Canford Heath and Talbot Heath, VJG.

Acarospora fuscata (Schrader) Th.Fr. Occasional on acid rock, sarsens or tombstones. 24.

Acarospora rufescens (Ach.) Krempelh. Rare on acid stone. 2. Child Okeford and Ibberton, PWJ in BM; Iwerne Minster, VJG.

Acarospora smaragdula (Wahlenb.) Massal. Scarce on granite or acid stone. 9.

Acrocordia conoidea (Fr.) Koerber. Occasional on hard limestone rock or walls, or on chalk pebbles 29. **BMH, DOR, RNG.**

Acrocordia gemmata (Ach.) Massal. Once frequent on elm, now less common on ash, oak or sycamore. 29. **BMH, DOR, RNG.**

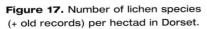

Figure 17. Number of lichen species (+ old records) per hectad in Dorset.

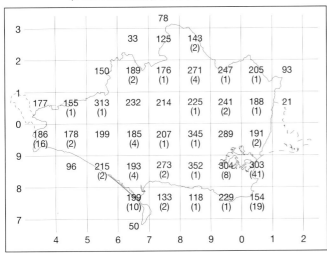

Acrocordia salweyi (Leighton ex Nyl.) A.L. Sm. Scarce on soft limestone or mortar near the coast. 10. **BM.**

Agonimia allobata (Stizenb.) P. James. A scarce old forest indicator on mossy bark. 8. Melbury Park, FR; Bracket's Coppice and Sydney Coppice, BE; Sherborne Park, VJG; Handcocks Bottom, FR; Hay Wood, BE; Bere Heath, BE: Morden Park, BE, VJG and NAS; Charborough Park and Kingston Lacy, BE; Langton West Wood, BE.

Agonimia octospora Coppins & P. James. A rare, old forest indicator on oak. 5. Melbury Park, FR; Oakers Wood, VJG; Woodbridge, BE; Whitehall, BE; Creech Grange, BE and VJG; Holt Forest, BE.

Agonimia tristicula (Nyl.) Zahlbr. Occasional on sheltered and shaded mossy limestone, or overgrowing gelatinous lichens, rarely on soil or bark. 24. **DOR, RNG.**

Amandinea lecideina (Mayrh. & Poelt) Scheidegger & Mayrh. Rare or overlooked on acid rock. 1. Agglestone, VJG.

Amandinea punctata (Hoffm.) Coppins & Scheidegger. Abundant on eutrophic bark and wood, rarely on flint. All grid squares. **BMH, DOR, RNG.**

Anaptychia ciliaris (L.) Koerber ex Massal. Once frequent on elm, now uncommon on large trees of ash, maple or sycamore, rarely on lime or oak. Once seen on limestone at Silton Ch. Often fertile. 30. **BMH, DOR, RNG.**

Anaptychia runcinata (With.) Laundon. On maritime boulders, acid rocks of stone circles, and rarely on ash or sycamore bark. 6. Valley of Stones, ! and BE; Portesham, !; Kingston Russell, !; Tyneham, PWJ; St Aldhelm's Head, BE and !; [Agglestone, EMH in **DOR**]. **RNG.**

Anisomeridium biforme (Borrer) R. Harris. Occasional on elder, elm, maple, sycamore or willow bark. 26. **BMH, DOR, RNG**.

Anisomeridium nyssaegeneum (Ellis & Everh.) R.C. Harris Local but overlooked on elder or eutrophic bark. 4. Charminster, VJG; White Nothe, !; Oakers Wood, VJG; Oakhills Coppice, VJG. **DOR.**

Arthonia anombrophila Coppins & P. James. Very rare on old oak bark. 1. Creech Grange, BE and VJG.

Arthonia astroidestra Nyl. A very rare old forest indicator. 1. Oakers Wood, DSR, also VJG in **DOR.**

Arthonia cinnabarina (DC.) Wallr. Occasional on oligotrophic bark of hazel, holly or oak in old woods. 20. **BMH, DOR, RNG.**

Arthonia didyma Koerber. Occasional on oligotrophic bark of hazel or oak in woods. 13. **BMH, DOR, RNG.**

Arthonia elegans (Ach.) Almq. Occasional on smooth bark of hazel or young oak in old woods. 17.

Arthonia endlicheri (Garov.) Oxmer. Rare on limestone. 2. East Portland undercliff, VJG in **DOR**; North of Blandford, PWJ.

Arthonia ilicina Taylor. Very rare on hazel bark. 1. East Burton, VJG in **DOR.**

Arthonia invadens Coppins on *Schismatomma quercicola*, Lulworth Park, BE.

Arthonia impolita (Hoffm.) Borrer. Occasional in dry bark crevices of large trees. 17. **DOR.**

Arthonia lapidicola (Taylor) Branth. & Rostrup. Local, on limestone stained with rust or lead salts, often on church windowsills. 22. **BM.**

Arthonia muscigena Th. Fr. Very rare, on elder bark. 1. Chapman's Pool, PWJ and VJG in **BM**.

Arthonia punctiformis Ach. Frequent on smooth oligotrophic bark of young trees or twigs. 16. **DOR, RNG.**

Arthonia radiata (Pers.) Ach. Abundant on smooth oligotrophic bark, especially on hazel in woods. 38 (not in SY66). **BMH, DOR, RNG.**

Arthonia spadicea Leighton. Occasional on shaded bark, often at the base of hazel or oak in woods. 23. **BMH, DOR, RNG.**

Arthonia vinosa Leighton. An old forest indicator, occasional on oak in old woods. 21. **BMH, DOR, RNG.**

Arthonia zwackhii Sandst. Very rare on bark. 1. Oakers Wood, DSR in **BM** and VJG in **E.**

Arthopyrenia antecellans (Nyl.) Arnold. Uncommon on beech, hazel or holly in old woods. 8. Hooke Park, FR and !; Melbury Park, FR; Bere Wood, BE; Oakers Wood, PWJ; Tyneham, FR; Rempstone, !; Delph Wood, VJG; Verwood, PWJ in **BM.**

Arthopyrenia lapponina Anzi. Occasional on smooth bark, especially hazel. 11. **DOR.**

Arthopyrenia platypyrenia (Nyl.) Arnold. Very rare on bark, a non-lichenized fungus. 1. Oakers Wood, ! in **E**, det. BJC.

Arthopyrenia punctiformis Massal. Occasional on smooth bark and twigs. 14. **BMH, DOR, RNG.**

Arthopyrenia ranunculospora Coppins & P. James. Local on smooth bark in old woods. 10.

Arthopyrenia viridescens Coppins. Rare, on hazel bark. 3. Tyneham Gwyle, Kings Wood and Great Wood, Creech, BE; Povington Wood, NAS; Studland, BE.

Aspilicia caesiocinerea (Nyl. ex Mahlbr.) Arnold. Uncommon on acid stone, sarsens or flint. 10. **DOR, NOT.**

Aspilicia calcarea (L.) Koerber. Abundant on limestone. All grid squares. **BMH, DOR, RNG.** The size of thalli on tombs at St Georges Ch, Portland, has been used to date rock exposures there. Many churches have thalli >200 mm diameter, and the largest noted is 437 mm on a 1670 tomb at Iwerne Courtney.

Aspilicia contorta (Hoffm.) Krempelh. Occasional on limestone, and rare on Chalk. 34. **DOR, RNG.**

Aspilicia subcircinata (Nyl.) Coppins. Scarce on limestone. 5. Chideock Ch, !; Milton Abbey, TWC and !; Iwerne Courtney Ch, TWC and VJG; West Orchard, on bridge !; [Corfe and Swanage, EMH in **DOR**].

Aspilicia sp. On Greensand wall. 1. Iwerne Courtney Ch, PWJ and VJG.

Bacidia absistens (Nyl.) Arnold. Very rare on bark of old trees. 1. Tyneham, SRD.

Bacidia arceutina (Ach.) Arnold. Scarce on smooth bark. 6. Oakers Wood, VJG; West Lulworth, !; Woolcombe, !; Hanford, !; Blandford, PWJ; Rempstone, !.

Bacidia arnoldiana Koerber. Rare on sheltered flint or limestone. 4. Tincleton, VJG; Tarrant Gunville, VJG; Corfe Mullen, VJG; Ballard Down, OLG.

Bacidia bagliettoana (Massal. & De Not.) Jatta. Uncommon on shallow calcareous soil. 5. Grove, BE; Durdle Door and Bindon Hill, !; Tyneham, FR; Fontmell Down, SRD; Ballard Down, EMH, also OLG. **DOR, NOT.**

Bacidia biatorina (Koerber) Vainio. Occasional on large ashes, elms or oaks in old woods. 15.

Bacidia caligans (Nyl.) A.L. Sm. Rare, on eutrophic elder bark. 1. Sydling St Nicholas, ! det BJC.

Bacidia delicata (Larbal. ex Leighton) Coppins. Scarce on eutrophic bark of elder, elm or maple. 5. On sheltered sarsen, Rye Water Farm, PWJ; Ringstead, BJC and FR; Hilton, VJG; Tollard Green, ! ; Studland wood, BE.

Bacidia friesiana (Hepp) Koerber. Rare on elder or elm bark. 3. Melbury Park; Ringstead, !; Brownsea Is, DSR and PWJ. **DOR.**

Bacidia herbarum (Stizenb.) Arnold. Rare on leached soil over Chalk. 2. Durdle Door, VJG; Bindon Hill, ! in **E**; Fontmell Down.

Bacidia incompta (Borrer ex Hooker) Anzi. Once occasional on elm, now rare on elder. 11. **RNG.**

Bacidia inundata (Fr.) Koerber. Rare or overlooked on rock in wet ravines. 1. Rye Water Farm, PWJ + BE, VJG and !.

Bacidia laurocerasi (Delise ex Duby) Zahlbr. Once occasional on elm, now on other eutrophic bark (ash, elder, sycamore; rarely on oak). 22. **DOR, RNG.**

Bacidia naegelii (Hepp) Zahlbr. Occasional on eutrophic bark of ash, elder, elm, maple, oak or sycamore. 13. **DOR, RNG.**

Bacidia phacodes Koerber. Mostly on shaded bark of ash and sycamore, once mainly on elm. 13. **DOR, RNG.**

Bacidia rubella (Hoffm.) Massal. Once frequent on elm, now less common on ash, maple or sycamore. 22. **DOR, RNG.**

Bacidia sabuletorum (Schreber) Lettau. Frequent on sheltered moss on limestone walls, or on shallow calcareous soil. 34. **BMH, DOR, RNG.**

Bacidia vezdae Coppins & P. James. Scarce on eutrophic bark of ash or willow. 7. Melbury Park, FR; Oakers Wood, VJG; Tyneham, FR; Handcocks Bottom, FR; Cranborne, VJG; Studland, VJG; Brownsea Is, PWJ. **DOR.**

Bacidia viridifarinosa Coppins & P. James. Rare on bark of old trees. 5. Ringstead, BJC and FR; Chapel Coppice, BE; Povington Wood, BE and FR; Bere Wood, BE; Langton West Wood and Linnings Copse, BE.

Bactrospora corticola (Fr.) Almq. Rare on dry oak bark in old woods. 3. Tyneham Gwyle, Woodstreet, Gough's Shoot and Froxen Copse, BE; Povington Wood, NAS.

Baeomyces roseus Pers. Uncommon on sheltered acid banks. 5. Puddletown Forest, !; Oakers Wood, VJG; Wool Heath, BE; Corfe Common, AEA Dunston; Hartland Moor, DSR; Brownsea Is, PWJ.

Baeomyces rufus (Huds.) Rebent. Occasional to frequent on dry, sandy or peaty banks, tolerant of some shade. 13. **DOR, RNG.**

Belonia nidarosiensis (Kindt.) P.M. Jorg. & Vezda. Locally dominant on sheltered, north-facing limestone rocks or church walls. 34.

(*Biatora cuprea* (Sommerf.) Fr. was reported in error from Abbotsbury (Holmes, 1906).)

Biatora epixanthoides (Nyl.) Diederich. A rare, sterile crust overgrowing mosses on bark. 1. Bracket's Coppice, VJG; Bakers Wood, BE.

Biatora sphaeroides (Dickson) Koerber. A scarce, old forest indicator, on sheltered ash, maple and oak bark. 9. **DOR.**

(*Biatora vernalis* (L.) Fr. has no certain record, though reported from the Melbury area, and occurring in SU10 in VC 11.)

Bryoria fuscescens (Gyelnik) Brodo & D. Hawksw. Rare on old oaks in woods. 4. Morden Decoy, DSR and FR, also on palings, BE; Earls Hill, Cranborne Chase, PWJ, probably in VC 8; [Bloxworth and Ballard Down, (Holmes, 1906)].

Buellia aethalea (Ach.) Th. Fr. Frequent on smooth, acid stone including sarsens, especially on granite tombs. 27. **DOR.**

Buellia disciformis (Fr.) Mudd. Uncommon on smooth bark of ash, oak or *Sorbus*. 5. Lyme Regis, GL; Monkton Wyld, !; Powerstock Common, !; Melcombe Park, !; Clenston, !; Fontmell Down, SRD. BMH, DOR, RNG.

Buellia griseovirens (Turner & Borrer ex Sm.) Almb. Occasional on the roots of old trees, or on wood. 13. **DOR.**

Buellia ocellata (Flotow) Koerber. Frequent on acid stone, sarsens, granite and flint. 26.

Buellia saxorum Massal. Rare on sarsens and puddingstones. 2. Valley of Stones, VJG; Portesham,! in **E**, also BE.

Buellia schaereri de Not. Scarce on pine bark or wood. 4. Melbury Park, VJG; Grange Heath, BE; Langton West Wood, !; Brownsea and Furzey Is, VJG.

Buellia stellulata (Taylor) Mudd. Uncommon on acid stone. 8. Bryanston, PWJ; Winspit, BJC; Cranborne Ch, VJG and !; Swanage, DL Hawksworth; Brownsea Is, PWJ; [Abbotsbury and Chesil Beach, (Watson, 1922)].

Buellia subdisciformis (Leighton) Jatta. Rare on Chert. 3. Portland, 1887, EMH; Church Ope Cove and East Weare, !; West Weare, BLS; Emmetts Hill, VJG. **DOR, NOT.**

Byssoloma subdiscordans (Nyl.) P. James. Very rare on elder bark. 1. White Nothe, VJG.

Calicium glaucellum Ach. Uncommon on dead wood. 10. **DOR, RNG.**

Calicium salicinum Pers. Occasional on dry, oligotrophic bark of ash, oak or yew, or on wood. 13. **DOR, RNG.**

Calicium viride Pers. Occasional to frequent in crevices of dry bark in woods, rarely on pine. 22. **RNG.**

Caloplaca alociza (Massal.) Migula. Rare, but locally abundant on exposed limestone. 2. Portland, VJG; Tilly Whim, !. **RNG.**

Caloplaca aurantia (Pers.) Steiner. Common on old limestone exposures and walls, especially near the coast. 35. **DOR, RNG.** A thallus 185 mm in diameter was seen at Dewlish Ch.

Caloplaca ceracea Laundon. Rare on Greensand. 1. Iwerne Courtney Ch, PWJ and VJG. Reports of this from stone at Pimperne Ch and Studland Ch, TWC need confirmation.

Caloplaca cerina (Ehrh. ex Hedw.) Th. Fr. Occasional on eutrophic bark of elder, elm, ash or sycamore, on *Suaeda vera* at Small Mouth (BE), rarely on wood, and on calcareous soil at the edges of cliffs at Durdle Door and Ballard Down, OLG. 20. **BMH, DOR, RNG.**

Caloplaca cerinella (Nyl.) Flagey. Rare on eutrophic bark of elm or elder. 2. Melbury Park, PWJ; Iwerne Courtney, PWJ. **BM.** A similar plant on elder bark at Abbotsbury and Tyneham Cap, BE, is probably an undescribed species in the *C. holocarpa* group.

Caloplaca chalybaea (Fr.) Muell. Arg. Rare on limestone. 4. Minterne Magna and Iwerne Minster, VJG; [Corfe and Swanage (Holmes, 1906)].

Caloplaca cirrochroa (Ach.) Th. Fr. Scarce on exposed limestone. 6. Portland Quarries (VJG, also BE and !); Dungy Head, BE, !; St Aldhelm's Head, BE; Corfe Castle ruins, !, *et al.*; Blackers Hole, BE; Durlston CP, VJG; Swanage Ch, TWC. **DOR.**

Caloplaca citrina (Hoffm.) Th. Fr. Abundant, but sterile, on sheltered calcareous rocks or mortared walls, sometimes extending over moss; also fertile in rain channels of eutrophic trees; BE noted an isidiate form on coastal limestone. All grid squares. **BMH, DOR, RNG.**

Caloplaca crenularia (With.) Laundon. Locally common on acid stone, especially on Greensand, also on sarsens and old flints, and extending to HWM on Portland. 16. **DOR, RNG.**

Caloplaca dalmatica (Massal.) H. Olivier. Common on dry, exposed, vertical limestone or chalk. 28. **BMH, DOR, RNG.**

Caloplaca decipiens (Arnold) Blomb. & Forss. Occasional on bird perching-rocks, eutrophic asbestos, concrete, or limestone, rarely on flint; mostly sterile, but fertile near Wool, !, and at Winspit, BE. 19. **DOR, RNG.**

[*Caloplaca ferruginea* (Huds.) Th. Fr. One record from beech bark at Uplyme in 1877 by GL, which may have been in VC 3.]

Caloplaca flavescens (Huds.) Laundon. Abundant on moderately eutrophic limestone, concrete, mortar and old brick; a depauperate form occurs on shallow calcareous soil near the coast, BE. Var. *brevilobata* (Nyl.) Wade was found at Gillingham. All grid squares. **BMH, DOR, RNG.**

Caloplaca flavorubescens (Huds.) Laundon. A nationally rare plant which is scarce on bark or wood. 5. Winterbourne Monkton, !; Dewlish Park, BE; Turners Piddle, !; Anderson, BJC and FR; Kingston, VJG; [Swanage and Studland (Holmes, 1906)].

Caloplaca flavovirescens (Wulfen) Dalla Torre & Sarnth. Occasional on sheltered limestone and vertical stone hedges, mostly near the coast. 12. **DOR.**

Caloplaca granulosa (Muell. Arg.) Jatta. Rare on sheltered limestone. 5. East Portland, VJG; Poyntington Ch, PWJ and VJG; Dungy Head, BE, !; Corfe Castle, VJG; St Aldhelm's Head, BE. **DOR.**

Caloplaca holocarpa (Hoffm.) Wade. A polymorphic species, common as an early colonist of limestone, and sometimes on bark or wood. All grid squares. **BMH, DOR, NOT, RNG.**

Caloplaca isidiigera Vezda. Scarce or overlooked on sheltered granite or acid stone. 5. Tincleton, VJG; Tarrant Gunville, !; Iwerne Courtney, Iwerne Minster and Okeford Fitzpaine, VJG.

Caloplaca lactea (Massal.) Zahlbr. Occasional on limestone, or on chalk pebbles. 13. **BMH, DOR, RNG.**

Caloplaca littorea Tavares. Rare on vertical faces of chert. 2. St Alhelm's Head, VJG, also BE and !; Ballard Head, !.

Caloplaca luteoalba (Turner) Th. Fr. Perhaps extinct; once rare on big elms at Herringston Park, Gillingham and Creech Grange, !. **DOR, RNG.**

Caloplaca marina (Wedd.) Zahlbr. ex Du Rietz. Locally frequent on boulders at HWM between Portland and Durlston Head, also on Brownsea Is. 9. **DOR, RNG.**

Caloplaca microthallina (Wedd.) Zahlbr. Rare on *Verrucaria maura* near HWM. 3. East Weares, !, also VJG; On rusty iron, Portland Breakwater, !; Fossil Forest, Gad Cliff and Emmetts Hill, BE. **E.**

Caloplaca obscurella (Lahm ex Koerber) Th. Fr. Scarce on eutrophic bark, usually elder. 7. Dottery, BE: Rampisham, !; Ringstead, !; Hog Hill, !; Binghams Melcombe, PWJ; Okeford Hill, !; Lytchett Matravers, BE; Tollard Green, !. **BM, BMH, DOR, RNG.**

Caloplaca ochracea (Schaerer) Flagey. Occasional on hard limestone, mainly coastal. 15. **BM, DOR.**

Caloplaca ruderum (Malbr.) Laundon. Rare on calcareous rock. 3. St Gabriels and Portland Bill, CJB Hitch; Piddlehinton, VJG..

Caloplaca saxicola (Hoffm.) Nordin. Occasional on exposed limestone, also on asbestos. 34. **BMH, DOR, RNG.** On eutrophic wooden bollards, Parkstone, VJG.

Caloplaca teicholyta (Ach.) Steiner. Common on limestone, asbestos and old brick, perhaps increasing. Mostly sterile, but fruiting at Tarrant Keyneston, VJG. 37. **DOR.**

Caloplaca thallincola (Wedd.) Du Rietz. Locally frequent on boulders near HWM between Portland and Swanage. 11. **BMH, DOR, RNG.**

Caloplaca ulcerosa Coppins & P. James. Local, but overlooked, on eutrophic bark of elder and big trees. 12. **DOR.**

Caloplaca variabilis (Pers.) Muell. Arg. Occasional on limestone. 14. **BM, DOR, RNG.**

Caloplaca verruculifera (Vainio) Zahlbr. Rare on rocks used as bird perches. 3. Portland, VJG; Between St Aldhelm's Head and Durlston Head, BE.

Caloplaca virescens (Sm.) Coppins. Very rare, at the base of an elm. 1. Fifehead Magdalen, ! det. BJC.

Candelaria concolor (Dickson) B. Stein. Frequent on eutrophic bark, rarely on eutrophic granite (at Anderson, !) or sarsens (Valley of Stones, BE); sometimes fertile. 32. **BMH, DOR, NOT, RNG.**

Candelariella aurella (Hoffm.) Zahlbr. Common on limestone, asbestos or concrete. 31. **BMH, DOR.**

Candelariella coralliza (Nyl.) Magn. Very local on sarsens and puddingstones. 1. Valley of Stones, !, also BE. **DOR.**

Candelariella medians (Nyl.) A.L. Sm. Common on bird perches, and on flat eutrophic limestone, mortar or asbestos, often in suburban sites. Usually sterile, but in fruit at Stourton Caundle, !. 31. **BMH, DOR, RNG.**

Candelariella reflexa (Nyl.) Lettau. Frequent on eutrophic bark, especially on elder. 27. **BMH, DOR, RNG.**

Candelariella vitellina (Hoffm.) Muell. Arg. Abundant on acid stone, brick or wood, occasional on cast iron. All grid squares. **BMH, DOR, RNG.**

Candelariella xanthostigma (Ach.) Lettau. Occasional on eutrophic bark of big trees in Parks. 15. **DOR.**

Catapyrenium cinereum (Pers.) Koerber. Rare on calcareous soil. 1. Corfe railway cutting, VJG.

Catapyrenium pilosellum Breuss. Rare on calcareous soil. 1. Ballard Down, OLG.

Catapyrenium psoromoides (Borr.) R. Sant. A very rare species. 1. On bark near Puddletown, BE.

Catapyrenium squamulosum (Ach.) Breuss. Occasional on shallow calcareous soil, especially on undercliffs and at Portland. 11. **BEL, BMH, DOR, RDG**.
Catillaria aphana (Nyl.) Coppins. Rare on chalk pebbles. 1. Ballard Down, OLG.
Catillaria atomarioides (Muell. Arg.) Kilias. Rare on flints. 2. Piddlehinton Ch, VJG; Godlingston Heath, VJG.
Catillaria atropurpurea (Schaerer) Th. Fr. A very local, old forest indicator, on old mossy oaks in woods or parks. 18. **DOR, RNG**.
Catillaria chalybeia (Borrer) Massal. Occasional on acid stone, especially on flint or chert, and on wooden posts in salt-marshes. 17. **DOR, RNG**.
Catillaria lenticularis (Ach.) Th. Fr. Occasional on limestone, rarely on old brick or weathered flint. 24. **DOR, RNG**.
Catillaria nigroclavata (Nyl.) Schuler. Very rare or extinct. 1. On elm bark, Winterbourne Came, !.
Catillaria pulverea (Borrer) Lettau. A rare, old forest indicator, on mossy trees. 4. On alder, Melbury Park, BE; Oakers Wood, VJG; Morden Decoy, FR; [Studland, EMH].
Celothelium ischnobelum (Nyl.) Aguirre. Rare on hazel bark in old woods. 2. Holt Forest, FR; Cranborne Chase, FR, perhaps in VC 8.
Cetraria chlorophylla (Willd.) Vainio. Occasional on oligotrophic branches in the canopy of big trees. 15. **BMH, DOR, RNG**.
Cetrelia olivetorum (Nyl.) Culb. & C. Culb. Very rare on bark. 1. On *Aesculus*, Melbury Park, FR, PWJ and SRD.
Chaenotheca brachypoda (Ach.) Tibell. Very rare on bark. 1. Melbury Park, FR, PWJ and SRD.
Chaenotheca brunneola (Ach.) Muell. Arg. Rare on dead wood of oak or pine. 10. Sadborough, ARV; Melbury Park, FR, PWJ and SRD; SherbornePark, VJG; Lulworth Park and Oakers Wood, VJG; Grange Heath and Whitehall, Steeple, BE; Morden Decoy, DSR; Oakhills Coppice, BE; Brownsea Is, VJG; Holt Forest, BE. **DOR.**
Chaenotheca chrysocephala (Turner ex Ach.) Th. Fr. Rare on dry conifer bark. 2. Sherborne Park, VJG; Wareham Forest, VJG.
Chaenotheca ferruginea (Turner ex Ach.) Mig. Local on dry, oligotrophic bark or wood of birch, maple, oak or pine in woods; often fertile, especially on pine wood or old palings. 15. **DOR, RNG.**
Chaenotheca furfuracea (L.) Tibell. Rare on bark. 2. Melbury Park, PWJ and !; Creech Grange, BE and VJG.
Chaenotheca hispidula (Ach.) Zahlbr. Scarce on dry, acid bark. 4. Melbury Park, BE; Oakers Wood, VJG; Henbury, BJC: Creech Grange, BE and VJG.
Chaenotheca stemonea (Ach.) Muell. Arg. Very rare on shaded bark. 1. Creech Grange, BE and VJG.
Chaenotheca trichialis (Ach.) Th. Fr. Rare on old oak bark. 2. Melbury Park and Chetnole Withybed, BE; The Oaks, Badbury, NAS.
Chaenothecopsis nigra Tibell. Rare. On stump, Sherborne Park, VJG in **E**; Creech Grange, BE and VJG. An as yet unidentified species occurred on algae on a stump in High Wood, Badbury, BE.

(*Chiodecton myrticola* Fee has been reported in error.)
Chromatochlamys muscorum (Fr.) Mayrh. & Poelt. Rare in the rain channel of eutrophic trees. 4. Forde Abbey, !; Melbury Park, FR,PWJ and SRD; Sherborne Park, VJG; in lichen heath on leached chalk soil, Bindon Hill, ! det. BJC.
Chrysothrix candelaris (L.) Laundon. Common in dry bark crevices of old trees, usually in woods, always sterile. All grid squares. **BMH, DOR, RNG**.
Chrysothrix chrysophthalma (P. James) P. James & Laundon (the segregate *C. flavovirens*, Tonsberg or pine). Occasional on dry oak bark or wood in old woods or parks. 9. **BMH, DOR, RNG**.
Cladonia arbuscula (Wallr.) Flotow. Locally frequent at the edges of bogs or on wet heaths in the Poole basin. 8. Extinct in the west, and perhaps decreasing. **BMH, DOR, NOT, RNG**.
Cladonia caespiticia (Pers.) Floerke. Scarce on mossy trees, shaded peat or wood. 7. Lamberts Castle, ARV; Hole Common and Wyld Warren, BE; Powerstock Common, BE; Oakers Wood, PWJ; Bere Wood and Bloxworth, BE; Grange Heath, BE; Great Ovens Hill, !; Holton Heath, BE; Brownsea Is, PWJ. **BM, DOR**.
Cladonia cariosa (Ach.) Sprengel
Rare on sandy soil. 2. Warmwell Heath, DJ Hill; Furzebrook, OLG and VJG.
Cladonia cervicornis (Ach.) Flotow
Uncommon on heaths and dunes in the Poole basin. 8. **BMH, DOR, RNG**. Subsp. *verticillata* (Hoffm.) Ahti is occasional in dry or moist acid soil, and extends to Hardown Hill in the west. 11. **BMH, DOR, NOT, RDG**.
Cladonia chlorophaea (Floerke ex Sommerf.) Sprengel. Very common on heaths and acid banks, also on ash beds, mossy walls and old wood. 33. **BMH, DOR**.
Cladonia ciliata Stirton var. *tenuis* (Floerke) Ahti. Locally frequent on heaths, dunes, old ash beds, or rarely on wood. 15. Mostly in the Poole basin, with outliers at Lamberts Castle, BE, Bindon Hill, !, and Fontmell Down, SRD. **BMH, DOR, RNG**.
Cladonia coccifera agg. (all or mostly *C. diversa* Asperges) Frequent on peaty soils and heaths in the west and in the Poole basin. 15. **DOR, RNG**.
Cladonia coniocraea (Floerke) Sprengel. Abundant on sheltered oligotrophic bark and wood, and around tree bases in woods. Sometimes on dunes, mossy walls, or among sarsen boulders. 34. **BMH, DOR**.
[*Cladonia convoluta* (Lam.) Cout. On soil, Ballard Down, EMH (Holmes, 1902).]
Cladonia crispata (Delise) Vainio var. *cetrariiformis*. Locally frequent on peaty heaths in the Poole basin, also at Hardown Hill. 12. **BMH, DOR, RNG**.
Cladonia digitata (L.) Hoffm. Occasional on oligotrophic bark of old trees, wood or peat banks. 12. **BMH, RNG**.
Cladonia fimbriata (L.) Fr. Common on wood or thatch, on fixed dunes, in lichen heaths over Chalk and among sarsen boulders. 32. **BMH, DOR, RNG**.
(*Cladonia firma* (Nyl.) Nyl. was reported from Agglestone, EMH in **NOT**, but lacks confirmation.)

307

Cladonia floerkeana (Fr.) Floerke. Common on peaty heaths, occasional on wood or old tile, and among sarsen boulders. 19. **BMH, DOR, RNG**.

Cladonia foliacea (Huds.) Willd. Very local on soil near the coast. 7. East Portland, !, also BE; Arish Mell and Bindon Hill, BE; West Lulworth, H Livens; Wealden cliffs, Worbarrow Bay, !; Hartland Moor, !; Studland Bay dunes, EMH, DSR, FR *et al.*; Durlston Head and Ballard Down, BE. **BEL, BMH, DOR, NOT, RNG**. This may be the *Lichen foliaceus* reported from Maiden Castle by R Pulteney in the 18th century.

Cladonia furcata (Huds.) Schrader. Locally frequent on heaths, sometimes on wood or old rail tracks. 20. **BMH, DOR, RNG**.

Cladonia glauca Floerke. Scarce on young heaths. 4. East of Red Bridge, Moreton, !; Bere Heath and Oak Hill, !; Furzebrook, VJG. **BMH, DOR, RNG**.

Cladonia gracilis (L.) Willd. Occasional on heaths in the Poole basin. 7. Black Down, !; Warmwell Heath, DJ Hill; Winfrith Heath, !; Throop Clump, !; Lower Hyde Heath, BE; Slepe Heath, VJG; Morden Decoy, DSR, !; Studland, EMH, also DSR; Brownsea Is, BE. **BMH, DOR, NOT, RNG, WRHM**.

Cladonia humilis (With.) Laundon. Locally frequent on heaths, sandy soils or dunes in the Poole basin, rare in the west (Marshwood, ARV). 8. **BMH, DOR, RNG**.

Cladonia incrassata Floerke. Local on sheltered peat or decaying *Molinia* tussocks in the Poole basin and the west. 6. Wyld Warren, BE; Woolsbarrow, BE; Rempstone and Newton Gully, BE; Hartland Moor and East Creech, DSR; Morden Decoy, !; Studland, DSR; Brownsea Is, PWJ. **BM, BMH, DOR, RNG**.

Cladonia macilenta Hoffm. Frequent on heaths, also on dunes and rarely on rotting wood. A form with squamular undersides yellow occurred at Sherborne Park. 20. **BMH, DOR, RNG**.

(*Cladonia mitis* Sandst. One doubtful report, from Hartland Moor, DSR; it might occur on shingle.)

Cladonia ochrochlora Floerke. Occasional on stumps, mostly in the west. 9. **DOR**.

Cladonia parasitica (Hoffm.) Hoffm. Local on old wood in parks or woods. 12. Powerstock Common, BE; Melbury Park, PWJ *et al.*; Sherborne Park, !; Lulworth Park, FR; Wool, BE; Oakers Wood, DSR, also VJG; Stock Gaylard Park, BE; Corfe, FR; Morden Decoy, FR and SRD; Morden Park, BE and !; Handcocks Bottom, VJG; Brownsea Is, BE. **DOR, RNG**.

(*Cladonia peziziformis* (With.) Laundon. The report from Lamberts Castle by ARV was not this, fide BJC.)

Cladonia pocillum (Ach.) Grognot. Frequent on shallow, calcareous soils, especially on Portland, and on mossy walls. 21. **BMH, DOR, RNG**.

Cladonia polydactyla (Floerke) Sprengel. Occasional on peat, wood or old thatch. 14. **BMH, DOR, RNG**.

Cladonia portentosa (Dufour) Coem. Locally dominant on heaths or young pine plantations in the west and the Poole basin. 24. Rarely in other habitats, e.g. on dunes at Studland, CCT; on Eocene sandstone, Canford, !; on bark, Holt, FR; on wood, Droop, ! and Canford Heath, VJG; in lichen heath over Chalk, Bindon Hill, !; fertile at Morden, DSR. **BMH, DOR, RNG**.

Cladonia pyxidata (L.) Hoffm. Occasional on dry soil, brick walls, tile or sandstone. 26. **BMH, DOR, NOT, RNG**.

Cladonia ramulosa (With.) Laundon. Occasional on sandy or peaty soil in the west or the Poole basin, rarely on bark. 14. **BMH, DOR, NOT, RNG**.

Cladonia rangiformis Hoffm. Locally frequent in short, calcareous turf, especially near the coast. 25. **BMH, DOR, NOT, RNG**.

Cladonia scabriuscula (Delise) Nyl. Uncommon in acid soils or on dunes. 5. Valley of Stones, !; Furzebrook, VJG; Canford Ch, !; Studland Bay, KL Alvin; St Leonards, BE.

Cladonia squamosa Hoffm. Frequent in woods, on peat or wood, among sarsens, on dunes and in calcareous lichen heaths. 22. Var. *subsquamosa* (Nyl. and Leighton) Vainio is less common than the type, sometimes on wet peat. **BMH, DOR, RNG**.

Cladonia strepsilis (Ach.) Grognot. Local on wet peat in heaths of the Poole basin. 9. **BMH, DOR, RNG**.

Cladonia subcervicornis (Vainio) Kernst. Scarce on peaty heaths. 2. Stoke and Warren Heaths, !. **BMH, DOR, RNG**.

Cladonia subrangiformis Sandst. Scarce in old calcareous turf. 7. Tout quarry, Portland, !; between Grove and Verne, BE; Bindon Hill, !; Fontmell Down, SRD; Little Down, Mr Warren; Breeze Hill, RM; Worth Matravers, CCT; Studland. **BMH, DOR, RNG**.

Cladonia subulata (L.) Weber ex Wigg. Occasional on sandy banks and heaths, less common on wood. 12. **BMH, DOR, RNG**.

Cladonia uncialis (L.) Weber ex Wigg. subsp. *biuncialis* (Hoffm.) Choisy. Locally frequent on wet or dry, bare places in heaths of the Poole basin. 11. **BMH, DOR, NOT, RNG**.

Clauzadea immersa (Hoffm.) Hafellner & Bellem. Uncommon on limestone or Chalk. 6. Gillingham; Middle Bottom and Bindon Hill, BE; Durdle Door, OLG; Winspit, BJC and FR; Ballard Head, OLG; near Poole, VJG.

Clauzadea metzleri (Koerber) Clauz. & Roux ex D Hawksw. Scarce on limestone or chalk pebbles. 4. Durdle Door, !, also OLG; Winspit, !; Durlston Head, VJG; Ballard Head, OLG. **DOR, RNG**.

Clauzadea monticola (Ach.) Hafellner & Bellem. Common on dry limestone or concrete. 33. **BMH, DOR, RNG**.

Cliostomum corrugatum (Ach.) Fr. A northern species, perhaps now extinct, on birch bark on Brownsea Is, DSR and PWJ.

Cliostomum griffithii (Sm.) Coppins. Abundant on dry, smooth bark, rarely on wood. 37. **BMH, DOR, RNG**.

Coelocaulon aculeatum (Schreber) Link. Locally abundant on nutrient-poor, acid soils, dunes or ball-clay waste, but damaged by trampling. 8. Seen in fruit at Morden Decoy and Studland. **BMH, DOR, NOT, RNG**.

Coelocaulon muricatum (Ach.) Laundon. In similar habitats to *C. aculeatum*, but less common. 5. Abbotsbury Castle, NCC; Warren Heath, !; Furzebrook, VJG; Morden Decoy, FR and SRD; Studland, EMH. **DOR, NOT**.

Collema auriforme (With.) Coppins & Laundon. Frequent

on horizontal limestone, such as chest tombs, also on chalk pebbles and asbestos roofs where rain lies. 36. **BMH, DOR, RNG**.

(*Collema bachmannianum* (Fink) Degel. was reported from Portland by VJG, but in error.)

Collema crispum (Huds.) Weber ex Wigg. Common on limestone and mortar, and seen on bark at Portland. 35. **BMH, DOR, RNG**.

Collema cristatum (L.) Weber ex Wigg. Scarce on limestone. 4. Beaminster, W Watson; Buckland Newton Ch, !; Corfe Castle, PWJ; Swanage, EMH. **DOR**.

[*Collema flaccidum* (Ach.) Ach. Among moss at Swanage, EMH.]

Collema fragile Taylor. A rare species, found on limestone on the undercliff, east Portland, VJG in **BM**.

Collema fragrans (Sm.) Ach. A rare plant of eutrophic bark, once on elm, now mainly on ash. 7. Marshwood, !; Forde Abbey, !; Melbury Park, !, also VJG; Dorchester, EMH; Came Park, !; Lulworth Park, !; Tyneham, SRD; Handcocks Bottom, VJG; Great Wood, Creech, BE and VJG. **BM, NOT**.

Collema furfuraceum (Arnold) Du Rietz. Scarce on eutrophic bark of ash or sycamore, once on elm. 7. **DOR, RNG**.

Collema fuscovirens (With.) Laundon. Occasional, or overlooked as *C. auriforme*, on horizontal limestone. 8.

Collema limosum (Ach.) Ach. Local on clay undercliffs. 2. Eype, !; Fortuneswell, DSR.

Collema nigrescens (Huds.) DC. Rare and decreasing on bark since the loss of elms. 4. Lyme Regis, GL, also ARV; Charmouth, EMH; Tyneham, SRD; Encombe, Mr Castell, also South Gwyle, BE; Chapmans Pool, !; Studland, EMH. **BM, DOR, E**.

Collema occultatum Bagl. Rare, on elm bark 1. Tyneham, FR.

Collema polycarpon Hoffm. Rare on limestone. 2. Melbury Park, PWJ and !; Tilly Whim, EMH in **NOT**.

Collema subflaccidum Degel. Scarce on ash or oak bark in woods, parks and waysides. 12. **DOR, RNG**.

Collema tenax (Swartz) Ach. Common on shallow calcareous soil and on old walls. 33. **BMH, DOR, RNG**. Var. *ceranoides* (Borrer) Degel. is somewhat less common. 25. **BMH, DOR, NOT, RNG**.

Cryptolechia carneolutea (Turner) Massal. Once scarce on elm bark, now rare on ash, oak or sycamore. 9. Melbury Park, PWJ and !; Church Ope Cove, HH Knight, also ! and VJG; Ringstead, !, also BJC and FR; Tyneham Gwyle, FR, also BE; Lulworth Park and Povington, BE; Chapmans Pool, VJG; Afflington Wood and Little Wood, Creech, BE; Kingston Lacy, BE; Studland, EMH, also UH Duncan. **BM, DOR, E, NOT, RNG, SLBI**.

Cyphelium inquinans (Sm.) Trevisan. Occasional on wood and the tops of old fence posts. 16. **DOR, RNG**.

Cyphelium sessile (Pers.) Trevisan. Occasional on *Pertusaria*, on oak or ivy bark. 13. **DOR, RNG**.

[*Degelia atlantica* (Degel.) P. Jorg. & P. James was on bark on Lyme Regis undercliff, GL, but probably in VC 3.]

(*Degelia plumbea* (Lightf.) P. Jorg. & P. James was reported from Bridport in error.)

Dermatocarpon miniatum (L.) Mann. Rare on horizontal limestone or conglomerate outcrops, or on limestone roofs.

5. Waddon Hill, BE; Abbotsbury Plains, !; Eggardon Hill, BE; West Weare, EMH, also TW Ottley; Milton Abbey, TWC and !. **DOR**.

[*Dictyonema interruptum* (Carm. ex Hook.) Parmasto. Very rare, on sheltered trees near Wareham, (Wright, 1890).]

Dimerella lutea (Dickson) Trevisan. Local in the rain channel of big trees, mostly ash, oak or sycamore. 19. **BMH, DOR, RNG**.

Dimerella pineti (Ach.) Vezda. Occasional on bark in deep shade, especially on *Salix*, also on ash, elm and oak. 19. **BMH, DOR, RNG**.

Diploicia canescens (Dickson) Massal. Abundant on dry limestone, and locally dominant on eutrophic bark, but uncommon in fruit; rarely on lead and glass of church windows. Thalli attain 220 mm in diameter at Kingston Lacy. All grid squares. **BMH, DOR, NOT, RNG**.

Diploschistes caesioplumbeus (Nyl.) Vainio. Rare on chert. 2. East Portland undercliff, VJG; Kingston Lacy, !.

[*Diploschistes gypsaceus* (Ach.) Zahlbr. On chalk, Ballard Down, (Holmes, 1906).]

Diploschistes muscorum (Scop.) R. Sant. Occasional on moss on old walls, also on *Cladonia* on oak at Melbury Park, BE; on soil, Grove, BE; on wooden rails, Affpuddle, VJG. 12. **BMH, DOR, RNG**.

Diploschistes scruposus (Schreber) Norman. Local on acid rocks, limestone, mortar, and old walls of brick or tile. 17. Thalli to 200 mm diameter at Wool Ch, !. **BMH, DOR, RNG**.

Diplotomma alboatrum (Hoffm.) Flotow. Common on dry, vertical limestone, also on dry eutrophic bark. 36. **BMH, DOR, RNG**.

Diplotomma epipolium (Ach.) Arnold. Rare or overlooked on calcareous stone. 5. West Weare, BE; on chalk pebbles, Durdle Door and Ballard Down, OLG; Emmetts Hill and Durlston Head, BE.

Dirina massiliensis Durieu & Mont. Very local on sheltered, vertical limestone near the sea. 3. East and West Weare, Church Ope Cove and Portland quarries, EMH, also BE, PWJ and VJG and !; Corfe Castle, EMH; St Aldhelm's Head, BE and !. **BM, DOR, NOT, RNG**. F. *sorediata* (Muell. Arg.) Tehler is locally dominant on the north side of limestone churches. 34. **BMH, DOR, RNG**.

Endocarpon sp. On Greensand wall, Iwerne Courtney Ch, VJG; the first British record of a species as yet undescribed.

Enterographa crassa (DC.) Fee. Common on smooth bark of old trees, often in deep shade. 36. BMH, DOR, RNG.

Enterographa sorediata Coppins & P. James. Rare on oak bark. 2. Melbury Park, BE, FR and NAS; Creech Grange, BE and VJG.

Eopyrenula grandicula Coppins. Overlooked on hazel, rarely on ash or oak bark. 6. Chapel Coppice, BE; Melbury Park, BE and FR; Oakers Wood, !; Hyde Ho, BE; Great Wood, Creech, BE; Langton West Wood, BE; Studland Wood, BE. **E**.

Evernia prunastri (L.) Ach. Common on exposed bark, branches and twigs; occasional on Greensand walls, acid standing-stones, lichen heaths on sand at Hartland Moor and the Wealden undercliff at Worbarrow Bay, !. Seen in fruit at Sadborow, ARV and Chetterwood, !. All grid squares but SY66. **BM, BMH, DOR, RNG**.

Fellhanera bouteillei (Desm.) Vezda. Rare on box twigs or leaves. 4. Sherborne Park, VJG; Milton Abbas, DSR; Bryanston, PWJ; Canford School, !. **DOR.**

Fuscidea cyathoides (Ach.) V. Wirth & Vezda. Scarce on flint or chert near the sea. 3. East Portland, VJG; Durdle Door and Ballard Down, OLG and VJG.

Fuscidea lightfootii (Sm.) Coppins & P. James. Widespread on smooth, oligotrophic twigs or bark of birch, hazel, oak or willow; often fertile. 15. **BMH, DOR, RNG.**

Graphina anguina (Mont.) Muell. Arg. Scarce or overlooked on smooth bark of hazel or holly. 10. **DOR, RNG.**

Graphis elegans (Borrer ex Sm.) Ach. Frequent on smooth, oligotrophic bark, especially hazel, but also beech, oak, pine and sycamore. 33. **BMH, DOR, RNG.**

Graphis scripta (L.) Ach. Frequent on smooth or rough, oligotrophic bark in woods, mostly on hazel but also on ash, beech, cherry, holly and oak. 34. **BMH, DOR, RNG.**

Gyalecta derivata (Nyl.) H. Olivier. Scarce on eutrophic bark of ash or elm. 7. Ashley Chase, !; Melbury Park, !; Pennsylvania Castle, Portland, !; Langton West wood, BE; Studland, EMH in **DOR**; Charborough Park, BE; Sutton Holms, NAS.

Gyalecta flotowii Koerber. Rare on eutrophic bark. 4. Melbury Park, FR; Portland, VJG; Studland, EMH; Cranborne, VJG.

Gyalecta jenensis (Batsch) Zahlbr. Occasional on rough, sheltered limestone, often in shade. 15. **DOR, RNG.**

Gyalecta truncigena (Ach.) Hepp. Once frequent on elm, now uncommon in rain channels of ash or maple. 23. **DOR, RNG.**

Gyalideopsis anastomosans P. James & Vezda. Scarce on smooth bark or wood. 8. Powerstock Common, BE and PWJ; Oakers Wood, VJG; Lulworth Park, VJG; Oakhills Coppice, VJG; Bryanston, PWJ; Iwerne Courtney, VJG; Delph Wood, VJG; Brownsea Is, VJG. **DOR.**

Haematomma ochroleucum (Necker) Laundon var. *porphyrium* (Pers.) Laundon. Locally frequent on sheltered acid stone or Greensand, fruiting at Stock Gaylard, Morden and Chettle; occasional, on oligotrophic bark in woods, fruiting in Povington Wood, BE. 15. **BMH, DOR, RNG.** Var. *ochroleucum* has only been seen at Iwerne Courtney Ch, VJG.

[*Heterodermia leucomelos* (L.) Poelt once occurred on cliffs, last seen in 1915. 3. Portland, R Paulson; south of Swanage, EMH; Ballard Down, W Johnson, also EMH. **BM, DOR, E, SLBI.**]

Heterodermia obscurata (Nyl.) Trevisan. Very rare on oak bark. 2. Melbury Park, OLG, also BE; [Shaftesbury, (Holmes, 1906)].

Hymenelia prevostii (Duby) Krempelh. Rare on calcareous stone or pebbles. 5. Portland, DSR; Durdle Door, OLG; Coombe Keynes Ch, !; Sandford Orcas Ch, VJG; Iwerne Courtney Ch, VJG; Edmondsham Ch, VJG.

Hyperphyscia adglutinata (Floerke) Mayrh. & Poelt. Frequent on eutrophic bark of elm or sycamore, occasional on eutrophic limestone or sarsens. 30. **BMH, DOR, RNG.**

Hypocenomyce caradocensis (Leighton ex Nyl.) P. James & G. Schneider. Rare on bark or palings. 2. Morden Bog, VJG; Holt Forest, !. **DOR.**

Hypocenomyce scalaris (Ach. ex Lilj.) M. Choisy.

Occasional on oligotrophic bark, especially pine, or on old wood; rarely on sandstone or Greensand. 20. **BMH, DOR, RNG.**

Hypogymnia physodes (L.) Nyl. Common on oligotrophic bark, twigs or wood, occasional on *Calluna* or *Ammophila*, and less common on sandstone or sandy cliffs. All grid squares but SY66. Rarely fertile, as at Sherborne Park, Norden, Shaftesbury and Arne. **BMH, DOR, RNG.**

Hypogymnia tubulosa (Schaerer) Havaas. Frequent on oligotrophic bark, usually on branches or twigs; rarely on peaty soil, sarsens or old tile. 31. **BMH, DOR, RNG.**

Icmadophila ericetorum (L.) Zahlbr. Rare on wet heaths in the Poole basin, mostly fertile. 3. Winfrith Heath, !; Hartland Moor, DSR; Coombe Heath, BE; Canford Heath, VJG.

Imshaugia aleurites (Ach.) S. Meyer. Uncommon on exposed pine bark or palings. 6. On oak, Melbury Park, FR, PWJ and SRD; Bovington, BE; Black Heath, Bere, and Wareham Forest, BE; Sandford, NAS; Morden decoy, DSR, also !; Morden Park, BE, also VJG; Stoney Down, BE; Brownsea Is, BE. **DOR.**

Japewia carrollii (Coppins & P. James) Tonsb. Rare on sallow bark in carr. 2. Powerstock Common, BE and PWJ; Oakers Wood, PWJ in **BM.**

Lecanactis abietina (Ach.) Koerber. Locally dominant on dry bark in woods, mainly oak, but also alder and maple, occasionally fertile 23. **BMH, DOR, RNG.**

Lecanactis amylacea (Ehrh. ex Pers.) Arnold. Rare on dry oak bark, sterile. 3. Melbury Park, FR, PWJ and SRD; Sherborne Park, VJG; Creech Grange, BE and VJG. Rare in Britain.

Lecanactis grumulosa (Dufour) Fr. Rare on dry, shaded coastal limestone. 3. Portland, EMH, also BE and VJG; Gad Cliff, BE; Corfe Castle, EMH. **BM, DOR, NOT.**

Lecanactis hemisphaerica Laundon. Very rare on sheltered rendered walls. 2. Winterbourne Came, VJG; Affpuddle Ch, PWJ and VJG in **BM**, also !.

Lecanactis lyncea (Sm.) Fr. Scarce on dry underhangs of old oak boughs. 14. **DOR, RNG.**

Lecanactis premnea (Ach.) Arnold. Occasional on dry bark of large, ancient oaks in woods and parks, once seen on sycamore and ivy. 19. **BMH, DOR, RNG.**

Lecanactis subabietina Coppins & P. James. Scarce on dry, shaded oak bark in old woods, especially in the south-east. 11. **DOR.**

Lecania aipospila (Wahlenb.) Th. Fr. Rare on limestone above HWM. 1. Kimmeridge, TW Ottley; Winspit, PWJ.

Lecania chlorotiza (Nyl.) P. James. Very rare on shaded bark. 1. Woolcombe, ! in **E**, det. BJC.

Lecania cuprea (Massal.) v.d.Boom & Coppins. Very rare under limestone overhang. 1. Bowers quarry, Portland, ! det. BJC.

Lecania cyrtella (Ach.) Th. Fr. Frequent on smooth eutrophic bark of ash, elder or elm. 22. **BMH, DOR, NOT, RNG.**

Lecania cyrtellina (Nyl.) Sandst. Rare on ash bark. 2. Minterne Magna, BE; Cranborne Chase, both in VC 8, and in Handcocks Bottom, BE.

Lecania erysibe (Ach.) Mudd. Frequent and variable, on limestone, mortar, old flints, cliff-top calcareous soil and rarely on wood. 22. **BMH, DOR, RNG**. f. *sorediata* Laundon was seen on Portland, TWC and !. A fertile form on shaded rock at Rye Water Farm, PWJ and !.

(*Lecania fuscella* (Schaerer) Koerber. Reports of *Lecanora syringea* Ach. (Holmes, 1906) might have been this, but it is now extinct in Britain.)

Lecania hutchinsiae (Nyl.) A.L. Sm. Rare on sheltered acid stone. 6. Tout Quarry, Portland, VJG; Cheselbourne Ch, !; Sturminster Marshall, Okeford Fitzpaine, Iwerne Courtney, Fontmell Magna, Tarrant Gunville and Edmondsham, VJG.

Lecania rabenhorstii (Hepp) Arnold. Rare on limestone. 1+. Piddlehinton Ch, PWJ and VJG; probably this at Portesham Ch and Studland Ch, TWC.

Lecania turicensis (Hepp) Muell. Arg. Rare on limestone. 2. Okeford Fitzpaine Ch, TWC and VJG; Winspit, PWJ.

Lecanora actophila Wedd. Rare on stone or wood near the sea. 4. On chert, east Portland, VJG; St Aldhelm's Head, BE and !; on chalk pebbles, Ballard Down, OLG; on wooden slipway, Poole Harbour, VJG.

Lecanora aghardiana Ach. Rare on limestone near HWM. 2. East Weare, ! det. A Fletcher; Fossil Forest, W Watson in **BM**.

Lecanora aitema (Ach.) Hepp. Local or overlooked, on young pine bark. 1. Woolsbarrow, BE, PWJ and VJG.

Lecanora albescens (Hoffm.) Branth & Rostrup. Frequent on limestone. 36. **BEL, DOR**.

Lecanora andrewii B. de Lesd. Very rare on chert. 2. Church Ope Cove, VJG in **DOR**; Grove, BE.

Lecanora campestris (Schaerer) Hue. Very common on eutrophic limestone, asbestos, Greensand and rarely on slate. 36. **BMH, DOR, RNG**. A thallus 137 mm in diameter was seen at Blandford Ch.

Lecanora carpinea (L.) Vainio. Occasional on smooth bark of ash or sycamore, rarely on lime or *Euonymus japonicus*. 20. **BMH, DOR, NOT, RNG**.

Lecanora chlarotera Nyl. Abundant on smooth bark or twigs, often with *Lecidella elaeochroma*; also on wood and rubber. 37. **BMH, DOR, RNG**.

Lecanora conferta (Duby ex Fr.) Grognot. Occasional on ironstone. 8. Almer, Tomson and Zelston, TWC and !; Belchalwell and Okeford Fitzpaine,VJG; Fontmell Magna,TWC; Steeple, VJG; Studland, TWC; Edmondsham, VJG.

Lecanora confusa Almb. Occasional on smooth bark, twigs and wooden posts. 27. **DOR, RNG**.

Lecanora conizaeoides Nyl. ex Crombie. Abundant on young, dry bark or wood, occasional on stone and brick, sometimes on cast iron or lead. 37. **BMH, DOR, RNG**.

Lecanora crenulata Hook. Common on limestone cliffs and the vertical, dry limestone of church walls. 36. **BMH, DOR**.

Lecanora dispersa (Pers.) Sommerf. Common on eutrophic rocks and walls, occasional on eutrophic bark such as elder or *Suaeda vera*, once on decaying *Armeria*, rarely on rubber. All grid squares. **BMH, DOR, NOT, RNG**.

Lecanora expallens Ach. Very common, but sterile, on smooth bark or wood. 38. **BMH, DOR**, RNG. The segregates *L. barkmaniana* Aptroot & Van Herk and *L. compallens* Van Herk & Aptroot occur, BE and VJG.

Lecanora fugiens Nyl. Occasional, but overlooked, on sarsens or granite. 5. Abbotsbury and Portesham, TWC and !; Portland, VJG; Preston, !; Milton-on-Stour, !. **DOR**.

Lecanora gangaleoides Nyl. Local on sarsens, sandstone or Greensand. 6. Valley of Stones and Portesham, !: Piddlehinton Ch, VJG; Shaftesbury, RH Bailey; St Aldhelm's Head, BE and !; Agglestone, VJG.

Lecanora helicopis (Wahlenb.) Ach. Rare on acid stone near the sea. 6. Chesil bank, (W Watson, 1922); Grove, ! det. A Fletcher; Lulworth, ! det. JR Laundon; Gad Cliff and Emmetts Hill, BE; on wooden groynes, Swanage, VJG. **DOR**.

Lecanora intricata (Ach.) Ach. Very rare on sandstone. 1. Lytchett St Mary Ch, VJG.

Lecanora intumescens (Rebent.) Rabenh. Rare on smooth bark or wood. 3. Hooke, BE; Tyneham, FR; Furzey Is, VJG.

Lecanora jamesii Laundon. Uncommon on smooth bark of sallow in carr, rarely on ash. 6. Mount Pleasant, !; Powerstock Common, W Sutcliffe, also BE and PWJ; Melbury Park, FR; Winterbourne Came, PWJ; Lulworth Park, FR; Oakers Wood, VJG; Morden Decoy, FR, also !; Morden Park, BE, NAS and VJG. **RNG**.

Lecanora muralis (Schreber) Rabenh. Locally frequent on eutrophic stone, including sarsens, slate, sandstone, limestone, concrete and asbestos, often on paving slabs. Rarely on bark, wood or cast iron. 31. **BMH, DOR, RNG**.

Lecanora orosthea (Ach.) Ach. Occasional on acid rock, sterile. 8. Valley of Stones, BLS; Tincleton, VJG; Bere Regis, VJG; Bryanston, PWJ; Iwerne Minster, VJG; Agglestone, EWJ, also !; Cranborne Ch, VJG and !. **NOT**.

Lecanora pallida (Schreber) Rabenh. Occasional on smooth bark of twigs in woods. 11. **DOR, RNG**.

Lecanora pannonica Szat. Very rare on sandstone, Iwerne Courtney Ch, VJG.

Lecanora piniperda Koerber. Rare on pine bark or wood. 2. Melbury Park, FR; Holme Priory, DSR.

Lecanora polytropa (Hoffm.) Rabenh. Widespread on acid stone; sarsens, granite, old brick or flint. 26.

Lecanora pruinosa Chaub. Local on old limestone walls. 8. Maiden Newton, !; Stock Gaylard Ch, BE and !; Came Ch, BE; Melcombe Horsey Ch, !; Hilton, Piddlehinton, Mappowder, Minterne Magna, West Stafford, Tarrant Gunville and Steeple, VJG.

Lecanora pulicaris (Pers.) Ach. Rare on smooth bark. 2. Melbury Park, VJG; Chetterwood, FR.

Lecanora quercicola Coppins & P. James. Very rare on oak bark. 1. Melbury Park, FR, PWJ and SRD, also !. **RNG**.

(*Lecanora rupicola* (L.) Zahlbr. A plant of acid rock, reported but not confirmed (Watson, 1953).)

Lecanora saligna (Schrader) Zahlbr. Rare on wood. 3. Studland, (EMH/,1906); Canford Heath and Bourne valley, VJG; On shed, Poole, VJG.

Lecanora sambuci (Pers.) Nyl. Rare on eutrophic bark. 2. On elm, Melbury Park, PWJ; Melbury Hill (L2).

Lecanora soralifera (Suza) Rasanen. Rare on acid stone. 2. Puddletown Ch and West Stafford Ch, VJG.

Lecanora sublivescens (Nyl. ex Crombie) AL Sm. Very rare on dry bark. 1. On ash, Melbury Park, FR, PWJ and !.

Lecanora sulphurea (Hoffm.) Ach. Frequent on acid stone, flint, chert, conglomerate, sandstone, ironstone and old brick, extending to HWM. 25. **BMH, DOR, NOT, RNG**.

Lecanora symmicta (Ach.) Ach. Occasional on oligotrophic bark or wooden fences. 9. **BMH, DOR, RNG**.

Lecanora varia (Hoffm.) Ach. Rare on old fence posts or timber. 4. Cerne Abbas, ARV; Hyde, VJG; La Lee Farm, !; Studland, EMH in **DOR**.

Lecidea doliiformis Coppins & P. James. Scarce on oligotrophic bark of oak or pine. 8. Melbury Park, NAS; Stoborough Heath, VJG; The Oaks, Badbury, NAS, also High Wood, BE; fertile at Brownsea Is, PWJ, also VJG; Canford Heath, VJG; Pug's Hole, Talbot Heath, VJG; Pergins Is, ! det. BJC; Holt Forest, BE. **BM, DOR, E**.

Lecidea fuscoatra (L.) Ach. Occasional on acid stone, roof tiles or old brick. 19. **DOR**.

(*Lecidea hypnorum* Lib. is a northern taxon, unlikely to be the correct name for reports of *L. sanguineoatra* and *L. templetonii* (Holmes, 1906).)

(*Lecidea lapicida* (Ach.) Ach. may have been found at Bloxworth and reported as *L. confluens* (Holmes, 1906).)

Lecidea lichenicola (A.L. Sm. & Ramsb.) D. Hawksw. Rare on chalk pebbles. 2. Durdle Door, OLG and VJG; below Hambury Tout, BE; Ballard Down, OLG.

Lecidea lithophila (Ach.) Ach. Very rare on flint. 1. Near the Agglestone, VJG. This taxon should probably be transferred to *Micarea*.

Lecidea turgidula Fr. Very rare on ironstone. 1. Oakers Wood, DSR and PWJ.

Lecidella elaeochroma (Ach.) Choisy. Abundant on smooth bark, twigs and occasional on wood. 38. **BMH, DOR, RNG**. f. *soralifera* (Erichsen) D. Hawksw. is less common. 8. A form with pale fruit was seen at Parnham.

Lecidella scabra (Taylor) Hertel & Leuckert. Common on acid rock, sarsens, sandstone, granite, chert, old brick or on timber 35. Mostly sterile. **BMH, DOR, RNG**.

Lecidella stigmatea (Ach.) Hertel & Leuckert. Common on eutrophic horizontal limestone or asbestos; once seen on rubber. 37. **BMH, DOR, RNG**.

Lecidella viridans (Flotow) Koerber. On Greensand. 1. Iwerne Courtney Ch, PWJ and VJG.

(*Lempholemma botryosum* (Massal.) Zahlbr., reported from Portland by EMH in **BM**, was probably another species in this genus.)

Lempholemma chalazanum (Ach.) B. de Lesd. Rare and needing confirmation, on mortar. 2. Corfe Castle, EMH in **DOR**; Brownsea Is, PWJ.

Lempholemma polyanthes (Bernh.) Malme. Rare on stone or mortar. 2. Tyneham, FR; Tarrant Gunville Ch, VJG det. BJC.

Lepraria eburnea Laundon. Rare or overlooked on mossy limestone. 1. Beaminster Ch, !.

Lepraria incana (L.) Ach. Abundant on dry, shaded bark and sheltered, shaded stone, both acid and calcareous. All grid squares. **BMH, DOR, RNG**.

Lepraria lesdainii (Hue) R. Harris. Occasional and probably under-recorded on sheltered, overhanging limestone, in shade. 9. **BMH, DOR, RNG**.

Lepraria lobificans Nyl. Occasional, but under-recorded, on north-facing limestone walls. 7. Stoke Abbott, BE and !; Sydling St Nicholas, ! det. BJC; Fontmell Magna, TWC; Wareham, TWC; Zelston, TWC; Lytchett chapel, VJG; Branksome, !. **BMH, DOR, RNG**.

Lepraria nivalis Laundon. Locally frequent on vertical shaded limestone. 1. Portland quarries, ! det. BJC, also VJG. **BMH, DOR, RNG**.

Leprocaulon microscopicum (Vill.) Gams ex D. Hawksw. Local on shaded Greensand or acid rock. 4. Bryanston, PWJ, also !; Blandford Forum Ch, !; Winspit, PWJ; Studland, EMH. **BM, DOR**.

Leproloma diffusum Laundon. Occasional on sheltered north-facing walls. 8. **DOR, RNG**.

Leproloma vouauxii (Hue) Laundon. Frequent on eutrophic bark in shade, or on sheltered stone. 21. **DOR, RNG**.

Leproplaca chrysodeta (Vainio ex Rasanen) Laundon. Occasional on shaded stone, usually in crevices or overhangs. 17. **DOR, RNG**.

Leproplaca xantholyta (Nyl.) Harm. Occasional on shaded limestone, extending to moss. 23. **DOR, RNG**.

Leptogium byssinum (Hoffm.) Zwackh. ex Nyl. Rare on shallow calcareous soil. 2. Portland quarries, DSR, also VJG; Ballard Down, OLG.

[*Leptogium corniculatum* (Hoffm.) Minks. On dunes, Studland, EMH in **NOT**; a recent report from Fontmell Down is doubtful.]

Leptogium gelatinosum (With.) Laundon. Common on sheltered, mossy limestone, shallow calcareous soil and tops of walls. 31. **BMH, DOR, RNG**.

Leptogium lichenoides (L.) Zahlbr. Scarce on mossy bark of old trees. 6. Melbury Park, FR *et al.*; Sherborne Park, VJG; Tyneham Great Wood, BE and FR; Povington Wood, BE; Oakers Wood, VJG; Handcocks Bottom, FR; Great Wood, Creech, BE. **DOR**. Early reports from walls are forms of the polymorphic *L. gelatinosum*.

Leptogium plicatile (Ach.) Leighton. Occasional on damp calcareous rocks and walls. 11. **DOR, NOT, RNG**.

Leptogium schraderi (Ach.) Nyl. Occasional on shallow calcareous soils and limestone walls. 14. **BMH, DOR, RNG**.

Leptogium subtile (Schrader) Torss. Rare on stumps. 1. Great Wood, Creech, VJG.

Leptogium tenuissimum (Dickson) Koerber. Probably this on shallow calcareous soil. 1. Tout quarry, Portland, ! det. BJC.

Leptogium teretiusculum (Wallr.) Arnold. On calcareous soil of undercliffs, old mortar, and sometimes on eutrophic bark of ash or elm. 26. **DOR, RNG**.

Leptogium turgidum (Ach.) Crombie. On mossy, calcareous walls, often hard to tell from *L. plicatile*. 13. **DOR, RNG**.

Leptorhaphis epidermidis (Ach.) Th. Fr. Scarce on birch bark. 4. Hurst Heath, !; Oakers Wood, DSR; Slepe Heath, VJG; Morden Decoy, FR and SRD. **DOR**.

Lichina confinis (Mueller) Agardh. Local on calcareous rock at HWM. 4. Portland Bill, !; Osmington Mills, EMH; Emmetts Hill, BE; St Aldhelm's Head, BE and !; Swanage, EMH. **DOR, NOT, RNG**.

Lichina pygmaea (Lightf.) Agardh. Occasional on vertical rock between tidemarks. 9. Lyme Regis, BE; west Portland, M Tarroway, !; east Portland, !; Bran Point, Ringstead !; Lulworth Cove, !; Mupe Bay, !; Gad Cliff, BE; Kimmeridge, !, also VJG; Chapmans Pool, !, also VJG; Brownsea Is; Swanage, EMH. **DOR, RNG**. A refuge for the tiny bivalve mollusc *Lasaea rubra*.

Lobaria amplissima (Scop.) Forss. Very rare on bark of big oaks. 3. Melbury Park, FR; Lulworth Park, with cephalodia, !; Studland, EMH. **DOR**.

Lobaria pulmonaria (L.) Hoffm. Scarce on large trees, mostly ash or oak, also on apple, elm and sycamore, sometimes fertile. 13. **BMH, DOR, RNG**. It has declined since the last century, but is still surviving well on many trees at Melbury Park, Minterne Park and Lulworth Park.

Lobaria scrobiculata (Scop.) DC. Very rare on a single oak. Morden decoy, FR and SRD.

Lobaria virens (With.) Laundon. Rare on oak or lime bark in old woods or parks, decreasing. 7. Hooke Park, BE; Melbury Park, !, also R Palmer; Povington Wood, BE, !, in fruit; west Great Wood, Creech, BE; Oakers Wood, VJG; Bloxworth, EMH; Creech Grange, DSR, also R Palmer and BE; Kingston Lacy, EW Jones (1942). **DOR**.

Loxospora elatina (Ach.) Massal. Rare on oak in old, humid woods or parks. 4. Melbury Park, FR and !; Lulworth Park, FR; west Great Wood, Creech, BE; Oakers Wood, FR; Morden Decoy, FR and SRD.

Macentina abscondita Coppins & Vezda. Very rare on elder bark. 1. Ringstead, VJG.

Macentina stigonemoides A. Orange. Occasional, or overlooked, on mossy elder bark. 7. Sydling St Nicholas, !; Affpuddle, !; Woolland, !;Minterne Magna, VJG; Tollard Green, !; Primrose Hill, !; Oakhills Coppice, VJG. **BMH, DOR, RNG**.

[*Megalaria grossa* (Pers. ex Nyl.) Hafellner. On bark of old trees. 1. Studland, EMH in **NOT**. It still occurs in Cranborne Chase near Tollard Royal, but in VC 8.]

Megalospora tuberculosa (Fee) Sipman. Very rare in old woods. 1. On ash, Woodcutts, FR.

Melaspilea lentiginosa (Lyell ex Leighton) Muell. Arg. A rare endophyte of *Phaeographis*. 1. Oakers Wood, DSR, also VJG.

Melaspilea ochrothalamia Nyl. Rare on smooth bark. 3. Oakers Wood, Delph Wood and Oakhills Coppice, VJG. **DOR**.

Micarea bauschiana (Koerber) V. Wirth & Vezda. Rare on dry stone. 2. Probably this, reported as *Lecidea sylvicola* var. *infidula* Nyl., at Abbotsbury, EMH; Iwerne Minster Ch, VJG.

Micarea botryoides (Nyl.) Coppins. Rare on shaded pebbles. 1. Wareham Forest, VJG.

Micarea coppinsii Tonsb. Rare on mossy humus in carr. 3. Oak Hill, !, det. BJC; VJG has found this to be widespread on *Calluna* stems at Woolsbarrow, Hartland Moor and Canford Heath.

Micarea denigrata (Fr.) Hedl. Frequent, but overlooked, on timber. 2. Melbury Park, FR, PWJ and SRD; Handcocks Bottom, VJG.

Micarea erratica (Koerber) Hertel, Rambold & Pietschm.

Common on exposed flint pebbles on heaths or in old pits. 12. **DOR, RNG**.

Micarea leprosula (Th. Fr.) Coppins & Fletcher. A northern taxon. 1. On clay, Furzebrook, VJG. **DOR**.

Micarea ligniaria (Ach.) Hedl. Occasional on wood, mosses or acid banks. 6. Hod Hill, TW Ottley; Worth Matravers, CCT; Agglestone, VJG; Brownsea and Furzey Is, VJG; Delph Wood, VJG. Var. *endoleuca* (Leighton) Coppins on ball clay, Corfe, BE det. BJC.

Micarea lithinella (Nyl.) Hedl. On flints, Agglestone, VJG.

Micarea melaena (Nyl.) Hedl. Very rare on wood. 1. Morden Decoy, VJG in **DOR**.

Micarea nitschkeana (Lahm. ex Rabenh.) Harm. Probably frequent on old, degenerate *Calluna* and wooden palings. 3. Woolsbarrow and Hartland Moor, BE, PWJ and VJG; On elder bark, Holt Heath, VJG.

Micarea peliocarpa (Anzi) Coppins & R. Sant. Rare or overlooked on plant debris. 2. Coombe Heath, BE det. BJC; Canford Heath, VJG.

Micarea prasina Fr. Occasional on sheltered bark of elder, pine and other trees. 17. **DOR**.

Micarea pycnidiophora Coppins & P. James. Very rare on holly bark. 1. Holt Forest, NAS, also BE.

Mniacea jungermanniae (Nees ex Fr.) Boud. Scarce, overgrowing mossy acid banks. 6. Puddletown Heath, VJG: Carey and Philliol's Heath, !; Green Pond, BE and VJG; Morden Decoy, !; Brownsea Is, !; Delph Wood, VJG. **BMH, DOR, RNG**.

Mycoblastus caesius (Coppins & P. James) Tonsb. Scarce on shaded bark. 5. Melbury Park, FR, PWJ and SRD; Morden Park, BE, NAS and VJG; Elder Moor, VJG; Oakhills Coppice, VJG; Studland Heath, VJG.

Mycoblastus sterilis Coppins & P. James. Occasional, but overlooked, on smooth bark, palings or timber in parks or sheltered swamps. 7.

Mycocalicium subtile (Pers.) Szat. Very rare on wood. 1. Melbury Park, FR, PWJ and SRD.

Mycoglaena myricae (Nyl.) R.C. Harris. Widespread on *Myrica gale* bark in bogs or carr. 5+. Hyde bog, BE; Oak Hill, !; Wytch and Shotover Moors, BE; Morden Bog, BE; Arrowsmith Coppice, NAS; Holton Heath, !: Studland Heath, BE.

Mycomicrothelia confusa D. Hawksw. Rare on sheltered bark of oak or holly. 2. Monkton Wyld, ! det. BJC; Furzey Is, VJG.

Mycoporum quercus (Massal.) Muell. Arg. Occasional on smooth oak bark or twigs. 10. **DOR**.

Nephroma laevigatum Ach. Very rare on sheltered bark. 2. On one oak, Ower, BE; [on apple, Lyme undercliff, GL, perhaps in VC 3].

Normandina pulchella (Borrer) Nyl. Frequent in rain channels or mossy bark of large trees, or rarely on thalli of *Parmelia* or *Peltigera*. 29. **BMH, DOR, RNG**.

Ochrolechia androgyna (Hoffm.) Arnold. Occasional on mossy, oligotrophic bark in old woods. 19. **BMH, DOR, RNG**.

Ochrolechia inversa (Nyl.) Laundon. A scarce, old forest indicator on bark in humid woods. 5. Melbury Park, BJC and FR; Deadmoor Common, DJ Hill, needing confirmation; Oakers Wood, VJG; Morden Decoy, FR and SRD; Morden Park, BE.

Ochrolechia parella (L.) Massal. Abundant on acid rock of all kinds and on old brick walls, less often on limestone boulders near the sea in Portland and Purbeck, and rarely on bark of ash, sycamore or on timber. 35. **BMH, DOR, RNG**. A thallus 220 mm in diameter was seen at Winterborne Kingston Ch.

Ochrolechia subviridis (Hoeg) Erichsen Frequent on bark of large trees, mostly oak, less often ash or elm, rarely on timber or soil in lichen heath. 24. Mostly sterile, but seen in fruit at Lower Strode. **BMH, DOR, RNG**.

(*Ochrolechia szatalaeensis* Vers. was reported from Bloxworth and Studland (Holmes, 1906), but was probably a form of *O. parella* growing on bark.)

Ochrolechia turneri (Sm.) Hasselrot. Occasional on bark and palings, always sterile. 21. **BMH, DOR, NOT, RNG**.

Omphalina umbellifera (L. ex Fr.) Quel. Uncommon on wet heaths. Not always lichenised: see Fungi, Hymenomycetes.

Opegrapha atra Pers. Frequent on dry, smooth bark, usually in shade. 34. **BMH, DOR, RNG**.

Opegrapha corticola Coppins & P James. Occasional on oak and lime bark, also seen on box, elm, hawthorn and maple. 26. **BMH, DOR, RNG**. This includes old corticolous reports of *O. gyrocarpa*.

Opegrapha dolomitica (Arnold) Clauz. & Roux. Rare on shaded stone. 3. West Portland, DSR, also VJG; Swanage, EMH; Agglestone, VJG. **DOR**.

Opegrapha fumosa Coppins & P. James. Very rare on oak bark. 1. Melbury Park, BE, FR and NAS.

Opegrapha gyrocarpa Flotow. Rare on acid rock overhangs. 1. Agglestone, VJG.

Opegrapha herbarum Mont. Occasional on dry, shaded bark. 14. **BM, DOR, RNG**.

Opegrapha mougeotii Massal. Scarce on shaded, sheltered limestone on undercliffs or old churches. 8. East Fleet, !; Poyntington, !; Portland, EMH; Durdle Pier, BE; East Chaldon, !; Milton Abbey, TWC and !; Fifehead Neville, !; Gad Cliff and St Aldhelm's Head, BE. **BMH, DOR, E, NOT, RNG**.

Opegrapha multipuncta Coppins & P. James. Rare or overlooked on sallow bark. 2. Rye Water Farm, BE and PWJ; Tyneham, BE, PWJ and VJG. Like *O. sorediifera* but C-.

Opegrapha niveoatra (Borrer) Laundon. Scarce on elder or yew bark. 7. Charminster, VJG; Trent, !; Minterne Magna, VJG; Milborne St Andrew, !; Oakers Wood, VJG; Handcocks Bottom, VJG; Edmondsham, VJG. **DOR, RNG**.

Opegrapha ochrocheila Nyl. Scarce on rough bark. 8. Burstock Down, BE; Melbury Park, FR; Fontmell Magna and Sutton Waldron, PWJ; Povington Wood, BE; Little Wood, Creech, ! and Great Wood, Creech, BE; Morden Park, BE, NAS and VJG; Langton West Wood, BE; Holt Forest, BE. **BM**.

Opegrapha parasitica (Massal.) H. Olivier A scarce parasite of *Verrucaria*. 7. Bowers quarry, Portland, !; Poxwell, !; Durdle Door, OLG; Corfe, VJG; Morden Ch, !; Durlston Head and Swanage, VJG. **DOR, E**.

Opegrapha prosodea Ach. Scarce on oak or yew bark. 4.

Whitehall and West Creech, BE; Knowlton and Wimborne St Giles, VJG.

Opegrapha rufescens Pers. Scarce on smooth, shaded, eutrophic bark. 6. Melbury Park, PWJ and !; Sydling St Nicholas, !; Fifehead Magdalen, !; Oakers Wood, VJG; Bryanston, PWJ; Oakhills Coppice, VJG. **DOR, RNG**.

Opegrapha saxatilis anct. non DC. Locally abundant on dry, sheltered limestone or mortar, especially near the coast. 31. **BM, BMH, DOR, NOT, RNG**.

Opegrapha saxigena Tayl. Rare on shaded chert. 1. Portland, VJG.

Opegrapha sorediifera P. James. Scarce on smooth, oligotrophic bark in old woods. 10. **DOR**. Some records may belong to *O. multipuncta*.

Opegrapha subelevata Nyl. A rare plant of limestone. 1. East Portland undercliff, VJG.

Opegrapha varia Pers. Common and variable on dry, shaded bark, especially ash, elm, maple and sycamore. 31. **BMH, DOR, RNG**.

Opegrapha vermicellifera (Kunze) Laundon. Occasional on dry, shaded bark, commonest on elm before 1970, now on many trees, especially yew. 21. **BMH, DOR, RNG**. Easily confused with *Lecanactis subabietina*.

(*Opegrapha viridis* (Ach.) Nyl. The specimen called this by EMH in **DOR**, from Studland, is *O. vulgata*.)

Opegrapha vulgata (Ach.) Ach. Abundant on smooth, dry bark, most common on hazel, sycamore or yew. 35. **BMH, DOR, RNG**.

Opegrapha xerica Torrente & Egea. Rare or overooked on dry, shaded bark in old woods. 6. Bloxworth Ch, !; Wood Street, BE, !; Povington Wood, BE, FR, NAS and !; Wytch Farm, BE and !; Afflington Wood, Langton West Wood and Studland Wood, BE; The Oaks, Badbury, BE;.

Opegrapha zonata Koerber (*Enterographa zonata* (Koerber) Kallsten). Very rare on granite. 1. Iwerne Courtney Ch, PWJ and VJG.

Pachyphiale carneola (Ach.) Arnold. An old forest indicator, usually in the rain channel of old oaks in woods; also on ash, beech, hazel, lime and maple. 18. **BMH, DOR, NOT, RNG**.

Pannaria conoplea (Ach.) Bory. A rare, old forest indicator, on oak bark. 2. Melbury Park, fertile, FR, also VJG; Woodcutts, FR; a report from Sutton Holms, T Hooker, needs confirmation.

Pannaria mediterranea C. Tav. A very rare, old forest indicator, on oak. 1. Chetterwood, FR and SRD.

(*Pannaria rubiginosa* (Ach.) Bory was reported from Studland (Holmes, 1906), but was most likely to have been *P. conoplea*.)

Parmelia acetabulum (Necker) Duby. Uncommon on eutrophic bark of ash, maple or sycamore, often high up. 7. Came Park, !; Kingston Maurward, DWT; Winterborne Kingston, !; Sturminster Newton, BE; Bryanston, PWJ; Pirford Bridge, JRW; Wareham, VJG. **DOR, RNG**.

Parmelia borreri (Sm.) Turner. Occasional on bark of large trees. 12. Usually sterile, but in fruit at Forston, BE. **DOR**.

Parmelia caperata (L.) Ach. Abundant on the bark of large trees, and luxuriant in humid sites, extending to the centre of Dorchester; occasional on old wood or thatch, acid stone, old brick and tile, on the Wealden sand undercliff at

Worbarrow Bay and the calcareous lichen heath on Bindon Hill. All grid squares but SY66. Mostly sterile, but in fruit in carr at Hurst Heath, Oakers Wood and Oak Hill; fruiting may be accelerated on dying substrates. **BMH, DOR, RNG**.

Parmelia conspersa (Ehrh. ex Ach.) Ach. Occasional on more or less horizontal acid stone, tile or slate, mainly in the north and west. 12. **BMH, DOR, RNG**.

Parmelia crinita Ach. Rare on oligotrophic bark of alder or oak. 3. Melbury Park, FR and !; Hay Wood and Minterne Park, BE: west Great Wood, Creech, BE, !. **BMH, DOR, RNG**.

Parmelia delisei (Duby) Nyl. Rare on sarsen stones, also on slate roofs. 7. Sadborow, ARV; Melbury Park, !; Portesham and Valley of Stones, !, also VJG; Piddlehinton, VJG; fertile at Turnworth, !; Morden, !. **BM, DOR, RNG**.

Parmelia disjuncta Erichsen. Very rare on slate roof. 1. Melbury Park, ! det. PWJ.

Parmelia elegantula (Zahlbr,) Szat. Occasional on horizontal branches of ash or sycamore. 10. **BMH, DOR, RNG**.

Parmelia exasperata De Not. Occasional on branches of ash, elm, lime or oak, and on wooden jetties in Poole Harbour. 9. **BMH, DOR, RNG**. Always fertile.

Parmelia exasperatula Nyl. Occasional, perhaps increasing, on horizontal branches with eutrophic bark. 10. **RNG**. Always sterile.

Parmelia glabratula (Lamy) Nyl. Very common on bark, twigs or wood, rarely on sarsens, as at Portesham. Mostly sterile, but in fruit at Tyneham, BE and on *Liriodendron* at Canford, !. All grid squares but SY66. **BMH, DOR, RNG**. Subsp. *fuliginosa* (Fr. ex Duby) Laundon, is frequent on acid stone, old brick and slate, or rarely on bark as at Lulworth Park, FR. 29. **BMH, DOR, RNG**.

Parmelia laciniatula (H. Olivier) Zahlbr. Occasional on branches of large trees, rarely on wood. 18. **BMH, DOR, RNG**.

Parmelia laevigata (Sm.) Ach. Rare on birch or sallow bark in carr. 5. Wytherston Marsh, BE; Oak Hill, SRD; Morden Decoy, FR and SRD; fertile, Corfe and Studland, EMH.

Parmelia loxodes Nyl. Rare on acid stone or slate. 4. Melbury Park, ! det. PWJ; Valley of Stones, BE, !; Okeford Fitzpaine Ch, VJG; Organford, VJG. **DOR**.

Parmelia minarum Vainio. A very rare plant, on oligotrophic bark. 2. On alder and willow, Powerstock Common, BE; on beech, Morden Park, BE.

Parmelia mougeotii Schaerer ex D. Dietr. Frequent on acid stone, sarsens, granite, tile and especially slate roofs, as a pioneering colonist. 28. **BMH, DOR, RNG**.

Parmelia omphalodes (L.) Ach. Very rare on sarsens. 1. Valley of Stones, !.

Parmelia pastillifera (Harm.) R. Schubert & Klem. Frequent on eutrophic branches, wood, stone or tile. 23. **BMH, DOR, NOT, RNG**. All old reports of *P. tiliacea* belong here.

Parmelia perlata (Huds.) Ach. Common, often co-dominant with *P. caperata* on the boles of large trees, sterile; occasional on acid stone, in lichen heath on Bindon Hill and on the Wealden sand undercliff at Worbarrow Bay. All grid squares but SY66. **BMH, DOR, RNG**.

Parmelia pulla Ach. Rare or inaccessible on old slate roofs, fertile. 2. Beaminster, BE and !; Hedge End, Turnworth, !.

Parmelia quercina (Willd.) Vainio. Scarce on eutrophic bark of ash, elder, elm or sycamore, rarely on oak, often high up; restricted to the coastal belt. 5. Abbotsbury, TDV Swinscow, !; Little Bredy, BE; Ringstead, !; Dungy Head, !; Tyneham Gwyle, BE and FR; Kingston, !, also VJG; South Gwyle, Encombe, BE. **BM, DOR, E, RNG**.

Parmelia reddenda Stirton. Occasional on oligotrophic bark, especially oak, in old woods. 21. **BMH, DOR, RNG**.

Parmelia reticulata Taylor. Occasional on oligotrophic bark in old woods. 17. **DOR, RNG**.

Parmelia revoluta Floerke. Frequent on oligotrophic bark in woods. 32. **BM, DOR, RNG**.

Parmelia saxatilis (L.) Ach. Common on oligotrophic bark or wood, and frequent on acid stone, sarsens or old brick; not seen fertile. 38. **DOR, RNG**.

[*Parmelia sinuosa* (Sm.) Ach. On bark, Studland, EMH (Holmes, 1906); also recorded from Bournemouth last century by A Bloxam in **NOT**, but in VC 11.]

Parmelia soredians Nyl. Scarce on bark, timber or acid stone in the south. 10. **DOR, RNG**.

Parmelia subaurifera Nyl. Common on horizontal branches or twigs. 36. **BMH, DOR, RNG**.

Parmelia subrudecta Nyl. Abundant on bark, occasional on wood, sterile. 37. **BMH, DOR, NOT, RNG**.

Parmelia sulcata Taylor. Abundant on ± eutrophic bark, on wood, and occasional on acid stone and the Wealden sand undercliff at Worbarrow Bay. 38. Rarely fertile. **BMH, DOR, RNG**.

Parmelia ulophylla (Ach.) F. Wilson. Recently found on bark, BE, and on Eocene rock, Godlingston, VJG.

Parmelia verruculifera Nyl. Frequent on acid stone, sarsens, old brick and slate roofs. 27. **BMH, DOR, RNG**.

Parmeliella jamesii S. Ahlner & P. Jorg. A very rare, old forest indicator, on mossy bark. 1. Melbury Park, FR.

Parmeliella triptophylla (Ach.) Muell. Arg. A very rare, old forest indicator. 2. On ash bark, Drew Coppice and Woodcutts, FR, also PWJ and !, on or over the county boundary; Studland, EMH in **DOR**.

Parmeliopsis ambigua (Wulfen) Nyl. Uncommon on oligotrophic bark. 4. Thorncombe Wood, !; On beech, Lulworth Park, !; Fontmell Down, SRJ; Brownsea Is, PWJ.

Parmeliopsis hyperopta (Ach.) Arnold. Rare on pine bark. 1. Brownsea Is, PWJ.

Peltigera canina (L.) Willd. Rare or overlooked in acid grassland. 1. Sandford, NAS; Middlebere Heath, BE.

Peltigera collina (Ach.) Schrader. A rare, old forest plant, on mossy bases of old trees. 3. Melbury Park, !; Chetterwood, FR and SRD; Cranborne Chase, FR. A Portland report (Holmes, 1906) was an error.

Peltigera didactyla (With.) Laundon. Occasional on dry, heathy tracks, fixed dunes, old rail tracks and shallow acid soils. 17. **BMH, DOR, RNG**.

Peltigera horizontalis (Huds.) Baumg. A scarce, old forest indicator on mossy bark or logs. 9. Powerstock Common, FR and !; Hooke Park, BE; Melbury Park, FR, !; Batcombe and Middlemarsh, BE; Sherborne Park, VJG; Melcombe Park, M Heath; Handcocks Bottom, VJG; Great Wood, Creech, BE; Harbins Park, BE. **DOR, RNG**.

Peltigera lactucifolia (With.) Laundon. Occasional on heaths, by sarsens, in dune slacks, rarely on logs. 19. **BMH, DOR, RNG**.

Peltigera membranacea (Ach.) Nyl. Occasional in woodland rides or sheltered grassland on acid soil, or on logs. 19. **DOR, RNG**.

Peltigera praetextata (Floerke ex Sommerf.) Zopf. Locally abundant on logs and tree bases in damp woods, rarely on walls or fixed dunes. 21. **BMH, DOR, RNG**.

Peltigera rufescens (Weiss) Humb. Scarce on shallow calcareous soil or asbestos roofs. 6. Verne, Portland, SL Jury; Glanville Wootton, CW Dale; Minterne Magna, !; Stourton Caundle, !; Blandford Camp, BE; Townsend, Swanage, BE; Brownsea Is, PWJ. **DOR, RNG**.

Pertusaria albescens (Huds.) M. Choisy & Werner. Common on oligotrophic bark in woods, and occasional on acid stone. Var. *corallina* (Zahlbr.) Laundon is equally common. 35. **BMH, DOR, RNG**.

Pertusaria amara (Ach.) Nyl. Common on oligotrophic bark in woods, occasional on acid stone or old brick. 34. **BMH, DOR, RNG**.

Pertusaria coccodes (Ach.) Nyl. Occasional on oak or sycamore bark in woods, and on timber. 21. **BMH, DOR, RNG**.

[*Pertusaria corallina* (L.) Arnold. Extinct on sandstone. 1. Agglestone, EMH in **DOR**; it still occurs at Hengistbury Head in VC 11, SRD.]

Pertusaria flavida (DC.) Laundon. An old forest indicator, locally frequent on big ashes and oaks. 13. **BMH, DOR, RNG**.

Pertusaria hemisphaerica (Floerke) Erichsen. Local on oligotrophic bark in woods. 23. **BMH, DOR, RNG**.

Pertusaria hymenea (Ach.) Schaerer. Common on oligotrophic bark in woods, and on stone at Stanbridge. 35. **BMH, DOR, RNG**.

Pertusaria leioplaca DC. Frequent on smooth oligotrophic bark, especially hazel, in woods. 28. **BMH, DOR, RNG**.

Pertusaria multipuncta (Turner) Nyl. Locally frequent on smooth, oligotrophic twigs or bark in woods. 14. **BMH, DOR, RNG**.

Pertusaria pertusa (Weigel) Tuck. Common on rough, oligotrophic bark in woods, occasional on wood, acid stone or old brick. 37. **BMH, DOR, RNG**.

Pertusaria pseudocorallina (Lilj.) Arnold. Rare on chert or acid conglomerate rock. 3. East Portland, VJG; Gad Cliff, BE; Agglestone, !, also VJG. **BMH, DOR, RNG**.

(*Pertusaria pustulata* (Ach.) Duby was reported from bark at Powerstock Common by M Sutcliffe; so rare a plant needs confirmation.)

Pertusaria velata (Turner) Nyl. A rare, old forest indicator on oak bark. 3. Melbury Park, BJC and FR, also BE; Hooke Park, FR and !; East Lulworth, EMH.

Petractis clausa (Hoffm.) Krempelh. Scarce, on chalk or limestone. 9. **BM, NOT**.

Phaeographis dendritica (Ach.) Muell. Arg. Occasional on smooth, oligotrophic bark, especially hazel, in woods. 23. **BMH, DOR, RNG**.

Phaeographis inusta (Ach.) Muell. Arg. Rare, on sallow bark. 2. Wicker Coppice, BE; Oakers Wood, VJG det. BJC.

Phaeographis lyellii (Sm.) Zahlbr. Rare on smooth, oligotrophic bark in woods. 4. Hooke Park, FR and !, also BE; Gore Wood, Middlemarsh, BE; Oakers Wood, DSR; Bere Wood, VJG; Affpuddle, PWJ; Oakhills Coppice, VJG. **BM**.

Phaeographis smithii (Leighton) B. de Lesd. Scarce on smooth, oligotrophic bark in woods. 10. **BMH, DOR, RNG**.

(*Phaeophyscia endophoenicea* (Harm.) Moberg has been reported from bark on the Lulworth estates on insufficient evidence.)

Phaeophyscia nigricans (Floerke) Moberg. Rare on hypereutrophic stone or wood. 5. Sadborow, ARV; Piddlehinton,VJG; Droop Ch, !; Branksome, VJG; Edmondsham, VJG.

Phaeophyscia orbicularis (Necker) Moberg. Abundant on eutrophic bark, and common on eutrophic stone, asbestos, concrete or slate. 38. Variable, with a dwarf black form in places; often fertile. **BMH, DOR, RNG**.

Phlyctis agelaea (Ach.) Flotow. Rare and declining on oligotrophic bark in old woods. 6. Melbury Park, FR; Oakers Wood, VJG; Fontmell Down, SRD; Slepe and Morden, DSR; Studland, EMH. **DOR**.

Phlyctis argena (Sprengel) Flotow. Abundant on bark, usually in woods, but always sterile. 35. **BMH, DOR, RNG**.

Phyllopsora rosei Coppins & P. James. Very rare on oak bark. 1. Melbury Park, FR, PWJ and SRD.

Physcia adscendens (Fr.) H. Olivier. Abundant on eutrophic bark or wood, but sterile; common on limestone or old brick, often fertile. All grid squares. **BMH, DOR, RNG**.

Physcia aipolia (Ehrh. ex Humb.) Fuernr. Frequent on bark or twigs of large trees in parks, once seen on limestone at Abbotsbury, TWC and !. 36. **BMH, DOR, RNG**.

Physcia caesia (Hoffm.) Fuernr. Frequent on eutrophic, horizontal stone, slate or tile, often fertile; rare on eutrophic bark at Anderson, and on lead and glass, Ibberton Ch, !. 33. Variable. **BMH, DOR**.

Physcia clementei (Sm.) Maas Geest. Scarce on eutrophic bark or limestone. 14. Stoke Abbott, BE and !; Little Bredy, BE; Pennsylvania Castle, Portland, !; Forston and Frome Whitfield bridge (fertile), BE; Yetminster Ch, !; Milborne St Andrew, BE: Droop, BE; Wood Street, BE, !; Tomson, BE; Spetisbury, BE; Charborough Park, BE; between Corfe and Kingston, VJG. **DOR**.

Physcia dubia (Hoffm.) Lettau. Occasional on horizontal, acid rock, slate, tile or eutrophic timber. 12. **BMH, DOR, RNG**.

Physcia semipinnata (Gmelin) Moberg. Scarce on twigs or wood, mostly near the coast, and rare on limestone. 9. West Weare, TW Ottley; East Weare, !; Came Park, FR and !; Oakers Wood, VJG; Arish Mell, BE: Emmetts Hill, BE; Durlston Head, ! and SRD; Swanage and Ballard Point, DL Hawksworth; Hamworthy, on jetties, VJG. **DOR, RNG**.

Physcia stellaris (L.) Nyl. Very rare. 1. Glanvilles Wootton Ch, VJG.

Physcia tenella (Scop.) DC. Abundant on eutrophic bark and twigs, sometimes fertile; less common on eutrophic horizontal limestone. 36. **BMH, DOR, RNG**.

Physcia tribacia (Ach.) Nyl. Occasional on large, free-standing trees, ash, cypress, oak or sycamore, less common on limestone or old tile. 18. **BMH, DOR, RNG**.

Physcia tribacioides Nyl. A scarce plant of eutrophic bark of free-standing ash, elm, maple, oak or sycamore. 10. Little Bredy, BE; South Perrott, BE; Toller Porcorum, BE; fertile at Ilsington !, and Herringston, BE; Whitcombe, VJG; Throop, BE; Muston and Tomson, BE; Creech Grange, BE and VJG; Corfe Mullen, BE.

Physconia distorta (With.) Laundon. Locally frequent on eutrophic bark, especially ash and elder; seen on stone at Sherford Bridge, BE. 32. **BMH, DOR, RNG**.

Physconia enteroxantha (Nyl.) Poelt. Uncommon on large, free-standing trees in parks. 4. Melbury Park, FR; Westrow Ho, !; Whatcombe Park, !; Canford Park, !. **BMH, DOR**.

Physconia grisea (Lam.) Poelt. Common on eutrophic bark, but sterile; equally common on eutrophic limestone, occasionally fertile, as at Silton Ch. 36. **BMH, DOR, RNG**.

Physconia perisidiosa (Erichsen) Moberg. Occasional on eutrophic bark, formerly on elm and now on ash or sycamore. 20. **BMH, DOR, RNG**.

Placynthiella icmalea (Ach.) Coppins & P. James. Occasional on elder bark, more common on timber roofs or fence posts and on acid soil. 14. **DOR**.

Placynthiella uliginosa (Schrader) Coppins & P. James. Locally frequent on peaty soil of heaths in the Poole basin, or on decaying moss. 14. **DOR**.

Placynthium nigrum (Huds.) Gray. Common on damp, horizontal limestone or asbestos, especially on or near ground-level; also on lichen heaths and at edges of cliffs. 38. **BMH, DOR, RNG**.

Placynthium tremniacum (Massal.) Jatta. Rare on mortar. 1. Brownsea Is, PWJ.

Platismatia glauca (L.) Culb. & C. Culb. Common on oligotrophic branches, often high up, in woods, occasional on timber or peaty soil. 29. **BMH, DOR**.

(*Poeltinula cerebrina* (DC.) Hafellner. A rare plant of limestone near the sea. 2. West Weare quarry, DSR, needing confirmation; Durlston Head, EMH in **DOR**, but this looks like *Opegrapha parasitica*, det. VJG.)

Polyblastia albida Arnold. Very rare on mortar. 1. North of Bryanston, PWJ; Corfe Castle, 1884, EMH in **NOT**, needing confirmation.

Polyblastia deminuta Arnold. Rare on chalk scree. 1. Church Knowle, DSR.

Polyblastia dermatodes Massal. Rare on chalk or mortar. 3. Batcombe, VJG; Five Maries and White Nothe, !; Brownsea Is, PWJ. **DOR**.

Polyblastia gelatinosa (Ach.) Th. Fr. Occasional on shallow calcareous soil. 6. Southwell, Verne and West Weare, BE; Durdle Door, OLG; Bindon Hill and Emmetts Hill, BE; Corfe, VJG; Ballard Down, EMH, also OLG. **DOR**.

Polysporina simplex (Davies) Vezda. Occasional on sarsens or rough granite. 20. **DOR**.

Porina aenea (Wallr.) Zahlbr. Frequent on smooth, shaded bark of hazel or maple in woods. 22. **BMH, DOR, RNG**.

Porina borreri (Trevisan) D. Hawksw. & P. James. Rare on bark. 2. Melbury Park, FR, PWJ and SRD; Affpuddle Ch, VJG. The specimen purporting to be this from Studland, EMH in **DOR**, was *P. aenea*, det. BJC.

Porina chlorotica (Ach.) Muell. Arg. Locally frequent on sheltered limestone or chalk, especially on Portland. 18. **DOR**.

Porina leptalea (Durieu & Mont.) A.L. Sm. Scarce on smooth bark, usually hazel, in woods. 10. Champernhayes, ! det. BJC; Melbury Park, !; Hurst Heath, !; Oakers Wood, PWJ; South Gwyle, Encombe, BE; Arne, !; Charborough Park, BE and !; on beech, Morden Park, BE; Oakhills Coppice, VJG; Holt Forest, BE. **BM, BMH, DOR, RNG**.

Porina linearis (Leighton) Zahlbr.) Occasional on sheltered limestone or chalk, mostly in the coastal belt. 9. **BMH, DOR, RNG**.

(*Porina mammillosa* (Th. Fr.) Vainio. A northern species, reported from Lulworth by N Stewart, but unlikely to be correct.)

Porpidia cinereoatra (Ach.) Hertel & Knopf. Rare on sarsen stones. 2. Valley of Stones and Portesham, ! and BE. **DOR**.

Porpidia crustulata (Ach.) Hertel & Knoph. Frequent on acid stone, sarsens, chert, flint or old brick. 25. **BMH, DOR, RNG**.

Porpidia macrocarpa (DC.) Hertel & Schwab. Probably common but overlooked on Eocene sandstone on heaths in the Poole basin, or on flint. 6. Chesil bank (Watson, 1922); Higher Hyde Heath, !; Slepe Heath, VJG; Arne and Hartland Moor, DSR; Lytchett Chapel, ! det. BJC; Agglestone, EMH (as *Lecidea contigua* in **DOR**); Newton Heath, TW Ottley; Brownsea Is, PWJ; Branksome Ch, !. **DOR, RNG**.

Porpidia platycarpoides (Bagl.) Hertel. Very rare on acid rock. 1. Agglestone, VJG.

Porpidia soredizoides (Lamy ex Nyl.) Laundon. Probably frequent on acid rock. 12.

Porpidia tuberculosa (Sm.) Hertel & Knoph. Abundant on sarsens, granite, sandstone, flint and slate. 33. **BMH, DOR, RNG**.

Protoblastenia calva (Dickson) Zahlbr. Occasional on exposed limestone. 11. **DOR, RNG**.

Protoblastenia incrustans (DC.) Steiner. Scarce on limestone. 7. Langton Herring Ch, DA Newman; Minterne Magna and Hilton Ch, VJG; Penns Weare, BE; Worbarrow Bay, DSR; Winspit, !; Corfe Mullen, VJG.

Protoblastenia rupestris (Scop.) Steiner. A common early colonist of limestone or concrete, rarely on weathered flint, in both exposed and shaded sites. 37. **BMH, DOR, RNG**. An albino form with pink fruit was found at Buckland Newton Ch, ! det. BJC in **E**.

[*Protoparmelia badia* (Hoffm.) Hafellner. On acid rock. 1. Ballard Down (Holmes, 1906).]

Pseudevernia furfuracea (L.) Zopf. Occasional, in small numbers, on oligotrophic bark or wood, sterile. 16. **BMH, DOR, RNG**.

[*Pseudocyphellaria aurata* (Ach.) Vainio. A very rare plant of bark, found in 1908 on apple at Lyme Regis undercliff, GL in **BM**, also A Lester in **NOT**, possibly in VC 3.]

Psilolechia leprosa Coppins & Purvis. Probably widespread but very local, on limestone stained with verdigris from nearby copper. 11. It secretes copper oxalate.

Psilolechia lucida (Ach.) M. Choisy. Locally abundant on vertical, acid stone or brick, rarely on shaded roots. 32. **BMH, DOR**.

Psora lurida (Ach.) DC. Rare on calcareous rock or soil. 2. Bowers quarry, Portland !; East Weare, VJG; Church Ope Cove, BE. **DOR.**

Psorotichia schaereri (Massal.) Arnold. Rare on shaded limestone. 3. Affpuddle Ch, VJG; Corfe railway cutting, VJG; Studland, EMH.

Pycnothelia papillaria Dufour. Local on heaths in the Poole basin, plus Hardown Hill. 9. **BMH, DOR, NOT, RNG.**

Pyrenocollema halodytes (Nyl.) R.C. Harris. Frequent between tide marks on limestone, chalk, barnacle and limpet shells. 9. **DOR, RNG.**

Pyrenocollema monense (Wheldon) Coppins. Rare on sheltered chalk pebbles. 3. Durweston Woods and Iwerne Courtney Ch, VJG; Rushmore, PWJ.

Pyrenocollema orustense (Erichsen) A. Fletcher. On flint pebbles, Chesil Bank, VJG.

Pyrenocollema saxicola (Massal.) Coppins. Occasional on sheltered limestone near the sea. 5. Portland Bill, PWJ; Portland quarries, !; East Weare, VJG; Winspit, PWJ, !; Brownsea Is, PWJ. **BM, BMH, DOR, RNG.**

Pyrenula chlorospila Arnold. Frequent on smooth bark in woods, most common in the wetter north and west. 30. **BMH, DOR, RNG.**

(*Pyrenula laevigata* (Pers.) Arnold. Two reports, one from (Watson, 1953) and the other from Corfe, need confirmation.)

Pyrenula macrospora (Degel.) Coppins & P. James. Frequent on smooth, shaded bark, mostly in woods in the wetter north and west. 32. **BMH, DOR, RNG.**

(*Pyrenula nitida* (Weigel) Ach. occurs in the New Forest, but all old records from VC 9 are *P. macrospora*.)

Pyrrhospora quernea (Dickson) Koerber. Common on oligotrophic bark of old trees, often fertile, rarely on wood. 36. **BMH, DOR, RNG.**

Ramalina calicaris (L.) Fr. Local on twigs and bark on trees in parks or sheltered valleys; fertile at Gillingham. 24. **BMH, DOR, RNG.**

Ramalina canariensis Steiner. Occasional on big trees of ash or sycamore, or on vertical, sheltered stone, sometimes forming a band down church towers. 33. **BMH, DOR, RNG.**

Ramalina farinacea (L.) Ach. Abundant on exposed bark and twigs, much less common on acid stone, also on a Wealden sand undercliff at Worbarrow Bay; always sterile. 37. **BMH, DOR, RNG.**

Ramalina fastigiata (Pers.) Ach. Abundant as an early colonist of exposed twigs, and on bark, always fertile. 37. **BMH, DOR, RNG.**

Ramalina fraxinea (L.) Ach. Frequent on big ashes and elms, perhaps more stunted than of old, but usually fertile. 28. **BMH, DOR, RNG.**

Ramalina lacera (With.) Laundon. Occasional on bark of wayside trees, less common on acid stone of old buildings such as Shapwick Ch, !, det. BJC, and Horton Tower, PWJ. 10. **BM, DOR, NOT.**

(*Ramalina pollinaria* (Westr.) Ach.; all early reports refer to other species; e.g Studland, 1884, EMH in **NOT**.)

(*Ramalina polymorpha* (Lilj.) Ach. has an old record from the Chesil beach which has never been confirmed (Watson, 1922).)

Ramalina siliquosa (Hudson) A.L. Sm. Local on sarsens, chert and old flint, or on limestone boulders and wall-tops near the sea; very rare on tombs later than megaliths. 13. Salway Ash Ch, !; Valley of Stones and Portesham, !; Grey Mare and her colts, !; Hell Stone, !; Chesil beach, !; West and East Weares, !: Poxwell Stone Circle, !; Gad Cliff, !; Chapmans Pool, FR; between Kingston and Houns Tout, !; Emmetts Hill and St Aldhelm's Head, !; Agglestone, EMH, also !, very stunted; Knowlton Ch, VJG. **BMH, DOR, RNG.** *R. cuspidata* reported from Portland in 1881 by EMH in **NOT** was probably this.

Ramalina subfarinacea (Nyl. ex Crombie) Nyl. Rare on acid stone. 3. Chesil beach, (Watson, 1922); East Weare, VJG; Shaftesbury Ch, RH Bailey.

Ramonia chrysophaea (Pers.) Vezda. Rare on oak bark. 4. Oakers Wood, VJG; Oakhills Coppice, VJG; Afflington Wood, BE; Holt Forest, BE.

Ramonia dictyospora Coppins. Rare on bark, and hard to see unless wetted. 2. Milborne Wood, BE; Holy Stream, BE.

Rhizocarpon concentricum (Davies) Beltr. Scarce on chert, flint or acid rocks. 9. **DOR.**

Rhizocarpon distinctum Th. Fr. Rare on sheltered acid stone. 1. Abbotsbury Ch, TWC and !.

Rhizocarpon geographicum (L.) DC. Abundant on old slate roofs, less common on sarsens, granite or tile. 28. **DOR, RNG.**

Rhizocarpon hochstetteri (Koerber) Vainio. Rare on a sarsen. 1. Portesham Ch, TWC and !.

Rhizocarpon obscuratum (Ach.) Massal. Abundant on all kinds of acid stone, especially on flints exposed on heaths or old quarries. 37. **BMH, DOR, RNG.**

Rhizocarpon oederi (Weber) Koerber. Rare on Eocene rock. 1. Newton Heath, TW Ottley and VJG.

Rhizocarpon richardii (Nyl.) Zahlbr. Scarce on chert or flint, mostly near the sea. 7. Chesil beach, (W. Watson, 1922), also !; West Weare, BLS; East Weare, VJG; Upwey, !; Gad Cliff, BE; Lytchett Manor, !; St Aldhelm's Head, BE and !. **DOR, RNG.**

Rhizocarpon umbilicatum (Ramond) Flagey. Rare on limestone or chalk pebbles. 3. Tincleton Ch, VJG; Fortuneswell, DSR; Durdle Door, OLG.

(*Rhizocarpon viridiatrum* (Wulfen) Koerber. A report from acid rock near Swanage needs confirming.)

Rinodina aspersa (Borrer) Laundon. Rare on flint pebbles on the Chesil bank, VJG. 1. A report from near Studland needs confirming.

Rinodina atrocinerea (Hook.) Koerber. Rare on sarsens. 2. Valley of Stones, VJG; Rye Water Farm, PWJ + BE, VJG and !.

Rinodina bischoffii (Hepp) Massal. Scarce on limestone or chalk near the sea. 5. Gad Cliff and Bindon Hill, BE; Durdle Door, OLG and VJG; Winspit, !; Durlston Head, EMH, also VJG; Ballard Down, OLG and VJG; Ulwell, VJG. **DOR, NOT.**

Rinodina efflorescens Malme. Rare on hazel bark. 2. Oakers Wood, ! det. BJC; Delph Wood, VJG.

Rinodina exigua Gray. Occasional on eutrophic bark, rarely on wood. 13. **BMH, DOR, RNG.**

Rinodina gennarii Bagl. Frequent on eutrophic stone, noted on limestone, sandstone, sarsens, chert and flint;

also a pioneer species on timber at Poole Harbour, VJG. 28. **BMH, DOR, RNG**.

Rinodina roboris (Dufour ex Nyl.) Ach. Frequent on rough bark of large trees, especially oak, in parks and woods. 31. **BMH, DOR, RNG**.

Rinodina sophodes (Ach.) Massal. Occasional but easily overlooked on twigs of oak or ash. 7. **BM, BMH, DOR, RNG**.

Rinodina teichophila (Nyl.) Arnold. Occasional on ironstone or golden limestone. 9.

[*Roccella fuciformis* (L.) DC. On coastal rock. 1. Portland (1859), GB Sowerby in **BM** det. JR Laundon.]

Roccella phycopsis Ach. Very local on cherty limestone boulders near the sea. 2. Portland, Lord Lewisham (1801), GB Sowerby (1859), and EMH (1881) in **BM**; Durdle Pier, J Carrington, *et al.*, !; Church Ope Cove and south, !, also BE. **BM, BMH, DOR, NOT, RNG**.

Sarcogyne regularis Koerber. Common on exposed limestone, concrete or mortar. 33. **BMH, DOR, RNG**.

Sarcopyrenia gibba (Nyl.) Nyl. Rare on horizontal stone. 4. Charmouth, CJB Hitch; Fontmell Magna Ch, TWC; Almer Ch, TWC and !; Tarrant Gunville Ch, VJG.

Sarcosagium campestre (Fr.) Poetsch. & Schied. Rare on shallow calcareous soil on Portland. 1. Tout and Trade quarries, VJG; Bowers quarry, !.

Schismatomma cretaceum (Hue) Laundon. Occasional on dry bark of ash, maple or oak. 25. **BMH, DOR, RNG**.

Schismatomma decolorans (Turner & Borrer ex Sm.) Clauz. & Vezda. Common in dry crevices of rough bark in woods; sterile. 34. **BMH, DOR, RNG**.

Schismatomma niveum D. Hawksw. & P. James. An old forest indicator, occasional on maple or oak bark. 16. **DOR**.

Schismatomma quercicola Coppins & P. James. Uncommon on alder, holly or oak bark in old woods or parks. 10. **DOR**.

Sclerophytum circumscriptum (Taylor) Zahlbr. Very rare in dry, cherty overhangs. 2. East Weare, VJG; St Aldhelm's Head, BE.

Scoliciosporum chlorococcum (Graewe ex Stenhammar) Vezda. Probably frequent, but overlooked, on eutrophic bark of elder, gorse, sycamore or willow, etc. 7+. **DOR**.

Scoliciosporum pruinosum (P James) Vezda. Rare on sheltered, smooth oak bark. 3. Oakers Wood, VJG; Chetterwood, FR and SRD; Delph Wood, VJG. **BM**.

Scoliciosporum umbrinum (Ach.) Arnold. Probably frequent, but overlooked, on acid stone. 12. **DOR**.

Solenospora candicans (Dickson) Steiner. Common on limestone, especially when horizontal, and abundant on Portland. 34. **BMH, DOR, NOT, RNG**.

Solenospora vulturiensis Massal. Rare on cherty rocks or soil near the sea. 5. Portland Bill, BE; East Weare, VJG; Durdle Pier, BE; Dungy Head, BE, ! ; St Aldhelm's Head, PWJ. **DOR**.

Sphaerophorus globosus (Huds.) Vainio. Rare. 1. In heathy plantation, Stony Down, SL Jury in **RNG**.

Sphinctrina turbinata (Pers.) De Not. An uncommon commensal on *Pertusaria pertusa* on old trees. 5. Melbury Park, VJG; Lulworth Park, VJG; Povington Wood, NAS; Charborough Park, !; Stanbridge, !; Yards Brake, Swanage, !. **DOR, RNG**.

Squamarina cartilaginea (With.) P. James. Locally common

on shallow, calcareous soil near the sea. 6. Portland Bill, !, also VJG; West Weare, ARV, !; East Weare, BE, !; Corfe Castle and Swanage, EMH; Durlston CP, BE. **BMH, DOR, RNG**.

Staurothele caesia (Arnold) Arnold. Very rare on limestone. 1. Portland, PWJ.

Staurothele hymenogonia (Nyl.) Th. Fr. Very rare on limestone. 1. Worbarrow bay, DSR and TDV Swinscow.

Staurothele rupifraga (Massal.) Arnold. Very rare on limestone. 1. Winspit, PWJ.

Stenocybe pullatula (Ach.) B Stein. Occasional on alder boughs in wet places. 8. **RNG**.

Stenocybe septata (Leighton) Massal. Occasional on holly twigs in woods or ravines. 24. **BMH, DOR, RNG**.

Stereocaulon pileatum Ach. Rare on brick walls, resistant to lead pollution. 2. Stourpaine Ch, VJG, !; Parkstone, VJG.

Sticta limbata (Sm.) Ach. Very local on mossy ash or oak bark. 5. Sadborow, ARV; Melbury Park, !; Batcombe Hill, !; Lulworth Park, FR and SRD, also VJG; West Creech, BE; Woodcutts, FR.

Sticta sylvatica (Huds.) Ach. Very local and rare, on mossy bark. 4. Sadborow, ARV; Minterne Park, BE; Lulworth Park, FR and SRD, also VJG; Woodcutts, FR.

Strangospora moriformis (Ach.) B. Stein. Locally common on old wood, sometimes in shade. 4. Hyde Golf Course, VJG; Arne, ! det. BJC; Sandbanks Road and Poole Park, VJG.

Strangospora ochrophora (Nyl.) R. Anderson. A scarce, old forest indicator on mossy bark. 6. Holwell Ch, ! det. A Orange; Oakers Wood, VJG; West Creech and Hill Bottom, BE; Chetterwood, VJG; Drew Coppice, FR.

Strigula jamesii (Swinscow) R.C. Harris. Rare on mortar. 1. Brownsea Is, PWJ, also VJG.

Strigula taylorii (Carroll ex Nyl.) R.C. Harris. Rare on smooth bark. 1. Creech Grange, BE and VJG.

(*Synalissa symphorea* (Ach.) Nyl. A specimen in **BM** from Portland was wrongly named, fide PWJ; there is a duplicate specimen in **NOT**.)

[*Teloschistes chrysophthalmus* (L.) Th. Fr. Very rare on sloe twigs. 1. Lyme undercliff, GL, perhaps in VC 3.]

Teloschistes flavicans (Sw.) Norman. Very local and decreasing on large, free-standing ashes or oaks; it was not uncommon in the 18th century (RP), but todays specimens are stunted (Gilbert, 1996b). 11. Monkton Wyld, GL; Sadborow, ARV; Bracket's Coppice, DHE Warren; Melbury Park, !, *et al.*; Bridport, FH Brightman; Hilfield, !; Batcombe and Hermitage, OLG and VJG; Glanvilles Wootton (1878) JC Dale; Greenhill Down, DHE Warren, !; Tyneham, BE; Lulworth Park, !, *et al.*; Bloxworth, EMH; Hyde Wood, !; Cranborne Chase (1889), Mr Berkeley; Ballard Down, EMH. **BM, DOR, RNG**.

Tephromela atra (Huds.) Hafellner ex Kalb. Common on acid stone, extending to HWM on Portland, also on old brick, and rarely on sycamore bark. 37. **BMH, DOR, RNG**.

Tephromela grumulosa (Pers.) Hafellner & Roux. Rare on acid rock. 1. Agglestone, VJG.

Thelidium decipiens (Nyl.) Krempelh. Scarce on limestone, mortar and chalk. 5. Melbury Ch, PWJ and !; north of Blandford, PWJ; Brownsea Is, PWJ; Tilly Whim, EMH; Ballard Down, OLG.

Thelidium incavatum Mudd. Scarce on limestone. 6. Portland, EMH; Piddlehinton Ch, VJG; Broomhill Bridge, DSR; St Aldhelm's Head, EMH; Church Knowle, DSR; Ballard Down, EMH; **DOR, NOT**.

Thelidium minutulum Koerber. Rare or overlooked on chalk pebbles. 3. Batcombe Ch and Iwerne Courtney Ch, VJG; Ballard Down, OLG.

Thelidium pyrenophorum (Ach.) Mudd. Rare on limestone. 1. Iwerne Courtney Ch, VJG.

Thelidium zwackhii (Hepp) Massal. On sheltered chalk. 2. Near HWM, Studland cliffs, EMH in **DOR**; Hambledon Hill and Compton Down, BE. Probably overlooked; found near Rushmore in VC 8 by PWJ.

Thelopsis rubella Nyl. A local, old forest indicator on bark of big oaks, rarely on lime. 12. **DOR, RNG**.

Thelotrema lepadinum (Ach.) Ach. An old forest indicator, locally frequent on bark, especially on holly, in woods. 21. **BMH, DOR, RNG**.

Tomasellia gelatinosa (Chevall.) Zahlbr. Scarce, or overlooked, on hazel bark or oak twigs in old woods. 9. Ashley Chase and Powerstock Common, !; Lower Kingcombe, BE; Hay Wood and Minterne Park, BE; Melcombe Park, BE; Povington Wood, FR; Tyneham Gwyle, BE; Oakers Wood, PWJ and VJG; Bere Wood, BE; Great Wood, Creech, BE; Woodcutts, FR, !; Studland, EMH, also BE. **DOR**.

Toninia aromatica (Sm.) Massal. Common in crevices of rough limestone or mortar. 38. **BMH, DOR, RNG**.

Toninia episema (Nyl.) Timdal. An occasional commensal on large plants of *Aspilicia calcarea*. 12.

Toninia lobulata (Sommerf.) Lynge. Rare on shallow calcareous soil. 2. St Gabriels, DERC; East Weare, ! det. BJC. **DOR**.

Toninia mesoidea (Nyl.) Zahlbr. Rare on limestone. 1. East Weare, VJG.

Toninia sedifolia (Scop.) Timdal. Locally frequent on shallow, calcareous soil, on the edge of cliffs and on wall-tops; most common on Portland. 12. **BMH, DOR, RNG**.

Toninia verrucarioides (Nyl.) Timdal. Rare on shallow, calcareous soil by tracks. 1. Portland, VJG.

Trapelia coarctata (Sm.) M. Choisy. Frequent on sarsens and on top of brick walls, often in shade and sterile. 20. **BMH, DOR, RNG**. A thallus 380 mm in diameter was seen at Hinton Martel Ch.

Trapelia corticola Coppins & P. James. Rare on oligotrophic bark in humid sites. 3. Melbury Park, BE, FR and NAS; Oakers Wood, VJG; Handcocks Bottom, VJG.

Trapelia involuta (Taylor) Hertel. Occasional or overlooked on acid stone. 4. Iwerne Courtney and Tarrant Gunville, VJG; Fontmell Magna Ch, TWC; Wareham, TWC; Furzey Is, VJG.

Trapelia obtegens (Th. Fr.) Hertel. Scarce on acid stone. 2. Wareham, TWC; on flints, Godlingston Heath, VJG.

Trapelia placodioides Coppins & P. James. Occasional on acid stone, flint or tile. 10.

Trapeliopsis flexuosa (Fr.) Coppins & P. James.Occasional to frequent on wood, also on ball-clay spoil. 7. Hole Common, BE; Hyde Ho, BE; Morden Bog, DSR; Norden, !; Brownsea and Furzey Is, VJG; South Haven peninsula, KL Alvin.

Trapeliopsis gelatinosa (Floerke) Coppins & P. James. Uncommon on acid soil or mossy bark. 2. Oakers Wood, PWJ; Studland Wood, BE.

(*Trapeliopsis glaucolepidea* (Nyl.) G. Schneider. Reported as doubtful (in Watson, 1953), has not been confirmed.)

Trapeliopsis granulosa (Hoffm.) Lumbsch. Locally dominant on peaty soil in heaths, and widespread on wood; on sandstone on Furzey Is, VJG. 26. **BMH, DOR, RNG**.

Trapeliopsis pseudogranulosa Coppins & P. James. Scarce on mossy bark. 1. Melbury Park, VJG.

[*Tylothallia biformigera* (Leighton) P. James & Kilias. On dry, acid, coastal rock. 1. Studland, EMH.]

Usnea articulata (L.) Hoffm. Locally abundant on twigs and branches of big trees in the north and west, less often on fence posts, *Calluna* stems, or hard, heathy soil (at Abbotsbury Castle). 17. First recorded by W. Stonestreet, Winterhays, 1717 in OXF. **BMH, DOR, RNG**.

Usnea ceratina Ach. Occasional to frequent on oligotrophic boles of big trees in woods; seen fertile at Slepe Copse and Morden Park, BE. 19. **BMH, DOR, RNG**.

Usnea cornuta Koerber. Frequent on oligotrophic bark, mostly in woods. 27. **BMH, DOR, RNG**.

Usnea esperantiana Clerc. Scarce on twigs or branches in woods or parks. 7. Monkton Wyld, !; Pilsdon Pen, !; Powerstock Common and Hooke Park, !; Stock Gaylard Park, !; Lulworth Park, FR; East Stoke, DSR; Morden Decoy, FR and SRD. **DOR**.

(*Usnea filipendula* Stirton, reported as *U. dasypoga* (Watson, 1953), probably does not occur.)

Usnea flammea Stirton. Rare on pine bark or wood. 2. Melbury Park, BE and PWJ; atypical in Morden Decoy wood, DSR det. PWJ; Morden Park, BE, NAS and VJG.

Usnea florida (L.) Weber ex Wigg. Very local and declining on sheltered twigs of birch, oak, sloe or *Sorbus* in humid sites, abundantly fertile. 13. **DOR**.

Usnea fragilescens Havaas ex Lynge. Scarce on twigs. 8. Abbotsbury Castle, !: Melbury Park, !; Batcombe Hill, !; Oakers Wood, PWJ and VJG; Tyneham, FR; Arne, DSR; Morden, DSR; Boys Wood, FR.

Usnea glabrescens (Nyl. ex Vainio) Vainio. Rare on oligotrophic bark. 3. Stock Gaylard, !; Lulworth and Slepe, DSR. A critical plant, whose old records need confirmation.

Usnea hirta (L.) Weber ex Wigg. Very rare on oligotrophic bark. 1. On birch, Cranborne Chase, PWJ in **BM**.

Usnea rubicunda Stirton. Frequent on oligotrophic bark in woods, mostly on ash or oak, also on alder and pine. 20. **BMH, DOR, RNG**.

Usnea subfloridana Stirton. Much the commonest *Usnea* in the county, on bark of boles and twigs, on wood, and once on acid stone in Coombe Wood; fertile in many humid sites. 36. Plants occur in central Dorchester. **BMH, DOR, RNG**.

(*Verrucaria aethiobola* Wahlenb. was reported from Tilly Whim by EMH (Holmes, 1906), but is doubtful.)

Verrucaria aquatilis Mudd. On acid rock in or by water. 2. Portland, VJG; Winterbourne Abbas, VJG.

Verrucaria baldensis Massal. Abundant on dry, exposed limestone. 37. **BMH, DOR, RNG**.

Verrucaria caerulea DC. Rare on limestone. 1. Seacombe, ! det. BJC; possibly this at Zelston Ch, TWC.

Verrucaria calciseda DC. Rare on limestone or Greensand. 3. Lyme Regis, GL; Cerne Abbas, ! det. TDV Swinscow; Swanage, EMH in **DOR**.

Verrucaria ditmarsica Erichsen. Rare on limestone or chalk at HWM. 2. White Nothe, ! det. A Fletcher; Winspit, BJC in **BM**.

Verrucaria dolosa Hepp. Locally abundant on shaded flints or chert. 7. Bowers quarry, Portland, !; Durdle Door, OLG; Durweston woods and Hod Hill, VJG; under yew, Hambledon Hill, !; Great Wood, Creech, BE and VJG; Cranborne Chase, FR, PWJ and !; Ballard Down,!, also OLG; Ulwell, VJG; Studland, EMH. **DOR, NOT**.

Verrucaria dufourii DC. On limestone, Portland, VJG. A specimen from Lyme, GL in **BM**, was not this, PWJ. A report from Tilly Whim (Holmes, 1906) is best treated as doubtful.

Verrucaria elaeomelaena (Massal.) Arnold. Perhaps not uncommon on calcareous pebbles in unpolluted streams (Gilbert, 1996a). Spring Bottom, Ringstead, ! in **E** det. BJC.

Verrucaria fusconigrescens Nyl. Rare on chert. 1. East Weare, VJG in **DOR**.

Verrucaria glaucina auct. Brit. Abundant on exposed limestone. 37. **BMH, DOR, RNG**. Records of *V. fuscella* (Turner) Winch and Thornhill are included in this taxon here.

Verrucaria hochstetteri Fr. Frequent on limestone and mortar. 37. **BMH, DOR, RNG**.

Verrucaria hydrela Ach. Locally abundant on pebbles in woodland streams in the west (Gilbert, 1996a). 6. Rye Water Farm, BE, PWJ, VJG and !. **BMH, DOR, RNG**.

Verrucaria macrostoma Dufour ex DC. Frequent, but often overlooked, on limestone. 17. Only recently split from *V. viridula*.

Verrucaria maura Wahlenb. Locally common on vertical stone cliffs between tide marks, but never as prominent as in west Britain. 9. **BMH, DOR, RNG**.

Verrucaria mucosa Wahlenb. Scarce on stone between tidemarks. 4. Blacknor, M Tarraway; Winspit, BJC; Durlston Head, KC; Studland, EMH.

Verrucaria muralis Ach. Abundant on exposed limestone, less common in shade. 38. **BMH, DOR, RNG**.

Verrucaria murina Leighton. Rare or overlooked on flint. 1. Ballard Down, OLG.

Verrucaria nigrescens Pers. Abundant on limestone, less common on weathered flint. 38. **BMH, DOR, RNG**.

Verrucaria papillosa Ach. Rare on Greensand. 1. Iwerne Courtney Ch, PWJ and VJG.

(*Verrucaria pinguicula* Massal. EMH's specimen from St Aldhelm's Head in **DOR** named *V. peloclita* needs redetermination.)

Verrucaria praetermissa (Trevisan) Anzi. On wet limestone pebbles. 1+. In a ravine, Rye Water Farm, PWJ + BE, VJG and !; Bracket's Coppice, VJG; probably this on damp, shaded limestone in a quarry near Kingston, ! det. BJC.

Verrucaria rheitrophila Zschacke. Occasional to frequent on submerged acid pebbles, especially in winterbournes (Gilbert, 1996a). 2+. Winterbourne Abbas, VJG; Stock Gaylard Park ?, !; Bere Regis, VJG.

Verrucaria simplex Mc Carthy. Rare on chalk pebbles. 1. Durdle Door, OLG.

Verrucaria striatula Wahlenb. Occasional on limestone between tidemarks. 2. East Weare, ! det. A Fletcher, also VJG; Chapmans Pool, ! det. TDV Swinscow; Winspit, PWJ. **DOR**.

Verrucaria viridula (Schrader) Ach. Common on eutrophic limestone, concrete, asbestos and old brick, often in shade. 38. **BMH, DOR, RNG**.

Vezdaea aestivalis (Ohl.) Tscherm.-Woess & Poelt. Perhaps overlooked as *Bacidia sabuletorum*, on decaying moss on walls. Fordington Ch, ! det. BJC.

Vezdaea leprosa (P. James) Vezda. Occasional on soil contaminated with zinc from gratings or wire fences. 2. Iwerne Courtney Ch, OLG and VJG; Wareham, VJG.

(*Vezdaea retigera* Poelt & Doebbeler, a northern species reported from near Lydlinch, needs confirmation.)

Wadeana dendrographa (Nyl.) Coppins & P. James. Scarce on shaded bark of large ashes or oaks. 13. Lyme Regis, GL; Hole Common, BE; Powerstock Common, FR and !; Melbury Park and High Stoy, BE; Bascombe Barn, Holworth, BE; Marlpits Wood, BE; Lulworth Park, BJC and FR; Tyneham, FR; Povington Wood and Lower Grounds, BE; west Great Wood, Creech, BE; Langton West Wood and Linnings Copse, BE; Steeple and Rempstone, BE; between Swanage and Studland, EMH; High Wood, Badbury, BE. **BM, E, NOT, RNG**.

Xanthoria calcicola Oxner. Frequent on exposed limestone, asbestos and concrete, rarely on old brick, but less common than *X. parietina*. 35. 20 cm diameter on Bindon Hill. **BMH, DOR**.

Xanthoria candelaria (L.) Th. Fr. Frequent on eutrophic bark at the base of trees, on wood, and on eutrophic horizontal stone. 23.

[*Xanthoria ectaneoides* (Nyl.) Zahlbr. was reported from rock near the Fleet (Watson, 1922), but needs modern confirmation for this north-western taxon.]

Xanthoria elegans (Link) Th. Fr. Uncommon on eutrophic stone, asbestos, granite or old brick. Sadborow, ARV; St Gabriels, DERC; Toller Porcorum Ch, !; Grove, !; Winterborne Stickland, !; Stourpaine Ch, VJG. **DOR**.

Xanthoria parietina (L.) Th. Fr. Abundant on eutrophic bark, twigs, stone, brick and slate, rarely on cast iron; especially common near the sea or by farms. All grid squares. Grey forms lacking parietin are frequent in shade. **BMH, DOR, RNG**.

Xanthoria polycarpa (Hoffm.) Th. Fr. ex Rieber. Frequent on eutrophic twigs, especially on elder; occasional on wood, stones used as bird perches and old rubber tyres. 35. **BMH, DOR, RNG**.

Xylographa vitiligo (Ach.) Laundon. Rare on wood. 1. Melbury Park, FR, PWJ and SRD.

Zamenhofia coralloidea (P. James) Clauz. & Roux. A scarce, old forest indicator on shaded bark. 9. Hooke Park, BE; Melbury Park, FR and !, also VJG; Lulworth Park, FR; Povington Wood, BE, FR and !; Chetterwood, !; Little Wood, Creech, !; Charborough Park, BE and !; Sutton Holms, NAS. **DOR, RNG**.

Zamenhofia rosei (Serusiaux) P. James. Rare on mossy oak bark. 3. Melbury Park, BE, !; Lulworth Park, NAS; Creech Grange, BE and VJG.

CHAPTER 9

DORSET FUNGI

Fungi are a diverse group of more than 100,000 species, all lacking chlorophyll and with haploid spores. They are all parasites or saprophytes, prefer acid habitats and are often treated as neither plants nor animals.

Dorset fungi have attracted little interest until this century. Early records include *Calocybe gambosa* at Weymouth (Stackhouse, 1801) and *Sparassis crispa* at Bradford Abbas (Lees, 1877), but the first and only county list was that of EF Linton in 1914–15; most records were by Linton, with a few by Mrs EW Baker, CB Green and Mrs Pringle. Almost nothing was published from then until the 1960s, and since that time the majority of records are lists from forays which were circulated privately to participants. For example, forays from the Kingcombe centre have added many records, particularly the British Mycological Society's foray in 1990, as a Memorial to Richard Jennings. By far the largest number of records have been assembled since 1971 by JG Keylock, and the Southern Fungus Recording Group which he initiated and led since 1986; the latter has included many well-known mycologists such as Gordon Dickson, Ann Leonard and PD Orton. Other known foray leaders or recorders are:
1960s: M Brett, RD Good (Good, 1961), M Holden, VT and CG Pickering (1962), British Mycological Society (Anon, 1960;1961), Canford School.
1970s: BA Bowen, J Edwards, JA Garfitt, KB Overton, A Simon, Bryanston School.
1980s: B Candy (Candy, 1982), A Horsfall, R Jennings, GF Le Pard, J Madgwick, CH Schofield, D Stephens, Borough of Poole Leisure Services.
1990s: P Austin, VL Breeze, B Collins, V Copp, C Cornell, B Eyles, J Fildes, PA Kauth, AJE Lyon, AA Wason.

The coverage of the county, mainly by forays, has been uneven (Figure 18). This is less a matter of chance than a result of the irregular distribution of old woods, especially those with acid soil which are rich in fungi. Figure 18 shows that seven 10 km × 10 km grid squares, mostly fragmentary, have very few records, while another seven have records of more than 300 Basidiomycetes. These seven, with their main sites of mycological interest, are:
SY59 Kingcombe and Powerstock Common
ST50 Bracket's Coppice and Melbury Park
SY79 Black Heath and Thorncombe Wood
SY89 Bere Forest and Oakers Wood
SY98 Arne Heath and Great Wood, Creech
SZ08 Brownsea Is and Studland Heath
SU01 Castle Hill Wood, Edmondsham and Garston Wood

As yet, relatively few fungi have been recorded from the old woods in Cranborne Chase (ST81 and 91). Because of

the large number of records from Avon Forest Park, SU10 in 'new' Dorset but strictly in VC 11, some of these are included in the lists below. As no Atlas of fungal records is available, it is hard to evaluate the rarity of most British fungi. Fortunately a scholarly account of the New Forest fungi, including *c.*2,100 species and *c.*500 lichens, has recently appeared (Dickson & Leonard, 1996). A comparison of this with the present work, which lists 1,496 fungi and 652 lichens, shows how few Ascomycetes and smaller fungi have been found in Dorset. The county lacks a resident expert, and many common species must await discovery. It is hoped that publication of how little is known in this field will stimulate further work.

Little can be said here about fungal ecology. Most of the larger species need undisturbed conditions, which in practice means that they are found in woods. Here their mycorrhizal associations with tree roots are a subject or current research; the absence of native beech in Dorset means that fungi associated with this tree are much less frequent than they are in the New Forest or the Chilterns.

Subjective observations suggest big differences in the fungal floras of woods on acid and calcareous soils. For example, calcareous woods such as Garston Wood, Creech Great Wood and Hendover Coppice, are richer than acid woods in species of *Lepiota*, but much poorer in species of *Boletus* and *Russula*. Plantations of conifers on heathland are already developing a characteristic association of fungi. Against this, the conversion of old grassland to arable or 'improved' grass must have caused a decline in pastoral species. There has been recent interest in assessing

Figure 18. Number of fungal species per hectad in Dorset: larger Basidcomycetes/Ascomycetes and smaller fungi.

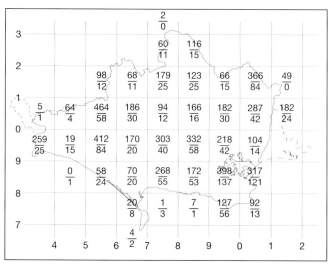

'Waxcap' grasslands from their CHEG profile, where C = no. of species of *Clavaria + Clavulinopsis*, H = no. of species of *Hygrocybe + Dermoloma*, E = no. of species of *Entoloma + Trichopilus* and G = no. of species of *Geoglossum, Microglossum, Spathularia + Trichoglossum* (Rotheroe, Newton, Evans & Feehan, 1996). At present the best Waxcap grasslands known here are at Kingcombe (C4, H12, E5, G1), Hog Cliff (C2, H13, E2, G0) and Brownsea Island (C6, H8, E1, G0), but further work is needed.

Changes in nomenclature and taxonomic classification have accelerated during the last few years, as has the continuing discovery or splitting of new taxa. The problem of arranging species in a county Flora is not simple. There is an accepted checklist of generic, but not of specific, fungal names (Hawksworth, Kirk, Sutton & Pegler, 1995). A revision of Ascomycete names is in progress, and a revision of the 1960 checklist of Agarics and Boleti (Dennis, Orton & Hora, 1960) would be welcome. The following *ad hoc* scheme for major groups has been adopted; within each group, species are listed alphabetically by genus (references for nomenclature are given in brackets). The number of known species in Dorset is given, with the corresponding number for the New Forest (Dickson & Leonard, 1996) given in square brackets.

ASCOMYCETES, omitting lichenised species (Dennis, 1968; Leonard, pers. comm.) – 270 species [461]
DEUTEROMYCETES or imperfect fungi, Coelomycetes and Hyphomycetes (Leonard, pers. comm.) – 64 species [234]
OOMYCETES and ZYGOMYCETES (Erikssen & Hawksworth, 1993; Hawksworth, Kirk, Sutton & Pegler, 1995) – 20 species [34]
BASIDIOMYCETES
UREDINIOMYCETES or rust fungi (Ellis, 1997; Wilson & Henderson, 1966) – 69 species [53]
USTILAGINOMYCETES or smut fungi (Mordue & Ainsworth, 1984) – 15 species [11]
HYMENOMYCETES (Agarics, Boletes and most Bracket Fungi; Bon, 1987; Dennis, Orton & Hora, 1960) – 1,007 species [1,280]
GASTEROMYCETES (Puffballs +) and TREMELLALES (Jelly Fungi; Bon, 1987) – 53 species [58]
(MYXOMYCETES are omitted; 115 species are listed for Dorset (Ing, 1968))

Included in the ensuing list are a few widely used English names, habitats where recorded, and localities only for rarer species. The number given under most taxa is the number of 10 km x 10 km grid squares from which it has been recorded (maximum 37), and is a measure of how widespread the taxon is known to be, at least for the more conspicuous fungi. Herbarium records are few and rarely quoted. Records from Brownsea Island are kept there by the Warden; they are a composite of his own observations with those of visitors, notably the British Mycological Society in the 1960s. Records attributed to JGK may be due to other members of the Southern Fungus Recording Group, especially for the commoner species. Both doubtful and purely VC 11 records are enclosed in brackets; neither are included in the total species count.

ASCOMYCETES

Abrothallus parmeliarum (Sommerf.) Nyl. Frequent on old thalli of *Parmelia*. 6.
Aleuria aurantia (Pers. ex Fr.) Fuckel. ORANGE-PEEL FUNGUS. Frequent on bare soil, either sandy or chalky, especially after disturbance. 15.
Aleuria luteonitens (Berk. & Broome) Gillet. On soil in pine wood, Ferndown, EFL.
Anisogramma vepris (Lacroix) Mesezhko. On bramble stems. 1. Ringstead, BAB.
Anthracobia macrocystis (Cooke) Boud. On burnt soil on heaths. 5. Powerstock Common, AL; Bracketts Coppice, JGK; Worgret, Encombe and Studland, CHS.
Anthracobia maurilabra (Cooke) Boud. Rare on burnt soil. 1. Rempstone Wood, BAB.
Anthracobia melaloma (Alb. & Schwein. ex Fr.) Boud. On burnt wood or soil. 3. Stourpaine and Creech, CHS; Garston Wood, !.
Apiocrea chrysosperma (Tul. & C. Tul.) P. Syd. & Syd. A yellow mould on dying *Boletus*. 7.
Ascobolus albidus P. Crouan. On dung. 1. Woolcombe, BAB.
Ascobolus furfuraceus Pers. ex Fr. On dung. 2. Kingcombe and Powerstock Common, JGK; Creech Grange, BAB; Wytch Farm, !.
Ascocoryne cylichnium (Tul.) Korf. Rare on logs. 1. Between Furzebrook and Norden Farm, BC.
Ascocoryne sarcoides (Jacq.) J.W. Groves & D.E. Wilson. Common on logs. 19.
Ascotremella faginea (Peck) Seavern. Rare on wood. 2. Hog Cliff, JGK and PDO; Kingston, BC.
Aulographum hederae Lib. Scarce on dead holly leaves. 1. Studland, BMS.
Bisporella citrina (Batsch ex Fr.) Korf & S.E. Carp. Common on dead wood. 15.
Bisporella sulfurina (Quel.) S.E. Carp. Occasional on wood. 3. Powerstock Common, JGK; Stubhampton Bottom, JGK; Arne, JGK.
Bulgaria inquinans (Pers.) Fr. Common on dead wood, especially ash. 17.
Calloria fusarioides (Berk.) Fr. On dead nettle stems. 3. Winterbourne Kingston, !; Worth and Studland, CHS.
Calosphaeria pulchella (Pers. ex Fr.) Schroet. On dead cherry wood. 1. Chamberlaynes, BAB.
Calycellina faginea (Schmidt & Arend.) Baralx. On beech leaves. 1. Powerstock Common, JGK.
Calycellina punctiformis (Grev.) Hoehn. On oak leaves. 2. Kingcombe, JGK; Edmondsham, JGK.
Catinella olivacea (Batsch ex Pers.) Boud. On wood. 2. Ringstead, BAB; The Oaks, AL.
Ceratostomella cirrhosa (Pers. ex Fr.) Sacc. On oak wood. 1. Ringstead, BAB.
Chaetosporella fusca (Fuckel) E. Muell. & C. Booth. 1. Hog Cliff, JGK and PDO.

Cheilymenea fimicola (De Not. & Bagl.) Dennis. On dung. 4. Winterborne Kingston, !; Worth and Slepe Heath, CHS.

Cheilymenea stercorea (F.H. Wigg ex Fr.) Boud. On dung. 1. Kingston, BAB.

Chlorencoelia versiformis (Pers.) J.F. Dixon. On rotten wood. 1. Wareham, CHS.

Chlorosplenium aeruginascens (Nyl.) P. Karst. Frequent on dead oak or hazel wood, which it stains blue-green. 20. Forms with discs drying yellow are *C. aeruginosum* (Fr.) De Not.

Ciboria batschiana (Zopf.) N.F. Buckw. Scarce on fallen acorns. Oakers Wood, JGK; Holt Forest, JGK.

Ciboria bolaris (Batsch) Fuckel. Rare. 1. Edmondsham, JGK, det. B Spooner.

Ciborinia hirtella (Boud.) R. Batra & Korf. On leaf litter. 1. Studland, BMS.

Cistella fugiens (Buckn.) Matheis. On rushes. 1. Holt Forest, AL.

Claviceps purpurea (Fr. ex Fr.) Tul. & C. Tul. Ergot. Frequent on seeds of *Molinia*, *Spartina* and other grasses. 9. The sclerotia contain several toxic alkaloids.

Clypeosphaeria notarisii Fuckel. On dead bramble stems. 1. Studland, BMS.

Coleroa robertiana (Fr. ex Fr.) E. Mueller. On *Geranium robertianum* leaves. 6.

Colpoma quercinum (Pers. ex Fr.) Wallr. On oak twigs. 1. Studland, BMS.

Coprobia granulata (Bull. ex Fr.) Boud. Common on cow dung. 12.

Cordyceps capitata (Holmskj. ex Fr.) Link. Rare on *Elaphomyces*. 1. Lyme Regis.

Cordyceps militaris (L. ex Fr.) Link. Occasional on pupae. 7. Hooke Park, Kingcombe and Powerstock Common, JGK; Bracketts Coppice, JGK; Piddles Wood, BAB; Lulworth Castle, CHS; Wareham forest, R Jennings.

Cordyceps ophioglossoides (Ehrh. ex Fr.) Link. Rare on *Elaphomyces*. 2. Puddletown Forest, BAB; Furzebrook, BC.

Creopus gelatinosus (Tode ex Fr.) Link. Scarce on wood. 3. Piddles Wood, JGK; Duncliffe Wood, JGK; Studland JGK.

Cryptodiaporthe aubertii (Westend.) Wehmeyer. Rare on *Myrica* twigs. 1. Morden Bog, BMS.

Cudoniella acicularis (Bull.) J. Schroet. On logs. 6. Arne, JGK; The Oaks, AL; Holt Forest, AL.

Cudoniella clavus (Alb. & Schwein.) Dennis. On rotting twigs in swamps. 1. Kingcombe, AL.

Cudoniella rubicunda (Rehm.) Dennis. Rare on fallen pine cones. 1. SU10.

Cyathicula coronata (Bull.) De Not. ex P. Karst. On dead grass stems. 2. Melbury Park; Edmondsham, JGK.

Cyathicula dolosella (P. Karst.) Dennis. 1. Kingcombe and Powerstock Common, JGK.

Cymadothea trifolii (Pers.) Wolf. Black blotch of clover. 1. Poxwell Big Wood, BAB.

Dacampiosphaeria rivana (De Not) D. Hawksw. On *Caloplaca teicholyta*, Wareham, VJG in **DOR**. 1.

Daldinia concentrica (Bolton ex Fr.) Ces. & De Not. Common on dead ash branches, rarely on birch or sycamore. 30.

Daldinia vernicosa (Schwein.) Ces. & De Not. Occasional on gorse stems. 3. Warmwell, BAB; Blacknoll, !; Brenscombe Hill, BAB; Furzebrook, BC.

Dasyscyphella nivea (Hedw. ex Fr.) Raitr. Frequent on oak or other wood. 9.

Dasyscyphus acerinus (Cooke & Ellis) Cash. On oak. 1. Kingcombe, JGK det. MC Clark; Ringstead, BAB.

Dematioscypha dematiicola (Berk. & Broome) Svrcek. On birch. 1. Studland, BMS.

Diaporthe eres Nitschke. On holly twigs. 2. Brenscombe Hill, BAB; Studland, BMS.

Diaporthe nobilis Sacc. & Speg. On *Laurus*. 1. Studland, BMS.

Diaporthe nucleata (Currey) Cooke. On dead *Ulex*. 1. Studland, BMS.

Diatrype bullata (H. Hoffm. ex Fr.) Fr. On sallow twigs. 1. Shire Rack, Salisbury NHS.

Diatrype disciformis (H. Hoffm. ex Fr.) Fr. Common on dead beech branches. 19.

Diatrype stigma (H. Hoffm. ex Fr.) Fr. Frequent on dead wood. 8.

Diatrypella favacea (Fr. ex Fr.) De Not. Scarce on alder twigs. 1. Bracket's Coppice, A Lyons.

Diatrypella quercina (Pers. ex Fr.) Cooke. On dead oak branches. 1. Edmondsham, EFL.

Diplocarpa bloxami (Berk.) Seaver. Rare on wood. 1. Wytherston Wood, BMS.

Discinella margarita Buckley. Scarce on dead *Molinia* 2. Bere Heath, CHS; Morden Bog, BMS.

Disciotis venosa (Pers.) Boud. On shaded soil. 4. Hendover Coppice, B Collins; Kingston, BC; Crichel Withybeds, EW Baker.

Ditopella ditopa (Fr.) Schroet. On alder twigs. 1. Sugar Hill, BMS.

Dumontinia tuberosa (Hedw.) Kohn. On tubers of *Anemone nemorosa*. 1. Edmondsham, EFL.

Durella commutata Fuckel. On wood. 1. Stoborough, CHS.

Elaphomyces granulatus Fr. ex Fr. Scarce in sandy woodland. 2. Lamberts Castle, JGK; Wareham, CHS.

Elaphomyces muricatus Fr. ex Fr. Scarce in sandy woodland. 2. Puddletown Forest, BAB; Brownsea Is, KC.

Encoelia furfuracea (Roth.) P. Karst. Scarce on branches of alder or hazel. 3. Puncknowle, BAB; Bracketts Coppice, JGK; Chamberlaynes, BAB.

Epichloe typhina (Pers. ex Fr.) Tul. On grass stems. 3. 51; Ringstead, BAB; Edmondsham, EFL.

Erysiphe cichoracearum DC. ex Merat. On dead *Eupatorium* stems. 1. Studland, BMS.

Erysiphe graminis DC. ex Merat. Powdery mildew of grasses and cereal crops, frequent.

Erysiphe polygoni DC. ex Merat. On dead *Brassica* stems. 1. Edmondsham, EFL.

Erysiphe ulmariae Desm. On *Filipendula*. 1. Lox Lane, Motcombe, !, det. J Manners.

Geoglossum cookeianum Nannf. Scarce in sandy turf. 1. Studland, CHS.

Geoglossum fallax Durand. Scarce in sandy turf. 1. Wareham Heath, CHS.

Geoglossum glutinosum Pers. ex Fr. Scarce in sandy turf. 3. Hartland Moor, CHS; Ferndown and Edmondsham, EFL.

(*Geoglossum starbaeckii* was reported from SY98, probably in error.)

Geopora arenosa (Fuckel) Ahmad. Rare in dune slacks. 1. Studland Bay, CHS, also BC.

Gloniopsis levantica (Schwein.) Underw. & Earle. On bramble stems etc. 4. SY67, 77, 97, SZ08.

Guignardia divieri (Vouaux) Sacc. On *Xanthoria*. 1. Swanage, DL Hawksworth.

Gyromitra esculenta (Pers. ex Fr.) Fr. Under conifers on sandy soil, fruiting in spring. 2. Stoborough, CHS; Brownsea Is, KC. It contains a toxin which breaks open red blood cells.

Helvella acetabulum (L.) Quel. Rare in woods on sandy soil. 1. Kingston, BC.

Helvella corium (Weberh.) Massee. Rare in woods. 1. Brownsea Is, KC.

Helvella crispa (Scop. ex Fr.) Fr. Occasional in woods. 14.

Helvella elastica Bull. Scarce in woods. 7. Bracket's Coppice, JGK; Wyke, BAB; Fifehead Wood, JGK; Poxwell Big Wood, BAB; Corfe Castle, CHS; also in ST81 and SU89.

Helvella lacunosa Afzel ex Fr. Occasional in woods or carr. 12.

Helvella leucomelaena (Pers.) Nannf. Rare in woods. 2. SU98; Brownsea Is, KC.

Helvella macropus (Pers. ex Fr.) P. Karst. Occasional in woods. 6. Powerstock Common, JGK; Bracket's Coppice, JGK and PDO; Thorncombe Wood, JGK; Hilfield, B Collins; Hurst Heath, BAB det. DA Reid; Corfe Castle and Studland, CHS; Edmondsham. JGK.

Humaria hemisphaerica (F.H. Wigg ex Fr.) Fuckel. On acid soil in woods. 7.

Hyaloscypha daedaleae Velen. On oak wood 1. Edmondsham, AL.

Hyaloscypha fuckelii Nannf. On wood. 2. Holt Forest, JGK; var. *alniseda* (Velen.) Hutinen, The Oaks, Badbury, AL.

Hyaloscypha herbarum Velen. On dead *Heracleum* stems. 1. Garston Wood, AL.

(*Hyaloscypha hyalina* (Pers. ex Fr.) Boud. On gorse wood, Studland, CHS; a *nomen ambiguum*.)

Hyaloscypha stevensonii (Berk. & Broome) Nannf. On pine wood. 2. Arne, JGK; Avon FP.

Hymenoscyphus albidus (Rob. ex Desm.) W. Phillips. On fallen ash petiole. 1. Hendover Coppice, JGK.

Hymenoscyphus calyculus (Sow.) W. Phillips. On dead twigs. 3. Hendover Coppice, JGK; Poxwell Big Wood, BAB; Castle Hill Wood, EFL.

Hymenoscyphus caudatus (P. Karst.) Dennis. On dead leaves. 4. Kingcombe, JGK; Stubhampton Bottom, JGK; Holt Forest, JGK; Edmondsham, AL.

Hymenoscyphus epiphyllus (Pers. ex Fr.) Rehm. On dead leaves. 1. Thorncombe Wood, JGK.

Hymenoscyphus fructigenus (Bull. ex Fr.) Gray. On fallen acorns. 10.

Hymenoscyphus herbarum (Pers. ex Fr.) Dennis. On bramble stems. 1. Kingcombe, JGK.

Hymenoscyphus phyllophilus (Desm.) O. Kuntze. On beech leaves. 1. Powerstock Common, JGK.

Hymenoscyphus pileatus (P. Karst.) O. Kuntze. On bracken stems. 2. Powerstock Common, JGK; The Oaks, AL.

Hymenoscyphus repandus (W. Phillips) Dennis. On nettle stems. 1. Edmondsham, JGK.

Hymenoscyphus scutula (Pers. ex Fr.) W Phillips. On herbaceous stems. 5. Kingcombe, Thorncombe Wood, Oakers Wood and Edmondsham, JGK; Kingston, BC.

Hypocrea pulvinata Fuckel. On old *Piptoporus betulinus* punks. 1. Furzebrook, BC.

Hypocrea rufa (Pers. ex Fr.) Fr. On old fungi, Studland, JGK.

Hypoderma commune (Fr.) Duby. Rare on dead herbaceous stems. 1. Chamberlaynes, BAB.

Hypoderma hederae De Not. On dead ivy leaves. 1. Church Ope Cove, ! det. AL.

Hypomyces aurantius (Pers. ex Fr.) Tul. & C. Tul. On bracket fungi. 2. Bracket's Coppice, BAB; Hogg Cliff Bottom, JGK.

Hypomyces rosellus (Alb. & Schwein. ex Fr.) Tul. On *Stereum hirsutum*. 2. Abbotsbury and Hyde Wood, BAB.

Hypoxylon cohaerens (Pers. ex Fr.) Fr. On dead beech wood. 1. Coombe Wood, BAB.

Hypoxylon fragiforme (Scop. ex Fr.) Kickx. Common on dead beech wood. 14.

Hypoxylon fraxinophyllum Pouzar. On dead ash wood 1. Kingston, CHS.

Hypoxylon fuscum (Pers. ex Fr.) Fr. On dead sticks, mostly hazel. 21.

Hypoxylon multiforme (Fr. ex Fr.) Fr. On dead birch twigs. 15.

Hypoxylon nummularium Bull. ex Fr. On dead beech twigs. 5.

Hypoxylon rubiginosum (Pers. ex Fr.) Fr. On dead wood. 3. Bracket's Coppice, JGK; Bere Heath, !; Brownsea Is, KC.

Hypoxylon serpens (Pers. ex Fr.) Fr. On dead wood. 2. Hendover Coppice, JGK; Creech, CHS.

Hysterium angustatum Alb. & Schwein. Common on shaded bark, often confused with *Opegrapha* spp. 7.

Hysterium pulicare Pers. ex Fr. On bark. 1. Kingcombe, JGK.

Hysterographium mori (Schwein.) Rehm. Rare on aspen bark. 1. Studland, BMS.

Inermisia fusispora (Berk.) Rifai. Rare on sandy soil with *Polytrichum*. 1. Slepe Heath, CHS.

Iodophanus carneus (Pers.) Korf. On dung. 1. Edmondsham, AL.

Lachenellula occidentalis (G.G. Hahn & Ayers) Dharne. On conifer bark. 2. Broadstone and Edmondsham, EFL.

Lachnum acutipilum (P. Karst.) P. Karst. On wood. 2. Piddles Wood, JGK; Studland, BMS.

Lachnum apalum (Berk. & Broome) Nannf. On dead rushes. 5. Kingcombe and Powerstock Common, AL; Holt Forest and Edmondsham, AL.

Lachnum controversum (Cooke) Rehm. On *Phragmites*. 1. Studland, BMS.

Lachnum diminutum (Rob.) Rehm. On dead rushes. 1. Edmondsham, AL.

Lachnum fugiens (Phill. Rehm. On dead rushes, Girdler's Copse, P. Austin.

Lachnum pulverulentum (Lib.) Sacc. On pine needles. 1. Ashley Heath.

Lachnum sulphureum (Pers.) P. Karst. On nettle or *Myrica* stems. 3. Kingcombe and Puddletown Forest, AL; Morden bog, BMS.

Lachnum virgineum (Batsch.) P. Karst. On beech mast, larch cones or gorse wood. 10.

Lamprospora sp. On burnt soil. 1. Studland, CHS; unlikely to be *L. astroidea* (Hazsl.) Boud., as reported.

Lanzia echinophila (Bull.) Korf. On chestnut husks. 2. Arne and Edmondsham, AL.

Lanzia luteovirescens (Rob. ex Desm.) Dumont & Korf. On *Acer* leaves. 2. Thorncombe Wood and Edmondsham, AL.

Lasiobolus ciliatus (J.C. Schmidt ex Fr.) Boud. On dung. 4. Kingcombe, JGK; Woolcombe, BAB; Kingston, BAB; Wytch Farm, !.

Lasiosphaeria ovina (Pers. ex Pers.) Ces. & De Not. On beech wood. 1. Hog Cliff, JGK.

Lasiosphaeria spermoides (H. Hoffm. ex Fr.) Ces. & De Not. On beech wood. 1. Clifton Wood, JGK.

Leotia lubrica (Scop. ex Fr.) Pers. Occasional in damp woods or hedges. 10.

Leptosphaeria acuta (Hoffm. ex Fr.) P. Karst. On nettle. 3. Woolcombe, BAB; The Oaks and Edmondsham, AL.

Leptosphaeria eustoma (Fuckel) Sacc. On *Carex* stems. 1. Sugar Hill, BMS.

Leptospora rubella (Pers. ex Fr.) Rabenh. On dead stems of umbellifers. 2. Woolcombe and Lulworth, BAB.

Leptotrochila repanda (Fr.) P. Karst. On leaves of *Galium saxatile*. 1. Morden bog, BMS.

Leucoscypha leucotricha (Alb. & Schwein. ex Fr.) Boud. On decaying leaves. 1. Brenscombe Hill, BAB det. DA Reid.

Lophiostoma angustilabrum (Berk. & Broome) Cooke. On dead gorse twigs. 1. Studland, BMS.

Lophiostoma caulium (Fr.) Ces. & De Not. On dead herbaceous stems. 1. Stoke Wake, ! det. BJC.

Lophodermium arundinaceum (Schrad.) Cheval. On dead reeds or sedge stems. 2. Sugar Hill and Studland, BMS.

Lophodermium pinastri (Schrad.) Cheval. On pine needles. 1. Morden bog, BMS.

Lophodermium seditiosum Minter, Stanley & Miller. 1. Powerstock Common, JGK.

Massaria inquinans (Tode ex Fr.) De Not. On dead sycamore branches. 1. Edmondsham, EFL.

Melastiza cornubiensis (Berk. & Broome) J. Moravec. On damp, sandy soil. 6. West Bay, !; Bracket's Coppice, JGK; Puddletown Forest, BAB; Binnegar, !; Worth, CHS.

Microglossum olivaceum (Pers. ex Fr.) Gillet. Rare in grassland. 1. Hartland Moor, CHS.

Microsphaera alphitoides Griffon & Maublanc. Powdery mildew of oak. Frequent. 4+.

Microsphaera euonymi-japonici Viennot-Bourzin. Powdery mildew of *Euonymus japonicus*. 1. 08.

Microsphaera grossulariae (Wallr.) Lev. Powdery mildew of gooseberry. 1+. Winterborne Kingston, !.

Microthyrium ilicinum (De Not.) Sacc. On *Quercus ilex* leaves. 1. Studland, BMS.

Microthyrium microscopicum Desm. On evergreen leaves. 2. Kingcombe, BMS; Morden bog, BMS.

Microthyrium versicolor (Desm.) Hoehn. On dead bramble stems. 1. Kingcombe, BMS.

Mitrophora semilibera (DC. ex Fr.) Lev. Spring fruiting in damp woods. 1. Hendover Coppice, B Collins.

Mitrula paludosa Fr. On wet leaves. 4. Kingcombe, JGK; Warmwell, BAB; Sugar Hill, BMS; Broadstone, Mr Deakin.

Mollisia amenticola (Sacc.) Rehm. On fallen catkins. 1. Kingcombe, JGK.

Mollisia cinerea (Batsch) P. Karst. Frequent on dead wood. 7.

Mollisia cinerella Sacc. On logs. 1. Milton Park Wood.

Mollisia discolor var. *longispora* Le Gal. On wood, Studland, AL. 1.

Mollisia ligni (Desm.) P. Karst. On dead wood. 2. Ringstead and Morden Park, BAB.

Morchella esculenta Pers. ex St Amans. COMMON MOREL. Spring fruiting in eutrophic soil. 8. Rew, HP Manton; Milborne Wood, HP Manton; Winterborne Kingston, !; Stroud Bridge, DAP; Wareham, AH; Witchampton, T Graveson; Durlston CP and Cranborne, BNSS.

Mycosphaerella iridis (Desm.) Schroeter. Probably this on *Iris foetidissima* leaves. 1. Durlston, ! det. J Manners.

Mycosphaerella punctiformis (Pers. ex Fr.) Starb. On fallen leaves. 1. Edmondsham, EFL.

Nectria cinnabarina (Tode ex Fr.) Fr. CORAL SPOT. Common on sticks. 22.

Nectria coccinea (Pers. ex Fr.) Fr. On sticks of ash, elder or *Myrica*. 4. Woolcombe, BAB; Bryanston, CHS; Morden bog, BMS; Edmondsham, EFL.

Nectria ditissima Tul & C. Tul. On apple twigs. 1. Edmondsham, EFL.

Nectria episphaeria (Tode ex Fr.) Fr. On *Xylaria*. 1. 60.

Nectria veuillotiana Roum. & Sacc. On twigs. 2. Winfrith Heath, CHS; Bryanston, BNHS.

Neobulgaria pura (Fr.) Petr. On beech logs. 7.

Neodasyscypha cerina (Pers. ex Fr.) Spooner. On wood. 1. Ringstead, BAB.

Nitschkia collapsa (Romell) Chenant. On wood. 1. Slepe Heath, CHS.

Octospora hetieri (Boud.) Dennis & Izerott. On burnt soil. 1. Brownsea Is, !.

Octospora rutilans (Fr.) Dennis & Izerott. On sandy soil with *Polytrichum*. 1. Stoborough, CHS.

Ophiostoma ulmi (Buisman) Nannf. Elm disease. 6+. Lethal since the late 1960s.

Orbilia curvatispora Boud. On dead *Heracleum* stems. 1. Garston Wood, AL.

Orbilia delicatula (P. Karst.) P. Karst. Frequent on wood. 7.

Otidea alutacea (Pers. ex Fr.) Massee. On soil in woods. 3. Kingcombe, JGK; Tincleton, CHS; Arne, JGK.

Otidea bufonia (Pers. ex Fr.) Boud. Rare in woods. 1. Birches Copse, JGK.

Otidea cochleata (L. ex Fr.) Fuckel. Rare in woods. 1. Wool, CHS.

Otidea leporina (Barsch.) Fuckel. On soil in woods. 4. Oakers Wood, JGK; Brownsea Is, KC; Broadstone, Canford School NHS; Holt Wood, EW Baker.

Otidea onotica (Pers. ex Fr.) Fuckel. On soil in woods. 3. Piddles Wood, JGK; Wool and Studland, CHS; Brownsea Is, KC.

Paraphaeosphaeria rusci (Wallr.) O. Erikss. On *Ruscus*. 2. Studland, BMS; Garston Wood, ! det. AL.

Passeriniella discors (Sacc. & Ellis) Apinis & Chesters. On *Ammophila* or *Spartina*. 1. Studland, BMS.

Peckiella viridis (Alb. & Schwein.) Sacc. On decaying *Lactarius*. 1. Furzebrook, BC.

Peziza acetabulum L. ex St Amans. Brownsea Is, B Collins det. JGK.

Peziza ammophila Dur. & Mont. On sandy cliffs. 1. Canford Cliffs, !.

Peziza ampelina Quel. On acid soil. 1. Kingcombe and Powerstock Common, JGK.

Peziza ampliata Pers. ex Fr. On rotten wood or thatch. 3. Tincleton, M Holden; Kingston, BC; Edmondsham, EFL.

Peziza badia Pers. ex Fr. On sandy soil. 12.

Peziza cerea Sow. ex Fr. On rotting debris. 3. Came Wood, !; Worth, CHS; Kingston, BC.

Peziza echinospora P. Karst. On burnt soil. 5. Bracket's Coppice, JGK; Hendover Coppice, B Collins; Wool, Encombe and Norden, CHS.

Peziza limosa (Grelet) Nannf. On wet mud. 2. Poxwell Big Wood and Brenscombe, BAB det. DA Reid.

Peziza micropus Pers. ex Fr. On rotten wood, once on a doormat. 2. Lulworth Park, BAB; Worth, CHS det. RWG Dennis.

Peziza petersii Berk. & Curtis. On burnt soil 2. Encombe, CHS; Brownsea Is, KC.

Peziza praetervisa Bres. On burnt soil. 1. Slepe Heath, CHS.

Peziza repanda Pers. ex Fr. On woodland soil or sawdust. 8.

Peziza saniosa Schrader ex Fr. Rare on woodland soil. 1. Garston Wood, AL.

Peziza sepiatra Cooke. Rare on soil. 1. Witchampton, EW Baker.

Peziza succosa Berk. On woodland soil. 6. Piddles Wood, JGK; Poxwell Big Wood and Rempstone, BAB; Kingston, BC; Broadstone and Cannon Hill, Canford School NHS.

Peziza varia (Hedw. ex Fr.) Fr. On soil or rotten wood. 2. Warmwell and Lulworth Park, BAB.

Peziza vesiculosa Bull. ex Fr. On rich soil or manure heaps. 9.

Peziziella alniella (Nyl.) Dennis. On alder cones or catkins. 3. Kingcombe and Powerstock Common, BMS; The Oaks, Badbury and Holt Forest, AL.

Peziziella campanuliformis (Fuckel) Dennis. On fern petioles. 2. Kingcombe, BMS; The Oaks, Badbury, AL.

Pezizella chryostigma (Fr.) Sacc. On bracken. 1. Kingcombe, BMS.

Pezizella roburnea Velen. 1. Kingcombe, BMS.

Pezizella rubescens Mouton. 1. Kingcombe, BMS.

Pezizella vulgaris (Fr.) Hoehn. On sticks. 1. Powerstock Common, BMS.

Phacidium multivalve (DC. ex Fr.) Schum. On dead holly leaves. 2. Kingcombe and Studland, BMS.

Phaeohelotium italicum (Sacc.) Dennis. 2. SY89, SZ08.

Phialina lachnobrachya (Desm.) Raitr. 1. Kingcombe, BMS.

Phialina pseudopuberula (Graddon) Raitr. 1. Kingcombe, BMS.

Phyllachora graminis (Pers. ex Fr.) Fuckel. On grass leaves. 3. Langdon Hill, BAB; Carey and Studland, BMS.

Phyllactinia guttata (Fr.) Lev. On hazel leaves. 1. Kingcombe, BMS.

Pleospora herbarum (Pers. ex Fr.) Rabenh. Common on herbaceous debris. 2. Winterborne Kingston, !; Edmondsham, EFL.

Poculum firmum (Pers. ex Fr.) Dumont. On dead oak twigs. 5. Hooke park, JGK; Bracket's Coppice, JGK and PDO; Lulworth Park, BAB; Holt Forest and Edmondsham, AL.

Poculum petiolorum (Rob.) Dumont & Korf. On fallen oak leaves. 2. Powerstock Common, JGK; Edmondsham, AL.

Poculum sydowianum (Rehm.) Dumont. On fallen oak leaves. 3. Kingcombe, BMS; The Oaks, Badbury and Edmondsham, AL.

Podosphaera clandestina (Wallr. ex Fr.) Lev. A mildew on rosaceous trees. 1. Studland, BMS.

Podosphaera leucotricha (Ellis & H.C. Evans) Salmon. Apple mildew. 1+. Carey, BMS.

Poronia punctata (L. ex Fr.) Fr. NAIL FUNGUS. Rare on horse dung. 1. Corfe Common, BC; Hartland Moor, JHSC.

Proliferodiscus pulveraceus (Alb. & Schwein. ex Fr.) Spooner. On dead twigs. 1. Kingcombe, BMS.

Protomyces macrosporus Ung. On *Aegopodium*. 1. Studland, BMS.

Pseudopeziza medicaginis (Lib.) Sacc. On *Medicago* leaves. 1. Kingcombe, BMS.

Pseudopeziza trifolii (Biv.-Bern.) Fuckel. Black blotch on *Trifolium* leaves. 1. Kingcombe, BMS.

Pseudoplectania nigrella (Pers. ex Fr.) Fuckel. Rare on coniferous litter. 1. Rempstone Wood, BAB.

Pyrenidium actinellum Nyl. On *Caloplaca teicholyta*. 1. Wareham.

Rhizina undulata Fr. ex Fr. On coniferous litter. 3. Powerstock Common, JGK; Rempstone Wood and Arne, BAB; Brownsea Is, KC.

Rhopographus filicinus (Fr. ex Fr.) Nitschke ex Fuckel. On bracken or ferns. 9.

Rhytisma acerinum (Pers. ex Fr.) Fr. The common Black spot of sycamore leaves. 23.

Rosellinia mammiformis (Pers. ex Fr.) Ces. & De Not. On ivy. 1. Fleet, BAB.

Rutstroemia fruticeti Rehm. 1. Piddles Wood, JGK.

Sarcoscypha coccinea (Fr.) Lambotte. On sticks, fruiting in spring. 19. The cup is normally red, but a form with an orange cup was seen at Kingston, !. Records may include *S. austriaca*, whose spores have truncate reather than rounded ends.

Sarea resinae (Fr.) Kuntze. On pine resin. 1. Brownsea Is, VJG in **DOR**.

Scutellinia scutellata (L. ex Fr.) Lambotte. On rotten wood. 14.

Scutellinia trechispora (Berk. & Broome) Lambotte. On damp soil. 1. Portland, BE.

Spathularia flava Pers. ex Fr. Under pines. 4. Powerstock Common, JGK; Piddles Wood, VL Breeze; Worgret Heath, CHS; Edmondsham, EFL.

Sphaerotheca pannosa (Wallr. ex Fr.) Lev. Mildew on *Rosa*. 2. Kingcombe, BMS; Carey, BMS.

Tapesia fusca (Pers. ex Fr.) Fuckel. On dead wood of birch or sallow. 4. Kingcombe, JGK; Piddles Wood, JGK; Morden bog, BMS; Birches Copse, JGK.

Taphrina betulina Rostrup. Forming Witches brooms on birch. 13.

Taphrina cerasi (Fuckel) Sadeb. Leaf-curl of cherry. 1+. Winterborne Kingston, !.

Taphrina deformans (Berk.) Tul. Leaf-curl of peach. 1. Edmondsham, EFL.

Taphrina pruni Tul. Leaf-curl of sloe. 1. Gains Cross, !.

Taphrina tosquinetii (Westend.) Magnus. On alder leaves. 1. Carey, BMS.

Tarzetta catinus (Holmsk. ex Fr.) Kork & J.K. Rogers. On soil in woods. 1. Bryanston, BMS.

Thuemenidium atropurpureum (Batsch ex Fr.) O. Kuntze. Rare on dunes. 1. Studland, CHS; Sandbanks, BAB.

Trichoglossum hirsutum (Pers. ex Fr.) Boud. In acid grassland. 3. Kingcombe, JGK; Creech and Hartland Moor, CHS; Studland, CHS.

Trichophaea hemisphaeroides (Mouton) Graddon. On burnt soil 1. Brownsea Is, KC.

Trochila craterium (DC.) Fr. On dead ivy leaves. 2. Kingcombe, BMS; Coombe Wood, BAB.

Trochila ilicina (Nees ex Fr.) Greenh. & Morgan-Jones. Common on dead holly leaves. 16.

Trochila laurocerasi (Desm.) Fr. On dead cherry-laurel leaves. 1. Edmondsham, EFL.

Tuber aestivum Vitt. Rare under beech. 1. Herringston Road, Dorchester, D Cove.

Tuber excavatum Vitt. Rare under beech. 1. Hod Wood, BAB.

Tuber rufum Pico ex Fr. Under soil in woods. 1. Rempstone Wood, BAB det. DA Reid.

Uncinula bicornis (Fr.) Lev. Sycamore mildew. 2. Kingcombe, BMS; Edmondsham, EFL.

Uncinula necator (Schwein.) Burrill. Vine mildew. 1+. Winterborne Kingston, !.

Unguicularia millepunctata (Lib.) Dennis. On wood. 1. Powerstock Common, JGK.

Ustulina deusta (Hoffm. ex Fr.) Lind. On beech stumps. 8.

Valsa ceratophora Tul. & C. Tul. On rose twigs. 1. Studland, BMS.

Venturia inaequalis (Cooke) G. Winter. Apple scab. 2+. Winterborne Kingston, !; Carey, BMS.

Venturia rumicis (Desm.) G. Winter. Abundant on dock leaves. 18+.

Verpa conica Swartz ex Pers. Uncommon in chalky hedgebanks or old pits. 3. Winterborne Kingston, !; Jerry's Hole, Fontmell, AH; Durlston and Herston, CHS;

Xylaria carpophila (Pers. ex Fr.) Fr. On fallen beech mast. 6. Lamberts Castle, JGK; Thorncombe Wood, JGK; Stubhampton Bottom, JGK; Creech Grange, BAB; The Oaks, Badbury, AL.

Xylaria hypoxylon (L. ex Fr.) Grev. Common on stumps. 30.

Xylaria longipes Nitschke. On branches of *Acer* or *Ulex*. 5. Puddletown Forest, AL; Kingston, BC; Edmondsham, JGK.

Xylaria polymorpha (Pers. ex Fr.) Grev. Frequent on stumps. 22.

DEUTEROMYCES
OR IMPERFECT FUNGI

Actinothyrium graminis Kunze. On *Juncus* or *Molinia*. Morden bog, BMS.

Alternaria solani Sorauer? Probably this on tomato. Edmondsham, EFL.

Ascochytulus symphoricarpi (Pass.) Died. On *Symphoricarpos*. Studland, BMS.

Ascodichaena rugosa Butin. Common on beech trunks. Warmwell, Lulworth, Oakers Wood and Morden, BAB.

Aspergillus flavus Link. On herbarium specimens. Edmondsham, EFL.

Aspergillus glaucus Link. On fruit. Edmondsham, EFL.

Aspergillus niger Tiegh. Black mould on fruit. Frequent, !.

Brachysporium masonii Hughes. Arne, JGK.

Botrytis cinerea Pers. Grey mould of lettuce etc. Frequent, EFL, also !.

Calcarisporium arbuscula Preuss. On *Lachnum virgineum*, Studland, AL.

Camarographum metableticum (Trail) Grove. On *Ammophila*. Studland, BMS.

Camarosporium caprifolii. On *Lonicera*. Studland, BMS.

Candida albicans (Robin) Berkh. THRUSH. Common in babies mouths etc.

Ceuthospora phacidioides Grev. On holly. Studland, BMS.

Ceuthospora rhododendri Grove. On *Rhododendron*. Morden bog, BMS.

Coleophoma cylindrospora (Desm.) Hoehn. On holly. Studland, BMS.

Colletotrichum trichellum (Fr.) Duke. On ivy. Studland, BMS.

(***Coniothecium betulinum*** Corda. A *nomen dubium*; on birch. Morden bog, BMS. Perhaps *Trimmatostroma betulinum* (Corda) Hughes.)

Coniothyrium ilicis A.L. Sm. & Ramsb. On holly. Studland, BMS.

Cryptosporium tami Grove. On *Tamus*. Studland, BMS.

Cryptosporium typhae. On *Typha*. Studland, BMS.

Dendryphion comosum Wallr. On dead *Urtica* or umbellifer stems. Puddletown Forest, AL; Studland, BMS.

Diplodina caricis Grove. On dead rushes. Studland, BMS.

Epicoccum nigrum Link. On *Spartina*. Studland, BMS.

Epidermophyton cruris (Castell.) Castell. & Chalm. ATHLETES FOOT. Common on human skin.

Gloeosporidella ribis (Lib.) Petrak. On *Ribes*. Sugar Hill, BMS.

Gloeosporidella rhododendri Briosi & Cavata. On *Rhododendron*. Ilsington, !; Sugar Hill, BMS.

Helminthosporium velutinum Link. On sycamore. Studland, BMS.

Hendersonia culmicola Sacc. On *Spartina*. Studland, BMS.

Hendersonia sarmentorum Westd. On *Laurus nobilis*. Studland, BMS.

Lichenoconium buelliae (Keissl.) D. Hawksw. On *Diploicia canescens*. Kimmeridge, ! det. DL Hawksworth.

Marssonina populi (Lib.) Magnus. On *Populus* x *canescens*. Studland, BMS.

Milospium graphideorum D. Hawksw. On *Lecanactis*. Melbury Park, FR; Wood Street, BE, !; Povington Wood, NAS.

Monostichella salicis (Westd.) Arx. On sallow. Abbotsbury (Anon, 1960).

Paecilomyces farinosus (Dicks. ex Fr.) A.H.S. Br. & G. Sm. Holt Forest, JGK.

Penicillium expansum Link. Blue mould of apples. Frequent, !.

Phoma statices Tassi. On *Limonium*. Studland, BMS.

Phoma typharum Sacc. On *Typha*. Studland, BMS.

Phomatospora dinemaspora Webster. On dead *Molinia* or wood. Morden bog and Studland, BMS.

Phomopsis conorum (Sacc.) Died. On spruce cones. Sugar Hill, BMS.

Phomopsis salicina (Westd.) Died. On *Salix*. Studland, BMS.

Phyllosticta typhina Sacc. & Mallar. On *Typha*. Studland, BMS.

Pollaccia radiosa (Lib.) Bald. & Cif. On poplar. Studland, BMS.

Psammina bommerae Rouss. & Sacc. On *Juncus effusus*. Studland, BMS.

Psammina stipitata D. Hawksw. On *Schismatomma*. Chettle, !.

Pycnostysanus azaleae (Peck) E.W. Mason. On *Rhododendron*. Arne, JGK; Sugar Hill and Morden bog, BMS.

Ramularia ari Fautr. On *Arum* leaves. Lower Bushey, ! det. J Manners.

Ramularia coleosporii Sacc. Abbotsbury (Anon, 1960).

Ramularia farinosa. On *Symphytum*. Sugar Hill, BMS.

Ramularia heraclei (Oudem.) Sacc. On *Heracleum*. Studland, BMS.

Ramularia obliqua. On dock. Studland, BMS. Perhaps *R. pratensis* Sacc.

Ramularia rubella (Bonord.) Nannf. On dock. Studland, BMS.

Rhizopus nigricans Ehrh. Frequent on decaying fruits, !.

Septoria euonymi Rabenh. On *Euonymus japonicus* leaves. Durlston, CBG (Dennis & Hubbard, 1962); Studland, BMS.

Septoria scabiosicola Desm. On *Knautia* leaves. Winterborne Kingston, ! det. J Manners.

Septoria stellariae (Rob.) Desm. On *Stellaria*. Sugar Hill, BMS.

Sphaeropsis sapinea (Fr.) Dyke & B. Sutton. On pine cones. Studland, BMS.

Sporidesmium lobatum Berk. & Broome. On Pine. Morden bog, BMS.

Sporobolomyces roseus Kluyver & van Niel. Frequent on cereal leaves. Sandbanks, BAB.

Stagonospora innumerosa (Desm.) Sacc. On rushes. Morden bog, BMS.

Stagonospora socia Grove. On rushes. Morden bog, BMS.

Stagonospora subseriata (Desm.) Sacc. On *Molinia*. Morden bog, BMS.

Torula herbarum (Pers.) Link. On nettle stems. Studland, BMS.

Trichothecium roseum (Pers.) Link. Apple rot. Frequent, EFL, also !.

Varicosporium elodeae W. Kegel. In water. Studland, BMS.

OOMYCETES AND ZYGOMYCETES

Achlya racemosa Hildebrand. In water. Carey, BMS.

Albugo candida (Pers.) O. Kuntze. On crucifers such as *Capsella*. Winterborne Kingston, !.

Albugo lepigoni (de Bary) O. Kuntze. On *Spergularia*. Lyme Regis, ! in **K**, det. BM Spooner; also HNR and WF (1884) in **K**.

Leptolegnia caudata de Bary. In water. Morden bog, BMS.

Mucor hiemalis Wehmer. A mould. Brownsea Is, KC.

Peronospora digitalis Gaum. On foxglove. Lulworth Castle, BMS.

Peronospora oerteliana Kuehn. On primroses. Carey, BMS.

Peronospora ranunculi Gaum. On buttercups. Carey, BMS.

Peronospora sordida Berk. On *Scrophularia nodosa*. Lulworth Castle, BMS.

Peronospora symphyti Gaum. On comfrey. Sugar Hill, BMS.

Peronospora trifoliorum de Bary. On *Trifolium dubium*. Sugar Hill, BMS.

Peronospora viciae (Berk.) Casp. On *Vicia sepium*. Rempstone Wood, BMS.

Phytophthora infestans (Mont.) de Bary. Blight of potato and tomato. Frequent, EFL, also !.

Pilaira anomala (Ces.) Schroeter. On dung. Bracket's Coppice and Morden Park, BAB.

Pilobolus crystallinus (Wiggers) Tode. On dung. Morden Park, BAB.

Plasmopara nivea (Unger) Schroeter. On umbellifers. Sugar Hill and Studland, BMS.

Plasmopara viticola (Berk. & Curt. ex de Bary) Berl. & de Toni. Vine Mildew. Winterborne Kingston, !.

Pythium undulatum H.E. Petersen. In water. Carey, BMS.

Saprolegnia paradoxa (a *nomen dubium*). In water. Studland, BMS.

Spinellus fusiger (Link) Tegh. On *Mycena*. Kingcome and Hooke Park, BMS.

BASIDIOMYCETES
UREDINIOMYCETES
OR RUST FUNGI

Coleosporium campanulae (Pers.) Lev. On *Campanula trachelium* leaves. Winterborne Kingston, !; Bushey, CHS. Perhaps a form of *C. tussilaginis*.

Coleosporium senecionis (Pers.) Fr. On *Senecio vulgaris*. Sugar Hill, BMS; Edmondsham, EFL.

Coleosporium tussilaginis (Pers.) Lev. On *Tussilago* or *Petasites* leaves. Forston and Binnegar, !; Bushey, CHS.

Cronartium flaccidum (Alb. & Schwein.) G. Winter. Abbotsbury (Anon, 1960).

Cumminsiella mirabilissima (Peck) Nannf. On *Mahonia*. Carey, BMS.

Kuehneola uredinis (Link) Arthur. On bramble stems. Ringstead, BAB; Worth, CHS.

Melampsora caprearum Thuem. Frequent on sallow. SY68, 78, 89, 97, 98.

Melampsora epitea Thuem. On sallow. Tatton Coppice, BAB; East Holme, CHS.

Melampsora euphorbiae Cast. On leaves of annual spurges. Edmondsham, EFL.

Melampsora hypericorum G. Winter. On *Hypericum* leaves. Creech Grange, BAB.

Melampsora populina Lev. On poplar leaves. Edmondsham, EFL.

Melampsora populnea (Pers.) P. Karst. On *Mercurialis* and *Populus* leaves. Shire Rack, Salisbury NHS; Studland, CHS.

Melampsorella symphyti Bubak. On comfrey leaves. Fleet, BAB; East Holme and Luckford Lake, CHS; Sugar Hill, BMS.

Melampsoridium betulinum (Fr.) Kleb. On birch leaves. Kingcombe, BMS; Rempstone Wood, BAB; Studland, CHS; Edmondsham, EFL.

Milesina dieteliana (Dyd.) Magnus. On *Polypodium*. Studland, BMS.

Milesina scolopendrii (Faull) D.M. Hend. On *Phyllitis*. Encombe, CHS.

Phragmidium bulbosum (Strauss) Schlecht. On bramble. Worth, CHS.

Phragmidium mucronatum (Pers.) Schlecht. On *Rosa*. Wyke Wood, BAB; Stourpaine, CHS; Bushey and Middlebere Heath, CHS; Edmondsham, EFL.

Phragmidium rubi-idaei (DC.) Karst. On raspberry. Worth, CHS.

Phragmidium sanguisorbae (DC.) Schroeter. On *Sanguisorba minor*. St Aldhelm's Head, CHS.

Phragmidium tuberculatum J. Mueller. On *Rosa*. Morcombelake, CHS.

Phragmidium violaceum (C.F. Schultz) G. Winter. Abundant on bramble leaves. 19.

Puccinia acetosae Koern. On *Rumex acetosa*. Monkton Wyld and Puncknowle, BAB; Carey, BMS.

Puccinia albescens Plowr. On *Adoxa*. Creech, CHS.

Puccinia adoxae DC. On *Adoxa*, usually commoner than *P. albescens*. Bere Wood, !.

Puccinia annularis (Strauss) Roehl. On *Teucrium scorodonia*. Slepe Heath, CHS.

Puccinia arenariae (Schum.) G. Winter. On *Silene*. Wyke and Lulworth, BAB; Corfe and Studland, CHS.

Puccinia brachypodii Otth. On *Anthoxanthum* or *Brachypodium*. Bushey, CHS, also BMS; Carey, BMS.

Puccinia brunellarum-moliniae Cruchet. On *Prunella*. Stubhampton Bottom, JGK.

Puccinia buxi DC. On *Buxus*. Bryanston, CHS; Carey, BMS; Edmondsham, EFL.

Puccinia caricina DC. On *Carex, Ribes* and *Urtica*. ST50, SY68, 78, 88, 98.

Puccinia chrysanthemi Roze. On *Dendranthema* leaves. Edmondsham, EFL.

Puccinia chrysosplenii Grev. On *Chrysosplenium*. Morden Park, BAB.

Puccinia circaeae Pers. On *Circaea* leaves. Puncknowle, BAB; Bracket's Coppice, JGK; Poxwell Big Wood, BAB; Creech Grange, BAB.

Puccinia coronata Corda. On grass leaves. Wyke, Ringstead, East Holme and Creech Grange, BAB.

Puccinia distincta McAlpine. On *Bellis perennis* leaves. Knighton pits and Weymouth, ! det. J Webster. A recent arrival from Australia.

Puccinia elymi West. On *Ammophila*. Studland dunes, BMS.

Puccinia graminis Pers. On *Berberis* or grasses. East Stoke and Worth, CHS.

Puccinia heraclei Grev. On hogweed. Poxwell Big Wood, BAB.

Puccinia hieracii Martinez. On *Leontodon* and *Taraxacum*. Carey, Sugar Hill and Studland, BMS.

Puccinia holcina Erikss. On *Holcus*. Kingcombe, BMS.

Puccinia iridis Rabenh. On *Iris foetidissima* Abbotsbury, BAB; Stourpaine, CHS.

Puccinia malvacearum Corda. On *Alcea* or *Malva*. SY67, 89, 97, SZ07, 08, SU01.

Puccinia menthae Pers. On mint. Ringstead, BAB; Luckford Lake, Worth, Creech and Bushey, CHS.

Puccinia obscura Schroeder. On *Luzula*. Hyde Wood, BAB; Studland, BMS.

Puccinia phragmitis (Schum.) Koern. On *Phragmites*. Abbotsbury (Anon, 1961); Ringstead, BAB; East Stoke, CHS.

Puccinia poarum Niels. On grasses or *Tussilago*. Worth, CHS.

Puccinia primulae Duby. On *Primula*. 68, 78, 89, 98, 08, 01.

Puccinia pulverulenta Grev. On *Epilobium hirsutum*. Worth, CHS.

Puccinia punctiformis (Strauss) Roehl. On *Cirsium arvense* +. SY48, 88, 97, 98, SZ08.

Puccinia recondita Rob. ex Desm. On *Anisantha, Bromus* or *Holcus*. SY58, 78, 89, 97, 98, 99, SZ08.

Puccinia sessilis Schroeter. On *Arum* or *Allium ursinum*. Oakers Wood, BAB; Stoborough and Bushey, CHS; Carey, BMS; Shire Rack, Salisbury NHS.

Puccinia smyrnii Biv.- Bernh. On alexanders. Portland Bill and Church Ope Cove, BAB, also CHS; Sandsfoot, !; Wareham, !.

Puccinia thesii Duby. Rare, on *Thesium*. Above Tilly Whim, CBG and RVS.

Puccinia veronicae Schroeter. On *Veronica montana*. Lulworth Castle, BAB; Rempstone Wood, BMS.

Puccinia violae DC. On violets. Puncknowle, BAB; Carey, BMS.

Tranzschelia pruni-spinosa (Pers.) Diet. On plum leaves. Edmondsham, EFL.

Triphragmium ulmariae (DC.) Link. On meadowsweet leaves. East Stoke, Wool and Hartland Moor, CHS; Carey, BMS.

Uromyces betae Kickx. On spinach leaves. Loders, R Jennings.

Uromyces dactylidis Otth. On *Ranunculus* or grasses. SY49, 58, 78, 89, 97, 98, ST50, 60.

Uromyces geranii (DC.) Fr. On *Geranium* leaves. Morcombelake, CHS; Studland, BMS.

Uromyces junci (Desm.) Tul. On *Pulicaria dysenterica*. Chapmans Pool, BMS.

Uromyces limonii (DC.) Berk. On sea lavender leaves. Studland, BMS.

Uromyces lineolatus (Desm.) Schroeter. On *Bolboschoenus*. Weymouth, CP Hurst (Hurst, 1923).

Uromyces minor Schroeter. Rare on *Trifolium dubium*. Worth, CHS.

Uromyces muscari (Duby) Graves. On bluebell leaves. Winterborne Kingston, ! det. J Manners; Studland, CHS, also BMS.

Uromyces polygoni -aviculare (Pers.) P. Karst. On knotgrass. Bridport (Anon, 1961); Kingcombe, BMS.

Uromyces valerianae Fuckel. On valerian leaves. Luckford Lake and Bushey, CHS.

Uromyces viciae-fabae (Pers.) de Bary. On broad bean leaves and pods. Winterborne Kingston, !.

USTILAGINOMYCETES OR SMUT FUNGI

Entyloma microsporum (Unger) Schroeter. On buttercups. Studland, BMS.

Entyloma serotinum Schroeter. On comfrey. (Mordue & Ainsworth, 1962)

Farysia thuemenii (A.A. Fisch. Waldh.) Nannf. On *Carex riparia*. Studland, BMS.

Melanotaenium endogenum (Unger) de Bary. Rare on *Galium*. Swanage, RVS.

Thecaphora deformans Dur. & Mont. On seeds of *Ulex minor*. (Mordue & Ainsworth, 1962).

Thecaphora seminis-convolvuli (Desm.) Liro. On seeds of *Calystegia* or *Convolvulus*. (Mordue & Ainsworth, 1962).

Urocystis anemones (Pers.) G. Winter. On *Ranunculus repens*. Ringstead, BAB; Studland, BMS.

Urocystis fischeri Koern. On *Carex flacca*. (Mordue & Ainsworth, 1962).

Ustilago cynodontis (Pass.) Henn. On *Cynodon*. Sandbanks, (Dennis & Leonard, 1996; Mordue & Ainsworth, 1962).

Ustilago hypodytes Fr. On *Ammophila*. Sandbanks, BAB.

Ustilago intermedia Schroeter. On *Scabiosa*. (Mordue & Ainsworth, 1962).

Ustilago marina Dur. de Mais. On *Eleocharis parvula*. Little Sea, EMH in **K**.

Ustilago segetum (Bull.) Roussel. Loose Smut of cereals. Worth, CHS; Edmondsham, EFL.

Ustilago succisae Magnus. On *Succisa*. Bracket's Coppice, JGK and PDO.

Ustilago violacea (Pers.) Roussel. On *Silene* anthers. Pilsdon Pen, BAB.

HYMENOMYCETES: AGARICS, BOLETI AND BRACKET FUNGI

Abortiporus biennis (Bull. ex Fr.) Singer. On buried wood. 6. Bride Head, MH; Kingcombe, Hooke Park, Melbury Park and Bracket's Coppice, JGK; Moigne Combe, BAB; Tincleton, MH.

Agaricus abruptibulbus Peck. In grass. 3. Newlands Batch, Bracket's Coppice and Foxhills, JGK.

Agaricus arvensis Schaeff. HORSE MUSHROOM. Occasional in pastures. 21.

Agaricus augustus Fr. In gardens or under trees. 5. Wootton Fitzpaine, BAB; Grays Wood, Piddlehinton and Race Down, !; Brownsea Is, KC.

Agaricus bernardii Quel. In pastures near the sea. 2. West Hill, Worth and Corfe Common, !.

Agaricus bisporus (Quel.) Sacc. CULTIVATED MUSHROOM. In manured grass. 5. Long Burton, BAB; Ringstead and Bere Regis, !; Bere Wood, JGK; Kingston, BC.

Agaricus bitorquis (Quel.) Sacc. In rich soil or verges. 3. Hurst Heath, !; Bryanston and Brownsea Is, BAB.

Agaricus bresadolianus Bohus. In calcareous grass. 2. Hog Cliff and Hendover Coppice, JGK.

Agaricus campestris L. ex Fr. FIELD MUSHROOM. Frequent in old pastures, especially when manured, but sporadic. 25.

Agaricus comtulus Fr. In grass or under oak. 2. Bere Heath, !; Stubhampton Bottom, JGK.

Agaricus excellens (Moeller) Moeller. In calcareous woods. 1. Oatclose Wood, !.

Agaricus fuscofibrillosus (Moeller) Pilat. In deciduous woods. 4. Newlands Batch, Kingcombe, Bracket's Coppice and Edmondsham, JGK.

Agaricus haemorrhoidarius Schulzer. In grassy places. 4. Kingcombe, AJEL; Portland, Hog Cliff and Hendover Coppice, JGK.

Agaricus impudicus (Rea) Pilat. In pastures near the sea. 1. Swyre, (SY58) JGK.

Agaricus langei (Moeller) Moeller. In deciduous woods. 4. Clifton Wood and Arne, JGK; Durlston CP and Brownsea Is, KC.

Agaricus lanipes Moeller & Schaeff. In woods. 1. Slepe Farm, M Tuck det. JGK.

Agaricus macrosporus (Moeller & Schaeff.) Pilat. A northern species, in pastures. 5. Powerstock Common, AJEL; Kingcombe, Hendover Coppice and Foxhills, JGK; Above Dancing Ledge, Corfe Common and Arne, !.

Agaricus placomyces Peck. Occasional in more or less wet woods. 8.

Agaricus porphyrizon Orton. In woods or hedgebanks. 3. Bracket's Coppice, Hog Cliff and Hendover Coppice, JGK.

Agaricus porphyrocephalus Moeller. In grassland near the sea. 1. Durlston CP, !.

Agaricus purpurellus (Moeller) Moeller. Under beech. 1. Bryanston, BNHS.

Agaricus romagnesi Wasser. In grass. 1. Hog Cliff, JGK.

Agaricus semotus Fr. Under trees. 3. Powerstock Common, JGK; Oakers Wood, BAB and !; The Oaks, Badbury, JGK.

Agaricus silvaticus Schaeff. ex Fr. Under pines. 14.

Agaricus silvicola (Vitt.) Peck. Frequent in woods or grassy rides. 21.

Agaricus subperonatus (J. Lange) Singer. By manure heaps. 1. Portland, !.

Agaricus vaporarius (Vitt.) Moser. In rich soils, once on fixed shingle. 5. Small Mouth, Bere Wood, Stokeford and Godlingston Heaths, !; Arne, JGK; Durlston Park, !.

Agaricus variegans Moeller. In grass or under trees. 5. Swyre, JGK; Bere Wood, Sherford Bridge and Arne, !; Bryanston, BNHS.

Agaricus xanthodermus Genev. YELLOW-STAINING MUSHROOM. 5. Bracket's Coppice, JGK; Bryanston, BNHS; Spyways Barn, BNSS; Avon FP, J Goss. Toxic.

Agrocybe cylindracea (DC. ex Fr.) Maire. On ash or willow stumps. 4. Kingcombe, JGK; Ringstead, BAB; Fontmell Magna, !; Swanage, CBG.

Agrocybe dura (Bolt. ex Fr.) Singer. In grass. 2. Bracket's Coppice, M Brett; Ringstead, BAB.

Agrocybe erebia (Fr.) Kuehn. In damp woods. 3. Hethfelton and Great Ovens Heath, !; Bryanston, BNHS.

Agrocybe praecox (Pers. ex Fr.) Fayod. In grass, fruiting in spring. 3. Hog Cliff, JGK; Athelhampton and Sovell Down, !.

Agrocybe semiorbicularis (Bull.) Fayod. In grass. 4. Bracket's Coppice, M Brett; Broadstone and Daggons, EFL; Castle Hill Wood, J Edwards.

Alboleptonia sericella (Fr. ex Fr.) Largent & Benedict.In grass. 3. Powerstock Common, JGK; Oakers Wood, BAB; St Giles Park, EFL.

Amanita citrina (Schaeff.) Gray. Frequent in deciduous woods, mostly var. *alba* (Gillet) Gillet. 17.

Amanita echinocephala (Vitt.) Quel. Rare in calcareous woods. 1. Reported from ST80.

Amanita excelsa (Fr.) Kumm. Occasional in damp, deciduous woods. 13.

Amanita fulva Schaeff. TAWNY GRISETTE. Frequent in deciduous woods on clay. 17.

Amanita inaurata Secr. In deciduous woods. 4. Kingcombe, Bracket's Coppice, Melbury Park and Hendover Coppice, JGK; Poxwell Big Wood, BAB.

(*Amanita lividopallescens* (Gilb.) Gilb. & Kuehn. Possibly this in Birches Copse, JGK.)

Amanita muscaria (L. ex Fr.) Hook. FLY AGARIC.Frequent under birch, rare under pine. 19. It contains toxic hallucinogens such as ibotenic acid and muscarine.

Amanita pantherina (DC. ex Fr.) Krombh. Uncommon in deciduous woods; toxic. 9.

Amanita phalloides (Vaill. ex Fr.) Link DEATH-CAP. In oak woods or grass. 21. Var. *alba* noted in Bracket's Coppice, JGK. It contains amatotoxins and phallotoxins, both dangerous cell poisons.

Amanita regalis (Fr.) Michael. Rare under spruce; toxic. 2. Powerstock Common and Bracket's Coppice, JGK.

Amanita rubescens (Pers. ex Fr.) Gray. THE BLUSHER. Common in woods. 20. Var. *annulosulphurea* found once in Bracket's Coppice, JGK.

Amanita solitaria (Bull. ex Fr.) Secr. Once reported from the edge of a wood at Sutton Holms, EFL; the plant might have been *A. echinocephala*.

Amanita submembranacea (Bon) Groeger. Under conifers. 2. Powerstock Common and Birches Copse, JGK.

Amanita vaginata (Bull. ex Fr.) Vitt. GRISETTE. Occasional in deciduous woods. 14.

Amanita virosa (Fr.) Bertol. DESTROYING ANGEL. Rare in deciduous woods on acid soil. 4. Sherford Bridge, AA Wason; Froxen Copse, KB Overton; Delph Wood, Canford School NHS; Castle Hill Wood, J Edwards. Equally toxic as *A. phalloides*.

Antrodia albida (Fr. ex Fr.) Donk. On wood. 2. Bracket's Coppice, M Brett; Holt Forest, JGK.

Antrodia albobrunnea (Romell) Ryvarden. On wood. 1. Bere Heath, !.

(*Antrodia serialis* (Fr.) Donk. On wood. 1. Avon FP, VC 11, J Banks.)

Antrodia xantha (Fr.) Ryvarden. On coniferous wood. 1. Reported from SY39.

Armillaria bulbosa (Barla) Kyle & Watling. Perhaps frequent on deciduous wood. 4. Kingcombe, Powerstock Common, Bracketts Coppice and Studland, JGK; Kingston, BC.

Armillaria mellea (Vahl ex Fr.) Kumm. HONEY FUNGUS. Common on deciduous wood. 30.

Armillaria ostoyae Romagn. On deciduous wood. 1. Birches Copse, JGK.

Armillaria tabescens (Scop. ex Fr.) Emel. On stumps. 3. Kingcombe, Bracket's Coppice and Piddles Wood, JGK.

Arrhenia acerosa (Fr. ex Fr.) Kuehn. On soil. 5. Powerstock Common, MH; Kingcombe and Piddles Wood, JGK; Lulworth Castle, BMS; Winterborne Kingston, !; Rempstone Wood, D Stephens.

Auriscalpium vulgare (Fr.) P. Karst. On fallen pine cones in the Poole basin. 8.

Baeospora myosura (Fr. ex Fr.) Singer. On fallen pine cones. 16.

Bjerkandera adusta (Willd. ex Fr.) P. Karst. On dead or living wood. 20.

Bjerkandera fumosa (Fr.) P. Karst. On dead wood. 1. Brownsea Is, KC.

Bolbitius vitellinus (Pers. ex Fr.) Fr. Frequent in manured grassland. 18.

Boletus: see also *Chalciporus*, *Leccinum*, *Suillus* and *Xerocomus*.

Boletus albidus Rocques. Rare in calcareous oak woods. 2. Kingcombe and Bracket's Coppice, AJEL.

Boletus appendiculatus Schaeff. ex Fr. Under oak. 2. Gallows Hill, !; Pamphill, J Fildes.

Boletus calopus Fr. In acid woodland soil. 3. Lamberts Castle, JGK; Bracket's Coppice, M Brett; Oakers Wood, BAB.

Boletus edulis (Bull.) Fr. PENNY BUN, CEP. Frequent in deciduous woods. 18.

Boletus erythropus (Fr. ex Fr.) Pers. Occasional in woods or acid banks. 9.

Boletus impolitus Fr. Uncommon under oak. 3. Kingcombe, JGK; Lytchett Minster and Delph Wood, Canford School NHS.

Boletus lanatus Rostk. Uncommon under oak. 2. Kingcombe and Powerstock Common, AJEL.

Boletus luridus Fr. Under oak. 6. Bracket's Coppice, J Madgwick; Hog Cliff, JGK; Oakers Wood, BAB; Bovington and Bramble Bottom, !; Furzebrook, BC; Morden Park, BNSS.

Boletus porosporus (Imler) Watling. In woods. 4. Hog Cliff, JGK; Westrow Ho, !; Hurst Heath, JGK; Rempstone Wood, D Stephens.

Boletus queletii Schulzer. Under oak or in grass. 3. Kingcombe and Oakers Wood, JGK; Upton CP, PBLS.

Boletus reticulatus (Schaeff.) Boud. In deciduous woods. 1. Post Green, !.

Boletus satanoides Smotl. Under oak. 1. Oakers Wood, JGK.

Bulbillomyces farinosus (Bres.) Juel. On dead twigs. 2. Oakers Wood, JGK; Studland, AL.

Byssomerulius corium (Fr.) Parmasto. On dead gorse and other twigs. 19.

Calocybe carnea (Bull. ex Fr.) Donk. In grassland. 5. Powerstock Common, AJEL; Kingcombe, Hog Cliff and Puddletown Forest, JGK; Broadwindsor and Ringstead, BAB.

Calocybe constricta (Fr.) Kuehn. In rich grassland. 1. Hog Cliff, JGK and PDO.

Calocybe gambosa (Fr.) Donk. ST GEORGES MUSHROOM. In rich grassland on more or less calcareous soil, fruiting in spring. 16. The first fungus to be recorded from Dorset (Stackhouse, 1801).

Calyptella campanula (Nees ex Pers.) Cooke. On dead foxglove stem. 1. The Oaks, Badbury, AL.

Calyptella capula (Holmsk. ex Pers.) Quel. On dead stems. 3. Kingcombe, Powerstock Common. Bracket's Coppice and Kingsettle Wood, JGK.

Cantharellula umbonata (Fr.) Singer. In acid woods. 1. Bloxworth, V Copp.

Cantharellus cibarius Fr. CHANTERELLE. On hard ground in woods on acid soils. 15.

Cantharellus cinereus (Pers.) Fr. In deciduous woods. 3. Hurst Heath !; Thorncombe Wood, JGK; Brenscome Hill, BAB.

Cantharellus infundibuliformis Fr. Occasional in woods. 12.

Ceriporia viridans (Berk. & Broome) Donk. On dead wood. 1. Hooke Park, JGK.

Cerrena unicolor (Fr.) Murrill. On beech. 1. Sutton Holms, WRL.

Chaetoporus euporus (P. Karst.) Boud. & Singer. 2. Bracket's Coppice and Hendover Coppice, JGK.

Chalciporus piperatus (Bull. ex Fr.) Battaille. In acid woods. 5. Powerstock Common, Hurst Heath and Avon FP, JGK; Tincleton, MH; Affpuddle, BAB; Canon Hill, Canford School NHS.

Chondrostereum purpureum (Fr.) Pouzar. Frequent on dead wood, including birch and yew, and the cause of 'Silverleaf' in plum. 19.

Chroogomphus rutilus (Schaeff. ex Fr.) Miller. Under pines. 6. Bracket's Coppice and Hendover Coppice, JGK; Hurst Heath, BAB, also JGK; Hethfelton, !; Avon FP, JGK.

Claudopus sericeonitidus (P.D. Orton) comb. nov. Brownsea Is, KC. 1.

Clavaria acuta (Fr.) Corner. In short turf. 5. Powerstock Common, Bracket's Coppice, Clifton Wood and Garston Wood, JGK; Brownsea Is, KC.

Clavaria argillacea Fr. In short turf. 2. Stonebarrow; Tincleton, MH.

Clavaria fumosa Fr. In short turf. 2. Parnham and Puddletown Forest, BAB.

Clavaria vermicularis (Fr.) Corner. In lawns. 3. Clifton Wood, JGK; Winterborne Kingston, !; Edmondsham, EFL.

Clavariadelphus contortus (Fr.) Corner. On birch wood. 3. Kingcombe, Powerstock Common, Bracketts Coppice and Kingsettle Wood, JGK. Perhaps a variety of *C. fistulosus*.

Clavariadelphus fistulosus (Fr.) Corner. On birch wood. 10.

Clavariadelphus junceus (Fr.) Corner. On leaf litter in woods. 5. Kingcombe, Powerstock Common and Hendover Coppice, JGK; Bracket's Coppice, M Brett; Brownsea Is, KC; Avon FP, J Goss.

Clavariadelphus pistillaris (Fr.) Donk. Rare under beech. 1. Bracket's Coppice, JGK.

Clavulicium delectabile (Jacks.) Hjortst. Rare on conifer logs. 1. Piddles Wood, JGK.

Clavulina cinerea (Fr.) Schroeter. In deciduous woods. 15.

Clavulina cristata (Fr.) Schroeter. Frequent in deciduous woods. 19.

Clavulina rugosa (Fr.) Schroeter. Under oak. 10.

Clavulinopsis corniculata (Fr.) Corner. In lawns. 10.

Clavulinopsis fusiformis (Fr.) Corner. In heathy turf or woodland. 8. Lamberts Castle, JGK; Bracket's Coppice, JGK; Duncliffe Wood, JGK; Morden Park, BNSS; Brownsea Is, KC; Birches Copse, Maldry Wood and Alderholt, EFL.

Clavulinopsis helvola (Fr.) Corner. In short turf. 9.

Clavulinopsis laeticolor (Berk. & Curtis) Petersen. Rare in short turf. 1. Powerstock Common, JGK.

Clavulinopsis luteoalba (Rea) Corner. In heathy turf. 1. Thorncombe Wood, JGK and !.

Clitocybe asterospora (J. Lange) Moser. 2. Upton CP, PBLS; Castle Hill Wood, J Edwards.

Clitocybe brumalis (Fr. ex Fr.) Quel. In acid conifer plantations. 1. Mount Pleasant, Horton, EFL.

Clitocybe candicans (Pers. ex Fr.) Kumm. In deciduous woods; toxic. 1. Edmondsham, EFL.

Clitocybe cerussata (Fr.) Gillet. In woods. 6. Bracket's Coppice, Hendover Coppice and Thorncombe Wood, JGK; Morden Park, BNSS; Stubhampton Bottom, V Copp; Edmondsham, EFL.

Clitocybe clavipes (Pers. ex Fr.) Kumm. Frequent in woods. 17.

Clitocybe dealbata (Sow. ex Fr.) Kumm. In short turf; toxic. 8. A source of muscarine.

Clitocybe diatreta (Fr. ex Fr.) Kumm. Rare in woods; toxic. 1. Kingcombe, JGK.

Clitocybe dicolor (Pers.) J. Lange. In deciduous woods. 10.

Clitocybe ditopus (Fr. ex Fr.) Gillet. Under conifers. 8.

Clitocybe fragrans (With. ex Fr.) Kumm. In grassy woods. 10.

Clitocybe geotropa (Bull.) Quel. Frequent under hazel in calcareous soil. 18.

Clitocybe gibba (Pers. ex Fr.) Kumm. Frequent in deciduous woods. 20.

Clitocybe harmajae Lam. In woods. 5. Bracket's Coppice, Clifton Wood, Hendover Coppice, Puddletown Forest and Kingsettle Wood, JGK.

Clitocybe hydrogramma (Bull. ex Fr.) Kumm. Under oak. 5. Powerstock Common and Bracket's Coppice, JGK; Great Wood, Creech, !; The Oaks, Badbury and Edmondsham, JGK.

Clitocybe langei Singer ex Hora. Under pine in wet woods. 4. Lamberts Castle and Stonebarrow, MH, also JGK; Clifton Wood, JGK; Bere Wood, BC; Delph Wood.

Clitocybe metachroa (Fr.) Kumm. In deciduous woods. 7. Powerstock Common, Hendover Coppice and Arne, JGK; Chetterwood, !; Birches Copse and Garston Wood, JGK; Broadstone, EFL.

Clitocybe nebularis (Batsch ex Fr.) Kumm. Common in deciduous woods. 26.

Clitocybe odora (Bull. ex Fr.) Kumm. Frequent in deciduous woods. 14.

Clitocybe phyllophila (Fr.) Kumm. Frequent in deciduous woods. 13.

Clitocybe quercina Pearson. Rare in deciduous woods. 1. Holt Forest, JGK.

Clitocybe rivulosa (Pers. ex Fr.) Kumm. In grassland; toxic. 7. Seven Ash Common, C Burchardt; Milborne Wood, !; Piddles Wood, JGK; Black Hill, Bere and Oatclose Wood, !; Chapmans Pool, CBG; Edmondsham, EFL, also JGK.

Clitocybe suaveolens (Schum. ex Fr.) Kumm. In deciduous woods. 3. Newlands Batch, !; Hendover Coppice, JGK; Bryanston, BNHS.

(*Clitocybe trullaeformis* (Fr.) P. Karst. Probably this in Puddletown Forest, !.)

Clitocybe vibecina (Fr.) Quel. In damp woods, often under pine. 10.

Clitocybula lacerata (Lasch.) Singer. Rare, on pine branches. 1. Wytherston Wood, MH det. **K**.

Clitopilus hobsonii (Berk. & Broome) Orton. On dead deciduous wood. 2. Bracket's Coppice and Piddles Wood, JGK.

Clitopilus prunulus (Scop. ex Fr.) Kumm. In woods on acid soil. 12.

Collybia butyracea (Bull. ex Fr.) Kumm. Common in woods. 23.

Collybia cirrhata (Schum. ex Fr.) Kumm. In woods. 9.

Collybia confluens (Pers. ex Fr.) Kumm. Frequent in woods. 20.

Collybia cookei (Bres.) Arnold. In woods, often on decaying fungi. 6. Wootton Hill, Kingcombe, Bracket's Coppice, Hendover Coppice, Piddles Wood and Garston Wood, JGK.

Collybia distorta (Fr.) Quel. Under conifers. 5. Powerstock Common, Hurst Heath, Thorncombe Wood, Arne and Studland, JGK.

Collybia dryophila (Bull. ex Fr.) Kumm. Common in woods. 24.

Collybia erythropus (Bull. ex Fr.) Kumm. Frequent in acid woods. 11.

Collybia fusipes (Bull. ex Fr.) Quel. Frequent on dead wood. 17.

Collybia impudica (Fr.) Singer. Rare under pines. 1. Kingcombe, JGK.

Collybia inodora (Pat.) Orton. Rare in woods. 2. Powerstock Common and Piddles Wood, JGK.

Collybia maculata (Alb. & Schwein ex Fr.) Kumm. Common in acid woods or heaths. 20.

Collybia peronata (Bolt. ex Fr.) Kumm. Frequent in woods. 18.

Collybia tuberosa (Bull. ex Fr.) Kumm. On other fungi in woods; once on dunes. 3. Piddles Wood, JGK; Upton CP, PBLS; Sandbanks, BAB.

Coltricia cinnamomea (Pers.) Murrill Rare on acid soil. 1. Broadstone, EFL.

Coltricia perennis (Fr.) Murrill. On acid soil, often under pine. 5. Moreton, AH; Rempstone Wood, BAB; Holton Lee, VL Breeze; Brownsea Is, KC; Avon FP, JGK.

Coniophora puteana (Schum. ex Fr.) P. Karst. On fallen oak wood. 8.

Conocybe aeruginosa Romagn. 1. Bracket's Coppice, H Baker det. **K**.

Conocybe ambigua (Kuehn.) Singer. 1. Hendover Coppice, JGK.

Conocybe appendiculata J. Lange & Kuehn. 1. Hendover Coppice, JGK.

Conocybe blattaria (Fr.) Kuehn. On fallen leaves. 1. Bryanston, BNHS.

Conocybe brunnea (J. Lange & Kuehn.) Watling. Under willow. 1. Poxwell Big Wood, BAB det. DA Reid.

Conocybe coprophila (Kuehn.) Kuehn. On dung. 1. Powerstock Common, JGK.

Conocybe filaris (Fr.) Kuehn. In woods. 1. Edmondsham, JGK.

Conocybe kuehneriana Singer. In grassland. 2. Hog Cliff and Studland, JGK.

(*Conocybe lactea* (J. Lange) Metrod. Possibly this on a verge near Poxwell, BAB.)

Conocybe mairei Kuehn. 1. Rempstone, BAB det. DA Reid.

Conocybe mesospora Kuehn. 1. Hog Cliff, JGK and PDO.

Conocybe pilosella (Pers. ex Fr.) Kuehn. On fallen twigs. 1. Hendover Coppice, B Collins det. JGK.

Conocybe pseudopilosella (Kuehn.) Kuehn. In short calcareous turf. 1. Above Durdle Door, !.

Conocybe pubescens (Gillet) Kuehn. On dung. 2. Kingston, BC; Pamphill, J Fildes.

Conocybe pygmaeoaffinis (Fr.) Kuehn. In mossy lawns. 1. Hendover Coppice, PDO.

Conocybe rickeniana Singer ex Orton. In grass. 1. Piddles Wood, JGK.

Conocybe rickenii (Schaeff.) Kuehn. On dung. 1. Pamphill, J Fildes.

(*Conocybe rugosa* (Peck) Watling. 1. Avon FP, VC 11, JGK.)

Conocybe semiglobata (Kuehn.) Kuehn. & Watling. In old turf. 1. Corfe Common, BNSS.

Conocybe subovalis (Kuehn.) Kuehn. & Romagn. In manured grass. 2. Came Wood, !; Hog Cliff, JGK and PDO.

Conocybe tenera (Schaeff. ex Fr.) Fayod. In grassland. 10.

Conocybe vexans Orton. 1. Hendover Coppice, B Collins det. JGK.

Coprinus acuminatus (Romagn.) Orton. On buried wood. 2. Kingcombe, Powerstock Common and Hooke Park, JGK.
Coprinus angulatus Peck. On burnt soil. 1. Bracket's Coppice, JGK.
Coprinus atramentarius (Bull. ex Fr.) Fr. In rich soil. 20.
Coprinus cinereus (Schaeff. ex Fr.) Gray. On dung or manured soil. 5. Rawlsbury Camp and Buddens Farm, !; Kingston, BC; Studland, CBG; Edmondsham, EFL.
Coprinus comatus (O.F. Muell. ex Fr.) Gray. Lawyers Wig or Shaggy Ink Cap. In rich soil. 26.
(*Coprinus congregatus* (Bull. ex St Amans) Fr. Probably this on dung at Ringstead, !.)
Coprinus cortinatus J. Lange. 1. Carey, BMS.
Coprinus disseminatus (Pers. ex Fr.) Gray. Frequent on dead wood. 20.
Coprinus domesticus (Bolt. ex Fr.) Gray. On rotten wood. 1. Duncliffe Wood, V Copp.
Coprinus ellisii Orton. 1. Piddles Wood, PDO.
Coprinus ephemerus (Bull. ex Fr.) Fr. 1. Kingston, BC.
Coprinus heptemerus M. Lange & A.H. Sm. On dung. 1. Kingston, BC.
Coprinus hiascens (Fr.) Quel. 2. Tatton Copse, BAB; Piddles Wood, JGK.
Coprinus impatiens (Fr.) Quel. On rich soil. 1. Almer, !; perhaps not distinct from *C. hiascens*.
Coprinus lagopides P. Karst. On burnt soil. 1. Bracket's Coppice, AJEL.
Coprinus lagopus (Fr.) Fr. In woods or gardens. 11.
Coprinus leiocephalus Orton. In deciduous woods. 6. Kingcombe, Powerstock Common. Hooke Park, Bracket's Coppice, Hendover Coppice and Fifehead Wood, JGK; Kingston and Furzebrook, BC.
Coprinus macrocephalus (Berk.) Berk. On fallen leaves. 1. Kingston, BC.
Coprinus micaceus (Bull. ex Fr.) Fr. Common on deciduous wood. 25.
Coprinus niveus (Pers. ex Fr.) Fr. Frequent on dung. 11.
Coprinus picaceus (Bull. ex Fr.) Gray. Under beech. 1. Whatcombe, BMS.
Coprinus plicatilis (Curtis ex Fr.) Fr. In short grass or woodland rides. 18.
Coprinus radians (Desm.) Fr. 1. Bride Head, MH.
Coprinus radiatus (Bolt. ex Fr.) Gray. On dung. 5. Wytherston Wood, BMS; Ringstead, BAB; Lydlinch; Kingston, BC; Edmondsham, JC Rayner.
Coprinus silvaticus Peck. 3. Yellowham Wood, J Edwards; Durlston CP, KC; St Giles Park, EFL.
Coprinus stercorarius (Bull.) Fr. On deer dung. 2. Kingston, BC; Edmondsham, JGK.
Coprinus sterquilinus (Fr.) Fr. On dung. 2. Turnworth, !; Kingston, BC.
Coprinus truncorum Schaeff. ex Fr. In woods. 4. Monkton Wyld and Poxwell Big Wood, BAB; Clifton Wood, JGK; Durlston CP, KC.
Coprinus urticicola (Berk. & Broome) Buller. 1. Poxwell Big Wood, BAB.
Coprinus xantholepis Orton. On dead thistle stems. 1. Piddles Wood, PDO.
Coprinus xanthothrix Romagn. In a garden. 1. Swanage, CBG.

Cortinarius: a genus containing undescribed species, and which may be split in future works.
Cortinarius acutus (Pers. ex Fr.) Fr. In wet woods. 5. Bovington, !; Brownsea Is, KC; Sutton Holms, St Giles Park and Daggons, EFL.
Cortinarius alboviolaceus (Pers. ex Fr.) Fr. In deciduous woods. 3. Bracket's Coppice, M Brett; Furzebrook, BC; Brownsea Is, KC.
Cortinarius anomalus (Fr. ex Fr.) Fr. In woods. 6. Lamberts Castle, MH; Bracket's Coppice, JGK; Puddletown Forest, PDO; 89; Furzebrook and Garston Wood, BC; Edmondsham, JGK.
Cortinarius anthracinus Fr. ex Fr. Under oak. 3. Hurst Heath and Birches Copse, JGK; Brownsea Is, KC.
Cortinarius armeniacus (Schaeff. ex Fr.) Fr. Under pine. 1. Birches Copse, EFL.
Cortinarius armillatus (Fr. ex Fr.) Fr. In deciduous woods. 1. Furzebrook, BC.
Cortinarius balteocumatilis Henry. Under beech. 2. Piddles Wood, JGK; perhaps this at Chetterwood.
Cortinarius basililaceus Orton. Under oak. 2. Kingcombe, JGK det. PDO; Edmondsham, JGK.
Cortinarius betuletorum (Moser) Moser. Under birch. 1. SY98.
Cortinarius bicolor Cooke. Under conifers. 1. Castle Hill Wood, EFL.
Cortinarius bolaris (Pers. ex Fr.) Fr. In woods. 2. Oakers Wood, JGK; Furzebrook, BC.
Cortinarius brunneus (Pers. ex Fr.) Fr. Under conifers 1. Castle Hill Wood, EFL.
Cortinarius caerulescens (Schaeff.) Fr. In deciduous woods. 1. Powerstock Common, JGK.
Cortinarius caesiocyaneus Britzelm. Rare, under beech. 1. Lamberts Castle, JGK.
Cortinarius caninus (Fr.) Fr. In deciduous woods. 4. Creech Grange, CBG; Sutton Holms, Birches Copse, Castle Hill Wood and Daggons, EFL.
Cortinarius causticus Fr. Under oak. 1. Oakers Wood, JGK.
Cortinarius cinnabarinus Fr. Under beech or hornbeam. 4. Bracket's Coppice, M Brett; Hooke Park, JGK; Bere Wood and Studland, BC; Holt Forest, EFL, also JGK.
Cortinarius cinnamomeoluteus Orton. In wet woods on sandy soil. 1. Studland, JGK.
Cortinarius cinnamomeus (L. ex Fr.) Fr. In woods. 10.
Cortinarius collinitus (Sow. ex Fr.) Fr. In woods. 1. Brownsea Is, KC.
Cortinarius cotoneus Fr. In woods. 1. Lamberts Castle, JGK.
Cortinarius croceus (Schaeff.) Bigeard. & Guill. Under conifers. 3. Puddletown Forest, PDO; Hurst Heath, JGK; Brownsea Is, KC.
Cortinarius decipiens (Pers. ex Fr.) Fr. In deciduous woods. 5. Bracket's Coppice, M Brett; Yellowham Wood, J Edwards; Furzebrook, BC; Rempstone, !; Brownsea Is, KC; Parkstone and Edmondsham, EFL.
Cortinarius delibutus Fr. Under birch. 4. Lamberts Castle, JGK; Thorncombe Wood, BAB and !; Birches Copse and Edmondsham, JGK.
Cortinarius durus Orton. 1. Bracket's Coppice, JGK and PDO.

Cortinarius elatior Fr. Under beech. 3. Bracket's Coppice, M Brett; Birches Copse and Castle Hill Wood, EFL.
Cortinarius ellisii Orton. 1. Piddles Wood, JGK and PDO.
Cortinarius emollitus Fr. 1. Birches Copse, JGK.
Cortinarius epsomiensis Orton. 1. Hogg Cliff, JGK and PDO.
Cortinarius flexipes (Pers. ex Fr.) Fr. In woods. 3. Bracket's Coppice, JGK; Furzebrook and Garston Wood, BC.
Cortinarius fulgens (Alb. & Schwein. ex Secr.) Fr. 1. Piddles Wood, JGK.
Cortinarius glandicolor (Fr.) Fr. 3. Stonebarrow, MH; Bracket's Coppice, M Brett; Brownsea Is, KC.
Cortinarius helvelloides (Fr.) Fr. Under alder. 2. Kingcombe and Bracket's Coppice, JGK.
Cortinarius hemitrichus (Pers. ex Fr.) Fr. Occasional under birch. 8.
Cortinarius hinnuleus Fr. Mostly under oak. 6. Kingcombe, JGK; Bracket's Coppice, M Brett; Melbury Park, BAB; Corfe, CBG; Furzebrook, BC; Brownsea Is, KC; Birches Copse, EFL.
Cortinarius hoeftii (Weinm.) Fr. Under oak. 1. Hurst Heath, PDO.
Cortinarius humicola (Quel.) Maire. In deciduous woods. 1. Lamberts Castle, JGK.
Cortinarius incisus (Pers. ex Fr.) Fr. 2. Broadstone and St Giles Park, EFL.
Cortinarius jubarinus Fr. Under pine. 2. Broadstone and Mannington, EFL.
Cortinarius largus Fr. In deciduous woods. 2. Powerstock Common, AJEL; Bracket's Coppice and Hooke Park, JGK.
Cortinarius lepidopus Cooke. In deciduous woods. 3. Kingcombe and Bracketts Coppice, JGK; Tincleton, MH.
Cortinarius malicorius Fr. A northern species of wet woods, under alder. 1. Lamberts Castle, JGK.
Cortinarius mucosus (Bull. ex Fr.) Cooke. Under birch or pine. 1. Furzebrook, BC.
Cortinarius myrtillinus Fr. Under beech. 1. Sutton Holms, EFL.
Cortinarius nemorensis (Fr.) J. Lange. In deciduous woods. 1. Bracket's Coppice, JGK.
Cortinarius obtusus (Fr.) Fr. Under conifers. 2. Bracket's Coppice, M Brett; Sandbanks, BAB.
Cortinarius ochroleucus (Schaeff. ex Fr.) Fr. In deciduous woods. 2. Ferndown and St Giles Park, EFL.
Cortinarius paleaceus (Weinm.) Fr. Under birch. 4. Lamberts Castle, MH; Bracket's Coppice, JGK; probably this at Oakers Wood, !; Upton Heath, Purbeck Fungus Group.
Cortinarius paleiferus Svrcek. In woods. 1. Kingcombe, JGK.
Cortinarius pholideus (Fr. ex Fr.) Fr. Under birch or oak. 2. Tincleton, MH; Brownsea Is, KC.
Cortinarius pseudosalor J. Lange. In deciduous woods. 2. Lamberts Castle, MH, also JGK; Oakers Wood, BAB, also JGK.
Cortinarius pulchellus J. Lange. In woods. 1. Kingcombe, JGK.
Cortinarius puniceus Orton. In deciduous woods. 6. Stonebarrow, Kingcombe, Bracket's Coppice, Hurst Heath and Holt Forest, JGK; Duncliffe Wood, V Copp.
Cortinarius purpurascens (Fr.) Fr. In deciduous woods. 5.

Powerstock Common, AJEL; Kingcombe and Bracket's Coppice, JGK; Pamphill, J Fildes; Brownsea Is, KC; Avon FP, JGK.
Cortinarius saniosus (Fr.) Fr. In wet, deciduous woods. 2. Bracket's Coppice, JGK; Warmwell, BAB.
Cortinarius saturninus (Fr.) Fr. 1. Witchampton, EW Baker.
Cortinarius semisanguineus (Fr.) Gillet. In acid woods. 7. Bracket's Coppice, M Brett; Puddletown Forest, BAB, also PDO; Hurst Heath, JGK; Trigon, AH; Furzebrook, BC; Brownsea Is, BAB.
Cortinarius subdelibutus Orton. In deciduous woods. 3 Powerstock Common, AJEL; Bracket's Coppice and Holt Forest, JGK.
Cortinarius subpurpurascens (Fr.) Kickx. 1. Bracket's Coppice, JGK; perhaps a form of *C. purpurascens*.
Cortinarius tabularis (Bull. ex Fr.) Fr. 1. Birches Copse, EFL.
Cortinarius torvus (Fr. ex Fr.) Fr. In oak woods. 4. Kingcombe, JGK; Morden Park, BNSS; Sutton Holms and Edmondsham, EFL.
Cortinarius triumphans Fr. Occasional in damp woods. 9.
Cortinarius trivialis J. Lange. In deciduous woods. 3. Warmwell, BAB; Furzebrook and Garston Wood, BC.
Cortinarius uliginosus Berk. In wet woods. 3. Powerstock Common, Studland and Holt Forest, JGK.
Cortinarius umbrinolens Orton. Under birch. 2. Holt Forest, Birches Copse and Edmondsham, PDO.
Cortinarius varius (Schaeff. ex Fr.) Fr. Under conifers. 1. Mannington, EFL.
Craterellus cornucopioides Pers. HORN-OF-PLENTY. In deciduous woods, mostly oak. 9.
Creolophus cirrhatus (Pers. ex Fr.) P. Karst. Rare on beech logs. 2. Lewesdon Hill, BAB det. DA Reid; Powerstock Common, JGK; Avon FP, J Goss.
Crepidotus amygdalosporus Kuehn. On twigs. 3. Puddletown Forest, Piddles Wood and Holt Forest, JGK.
Crepidotus applanatus (Pers. ex Pers.) Kumm. On dead beech wood. 3. Bryanston, BNHS; Holt Forest, JGK; Edmondsham, EFL.
Crepidotus autochthonus J Lange. 2. Bracket's Coppice and Piddles Wood, JGK.
Crepidotus calolepis (Fr.) Sing. 1. Bracket's Coppice, JGK; hardly distinct from *C. mollis*.
Crepidotus cesati (Rabenh.) Sacc. 6. Kingcombe, Powerstock Common, Bracket's Coppice, Hooke Park, Hendover Coppice, Piddles Wood, The Oaks and Holt Forest, JGK.
Crepidotus luteolus (Lambotte) Sacc. 1. Hendover Coppice, JGK.
Crepidotus mollis (Schaeff. ex Fr.) Kumm. On dead twigs. 16.
Crepidotus pubescens Bres. 1. Oakers Wood.
Crepidotus subsphaerosporus (J. Lange) Kuehn. & Romagn. 5. Bracket's Coppice, Hendover Coppice, Fifehead Wood, Kingsettle Wood and Holt Forest, JGK.
Crepidotus subtilis Orton. 2. Hendover Coppice and Oakers Wood, PDO.
Crepidotus variabilis (Pers. ex Fr.) Kumm. On dead twigs or herb stems. 16.
Cylindrobasidium evolvens (Fr. ex Fr.) Juel. On dead

twigs. 5. Powerstock Common, JGK; Blue Pool, !; Kingston, !; Brownsea Is, KC; Castle Hill Wood, EFL, also JGK.

Cystoderma amianthemum (Scop. ex Fr.) Fayod. On mossy soil in acid woods. 6. Wootton Hill, Lamberts Castle, Hurst Heath, Thorncombe Wood and Oakers Wood, JGK; Horton and Edmondsham, EFL.

Cystoderma jasonis (Cooke & Massee) Harmaja. On mossy heaths. 2. Kingcombe, JGK; Hyde Heath, !.

Daedalea quercina Fr. Frequent on oak stumps. 18.

Daedaleopsis confragosa (Bolt. ex Fr.) Schroeter. Common on live or dead sallow or other deciduous trees. 28.

Datronia mollis (Sommerf. ex Fr.) Donk. On dead deciduous wood. 6. Lamberts Castle, Wootton Hill, Bracket's Coppice, Hog Cliff and Thorncombe Wood, JGK; Kingston, BC; Broadstone, EFL.

Dermoloma cuneifolium (Fr. ex Fr.) Singer. In short turf. 2. Lamberts Castle and Kingcombe, JGK.

Drosella fracida (Fr.) Singer. In heathy woods. 1. Furzebrook, BC.

Entoloma; see also *Leptonia* and *Trichopilus* (Orton, 1991)

Entoloma bisporigerum (Orton) Noordel. 1. Kingcombe, JGK and PDO.

Entoloma bloxamii (Berk. & Broome) Sacc. In calcareous woods. 2. Hog Cliff, JGK; Kingston, BC.

Entoloma costatum (Fr.) Kumm. In pasture. 1. Near Birches Copse, EFL.

Entoloma eulividum Noordel. Under oak. 5. Bracket's Coppice, M Brett; Oakers Wood, JGK; Norden, BC; Sutton Holms and Edmondsham, EFL.

Entoloma lividoalbum Kuehn. & Romagn.) Kub. In woods. 1. Fifehead Wood, JGK.

Entoloma majaloides Orton. In woods. 2. Kingcombe and Hendover Coppice, JGK and PDO.

Entoloma nidorosum (Fr.) Quel. Occasional in damp, deciduous woods. 8.

Entoloma pernitrosum (Orton) Orton. 1. Bracket's Coppice, JGK and PDO.

Entoloma politum (Pers. ex Fr.) Donk. In grassy woods. 2. Bracket's Coppice and Holt Forest, JGK.

Entoloma prunuloides (Fr.) Quel. In grassland. 3. Hog Cliff, JGK and PDO; Poxwell Big Wood, BAB det. DA Reid; Edmondsham, EFL.

Entoloma rhodopolium (Fr.) Kumm. Occasional under oak. 8.

Entoloma sericatum (Britzelm.) Sacc. On buried wood. 1. Kingcombe, JGK and PDO.

Exobasidium vaccinii (Fuckel) Woron. On *Rhododendron* leaves. 2. Creech Grange and Morden Park, BAB.

Fistulina hepatica Schaeff. ex Fr. POOR-MAN'S BEEFSTEAK. Frequent on live oak boles. 21.

Flagelloscypha minutissima (Burt.) Donk On dead wood. 2. Hog Cliff and Garston Wood, JGK.

Flammulaster granulosus (J. Lange) Watling. On soil. 1. Badbury Rings, !.

Flammulaster novasilvensis Orton. Under oak. 1. Studland, JGK and PDO.

Flammulina velutipes (Curtis ex Fr.) P. Karst. Frequent on dead wood of gorse and other trees. 17.

Flocculina limulata (Fr. ex Weinm.) Orton. Under beech. 1. Bryanston, BNHS.

Fomes fomentarius (Fr.) Kickx. On dead beech wood. 1. Witchampton, EW Baker.

Galerina atkinsoniana A.H. Sm. & Orton. 1. Lamberts Castle.

Galerina autumnalis (Peck) A.H. Sm. & Singer. Frequent on dead deciduous wood. 19.

Galerina hypnorum (Schrank ex Fr.) Kuehn. Occasional on mossy banks in woods. 11.

(***Galerina mniophila*** (Lasch) Kuehn. In woods. 1. Avon FP, VC 11, I Cross.)

Galerina mycenopsis (Fr. ex Fr.) Kuehn. In woods. 2. Park Wood, Lulworth and Garston Wood, BAB.

Galerina paludosa (Fr.) Kuehn. Among *Sphagnum*. 4. Pilsdon Pen, BAB; Oakers bog, !; Furzebrook, BC; Arne, JGK; Studland, BMS.

Galerina praticola (Moeller) Orton. In wet grassland. 1. Quarleston, !.

Galerina stylifera (Atk.) A.H. Sm. & Singer. Rare on pine litter. 2. Kingcombe and Hurst Heath, JGK.

Galerina vittaeformis (Fr.) Moser. On acid mossy banks. 2. Bracket's Coppice and Hurst Heath, JGK.

Ganoderma applanatum (Pers.) Pat. On live beech boles, early records lumped with *G. australe*. 19.

Ganoderma australe (Fr.) Pat. Frequent on live beech or oak boles. 21.

Ganoderma lucidum (Fr.) P. Karst. On elm boles or stumps. 1. Piddles Wood, JGK.

Ganoderma pfeifferi Bres. On beech boles. 1. Kingston, BC.

Ganoderma resinaceum Boud. ex Pat. On oak boles. 1. Furzebrook, BC.

Gomphidius glutinosus (Schaeff. ex Fr.) Fr. Under conifers. 5. Melbury Park, BAB; Stubhampton Bottom, JGK; Furzebrook, BC; Fitzworth Heath, CBG; Avon FP, JGK.

Gomphidius roseus (Fr.) P. Karst. Under pines. 4. Bracket's Coppice, M Brett; Puddletown Forest, BAB; Moreton, JGK; Stoborough, AH; Furzebrook, BC.

Grifola frondosa (Fr.) Gray. At the base of living oaks. 8. Kingcombe, Bracket's Coppice, Hog Cliff, Thorncombe Wood, Piddles Wood, Duncliffe Wood and Edmondsham, JGK; Sherford Bridge, AA Wason.

Gymnopilus hybridus (Fr. ex Fr.) Singer. On dead wood. 4. Hooke Park, Arne and Kingsettle Wod, JGK; Ferndown, EFL.

Gymnopilus junonius (Fr.) Orton. Frequent on dead wood and stumps. 18

Gymnopilus penetrans (Fr. ex Fr.) Murrill. Frequent on dead conifer wood. 17.

Gymnopilus sapineus (Fr.) Maire. On dead conifer wood. 3. Brownsea Is, KC; Branksome Park, EW Baker; Broadstone and Mannington, EFL.

Gyrodon lividus (Bull. ex Fr.) Secr. Rare under alder. 1. Bracket's Coppice, JGK det. DA Reid.

Gyroporus castaneus (Bull. ex Fr.) Quel. Under oak. 2. Lewesdon Hill and Brenscombe Hill, BAB.

Gyroporus cyanescens (Bull. ex Fr.) Quel. Under birch. 1. Canford Heath, Canford School NHS.

Hapalopilus nidulans (Pers. ex Fr.) P. Karst. On dead deciduous branches. 6. Bracket's Coppce, JGK; Wootton

Hill and Piddles Wood, JGK; Talbots Wood, CBG; Organford and Badbury Rings, !.

Hebeloma anthracophilum Maire. On burnt soil. 4. Powerstock Common, Bracket's Coppice, Hendover Coppice and Arne, JGK.

Hebeloma crustuliniforme (Bull.) Quel. Frequent in deciduous woods. 16.

Hebeloma fastibile (Pers. ex Fr.) Kumm. 2. Bushey, CBG; Edmondsham, EFL.

Hebeloma leucosarx Orton. In damp, deciduous woods. 6. Kingcombe, Bracket's Coppice, Puddletown Forest, Hurst Heath, Oakers Wood and Studland, JGK.

Hebeloma longicaudum (Pers. ex Fr.) Kumm. 6. Lamberts Castle, MH; Bracket's Coppice and Piddles Wood, JGK; Oakers Wood and Garston Wood, BC; Brownsea Is, KC.

Hebeloma mesophaeum (Pers.) Quel. Mostly under birch. 7. Lamberts Castle, VL Breeze; Kingcombe, JGK; Yellowham Wood, J Edwards; Furzebrook, BC; Branksome and Edmondsham, EFL; Avon FP, JGK.

Hebeloma pumilum J. Lange. In wet woods. 2. Bracket's Coppice, MH; Piddles Wood, JGK.

Hebeloma pusillum J. Lange. 2. Kingcombe and Arne, JGK.

Hebeloma radicosum (Bull. ex Fr.) Ricken. On buried wood. 3. Kingcombe, Bracket's Coppice and Arne, JGK.

Hebeloma sacchariolens Quel. Occasional in damp woods. 10.

Hebeloma sinapizans (Paulet ex Fr.) Gillet. In deciduous woods. 8.

Hebeloma sinuosum (Fr.) Quel. In deciduous woods. 4. Wootton Hill, Kingcombe, Powerstock Common, Bracket's Coppice and Hooke Park, JGK; Durlston CP, KC.

(*Hebeloma truncatum* (Schaeff. ex Fr.) Quel. Probably this at Arne and Rempstone, !.)

Hemimycena delectabilis (Peck) Singer. On pine litter. 2. Hendover Coppice, JGK; Garston Wood, AL.

Hemimycena lactea (Pers. ex Fr.) Singer. On pine litter. 5. Bracket's Coppice, M.Brett; Hendover Coppice, JGK; Oakers Wood, !; Ferndown, Mrs Pringle; Castle Hill Wood, EFL.

Hemimycena tortuosa (Orton) Redhead. On deciduous litter. 7. Kingcombe, Powerstock Common, Bracket's Coppice, Hilfield, Fifehead Wood and Edmondsham, JGK; Carey, BMS.

Henningsomyces candidus (Pers. ex Schleich.) O. Kuntze On dead twigs. 2. Studland, AL and JGK; Edmondsham, JGK.

Hericium coralloides (Scop. ex Fr.) Pers. On dead deciduous wood. 2. Upton CP, PBLS; Brownsea Is, KC.

Hericium erinaceum (Bull. ex Fr.) Pers. Rare on dead beech wood. 1. Oatclose Wood, !.

Heterobasidion annosum (Fr.) Bref. On living pine trees. 19.

Hohenbuehelia geogenia (DC. ex Fr.) Singer. Under chestnut. 1. Melbury Park, !.

Hydnellum ferrugineum (Fr. ex Fr.) P. Karst. Rare on conifer wood. 1. Edmondsham, EFL.

Hydnellum scrobiculatum (Fr.) P. Karst. Rare under pine. 2. Moreton, JGK; Pamphill, !.

Hydnellum spongiosipes (Peck) Pouzar. Rare under oak. 1. Furzebrook.

Hydnum repandum L. ex Fr. Occasional in woods, often in deep shade. 17.

Hydnum rufescens Fr. In deciduous woods. 4. Powerstock Common, Bracket's Coppice and Oakers Wood, JGK; Cole Wood, !.

Hygrocybe brevispora Moeller. 1. Kingcombe, JGK.

Hygrocybe calyptraeformis (Berk. & Broome) Fayod. In grass. 2. Stonebarrow; Kingcombe, JGK.

Hygrocybe ceracea (Wulfen ex Fr.) Kumm. In grass. 4. Newlands Batch and Nettlecombe, JGK; Morden Park, BNSS; Daggons, EFL.

Hygrocybe citrina (Rea) J. Lange. In grass. 2. Lamberts Castle, MH; Edmondsham.

Hygrocybe coccinea (Schaeff. ex Fr.) Kumm. In short, neutral or acid turf. 12.

Hygrocybe colemanniana (Berk. & Broome) Orton & Watling. In acid turf. 2. Puddletown Forest, BAB; Corfe Common, !.

Hygrocybe conica (Scop. ex Fr.) Kumm. Frequent in short turf, (including *H. nigrescens*). 19.

Hygrocybe cystidiata Arnolds. In short turf. 4. Brownsea Is, KC; Crichel and Daggons, EFL.

Hygrocybe flavescens (Kauffman) Singer. In short, acid or calcareous turf. 5. Newlands Batch, Powerstock Common, Kingcombe, Portland and Edmondsham, JGK; Chetterwood, !.

Hygrocybe fornicata (Fr.) Singer. A northern species, in grass. 1. Kingcombe, JGK.

Hygrocybe glutinipes (J. Lange) Haller. In short turf. 1. Lamberts Castle, JGK.

Hygrocybe insipida (J. Lange) Moser. In short turf. 2. Hog Cliff, JGK; Stroud Bridge, !.

Hygrocybe laeta (Pers. ex Fr.) Kumm. In acid turf. 4. Stonebarrow,, Kingcombe and Studland, JGK; Brownsea, KC; Canon Hill, Canford School NHS.

Hygrocybe miniata (Fr.) Kumm. In acid turf. 6. Thorncombe Wood, JGK; Wareham Common, MHL; Morden Park, BNSS; Quince Hill, Corfe Common, and Nine Barrow Down, CBG; Edmondsham, EFL.

Hygrocybe nitrata (Quel.) Wuensche. In acid pasture. 1. Arne, !.

Hygrocybe ochraceopallida Orton. In short turf. 2. Kingcombe and Holt Forest, JGK and PDO.

Hygrocybe persistens (Britzelm.) Singer. In short turf. 4. Winfrith Heath and Bloxworth, !; Kingwood Down, D Stephens; Upton Heath, Purbeck Fungus Group.

Hygrocybe pratensis (Pers. ex Fr.) Murrill. In short turf, often on calcareous soil. 12. Var. *pallida* (Cooke) Arnolds. Hog Cliff, JGK; Binnegar, !.

Hygrocybe psittacina (Schaeff. ex Fr.) Wuensche. In acid or neutral turf. 12.

Hygrocybe punicea (Fr.) Kumm. Occasional in grass or woodland rides. 9. The only recent records are from Stonebarrow, Kingcombe and Hog Cliff, JGK and Loscombe, DL Thomas.

Hygrocybe quieta (Kuehn.) Singer. In short turf. 5. Newlands Batch,, Melbury Park, Clifton Wood and Hog Cliff, JGK; Furzebrook, BC.

Hygrocybe reai (Maire) Maire. In grass. 1. Hog Cliff, JGK and PDO.

Hygrocybe reidii Kuehn. In grass. 1. Hog Cliff, JGK and PDO.

Hygrocybe russocoriacea (Berk. & Mill.) Orton & Watling In short turf. 2. Hog Cliff and Castle Hill Wood, JGK.

Hygrocybe splendidissima (Orton) Svrcek. In grass. 1. Hog Cliff, JGK and PDO.

Hygrocybe strangulata (Orton) Svrcek. In acid turf. 4. Lamberts Castle, Kingcombe and Puddletown Forest, JGK; Tincleton, MH; Kingswood Down, D Stephens.

Hygrocybe streptopus (Fr.) Bon. In calcareous ride. 1. High Wood, Badbury, !.

Hygrocybe subminutula Murrill. In short turf. 1. Kingcombe, JGK.

Hygrocybe unguinosa (Fr.) P. Karst. In short turf. 2. Stonebarrow and Kingcombe, JGK.

Hygrocybe virginea (Wulfen ex Fr.) Orton & Watling. In short turf, often on calcareous soil. 15.

Hygrocybe vitellina (Fr.) P. Karst. Under oak. Furzebrook, BC; Pamphill, !.

Hygrophoropsis aurantiaca (Wulfen ex Fr.) Maire. Common in woods. 20.

Hygrophorus cossus (Sow. ex Berk.) Fr. Under beech, Creech Hill Wood, EFL; this might have been *H. discoxanthus*.

Hygrophorus discoxanthus (Fr.) Rea. Under beech. 2. Stubhampton Bottom, JGK; Brenscombe Wood.

Hygrophorus eburneus (Bull. ex Fr.) Fr. Under beech. 6. Kingcombe, Bracket's Coppice and Clifton Wood, JGK; Sherborne Park, !; Whatcombe, BMS; Garston Wood, BC.

Hygrophorus hypothejus (Fr. ex Fr.) Fr. Occasional under conifers on acid soil. 11.

Hygrophorus leucophaeus (Scop. ex Fr.) P. Karst. Under beech. 1. Hod Wood, !.

Hygrophorus persoonii Arnolds. In deciduous woods. 3. Bracket's Coppice, M Brett; Piddles Wood, JGK; Castle Hill Wood, J Edwards.

Hymenochaete corrugata (Fr.) Lev. Occasional on dead wood. 8.

Hymenochaete rubiginosa (Dicks. ex Fr.) Lev. Frequent on dead wood, mostly oak. 16.

Hyphoderma radula (Fr.) Donk. On twigs. 1. Oakers Wood, JGK.

Hyphoderma setigerum (Fr.) Donk. On deciduous twigs. 1. Oakers Wood.

Hyphodontia alutaria (Burt.) Erikss. On pine twigs. 2. High Wood, Badbury, AL; Avon FP, JGK.

Hyphodontia crustosa (Fr.) Erikss. On twigs. 1. Bracket's Coppice, JGK det. DA Pegler.

Hyphodontia sambuci (Pers.) Erikss. On dead elder wood, frequent. 10.

Hypholoma capnoides (Fr. ex Fr.) Kumm. On dead conifer wood. 6. Bracket's Coppice and Moreton, JGK; Mannington, Edmondsham and Daggons, EFL; Avon FP, C Edwards.

Hypholoma elongatum (Pers. ex Fr.) Ricken. Among *Sphagnum*. 3. Tadnoll, !; Furzebrook, BC; Studland, BMS.

Hypholoma epixanthum (Fr.) Quel. On dead beech wood. 1. Ferndown, EFL det. JF Rayner.

Hypholoma ericaceum (Pers. ex Fr.) Kuehn. In wet acid soil. 2. Lamberts Castle, MH, also JGK; Studland Heath, CBG.

Hypholoma fasciculare (Huds. ex Fr.) Kumm. SULPHUR-TUFT. Very common on dead wood. 30.

Hypholoma marginatum (Pers. ex Fr.) Schroeter. On wood or pine cones. 3. Cowpound Wood and Binnegar, !; Edmondsham, JGK.

Hypholoma radicosum J. Lange. A northern species, on conifer stumps. 1. Brownsea Is, KC.

Hypholoma sublateritium (Fr.) Quel. Frequent on dead wood. 17.

Hypholoma udum (Pers. ex Fr.) Kuehn. In wet acid soil. 2. Thorncombe Wood, BAB; Sares Wood, !.

Hypochnicium bombycinum (Sommerf. & Fr.) Erikss. On sallow twigs. 1. Castle Hill Wood, J Edwards.

Incrustoporia semipileata (Peck) Donk. On dead deciduous twigs. 3. Woodstreet and Stubhampton Bottom, !; Sherford Bridge, AA Wason.

Inocybe aghardii (Lund) Orton. Under willow. 1. Studland, JGK.

Inocybe asterospora Quel. Occasional in woods. 9.

Inocybe bongardii (Weinm.) Quel. In woods on clay. 4. Powerstock Common and Hooke Park, JGK; Bryanston, BNHS; Kingston, BC.

Inocybe brunneoatra (Heim.) Orton. In woods. 2. Bracket's Coppice, JGK; Bryanston, BNHS.

Inocybe calospora Quel. Rare in wet woods. 1. Hooke Park.

Inocybe casimiri Vel. In wet woods. 3. Bracket's Coppice, J Madgwick; Carey and Sugar Hill, BMS; Brownsea Is, KC.

Inocybe cervicolor (Pers. ex Fr.) Quel. In woods. 1. Bryanston, BNHS.

Inocybe cincinnatula Kuehn. Rare in woods. 1. Bracket's Coppice, JGK.

Inocybe cookei Bres. In woods on neutral or calcareous soil. 3. Wyke Wood, BAB; Kingston, BC; Edmondsham, EFL.

Inocybe descissa (Fr.) Quel. Rare in woods. 1. Powerstock Common, JGK.

Inocybe dulcamara (Alb. & Schwein. ex Pers.) Kumm. On bare soil in shade. 4. Monkton Wyld, BAB; Corfe Common, !; Arne, JGK; Pamphill, J Fildes; Avon FP, J Goss.

Inocybe eutheles (Berk. & Broome) Quel. In woods. 5. Puddletown Forest, BMS; Hethfelton, !; Oakers Wood and Moreton, JGK; Avon FP, JGK.

Inocybe fastigiata (Schaeff. ex Fr.) Quel. Occasional in deciduous woods. 9.

Inocybe flocculosa (Berk.) Sacc. In deciduous woods. 5. Bracket's Coppice, JGK; Moigne Combe and Ringstead, BAB det. DA Reid; Bryanston, BAB; Studland, JGK; Brownsea Is, KC.

Inocybe geophylla (Sow. ex Fr.) Kumm. Common in woods, with var. *lilacina* (Fr.) Gillet. 22.

Inocybe griseolilacina J. Lange. In deciduous woods. 3. Bracket's Coppice, JGK; Poxwell Big Wood, BAB; Fontmell Down, DWT.

Inocybe haemacta (Berk. & Cooke) Sacc. In woods; toxic. 1. Hendover Coppice, JGK.

Inocybe hystrix (Fr.) P. Karst. In wet woods. 3. Kingston, BC; Holt Forest and Edmondsham, JGK.

Inocybe jurana Pat. In woods. 3. Parnham and Poxwell Big Wood, BAB; Kingston, BC.

Inocybe lacera (Fr.) Kumm. Occasional in wet woods, heaths or bogs. 9.

Inocybe lanuginella (Schroeter) Konrad & Maubl. In deciduous woods. 2. Hurst Heath and Puddletown Forest, BAB det. DA Reid.

Inocybe lanuginosa (Bull. ex Fr.) Kumm. In wet woods. 3. Puddletown Forest and Arne, JGK; Furzebrook, BC; Brownsea Is, KC.

Inocybe longicystis Atk. In acid woods. 1. Thorncombe Wood, JGK.

Inocybe maculata Boud. In damp, deciduous woods. 5. Bracket's Coppice, M Brett; Hendover Coppice, JGK; Thorncombe Wood, BAB; Piddle Wood, Bere, !; Bryanston, BNHS.

Inocybe microspora J. Lange. In acid woods. 1. Moigne Combe, BAB det. DA Reid.

Inocybe mixtilis (Britzelm.) Sacc. Under conifers. 2. Hog Cliff, JGK and PDO; Hurst Heath, BAB det. DA Reid.

Inocybe napipes J. Lange. Occasional in wet deciduous woods. 8.

Inocybe nitidiuscula (Britzelm.) Sacc. Under pine. 3. Hethfelton, !; Sutton Holms and Edmondsham, EFL (as *I. scabella*).

Inocybe obscura (Pers. ex Pers.) Gillet. In deciduous woods. 2. Great Wood, Creech, !; Brownsea Is, KC.

Inocybe perlata (Cooke) Sacc. In woods. 1. Bracket's Coppice, JGK.

Inocybe petiginosa (Fr. ex Fr.) Gillet. In deciduous woods. 2. Kingcombe, Powerstock Common and Bracket's Coppice, JGK.

Inocybe posterula (Britzelm.) Sacc. In acid woods. 2. Moreton, JGK; Brownsea Is, KC.

Inocybe praetervisa Quel. In deciduous woods. 1. Kingston, BC.

Inocybe pusio P. Karst. In deciduous woods. 3. Hendover Coppice, JGK; Bryanston, BNHS; Kingston, BC.

Inocybe pyriodora (Pers. ex Fr.) Kumm. In woods on clay. 3. Melbury Park and Clifton Wood, JGK; Poxwell Big Wood, BAB.

Inocybe sambucina (Fr.) Quel. In acid woods. 1. Puddletown Forest, JGK.

Inocybe squamata J. Lange. 1. Brownsea Is, KC.

Inocybe terrigena (Fr.) Kuehn. Under conifers. 1. Studland, BC.

Inocybe umbrina Bres. In deciduous woods. 2+. Hooke Park, JGK; Brenscombe Wood, BAB det. DA Reid; probably this at Moreton Sta, Stoke Wake and Chetterwood, !.

Inonotus dryadeus (Fr.) Murrill. On living oaks, rarely on beech or lime. 9.

Inonotus hispidus (Fr.) P. Karst. On living ashes, once also on elms. 8.

Inonotus radiatus (Fr.) P. Karst. On living deciduous trees, especially alder. 9.

Ischnoderma benzoinum (Wahl. ex Fr.) P. Karst. On pine wood. 3. Wootton Hill; Tincleton, MH det. **K**; Stubhampton Bottom, JGK.

Junghuhnia nitida (Pers. ex Fr.) Ryvarden. On beech wood. 1. High Wood, Badbury, G Dickson.

Kuehneromyces mutabilis (Schaeff. ex Fr.) Singer & A.H. Sm. Common on deciduous wood. 18.

Laccaria amethystea (Bull.) Murrill. Common in woods. 24.

Laccaria bicolor (Maire) Orton. In woods on acid soil. 8.

Laccaria laccata (Scop. ex Fr.) Cooke. Very common and variable in woods. 28.

Laccaria proxima (Boud.) Pat. In wet woods or among *Sphagnum*. 3. Hurst Heath and Puddletown Forest, JGK; Studland Heath, BAB.

Lachnella villosa (Pers.) Gillet. 1. Kingcombe, JGK.

Lachrymaria pyrotricha (Holmsk. ex Fr.) Konrad & Maubl. On conifer stumps. 1. Bracket's Coppice, JGK.

Lactarius acerrimus Britzelm. In damp oak woods. 2. Bracket's Coppice and Fifehead Wood, JGK.

Lactarius aquifluus Peck. Under pine or birch. 3. Crossways, !; Froxen Copse, VJG; Avon FP, J Goss.

Lactarius azonites Bull. ex Fr. Under oak. 2. Powerstock Common and Bracket's Coppice, JGK.

Lactarius bertillonii (Neuh. ex Schaefer) Bon. In deciduous woods. 4. Powerstock Common, Bracket's Coppice, Piddles Wood and Foxhills, JGK.

Lactarius blennius (Fr. ex Fr.) Fr. Frequent under beech. 19.

Lactarius camphoratus (Bull. ex Fr.) Fr. Frequent in woods on clay. 15.

Lactarius chrysorheus Fr. Under oak. 11.

Lactarius circellatus Fr. Rare, under hornbeam. 1. Bracket's Coppice, JGK.

Lactarius controversus (Fr. ex Fr.) Fr. Under willow. 4. Powerstock Common, Oakers Wood and Duncliffe Wood, JGK; Studland, BC.

Lactarius decipiens Quel. In deciduous woods. 2. Hurst Heath, BAB; Holt Forest, AL.

Lactarius deliciosus (L. ex Fr.) Gray. Under pine. 14.

Lactarius deterrimus Groeger. Occasional under pine or spruce. 8.

Lactarius fluens Boud. Under beech. 1. Oatclose Wood, !.

Lactarius fuliginosus (Fr. ex Fr.) Fr. In deciduous woods. 3. Bracket's Coppice and Oakers Wood, JGK; Castle Hill Wood, EFL.

Lactarius fulvissimus Romagn. In deciduous woods. 3. Kingcombe, Powerstock Common and Bracket's Coppice, JGK; Kingston, BC.

Lactarius glyciosmus (Fr. ex Fr.) Fr. Under birch. 13.

Lactarius hepaticus Plowr. ex Boud. Under conifers. 11.

Lactarius hysginus (Fr. ex Fr.) Fr. In damp woods. 2. Thorncombe Wood and Oakers Wood, JGK.

Lactarius insulsus (Fr.) Fr. Under oak. 1. Sutton Holms, EFL; it differs little from *L. acerrimus*.

Lactarius lacunarum Romagn. ex Hora. Beside ponds. 2. Bracket's Coppice and Studland, JGK.

Lactarius lilacinus (Lasch) Fr. Under alder. 1. Hartland Bog, BC.

Lactarius mitissimus (Fr.) Fr. In woods. 5. Poxwell Big Wood, BAB; Arne, JGK; Furzebrook, BC; Witchampton, Sutton Holms and Castle Hill Wood, EFL.

Lactarius obnubilis Lasch ex Fr. Under alder. 3. Kingcombe and Bracket's Coppice, JGK; Oakers Wood, !.

Lactarius pallidus (Pers. ex Fr.) Fr. Under beech. 2. Champernhayes, JGK; Castle Hill Wood, EFL.

Lactarius piperatus (Scop. ex Fr.) Gray. In deciduous woods. 7. Powerstock Common, JGK, also BMS; Bracket's Coppice, M Brett; Piddles Wood, BAB; Oakers Wood, !; Sherford Bridge, AA Wason; Durlston CP, KC; Birches Copse, DWT.

Lactarius pubescens (Schrad. ex Fr.) Fr. Occasional under birch. 8.

Lactarius pyrogalus (Bull. ex Fr.) Fr. Under hazel. 17.

Lactarius quietus (Fr.) Fr. Frequent under oak. 24.

Lactarius resimus (Fr.) Fr. Black Hill, Bere, !. 1.

Lactarius rufus (Scop. ex Fr.) Fr. Frequent under pine. 16.

Lactarius subdulcis (Pers. ex Fr.) Gray. Frequent in deciduous woods. 20.

Lactarius subumbonatus Lindgr. In wet deciduous woods. 9.

Lactarius tabidus Fr. Frequent in wet deciduous woods. 16.

Lactarius torminosus (Schaeff. ex Fr.) Gray. Frequent under birch. 15.

(*Lactarius trivialis* (Fr. ex Fr.) Fr. Probably this in wet scrub on sand, Knighton Pits, !.)

Lactarius turpis (Weinm.) Fr. Frequent in deep shade in acid woods. 13.

Lactarius uvidus (Schrad. ex Fr.) Fr. In damp woods. 4. Bracket's Coppice, Clifton Wood, Oakers Wood and Holt Forest, JGK.

Lactarius vellereus (Fr.) Fr. Frequent in deciduous woods. 12.

Lactarius vietus (Fr.) Fr. Frequent in wet birch woods. 13.

Lactarius volemus (Fr.) Fr. Under oak. 4. Bracket's Coppice, M Brett; Froxen Copse, VJG; Brownsea Is, KC; Edmondsham, EFL.

Lactarius zonarius (Bull. ex St Amans) Fr. In deciduous woods. 4. Bracket's Coppice and Clifton Wood, JGK; Tatton Copse and Poxwell Big Wood, BAB.

Laetiporus sulphureus (Fr.) Murrill. Occasional on live deciduous trees or yew. 15.

Leccinum carpini (Schulz.) Moser. Under hazel. 3. Powerstock Common and Bracket's Coppice, JGK: Poxwell Big Wood, BAB.

Leccinum holopus (Rostk.) Watling. In wet birch woods. 2. Furzebrook, BC; Studland dunes, !.

Leccinum melaneum (Smotl.) Pilat & Dermek. Under birch. 1. Studland, JGK.

Leccinum quercinum (Pilat) Green & Watling. Under oak. 1. Oakers Wood, BC, also BAB and JGK.

Leccinum rigidipes Orton. Under birch. 6. Wootton Hill, Kingcombe, Melbury Park, Bracket's Coppice, Oakers Wood, Moreton, Arne and Studland, JGK.

Leccinum roseofracta (Watling) Bon. Under birch. 1. Brownsea Is, KC.

Leccinum scabrum (Fr.) Gray. Frequent under birch. 20.

Leccinum variicolor Watling. Occasional under birch. 10.

Leccinum versipelle (Fr. & Hoek) Snell. Under birch. 8.

Lentinellus cochleatus (Pers. ex Fr.) P. Karst. On dead wood. 7. Bracket's Coppice, Clifton Wood and Hurst Heath, JGK; Milborne Wood, !; Furzebrook, BC; Edmondsham, EFL.

Lentinus lepideus (Fr. ex Fr.) Fr. On dead pine wood. 1. Castle Hill Wood, J Edwards.

Lenzites betulina (Fr.) Fr. On dead wood, mostly birch. 8.

Lepiota: see also *Leucoagaricus*, *Macrolepiota* and *Melanophyllum*.

Lepiota adulterina Moeller. 1. Hogg Cliff, JGK and PDO.

Lepiota alba (Bres.) Sacc. In saline turf. 1. Greenlands Farm, Studland, !.

Lepiota aspersa (Pers. ex Fr.) Quel. Under beech on calcareous soil. 3. Came Wood, !; Warmwell, BAB; Hog Cliff, Hendover Copse and Stubhampton Bottom, JGK.

Lepiota bucknallii (Berk. & Broome) Sacc. In rich, shaded soil. 2. Hendover Coppice, JGK; Bryanston, BNHS.

Lepiota castanea Quel. In deciduous woods. 4. Stoke Wake !; Oakers Wood and Edmondsham, JGK; Garston Wood, BC.

Lepiota clypeolaria (Bull. ex Fr.) Kumm. In deciduous woods. 4. Lamberts Castle, JGK; Bracket's Coppice, M Brett; Came Wood, BAB; Brownsea Is, KC.

Lepiota cristata (Fr.) Kumm. Frequent on rich soil of woods and lawns. 20.

Lepiota eriophora Peck. In damp woods. 3. Bracket's Coppice and Hendover Coppice, JGK; Bryanston, BNHS.

Lepiota excoriata (Schaeff. ex Fr.) Kumm. In woods or rich grassland. 4. Ratcliffe Wood, ! det. JGK; Scratchy Bottom, !; Pamphill, J Fildes; Durlston CP, KC.

Lepiota felina (Pers ex Fr.) P. Karst. Under conifers. 1. Red Bridge, Moreton, AH det. BC.

Lepiota fulvella Rea. In calcareous woods. 5. Powerstock Common, Bracket's Coppice, Hendover Coppice and Garston Wood, JGK; Bryanston, BNHS.

Lepiota fuscovinacea Moeller & J. Lange. In acid woods. 2. Bracket's Coppice, M Brett; Puddletown Forest, BMS.

Lepiota georginae (WG Smith) Sacc. 1. Rempstone Wood, BAB det. DA Reid.

Lepiota grangei (Eyre) J. Lange. 1. Stubhampton Bottom, JGK.

Lepiota holosericea (Fr.) Gillet. In grassland. 1. Nine Barrow Down, CBG.

Lepiota ignivolvata Joss. Rare in woods. 2. Oakers Wood, JGK; Chetterwood, !.

Lepiota pseudohelvola Kuehn. ex Hora. In deciduous woods; toxic. 1. Broadley Wood, !.

Lepiota serena (Fr.) Sacc. In woods. 2. Puddletown Forest, BAB; probably this at Stoke Wake, !; Edmondsham, EFL.

Lepiota sistrata (Fr.) Quel. Occasional in hedgerows. 8.

Lepiota subgracilis Kuehn. In damp, deciduous woods. 4. Kingcombe, Bracket's Coppice and Edmondsham, JGK; Greenhill Down, JGK det. DA Reid.

Lepiota ventriosospora Reid. In woods. 5. Kingcombe, Bracket's Coppice, Hendover Coppice and Oakers Wood, JGK; Oatclose Wood, !.

Lepista gilva (Pers. ex Fr.) Pat. In woods. 1. Bryanston, BNHS.

Lepista inversa (Scop.) Pat. Common in grassy woods. 24.

Lepista irina (Fr.) Bigelow. In woods. 1. Black Down, Portesham, B Eyles.

Lepista luscina (Fr. ex Fr.) Singer. In grassland. 2. Swyre and Hog Cliff, JGK.

Lepista nuda (Bull. ex Fr.) Cooke. WOOD BLEWITS. In rich humus in woods, or on fixed shingle. 25.

Lepista saeva (Fr.) Orton. BLEWITS. In manured grass or gardens. 7. Portland, JGK; Puddletown Forest, BAB; Swanage, Corfe and Nine Barrow Down, CBG; Durlston CP, KC; Edmondsham, EFL.

Lepista sordida (Fr.) Singer. In rich humus in woods. 9.

Leptonia asprella (Fr. ex Fr.) Kumm. 2. Oakers Wood and Arne, JGK.

Leptonia atrocoerulea (Noordel.) Orton. In grass. 1. Edmondsham, JGK and PDO.

Leptonia chalybaea (Fr. ex Fr.) Kumm. In grass. 2. Stonebarrow and Kingcombe, JGK.

Leptonia corvina (Kuehn.) Orton. 3. Hog Cliff and Oakers Wood, JGK; Red Bridge, Moreton, AH.

Leptonia dichroa (Pers.) comb. nov. On dead beech wood. 1. Bere Wood, BC.

Leptonia euchroa (Pers. ex Fr.) Kumm. Under alder. 3. Hendover Copse and Oakers Wood, JGK; Norden, BC.

Leptonia exilis (Fr. ex Fr.) Orton. In grass. 1. Hog Cliff, JGK and PDO.

Leptonia incana (Fr.) Gillet. 2. Hog Cliff and Hendover Coppice, JGK and PDO.

Leptonia lampropus (Fr. ex Fr.) Quel. In grass. 3. Kingcombe, JGK; Sutton Holms and Edmondsham, EFL.

Leptonia lazulina (Fr.) Quel. In short grass. 1. Upton Heath, Purbeck Fungus Group.

Leptonia lividocyanula (Kuehn.) Orton. 1. Kingcombe, PDO.

Leptonia serrulata (Fr. ex Fr.) Kumm. In grass. 4. Greenhill Down, JGK; Bovington; Oakers Wood; Arne, JGK.

Leptonia turci Bres. 1. Hog Cliff, JGK and PDO.

Leucoagaricus badhamii (Berk.) Singer. In calcareous woods. 3. Sandford Orcas, MC Sheahan; Great Wood, Creech and Chetterwood, !.

Leucoagaricus leucothites (Vitt.) Wasser. In grass. 6. Bride Head, MH; saline turf below Grove, !; Ringstead, BAB det. DA Reid; Knighton pits, !; Holwell, VL Breeze; Charborough Park, !; Badbury Rings, JGK.

Leucocoprinus brebissonii (Godey) Locq. Occasional in woods. 13.

Leucopaxillus giganteus (Sow. ex Fr.) Singer. In acid woods. 4. Bere Wood, !; by fishing lakes N of Holton Heath, !: Brownsea Is, KC; Holt Forest, JGK.

Limacella lenticularis (Lasch.) Earle. In calcareous woods. 1. Castle Hill Wood, J Edwards.

Limacella vinosorubescens Forrer. Rare in calcareous woods. 1. Garston Wood, JGK.

Lyophyllum anthracophilum (Lasch.) M. Lange & Sivertsen On bonfire-sites. 5. Powerstock Common, Hendover Coppice, Arne and Studland, JGK; Poxwell Big Wood, BAB; Brownsea Is, KC.

Lyophyllum connatum (Schum. ex Fr.) Singer. In woodland rides. 2. Bracket's Coppice, M Brett; Brownsea Is, KC.

Lyophyllum decastes (Fr. ex Fr.) Singer. Occasional in woods and gardens. 9.

Lyophyllum leucophaeatum (P. Karst.) P. Karst. In woods. 2. Oakers Wood, JGK; Brownsea Is, KC.

Macrocystidia cucumis (Pers. ex Fr.) Heim. At edges of calcareous woods. 2. Hendover Coppice and Stubhampton Bottom, JGK.

Macrolepiota gracilenta (Fr.) Kumm. In grassy calcareous woods. 2. Hendover Coppice, JGK; Garston Wood, !.

Macrolepiota konradii (Huijsman ex Orton) Moser. In grassland. 4. Kingcombe, Morden Park and Garston Wood, JGK; Rempstone Wood, !.

Macrolepiota mastoidea (Fr.) Singer. In deciduous woods. 3. Kingcombe and Hog Cliff, JGK; Sherford Bridge, AA Wason.

Macrolepiota procera (Scop. ex Fr.) Singer. PARASOL MUSHROOM. Widespread in open woods, hedgebanks or chalk turf. 22.

Macrolepiota rhacodes (Vitt.) Singer. Mostly under conifers. 20.

Marasmiellus candidus (Bolt.) Singer. On fallen twigs. 2. Hendover Coppice, JGK; Puddletown Forest, AL.

Marasmiellus ramealis (Bull. ex Fr.) Singer. On fallen twigs or dead bramble stems. 19.

Marasmiellus vaillantii (Pers. ex Fr.) Singer. On fallen leaves, twigs or grass. 4. Canford Park, Canford School NHS; Studland, AL and JGK; Holt Forest, AL; Sutton Holms and Edmondsham, EFL.

Marasmius; see also *Marasmiellus* and *Setulipes*.

Marasmius bulliardiii Quel. On litter. 1. Hog Cliff, JGK and PDO.

Marasmius calopus (Pers. ex Fr.) Fr. On dead herbaceous stems. 3. Thorncombe Wood, JGK; Park Wood, Lulworth, BAB; Daggons Road Sta, EFL.

Marasmius cohaerens (Pers.) Cooke & Quel. In woods. 4. Kingcombe, Powerstock Common, Bracket's Coppice, Hendover Coppice and Piddles Wood, JGK.

Marasmius epiphylloides (Rea) Sacc. & Trotter. On dead ivy leaves. 2. Oakers Wood and Holt Forest, JGK.

Marasmius epiphyllus (Pers. ex Fr.) Fr. On dead leaves. 4. Kingcombe, Powerstock Common and Clifton Wood, JGK; Kingston, BC; Edmondsham, EFL.

Marasmius hudsonii (Pers. ex Fr.) Fr. Scarce in acid woods. 4. Lamberts Castle and Kingcombe, JGK; Corfe Hills, FN Brook; Holt Forest, EW Baker, also P Austin.

Marasmius lupuletorum (Weinm.) Bres. 2. Bracket's Coppice and Hendover Coppice, JGK.

Marasmius oreades (Bolt. ex Fr.) Fr. FAIRY-RING MUSHROOM. Common in grassland. 22.

Marasmius recubans Quel. Under birch or sallow. 5. Bracket's Coppice, Hendover Coppice, Moreton and Holt Forest, JGK; Garston Wood, AL.

Marasmius rotula (Scop. ex Fr.) Fr. Frequent on litter. 14.

Marasmius scorodonius (Fr.) Fr. Rare on litter. 1. Stubhampton Bottom, !.

Marasmius torquescens Quel. On dead leaves. 1. Hendover Coppice, JGK.

Marasmius undatus (Berk.) Fr. On dead bracken. 3. Moreton, Arne and Studland, JGK; Brownsea Is, KC.

Marasmius wynnei Berk. & Broome. In deciduous woods. 4. Bracket's Coppice, Hendover Coppice and Thorncombe Wood, JGK; Castle Hill Wood, J Edwards.

Megacollybia platyphylla (Pers. ex Fr.) Kotl. & Pouzar. Occasional in woods. 13.

Melanoleuca arcuata (Fr.) Singer. 3. Hog Cliff, JGK; Arne and Chetterwood, !.

Melanoleuca cognata (Fr.) Konrad & Maubl. In woods, often fruiting in spring. 2. Bracket's Coppice and Avon FP, JGK.

Melanoleuca exscissa (Fr.) Singer. In woods. 1. Bryanston, BNHS.

Melanoleuca grammopodia (Bull. ex Fr.) Pat. In woods, once in dunes. 5. Bushey, CBG; Scrubbity Barrows, !; Studland, JGK; Branksome Park and Edmondsham, EFL.

Melanoleuca melaleuca (Pers. ex Fr.) Singer. In deciduous woods or in grassland. 12.

Melanoleuca paedida (Fr.) Kuehn. & Muc. 1. Peveril Downs and Swanage, CBG.

Melanoleuca subpulverentula (Pers.) Singer. Under pine. 1. Lamberts Castle, JGK.

Melanophyllum eyrei (Mass.) Sing. Under oak. 1. Bracket's Coppice, JGK.

Melanophyllum haematospermum (Bull. ex Fr.) Kreis. In rich soil. 1. Hendover Coppice, JGK.

Meripilus giganteus (Pers. ex Fr.) P. Karst. Occasional on beech stumps. 12.

Meruliopsis taxicola (Pers.) Boud. On conifer stumps. 1. Ferndown, EFL.

Micromphale foetidum (Fr.) Singer. On dead branches. 2. Hendover Coppice, JGK; Yellowham Wood, J Edwards.

Mycena: see also *Hemimycena* and *Rickenella*.

Mycena acicula (Schaeff. ex Fr.) Kumm. Among moss. 4. Bracket's Coppice and Garston Wood, JGK; Durlston CP, KC; Canford Heath, BNSS.

Mycena adscendens (Lasch.) Maas Geest. On dead deciduous wood. 3. Powerstock Common, Bracketts Coppice and Kingsettle Wood, JGK.

Mycena aetites (Fr.) Quel. Occasional on lawns. 8.

Mycena alba (Bres.) Kuehn. 3. Bracket's Coppice, JGK; Oakers Wood, BC; Brownsea Is, KC.

Mycena amicta (Fr.) Quel. In woodland litter. 6. Bracket's Coppice, Piddles Wood, Moreton, Kingsettle Wood and Arne, JGK; Edmondsham, EFL.

Mycena arcangeliana Bres. Occasional on dead twigs. 9.

Mycena bulbosa (Cejp.) Kuehn. On dead rush stems. 1. Kingcombe and Powerstock Common, JGK.

Mycena capillaripes Peck. Under pines. 5. Hurst Heath, BAB; Hethfelton and Sares Wood, !; Bryanston, BNHS; Brownsea Is, BMS.

Mycena capillaris (Schum. ex Fr.) Kumm. Occasional on fallen beech leaves. 8.

Mycena chlorantha (Fr. ex Fr.) Kumm. In dune slacks. 1. Studland dunes, JGK.

Mycena cinerella P. Karst. On woodland litter. 6. Powerstock Common, Puddletown Forest and Avon FP, JGK; Hurst Heath, BAB; Bryanston, BNHS; Brownsea Is, KC.

Mycena citrinomarginata Gillet. 1. Hog Cliff, JGK.

Mycena clavicularis (Fr.) Gillet. 1. Wootton Hill, JGK.

Mycena clavularis (Batsch. ex Fr.) Secr. On pine litter. 5. Kingcombe, Powerstock Common and Bracket's Coppice, JGK; Ringstead and Studland, BAB; Broadstone, EFL.

Mycena corticola (Pers. ex Fr.) Gray. On mossy bark. 3. Lulworth Park, BAB; Sutton Holms and Edmondsham, EFL.

Mycena crispula (Quel.) Kuehn. 1. Hendover Coppice, JGK.

Mycena epipterygia (Scop. ex Fr.) Gray. Frequent under conifers. 15.

Mycena erubescens Hoehnel. In mossy litter. 2. Bracket's Coppice and Arne, JGK.

Mycena filopes (Bull. ex Fr.) Kumm. Occasional in mossy litter. 11.

Mycena flavescens Vel. In mossy pine litter. 1. Edmondsham, EFL, as *M. luteoalba*.

Mycena flavoalba (Fr.) Quel. In lawns. 5. Kingcombe, Hog Cliff and Castle Hill Wood, JGK; Bracket's Coppice, M Brett; Coombe Bottom, !.

Mycena fusconigra Orton. 1. Kingcombe, JGK and PDO.

Mycena galericulata (Scop. ex Fr.) Gray. Common on stumps and dead wood. 27.

Mycena galopus (Pers. ex Fr.) Kumm. Common in woods, with vars. *candida* and *nigra*. 25.

Mycena haematopus (Pers ex Fr.) Kumm. Frequent on dead wood. 18.

Mycena hiemalis (Osbeck ex Fr.) Quel. On mossy bark. 2. Bracket's Coppice, JGK; Brenscombe Hill, BAB.

Mycena inclinata (Fr.) Quel. Common on dead oak wood. 22.

Mycena integrella (Pers. ex Fr.) Gray. 1. Fifehead Wood, JGK.

Mycena leptocephala (Pers. ex Fr.) Gillet. Occasional in pine litter. 11.

Mycena leucogala (Cooke) Sacc. In woods, probably a form of *M. galopus*. 9.

Mycena lineata (Bull. ex Fr.) Kumm. In woods. 1. Poxwell Big Wood, BAB.

Mycena maculata P. Karst. In oak litter. 2. Bracket's Coppice and Arne, JGK.

Mycena megaspora Kauffmann. In wet heaths or *Sphagnum* bogs. 3. Winfrith Heath, !; Furzebrook, BC; Avon FP.

Mycena metata (Fr. ex Fr.) Kumm. Occasional under conifers. 10.

Mycena mucor (Batsch ex Fr.) Gillet. In acid woods. 1. Thorncombe Wood, JGK.

Mycena olida Bres. Occasional on bark or in deciduous litter. 11.

Mycena olivaceomarginata (Massee) Massee. In grassland. 3. Kingcombe and Hog Cliff, JGK; Black Down, Portesham.

Mycena pearsoniana Dennis ex Singer. In pine litter. 2. Stubhampton Bottom and Edmondsham, JGK.

Mycena pelianthina (Fr.) Quel. In deciduous litter. 6. Powerstock Common, Hooke Park, Bracket's Coppice, Hendover Coppice, The Oaks at Badbury, Holt Forest, Garston Wood and Edmondsham, JGK.

Mycena polyadelpha (Lasch.) Kuehn. On dead oak leaves. 4. Kingcombe, Bracket's Coppice and Fifehead Wood, JGK; Durlston CP, KC.

Mycena polygramma (Bull. ex Fr.) Gray. Common on deciduous stumps. 21.

Mycena pseudocorticola Kuehn. On mossy bark. 4. Newlands Batch, Powerstock Common and Bracket's Coppice, JGK; Woolland, !.

Mycena pudica Hora. 1. Kingston, BC.

Mycena pullata (Berk. & Cooke) Sacc. On dead leaves. 1. Broadstone, EFL.

Mycena pura (Pers. ex Fr.) Kumm. Common in woods. 24.

Mycena rorida (Scop. ex Fr.) Quel. Occasional on dead twigs. 8.

Mycena rosea (Bull.) Grauberg. Under beech. 2. Powerstock Common, JGK; Ruins plantation, !. Probably a variety of *M. pura*, and under-recorded.

Mycena sanguinolenta (Alb. & Schwein. ex Fr.) Kumm. Frequent in woodland litter. 14.

Mycena sepia J Lange. In woods. 3. Brenscombe Hill, BAB; Brownsea Is, KC; St Leonards, AL.

Mycena seynii Quel. On fallen pine cones. 2. Brownsea Is, KC; Avon FP, JGK.

Mycena speirea (Fr. ex Fr.) Gillet. Frequent on bark or woodland litter. 15.

Mycena stipata Maas Geest & Schroebel. On dead deciduous twigs. 11.

Mycena stylobates (Pers. ex Fr.) Kumm. On dead twigs. 8.

Mycena tintinnabulum (Fr.) Quel. On deciduous stumps. 2. Swanage, CBG; Castle Hill Wood, J Edwards.

Mycena viscosa (Secr.) Maire. Under conifers. 2. Powerstock Common and Bracket's Coppice, JGK.

Mycena vitilis (Fr.) Quel. Frequent in deciduous litter. 19.

Mycena vulgaris (Pers. ex Fr.) Kumm. Under conifers. 2. Park Wood, Lulworth, BAB; Avon FP, J Banks.

Mycoacia fuscoatra (Fr.) Donk. On dead wood. 1. Bracket's Coppice, JGK.

Mycoacia uda (Fr.) Donk. On dead wood. 3. Kingcombe and Piddles Wood, JGK; Whatcombe, BMS.

Myxomphalia maura (Fr.) Hora. On burnt soil. 4. Powerstock Common, AJEL; Cowpound Wood and High Wood, Badbury, !; Arne, JGK; Furzebrook, BC.

Naucoria celluloderma Orton. Under alder. 1. Bracket's Coppice, JGK and PDO.

Naucoria escharoides (Fr. ex Fr.) Kumm. Under alder. 7. Kingcombe, Powerstock Common, Bracket's Coppice, Hooke Park, Melbury Park, Hilfield and Arne, JGK; Peggs Farm, !; Brownsea Is, KC; Edmondsham, EFL.

Naucoria scolecina (Fr.) Quel. Rare under alder. 2. Kingcombe and Melbury Park, JGK.

Naucoria striatula Orton. Under alder. 4. Monkton Wyld, BAB; Kingcombe, Bracket's Coppice and Arne, JGK.

Nolanea ameides Berk. & Bres. 1. Kingcombe, AJEL.

Nolanea cetrata (Fr. ex Fr.) Kumm. Mostly under conifers, fruiting in winter. 3. Bracket's Coppice, AJEL; Bere Wood, BC; Brownsea Is, KC.

Nolanea clandestina (Fr. ex Fr.) Kumm. In grass. 1. Hog Cliff, JGK and PDO.

Nolanea hebes (Romagn.) Orton. Under oak or pine. 3. Powerstock Common, JGK and PDO; Bere Heath and Rempstone Wood, !.

Nolanea lucida Orton. 1. Kingcombe, PDO.

Nolanea mammosa (L. ex Fr.) Quel. 1. Brenscombe Wood, BAB det. DA Reid

Nolanea papillata Bres. 1. Thorncombe Wood, JGK.

Nolanea rhombispora (Kuehn. & Bours.) Orton. 1. Kingcombe, JGK and PDO.

Nolanea sericea (Bull.) Orton. Occasional in grass. 9.

Nolanea solstitialis (Fr.) Orton. In grass. 1. Broadstone, EFL.

Nolanea staurospora Bres. Occasional in grass. 11.

Nolanea tenuipes Orton. 1. Bracket's Coppice, JGK and PDO.

Nyctalis asterophora Fr. On dying *Russula nigricans*. 8.

Nyctalis parasitica (Bull. ex Fr.) Fr. On dying *Lactarius* or *Russula*. 3. Lamberts Castle, MH; Kingcombe, AJEL; Powerstock Common, JGK; Bracket's Coppice, J Madgwick.

Omphalina chlorocyanea (Pat.) Singer. 1. Puddletown Forest.

Omphalina obscurata Reid. In acid lawns. 1. Hog Cliff, JGK; probably this at Cripplestyle, !.

Omphalina postii (Fr.) Singer. In acid lawns. 1. Lamberts Castle, VL Breeze.

Omphalina pyxidata (Bull. ex Fr.) Quel. In acid lawns. 2. Buddens Farm, Bere, !; Castle Hill Wood, J Edwards.

Omphalina sphagnicola (Berk.) Moser. In *Sphagnum* bogs. 1. Slepe Heath, AH det. BC.

Omphalina umbellifera (L. ex Fr.) Quel. In wet heaths; may be lichenized. 3. Lamberts Castle and Puddletown Forest, BAB; Boswells plantation, !; Sutton Holms, WRL.

Omphalina umbratilis (Fr.) Quel. 1. Broadstone, EFL.

Oudemansiella mucida (Schrad. ex Fr.) Hoehnel. On dead beech wood. 16.

Oudemansiella radicata (Relhan ex Fr.) Singer. Frequent under beech. 23.

Oxyporus populinus (Schum. ex Fr.) Donk. On wood. 1. Great Wood, Creech, !.

Panaeolus antillarum (Fr.) Dennis. On dung. 1. Studland, CBG.

Panaeolus ater (J.Lange) Kuehn. & Romagn. 2. Came Wood, !; Upton CP, PBLS.

Panaeolus campanulatus (Bull. ex Fr.) Quel. On dung or manured grass. 6. Lamberts Castle, MH, also JGK; Hod Hill and Corfe Common, !; Morden Park, BNSS; Kingswood, D Stephen; Castle Hill Wood, J Edwards.

Panaeolus fimicola (Fr.) Quel. In calcareous pastures. 4. Doghouse Hill, Rawlsbury Camp, Ruins Plantation and Badbury Rings, !.

Panaeolus foenesecii (Pers. ex Fr.) Maire. In short manured turf. 12.

Panaeolus papilionaceus (Bull. ex Fr.) Quel. In calcareous turf. 1. Portland, JGK.

Panaeolus rickenii Hora. Frequent in grassy woodland rides. 14.

Panaeolus semiovatus (Sow. ex Fr.) Lindell. Occasional on dung. 9.

Panaeolus sphinctrinus (Fr.) Quel. Frequent on dung or manured grassland. 16.

Panaeolus subbalteatus (Berk. & Broome) Sacc. In sandy turf. 2. Buddens Farm, Bere, !: Kingston, BC.

Panellus mitis (Pers. ex Fr.) Singer. On dead conifer wood. 3. Yellowham Wood, J Edwards; on gorse, Black Hill, Bere !; Brownsea Is, KC; Avon FP, J Goss.

Panellus serotinus (Schrad. ex Fr.) Kuehn. On dead deciduous wood. 4. Bracket's Coppice, Hog Cliff and Thorncombe Wood, JGK; Bryanston, BNHS.

Panellus stipticus (Bull. ex Fr.) P. Karst. Frequent on dead deciduous wood. 14.

Panus torulosus (Pers. ex Fr.) Fr. On dead deciduous wood. 2. Oakers Wood, BAB; Brownsea Is, KC.

Paxillus atrotomentosus (Batsch ex Fr.) Fr. On dead pine wood. 5. Wootton Hill, JGK; Charlton Marshall, JV Sullivan; Furzebrook, BC; Brownsea Is, KC; Avon FP, GF Le Pard.
Paxillus involutus (Batsch ex Fr.) Fr. Common in woods; toxic. 27.
Paxillus panuoides (Fr. ex Fr.) Fr. On dead pine wood. 2. Brownsea Is, KC; Ferndown, EFL.
Paxillus rubicundus Orton. Under alder. 1. Bracket's Coppice, JGK and PDO.
Peniophora incarnata (Pers. ex Fr.) P. Karst. Probably widespread on dead gorse.
Peniophora limitata (Fr.) Cooke. On dead ash branches. 1. Castle Hill Wood, JGK.
Peniophora quercina (Fr.) Cooke. On dead branches of oak or beech. 12.
Phaeolus schweinitzii (Fr.) Pat. Occasional on living conifers. 8.
Phaeomarasmius erinaceus (Fr.) Kuehn. On deciduous wood. 1. Kingcombe, JGK.
Phanerochaete sanguinea (Fr.) Pouzar. Staining gorse wood red. 1. Abbotsbury Castle, ! det. JGK.
Phellinus contiguus (Fr.) Pat. On wood. 1. Bracket's Coppice, JGK det. DN Pegler.
Phellinus ferreus (Pers.) Bourdot & Galzin. On deciduous twigs. 7. Kingcombe, Powerstock Common. Bracket's Coppice, Clifton Wood, Thorncombe Wood, Castle Hill Wood and Garston Wood, JGK; Holt Forest, AL.
Phellinus ferruginosus (Fr.) Pat. Mostly on beech twigs. 4. Bracket's Coppice and Greenhill Down, JGK; High Wood, Badbury, AL; Edmondsham, EFL.
Phellinus igniarius (Fr.) Quel. Occasional on willow trunks. 12.
Phellinus laevigatus (Fr.) Bourdot & Galzin. On deciduous wood. 1. Warmwell, BAB.
Phellinus pomaceus (Fr.) Maire. On rosaceous trees. 4. Bracket's Coppice and Greenhill Down, JGK; Oakers Wood and Chamberlaynes, BAB; Langton Matravers, RE Stebbings.
Phellinus ribis (Schum. ex Fr.) P. Karst. On *Ribes* bushes. 2. Winterborne Kingston, !; Edmondsham, EFL.
Phellinus trivialis (Bres.) Kreisel. On sallow. 1. Warmwell, BAB.
Phellodon niger (Fr. ex Fr.) P. Karst. Rare under oak. 1. Moreton and Oakers Wood, JGK.
Phlebia cornea Parmasto. On dead branches. 1. Edmondsham, JGK.
Phlebia radiata Fr. Frequent on dead bark. 14.
Phlebiopsis gigantea (Fr.) Juel. On pine stumps. 2. Branksome Park and Edmondsham, EFL.
Phleogena faginea (Fr. ex Fr.) Link. On beech wood. 1. Duncliffe Wood, JGK.
Pholiota adiposa (Fr.) Kumm. On deciduous trees. 2. Upton CP, Poole Borough Council; Castle Hill Wood, J Edwards.
Pholiota alnicola (Fr.) Singer. On deciduous trees, mostly birch. 5. Powerstock Common, JGK; Melbury Park, BAB; Furzebrook, BC; Upton CP, PBLS; St Leonards, AL.
Pholiota apicrea (Fr.) Moser. On pine wood. 2. Ringstead, BAB; Yellowham Wood, J Edwards.

(*Pholiota astragalina* (Fr.) Singer. On conifer wood, a northern species. 1. Avon FP, VC 11.)
Pholiota cerifera (P Karst.) P. Karst. On live or dead beech trunks. 3. Powerstock Common, JGK; Stock Gaylard Park, !; Bryanston, BNHS.
Pholiota flammans (Fr.) Kumm. A northern species on conifer wood. 2. Puddletown Forest, BAB; Hooke Park, JGK.
Pholiota gummosa (Lasch.) Singer. On woody debris. 7. Kingcombe and Bracket's Coppice, JGK; Oakers Wood, BC; Bagber and Lydlinch Common, !; Kingston and Furzebrook, BC; Studland, JGK.
Pholiota highlandensis (Peck) A.H. Sm. & Hesler. Occasional on burnt soil or wood. 8.
Pholiota inaurata (Fr.) J. Lange. On dead birch wood. 1. Bracket's Coppice, JGK.
Pholiota lenta (Pers. ex Fr.) Singer. On woody debris. 2. Bride Head, MH; Cranborne, EFL.
Pholiota lubrica (Pers. ex Fr.) Singer. On wood. 1. Kingcombe, JGK.
Pholiota lucifera (Lasch.) Quel. On buried wood. 2. Bracket's Coppice, M Brett.
Pholiota myosotis (Fr. ex Fr.) Singer. In wet heaths or bogs. 1. Studland Heath, BC.
Pholiota perscina Orton. 1. Bracket's Coppice, PDO.
Pholiota squarrosa (Pers. ex Fr.) Kumm. On live or dead deciduous trees, mostly beech. 14.
Phylloporus rhodoxanthus (Schwein.) Bres. Under beech. 1. Hog Cliff, JGK.
Physisporinus sanguinolentus (Fr.) Pilat. On dead wood. 6. Powerstock Common and Thorncombe Wood, JGK; Ringstead, Park Wood near Lulworth and Morden Park, BAB; Moors River.
Physisporinus vitreus (Fr.) P. Karst. On beech wood. 2. Piddles Wood, JGK; Castle Hill Wood, EFL.
Piptoporus betulinus (Bull. ex Fr.) P. Karst. A common parasite of birch. 24.
Pleurotellus graminicola Fayod. In herbaceous litter. 1. Bracket's Coppice, JGK.
Pleurotellus porrigens (Pers. ex Fr.) Kuehn. & Romagn. On pine trunks. 1. Crichel, EW Baker.
Pleurotus cornucopiae (Paulet ex Pers.) Rolland. Occasional on deciduous trees. 8.
Pleurotus dryinus (Pers. ex Fr.) Kumm. On deciduous boles. 1. Tincleton, MH.
Pleurotus lignatilis (Pers. ex Fr.) Kumm. On dead beech wood. 1. Piddletrenthide, !.
Pleurotus ostreatus (Jacq. ex Fr.) Kumm. OYSTER FUNGUS. Frequent on deciduous stumps. 22.
Pleurotus pulmonarius (Fr.) Quel. On beech or elm stumps. 4. Kingcombe, Powerstock Common, Hooke Park, Clifton Wood and Hog Cliff, JGK.
Pleurotus ulmarius (Bull. ex Fr.) Kumm. On elm boles. 2. Kingston, BC; Crichel, EFL.
Pluteolus aleuriatus (Fr. ex Fr.) P. Karst. 1. Bracket's Coppice, JGK.
Pluteus atromarginatus (Konrad) Kuehn. On stumps. 1. Lulworth Castle, BAB.
Pluteus aurantiorugosus (Trog) Sacc. On elm stumps. 2. Hendover Coppice, JGK; 79.

Pluteus boudieri Orton. On beech stumps 1. Kingston, BC.

Pluteus cervinus (Schaeff. ex Fr.) Kumm. Common on deciduous logs. 26.

Pluteus cinereofuscus J. Lange. On rotten wood. 1. Bracket's Coppice, M Brett, also J Madgwick.

Pluteus galeroides Orton. On beech wood. 4. Kingcombe, Hog Cliff, Thorncombe Wood and Fifehead Wood, JGK.

Pluteus griseoluridus Orton. On wood of alder or sallow. 2. Hendover Coppice and Arne, JGK.

Pluteus griseopus Orton. On beech wood. 3. Rapehole Coppice, Hog Cliff and Oakers Wood, JGK.

Pluteus hispidulus (Fr. ex Fr.) Gillet. On beech wood. 1. Piddles Wood, JGK.

Pluteus luteovirens Rea. On beech wood. 1. Hendover coppice, JGK.

Pluteus minutissimus Maire. On beech wood. 2. Hendover Coppice, JGK; Poxwell Big Wood, BAB det. DA Reid.

Pluteus nanus (Pers. ex Fr.) Kumm. On dead wood. 1. Hendover Coppice, JGK.

Pluteus pellitus (Pers.ex Fr.) Kumm. Rare on wood. 1. Hog Cliff, JGK.

Pluteus phlebophorus (Ditmar ex Fr.) Kumm. On dead wood. 1. Hendover Coppice, JGK.

(*Pluteus plautus* (Weinm.) Gillet. Possibly this on dead wood at Hod Wood, !.)

Pluteus romellii (Britzelm.) Lapl. On deciduous wood. 6. Bracket's Coppice, Hooke Park, Hendover Coppice and Piddles Wood, JGK; Poxwell Big Wood, BAB; Bryanston, BNHS; Kingston, BC.

Pluteus salicinus (Pers. ex Fr.) Kumm. On deciduous wood. 12.

Pluteus semibulbosus (Lasch.) Gillet. On deciduous wood. 1. Great Wood, Creech, !.

Pluteus thomsonii (Berk. & Broome) Dennis. On deciduous wood. 3. Poxwell Big Wood, BAB; Bryanston, BNHS; Furzebrook, BC.

Pluteus umbrosus (Pers. ex Fr.) Kumm. On deciduous wood. 5. Powerstock Common and Hendover Coppice, JGK; Bere Heath, !; Furzebrook, BC; Studland Heath, JGK.

Pluteus xanthophaeus Orton. On beech wood. 2. Ringstead, BAB det. DA Reid; Piddles Wood, JGK.

Podoscypha multizonata (Berk. & Broome) Pat. Under oak. 1. Arrowsmith lane, Canford Heath, !.

Polyporus brumalis Fr. On deciduous wood. 6. Stonebarrow, Hendover Coppice and Studland, JGK; Kingston, BC; Rushmore, BAB; Castle Hill Wood, J Edwards.

Polyporus ciliatus Fr. On birch wood. 3. Powerstock Common, JGK; Warmwell, !; Bourton, !.

Polyporus nummularius (Bull.) Pers. On wood 2. Puddletown Forest, BAB; Studland, JGK det. DA Reid. Perhaps a small form of *P. varius*.

Polyporus rostkovii Fr. On stumps. 1. Bryanston, BNHS.

Polyporus squamosus Fr. Frequent on deciduous wood. 20.

Polyporus tuberaster Jacq. ex Fr. On gorse wood. 3. Warmwell Heath, BAB; Furzebrook, BC; Avon FP, JGK.

Polyporus varius Fr. Frequent on deciduous wood. 17.

Poria gilvescens Bres. On beech twigs. 3. Powerstock Common and Foxhills, JGK; Westrow Ho, !.

Porphyrellus pseudoscaber (Secr.) Singer. In woods. 2. Hendover Coppice, JGK; Bere Wood and Bloxworth Heath, !.

Porpomyces mucidus (Pers. ex Fr.) Juel. On dead branches. 2. Holt Forest and Avon FP, JGK.

Psathyrella albidula (Romagn.) Moser. 1. Kingston, BC.

Psathyrella ammophila (Dur. & Lev.) Orton. On dunes. 1. Sandbanks, BAB.

Psathyrella artemisiae (Pass.) Konrad & Maubl. Under beech. 1. Champernhayes.

Psathyrella atomata (Fr.) Quel. In grassland. 2. Swanage, CBG; St Giles Park, EFL.

Psathyrella candolleana (Fr.) Maire. In rich soil. 14.

Psathyrella conopilea (Fr.) Pearson & Dennis. In rich soil. 3. Bracket's Coppice, J Madgwick; Park Wood, Lulworth, BAB; Upton CP, PBLS.

Psathyrella cortinarioides Orton. 1. Bracket's Coppice, PDO.

Psathyrella cotonea (Quel.) Konrad & Maubl. On stumps. 1. Kingcombe, JGK.

Psathyrella gossypina (Bull. ex Fr.) Pearson & Dennis. In woods. 1. Piddles Wood, JGK.

Psathyrella gracilis (Fr.) Quel. On buried twigs. 6. Bracket's Coppice, J Madgwick; Came Wood and Puddletown Forest, !; Herston, CBG; Sutton Holms and Edmondsham, EFL.

Psathyrella hirta Peck. On dung. 2. Kingcombe, JGK; Kingston, BC.

Psathyrella hydrophila (Bull. ex Fr.) Maire. Frequent on deciduous stumps. 14.

Psathyrella lachrymabunda (Bull. ex Fr.) Moser. Frequent in rich soil. 15.

Psathyrella leucotephra (Berk. & Broome) Orton. In rich soil. 2. Hendover Coppice, JGK; Bryanston, BNHS.

Psathyrella marcesibilis (Britzelm.) Singer. 2. Ringstead, BAB det. DA Reid; Oatclose Wood, !.

Psathyrella microrhiza (Lasch.) Konrad & Maubl. 3. Bracket's Coppice and Hog Cliff, JGK; Kingston, BC.

Psathyrella multipedata (Peck) A.H. Sm. On deciduous stumps. 3. Hooke Park and Bracket's Coppice, JGK; Post Green and Pamphill, !.

Psathyrella obtusata (Fr.) A.H. Sm. On dead wood. 2. Bracket's Coppice, J Madgwick; Upton CP, PBLC.

Psathyrella pennata (Fr.) Pearson & Dennis. On burnt soil. 3. Powerstock Common, Bracket's Coppice and Hendover Coppice, JGK.

Psathyrella piluliformis (Bull. ex Merat) Orton. On deciduous stumps. 2. Bracket's Coppice and Clifton Wood, JGK.

Psathyrella sarcocephala (Fr. ex Fr.) Singer. On deciduous stumps. 1. Bracket's Coppice, JGK.

Psathyrella spadicea (Schaeff.) Singer. On rotten wood. 1. Sutton Holms, EFL.

Psathyrella spadiceogrisea (Fr.) Maire. In deciduous woods, fruiting in spring. 4. Eggardon Hill, Mr Douglas; Poxwell Big Wood, BAB; Red Bridge, Moreton, AH; Kingston, BC; Durlston CP, KC.

Psathyrella squamosa (P. Karst.) Moser. In deciduous woods. 2. Lamberts Castle, MH; Oakers Wood, JGK.

Pseudoclitocybe cyathiformis (Bull. ex Fr.) Singer. Occasional in deciduous woods. 8.

Pseudocraterellus sinuosus (Fr.) Reid. Under birch. 2. Bracket's Coppice and Piddles Wood, JGK.

Psilocybe apelliculosa Orton. In grass. 1. Hog Cliff, JGK and PDO.

Psilocybe coprophila (Bull. ex Fr.) Quel. On dung. 3. Lamberts Castle, JGK; Rawlsbury Camp, !; Grange Arch, !.

Psilocybe crobula (Fr.) M. Lange. On fallen twigs. 1. Thorncombe Wood, JGK.

Psilocybe merdaria (Fr.) Ricken. On dung 3. Corfe Common and Nine Barrow Down, CBG; Birches Copse, EFL.

(*Psilocybe montana* (Pers. ex Fr.) Orton. 1. Avon FP, VC 11, J Goss.)

Psilocybe phyllogena (Peck) Peck. 1. Kingcombe, JGK.

Psilocybe semilanceata (Fr.) Kumm. MAGIC MUSHROOM. Frequent in old pastures. 14. It contains the toxic hallucinogens psilocin and psilocybin.

Pulcherricium caeruleum (Fr.) Parmasto. A deep blue crust on hazel sticks. 3. Bracket's Coppice, JGK; Winterborne Kingston, !; Zigzag Hill, F Preedy.

Pterula multifida Fr. Rare under oak. 1. Bracket's Coppice, JGK.

Ramaria botrytis (Fr.) Ricken. In deciduous woods. 2. Yellowham Wood, J Edwards; Durlston CP, KC.

Ramaria decurrens (Pers.) R.H. Petersen. In acid soil. 1. Stoborough, AH det. P Roberts.

Ramaria eumorpha (P. Karst.) Corner. On pine needle litter. 1. Arne, BAB.

Ramaria formosa (Fr.) Quel. In deciduous woods. 1. Brownsea Is, KC.

Ramaria ochraceovirens (Jungh.) Donk. On coniferous litter. 1. Clifton Wood, JGK.

Ramaria stricta (Fr.) Quel. Occasional on dead wood or litter. 10.

Ramicola haustellaris (Fr.) Watling. On wood. 1. Kingsettle Wood, PDO.

Resupinatus applicatus (Batsch ex Fr.) Gray. On dead twigs. 3. Winterborne Kingston, !; Coombe Bottom, Hambledon, !; Edmondsham, JGK.

Resupinatus trichotis (Pers.) Singer. On dead twigs. 1. Piddles Wood, JGK.

Rhodotus palmatus (Bull. ex Fr.) Maire. On deciduous stumps, mainly elm. 8. Hog Cliff, Piddles Wood and Kingsettle Wood, JGK; Ratcliffe Wood, !; Upton CP, PBLS; Kingston, BC; The Oaks, Badbury, JGK; Pamphill, J Fildes; Studland, BC, also JGK; Brownsea Is, KC.

Rickenella fibula (Bull. ex Fr.) Raith. In mossy, acid lawns. 15.

Rickenella setipes (Fr. ex Fr.) Raith. In short turf. 4. Oakers Wood, Arne and Edmondsham, JGK; Furzebrook, BC; Brenscombe Hill, BAB; Brownsea Is, KC; Avon FP, JGK.

Rigidiporus ulmarius (Fr.) Imazeki. On dead elm logs. 3. Woolland, !; Studland, CBG; Brownsea Is, KC; Canford, Canford School NHS.

Russula adusta (Pers. ex Fr.) Fr. In deciduous woods. 4. Kingcombe, AJEL; Brownsea Is, KC; Edmondsham and Daggons, EFL.

Russula aeruginea Lindblad. Occasional in woods. 9.

Russula albonigra (Krombh.) Fr. In deciduous woods. 1. Bracket's Coppice, JGK.

Russula aquosa Leclair. In *Sphagnum* under birch. 1. Furzebrook, BC.

Russula betularum Hora. Occasional under birch. 11.

Russula brunneoviolacea Crawshay. Under oak. 2. Powerstock Common, JGK; Lulworth Castle, BMS.

Russula caerulea (Pers.) Fr. Occasional under pine. 8.

Russula claroflava Grove. Occasional in wet woods of birch or pine. 10.

Russula cyanoxantha (Schaeff.) Fr. Common in deciduous woods. 22. Var. *peltereaui* was noted from Powerstock Common and Oakers Wood.

Russula delica Fr. In deciduous woods. 7. Bracket's Coppice, M Brett; Powersock Common and Melbury Park, JGK; Affpuddle, Poxwell Big Wood, Knighton, Puddletown Forest and Hurst Heath, BAB; Bere Wood, !; Furzebrook, BC; Avon FP, GF Le Pard.

Russula densifolia (Secr.) Gillet. In deciduous woods. 13.

Russula emetica (Schaeff. ex Fr.) Pers. Frequent under pine. 15.

Russula emeticella (Singer) Hora. Under oak. 6. Kingcombe, Piddles Wood, Oakers Wood, Moreton and Studland, JGK; Trigon, AH; Furzebrook, BC.

Russula farinipes Rommell. Under oak. 2. Poxwell Big Wood and Rempstone Wood, BAB.

Russula fellea (Fr.) Fr. Frequent under beech. 19.

Russula firmula Schaeff. Under conifers. 1. Bracket's Coppice, JGK.

Russula foetens (Pers. ex Fr.) Fr. Occasional in damp woods. 9.

Russula fragilis (Pers. ex Fr.) Fr. Frequent in deciduous woods. 15.

Russula gracillima Schaeff. In wet birch woods. 4. Powerstock Common and Bracket's Coppice, JGK; Hurst Heath, BAB; Avon FP, C Edwards.

Russula grata (Britzelm.) Kuehn. & Romagn. Under birch or pine. 3. Hog Cliff, JGK; Oakers Wood, BAB; Furzebrook, BC.

Russula grisea (Pers. ex Secr.) Fr. In deciduous woods. 4. Affpuddle, Hyde Wood, Moreton and Poxwell Big Wood, BAB; Brownsea Is, KC.

Russula heterophylla (Fr.) Fr. In deciduous woods. 5. Puncknowle, BAB; Kingcombe, Bracket's Coppice and Thorncombe Wood, JGK; Sutton Holms, EFL.

Russula illota Romagn. Under oak. 1. Kingcombe, PDO.

Russula integra Fr. Under conifers. 1. Maldry Wood and Sutton Copse, EFL.

Russula knauthii (Singer) Hora. In wet oak woods. 1. Poxwell Big Wood, BAB.

Russula krombholzii Shaffer. Frequent under oak. 19.

Russula laccata Huijsman. In dunes. 1. South Haven, JGK.

Russula lepida Fr. In deciduous woods. 6. Eight Acre Copse, A Simon; Powerstock Common, MH; Bracket's Coppice, M Brett; Trigon, DWT; Studland, BC; Brownsea Is, KC; Daggons, EFL.

Russula lutea (Huds. ex Fr.) Gray. In deciduous woods. 3. Puddletown Forest, BMS; Oakers Wood, BAB and !; St Leonards, AL.

Russula luteotacta Rea. In deciduous woods. 6. Kingcombe, Powerstock Common, Bracket's Coppice, Oakers Wood and Moreton, JGK; Tatton Coppice, Poxwell Big Wood, Moigne Combe and Brenscombe Hill, BAB.

Russula mairei Singer. Frequent under beech. 17.

Russula nauseosa (Pers. ex Secr.) Fr. A northern species, under pine. 2. Parnham and Melbury Park, BAB.

Russula nigricans (Bull.) Fr. Frequent in deciduous woods. 21.

Russula nitida (Pers. ex Fr.) Fr. In deciduous woods. 6. Lamberts Castle, VL Breeze; Stonebarrow, MH; Bracket's Coppice, JGK; Hurst Heath, BAB; Arne, !; Studland, JGK; St Leonards, AL.

Russula ochroleuca (Pers.) Fr. Common in woods. 29.

Russula odorata Romagn. In woods. 2. Bracket's Coppice and Piddles Wood, PDO.

Russula olivacea (Schaeff.) Pers. In acid woods. 3. Bracket's Coppice M Brett; Canford Heath, Canford School NHS; Castle Hill Wood, J Edwards.

Russula parazurea Schaeff. Mostly under oak. 4. Powerstock Common, JGK; Hurst Heath and Creech Grange, BAB; Norden, BC.

Russula pectinata (Bull. ex St Amans) Fr. In woods. 1. Rempstone Wood, BAB.

Russula pseudointegra Arn. & Goris. In deciduous woods. 3. Powerstock Common, JGK; Wyke Wood and Poxwell Big Wood, BAB.

Russula puellaris Fr. In deciduous woods. 5. Kingcombe and Powerstock Common, JGK; Bracket's Coppice, M Brett; Hurst Heath and Ilsington, BAB; Castle Hill Wood, EFL.

Russula pulchella Borsz. In sandy soil. 1. Gallows Hill, !.

Russula queletii Fr. In woods, on acid or calcareous soils. 5. Lamberts Castle, Powerstock Common and Stubhampton Bottom, JGK; Furzebrook, MHL; Brownsea Is, KC.

Russula sanguinea (Bull.) Fr. In acid woods. 8. Lamberts Castle, VL Breeze; Powerstock Common, JGK; Hurst Heath and Ilsington Wood, BAB; Red Bridge, Moreton and Trigon, AH; Hethfelton and Bere Heath, !; Moreton and Oakers Wood, JGK; Furzebrook, BC.

Russula sardonia Fr. Frequent under pine on acid soil. 18.

Russula sororia (Fr.) Romell. Mostly under oak. 5. Kingcombe, AJEL; Red Bridge, Moreton, AH; Thorncombe Wood, JGK; Furzebrook, BC; Holton Lee, VL Breeze.

Russula turci Bres. Rare under pine. 1. Powerstock Common, JGK.

Russula velenovski Melzer & Zvara. Under oak on acid soil. 2. Oakers Wood, BAB; Gallows Hill, !; Furzebrook, BC.

Russula versicolor Schaeff. Under birch. 3. Bracket's Coppice, M Brett; Melcombe, BAB; Furzebrook, BC; Arne, JGK.

Russula vesca Fr. Frequent in deciduous woods on acid soil. 15.

Russula veternosa Fr. In deciduous woods. 2. Bracket's Coppice, JGK; Quince Hill Wood, CBG.

Russula violacea Quel. In deciduous woods. 1. Chetterwood, !.

Russula violeipes Quel. In deciduous woods. 3. Hooke Park, JGK; Piddles Wood, BAB, also JGK; Affpuddle, BAB.

Russula virescens (Schaeff.) Fr. In deciduous woods on

sandy soil. 4. Eight Acre Copse, A Simon; Lulworth Park, BAB; Brownsea Is, KC; Birches Copse, EFL.

Russula xerampelina (Schaeff.) Fr. Occasional under conifers. 10.

Sarcodon imbricatum (L. ex Fr.) P. Karst. Rare under conifers. 1. Moreton and Oakers Wood, JGK.

Schizophyllum commune (L.) Fr. On dead deciduous wood. 2. Kingcombe, JGK; Brownsea Is, KC.

Schizopora paradoxa (Schrad. ex Fr.) Donk. Occasional on dead twigs. 12.

Serpula lachrymans (Wulf.) Schroet. DRY ROT. On wood. 5. Poxwell, Ringstead and Lulworth Park, BAB; Brownsea Is, KC; Shapwick and Edmondsham, EFL.

Setulipes androsaceus (L. ex Fr.) Antonin. On pine litter or heather stems. 13.

Setulipes quercophilus (Pouz.) Antonin. On deciduous leaves. 3. Moreton, Oakers Wood and Holt Forest, JGK; Fifehead Wood, AL.

Simocybe centunculus (Fr.) Singer. On dead deciduous wood. 3. Bracket's Coppice, AJEL; Bryanston, BNHS; Holt Forest, PDO.

Simocybe sumptuosa (Orton) Singer. On dead beech wood. 2. Powerstock Common and Bracket's Coppice, JGK and PDO.

Skeletocutis amorpha (Fr.) Kotl. & Pouzar. On dead pine wood. 6. Hurst Heath, Arne, Castle Hill Wood and Avon FP, JGK; Rempstone, CBG; Ferndown and Daggons, EFL.

Skeletocutis nivea (Jungh.) Keller. On dead deciduous wood. 4. Kingcombe, Bracket's Coppice and Arne, JGK; Edmondsham, EFL.

Sparassis crispa (Wulf. ex Fr.) Fr. CAULIFLOWER FUNGUS. Single plants on pine stumps. 15.

Sparassis laminosa Fr. Under pine. 1. Ferndown, EFL.

Spongipellis spumeus (Fr.) Pat. On hardwood stump, Studland, JGK det. G Dickson. 1.

Steccherinum fimbriatum (Pers. ex Fr.) Erikss. 1. High Wood, Badbury, AL.

Steccherinum ochraceum (Pers ex Fr.) Gray. On dead beech twigs 2. Birches Copse and Edmondsham, EFL.

Stereum gausapatum (Fr.) Fr. On dead oak wood. 10.

Stereum hirsutum (Willd. ex Fr.) Gray. Very common on dead wood. 31.

Stereum rameale (Pers. ex Fr.) Burt. On dead wood. 12.

Stereum rugosum (Pers. ex Fr.) Burt. Common on live or dead wood. 20

Stereum sanguinolentum (Alb. & Schwein ex Fr.) Fr. On conifer wood. 6. Powerstock Common, JGK; Ringstead, Lulworth Park and Morden Park, BAB; Avon FP, GF Le Pard.

Stereum subtomentosum Pouzar. On dead wood. 1. Edmondsham, EFL.

Strobilurus esculentus (Wulf. ex Fr.) Singer. On fallen spruce cones. 1. Foxhills, JGK.

Strobilurus tenacellus (Pers. ex Fr.) Singer. On fallen pine cones. 6. Morden Bog, Carey, Sugar Hill and Studland, BMS; Witchampton and Ferndown, EFL; Avon FP, JGK.

Stropharia aeruginosa (Curtis ex Fr.) Quel. In birch woods. 9.

Stropharia caerulea Kreissel. In rich soil. 6. Stonebarrow, Kingcombe, Bracket's Coppice, Hendover Coppice, Oakers Wood and Castle Hill Wood, JGK.

Stropharia coronilla (Bull. ex Fr.) Quel. In sandy lawns. 1. Sandford, !.

Stropharia hornemannii (Fr. ex Fr.) Lundell & Nannf. Rare in humus-rich soil. 1. Lytchett Minster, Canford School, NHS.

Stropharia inuncta (Fr.) Quel. In old pasture. 2. Hog Cliff, JGK; Edmondshsam, EFL.

Stropharia pseudocyanea (Desm.) Morgan. In grass. 4. Lamberts Castle, Hog Cliff, Hendover Coppice and Edmondsham, JGK. Perhaps a form of *S. aeruginosa*.

Stropharia semiglobata (Batsch ex Fr.) Quel. Frequent in old, manured pastures. 17.

Suillus aeruginascens(Opat) Snell. Under larch. 3. Moigne Combe; Furzebrook, BC; Edmondsham, EFL.

Suillus bovinus (L. ex Fr.) Kuntze. Frequent under pine. 15.

Suillus citrinovirens Watling. 1. Furzebrook, BC.

Suillus collinitus (Fr.) Kuntze In acid woods. 2. Moreton Sta, !; Moreton, JGK.

Suillus flavidus (Fr.) Singer. Rare under pine. l. Moreton, JGK.

Suillus granulatus (L. ex Fr.) Kuntze. Frequent under pine. 15.

Suillus grevillei (Klotsch. ex Fr.) Singer. Occasional under larch. 13.

Suillus luteus (L. ex Fr.) Gray. Under pine. 8.

Suillus variegatus (Fr.) Kuntze. Under conifers. 13.

Tephrocybe atrata (Fr.) Donk. In burnt soil. 3. Powerstock Common and Bracket's Coppice, JGK; Brownsea Is, KC.

Tephrocybe ellisii Orton. 3. Puddletown Forest, Holt Forest and Edmondsham, PDO.

Tephrocybe ferruginella (Pearson) Orton. 1. Hooke Park, S Berridge det. PDO.

Tephrocybe graminicola Bon. In calcareous turf. 1. Badbury Rings, !.

Tephrocybe palustris (Peck) Donk. Among *Sphagnum* in bogs. 4. Morden Bog, Stroud Bridge and Holton Lee, !; Furzebrook, BC; Studland Heath, BMS, also !.

Tephrocybe rancida (Fr.) Donk. In deciduous woods. 4. Bracket's Coppice, JGK; Milborne Wood, Stoke Wake and Chetterwood, !.

Thelephora penicillata. Corner. In acid woodland. 1. Bracket's Coppice, JGK.

Thelephora spiculosa (Fr.) Burt. In acid litter. 1. Hooke Park, JGK.

Thelephora terrestris (Ehrh.) Fr. Frequent on pine litter or along heath tracks. 15.

Tomentella crinalis (Fr.) Larsen. On dead wood. 1. Thorncombe Wood, JGK.

Tomentella fimbriata Christ. On dead wood. 1. St Giles Park, EFL.

Tomentellopsis echinospora (Ellis) Hjortst. On dead wood. 1. Bracket's Coppice, JGK.

Trametes gibbosa (Pers. ex Fr.) Fr. On dead wood, mostly beech. 14.

Trametes pubescens (Fr.) Pilat. On dead wood. 3. Hooke Wood, Fifehead Wood and Holt Forest, JGK.

Trametes versicolor (L. ex Fr.) Pilat. Very common on dead wood. 31.

Trechispora farinacea (Pers. ex Fr.) Lib. On dead wood. 1. Bracket's Coppice, JGK.

Trechispora fastidiosa (Pers. ex Fr.) Lib. On dead wood. 1. Garston Wood, AL.

Trechispora mollusca (Pers. ex Fr.) Lib. On dead wood. 3. Hurst Heath, JGK; Bovington and Broadley Wood, !.

Trichaptum abietinum (Fr.) Ryvarden. Frequent on conifer wood. 16.

Tricholoma: see also *Calocybe* and *Lepista*.

Tricholoma acerbum (Bull. ex Fr.) Quel. Rare in woods. 2. Oakers Wood, JGK; Castle Hill Wood, EFL.

Tricholoma albobrunneum (Pers. ex Fr.) Kumm. In woods. 3. Newlands Batch, JGK; Morden Park, BNSS; Castle Hill Wood, EFL; closely related to *T. ustaloides*.

Tricholoma album (Fr.) Kumm. In oak woods on acid soil. 6. Bracket's Coppice, JGK; Oakers Wood; Bryanston, BNHS; Morden Park, BNSS; Knoll Farm, !; Brownsea Is, KC; Edmondsham, EFL.

Tricholoma atrosquamosum (Chevall.) Sacc. Rare under oak. 1. Furzebrook, BC.

Tricholoma cingulatum (Almfelt) Jacobasch. In wet woods under birch or sallow. 4. Hendover Coppice, Clifton Wood and Arne, JGK; Furzebrook, BC; Avon FP.

Tricholoma columbetta (Fr.) Kumm. In oak woods on sandy soil. 3. Westrow Ho and Oatclose Wood, !; Arne.

Tricholoma equestre (L. ex Fr.) Kumm. In woods. 3. Furzebrook, BC; Broadstone and Daggons, EFL.

Tricholoma fulvum (DC. ex Fr.) Sacc. Occasional under birch. 14.

Tricholoma imbricatum (Fr. ex Fr.) Kumm. Occasional under pine. 11.

Tricholoma lascivum (Fr.) Gillet. In wet, deciduous woods. 7. Newlands Batch, Kingcombe, Bracket's Coppice, Greenhill Down and Garston Wood, JGK; Piddles Wood, D Stephens; Kingston, BC; Edmondsham, EFL.

(*Tricholoma leucocephalum* (Fr.) Quel. Possibly this at Chetterwood, !.)

Tricholoma portentosum (Fr.) Quel. Under pine. 3. Lamberts Castle, MH, also JGK; Furzebrook, BC; Edmondsham, EFL.

Tricholoma psammopus (Kalchbr.) Quel. Under conifers. 2. Powerstock Common, JGK; Puddletown Forest, BAB.

Tricholoma resplendens (Fr.) P. Karst. In deciduous woods. 1. Bracket's Coppice, PDO.

Tricholoma saponaceum (Fr.) Kumm. In deciduous woods. 5. Powerstock Common and Edmondsham, JGK; Sherborne Park, Westrow Ho and Hethfelton, !; Sutton Holms and Castle Hill Wood, EFL.

Tricholoma scalpturatum (Fr.) Quel. Occasional under oak. 8.

Tricholoma sciodes (Pers.) Martin. Under beech. Stonebarrow, MH, also JGK; Edmondsham, EFL.

Tricholoma sejunctum (Sow. ex Fr.) Quel. Rare in deciduous woods. 2. Bracket's Coppice, JGK; Oakers Wood, BC.

Tricholoma stiparophyllum (Lund.) P. Karst. In woods, closely related to *T. album*. 4. Powerstock Common, Bracket's Coppice, Thorncombe Wood and Garston Wood, JGK.

Tricholoma sudum (Fr.) Quel. In woods. 1. Durlston Park, KC.

Tricholoma sulphureum (Bull. ex Fr.) Quel. Widespread in woods on acid soil. 16.

Tricholoma terreum (Schaeff. ex Fr.) Kumm. In woods, often under pine. 13.

Tricholoma ustale (Fr. ex Fr.) Kumm. In deciduous woods. 4. Lamberts Castle, JGK; Powerstock Common, MH, also BMS; Thorncombe Wood and Holt Forest, JGK; Puddletown Forest, BAB.

Tricholoma ustaloides Romagn. In deciduous woods on clay. 3. Wootton Hill, Kingcombe, Powerstock Common and Birches Copse, JGK.

Tricholoma vaccinum (Pers. ex Fr.) Kumm. Under pines. 1. Thorncombe Wood, BAB.

Tricholoma virgatum (Fr. ex Fr.) Kumm. Under pines. 4. Lamberts Castle and Bracket's Coppice, JGK; Brownsea Is, KC; Daggons, EFL.

Tricholomopsis rutilans (Schaeff. ex Fr.) Singer. PLUMS AND CUSTARD. Frequent in woods. 21.

Trichopilus jubatus (Fr.) comb. nov. On heaths. 1. Edmondsham, EFL.

Trichopilus porphyrophaeus (Fr.) comb. nov. In grassland. 7. Lamberts Castle, MH; Powerstock Common and Greenhill Down, JGK; Melbury Park, BAB; Bindon Abbey, !; Upton CP, PBLS; Durlston CP, KC.

Tubaria conspersa (Pers. ex Fr.) Fayod. On woody debris. 7. Bracket's Coppice, Moreton, Arne, Studland, Holt Forest and Edmondsham, JGK; Hurst Heath, BAB.

Tubaria furfuracea (Pers. ex Fr.) Gillet. On woody debris. 15.

(*Tubaria hiemalis* Romagn. On woody debris. 1. Avon FP, VC 11, AL.)

Tubaria romagnesiana Arnolds. Under gorse. 1. Stonebarrow, JGK.

Tubulicrinis glebulosus (Bres.) Donk. On dead wood 1. Bracket's Coppice, JGK.

Tylopilus felleus (Fr.) P. Karst. In acid deciduous woods. 8.

Typhula erythropus Pers. ex Fr. On leaf litter. 5. Kingcombe, Powerstock Common, Bracket's Coppice, Fifehead Wood, Holt Forest and Edmondsham, JGK.

Typhula phacorrhiza Fr. On leaf litter in a glasshouse. 1. Edmondsham, EFL.

Typhula quisquiliaris (Fr.) Corner. On dead bracken stems. 4. Bracket's Coppice, Arne and Holt Forest, JGF; Puddletown Forest, AL.

Typhula sclerotioides Pers. ex Fr. On dead twigs. 1. Hendover Coppice, JGK.

Typhula setipes (Grev.) Berthier. On dead alder twig. 1. Kingcombe, JGK.

Typhula uncialis (Grev.) Berthier. On dead twigs. 2. Kingcombe, BMS; Fifehead Wood, JGK.

Tyromyces caesius (Fr.) Murrill. On rotten wood. 10.

Tyromyces chioneus (Fr.) P. Karst. On dead wood. 7. Lewesdon Hill, BAB; Kingcombe, Bracket's Coppice, Melbury Park, Clifton Wood, Piddles Wood and Holt Forest, JGK; The Oaks, Badbury, AL.

Tyromyces fissilis (Berk. & Curtis) Donk. On dead wood. 1. Fifehead Wood, G Dickson.

Tyromyces gloeocystidiatus Kotl. & Pouzar. On dead wood. 3. Puddletown Forest, BAB; Branksome Park and Ferndown, EFL.

Tyromyces lacteus (Fr. ex Fr.) Murrill. On dead wood. 9.

Tyromyces stipticus (Pers. ex Fr.) Kotl. & Pouzar. On dead conifer wood. 14.

Tyromyces subcaesius David. On dead deciduous wood. 11.

Tyromyces tephroleucus (Fr.) Donk. On dead wood. 1. Creech Grange, BAB.

Volvariella bombycina (Schaeff. ex Fr.) Singer. On live or dead wood. 3. Bride Head, MH; Kingcombe, J Harris; Lulworth Park, !.

Volvariella speciosa (Fr. ex Fr.) Singer. Locally common on rich calcareous soil. 9.

Volvariella taylorii (Berk. & Bres.) Singer. In limestone pasture. 1. Portland, JGK.

Vuilleminia comedens (Nees. ex Fr.) Maire. On dead deciduous branches. 2. Kingcombe and Thorncombe Wood, JGK.

Xerocomus badius (Fr. ex Fr.) Gillet. Frequent in woods, mostly under conifers. 19.

Xerocomus chrysenteron (Bull.) Quel. Frequent, mainly in deciduous woods. 21.

Xerocomus parasiticus (Bull. ex Fr.) Quel. Occasional on dying *Scleroderma citrinum*. 10.

Xerocomus pruinatus (Fr.) Quel. Under oak. Poxwell Big Wood, BAB; Bere Heath, !; Pamphill, J Fildes.

Xerocomus pulverulentus (Opat) Moser. Under birch or pine. 2. Hog Cliff, JGK; Hendover Coppice, B Collins.

Xerocomus rubellus (Krombh.) Moser. Under oak. 1. Furzebrook, BC.

Xerocomus spadiceus (Fr.) Quel. In acid woods. 1. Furzebrook, BC.

Xerocomus subtomentosus (L.) Quel. Occasional in deciduous woods. 12.

GASTEROMYCETES AND TREMELLALES

Astraeus hygrometricus (Pers.) Morg. On sandy soil. 1. Sherborne, Mrs Sidaway (Good, 1961–63).

Auricularia auricula-judae (Bull. ex Fr.) Wellst. JEW'S-EAR. Very common on elder, rare on other trees. 27.

Auricula mesenterica (Dicks. ex Fr.) Pers. Occasional on dead wood, mostly elm. 17.

Bovista dermoxantha (Vitt.) De Toni. Scarce on sandy heaths. 3. Yellowham Wood, J Edwards; Durlston CP and Brownsea Is, KC.

Bovista nigrescens (Pers.) Pers. On heaths. 6. Puddletown Forest and Rushmore, BAB; Oatclose Wood, !; Fontmell Down, DWT; Froxen Copse, KB Overton; Avon FP, J Goss.

Bovista plumbea (Pers.) Pers. On sandy lawns. 6. Loscombe, DL Thomas; Hog Cliff, JGK; Ringstead, BAB det. DA Reid; Brownsea Is, KC; Cranborne, EFL; Avon FP, J Goss.

Calocera cornea (Batsch ex Fr.) Fr. Frequent on deciduous logs. 18.

Calocera glossoides (Pers. ex Fr.) Fr. Rare on dead oak wood. 1. Holt Forest, G Dickson.

Calocera pallidospathulata Reid. Rare on dead conifer branches. 2. Bracket's Coppice, JGK; Studland, JGK.

Calocera viscosa (Pers. ex Fr.) Fr. Common on dead conifer wood. 22.

Calvatia gigantea (Batsch ex Pers.) Lloyd. GIANT PUFFBALL. Sporadic on rich, manured soil. 19.

Clathrus ruber Pers. Sporadic on rich soil near the coast. 5. Lyme Regis, M Brett (Good, 1961; 1962; 1963); Bridport, RDG (Good, 1961; 1962; 1963); West Bay, D Kempe; Dorchester, ! *et al.*; Durlston CP, AM and P de S Barrow. A Mediterranean species at the northern edge of its range.

Cyathus olla (Batsch) Pers. Rare, on sticks. 1. Roadside, Studland, JGK.

Cyathus striatus (Huds.) Pers. Rare, on sticks or stumps. 2. Bracket's Coppice, JGK; Maldry Wood, EFL.

Dacrymyces stillatus (Nees.) Fr. Common on dead wood. 24.

Exidia glandulosa (Buill.) Fr. Occasional on dead wood, mostly oak. 12.

Exidia plana (Wigg. ex Schleich.) Donk. Rare on dead wood. 1. Furzebrook, BC.

Exidia recisa (Ditm. ex Gray) Fr. Rare on dead wood. 1. Kingston Lacy, Canford School, NHS.

Exidia thuretiana (Lev.) Fr. Occasional on dead deciduous branches. 9.

Geastrum fimbriatum Fr. Rare on woodland soil. 2. Hendover Coppice, JGK; Poole, C Pike det. JGK.

Geastrum fornicatum (Huds.) Fr. 1. Witchampton, EFL.

Geastrum lageniforme Vitt. 1. Witchampton, EW Baker.

Geastrum pectinatum Pers. In sandy soil under pine. 1. Canford Cliffs, C Pike, det. JGK.

Geastrum rufescens (Pers.) Fr. In woods. 3. Bryanston, BNHS; Witchampton, EW Baker; Rushmore, BAB.

Geastrum striatum DC. Rare in woods. 2. Charmouth, RDG, as *G. bryantii* (Good, 1961; 1962; 1963); Ringstead, BAB det. DA Reid.

Geastrum triplex Jungh. EARTH-STAR. Local on rich soil in woods. 13.

Handkea excipuliformis (Bull. ex Pers.) Kreisel. Occasional in sandy woods or turf. 18.

Handkea utriformis (Bull. ex Pers.) Kreisel. In sandy turf in the south. 9.

Lycoperdon echinatum Pers. In rich woodland soil. 7. Eight Acre Copse, A Scrion; Bracket's Coppice, M Brett; Bryanston Hangings, BNHS; Creech; Holt Forest, JGK; Garston Wood, BC.

Lycoperdon mammiforme Pers. On soil in calcareous woods. 2. Hendover Coppice, JGK; Garston Wood, BC, also JGK.

Lycoperdon molle Pers. Occasional on acid soil in woods or under conifers. 8.

Lycoperdon nigrescens Pers. Occasional in woods or conifer plantations. 14.

Lycoperdon perlatum Pers. Frequent on acid soil in woods. 25.

Lycoperdon pyriforme Schaeff. ex Pers. Common on rotten wood in woods. 26.

Lycoperdon spadiceum Pers. On sandy or calcareous lawns, or on dunes. 5. Hog Cliff, JGK; The Warren at Durdle Door, Hethfelton and Durlston CP, !; Studland dunes, JGK.

Lycoperdon umbrinum Pers. On acid soil in deciduous woods. 6. Kingcombe, Powerstock Common, Hendover Coppice, Puddletown Forest, Holt Forest, Edmondsham and Avon FP, JGK.

Mutinus caninus (Huds. ex Pers.) Singer. DOG'S STINKHORN. Local on rich soil in woods. 16.

Myxarium nucleatum Wallr. On fallen deciduous branches. 4. Bracket's Coppice, Hog Cliff, JGK; Oakers Wood; Townsend, Swanage, WGT.

Phallus impudicus (L.) Pers. STINKHORN. Frequent in rich soil in woods and hedgebanks. 28. Var. *togatus* (Kalchbr.) Costant & Dufour was found at Hillcombe Coppice, Durweston by CJ Cornell.

Pisolithus tinctorius (Pers.) Desv. Very rare in sandy soil. 2. Black Hill, Bere, !; Greenland, SFRG.

Pseudohydnum gelatinosum (Scop. ex Fr.) P. Karst. Rare on conifer logs. 2. Oakers Wood; Furzebrook, BC.

Rhizopogon luteolus Fr. On sandy soil under pines. 6. Puddletown Forest, BAB; Oakers Wood, Moreton and Studland, JGK; Furzebrook, BC; Colehill, EW Baker; Daggons, EFL.

Scleroderma areolatum Ehrenb. On soil in woods. 6. Kingcombe and Hog Cliff, JGK; Fifehead Wood, JGK; Oakers Wood and Broadley Wood, !; Holt Forest, G Dickson.

Scleroderma bovista Fr. On soil in woods. 3. Melbury Park, JGK; Woolsbarrow, !; Holt Forest, G Dickson.

Scleroderma citrinum Pers. ex Pers. EARTHBALL. Common in woods, hedgebanks and heaths, extending to County Hall lawns in Dorchester. 23.

Scleroderma verrucosum (Bull. ex Pers.) Pers. Frequent on soil in woods. 16.

Sebacina incrustans (Fr.) Tul. On soil or sticks in hazel copses. 6. Bracket's Coppice, JGK; Wyke Wood, Poxwell Big Wood and Moigne Combe, BAB; Winterborne Kingston, !; Ferndown, Mrs Pringle; St Giles Park, EFL.

Sphaerobolus stellatus (Tode) Pers. Rare on fallen twigs. 2. Powerstock Common, JGK; Edmondsham, EFL.

Tremella encephala Pers. ex Fr. On *Stereum*, on fallen pine branches. 1. Studland, JGK.

Tremella foliacea Pers. ex Fr. On dead conifer branches. 8.

Tremella lutescens Pers. ex Fr. On dead branches. 1. Stonebarrow, JGK.

Tremella mesenterica Retz. ex Hook. JELLY FUNGUS. On *Peniophora*, mostly on dead gorse. 29.

Vascellum pratense (Pers. ex Pers.) Kreisel. Frequent in short, calcareous or sandy turf. 15.

CHAPTER 10
DORSET ALGAE

DORSET CHAROPHYTES OR STONEWORTS

An isolated group of about 400 species of Green Algae with notably large cells. Charophytes have been present in the county for a very long time. Their fossils are the subject of a monograph by TM Harris (1939). Modern Stoneworts occur in freshwater and can become dominant in shallow or deep water; they often prefer new ponds which are low in phosphate. Species of *Nitella* and *Nitellopsis* prefer acid water, *Lamprothamnium papillosum* needs brackish water, and species of *Chara* tolerate alkaline or sulphate-rich water from which they secrete a 'skeleton' of calcium carbonate. Few species occur in Dorset, and fewer still have been recorded this century. A detailed Stonewort Flora of the county is being prepared by N.F. Stewart, who has kindly let me see his work. In the list below, species are listed in alphabetical order, with a list of recent localities for the less common plants, plus all 10 km × 10 km grid squares where they have been found.

Note that very little work has been done on other freshwater Green Algae, most of which are microscopic. A species of *Enteromorpha* becomes common in the River Stour in the summer, and species of *Cladophora* and *Spirogyra* often clog garden ponds. Four hundred and twenty-three species of freshwater Green Algae, mostly desmids and diatoms, have been found in south Hampshire (Woodhead & Tweed, 1946–47). Fossils of calcareous algae occur in the Lower Purbeck (Jurassic) beds of Portland and Purbeck; at that time the region was covered by a large freshwater lake (House, 1968).

CHAROPHYCACEAE

[*Chara aspera* Deth. ex Willd. South of Wareham, also at Studland and Little Sea, but not since 1902; 98, 08; specimens in **BM, DOR, E, NMW**.]
[*Chara baltica* Bruz. Studland Heath (98, 1870, H Groves and J Groves).]
[*Chara canescens* Desv. & Lois. Hamworthy clay-pits (98, 1902, GR Bullock-Webster in BM); Little Sea (08, 1865–85, many botanists in **BM, DOR, E**).]
(*Chara contraria* records are subsumed under *C. vulgaris*.)
Chara globularis Thuill. Most, if not all, old records would now be called *C. virgata*. R. Lydden at Marnhull (71, 1995, DAP and RF).
(*Chara hispida* L. Reports from Canford and Wareham Heaths (1799, RP) were probably not this.)
Chara virgata Kuetz. Occasional in ditches or clay-pits, often in acid water. Powerstock Common (59, 1996,

B Gale and DAP); Lake near Coombe Keynes (88, 1996, FAW); Hyde Ho (88, 1992, ! in **RNG**); Furzebrook (98, 1996, NFS in **DOR**); The Moors (98, 1990, AJB, DAP and RMW); [Morden Park Lake (99, 1890, JCMP in **DOR**)]; Little Sea (08, 1977, JA Moore and SRD). There are many old records, and specimens in **BM, BMH, DOR, RNG**.
Chara vulgaris L. Occasional, and the most frequent species, recorded from the following: 39 (1947), 48, 58, 50, [66], 67, 69, 77, [78], 71, 87, [89], 82, 97, 98, 99, [90], 07, 08 (1978), 09, 00 and 10. Both var. *longibracteata* (Kuetz.) J Groves and Bullock-Webster and var. *papillata* Wallr. ex A. Braun are found. There are specimens in **BM, BMH, DOR, RNG**.
Lamprothamnium papulosum (Wallr.) J. Groves. A national rarity which is still locally abundant in the Fleet but extinct near Poole. The Fleet (58, 67 and 68, 1887, SMP and WBB in **BM**, *et al.* in **BM, CGE, DOR, E, RNG, SLBI**); Hamworthy, in tidal clay-pits (99, 1899, LVL in **BM**, *et al.*, last seen 1904).
Nitella opaca (Bruz.) Agardh. Rare. [Coombe Keynes Lake (88, 1891, JCMP)]; ditches at Holme Lane and Redcliff (98, 1996, NFS in **DOR**); [Little Sea (08, 1903, GR Bullock-Webster in **BM**); Creekmoor (09, 1900, EFL); Woodland Park (00, 1927, LBH in **BM**)].
(*Nitella gracilis* (Smith) Agardh. No Dorset records, but seen at Bransgore (l9, 1979, JA Moore) in VC 11.)
Nitella translucens (Persoon) Agardh. Scarce in acid ditches and flooded pits. Povington ranges (88, 1994, BE, CDP and DAP); Breach Pond (98, 1996, NFS); Furzebrook (98, 1986, AM Paul and JM Camus, and 1996, NFS in **DOR**); The Moors (98, 1990, BE, CDP and DAP); Ham Pond (99, 1988, CDP and NFS); Little Sea (08, 1962, AC Jermy in **BM**, *et al.*, and 1995, BE and DAP). There are specimens in **BM, BMH, CGE, DOR, RNG**.
Nitellopsis obtusa (Desv.) J. Groves. In the narrows of Little Sea (08, 1978, SRD), needing confirmation.
([*Tolypella glomerata* (Desv.) Leonh. In water-meadows at Burton (19, 1893, F Townsend), in VC 11]).

MARINE ALGAE

Introduction
Collection and identification of living seaweeds dates back to 1690 (Carter, 1957), but the earliest collectors, such as S Goodenough (1795), W Hudson (1762) and R Pulteney, only noted a few species from Weymouth and Portland (Hutchins, 1796). When King George III and his court began to pay regular summer visits to Weymouth, between 1789 and 1805, phycologists followed in their wake. These

included T Velley (1795); in **BM** and **LIV**), J Stackhouse (1795–1801), LW Dillwyn (1802), D Turner (1802) and E Forster (in **BM**). Lyme Regis was another location for amateur collectors, for example CD Pigott around 1798 and later T Walker (1884). WB Barrett listed 67 taxa from Weymouth and Portland under their 18th century names (WB Barrett, *c.*1856).

In the 19th century, seaweed-collecting became popular, but many of the extant herbaria lack dates, localities and sometimes collectors names: examples are those of Mrs Gray (*c.*1835 in **DOR**), Mrs HM Nelson (1849 in **DOR**) and Miss Tyrell *et al.*. (1859–65), and see also Berkeley, 1833; Harvey, 1846; Gatty, 1872. At the turn of the century extensive collections were made by EAL Batters in **BM**, EM Holmes in **BMG, CGE** and **NGM** and AD Cotton (1907; 1908; 1914). Holmes and Batters work resulted in many new county records (Batters, 1902; Batters & Holmes, 1891). Herbaria said to contain Dorset Algae but which have not been systematically searched include **BM, CLR, FKE, GLAM, PMH** and **WRN**. A small collection by Mrs L Ogilvy-Morris and donated to Poole public library could not be traced in 1994.

Twentieth century phycologists added little until recently. **RDG** has small collections made by Dr and Mrs Hurry at Swanage in 1907 and by SO Ridley at Swanage and Weymouth in 1926–27, while VM Grubb studied algal ecology at Swanage in 1936. Then Miss EM Burrows retired to Dorset from 1973 until her death in 1987, and added much to our knowledge (Burrows, 1964; 1981). Her beautiful specimens in **DOR** and **LIV**, and her numerous records in DERC (referred to as EMB), form the most important source for the list of species below. Other recent collectors include CI Dickinson in **CGE**, HJM Bowen in **RNG**, C Maggs (Maggs & Hommersand, 1993) and I Tittley (1988).

Recent taxonomic activity has allowed a better understanding of algal species, especially in the Red Algae, where juvenile and adult forms may look quite different.

It has lumped some older taxa, split others and changed a lot of familiar names, and the work is still going on (Dixon & Irvine, 1977; Fletcher, 1987; Irvine, 1983; Maggs & Hommersand, 1993). The distribution of some species in Britain is mapped in Norton's Atlas (1985). Burrows (1964) commented that no recent list of Dorset Algae was available, in a paper recording 135 species (22 Green, 44 Brown and 69 Red). The list below, excluding a few doubtful records, brings the number of species to 375 (67 Green, 2 Yellow-green, 114 Brown and 192 Red). Many of these are based on old records, and much remains to be done in this group.

Climatic and Edaphic Factors

Temperature is assumed to be important, but there is little variation in mean temperatures along the coastline beween Lyme Regis and Poole. Temperatures further west in Devon and Cornwall differ significantly, e.g.

1. The July mean of 16.5°C is slightly higher than that further west.
2. The January mean of 5.6–6.2°C is colder than that of 6.7°C in S. Devon and 7.5°C in W. Cornwall.
3. The February minimum of 2.7–3.3°C is colder than that of 3.9°C in S. Devon and 7.5°C in W. Cornwall. Since Devon and Cornwall have richer algal floras than Dorset, this suggests that winter temperatures are a factor limiting algal distribution.

The variability of edaphic factors, such as salinity and the nature of littoral and sublittoral substrates, must contribute to the observed richness of the algal flora. The coastline includes rock, shingle, sand and mud, distributed broadly as follows.

Much of the coastline consists of limestone rock descending more or less vertically into the sea, or forming a bouldery beach, as at Charmouth and Eype (SY39, 49), Portland (SY66, 67, 77), Redcliff Point (SY78) and between Durdle Door and Peveril Point (SY87, 88, 97 and SZ07). Vertical Chalk cliffs, whose algae have been studied (Tittley, 1988), are exposed between White Nothe and Lulworth (SY77, 87), at Mupe Bay (SY87) and between Ballard Head and Old Harry (SZ08). Extensive, near-horizontal limestone reefs, with tidal rock-pools, often accompany rocky shorelines, as at Lyme Regis (SY39), West Bay (SY49), Portland Harbour (a very rich area; SY67), between Redcliff Point and Ringstead Bay (SY78), Broad Bench and Kimmeridge (SY87, 97), Chapmans Pool (SY97) and Peveril Point (SZ07). Kimmeridge Bay is a marine nature reserve and comparatively well-explored. The large rock-pool at Dancing Ledge (SY97) is largely man-made.

Long stretches of beach consist of shingle, exposed to winter gales, and a poor habitat for algae. The shingle is mostly of rounded flint pebbles, as between Lyme Regis and the Portland end of the Chesil bank (SY39, 49, 58, 67, 68), between Weymouth and Durdle Door (SY78, 88) and between Worbarrow Bay and Kimmeridge (SY87, 97). The eastern part of the Fleet has a shingle bottom between

Figure 19. Number of marine algae per hectad in Dorset.

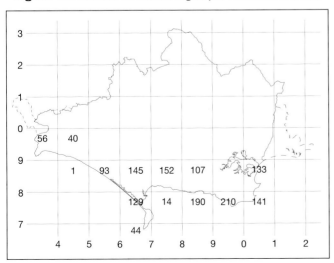

0.3 and 1.5 m below low water mark, and is subject to rapid tidal flow (Robinson, 1981).

Only beaches facing east have fine sands and a low gradient. They include Weymouth (SY67), Swanage, Studland and Sandbanks (SZ07, 08). Both the north and south-west corners of Studland Bay have sandstone rocks overlain by sand on the beach, with specialised algae.

Extensive muddy shores occur along the edges of the Fleet and in Poole Harbour. The west end of the Fleet at Abbotsbury is scarcely tidal and has a salinity of only 1.2 ± 0.7% (Whittaker, 1978; Burrows, 1981). Radipole Lake (SY67, 68) is a brackish reedswamp on mud. Poole Harbour (SY98, 99 and SZ08, 09) covers a large area of mud, mostly 0.3–1.7 m below low water mark, and has been neglected by phycologists.

The remainder of this chapter lists Green, Brown and Red Algae in alphabetical order by genus. Brief notes on abundance, habitat, occurrence in grid Squares (without their qualifying pair of letters), known herbarium material and/or references are appended.

CHLOROPHYCOTA

Green Algae are a diverse group of about 5,000 diploid species. They contain chlorophyll, produce haploid spores, and are the probable ancestors of higher plants and perhaps also of Brown Algae.

Acrochaete repens Pringsheim. A rare endophyte on *Chorda* or *Fucus*. 97. (Burrows, 1991).
Blidingia marginata (J. Ag.) P. Dangeard. On littoral rocks. 39, 49, 58, 67, 68, 78, 97, 07, 08. (Burrows, 1991).
Blidingia minima (Naeg.) Kylin. On littoral rocks. All grid squares. (Norton, 1985; Burrows, 1991).
Bolbocoleon piliferum Pringsheim. A scarce epiphyte or endophyte on *Chorda*. 58, 68, 78, 97. (EMB).
Bryopsis hypnoides Lamour. Scarce in rock-pools near LWM. 67, 78, 87, 97, 08; **DOR, LIV**. (Burrows, 1991).
Bryopsis plumosa (Huds.) C. Ag. In littoral rock-pools. 67, 78, 87, 88, 97, 07, 08; **DOR, LIV**. (Cotton, 1908a, b, 1914; Norton, 1985; Burrows, 1991).
Capsosiphon fulvescens (C. Ag.) Setchell & Gardner. Scarce in brackish water. 78, 97. (EMB).
Chaetomorpha linum (O. F. Muell.) Kuetz. In sandy,muddy or brackish pools. 39, 58, 67, 68, 78, 87, 97, 07, 08; **DOR, LIV**. (Norton, 1985; Burrows, 1991).
Chaetomorpha mediterranea (Kuetz.) Kuetz. In pools. 68, 78, 87, 88, 97. (Burrows, 1991).
Chaetomorpha melagonium (Weber & Mohr) Kuetz. On rock and below LWM. 39, 49, 78, 87, 97, 07, 08; **LIV**. (Norton, 1985; Burrows, 1991).
Chlorochytrium cohnii Wright. Endophytic in diatoms. 87, 97. (Burrows, 1991).
Chlorochytrium moorei Garch. Endophytic in *Blidingia*. 39, 49, 67, 68, 78, 87, 88, 97, 08; **LIV**. (Burrows, 1991).
Cladophora aegagrophila (L.) Rabenhorst. Scarce in the Fleet. 58, 68. (Whittaker, 1978).

Cladophora albida (Huds.) Kuetz. Littoral, and brackish water. 58, 68, 78, 87, 88, 97; **DOR, LIV**. (Burrows, 1991). Perhaps the *Fucus albidus* reported last century from Weymouth and Portland (Barrett, *c.*1856).
Cladophora battersii Van der Hoek. Very rare, with *Zostera*. 67. (Burrows, 1981).
Cladophora coelothrix Kuetz. Rare in shade, in the lower littoral. (Burrows, 1991).
Cladophora dalmatica Kuetz. In littoral and brackish waters. (Burrows, 1991).
Cladophora globulina (Kuetz.) Kuetz. Rare in brackish water. 08. (Burrows, 1991).
Cladophora hutchinsiae (Dillw.) Kuetz. In littoral pools. 39, 58, 68, 78, 87, 97, 07, 08; **DOR**. (Hudson, 1762; Norton, 1985; Burrows, 1991).
Cladophora laetevirens (Dillw.) Kuetz. In rock-pools. 58, [67], 68, 87, 88, 97; **DOR**. (Burrows, 1991).
Cladophora pellucida (Huds.) Kuetz. In more or less shaded rock-pools. 39, 58, 66, 67, 68, 87, 88, 97, 07; **DOR**. (Norton, 1985, Burrows, 1991).
Cladophora prolifera (Roth.) Kuetz. Rare in shade in the west Fleet. 58, 68. (Burrows, 1981).
Cladophora retroflexa (Bonnem.) Van der Hoek. Rare in shelter in the Fleet. 58, 68. (Burrows, 1981).
Cladophora rupestris (L.) Kuetz. Common, littoral and sublittoral. All grid squares; **DOR, LIV, RDG, RNG**. (Burrows, 1964, 1991; Norton 1985).
Cladophora sericea (Huds.) Kuetz. In more or less brackish pools. 78, 88, 97, 08. (Norton, 1985; Burrows, 1991).
Cladophora vagabunda (L.) Van der Hoek. Locally dominant under *Zostera* in the west Fleet. 58, 68. (Burrows, 1981).
(*Cladophora fracta* and **C. glomerata** occur in fresh water in the county; the modern name of **C. sonderae** Kuetz., reported in (Newton, 1931), has not been traced.)
Codium adhaerens C. Ag. Scarce on submerged rock. 67, 87; **DOR**. (Batters & Holmes, 1891; Burrows 1991).
Codium fragile (Suringer) Hariot. On littoral and submerged rock. 58, 67, 68, 78, 87, 88, 98, 07, 08; **LIV**. (Norton, 1985; Burrows, 1991). Var. *tomentosoides* in the Fleet (Whittaker, 1978). The chloroplasts of this plant are retained by the sea slug *Elysia viridis* which feeds on it.
Codium tomentosum Stackh. Decreasing, perhaps extinct. 66, 88, 97, 08, 09; **DOR, LIV**. (Burrows, 1964, 1991; Norton, 1985).
Codium vermilara (Olivi) Chiage. A rare south-western species of submerged rock. 87, 97, 08; **DOR**. (Burrows, 1991).
Derbesia marina (Lyngb.) Solier. A rare epiphyte in moving water. 87, 97. (Burrows, 1991).
Derbesia tenuissimum (De Not.) Crouan fr. Rare in sublittoral pools. [07]. (Cotton, 1908a, b, 1914; Burrows, 1991).
Enteromorpha clathrata (Roth) Greville. An epiphyte, or on littoral or sublittoral rock. 67, 78, 87, 88, 97, 07, 08; **LIV**. (Burrows, 1964; 1991).
Enteromorpha flexuosa (Wulfen ex Roth) J. Ag. Locally frequent, also in fresh water. 58, 68, 97. (Burrows, 1981, 1991).
Enteromorpha intestinalis (L.) Link. Common, locally abundant on beaches especially inside Poole Harbour, also in fresh water; an indicator of eutrophication. All grid

354

squares; **DOR, LIV, RNG**. (Burrows, 1964; Norton, 1985; Withering 1801).

Enteromorpha linza (L.) J. Ag. On rock or in pools near LWM. 67, 68, 78, 87, 88, 97, 08, 09; **DOR, LIV**. (Burrows, 1964; Norton, 1985).

Enteromorpha prolifera (O.F. Muell.) J. Ag. In sheltered sublittoral and brackish water, locally dominant in the Fleet. 49, 58, 67, 68, 78, 87, 88, 97, 07, 08; **LIV**. (Burrows, 1991; Norton, 1985).

Entocladia flustrae (Reinke) Taylor. An epiphyte, especially on Bryozoa. 78, 88. (Burrows, 1991).

Entocladia leptochaete (Huber) Burrows. A rare endophyte in other algae. (Burrows, 1991).

Entocladia perforans (Huber) Levring. An endophyte in *Zostera* and mollusc shells. 58, 66, 67, 68, 78, 87, 97, 08. (Burrows, 1991; Norton, 1985).

Entocladia viridis Reinke. An endophyte in algae or *Zostera*. 58, 68, 87, 97. (Burrows, 1991).

Eugomontia sacculata Kornmann. Rare, in dead mollusc shells. 67. (Whittaker, 1978).

(*Gormontia polyrhiza* (Lagenheim) Bornet & Flahault. This lives in dead mollusc shells; no certain record.)

Microspora filiculinae P. Dangeard. Rare in the Fleet. 68. (WF Farnham, Burrows, 1991)

Monostroma grevillei (Thur.) Wittrock. An ectophyte or shell-borer. 67, 78, 87, 88, 97. (Burrows, 1991).

Monostroma oxyspermum (Kuetz.) Doty. Rare in brackish water in the west Fleet. 58, 68. (Burrows, 1981).

Ochlochaete hystrix Thwaites. An epiphyte on *Zostera* in brackish water. 58, 68, [08]. (Burrows, 1981; Harvey, 1846).

Ostreobium quekettii Bornet & Flahault. Boring into mollusc shells. (Burrows, 1991).

Percursaria percursa (C. Ag.) Reav. A tiny ephemeral on brackish mud, locally dominant at Small Mouth. 58, 67, 68, 87, 08; **DOR**. (Norton, 1985).

Phaeophila dendroides (Gouan) Batters. An endophyte in algae, barnacle and mollusc shells. 58, 68, 97. (Whittaker, 1978).

Pilinia ramosa Kuetz. Rare on chalk in a cave. 08. (Burrows, 1991, Tittley, 1988).

Prasiola stipitata Suhr ex Jessu. Ephemeral on rock in the upper littoral. 58, 67, 78, 87. (Burrows, 1981, 1991).

Pringsheimiella scutata (Reinke) Marchew. An epiphyte on algae or *Zostera*. (Burrows, 1991).

[*Protococcus marinus* Kuetz. Among other algae. (Newton, 1931).]

Pseudendoclonium submarinum Wille. Local in the upper sublittoral. 58, 68, 87, 08. (Burrows, 1981, 1991; Tittley 1988).

[*Pseudopringsheimia confluens* (Rosenv.) Wille. A rare epiphyte on *Laminaria* (Batters & Holmes, 1891).]

Rhizoclonium tortuosum (Dillw.) Kuetz. Locally dominant on mud in the upper littoral. 58, 67, 68, 78, 87, 88, 08, 09; **DOR, LIV**. (Burrows, 1991; Norton, 1985).

Spongomorpha aeruginosa (L.) Van der Hoek. A sublittoral epiphyte. 58, 68, 78, 87, 88, 97, 07, 08; **DOR, LIV**. (Burrows, 1991).

Spongomorpha arcta (Dillw.) Kuetz. On littoral rock. 67, 78, 87, 88, 97, 07, 08; **LIV**. (Burrows, 1964; Norton, 1985).

Tellamia contorta Batters. On shells of living *Littorina* molluscs. 78, 97, 07 (Batters, 1902; Norton, 1985).

Ulothrix flacca (Dillw.) Thuret. On littoral rocks and shells.

58, 66, 67, 68, 77, 78, 87, 88, 97; **LIV**. (Burrows, 1964, Norton, 1985).

Ulothrix implexa (Kuetz.) Kuetz. In sheltered water, the Fleet. 58, 67, 68. (Burrows, 1981).

Ulothrix speciosa (Carmich.) Kuetz. Rare on sheltered beaches. 97. (Burrows, 1991).

Ulva lactuca L. Abundant on stony, sandy or muddy substrates below LWM, notably inside Poole Harbour; perhaps an indicator of eutrophication. All grid squares; **DOR, LIV, RDG, RNG**. (Burrows, 1964; Norton 1985).

Ulva rigida C. Ag. On rock, wood or as an epiphyte. 58, 68, 78, 87, 88, 97, 07, 08. (Norton, 1985).

Ulvella lens Crouan fr. Rare on stones or shells, or an an epiphyte on *Zostera*. 58, 68, 97. (Burrows, 1981; Norton, 1985).

Urospora penicilliformis (Roth) Aresch. On littoral rock. 58, 67, 68, 78, 87, 97, 07, 08; **LIV**. (Burrows, 1964).

Urospora wormskioldii (Mertens) Rosenv. Rare on chalk in the upper littoral. 87, 08. (Tittley, 1988).

Wittrockiella paradoxa Wille. Rare in salt-marshes. (Burrows, 1991).

XANTHOPHYCOTA

Yellow-green Algae are a group of about 300 species, mostly unicellular and in fresh water.

[*Vaucheria dichotoma* Lyngb. var. *submarina* C. Ag. (Berkeley, 1833).]

Vaucheria thuretii Woron. Muddy shores, needing confirmation. (N Chadwick in DERC).

Vaucheria sp.
The Fleet. 67. (Whittaker, 1978).

CHROMOPHYCOTA

Brown Algae are a group of about 1,700 species, containing chlorophyll and the brown fucoxanthin, with haploid spores; many accumulate iodine, which can be seen as a violet vapour when dried algae are burned.

Acinetospora crinita (Carmich.) Kornm. An endophyte in *Enteromorpha*. [07], 08 (Tittley, 1988; Walker, 1884).

Acrothix gracilis Kylin. An epiphyte in the Fleet. 67. (Whittaker, 1978).

[*Alaria esculenta* Grev. Two old and doubtful reports of this northern species, which grows on rocks below LWM. [67]. (Stackhouse, 1795; Barrett, *c.*1856).]

Arthrocladia villosa (Huds.) Duby. Rare on stones or shells below LWM. 87, 08; **DOR, LIV**. (Burrows, 1964; Cotton, 1908a, 1908b, 1914).

Ascophyllum nodosum Le Jol. Abundant on littoral rock. All grid squares but 58; **DOR, LIV**. (Burrows, 1964; Norton, 1985; Withering, 1801).

Asperococcus fistulosus (Hudson) Hooker. An epiphyte on *Fucus*. 67, 78, 87, 88, 97; **LIV**. (Burrows, 1964; Fletcher, 1987).

Asperococcus scaber Kuckuck. Rare on stones in pools. (Batters, 1902; Fletcher, 1987).

Asperococcus turneri (Smith) Hooker. An epiphyte. 67, 87, 88, 97: **DOR**. (Whittaker, 1978).

Bifurcaria bifurcata Ross. On rock. [66], 87, [07], 08; **DOR, LIV, RNG**. (Cotton, 1908a, b, 1914; Withering, 1801).

Buffhamia speciosa Batters. A rare epiphyte on *Mesogloia*. **BM**. (Batters, 1902).

Carpomitra costata (Stackh.) Batters. On sand-covered rock. 78. (J Sanders in DERC).

Chilionema foecundum (Stroemfelt) Fletcher. A rare epiphyte. (Batters, 1902; Fletcher, 1987).

Chilionema hispanicum (Sauvageau) Fletcher. A rare epiphyte on Brown Algae. (Batters, 1902; Fletcher, 1987).

Chilionema ocellatum (Kuetz.) Kuckuck. An epiphyte on *Palmaria*. 58, 68, 87, 07, 08. (Fletcher, 1987).

Chilionema reptans (Crouan) Sauvageau. A rare epiphyte on *Fucus*. (Fletcher, 1987).

Chorda filum Lamour. Frequent on sublittoral rock. 39, 67, 78, 87, 88, 97, 98, 07, 08, 09; **DOR, LIV, RDG**. (Burrows, 1964; Norton, 1985).

Chordaria flagelliformis C. Ag. On littoral and sublittoral rock. 68, 78, 87, 88, 97; **DOR, LIV**. (Burrows, 1964).

Cladosiphon contortus (Thuret) Kylin. On *Zostera* in the west Fleet. 58, 68. (Whittaker, 1978).

Cladosiphon zosterae (J. Ag.) Kylin. On *Zostera* in the west Fleet. 58, 68. (Burrows, 1981).

Cladostephus spongiosus J. Ag. On rock near LWM. All grid squares but 58; **DOR, LIV, RDG**. (Burrows, 1964; Norton, 1985).

Colpomenia peregrina (Sauvageau) Hamel. An alien epiphyte first recorded in 1908. 58, 66, 67, 68, 78, 87, 97,[07,08]; **DOR, LIV**. (Burrows, 1991; Cotton, 1908a, b, 1914; Norton, 1985).

(*Compsonema saxicolum*, a crust on rocks or limpets recorded from 67, 07 and 08, is now thought to be a phase of either *Petalonia* or *Scytosiphon* sp.)

Corynophloea crispa (Harvey) Kuckuck. A rare epiphyte on *Chondrus*. (Cotton, 1908a, b, 1914; Fletcher, 1987).

Cutleria multifida (Smith) Grev. Frequent on sublittoral rock. All grid squares; **LIV**. (Burrows, 1991; Cotton, 1908a, b, 1914; Fletcher, 1987; Hudson, 1762; Norton, 1985).

Cylindrocarpus microscopicus Crouan. An endophyte of *Gracilaria* in the Fleet. 58, 68. (Fletcher, 1987; Whittaker, 1978).

Cystoseira baccata (S. Gmel.) Silva. On sublittoral rock. 66, 67, 68, 78, 87, 88, 97, 07, 08; **DOR, LIV, RNG**. (Burrows, 1991; Norton, 1985).

Cystoseira foeniculacea (L.) Grev. On rock. 58, [67], 68, 78, 87, 97. (Hudson, 1762; ; Norton, 1985; Stackhouse, 1795; Barrett, c.1856).

Cystoseira nodicaulis (With.) Roberts. In rock pools or sublittoral. 67, 78, 87, 88, 97, 07, 08; **DOR**. (Norton, 1985).

Cystoseira tamariscifolia (Hudson) Papenf. On sublittoral rock. 67, 78, 87, 88, 97, 07, 08; **DOR, LIV, RDG, RNG**. (Burrows, 1991; Norton, 1985).

Desmarestia aculeata (L.) Lamour. On sublittoral rock. 39, [66], 67, 68, 78, 87, 88, 97, 07, 08; **LIV, RNG**. (Burrows, 1991; Norton, 1985; Stackhouse, 1795; Withering, 1801). The sap of this and other *Desmarestia* species contains sulphuric acid.

Desmarestia ligulata (Lightf.) Lamour. On sublittoral rock. 67, 68, 78, 87, 88, 97, 07, 08, 09; **DOR, LIV**. (Burrows, 1991; Goodenough, 1795; Norton, 1985).

Desmarestia viridis O.F. Muell. On rock, especially at Portland. 66, 67, 77, 97, 07; **DOR**.

Dictyopteris membranacea Batters. A southern species on submerged rock. 39, 67, 78, 87, 88, 97, 07, 08; **DOR, LIV**. (Burrows, 1991; Cotton 1908a, b, 1914; Norton, 1985).

Dictyosiphon foeniculaceus Grev. On sublittoral rock,or an epiphyte. 66,68,78,97,07; **LIV**. (Burrows, 1991).

Dictyota dichotoma Lamour. Common on rock above or below LWM. All grid squares; **DOR, RDG, LIV**. (Burrows, 1991; Norton, 1985).

Ectocarpus fasciculatus Harvey. A sublittoral epiphyte. 67, 87, 88, 97, 07, 08: **DOR, LIV**. (Burrows, 1991).

Ectocarpus siliculosus (Dillw.) Lyngb. A frequent sublittoral epiphyte on *Zostera* or algae. 58,67,68,78,87,88,97,07, 08; **DOR, LIV**. (Burrows, 1991; Norton, 1985). A revision of *Ectocarpus* and *Streblonema* has not been published. There are records of *E.stilophorae* Crouan from Portland (**LIV**; Newton, 1931), *E. reinboldii* Reinke and *E. lebelli* Crouan (Batters & Holmes, 1891; Newton, 1931), but their modern names have not been traced; some may now be in *Polytretus*.

Elachista flaccida (Dillw.) Aresch. A sublittoral epiphyte on *Cystoseira*. 67, 78, 87, 88, 97; **DOR, LIV**. (Burrows, 1991).

Elachista fucicola (Velley) Aresch. An epiphyte on *Fucus*. 58, 67, 68, 78, 87, 88, 97, 08, 09; **LIV**. (Burrows, 1991; Norton, 1985; Velley 1795).

(*Elachista scutulata* (Smith) Duby. An epiphyte of *Himanthalia*, with no certain record.)

Elachista stellaris Aresch. A scarce epiphyte. (Fletcher, 1987).

Eudesme virescens (Carmich.) J. Ag. Uncommon on sand-covered littoral rock. 97, 08; **LIV**. (Burrows, 1991).

Feldmannia globifera (Kurtz.) Hamel. A rare epiphyte on *Palmaria*. (Batters & Holmes, 1891).

Feldmannia simplex (Crouan fr.) Hamel. A scarce epiphyte on *Codium*. 07, 08. (Batters & Holmes, 1891).

Fucus ceranoides L. On rock, often in brackish water. 66, 08, 09; **DOR, RDG**. (Norton, 1985).

Fucus serratus L. Abundant on littoral rock. All grid squares; **DOR, LIV, RDG**. (Burrows, 1991; Norton, 1985).

Fucus spiralis L. Widespread on rock in the upper littoral. All grid squares; **DOR, LIV**. (Burrows, 1991; Norton, 1985).

Fucus vesiculosus L. Abundant on littoral rock. All grid squares; **DOR, LIV**. (Burrows, 1991; Norton, 1985).

Giffordia granulosa (Sm.) Hamel. An epiphyte on *Corallina*. 67, 87, 07, 08; **DOR**. (Whittaker, 1978).

[*Giffordia mitchelliae* (Harvey) Hamel. A rare epiphyte. (Newton, 1931).]

Giraudia sphacelarioides Derb. & Sol. An epiphyte on *Zostera* in the west Fleet. 58, 68. (Burrows, 1981).

Halidrys siliquosa Lyngb. Frequent on submerged rock. All grid squares but 58; **DOR, RDG, LIV**. (Burrows, 1991; Norton, 1985; Stackhouse, 1795–1801).

Halopteris filicina (Grat.) Kuetz. On rock, or epiphytic. 39, 49, 68, 78, 87, 97, 08; **DOR, RDG**. (Norton, 1985).

Halopteris scoparia (L.) Sauvageau. On rock, or an epiphyte. 39, 49, 67, 68, 78, 87, 88, 97, 07; **DOR, LIV, RDG**. (Burrows, 1991; Norton, 1985; Walker, 1884).

Halothrix lumbricalis (Kuetz.) Reinke. A rare epiphyte on *Zostera*. [67]. (Batters, 1902; Fletcher, 1987; Barrett, *c.*1856).

Hecatonema maculans (Collins) Sauvageau. An epiphyte, confused with *Chilionema ocellatum*. 58, 68, 87, 07, 08. (Fletcher, 1987; Whittaker, 1978).

Himanthalia elongata (L.) S.F. Gray. On submerged rock. 67, 77, 78, 87, 88, 97, 98, 07, 08; **LIV, RDG**. (Burrows, 1991; Norton, 1985).

Hincksia sandrianus (Zanard.) P. Silva. A rare epiphyte. [08]. (Cotton, 1908a, b, 1914).

Laminaria digitata Lamour. Abundant on submerged rock. 39, 66, 67, 68, 77, 78, 87, 88, 97, 07, 08, 09; **DOR, LIV, RDG**. (Burrows, 1991; Norton, 1985; Stackhouse, 1795–1801). Along with other Brown Algae, this concentrates iodine from sea water, so that dry fronds emit a violet vapour of the element when burned.

Laminaria hyperborea Foslie. Abundant on submerged rock, in deeper water than *L. digitata*. All grid squares but 58; **LIV**. (Burrows, 1991; Norton, 1985).

Laminaria saccharina Lamour. Abundant below LWM. All grid squares; **DOR, LIV, RDG, RNG**. (Burrows, 1991; Norton, 1985; Whittaker, 1978).

Laminariocolax tomentosoides (Farlow) Kylin. Epiphytic or parasitic on *Laminaria saccharina*. [67], 78, 87, 97. (EMB).

Leathesia difformis (L.) Aresch. Epiphytic on *Corallina* in rock pools. 66, 67, 77, 78, 87, 88, 97, 07, 08; **DOR, LIV, RNG**. (Burrows, 1991; Cotton, 1908a, b, 1914; Norton, 1985).

Leblondiella densa (Batters) Hamel. A rare epiphyte on *Zostera*. (Fletcher, 1987).

Litosiphon laminariae (Lyngb.) Harvey. An epiphyte on *Chorda* or *Laminaria*. 58, 68, 78, 87, 88, 97, 07; **LIV**. (Burrows, 1991; Fletcher, 1987; Norton, 1985).

Mesogloia vermiculata Le Jol. On rock, or epiphytic on *Cystoseira*. 66, 78, 87, 88, 97; **LIV**. (Burrows, 1991).

[*Microcoryne ocellata* Stroemfelt. A rare epiphyte on *Chorda*. 67. (Batters, 1902; Fletcher, 1987).]

Microsporangium globosum Reinke. A scarce epiphyte on *Fucus*. 07. (Cotton, 1907; Fletcher, 1987).

Myriactula chordae (Aresch.) Levring. A rare epiphyte. (Fletcher, 1987).

Myriactula clandestina (Crouan) J. Feldmann. An epiphyte on *Fucus*. (Fletcher, 1987).

Myriactula rivulariae (Suhr) J. Feldmann. An epiphyte on *Cystoseira* in the Fleet. 58, 68, 97. (Whittaker, 1978, EMB).

Myriactula stellulata (Harvey) Levring. A rare epiphyte on *Dictyota*. 97. (EMB).

[*Myriocladia tomentosa* Crouan. A rare, deep-water plant. 66. (Newton, 1931; Barrett *c.*1856).]

Myrionema corunnae Sauvageau. An epiphyte on *Laminaria* blades. (Fletcher, 1987).

(*Myrionema magnusii* (Sauvageau) Loiseaux. An epiphyte on *Zostera* with no certain record.)

Myrionema papillosum Sauvageau. A rare epiphyte on *Laminaria saccharina*. 07, 08. (Fletcher, 1987).

Myrionema strangulans Grev. An epiphyte on *Enteromorpha* and *Ulva*. 58, 67, 68, 78, 87, 88, 97, 07, 08; **DOR, LIV**. (Burrows, 1991; Norton, 1985).

Myriotrichia clavaeformis Harvey. An epiphyte or endophyte on *Zostera* or algae. 78. (EMB).

Padina pavonia Gaillon. On rock below LWM. 39, 49, 67, 68, 78, 87, 97, [07]; **DOR, LIV**. (Burrows, 1991; Norton, 1985; Walker, 1884).

Pelvetia canaliculata Decaisne & Thur. Local, on rock near HWM. 67, 68, 78, 87, 88, 97, 07, 08; **LIV**. (Burrows, 1991; Cotton, 1908a, b, 1914; Norton, 1985).

Petalonia fascia (O.F. Mueller) O. Kuntze. Uncommon on rock. 97, 07, 08. (EMB).

Petalonia filiformis (Batters) O. Kuntze. Rare on rock. 08. (Tittley, 1988).

Petalonia zosterifolia (Reinke) O. Kuntze. Uncommon on vertical rock. 97, 07, 08. (EMB).

Petrodermia maculiforme (Wollny) Kuckuck. Scarce on rock. 07, 08. (Tittley, 1988).

Petrospongium berkeleyi (Grev.) Naegeli. Uncommon on rock. 87, 88, [07]. (Cotton, 1908a, b, 1914; Fletcher, 1987).

Phaeostroma pustulosum Kuckuck. An epiphyte on *Zostera* in the west Fleet. 58, 68. (Burrows, 1991; Whittaker, 1978).

Pilayella littoralis Kjellm. On rock, or an epiphyte. 39, 67, 68, 78, 87, 97, 07, 08; **DOR, LIV**.

Pogotrichum filiforme Reinke. An epiphyte on *Laminaria*. **DOR**. (Fletcher, 1987).

Protoectocarpus speciosus (Borg.) Kuckuck. A rare epiphyte. (Withering, 1801).

Pseudolithoderma extensum (Crouan) Lund. Scarce on rock. 87, 97. (EMB).

Pseudolithoderma roscoffense Loiseaux. On sheltered flints. (Fletcher, 1987).

Punctaria latifolia Grev. A scarce epiphyte. 58, 68, 78. (EMB).

Punctaria plantaginea (Roth) Grev. Mostly on rock. 87, 88. (Fletcher, 1987).

Punctaria tenuissima (C. Ag.) Grev. An epiphyte in the littoral zone. (Fletcher, 1987).

Ralfsia verrucosa (Aresch.) J. Ag. A crust on rock. 87, 07, 08. (EMB).

Saccorhiza polyschides (Lightf.) Batters. On submerged rock. 48, 49, [66], 67, 78, 87, 88, 97, 07; **DOR, LIV**. (Burrows, 1991; Norton, 1985; Barrett *c.*1856).

Sargassum muticum (Yendo) Fensholt. A Japanese alien, first seen in 1973, spreading fast. 67, 77, 87, 97, 08; **RNG**. (Norton, 1985).

Sauvageaugloia griffithsiana (Griff.) Hamel. In rock pools. [07]; Mrs Gray in **DOR**. (Cotton, 1908a, b, 1914).

Scytosiphon lomentaria (Lyngb.) Link. On rock, to below LWM. 39, 66, 67, 68, 78, 87, 88, 97, 07, 08; **LIV**. (Norton, 1985).

[*Sorocarpus uvaeformis* Pringsh. A rare epiphyte. (Newton, 1931).]

Sphacelaria cirrosa (Roth) C. Ag. A littoral and sublittoral epiphyte. 67, 78, 87, 88, 97, 07, 08; **DOR, LIV**. (Burrows, 1991).

Sphacelaria furcigera Kuetz. A rare epiphyte. 68, 78, 97, [08]. (Burrows, 1964; Cotton, 1908a, b, 1914; Newton, 1931).

Sphacelaria plumula Zanard. Rare, submerged with *Zostera*. 78. (Norton, 1985).

Sphacelaria radicans (Dillw.) C. Ag. On sand-covered littoral rocks. 78, 88, 08; **LIV**. (Tittley, 1988).

Spongonema tomentosum (Huds.) Kuetz. An epiphyte on littoral Brown Algae. 58, 68, 87, 88, 97; **DOR, LIV**.

Sporochnus pedunculatus (Huds.) C. Ag. Rare, on stones. [66,

67], 87, 07, [08]; **DOR**. (Berkeley, 1833; Cotton, 1908a, b, 1914; Hudson, 1762; Stackhouse, 1795–1801, Barrett, c. 1856).

Stilophora rhizodes J. Ag. A rare epiphyte near LWM. 97; **DOR**. (Batters, 1902; Norton, 1985).

Stragularia clavata (Harvey) Hamel. Mostly on rock. 87, 07, 08. (Fletcher, 1987).

Stragularia spongiocarpa (Batters) Hamel. In sandy rock pools. 87, 97, 07, 08. (Fletcher, 1987; Norton, 1985).

Streblonema aequale Oltm. A rare endophyte in *Chorda*. 97. (EMB).

[*Streblonema intestinum* (Reinsch.) Batters. An endophyte in *Brongnartiella* 67. (Newton, 1931).]

[*Streblonema stilophorae* (Crouan fr.) Hamel. A parasite on *Stilophora*. 97. (Newton, 1931).]

[*Streblonema tenuissimum* Hauck. An endophyte in *Sauvageaugloia*. (Newton, 1931).]

(*Streblonema volubile* (Crouan fr.) Thur. An endophyte in *Dudresnaya*, modern name untraced (Newton, 1931).)

[*Strepsithalia buffhamiana* Batters. A rare endophyte in *Mesogloia*. (Newton, 1931).]

Striaria attenuata (Grev.) Grev. At the base of *Zostera* in the west Fleet. 58, 68. (Burrows, 1964; Newton, 1931).

Taonia atomaria J. Ag. A south-western species on submerged rock. 39, 49, 87, 97, 07; **DOR, LIV**. (Cotton, 1908a, b, 1914; Norton, 1985; Walker, 1884).

Tilopteris mertensii (Turner) Kuetz. Rare on stones. [07]. (Cotton, 1908a, b, 1914).

(*Ulonema rhizophorum* Foslie. An epiphyte on *Dumontia*, with no certain record.)

Zanardinia prototypus (Nardo) Nardo. Rare on rock. 87, 97; **DOR**. (Norton, 1985).

RHODOPHYCOTA

Red Algae are a diverse group of 2,500 species, often with alternate haploid and diploid generations. They contain chlorophyll, red phycoerythrin and blue phycocyanin; the latter also occurs in Blue-green Algae from which Red Algae may have evolved. Some species contain chloroform, bromoform and other toxic compounds with strong tastes.

Acrochaetium endozoicum Hamel. A rare endophyte in the Bryozoan *Alcyonidium*. 87; **DOR**.

(*Acrochaetium minutum* (Suhr) Hamel was reported in (Batters & Holmes, 1891) but later excluded (Dixon & Irvine, 1977).)

Acrosorium venulosum (Zanard.) Kylin. On rock or corallines. (Maggs & Hommersand, 1993).

Agardhiella subulata (C. Ag.) Kraft & Wynne. On a beach north of Arne, ! det. C Maggs. 98.

Aglaothamnion byssoides (Arn.) Halos & Rueves. A scarce sublittoral epiphyte. 97, 07; **LIV**. (Maggs & Hommersand, 1993).

Aglaothamnion feldmanniae Halos. On stones or shells. (Maggs & Hommersand, 1993).

Aglaothamnion hookeri (Dillw.) Maggs & Homm. On chalk. (Tittley, 1988).

Aglaothamnion pseudobyssoides (Crouan fr.) Halos. On rock, or an epiphyte. (Maggs & Hommersand, 1993).

(*Aglaothamnion roseum* (Roth) Maggs & Halos. An epiphyte with no certain record.)

Aglaothamnion tripinnatum (C. Ag.) Feldm.-Maz. On rock, or an epiphyte. (Maggs & Hommersand, 1993).

Ahnfeltia plicata (Huds.) Fr. In rock pools near LWM. 39, 49, 67, 78, 87, 88, 97, 07, 08; **DOR, LIV, RNG**. (Burrows, 1991; Norton, 1985; Stackhouse, 1795– 1801).

[*Anotrichium barbatum* (C. Ag.) Naegeli. A rare epiphyte. (Maggs & Hommersand, 1993).]

Anotrichium furcellatum (J. Ag.) Baldock. An epiphyte found in 1976. (Maggs & Hommersand, 1993).

Antithamnion cruciatum (C. Ag.) Naegeli. A scarce epiphyte. 97. (EMB).

Antithamnion villosum (Kuetz.) Athan. A scarce epiphyte. (Maggs & Hommersand, 1993).

Antithamnionella spirographidis (Schiff.) Woll. Rare in rock-pools. 97. (EMB).

Apoglossocolax pusilla Maggs & Homm. A hemiparasite on *Apoglossum*. 08; **BM**. (Norton, 1985).

Apoglossum ruscifolium (Turner) J. Ag. Epiphytic on *Laminaria*. 68, 78, 87, 88, 97, 07, 08, 09; **DOR, LIV**. (Burrows, 1991; Norton, 1985).

Asparagopsis armata Harvey. A sublittoral, perhaps alien, epiphyte. 78, 87, 88, 97, 07, 08; **DOR**.

Atractophora hypnoides Crouan. An epiphyte on Corallines. [07]. (Cotton, 1908a, b, 1914; Dixon & Irvine, 1977).

Audouinella bonnemaisoniae (Batters) Dixon. An endophyte in *Bonnemaisonia*. 97. (EMB).

Audouinella caespitosa (J. Ag.) Dixon. An epiphyte, or on mollusc shells. (Dixon & Irvine, 1977).

Audouinella daviesii (Dillw.) Woelk. An epiphyte on *Palmaria*. 68, 78, 87, 88, 97; **BM, DOR**. (Batters, 1902).

Audouinella floridula (Dillw.) Woelk. A littoral sand-binder. 39, 49, 67, 68, 87, 07, 08. (Norton, 1985).

Audouinella purpurea (Lightf.) Woelk. On rock, or epiphytic on *Laminaria*. 58, 68, 87, 88, 08. (EMB).

Audouinella thuretii (Bornet) Woelk. An epiphyte in the Fleet. 67. (Whittaker, 1978).

[*Audouinella trifila* (Buffham) Dixon. A rare epiphyte. 07. (Newton, 1931).]

Audouinella virgatula (Harvey) Dixon. An epiphyte on *Zostera* in The Fleet. 58, 68. (Burrows, 1981).

Bangia atropurpurea (Roth) C. Ag. Scarce on rock at LWM. 97, 08. (Tittley, 1988).

Boergeseniella fruticulosa (Wulfen) Kylin. A sublittoral epiphyte. 67, 97, 07; **LIV**. (Maggs & Hommersand, 1993).

Boergeseniella thuyoides (Harvey) Kylin. On rock or corallines. 87. (10.21).

Bonnemaisonia asparagoides (Woodw.) C. Ag. On rock or shell. 68, 78, 97, 07; **DOR, RNG**. (Cotton, 1908a, b; 1914, EMB).

Bonnemaisonia hamifera Hariot. A Japanese alien since 1890; one phase has been reported as *Trailliella intricata* Bat. 67, 68, 78, 87, 97, [07,08]. (Batters & Holmes, 1891; Cotton, 1908a, b, 1914).

Brongniartella byssoides (Good. & Woodw.) Schmitz. A

sublittoral epiphyte, or on rock. 67, 68, 78, 87, 88, 97, 07, 08; **DOR, LIV**. (Burrows, 1991).
Calliblepharis ciliata (Huds.) Kuetz. Frequent on sublittoral rock. All grid squares; **DOR, LIV**. (Burrows, 1991; Norton, 1985).
Calliblepharis jubata (Good. & Woodw.) Kuetz. A sublittoral epiphyte. [66], 67, 68, 78, 87, 88, 97, 07, 08; **DOR, LIV**. (Burrows, 1991; Norton, 1985; Barrett, *c.*1856).
Callithamnion corymbosum (J.E. Sm.) Lyngb. Scarce, usually epiphytic. 87, 97; **DOR**. (Walker, 1884).
Callithamnion tetragonum (With.) S.F. Gray. Epiphytic on *Laminaria*.. [66, 67], 68, 78, 87, 97, 07, 08; **DOR, LIV**. (Burrows, 1964; Cotton, 1908a, b, 1914, Withering, 1801; Barrett, *c.*1856).
(*Callithamnion tetricum* (Dillw.) S.F. Gray from vertical rock, has no certain record.)
(*C. rabenhorstii* Crouan was recorded (Newton, 1931), but its modern name is untraced.)
Callocolax neglectus Schmitz ex Batters. A sublittoral epiphyte or parasite on *Callophyllis*. 78, 87, 88, 97, 07, 08; **BM, LIV**. (Burrows, 1991; Norton, 1985).
Callophyllis laciniata (Huds.) Kuetz. An epiphyte on *Laminaria*. 67,78,87,88,97,07,08,09; **DOR, LIV, RDG, RNG**. (Burrows, 1991; Norton, 1985).
[*Calosiphonia vermicularis* (J. Ag.) Schmitz. Rare on sheltered stone. 67. (Batters, 1902).]
Catenella caespitosa (With.) L. Irvine. Littoral, including the Fleet. 58, 68, 97, 08; **LIV**. (Hudson, 1762).
Ceramium ciliatum (Ellis) Ducluz. On littoral rock or in *Zostera* beds. 58, 68, 78, 87, 88, 97; **DOR, LIV**. (Burrows, 1991).
Ceramium diaphanum (Lightf.) Roth. A sublittoral epiphyte. 58, 68, 78, 87, 88, 97, 08; **DOR, LIV**. (Burrows, 1991).
Ceramium echionotum J. Ag. Littoral, on rock or epiphytic. 58, [67], 68, 87, 88, 97; **DOR, LIV** (Burrows, 1991).
Ceramium flaccidum (Kuetz.) Ardis. Littoral, on rock or epiphytic. 78, 87, 88. (Maggs & Hommersand, 1993).
Ceramium gaditanum (Clem.) Clem. Scarce between tidemarks. 78; **LIV**. (Maggs & Hommersand, 1993).
Ceramium nodulosum (Lightf.) Ducluz. A frequent epiphyte. All grid squares; **DOR, LIV, RNG**. (Burrows, 1981; Norton, 1985).
(*Ceramium pallidum* (Naegeli ex Kuetz.) Maggs & Homm. No certain record.)
Ceramium securidatum Lyngb. An epiphyte, recently found on a beach north of Arne, ! det. C Maggs. 98. (Maggs & Hommersand, 1993).
Ceramium shuttleworthianum (Kuetz.) Silva. On littoral rock. 68, 78, 87, 88, 97; **LIV**. (Burrows, 1991).
Ceramium siliquosum (Kuetz.) Maggs & Homm. An epiphyte on small algae. (Maggs & Hommersand, 1993).
Ceramium strictum Harvey. Among *Zostera* in the Fleet. 58, 68; **DOR**. (Burrows, 1981).
Champia parvula (C. Ag.) Harvey. A scarce epiphyte in rock pools. 97; **DOR**. (Norton, 1985).
Chondria capillaris (Huds.) M. Wynne. On sheltered sublittoral pebbles. 58, [67], 68, 87, 97, 07; **LIV**. (Cotton, 1908a, b, 1914; Goodenough, 1795).

Chondria dasyphylla (Woodw.) C. Ag. Frequent on sublittoral stones. All grid squares; **LIV**. (Greville, 1830).
Chondrus crispus Stackh. Abundant on rock near LWM, and in brackish water. All grid squares; **DOR, LIV, RDG, RNG**. (Turner, 1802).
Choreocolax polysiphoniae Reinsch. A rare parasite on *Polysiphonia*. 07, [08]; **BM**. (Norton, 1985).
Choreonema thuretii Schmitz. A parasite on *Corallina*. 97, [07]. (Cotton, 1908a, b, 1914; EMB).
[*Chroodactylon ornatum* (C. Ag.) Basson. On *Cladophora* in brackish pools. (Gatty, 1872; Harvey, 1846).]
Chylocladia verticillata (Lightf.) Bliding. On sublittoral rock or epiphytic. 58, 67, 68, 78, 87, 88, 97, 07, 08; **DOR, LIV**. (Stackhouse, 1795–1801).
(*Conferva intertexta* has been recorded, but the name is of uncertain application (Maggs & Hommersand, 1993).)
Corallina officinalis L. Abundant in rock pools near LWM. All grid squares but 09; **DOR, LIV, RNG**. (Stackhouse, 1795–1801).
Cordylecladia erecta (Grev.) J. Ag. An epiphyte in sandy places. 58, 68, 78, 97, 07, 08. (EMB).
Crouania attenuata (C. Ag.) J. Ag. A rare epiphyte. 87, 97. (Norton, 1985).
Cruoria pellita (Lyngb.) Fr. A crustose epiphyte on corallines. 67, 87, 97. (Stackhouse, 1795–1801).
Cryptopleura ramosa (Huds.) Kylin ex Newton. On sublittoral rock or epiphytic. 67, 68, 78, 87, 88, 97, 07, 08; **DOR, LIV**. (Burrows, 1991; Norton, 1985).
Cystoclonium purpureum (Huds.) Batters. Frequent on more or less shaded rock near LWM. All grid squares; **DOR, LIV**. (Burrows, 1991; Norton, 1985).
[*Dasya corymbifera* J. Ag. was reported in 1861 by M Gray in **CGE**, and from Studland by AD Cotton (1908a, b; 1914), but is now confined to the Channel Isles (Maggs & Hommersand, 1993).]
Dasya hutchinsiae Harvey. On rock, or epiphytic on corallines. (Maggs & Hommersand, 1993).
Dasya punicea Menegh. ex Zanard. On friable rock. [08]. (Cotton, 1908a, b, 1914; Newton, 1931).
Delesseria sanguinea (Huds.) Lamour. On submerged rock. 66, 67, 77, 78, 87, 88, 97, 07, 09; **DOR, LIV**. (Burrows, 1991; Norton, 1985).
Dermatolithon litorale (Suneson) Lemoine. Scarce on rock. 97. (EMB).
Dilsea carnosa (Schmidd.) O. Kuntze. Frequent on sublittoral rock. All grid squares but 09; **DOR, LIV, RNG**. (Burrows, 1991; Norton, 1985).
Drachiella heterocarpa (Chauvin ex Duby) Maggs & Homm. On rock to 30 m deep. (Maggs & Hommersand, 1993).
Dudresnaya verticillata (With.) Le Jol. Scarce on stones. 87, 97, [07, 08]; **DOR**. (Cotton, 1908a, b, 1914; EMB).
Dumontia contorta (S.G. Gmel.) Rupr. In littoral rock pools. 66, 67, 68, 78, 87, 88, 97, 07; **LIV**. (Norton, 1985).
Erythrodermis traillii (Holmes ex Batters) Guiry & Garbary. Rare on shaded rock. 97 (EMB).
Erythroglossum laciniatum (Lightf.) Maggs & Homm. On rock. 78,07,08 (Maggs & Hommersand, 1993).
Erythropeltis discigera Schmitz. An epiphyte on the bryozoan *Flustra*. 08; **DOR**.

Erythrotrichia carnea J. Ag. An epiphyte in rock-pools. 67, 78, 87, 88, 08. (EMB).

[*Erythrotrichia reflexa* Thuret. A rare epiphyte. (Newton, 1931).]

Erythrotrichia welwitschii Batters. A rare epiphyte on *Ralfsia*. [08]. (Batters, 1902; Cotton, 1908a, b; 1914).

Fosliella farinosa (Lamour.) Howe. A rare epiphyte on *Zostera*. 58, 68, 97. (Burrows, 1981).

Furcellaria lumbricalis (Huds.) Lamour. Frequent on rock near LWM. All grid squares but 58; **DOR, LIV, RNG**. (Burrows, 1991; Norton, 1985).

Gastroclonium ovatum (Huds.) Papenf. On sublittoral rock or corallines. 67, 78, 87, 88, 97, 07, 08; **DOR, LIV**. (Turner, 1802).

Gelidium latifolium (Grev.) Bornet & Thuret. On littoral rock. 67, 87, 88, 97, 07, 08; **DOR, LIV**. (Burrows, 1991).

Gelidium pusillum (Stackh.) Le Jol. On rock near LWM. 58, 68, 87, 97, 08; **LIV**. (Newton, 1931).

(A report of *Gelidium sesquipedale* (Clemente) Thuret from Small Mouth in DERC was an error.)

Gigartina acicularis (Roth) Lamour. Occasional on rock. 58, 68, 78, 87, 88, 97, 07; **LIV**. (Burrows, 1991; Norton, 1985).

Gigartina teedii (Roth) Lamour. Rare on sheltered rock. 68, 78. (Irvine, 1983; Norton, 1985).

Gloiosiphonia capillaris (Huds.) Carmich. Rare on rock. 67, 78. (Whittaker, 1978).

Gonimophyllum buffhamii Batters. A hemiparasite on *Cryptopleura*. (Maggs & Hommersand, 1993).

Gracilaria bursa-pastoris (S.G. Gmel.) Silva. Occasional on rock. 58, 66, 67, 68, 77, 78, 07, 08. (Norton, 1985).

Gracilaria foliifera (Forskal) Boerg. Scarce on rock, and in the Fleet. 58, 68, 97. (Whittaker, 1978).

Gracilaria verrucosa (Huds.) Papenf. Frequent on sheltered sublittoral rock. 39, 49, 58, 67, 68, 78, 97, 07, 08, 09; **DOR, LIV, RNG**. (Burrows, 1991).

Gracilariopsis sp. On a beach north of Arne, ! det. C Maggs. 98.

Grateloupia dichotoma J. Ag. On vertical rock. 68, 78, 97. (Norton, 1985).

Grateloupia filicina (Lamour.) C. Ag. Uncommon on rock. 58, 67, 87, 97; **RNG**. (Burrows, 1991; Norton, 1985).

Griffithsia corallinoides (L.) Batters. An epiphyte. 67, 68, 78, 87, 07, 08; **DOR**.

Griffithsia devoniensis Harvey. On mud or stones. (Maggs & Hommersand, 1993).

Gymnogongrus crenulatus (Turner) J. Ag. In rock pools. 68, 78, 87, 97; **DOR**.

Gymnogongrus griffithsiae (Turner) Martins. In rock pools. 78, 97; **RNG**.

Haematocelis rubens J. Ag. Epiphytic on *Laminaria*. (Irvine, 1983).

Halarachnion ligulatum (Woodw.) Kuetz. On stones in deep water. [67], 68, 78, 87, 97, 07, 08; **DOR**. (Cotton, 1908a, b, 1914; Barrett, *c*.1856).

Haliptylon squamatum (L.) Johansen, Irvine & Webster. A coralline near LWM. 87, 97; **DOR**.

Halopitys incurvus (Huds.) Batters. Frequent on rock. 39, 58, 67, 68, 78, 87, 88, 97, 07, 08; **DOR, LIV, RDG**. (Burrows, 1991; Cotton, 1908a, b, 1914; Norton, 1985).

Halurus equisetifolius (Lightf.) Kuetz. Occasional on sublittoral rock. 67, 78, 87, 88, 97, 07, 08, 09; **DOR, LIV**. (Burrows, 1991; Norton, 1985).

Halurus flosculosus (Ellis) Maggs & Homm. Epiphytic on *Laminaria*. 66, 68, 78, 87, 97, 07, 08; **DOR, LIV**.

(*Halymenia latifolia* was reported in (Berkeley, 1833), probably in error for *Halarachnion* (Irvine, 1983).)

Haraldiophyllum bonnemaisonii (Kylin) A. Zinova. On sublittoral rock or epiphytic. 66, 78, 87, 07; **DOR, RNG**. (Dixon & Irvine, 1977; Norton, 1985).

Helminthocladia calvadosii (Lamour.) Setchell. Rare on sublittoral gravel or shells. 87. (EMB).

Helminthora divaricata (C. Ag.) J. Ag. Rare in rock pools. 97; **LIV**. (Burrows, 1991).

Heterosiphonia plumosa (Ellis) Batters. Common on sublittoral stone, or an epiphyte. 39, 49, 67, 78, 87, 97, 07, 08; **DOR, LIV**. (Burrows, 1991; Norton, 1985).

Hildenbrandia crouanii J. Ag. Rare, coating stones. 67. (Norton, 1985).

Hildenbrandia rubra (Sommerf.) Menegh. Coating pebbles in littoral pools. 58, 68, 87, 07, 08. (Burrows, 1964).

Holmesella pachyderma (Reinsch) Sturch. A parasite on *Gracilaria*. 67, 07, 08. (Whittaker, 1978).

Hypoglossum hypoglossoides (Stackh.) Collins & Harvey. On rock, or an epiphyte. [67], 68, 78, 87, 88, 97, 07, 08; **LIV, RDG**. (Stackhouse, 1795–1801; Barrett, *c*.1856).

Jania corniculata (L.) Lamour. A rare epiphyte. [67], 97; **DOR**. (EMB).

Jania rubens (L.) Lamour. An epiphyte on the shore and below it. 67, 78, 87, 88, 97, 07, 08; **DOR, LIV**. (Burrows, 1991).

Kallymenia reniformis (Turner) J. Ag. A rare epiphyte on *Laminaria*. 87. (Norton, 1985).

Laurencia pinnatifida (Huds.) Lamour. Frequent to locally dominant in rock pools near LWM. 39, 58, 66, 67, 68, 77, 78, 87, 88, 97, 07, 09; **DOR, LIV, RNG**. (Burrows, 1991; Norton, 1985; Stackhouse, 1795–1801).

[*Laurencia pyramidalis* Bory ex Kuetz. A rare epiphyte in rock pools. (Batters, 1902).]

Lithophyllum incrustans Philippi. On rock in the Fleet, etc. 58, 67, 68, 97. (Whittaker, 1978).

Lomentaria articulata (Huds.) Lyngb. On rock, or an epiphyte. 39, 68, 78, 87, 88, 97, 07, 08; **DOR, LIV**. (Withering, 1801).

Lomentaria clavellosa (Turner) Guillon. On rock, or an epiphyte. 87, 97, 07, 08, 09. (Norton, 1985).

Lomentaria orcadensis (Harvey) Collins. An epiphyte on *Laminaria*. 97, 07. (EMB, KC in DERC).

Lophosiphonia reptabunda (Suhr) Kylin. Rare on sheltered rock. (Maggs & Hommersand, 1993).

Mastocarpus stellatus (Stackh.) Guiry. Common on rock near LWM. All grid squares but 58; **DOR, RDG**. (Burrows, 1991; Norton, 1985; Stackhouse, 1795–1801). This is a potential source of agar.

Melobesia membranacea (Esper.) Lamour. An epiphyte or on shells. 39, 67, 68, 78, 87, 88, 97, 07. (Norton, 1985).

Membranoptera alata (Huds.) Stackh. On rock, or on *Laminaria*. 58, 67, 68, 78, 87, 88, 97, 07, 08; **DOR, LIV, RDG**. (Burrows, 1991; Norton, 1985).

Meredithia microphylla J. Ag. On vertical sublittoral rock. 67, 78, 87, 88, 97. (EMB).

Mesophyllum lichenoides (Ellis) Lemoine. Rare, on *Corallina* in rock pools. 97. (Norton, 1985).

Monosporus pedicillatus (J.E. Sm.) Solier. An epiphyte. 58, 67, 68, 78, 87, 88, 97, [07], 08; **DOR, LIV**. (Cotton, 1908a, b, 1914; Norton, 1985).

Naccaria wiggii (Turner) Endl. A scarce epiphyte. 87, 97, [07,08]; **DOR**. (Cotton, 1908a, b, 1914; Norton, 1985).

Nemalion helminthoides (Velley) Batters. On rock or mussel shells. [66], 67, 87, 97, 07; **DOR, LIV**. (Burrows, 1991; Cotton, 1908a, b, 1914; Velley, 1795).

Nitophyllum punctatum (Stackh.) Grev. On sublittoral rock, or an epiphyte. 66, [67], 87, 97, 07, 08; **DOR, LIV**. (Burrows, 1991; Stackhouse, 1795–1801; Barrett, *c.*1856).

Osmundea hybrida (DC.) Nam *et al.* Occasional on rock. 67, 68, 78, 87, 88, 97, 07, 08; **LIV**. (Burrows, 1991).

Osmundea pinnatifida (Huds.) Nam *et al.* Frequent to locally dominant in rock pools near LWM. 39, 58, 66, 67, 68, 77, 78, 87, 88, 89, 97, 07, 09; **DOR, LIV, RNG**. (Burrows, 1991; Norton, 1985; Stackhouse, 1795–1801).

Osmundea truncata (Kuetz.) Nam *et al.* An epiphyte on *Fucus*. (Maggs & Hommersand, 1993).

Palmaria palmata (L.) O. Kuntze. A common sublittoral epiphyte. All grid squares but 58; **DOR**. (Norton, 1985).

Peyssonnelia atropurpurea Crouan. An epiphyte on corallines. 97 (EMB).

Peyssonnelia dubyi Crouan. An epiphyte on corallines or holdfasts. 67, 87. (EMB).

Peyssonnelia harveyana J. Ag. An epiphyte on corallines or holdfasts. 87. (EMB).

Peyssonnelia immersa Maggs & Irvine. On shells in shelter. (Irvine, 1983).

Phycodrys rubens (L.) Batters. A frequent epiphyte on *Laminaria*. 66, 67, 68, 78, 87, 88, 97, 07, 08, 09; **DOR, LIV**. (Hudson, 1762).

Phyllophora crispa (Huds.) Dixon. On shaded, sublittoral rock. 58, 67, 68, 78, 87, 97, 07, 08, 09; **DOR, LIV, RDG, RNG**. (Burrows, 1991).

Phyllophora pseudoceranoides (S.G. Gmel.) Newroth & Tayl. On sublittoral rock. [67], 68, 78, 87, 97, 07, 08; **DOR, LIV**. (Burrows, 1964; Barrett, *c.*1856).

Phyllophora sicula (Kuetz.) Gairy & Irvine. Occasional in rock-pools near LWM. 58, 68, 78, 87, 97, 07, 08; **RNG**. (Norton, 1985).

Phymatolithon lenormandii (Aresch.) Adey. On vertical chalk, or in rock-pools, near LWM. 39, 67, 87, 97, 08. (Norton, 1985, EMB).

Pleonosporium borreri Naeg. Rare on muddy rocks at Swanage. [07]. (Cotton, 1908a, b, 1914).

Plocamium cartilagineum (L.) Dixon. A frequent sublittoral epiphyte, or on rock. All grid squares; **DOR, LIV, RNG** (Burrows, 1991; Norton, 1985).

Plumaria plumosa (Huds.) O. Kuntze. On littoral rock. 66, 67, 77, 78, 87, 88, 97, 07, 09; **DOR, LIV**. (Burrows, 1991; Norton, 1985).

Pneophyllum confervicolum (Kuetz.) Chamberlain. Epiphytic on *Laurencia*. 78, 87, 88, 97; **LIV**.

Pneophyllum lejolisii (Rosan.) Chamberlain. In shelter, in the Fleet. 58, 68. (Whittaker, 1978).

Pneophyllum limitatum (Foslie) Chamberlain. A scarce calcareous species. 78, 88, 97. (EMB).

Polyides rotunda (Huds.) Grev. On littoral rock. 39, 49, 67, 87, 97, 07, 08; **DOR, LIV**. (Burrows, 1991).

Polyneura bonnemaisonii (C. Ag.) Maggs & Homm. On rock, or epiphytic. 09. (Maggs & Hommersand, 1993; N Chadwick in DERC).

Polysiphonia atlantica Kapr. & J. Norris. Rare on rock or corallines. 97. (Maggs & Hommersand, 1993).

Polysiphonia brodiaei (Dillw.) Sprengel. Occasional on rock or corallines near LWM. 67, 68, 78, 87, 88, 97, 07; **DOR, LIV**. (Burrows, 1991).

Polysiphonia ceramiaeformis Crouan fr. A rare epiphyte on *Ulva*. [07]. (Cotton, 1908a, b, 1914; Maggs & Hommersand, 1993).

Polysiphonia denudata (Dillw.) Grev. On sheltered rock. **DOR**. (Velley, 1795).

Polysiphonia devoniensis Maggs & Homm. On sand-covered rock. (Maggs & Hommersand, 1993).

Polysiphonia elongata (Huds.) Sprengel. On rock or shells. 66, 67, 97, 07, 08; **DOR, RNG**.

Polysiphomia fibrillosa (Dillw.) Sprengel. On rock, or epiphytic. [39], 67, 78, 87, 97, 08; **DOR, LIV**. (Maggs & Hommersand, 1993; Walker, 1884).

Polysiphonia fucoides (Huds.) Grev. Frequent on littoral rock and among *Zostera*. 58, 67, 68, 78, 87, 88, 97, 07, 08; **DOR, LIV**. (Burrows, 1991).

[*Polysiphonia harveyi* Bailey. A rare epiphyte. 67. (Cotton, 1908a, b, 1914; Maggs & Hommersand, 1993).]

Polysiphonia lanosa (L.) Tandy. A frequent epiphyte on *Ascophyllum*. 66,67,68,78,87,88, 97,07,08; **DOR, LIV**. (10.6, 10.23).

Polysiphonia nigra (Huds.) Batters. On littoral pebbles and rock. 78,87,88,97; **DOR, LIV**. (10.6).

Polysiphonia simulans Harvey. An epiphyte on *Chondrus* or *Laminaria*. [07,08]. (10.9, 10.21).

Polysiphonia stricta (Dillw.) Grev. Frequent, often with *Zostera*. 58, 67, 68, 78, 87, 88, 97, 07; **DOR, LIV**. (Burrows, 1991; Norton, 1985).

[*Polysiphonia subulifera* (C. Ag.) Harvey. On rock, or epiphytic. (Gatty, 1872; Harvey, 1846).]

Porphyra linearis Grev. Frequent on littoral rock. 58, 67, 78, 87, 88, 97, 07, 08, 09; **LIV**. (Burrows, 1991; Norton, 1985).

Porphyra purpurea (Roth) C. Ag. Scarce on rock. 58, 68, 87. (EMB).

Porphyra umbilicalis J. Ag. Abundant on littoral rock. All grid squares but 58; **DOR, LIV**. (Burrows, 1991; Norton, 1985).

Pterocladia capillacea (S.G. Gmel.) Bornet & Thuret. Occasional on rock near LWM. 77, 78, 87, 97, 07, 08; **DOR**. (Norton, 1985).

Pterosiphonia parasitica (Huds.) Falk. A rare epiphyte on corallines. 87. (EMB).

Pterothamnion crispum (Ducluz) Naegeli. A rare epiphyte. (Maggs & Hommersand, 1993).

Pterothamnion plumula (Ellis) Naegeli. In the Fleet. 67. (Whittaker, 1978).

Ptilothamnion pluma (Dillw.) Thuret. An epiphyte on *Laminaria*. 97, [07, 08]; **RNG**. (Berkeley, 1833; Cotton, 1908a, b, 1914).

Ptilothamnion sphaericum (Crouan fr.) Maggs & Homm. Rare, on rock. (Maggs & Hommersand, 1993).

Radicilingua thysanorhizans (Holmes) Papenf. On unstable sand. (Maggs & Hommersand, 1993).

Rhodomela confervoides (Huds.) Silva. Occasional on sublittoral rock. 39, 49, [66], 67, 78, 87, 88, 97, 07; **LIV, RNG**. (Turner, 1802; Barrett, *c*.1856).

(*Rhodomela lycopodioides* (L.) C. Ag. Reported as an epiphyte on *Laminaria*, but very unlikely as it is a northern plant. [39], 97. (Walker, 1884, EMB).)

Rhodophyllis divaricata (Stackh.) Papenf. On rock, or epiphytic. 58, 68, 87, 97; **RNG**. (EMB).

Rhodophysema elegans (Crouan) Dixon. A rare crust on rock. 67, 07, 08. (EMB).

Rhodophysema georgii Batters. An epiphyte on *Zostera*. (Dixon & Irvine, 1977; Newton, 1931).

Rhodymenia pseudopalmata (Lamour.) Silva. On rock, or epiphytic. 39, 87, 88, 97, 08; **DOR, RNG**. (Burrows, 1991; Norton, 1985).

(*Rhodymenia diffusa*, reported (Cotton, 1908a, b, 1914), is a name of uncertain application.)

Schizymenia dubyi (Chauvin ex Duby) J. Ag. A scarce, sublittoral epiphyte. 87,88,97; **LIV**. (10.4).

Schmitziella endophloea Born. & Batters. An endoparasite in *Cladophora*. 58, 68, 78, 87, 97. (EMB).

Schottera nicaeensis (Lamour. ex Duby) Guiry & Holl. Scarce in rock pools near LWM. 58, 68, 87. (Norton, 1985).

Scinaia forcellata Bivona. Scarce in rock pools. 07,08. (Cotton, 1908a, b, 1914; Norton, 1985).

Scinaia turgida Chemin. On sublittoral rock. (Dixon & Irvine, 1977).

Solieria chordalis (C. Ag.) J. Ag. Submerged in the Fleet. 58, 67, 68. (Burrows, 1981).

Spermothamnion repens (Dillw.) Rosenv. Epiphytic on *Laminaria*. 87, 97, 07, 08; **LIV**. (Burrows, 1964).

Spermothamnion strictum (C. Ag.) Ardis. Epiphytic on *Laminaria*. 67. (Maggs & Hommersand, 1993).

Sphaerococcus coronopifolius Stackh. On sublittoral rock. [67], 87, 88, 97, 07, 08; **DOR**. (Cotton, 1908a, b, 1914; Norton, 1985).

Sphondylothamnion multifidum (Huds.) Naegeli. Frequent on subtidal rock. 58, 67, 68, 78, 87, 97, 07, 08; **DOR**. (Norton, 1985; Withering, 1801).

Spyridia filamentosa (Wulfen) Harvey. On sublittoral rock. 58, 68, 97, 07; **DOR, LIV**. (Norton, 1985; Walker, 1884).

Stenogramme interrupta (C. Ag.) Montagne ex Harvey. Scarce on sheltered sublittoral gravel. 87; **DOR**. (Norton, 1985).

Titanoderma cystoseirae (Hauck) Huve. Epiphytic on *Cystoseira*. 67. (Whittaker, 1978).

CHRYSOPHYCOTA

Four species of Golden-brown Algae have been recorded from vertical chalk, 78, 08; *Apistonema carterae, Chrysotila lamellosa, C. litorale* and *Thallochrysis litoralis* (Tittley, 1988).

MYXOPHYCOTA

The taxonomically difficult group of Blue-green Algae is not treated here. *Nostoc* is fairly common in bare, dry places inland, and a few marine species are recorded (Burrows, 1981; Cotton, 1908a, b, 1914; Newton, 1931; Tittley, 1988).

REFERENCES

Agricultural Statistics, 1866–1966, MOAFF Yearbooks.

Aitken G and N. 1982. *Proc. Dorset Nat. Hist. Archaeol. Soc.* **104**: 93–126.

Aiton W. 1789. *Hortus Kewensis*. Nicol G. London.

Allen DE and Randall RD. 1995. *Watsonia*. **20**: 407.

Allen MJ. 1997. *In:* Smith RJC. *et al*. Wessex Archaeol. Rept. **11**: 276–283.

Amherst A. 1896. *A History of Gardening in England*.

Anon. 1960. News Bull. Brit. Mycol. Soc. **16**: 4.

Anon. 1960. News Bull. Brit. Mycol. Soc. **17**: 4.

Anon. 1994. *The Dorset Coast today*. Dorset County Council.

Appleyard J. 1953. *Proc. Brit. Bryol. Soc.* 336.

Arkell WJ. 1947. *Geology of the country around Weymouth, Swanage, Corfe and Lulworth*. HMSO.

Armitage PD, Blackburn JH and Symes KL. 1996. *Proc. Dorset Nat. Hist. Archaeol. Soc.* **118**: 125.

Ashbee P and Dimbleby GW. 1958. *Proc. Dorset Nat. Hist. Archaeol. Soc.* **80**:146.

Ashbee P and Dimbleby GW. 1958. *Ibid*. **80**: 39–50.

Avery BW. 1980. *Soil Classification for England and Wales*. Soil Survey Tech. Monograph No.14. Harpenden.

Bales HGA and Copeland W, eds. 1997. *The Natural History of Dorset*. Dovecote Press. Wimborne.

Bandulska H. 1923. *J Linn. Soc. Bot.* **46**: 241.

Bandulska H. 1926 *J Linn. Soc. Bot.* **47**: 383.

Bandulska H. 1928. *J Linn. Soc. Bot.* **48**: 139.

Barrett WB. *c.*1856. Flora Weymouthiensis. Ms in Weymouth Public Library.

Bates HGA and Copland W, eds. 1997. *The Natural History of Dorset*. Dovecot Press. Wimborne.

Batters EAL. 1902. *J Bot.* **40**: Suppl. 1–107.

Batters EAL and Holmes EM. 1891. *Ann. Bot.* **5**: 63–107.

Battrick J and Lawson G. 1978. *Brownsea Islander*. Poole Historical Trust.

Bell M. 1981. *In:* Jones H and Dimbleby GW, eds. The Environment of Man from the Iron Age to the Anglo-Saxon period. *Brit. Archaeol. Record*. **87**: 75–91. Oxford University Press.

Berkeley MJ. 1833. *Gleanings of British Algae*.

Bettey JH. 1974. *Dorset*. David & Charles.

Bettey JH. 1980. *The Landscape of Wessex*. Moonraker Press. Bradford-on-Avon.

Blackburn JH, Armitage PD and Symes KL. 1997. *Proc. Dorset Nat. Hist. Archaeol. Soc.* **119**: 141.

Blockeel TL and Long DC. 1998. *A check-list and Census Catalogue of British and Irish Bryophytes*. Brit. Bryol. Soc. Cardiff.

Bon M. 1987. *The Mushrooms and Toadstools of Britain and North-western Europe*. Hodder & Stoughton.

Bowen HJM. 1976. *Lichenologist*. **8**: 1–33.

Bruce P. 1996. *Inshore along the Dorset coast*. 2nd ed. gives a good description of coastal topography between Portland and Bournemouth from a sailor's viewpoint, with many excellent colour photographs, mostly from the air.

Bullock JM, Connor J, Carrington S and Edwards RJ. 1998. *Watsonia*. **22**:143.

Burrows EM. 1964. *Br. Phycol. Bull.* **2**: 364–8.

Burrows EM. 1981. *In:* M Ladle ed. *The Fleet and Chesil Beach*. 39.

Burrows EM. 1991. *Seaweeds of the British Isles*. Vol. II. Brit. Mus. Nat. Hist.

Burt R. 1987. A detailed study of *Ophrys sphegodes*. Dorset County Council.

Byfield AJ and Pearman DA. 1994. *Dorset's disappearing heathland flora*. Plantlife and RSPB.

Cameron N and Scaife RG. 1988. *In:* Cox P. *Proc. Dorset Nat. Hist. Archaeol. Soc.* **110**: 65–70.

Candy B. 1982. Proc. Dorset Nat. Hist. Archaeol. Soc. **104**: 202.

Carruthers WJ. 1991. *In:* Cox PW and Hearne C, eds. *Dorset Nat. Hist. Archaeol. Monogr.* **9**: 203–209.

Carruthers WJ. 1991. *In:* Hearne CM and Smith RJC. *Proc. Dorset Nat. Hist. Archaeol. Soc.* **113**: 97.

Carruthers WJ. 1991. *In:* Woodward PJ. *Dorset Nat. Hist. Archaeol. Soc. Monog.* **8**: 113–114.

Carruthers WJ. 1992. *In:* Smith RJC, *et al*. *Proc. Dorset Nat. Hist. Archaeol. Soc.* **114**: 40–41.

Carruthers WJ. 1995. *In:* Harding PA, et al. *Ibid*. **17**: 86–89.

Carter PW. 1957. *Proc. Dorset Nat. Hist. Archaeol. Soc.* **79**: 73–98.

Casey H and Clarke RT. 1986. *IAHS Publ.* **No.157**: 257.

Casey H, Clarke RT and Smith SM. 1993. *Chem. and Ecol.* **8**: 105.

Casey H and Norton PVR. l973. *Freshw. Biol.* **3**: 317.

Chafin W. 1818. *Anecdotes and History of Cranbourne Chase*.

Chapman SB. 1975. *J. Ecol.* **63**: 809–824.

Chapman SB, Rose RJ and Clarke RT. 1989. *J. Appl. Ecol.* **26**: 1059–1072.

Chapman SB and Rose RJ. 1994. *Watsonia*. **20**: 89–95.

Chatwin CP. 1960. *The Hampshire basin and adjoining areas*. HMSO.

Claridge J. 1793. *A General View of the Agriculture of the County of Dorset*.

Clifton-Taylor A. 1972. *In:* Pevsner N and Nairn J, eds. *The buildings of England, Dorset*. Penguin.

Colborne GJN and Staines SJ. 1987. *Soil Survey Record No. 11. Soils in Somerset 1*. Harpenden.

Cotton AD. 1907. *J Bot.* **45**: 368–373.

Cotton AD. 1908. *J Bot.* **46**. 82.

Cotton AD. 1908. *J Bot.* **46**. 329; see also Cotton AD. 1914. *In:* Morris D, ed. *Natural History of Bournemouth and district*. Bournemouth Nat. Sci. Soc. 190–198.

Cotton AD. 1914. *In:* Morris D, ed. *Natural History of Bournemouth and district*. Bournemouth Nat. Sci. Soc. 190–198.

Cunliffe B. 1974. *Iron Age Communities in Britain*. Routledge and Kegan Paul.

Cunliffe B. 1978. *Hengistbury Head*. P Elek. London.

Dale CW. 1883. *Proc. Dorset Nat. Hist. Archaeol. Soc.* **5**: 151.

Davies GM. 1935. *The Dorset Coast, a Geological Guide*. T Murby.

Dennis RWG. 1968. *British Ascomycetes*. J.Cramer.

Dennis RWG, Orton PD and Hora FB. 1960. *New Checklist of British Agarics and Boleti*. Cambridge.

Dennis RWG and Hubbard CE. 1962. *Kew Bull.* **15**: 379.

Dickson G and Leonard AC. 1996. *Fungi of the New Forest*. English Nature.

Digest of Agricultural Census Statistics. 1991. HMSO.

Dillwyn LW. 1802. *British Confervae*.

Dimbleby GW. 1952. Also Orr MY. *In:* Case H. *Proc. Prehist. Soc.* **18**: 158–159.

Dimbleby GW. 1964. *In:* Fields NH, et al. *Ibid*. **30**: 360.

Diver C. 1933. *Geog.J.* **81**: 404.

Dixon PS and Irvine LM. 1977. *Seaweeds of the British Isles*, Vol. I, part 1. Brit. Mus. Nat. Hist.

Dorset Fed. Women's Institutes. 1990. *Hidden Dorset*. Countryside Books. Newbury.

Dudman AA and Richards AJ. 1997. *Dandelions of Great Britain and Ireland*. BSBI.

Dunston AEA. 1943. *Proc. Dorset Nat. Hist. Archaeol. Soc.* **65**: 130.

Drew CD. 1929. *Proc. Dorset Nat. Hist. Antiq. Field Club.* **51**: 232.

Drew CD. Undated. *Medieval Index*. Dorchester County Museum.

Ede J. 1988. *In:* Howard S. *Proc. Dorset Nat. Hist. Archaeol. Soc.* **110**: 101–105.

Ede J. 1988. *In:* Maynard D.*Ibid.* **110**: 93–94.

Ede J. 1989. *In:* Addison P. *Ibid.* **111**: 28.

Ede J. 1991. Also Gale R. *In:* Heaton MJ.*Ibid.* **114**: 121–122.

Ede J. 1992. *In:* Coe D and Hawkes JW. *Ibid.* **114**: 142–143.

Ede J. 1993. *In:* Smith RJC. *Wessex Archaeol. Rept.* **4**: 73–77.

Edwards B. 1997. *Bryophyte survey of Poole basin Mires.* DERC. Dorset County Council.

Edwards B. 1998. *Survey of Chalk Grassland in Dorset.* Dorset County Council.

Edwards B. 1997. *A survey of Early Gentian in Dorset.* And Supplement 1998. DERC report for Plantlife.

Ellis MB and Ellis JP. 1997. *Microfungi on land plants.* Richmond.

Erikssen OE and Hawksworth DL. 1993. *Outline of the Ascomycetes.*

Evans AM and Jones MK. 1979. *In:* Wainwright GW. *Dept. Environ. Archaeol. Rept.* **10**: 172–175.

Everett A. l994. *Mitchell's Materials.* Longmans.

Findlay DC, Colborne GJN, et al. 1984. *Soil Survey of England and Wales, Bulletin No.14 and map: Soils and Land Use in South West England.* Harpenden.

Fitzgerald R. 1988. *CSD Confidential Report 1061.* NCC.

Fletcher RL. 1987. *Seaweeds of the British Isles, Vol. III.* Brit. Mus. Nat. Hist.

Francis JE. 1983. *Palaeontology.* **26**: 277.

Gale R. 1990. *In:* Papworth M. *Proc. Dorset Nat. Hist.Archeol. Soc.* **112**: 129.

Gale R. 1991. *In:* Sharples NM. *Maiden Castle.* HBMC Archaeol. Rept. **19**: 125–129.

Galpine JK, ed. 1983. *The Georgian Garden.*

Gatty A. 1872. *British Seaweeds.*

Gilbert OL. 1993. The lichens of Chalk Grassland. *Lichenol.* **25**: 379–414.

Gilbert OL. 1993. *Lichenologist.* **25**: 379.

Gilbert OL. 1996. *Lichenologist.* **28**: 145.

Gilbert OL. 1996. *Lichenologist.* **28**: 493.

Girling MA. 1988. *In:* Jones M, ed. *Archaeology and the Flora of the British Isles.* OUCA Monog. **14**: 34–38.

Good RD. 1948. *A Geographical Handbook of the Dorset Flora.* Dorset County Museum.

Good RD. 1966. *The old roads of Dorset.* H.G.Commin. Bournemouth.

Good RD. 1961. *Proc. Dorset Nat. Hist. Archaeol. Soc.* **83**: 35.

Good RD. 1962. *Proc. Dorset Nat. Hist. Archaeol. Soc.* **84**: 37.

Good RD. 1963. *Proc. Dorset Nat. Hist. Archaeol. Soc.* **85**: 39.

Goodenough S. 1795. *British Fuci.*

Graveson AW. 1915. Ms notes in **DOR**.

Gray St. GH. 1918. *Proc. Dorset Nat. Hist. Antiq. Field Club.* **38**: 68–73.

Green CB. 1915. *Proc. Bournemouth Nat. Sci. Soc.* **7**: 46.

Green FJ. 1978. *In:* Hinton DA and Horsey I. *Proc. Dorset Nat. Hist. Archaeol. Soc.* **100**: 125.

Green FJ. 1980. *In:* Collins J and Field M. *Ibid.* **102**: 88.

Green FJ. 1980. Also Haskins L *In:* Horsey I and Shackley M. *Ibid.* **102**: 41.

Green FJ. 1980. *In:* Woodward PJ. *Ibid.* **102**: 101.

Green FJ. 1983. *In:* Jarvis K. *Dorset Nat. Hist. Archaeol. Soc. Monog.* **5**: 98.

Green M and Allen MJ. 1997. *Oxford J. Archaeol.* **16**: 121–132.

Greville RK. 1830. *Algae Britanniae.*

Groube LM and Bowden MCB. *In:* Bradley R, ed. 1982. *The archaeology of rural Dorset.* Dorset Nat. Hist. Archaeol. Soc. **14**.

Grove WB. 1935–37. *British Stem- and Leaf-fungi.* Cambridge.

Harris TM. 1939. *British Purbeck Charophyta.* Brit. Mus. Nat. Hist.

Harvey WH. 1846. *Phycologia Britannia.*

Haskins LA. 1980. *The Vegetational History of South-east Dorset.* Ph.D.Thesis. University of Southampton.

Haskins LE. 1992. *In:* Jarvis K. *Proc. Dorset Nat. Hist. Archaeol. Soc.* **114**: 95.

Hawksworth DL, Kirk PM, Sutton BC and Pegler DN. 1995. *Dictionary of the Fungi, 8th ed.* CAB International, Wallingford.

Helbaek H. 1952. *Proc. Prehist. Soc.* **18**: 194–233.

Helbaek H. 1956. *In:* Calkin JB. *Proc. Dorset Nat. Hist. Archaeol. Soc.* **88**: 49–51.

Hill MO. 1970. *Trans. Brit. Bryol. Soc.* **6**; 212–214.

Hill MO. 1977. *Bull. Brit. Bryol. Soc.* **30**; 6–7.

Hill MO. 1995. *Bull. Brit. Bryol. Soc.* **65**; 4–7.

Hill MO, Preston CD and Smith AJE. 1991–94. *Atlas of the Bryophytes of Britain and Ireland,* Vols. 1–3. Cambridge University Press.

Hinton P. 1999. *In:* Hearne C. *in press.*

Hinton P. *In:* Hinton P and Peacock DPS. *in press.*

Hinton P. *In:* Rawlings M. *to be published.*

Holmes EM. 1906. *unpublished MS in the Natural History Museum.* London.

Holmes NTH. 1983. *Classification of British rivers according to their Flora.* NCC.

Holmes NT. 1985. *Survey of the Fleet.* Report 648. NCC.

Hooker T. 1993. *The Trees of Canford Park.* Eureka.

Horsfall A. 1991. *Names of Wild Flowers in Dorset.*

Horsfall A. 1996. *Proc. Dorset Nat. Hist. Archaeol. Soc.* **118**: 1.

Horsfall A. 1997. *Proc. Dorset Nat. Hist. Archaeol. Soc.* **119**: 59.

House MR. 1968. *Proc. Dorset Nat. Hist. Archaeol. Soc.* **89**: 42.

Hudson W. 1762. *Flora Anglica.*

Hurst CP. 1923. *Proc. Dorset Nat. Hist. Antiq. Field Club.* **49**: 207.

Hutchins J. 1796. *History of Dorset.* 2nd ed.

Hutchins J. 1803. *History of the antiquities of the County of Dorset.* Second edition. Westminster.

Hutchins MJ. 1987a. *J.Ecol.* **75**: 711.

Hutchins MJ. 1987b. *J.Ecol.* **75**: 729.

Hyland P. 1989. *Purbeck, the ingrained island.* Dovecot Press. Wimborne.

Ilchester M. 1899. *Abbotsbury Garden Catalogue.*

Ing B. 1968. *Census catalogue of Myxomycetes.*

Irvine LM. 1983. *Seaweeds of the British Isles,* Vol. I, part 2A. Brit. Mus. Nat. Hist.

Jarvis K. 1993. *Proc. Dorset Nat. Hist. Archaeol. Soc.* **114**: 89.

Jenkinson MN. 1991. *Wild Orchids of Dorset.* Orchid Sundries, Gillingham.

Jenkinson MN. 1992. *Proc. Hampsh. Field Club Archaeol. Soc.* **47**: 225.

Jones AW and Roy J. 1996. *Hort. Soc.* **121**: 789.

Jones CA. 1973. *The Conservation of Chalk Downland in Dorset.* Dorset County Council.

Jones G and Legge A. 1987. *Antiquity.* **61**: 452–5.

Jones G and Legge A. *In:* Mercer R and Healy F. *Hambledon Hill.* Eng. Nature Archaeol. Monog. in press.

Jones G. 1986. *In:* Hurst JG and Wacker JS. *Proc. Dorset Nat. Hist. Archaeol. Soc.* **108**: 78.

Jones J and Straker V. 1993. *In:* Woodward PJ, et al. *Dorsetr Nat. Hist. Archaeol. Soc. Monog.* **12**: 349–350.

Jones M. 1980. *Proc. Prehist. Soc.* **46**: 61–63.

Jones NGD. 1981. *A History of English Forestry.* Blackwell.

Lees E. 1877. *Proc. Dorset Nat. Hist. Antiq. Field Club.* **1**: 40.

Letts JB. 1997. *In:* Smith RJC. *Wessex Archaeol. Rept.* **11**: 267–270.

Limbrey S. 1977. *In:* Gelling PS. *Proc. Prehist. Soc.* **43**: 277.

Limbrey S. 1975. *Soil Science and Archaeology.* London.

Linton EF. 1914. *Proc. Dorset Nat. Hist. Antiq. Field Club.* **35**: 141.

Linton EF. 1915. *Proc. Dorset Nat. Hist. Antiq. Field Club.* **36**: 148.

Lock M. 1998. *Supplement to 5.3.* Dorset Environmental Records Centre.

Lucking JH. 1982. *Dorset Railways.* Dovecot Press. Wimborne

Maby CJ. 1936. *In:* Drew CD andPiggott S. *Ibid.* **2**: 94.

MacFadyen WA. 1970. *Geological highlights of the West Country.* Butterworth.

Maggs CA and Hommersand M. 1993. *Seaweeds of the British Isles,* Vol. I, part 3A . Brit. Mus. Nat. Hist.

Mahon A and Pearman DA, eds. 1993. *Endangered Wildlife in Dorset.* Dorset Environmental Records Centre.

Mansel JC. 1886. *Proc. Dorset Nat. Hist. Antiq. Field Club.* **7**: 109–113.

Mansel-Pleydell JC. 1895. *The Flora of Dorsetshire.* Dorchester. (2nd ed) Privately printed.

Marker AFH, Casey H and Rother JA. 1984. *Verh. Internat. Verein. Limnol.,* **22**: 1949.

Melville RV and Freshey EC. 1982. *The Hampshire basin and adjoining areas*. HMSO.

Mills AD. 1986. *Dorset Place Names*. Ensign, Southampton.

Monk M. 1977. *In:* Hinton DA and Hodges R. *Proc. Dorset Nat. Hist. Archaeol. Soc.* **99**: 76.

Monk M. 1987. *In:* Green C.S. *Dorset Nat. Hist. Archaeol. Soc. Monog*. **7**: 132–137.

Moore NW. 1962. *J. Ecol.* **50**: 369–391.

Mordue JEM and Ainsworth GC. 1984. *Ustilaginales of the British Isles*. Commonw. Mycol. Inst.

Morgan GC. 1979. *In:* Wainwright GH. *Res. Rept. Comm. Soc. Antiq*. London, **37**: 253.

Morgan GC. 1980. *In:* Wacher JS. *Proc. Dorset Nat. Hist. Soc*. **102**: 42.

Newton L. 1931. *A Handbook of British Seaweeds*. Brit. Mus. Nat. Hist.

Norton TA . 1985. *Provisional Atlas of the Marine Algae of Britain and Ireland*. Inst. Terrest. Biol.

Nye S and Jones M. 1987. *In:* Cunliffe B. *Hengistbury Head, Vol 1*. 323–328

Orr MY. *In:* Case H. *Proc. Prehist. Soc.* **18**: 158–159.

Orton PD. 1991. *Mycologist* **5**: 123–138

Orton PD. 1991. *Mycologist* **5**: 172–176

Pahl J. 1960. *Proc. Dorset Nat. Hist. Archaeol. Soc.* **82**: 143–154.

Palmer C and Jones M. 1991. *In:* Sharples NM. *Maiden Castle*. HBMC *Archaeol. Rept*. **19**: 129–139.

Paterson A. 1978. *The Gardens of Britain, 2*. Batsford.

Pearman DA. 1994. *Sedges and their allies in Dorset*. DERC. Dorchester.

Perkins JW. l977. *Geology explained in Dorset*. David and Charles.

Perring FH. 1996. *The Garden*. **121**: 360.

Pickering VT and CG. 1962. *Proc. Bournemouth Nat. Sci. Soc.* **52**: 14.

Pickess B and Giavarini VJ. 1981. *Proc. Dorset Nat. Hist. Archaeol. Soc.* **103**: 107.

Porley RD and Ulf-Hansen P. 1991. *Proc. Dorset Nat. Hist. Archaeol. Soc*. **113**: 161.

Preece RC. 1980. *J. Archaeol. Sci.* **7**: 345–62.

Prentice HC. and IC. 1983. *Proc. Dorset Nat. Hist. Archaeol. Soc.* **105**: 127.

Preston CD. 1995. *Watsonia*, **20**: 255–262.

Preston CD and Pearman DA. 1998. *Watsonia*, **22**: 163–172.

Pulteney R. 1796. *In:* Hutchins J. *History of Dorset, 2nd ed*.

Purvis OW, Coppins BJ, Hawksworth DL, James PW and Moore DM. 1992. *The Lichen Flora of Great Britain and Ireland*. Natural History Museum. London.

Purvis OW, Coppins BJ and James PW. 1994. *Checklist of lichens of Great Britain and Ireland*, Brit. Lichen Soc.

Putnam W. 1984. *Roman Dorset*. Dovecot Press

Rackham O. 1986. *The History of the Countryside*. Dent.

Raybould AF. 1997. *Proc. Dorset Nat. Hist. Archaeol. Soc.* **119**: 147.

Read WJ. 1950–1. *Proc. Bournemouth Nat. Sci. Soc.* **41**: 56.

Reid C. 1896. *Proc. Dorset Nat. Hist. Archaeol. Soc.* **17**: 67–75.

Robinson D and Rasmussen P. 1989. *In:* Milles A, et al. eds. *The Beginnings of Agriculture*. BAR Internat. Ser. **496**: 149–163.

Robinson IS. 1981. *In:* Ladle M, ed. *The Fleet and Chesil Beach*. **39**: 33.

Robinson KL. 1948. *The Soils of Dorset*. Chapter 2 *In:* Good R. *A Geographical Handbook of the Dorset Flora*. Dorset Nat. Hist. Archaeol. Soc.

Rodwell JS, ed. 1992. *British Plant Communities*. Vol.3, *Grassland and Montane Communities*. Cambridge.

Rodwell JS, ed. 1991. *British Plant Communities*. Vol. 1, *Woodland and Scrub*. Cambridge.

Rodwell JS, ed. 1995. *British Plant Communities*. Vol 2, Cambridge.

Rodwell JS, ed. 1995. *British Plant Communities*. Vol 4, *Aquatic Communities*. Cambridge.

Rose F. 1992. *In:* Bates JW and Farmer AM, eds. *Temperate Forest Management; its effects on bryophyte and lichen flora and habitats*, pp.211–233

Rose RJ, Bannister P and Chapman SB. 1996. *J. Ecol.* **84**: 617–628.

Rose RJ, Webb NR, Clarke RT and Traynor CH. In press. *Biol. Conserv*.

Rotheroe M, Newton A, Evans S and Feehan J. 1996. *Mycologist* **10**.

Royal Commission on Historic Monuments. 1970. *Vol.3, part 2*. 318.

Ryley C. 1995. *Roman Plants Guide, Fishbourne Palace*. Sussex Archaeol. Soc.

Sanecki KN. 1994. *History of the English Herb Garden*. Ward Lock

Scaife RG. 1991. *In:* Cox PW and Hearne C, eds. *Dorset Nat. Hist. Archaeol. Soc. Monogr*. **9**: 180–197.

Scaife RG. 1993. *In:* Papworth M. *Proc. Dorset Nat. Hist. Archaeol. Soc.* **115**: 51–62.

Shuttleworth BM. 1984. *A Gazetteer of Dorset Place-names*. Dorset Environmental Records Centre.

Simons MA. 1998. *Dorset Data Book*. County Hall. Dorchester.

Smith AJE. 1990. *The Liverworts of Britain and Ireland*. Cambridge University Press.

Smith AJE. 1978. *The Moss Flora of Britain and Ireland*. Cambridge University Press.

Spencer JW. 1988. *Inventory of Ancient Woodland – Dorset*. NCC Report

Stace CA, ed. 1975. *Hybridisation and the Flora of the British Isles*. Academic Press.

Stace CA. 1997. *New Flora of the British Isles, 2nd ed.* Cambridge

Stackhouse J. 1795–1801. *Nereis Britannica*. J White. London.

Stackhouse J. 1801. *In:* Withering W. *A systematic arrangement of British Plants*. **4**: 222.

Staines SJ. l991. *In:* Sharples NM, ed. *Maiden Castle excavations and Field Survey 1985–6*. English Heritage Archaeol. Rep. No.**19**.

Staines SJ. 1993. *In:* Woodward PJ, et al. eds. *Dorset Nat. Hist. Archaeol. Soc. Monogr. No.***12**.

Staines SJ. 1997. *In:* Smith RJC, et al. eds. *Wessex Archaeol. Rep. No.***11**.

Stern RC. 1982. *Proc. Dorset Nat. Hist. Archaeol. Soc.* **104**: 202.

Straker V. 1997. *In:* Allen MJ, Smith RJC, et al. *Wessex Archaeol. Rpt.* **11**: 184–190.

Talbot JDR, House WA and Pethybridge AD. 1990. *Water Res*. **24**: 1295.

Taylor I. 1765. *Map of the County of Dorset*.

Thacker C. 1979. *The History of Gardens*. Croom Helm. London.

Tittley I. 1988. *Chalk cliffs and algal communities outside S.E.England*. NCC Report **878**. Brit. Mus. Nat. Hist.

Tomlinson PR. 1995. *In:* Higgins D. *Proc. Dorset Nat. Hist. Archaeol. Soc.* **117**: 146.

Turner D. 1802. *Synopsis of the British Fuci*.

Udal JS. 1889. *Proc. Dorset Nat. Hist. Antiq. Field Club.* **10**:19.

Velley T. 1795. *Marine Plants*. J White. London.

Walker TW. 1884. *J Bot.* **22**.

Waton PV and Barber KE. 1987. *In:* Barber KE, ed. *Wessex and the Isle of Wight*. Quatern. Res. Ass. Cambridge.

Watson W. 1953. *Census Catalogue of British Lichens*. Cambridge University Press.

Watson W. 1922. *J. Ecol.* **10**: 255–6.

Webb NR. 1990. *Biol. Conserv.* **51**: 273–286.

Westlake DF. 1968. *Proc. Dorset Nat. Hist. Archaeol. Soc.* **90**:155.

Whittaker JE. 1978. *Proc. Dorset Nat. Hist. Archaeol. Soc.* **100:** 73–79.

Wigginton M. 1987. *Dorset Chalk Grassland Survey 1983/4*. NCC Report.

Williams A. 1950. *Proc. Dorset Nat. Hist. Archaeol. Soc.* **72**: 20.

Wilson JD, Arroyo BE and Clark SC. 1997. *The diet of birds of lowland farmland*. EN report. Dept. Zoology. University of Oxford.

Wilson M and Henderson DM. 1966. *British Rust Fungi*. Cambridge.

Wilson V, et al. 1958. *Geology of the country around Bridport and Lyme Regis*. HMSO.

Withering W. 1801. *A systematic arrangement of British plants*.

Woodhead FA. 1994. *The Flora of the Christchurch area*. Blackmore Press. Shaftesbury.

Woodhead FA and Clement EJ. 1997. *BSBI News*. **76**:56.

Woodhead N and Tweed RD. 1946–7. *Proc. Bournemouth Nat. Sci. Soc.* **37**: 50–67.

Woodward PJ. 1991. *Dorset Nat. Hist. Archaeol. Soc. Monog.* **8**: Appendix 3, 170–172.

Wright CH. 1890. *J. Roy. Microscop. Soc.* 647.

Young D. 1972. *Proc. Dorset Nat. Hist. Archaeol. Soc.* **93**: 213–242.

INDEX